# Name Reactions

Jie Jack Li

# Name Reactions

## A Collection of Detailed Mechanisms and Synthetic Applications

Fifth Edition

 Springer

Jie Jack Li
Department of Chemistry
College of Arts and Sciences
University of San Francisco
San Francisco
CA, USA

ISBN 978-3-319-37431-4          ISBN 978-3-319-03979-4 (eBook)
DOI 10.1007/978-3-319-03979-4
Springer Cham Heidelberg New York Dordrecht London

1007393261

© Springer International Publishing Switzerland 2014
Softcover reprint of the hardcover 5th edition 2014

Printed on acid-free paper

Springer is part of Springer Science+Business Media (www.springer.com)

*To Prof. Claire Castro*

Kurt Alder
1902–1958
Nobel Prize, 1950

Eduard Buchner
1860–1917
Nobel Prize, 1907

Adolf von Baeyer
1835–1917
Nobel Prize, 1905

Elias James Corey
1928–
Nobel Prize, 1990

Derek H. R. Barton
1918–1999
Nobel Prize, 1969

Otto Paul Hermann Diels
1876–1954
Nobel Prize, 1950

Emil Fischer
1852–1919
Nobel Prize, 1902

Robert Robinson
1886–1975
Nobel Prize, 1947

Otto Wallach
1847–1931
Nobel Prize, 1910

Victor Grignard
1871–1935
Nobel Prize, 1912

Hermann Staudinger
1881–1965
Nobel Prize, 1953

Georg Wittig
1897–1987
Nobel Prize, 1979

Karl Ziegler
1898–1973
Nobel Prize, 1963

# Preface

Four years have gone by since the fourth edition was published and much has happened since then. Professionally, I have moved from industry to academia to teach organic and medicinal chemistry. This change is reflected in my choice to include most of the basic name reactions so that this book will be useful for my undergraduate students. I have also had the opportunity to make corrections to several quinoline- and isoquinoline-related mechanisms. In addition, new name reactions have emerged, and new references appeared for old name reactions. I have added 27 new name reactions to reflect the latest developments in organic chemistry and updated synthetic applications for each old name reaction. By popular demand, a brief biographical description of the inventor of nearly *every* name reaction has been added to this edition.

As in previous editions each reaction is delineated by its detailed step-by-step, electron-pushing mechanism, supplemented with the original and the latest references, especially review articles. Now, with the addition of many synthetic applications, this book is not only an indispensable resource for senior undergraduate and graduate students for learning mechanisms and the synthetic utility of name reactions and preparing for their exams, but it is also a good reference book for all organic chemists in both industry and academia.

I wish to thank Dr. Jonathan W. Lockner at Scripps Research Institute and Dr. Jun Cindy Shi of Bristol-Myers Squibb for their help in preparing and proofreading the manuscript. I also wish to thank Prof. Neil K. Garg at UCLA and his students, Grace Chiou, Adam Goetz, Liana Hie, Dr. Travis McMahon, Tejas Shah, Noah Fine Nathel, Joel M. Smith, Amanda Silberstein, and Evan D. Styduhar for proofreading the final version of the manuscript. Their knowledge and input have tremendously enhanced the quality of this book. Any remaining errors are, of course, solely my own responsibility.

As always, I welcome your critique! Please send your comments to this email address: lijiejackli@gmail.com.

October 2013                                                  Jie Jack Li
San Francisco, CA

# Table of Contents

# Abbreviations and Acronyms

| | |
|---|---|
| ◯— | Polymer support |
| Δ | Solvent heated under reflux |
| (DHQ)$_2$-PHAL | 1,4-*bis*(9-*O*-Dihydroquinine)-phthalazine |
| (DHQD)$_2$-PHAL | 1,4-*bis*(9-*O*-Dihydroquinidine)-phthalazine |
| [bimim]Cl•2AlCl$_3$ | 1-Butyl-3-methylimidazolium chloroaluminuminate |
| 3CC | Three-component condensation |
| 4CC | Four-component condensation |
| 9-BBN | 9-Borabicyclo[3.3.1]nonane |
| A | Adenosine |
| Ac | Acetyl |
| ADDP | 1,1′-(azodicarbonyl)dipiperidine |
| AIBN | 2,2′-azobisisobutyronitrile |
| Alpine-borane® | *B*-isopinocampheyl-9-borabicyclo[3.3.1]-nonane |
| AOM | *p*-Anisyloxymethyl = *p*-MeOC$_6$H$_4$OCH$_2$- |
| Ar | Aryl |
| B: | Generic base |
| BINAP | 2,2′-*bis*(Diphenylphosphino)-1,1′-binaphthyl |
| Bn | Benzyl |
| Boc | *tert*-Butyloxycarbonyl |
| BQ | Benzoquinone |
| BT | Benzothiazole |
| Bz | Benzoyl |
| Cbz | Benzyloxycarbonyl |
| CuTC | Copper thiophene-2-carboxylate |
| DABCO | 1,4-Diazabicyclo[2.2.2]octane |
| dba | Dibenzylideneacetone |
| DBU | 1,8-Diazabicyclo[5.4.0]undec-7-ene |
| DCC | 1,3-Dicyclohexylcarbodiimide |
| DDQ | 2,3-Dichloro-5,6-dicyano-1,4-benzoquinone |

| | |
|---|---|
| *de* | Diastereoselctive excess |
| DEAD | Diethyl azodicarboxylate |
| DIAD | Diisopropyl azodidicarboxylate |
| DIBAL | Diisobutylaluminum hydride |
| DIPEA | Diisopropylethylamine |
| DMA | *N*,*N*-dimethylacetamide |
| DMAP | 4-*N*,*N*-dimethylaminopyridine |
| DME | 1,2-Dimethoxyethane |
| DMF | *N*,*N*-dimethylformamide |
| DMFDMA | *N*,*N*-dimethylformamide dimethyl acetal |
| DMS | Dimethylsulfide |
| DMSO | Dimethylsulfoxide |
| DMSY | Dimethylsulfoxonium methylide |
| DMT | Dimethoxytrityl |
| DPPA | Diphenylphosphoryl azide |
| dppb | 1,4-*bis*(Diphenylphosphino)butane |
| dppe | 1,2-*bis*(Diphenylphosphino)ethane |
| dppf | 1,1′-*bis*(Diphenylphosphino)ferrocene |
| dppp | 1,3-*bis*(Diphenylphosphino)propane |
| *dr* | Diastereoselctive ratio |
| DTBAD | Di-*tert*-butylazodicarbonate |
| DTBMP | 2,6-Di-*tert*-butyl-4-methylpyridine |
| E1 | Unimolecular elimination |
| E1cB | 2-Step, base-induced β-elimination *via* carbanion |
| E2 | Bimolecular elimination |
| EAN | Ethylammonium nitrate |
| EDCI | 1-Ethyl-3-(3-dimethylaminopropyl)carbodiimide |
| EDDA | Ethylenediamine diacetate |
| *ee* | Enantiomeric excess |
| Ei | Two groups leave at about the same time and bond to each other as they are doing so |
| Eq | Equivalent |
| Et | Ethyl |
| EtOAc | Ethyl acetate |
| HMDS | Hexamethyldisilazane |
| HMPA | Hexamethylphosphoramide |
| HMTA | Examethylenetetramine |
| HMTTA | 1,1,4,7,10,10-Hexamethyltriethylenetetramine |
| IBX | *o*-iodoxybenzoic acid |
| Imd | Imidazole |
| KHMDS | Potassium hexamethyldisilazide |
| LAH | Lithium aluminum hydride |
| LDA | Lithium diisopropylamide |
| LHMDS | Lithium hexamethyldisilazide |

| | |
|---|---|
| LTMP | Lithium 2,2,6,6-tetramethylpiperidide |
| M | Metal |
| *m*-CPBA | *m*-chloroperoxybenzoic acid |
| MCRs | Multicomponent reactions |
| Mes | Mesityl |
| MPM | Methyl phenylmethyl |
| MPS | Morpholine-polysulfide |
| Ms | Methanesulfonyl |
| MTBE | Methyl tertiary butyl ether |
| MVK | Methyl vinyl ketone |
| MWI | Microwave irradiation |
| NBS | *N*-bromosuccinimide |
| NCS | *N*-chlorosuccinimide |
| NIS | *N*-iodosuccinimide |
| NMP | 1-Methyl-2-pyrrolidinone |
| Nos | Nosylate (4-nitrobenzenesulfonyl) |
| *N*-PSP | *N*-phenylselenophthalimide |
| *N*-PSS | *N*-phenylselenosuccinimide |
| Nu | Nucleophile |
| PCC | Pyridinium chlorochromate |
| PDC | Pyridinium dichromate |
| PE | Premature ejaculation |
| Piv | Pivaloyl |
| PMB | *para*-Methoxybenzyl |
| PPA | Polyphosphoric acid |
| PPTS | Pyridinium *p*-toluenesulfonate |
| PT | Phenyltetrazolyl |
| PyPh$_2$P | Diphenyl 2-pyridylphosphine |
| Pyr | Pyridine |
| Red-Al | Sodium *bis*(methoxy-ethoxy)aluminum hydride |
| Red-Al | Sodium *bis*(methoxy-ethoxy)aluminum hydride (SMEAH) |
| Salen | *N,N'*-disalicylidene-ethylenediamine |
| SET | Single electron transfer |
| SIBX | Stabilized IBX |
| SM | Starting material |
| SMEAH | Sodium *bis*(methoxy-ethoxy)aluminum hydride |
| S$_N$1 | Unimolecular nucleophilic substitution |
| S$_N$2 | Bimolecular nucleophilic substitution |
| S$_N$Ar | Nucleophilic substitution on an aromatic ring |
| SSRI | Selective serotonin reuptake inhibitor |
| TBABB | tetra-*n*-butylammonium bibenzoate |
| TBAF | tetra-*n*-butylammonium fluoride |
| TBAO | 1,3,3-Trimethyl-6-azabicyclo[3.2.1]octane |
| TBDMS | *tert*-Butyldimethylsilyl |

| | |
|---|---|
| TBDPS | *tert*-Butyldiphenylsilyl |
| TBS | *tert*-Butyldimethylsilyl |
| *t*-Bu | *tert*-Butyl |
| TDS | Thexyldimethylsilyl |
| TEA | Triethylamine |
| TEMPO | 2,2,6,6-Tetramethylpiperidinyloxy |
| TEOC | Trimethysilylethoxycarbonyl |
| Tf | Trifluoromethanesulfonyl (triflyl) |
| TFA | Trifluoroacetic acid |
| TFAA | Trifluoroacetic anhydride |
| TFP | Tri-2-furylphosphine |
| THF | Tetrahydrofuran |
| TIPS | Triisopropylsilyl |
| TMEDA | $N,N,N',N'$-tetramethylethylenediamine |
| TMG | 1,1,3,3-Tetramethylguanidine |
| TMP | Tetramethylpiperidine |
| TMS | Trimethylsilyl |
| TMSCl | Trimethylsilyl chloride |
| TMSCN | Trimethylsilyl cyanide |
| TMSI | Trimethylsilyl iodide |
| TMSOTf | Trimethylsilyl triflate |
| Tol | Toluene or tolyl |
| Tol-BINAP | 2,2'-*bis*(di-*p*-tolylphosphino)-1,1'-binaphthyl |
| TosMIC | (*p*-tolylsulfonyl)methyl isocyanide |
| Ts | Tosyl |
| TsO | Tosylate |
| UHP | Urea-hydrogen peroxide |

# Alder ene reaction

The Alder ene reaction, also known as the hydro-allyl addition, is addition of an enophile to an alkene (ene) *via* allylic transposition. The four-electron system including an alkene π-bond and an allylic C–H σ-bond can participate in a pericyclic reaction in which the double bond shifts and new C–H and C–C σ-bonds are formed.

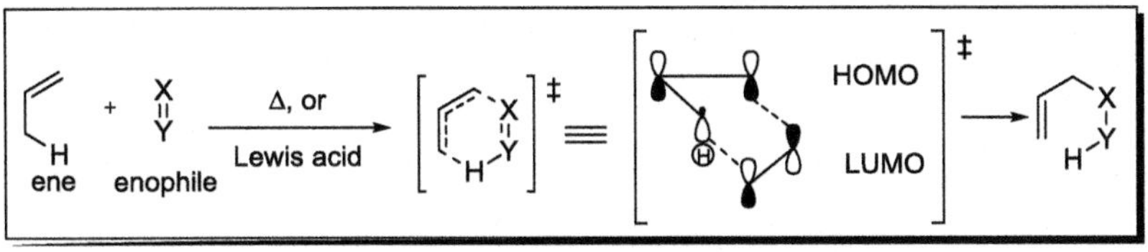

X=Y: C=C, C≡C, C=O, C=N, N=N, N=O, S=O, *etc.*

Example 1[5]

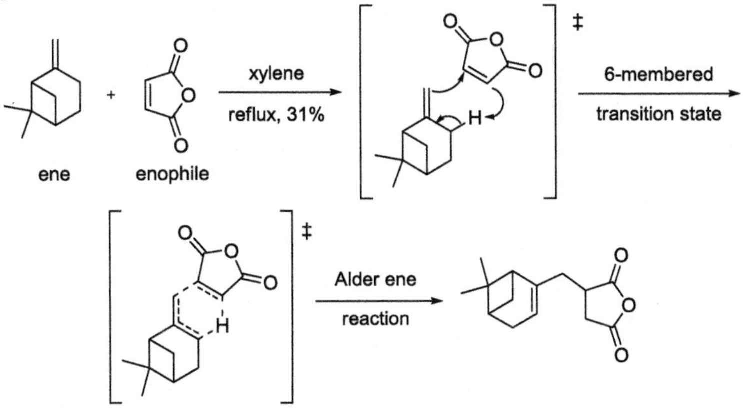

Example 2[7]

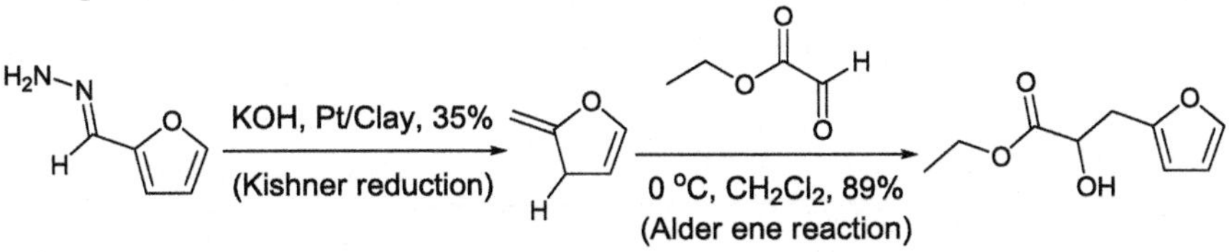

Example 3, Intramolecular Alder ene reaction[8]

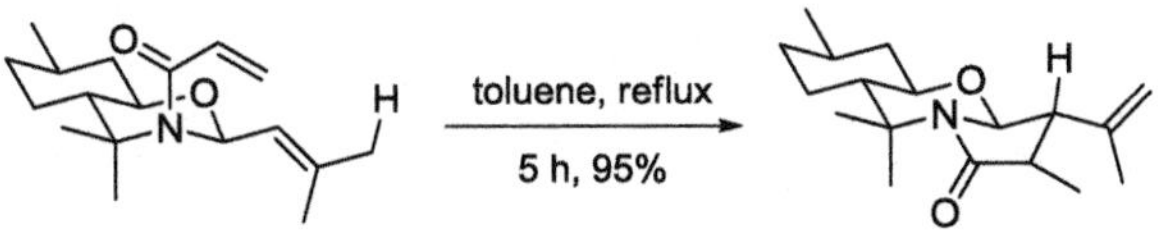

Example 4, Cobalt-catalyzed Alder ene reaction[9]

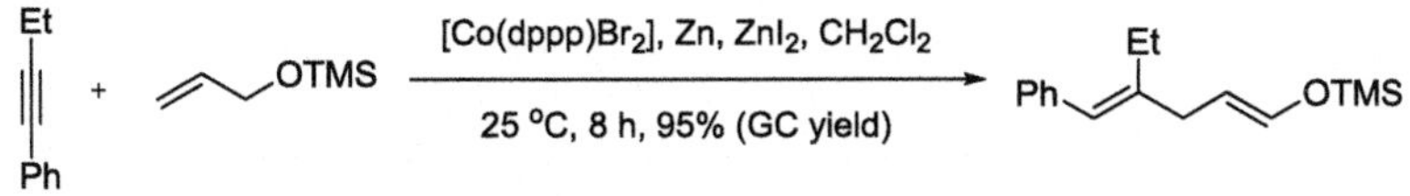

J.J. Li, *Name Reactions: A Collection of Detailed Mechanisms and Synthetic Applications*, DOI 10.1007/978-3-319-03979-4_1, © Springer International Publishing Switzerland 2014

Example 5, Nitrile Alder ene reaction[10]

Example 6[11]

Example 7[13]

**References**

1.  Alder, K.; Pascher, F.; Schmitz, A. *Ber.* **1943**, *76*, 27–53. Kurt Alder (Germany, 1902–1958) shared the Nobel Prize in Chemistry in 1950 with his teacher Otto Diels (Germany, 1876–1954) for the development of the diene synthesis.
2.  Oppolzer, W. *Pure Appl. Chem.* **1981**, *53*, 1181–1201. (Review).
3.  Johnson, J. S.; Evans, D. A. *Acc. Chem. Res.* **2000**, *33*, 325–335. (Review).
4.  Mikami, K.; Nakai, T. In *Catalytic Asymmetric Synthesis;* 2nd edn.; Ojima, I., ed.; Wiley–VCH: New York, **2000**, 543–568. (Review).
5.  Sulikowski, G. A.; Sulikowski, M. M. *e-EROS Encyclopedia of Reagents for Organic Synthesis* **2001**, Wiley: Chichester, UK.
6.  Brummond, K. M.; McCabe, J. M. *The Rhodium(I)-Catalyzed Alder ene Reaction.* In *Modern Rhodium-Catalyzed Organic Reactions* **2005**, 151–172. (Review).
7.  Miles, W. H.; Dethoff, E. A.; Tuson, H. H.; Ulas, G. *J. Org. Chem.* **2005**, *70*, 2862–2865.
8.  Pedrosa, R.; Andres, C.; Martin, L.; Nieto, J.; Roson, C. *J. Org. Chem.* **2005**, *70*, 4332–4337.
9.  Hilt, G.; Treutwein, J. *Angew. Chem. Int. Ed.* **2007**, *46*, 8500–8502.
10. Ashirov, R. V.; Shamov, G. A.; Lodochnikova, O. A.; Litvynov, I. A.; Appolonova, S. A.; Plemenkov, V. V. *J. Org. Chem.* **2008**, *73*, 5985–5988.
11. Cho, E. J.; Lee, D. *Org. Lett.* **2008**, *10*, 257–259.
12. Curran, T. T. *Alder Ene Reaction.* In *Name Reactions for Homologations-Part II*; Li, J. J., Ed.; Wiley: Hoboken, NJ, **2009**, pp 2–32. (Review).
13. Trost, B. M.; Quintard, A. *Org. Lett.* **2012**, *14*, 4698–4670.
14. Karmakar, R.; Mamidipalli, P.; Yun, S. Y.; Lee, D. *Org. Lett.* **2013**, *15*, 1938–1941.

## Aldol condensation

The aldol condensation is the coupling of an enolate ion with a carbonyl compound to form a β-hydroxycarbonyl, and sometimes, followed by dehydration to give a conjugated enone. A simple case is addition of an enolate to an **ald**ehyde to afford an alcohol, thus the name **aldol**.

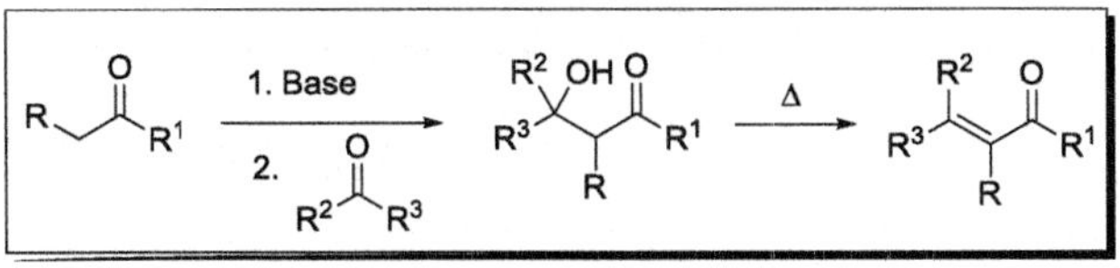

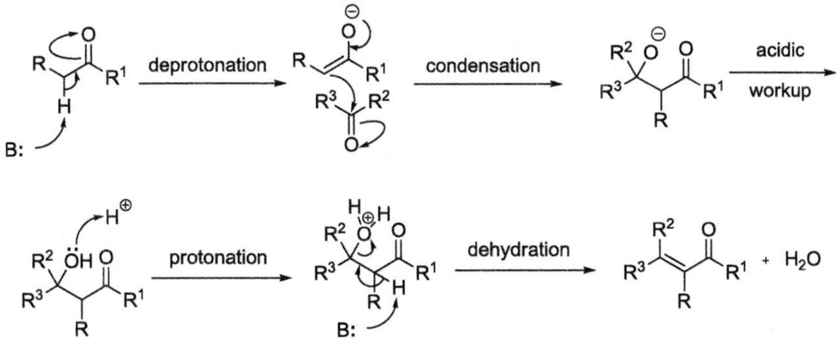

Example 1[3]

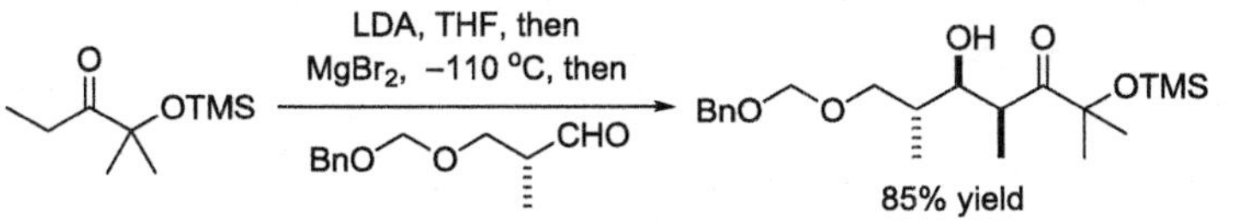

85% yield

Example 2[8]

22% of 6S,7R-diastereomer
and 10% recovered SM

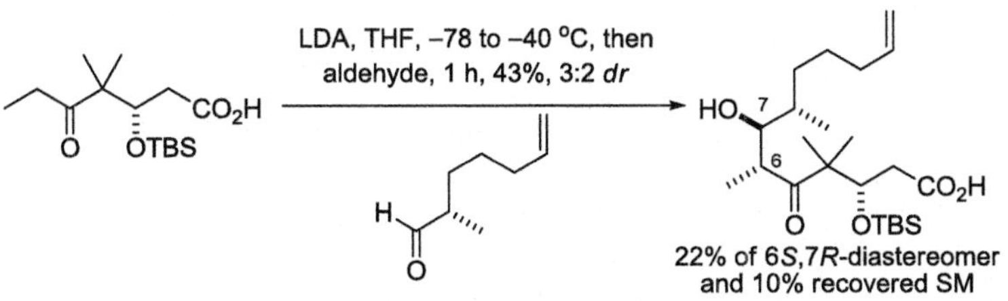

J.J. Li, *Name Reactions: A Collection of Detailed Mechanisms and Synthetic Applications*,
DOI 10.1007/978-3-319-03979-4_2, © Springer International Publishing Switzerland 2014

Example 3, Enantioselective Mukaiyama aldol reaction[10]

Example 4, Intermolecular aldol reaction using organocatalyst[12]

Example 5, Intramolecular aldol reaction[13]

Example 6, Intramolecular vinylogous aldol reaction[15]

## References

1.  Wurtz, C. A. *Bull. Soc. Chim. Fr.* **1872,** *17,* 436–442. Charles Adolphe Wurtz (1817–1884) was born in Strasbourg, France. After his doctoral training, he spent a year under Liebig in 1843. In 1874, Wurtz became the Chair of Organic Chemistry at the Sorbonne, where he educated many illustrous chemists such as Crafts, Fittig, Friedel, and van't Hoff. The Wurtz reaction, where two alkyl halides are treated with

sodium to form a new carbon–carbon bond, is no longer considered synthetically useful, although *the aldol reaction* that Wurtz discovered in 1872 has become a staple in organic synthesis. Alexander P. Borodin is also credited with the discovery of the aldol reaction together with Wurtz. In 1872 he announced to the Russian Chemical Society the discovery of a new by-product in aldehyde reactions with properties like that of an alcohol, and he noted similarities with compounds already discussed in publications by Wurtz from the same year.

2.   Nielsen, A. T.; Houlihan, W. J. *Org. React.* **1968,** *16*, 1–438. (Review).

3.   Still, W. C.; McDonald, J. H., III. *Tetrahedron Lett.* **1980,** *21*, 1031–1034.

4.   Mukaiyama, T. *Org. React.* **1982,** *28*, 203–331. (Review).

5.   Mukaiyama, T.; Kobayashi, S. *Org. React.* **1994,** *46*, 1–103. (Review on tin(II) enolates).

6.   Johnson, J. S.; Evans, D. A. *Acc. Chem. Res.* **2000,** *33*, 325–335. (Review).

7.   Denmark, S. E.; Stavenger, R. A. *Acc. Chem. Res.* **2000,** *33*, 432–440. (Review).

8.   Yang, Z.; He, Y.; Vourloumis, D.; Vallberg, H.; Nicolaou, K. C. *Angew. Chem. Int. Ed.* **1997,** *36*, 166–168.

9.   Mahrwald, R. (ed.) *Modern Aldol Reactions,* Wiley–VCH: Weinheim, Germany, **2004.** (Book).

10.  Desimoni, G.; Faita, G.; Piccinini, F.; Toscanini, M. *Eur. J. Org. Chem.* **2006,** 5228–5230.

11.  Guillena, G.; Najera, C.; Ramon, D. J. *Tetrahedron: Asymmetry* **2007,** *18*, 2249–2293. (Review on enantioselective direct aldol reaction using organocatalysis.)

12.  Doherty, S.; Knight, J. G.; McRae, A.; Harrington, R. W.; Clegg, W. *Eur. J. Org. Chem.* **2008,** 1759–1766.

13.  O'Brien, E. M.; Morgan, B. J.; Kozlowski, M. C. *Angew. Chem. Int. Ed.* **2008,** *47*, 6877–6880.

14.  Trost, B. M.; Brindle, C. S. *Chem. Soc. Rev.* **2010,** *39*, 1600–1632. (Review).

15.  Gazaille, J. A.; Abramite, J. A.; Sammakia, T. *Org. Lett.* **2012,** *14*, 178–181.

16.  Esumi, T.; Yamamoto, C.; Tsugawa, Y.; Toyota, M.; Asakawa, Y.; Fukuyama Y. *Org. Lett.* **2013,** *15*, 1898–1901.

## Algar–Flynn–Oyamada Reaction

Conversion of 2′-hydroxychalcones to 2-aryl-3-hydroxy-4$H$-1-benzopyran-4-ones (flavonols) by an oxidative cyclization.

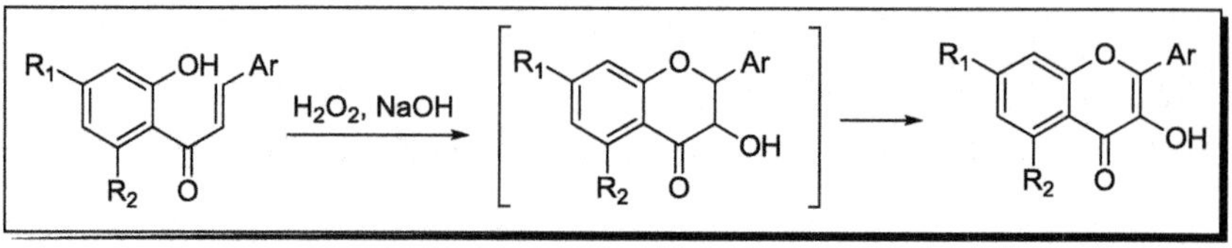

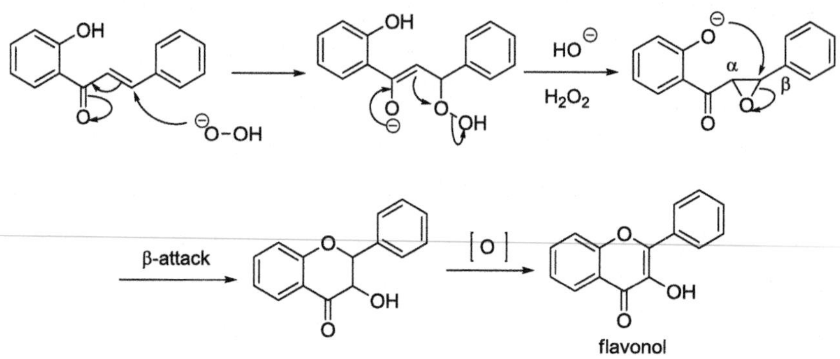

A side reaction:

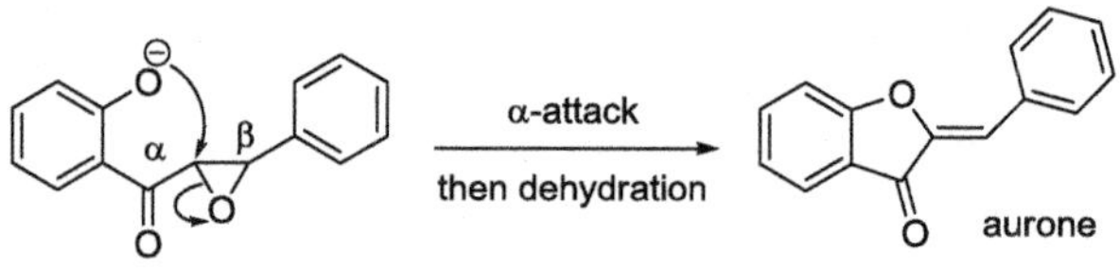

Example 1[5]

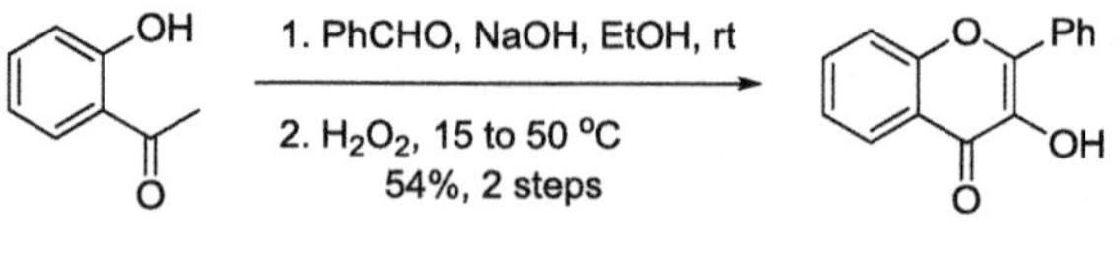

Example 2[5]

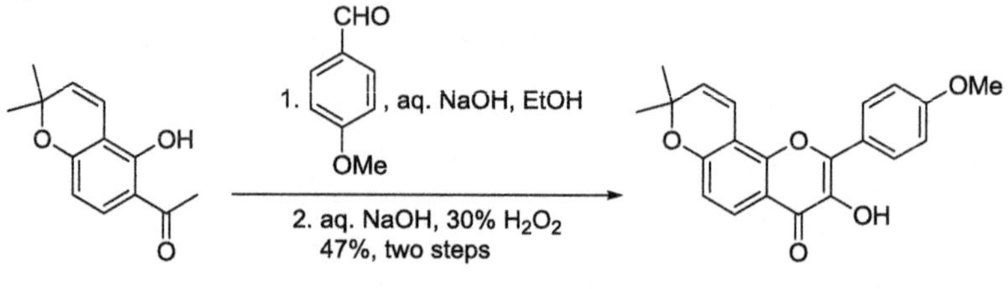

J.J. Li, *Name Reactions: A Collection of Detailed Mechanisms and Synthetic Applications*, DOI 10.1007/978-3-319-03979-4_3, © Springer International Publishing Switzerland 2014

Example 3, The side reaction dominated to give the aurone derivative[9]

Example 4[12]

Example 5, The side recation dominated to give the aurone derivative[13]

## References

1.  Algar, J.; Flynn, J. P. *Proc. Roy. Irish Acad.* **1934**, *B42*, 1–8. Algar and Flynn were Irish chemists.
2.  Oyamada, T. *J. Chem. Soc. Jpn* **1934**, *55*, 1256–1261.
3.  Oyamada, T. *Bull. Chem. Soc. Jpn.* **1935**, *10*, 182–186.
4.  Wheeler, T. S. *Record Chem. Progr.* **1957**, *18*, 133–161. (Review).
5.  Smith, M. A.; Neumann, R. M.; Webb, R. A. *J. Heterocycl. Chem.* **1968**, *5*, 425–426.
6.  Wagner, H.; Farkas, L. In *The Flavonoids*; Harborne, J. B.; Mabry, T. J.; Mabry H., Eds.; Academic Press: New York, **1975**, *1*, pp 127–213. (Review).
7.  Wollenweber, E. In *The Flavonoids: Advances in Research*; Harborne, J. B.; Mabry, T. J., Eds; Chapman and Hall: New York, **1982**, pp 189–259. (Review).
8.  Wollenweber, E. In *The Flavonoids: Advances in Research since 1986*; Harborne, J. B., Ed.; Chapman and Hall: New York, **1994**, pp 259–335. (Review).
9.  Bennett, M.; Burke, A. J.; O'Sullivan, W. I. *Tetrahedron* **1996**, *52*, 7163–7178.
10. Bohm, B. A.; Stuessy, T. F. *Flavonoids of the Sunflower Family (Asteraceae)*; Springer-Verlag: New York, **2000**. (Review).
11. Limberakis, C. *Algar–Flynn–Oyamada Reaction*. In *Name Reactions in Heterocyclic Chemistry*; Li, J. J., Ed.; Wiley: Hoboken, NJ, **2005**, pp 496–503. (Review).
12. Li, Z.; Ngojeh, G.; DeWitt, P.; Zheng, Z.; Chen, M.; Lainhart, B.; Li, V.; Felpo, P. *Tetrahedron Lett.* **2008**, *49*, 7243–7245.
13. Zhao, X.; Liu, J.; Xie, Z.; Li, Y. *Synthesis* **2012**, *44*, 2217–2224.

# Allan–Robinson reaction

Synthesis of flavones or isoflavones by the treatment of of *o*-hydroxyaryl ketones with aromatic aldehydes in the presence of a base. *Cf.* Kostanecki reaction.

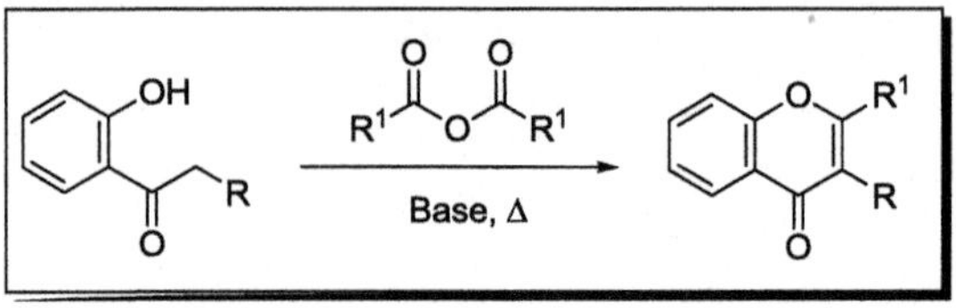
Base, Δ

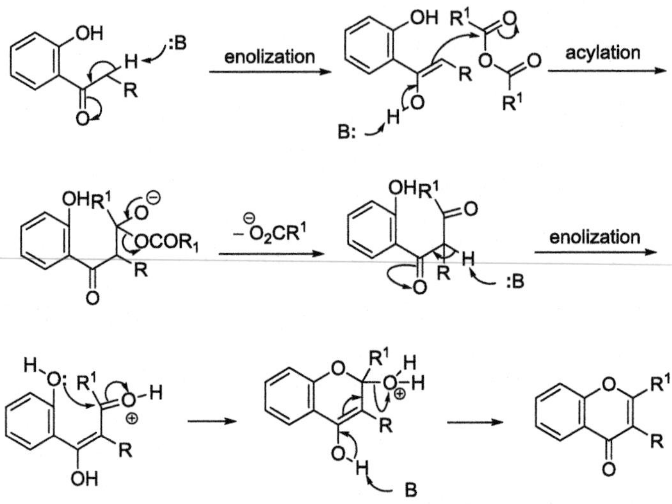
enolization
acylation
enolization

Example 1[6]

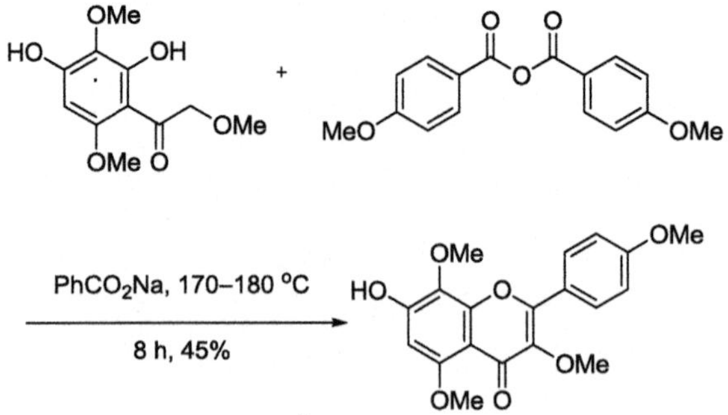
PhCO2Na, 170–180 °C
8 h, 45%

Example 2, Non-aromatic anhydride[9]

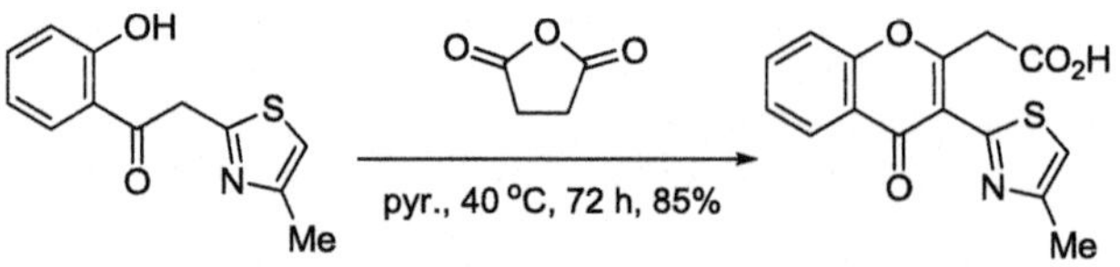
pyr., 40 °C, 72 h, 85%

J.J. Li, *Name Reactions: A Collection of Detailed Mechanisms and Synthetic Applications*, DOI 10.1007/978-3-319-03979-4_4, © Springer International Publishing Switzerland 2014

Example 3, Non-aromatic anhydride[10]

Et₃N, reflux, 12 h, 87%

Example 4, Acid chloride in place of anhydride[10]

pyr., rt
1–3 h, 61%

**References**
1. Allan, J.; Robinson, R. *J. Chem. Soc.* **1924**, *125*, 2192–2195. Robert Robinson (United Kingdom, 1886–1975) won the Nobel Prize in Chemistry in 1947 for his studies on alkaloids. However, Robinson himself considered his greatest contribution to science was that he founded the qualitative theory of electronic mechanisms in organic chemistry. Robinson, along with Lapworth (a friend) and Ingold (a rival), pioneered the arrow pushing approach to organic reaction mechanism. Robinson was also an accomplished pianist. James Allan, his student, also coauthored another important paper with Robinson on the relative directive powers of groups for aromatic substitution.
2. Széll, T.; Dózsai, L.; Zarándy, M.; Menyhárth, K. *Tetrahedron* **1969**, *25*, 715–724.
3. Wagner, H.; Maurer, I.; Farkas, L.; Strelisky, J. *Tetrahedron* **1977**, *33*, 1405–1409.
4. Dutta, P. K.; Bagchi, D.; Pakrashi, S. C. *Indian J. Chem., Sect. B* **1982**, *21B*, 1037–1038.
5. Patwardhan, S. A.; Gupta, A. S. *J. Chem. Res., (S)* **1984**, 395.
6. Horie, T.; Tsukayama, M.; Kawamura, Y.; Seno, M. *J. Org. Chem.* **1987**, *52*, 4702–4709.
7. Horie, T.; Tsukayama, M.; Kawamura, Y.; Yamamoto, S. *Chem. Pharm. Bull.* **1987**, *35*, 4465–4472.
8. Horie, T.; Kawamura, Y.; Tsukayama, M.; Yoshizaki, S. *Chem. Pharm. Bull.* **1989**, *37*, 1216–1220.
9. Poyarkov, A. A.; Frasinyuk, M. S.; Kibirev, V. K.; Poyarkova, S. A. *Russ. J. Bioorg. Chem.* **2006**, *32*, 277–279.
10. Peng, C.-C.; Rushmore, T.; Crouch, G. J.; Jones, J. P. *Bioorg. Med. Chem. Lett.* **2008**, *16*, 4064–4074.
11. Levchenko, K. S.; Semenova, I. S.; Yarovenko, V. N.; Shmelin, P. S.; Krayushkin, M. M. *Tetrahedron Lett.* **2012**, *53*, 3630–3632;

# Arndt–Eistert homologation

One-carbon homologation of carboxylic acids using diazomethane.

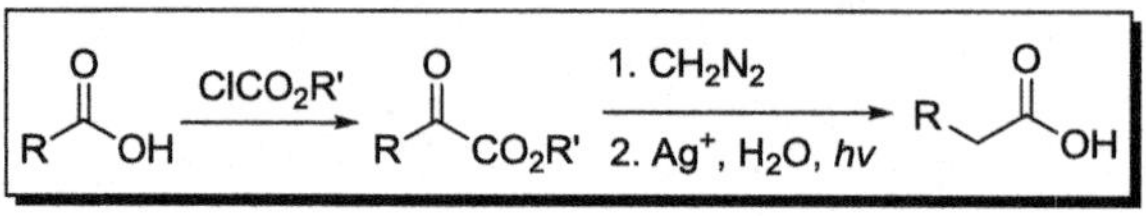
ClCO2R'
1. CH2N2
2. Ag+, H2O, hv

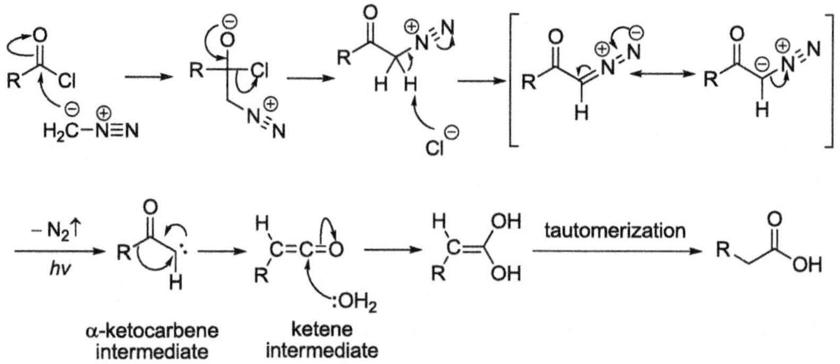
– N2↑
hv
tautomerization
α-ketocarbene intermediate
ketene intermediate

Example 1[7]

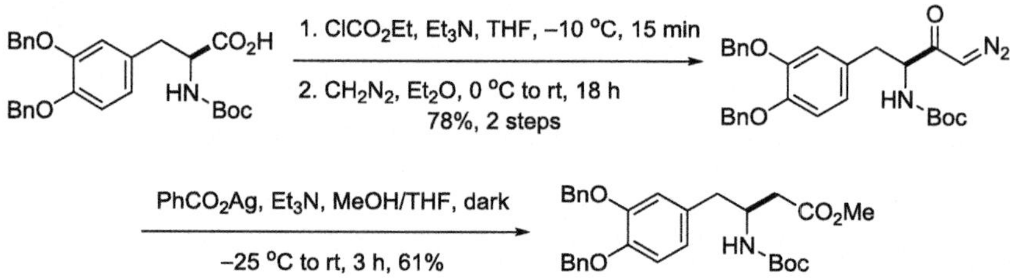
1. ClCO2Et, Et3N, THF, –10 °C, 15 min
2. CH2N2, Et2O, 0 °C to rt, 18 h
78%, 2 steps
PhCO2Ag, Et3N, MeOH/THF, dark
–25 °C to rt, 3 h, 61%

Example 2, An interesting variation[9]

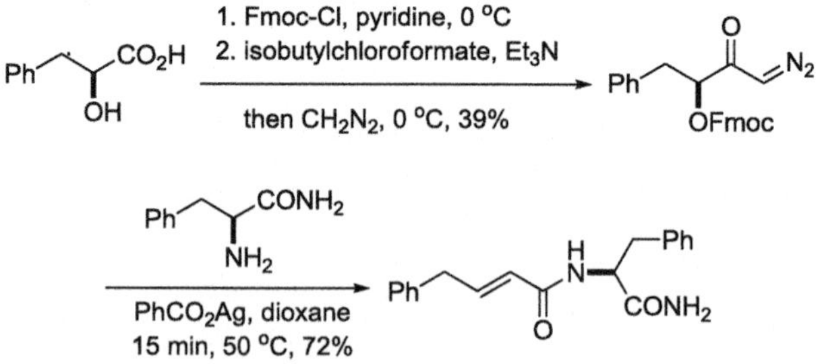
1. Fmoc-Cl, pyridine, 0 °C
2. isobutylchloroformate, Et3N
then CH2N2, 0 °C, 39%
PhCO2Ag, dioxane
15 min, 50 °C, 72%

Example 3[10]

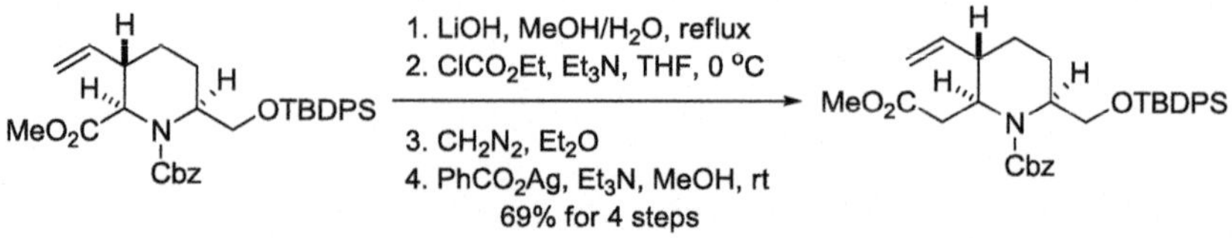
1. LiOH, MeOH/H2O, reflux
2. ClCO2Et, Et3N, THF, 0 °C
3. CH2N2, Et2O
4. PhCO2Ag, Et3N, MeOH, rt
69% for 4 steps

J.J. Li, *Name Reactions: A Collection of Detailed Mechanisms and Synthetic Applications*, DOI 10.1007/978-3-319-03979-4_5, © Springer International Publishing Switzerland 2014

Example 4[10]

## References

1. Arndt, F.; Eistert, B. *Ber.* **1935**, *68*, 200–208. Fritz Arndt (1885–1969) was born in Hamburg, Germany. He discovered the Arndt–Eistert homologation at the University of Breslau where he extensively investigated the synthesis of diazomethane and its reactions with aldehydes, ketones, and acid chlorides. Fritz Arndt's chain-smoking of cigars ensured that his presence in the laboratories was always well advertised. Bernd Eistert (1902–1978), born in Ohlau, Silesia, was Arndt's Ph.D. student. Eistert later joined I. G. Farbenindustrie, which became BASF after the Allies broke up the conglomerate following WWII.
2. Podlech, J.; Seebach, D. *Angew. Chem. Int. Ed.* **1995**, *34*, 471–472.
3. Matthews, J. L.; Braun, C.; Guibourdenche, C.; Overhand, M.; Seebach, D. In *Enantioselective Synthesis of β-Amino Acids* Juaristi, E. ed.; Wiley-VCH: Weinheim, Germany, **1996**, pp 105–126. (Review).
4. Katritzky, A. R.; Zhang, S.; Fang, Y. *Org. Lett.* **2000**, *2*, 3789–3791.
5. Vasanthakumar, G.-R.; Babu, V. V. S. *Synth. Commun.* **2002**, *32*, 651–657.
6. Chakravarty, P. K.; Shih, T. L.; Colletti, S. L.; Ayer, M. B.; Snedden, C.; Kuo, H.; Tyagarajan, S.; Gregory, L.; Zakson-Aiken, M.; Shoop, W. L.; Schmatz, D. M.; Wyvratt, M. J.; Fisher, M. H.; Meinke, P. T. *Bioorg. Med. Chem. Lett.* **2003**, *13*, 147–150.
7. Gaucher, A.; Dutot, L.; Barbeau, O.; Hamchaoui, W.; Wakselman, M.; Mazaleyrat, J.-P. *Tetrahedron: Asymmetry* **2005**, *16*, 857–864.
8. Podlech, J. In *Enantioselective Synthesis of β-Amino Acids (2nd Edn.)* Wiley: Hoboken, NJ, **2005**, pp 93–106. (Review).
9. Spengler, J.; Ruiz-Rodriguez, J.; Burger, K.; Albericio, F. *Tetrahedron Lett.* **2006**, *47*, 4557–4560.
10. Toyooka, N.; Kobayashi, S.; Zhou, D.; Tsuneki, H.; Wada, T.; Sakai, H.; Nemoto, H.; Sasaoka, T.; Garraffo, H. M.; Spande, T. F.; Daly, J. W. *Bioorg. Med. Chem. Lett.* **2007**, *17*, 5872–5875.
11. Fuchter, M. J. *Arndt–Eistert Homologation.* In *Name Reactions for Homologations-Part I*; Li, J. J., Ed.; Wiley: Hoboken, NJ, **2009**, pp 336–349. (Review).
12. Saavedra, C. J.; Boto, A.; Hernández, R. *Org. Lett.* **2012**, *14*, 3542–3545.

## Baeyer–Villiger oxidation

General scheme:

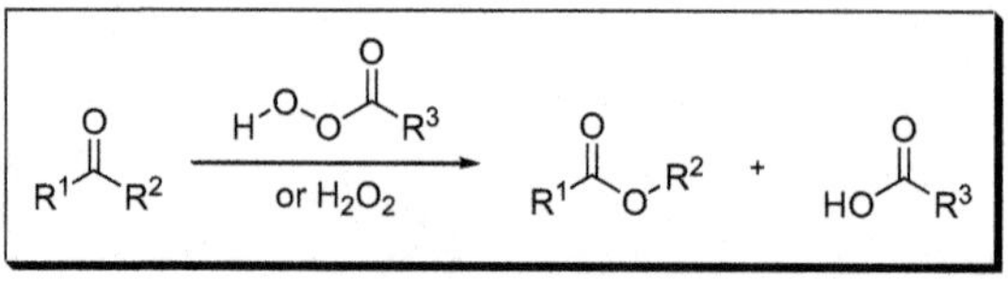

The most electron-rich alkyl group (more substituted carbon) migrates first. The general migration order: tertiary alkyl > cyclohexyl > secondary alkyl > benzyl > phenyl > primary alkyl > methyl >> H.

For substituted aryls:

$p$-MeO-Ar > $p$-Me-Ar > $p$-Cl-Ar > $p$-Br-Ar > $p$-O$_2$N-Ar

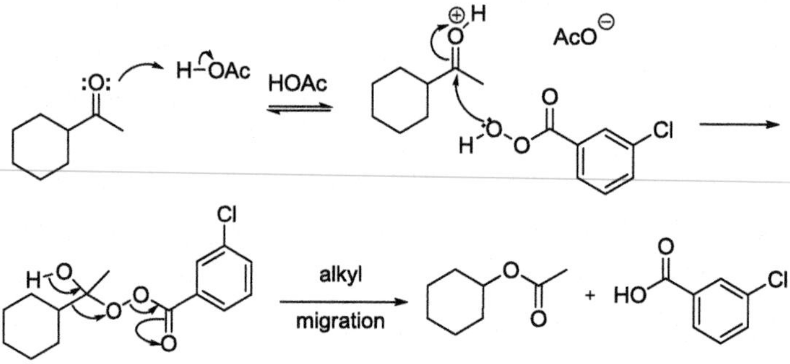

Example 1[4]

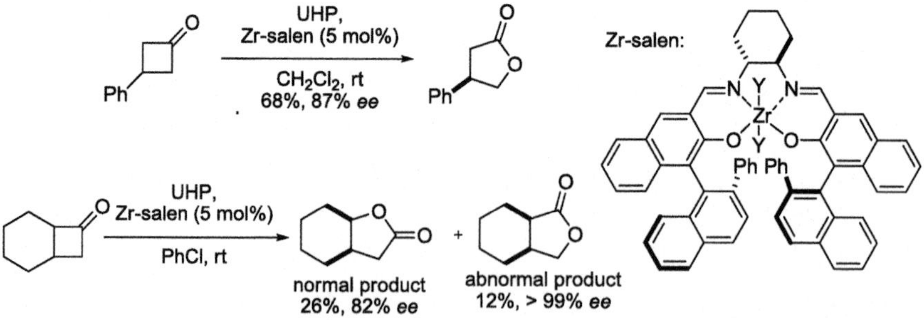

UHP = Urea-hydrogen peroxide complex

Example 2, Chemoselective over lactam[5]

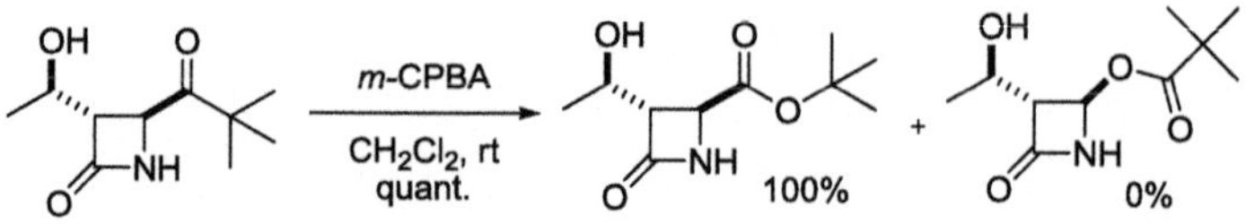

J.J. Li, *Name Reactions: A Collection of Detailed Mechanisms and Synthetic Applications*, DOI 10.1007/978-3-319-03979-4_6, © Springer International Publishing Switzerland 2014

Example 3, Chemoselective over lactone[6]

*m*-CPBA, NaHCO$_3$

CH$_2$Cl$_2$, 0 °C, 4 h
60%

Example 4, Chemoselective over ester[8]

*m*-CPBA, CF$_3$SO$_3$H

CH$_2$Cl$_2$, 45 min, 90%

## References

1. v. Baeyer, A.; Villiger, V. *Ber.* **1899**, *32*, 3625–3633. Adolf von Baeyer (1835–1917) was one of the most illustrious organic chemists in history. He contributed to many areas of the field. The Baeyer–Drewson indigo synthesis made possible the commercialization of synthetic indigo. Another one of Baeyer's claim of fame is his synthesis of barbituric acid, named after his then girlfriend, Barbara. Baeyer's real joy was in his laboratory and he deplored any outside work that took him away from his bench. When a visitor expressed envy that fortune had blessed so much of Baeyer's work with success, Baeyer retorted dryly: "Herr Kollege, I experiment more than you." As a scientist, Baeyer was free of vanity. Unlike other scholastic masters of his time (Liebig for instance), he was always ready to acknowledge ungrudgingly the merits of others. Baeyer's famous greenish-black hat was a part of his perpetual wardrobe and he had a ritual of tipping his hat when he admired novel compounds. Adolf von Baeyer received the Nobel Prize in Chemistry in 1905 at age seventy. His apprentice, Emil Fischer, won it in 1902 when he was fifty, three years before his teacher. Victor Villiger (1868–1934), born in Switzerland, went to Munich and worked with Adolf von Baeyer for eleven years.
2. Krow, G. R. *Org. React.* **1993**, *43*, 251–798. (Review).
3. Renz, M.; Meunier, B. *Eur. J. Org. Chem.* **1999**, *4*, 737–750. (Review).
4. Wantanabe, A.; Uchida, T.; Ito, K.; Katsuki, T. *Tetrahedron Lett.* **2002**, *43*, 4481–4485.
5. Laurent, M.; Ceresiat, M.; Marchand-Brynaert, J. *J. Org. Chem.* **2004**, *69*, 3194–3197.
6. Brady, T. P.; Kim, S. H.; Wen, K.; Kim, C.; Theodorakis, E. A. *Chem. Eur. J.* **2005**, *11*, 7175–7190.
7. Curran, T. T. *Baeyer–Villiger Oxidation.* In *Name Reactions for Functional Group Transformations*; Li, J. J., Ed.; Wiley: Hoboken, NJ, **2007**, pp 160–182. (Review).
8. Demir, A. S.; Aybey, A. *Tetrahedron* **2008**, *64*, 11256–11261.
9. Zhou, L.; Liu, X.; Ji, J.; Zhang, Y.; Hu, X.; Lin, L.; Feng, X. *J. Am. Chem. Soc.* **2012**, *134*, 17023–17026. (Desymmetrization and Kinetic Resolution).
10. Itoh, Y.; Yamanaka, M.; Mikami, K. *J. Org. Chem.* **2013**, *78*, 146–153.

## Baker–Venkataraman rearrangement

Base-catalyzed acyl transfer reaction that converts α-acyloxyketones to β-diketones.

Example 1, Carbamoyl Baker–Venkataraman rearrangement[5]

Example 2, Carbamoyl Baker–Venkataraman rearrangement, followed by cyclization[6]

J.J. Li, *Name Reactions: A Collection of Detailed Mechanisms and Synthetic Applications*,
DOI 10.1007/978-3-319-03979-4_7, © Springer International Publishing Switzerland 2014

Example 3, Baker–Venkataraman rearrangement[9]

Example 4, Baker–Venkataraman rearrangement[10]

## References

1. Baker, W. *J. Chem. Soc.* **1933,** 1381–1389. Wilson Baker (1900–2002) was born in Runcorn, England. He studied chemistry at Manchester under Arthur Lapworth and at Oxford under Robinson. In 1943, Baker was the first to confirm that penicillin contained sulfur, of which Robinson commented: "This is a feather in your cap, Baker." Baker began his independent academic career at University of Bristol. He retired in 1965 as the Head of the School of Chemistry. Baker was a well-known chemist centenarian, spending 47 years in retirement!
2. Mahal, H. S.; Venkataraman, K. *J. Chem. Soc.* **1934,** 1767–1771. K. Venkataraman studied under Robert Robinson Manchester. He returned to India and later arose to be the Director of the National Chemical Laboratory at Poona.
3. Kraus, G. A.; Fulton, B. S.; Wood, S. H. *J. Org. Chem.* **1984,** *49*, 3212–3214.
4. Reddy, B. P.; Krupadanam, G. L. D. *J. Heterocycl. Chem.* **1996,** *33*, 1561–1565.
5. Kalinin, A. V.; da Silva, A. J. M.; Lopes, C. C.; Lopes, R. S. C.; Snieckus, V. *Tetrahedron Lett.* **1998,** *39*, 4995–4998.
6. Kalinin, A. V.; Snieckus, V. *Tetrahedron Lett.* **1998,** *39*, 4999–5002.
7. Thasana, N.; Ruchirawat, S. *Tetrahedron Lett.* **2002,** *43*, 4515–4517.
8. Santos, C. M. M.; Silva, A. M. S.; Cavaleiro, J. A. S. *Eur. J. Org. Chem.* **2003,** 4575–4585.
9. Krohn, K.; Vidal, A.; Vitz, J.; Westermann, B.; Abbas, M.; Green, I. *Tetrahedron: Asymmetry* **2006,** *17*, 3051–3057.
10. Yu, Y.; Hu, Y.; Shao, W.; Huang, J.; Zuo, Y.; Huo, Y.; An, L.; Du, J.; Bu, X. *E. J. Org. Chem.* **2011,** 4551–4563.

## Bamford–Stevens reaction

The Bamford–Stevens reaction and the Shapiro reaction share a similar mechanistic pathway. The former uses a base such as Na, NaOMe, LiH, NaH, NaNH$_2$, heat, *etc.*, whereas the latter employs bases such as alkyllithiums and Grignard reagents. As a result, the Bamford–Stevens reaction furnishes more-substituted olefins as the thermodynamic products, while the Shapiro reaction generally affords less-substituted olefins as the kinetic products.

In protic solvent (S–H):

In aprotic solvent:

Example 1, Tandem Bamford–Stevens/thermal aliphatic Claisen rearrangement sequence[6]

The starting material *N*-aziridinyl imine is also known as Eschenmoser hydrazone.

Example 2, Thermal Bamford–Stevens[6]

J.J. Li, *Name Reactions: A Collection of Detailed Mechanisms and Synthetic Applications*, DOI 10.1007/978-3-319-03979-4_8, © Springer International Publishing Switzerland 2014

Example 3[7]

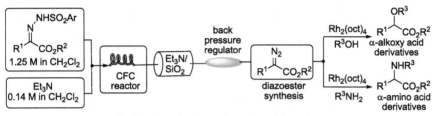

Example 4[8]

Example 5, Diazoesters from arylsulfonylhydrazones by means of in-flow Bamford-Stevens reactions[13]

CFC = Continuous-Flow Centrifugation

## References

1.  Bamford, W. R.; Stevens, T. S. M. *J. Chem. Soc.* **1952**, 4735–4740. Thomas Stevens (1900–2000), another chemist centenarian, was born in Renfrew, Scotland. He and his student W. R. Bamford published this paper at the University of Sheffield, UK. Stevens also contributed to another name reaction, the McFadyen–Stevens reaction.
2.  Felix, D.; Müller, R. K.; Horn, U.; Joos, R.; Schreiber, J.; Eschenmoser, A. *Helv. Chim. Acta* **1972**, *55*, 1276–1319.
3.  Shapiro, R. H. *Org. React.* **1976**, *23*, 405–507. (Review).
4.  Adlington, R. M.; Barrett, A. G. M. *Acc. Chem. Res.* **1983**, *16*, 55–59. (Review on the Shapiro reaction).
5.  Chamberlin, A. R.; Bloom, S. H. *Org. React.* **1990**, *39*, 1–83. (Review).
6.  Sarkar, T. K.; Ghorai, B. K. *J. Chem. Soc., Chem. Commun.* **1992**, *17*, 1184–1185.
7.  Chandrasekhar, S.; Rajaiah, G.; Chandraiah, L.; Swamy, D. N. *Synlett* **2001**, 1779–1780.
8.  Aggarwal, V. K.; Alonso, E.; Hynd, G.; Lydon, K. M.; Palmer, M. J.; Porcelloni, M.; Studley, J. R. *Angew. Chem. Int. Ed.* **2001**, *40*, 1430–1433.
9.  May, J. A.; Stoltz, B. M. *J. Am. Chem. Soc.* **2002**, *124*, 12426–12427.
10. Zhu, S.; Liao, Y.; Zhu, S. *Org. Lett.* **2004**, *6*, 377–380.
11. Baldwin, J. E.; Bogdan, A. R.; Leber, P. A.; Powers, D. C. *Org. Lett.* **2005**, *7*, 5195–5197.
12. Humphries, P. *Bamford–Stevens Reaction.* In *Name Reactions for Homologations-Part II*; Li, J. J., Ed.; Wiley: Hoboken, NJ, **2009**, pp 642–652. (Review).
13. Bartrum, H. E.; Blakemore, D. C.; Moody, C. J.; Hayes, C. J. *Chem. Eur. J.* **2011**, *17*, 9586–9589.

## Baran reagents

Zinc bis(alkanesulfinate) salts, which permit direct C–H functionalization of heteroarenes. Several of these reagents are now commercially available.[6,7]

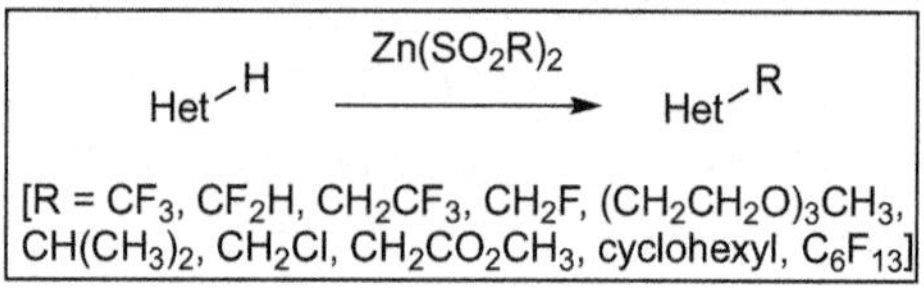
Zn(SO2R)2
Het–H
Het–R
[R = CF3, CF2H, CH2CF3, CH2F, (CH2CH2O)3CH3, CH(CH3)2, CH2Cl, CH2CO2CH3, cyclohexyl, C6F13]

A proposed mechanism for $CF_3$ radical generation is shown below.[5] Two regimes of differing reaction rates have been observed, and this tentative mechanism is still under study.[10]

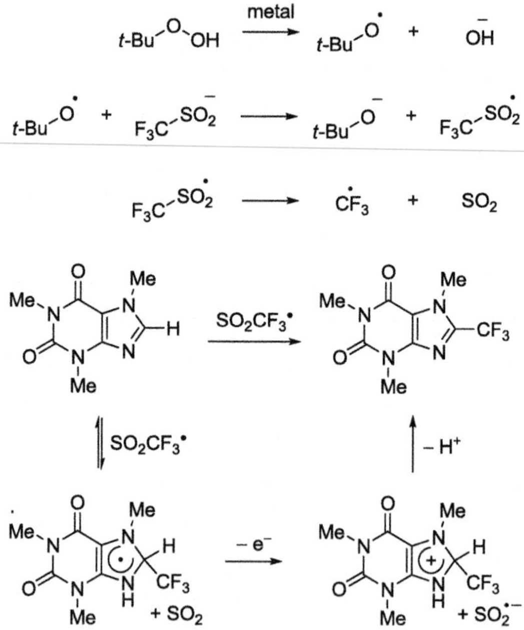
metal
t-Bu
OH
t-Bu
F3C
SO2
CF3
SO2
SO2CF3•
– H+
– e–
+ SO2

Example 1, Difluoromethylation of caffeine[6]

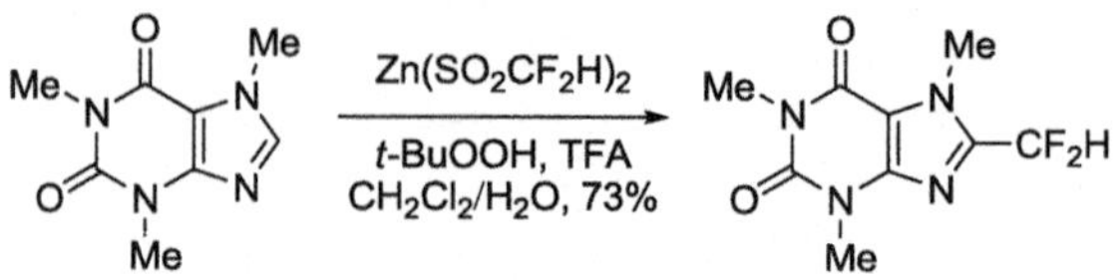
Zn(SO2CF2H)2
t-BuOOH, TFA
CH2Cl2/H2O, 73%
CF2H

J.J. Li, *Name Reactions: A Collection of Detailed Mechanisms and Synthetic Applications*, DOI 10.1007/978-3-319-03979-4_9, © Springer International Publishing Switzerland 2014

Example 2, sequential functionalization of dihydroquinine[7]

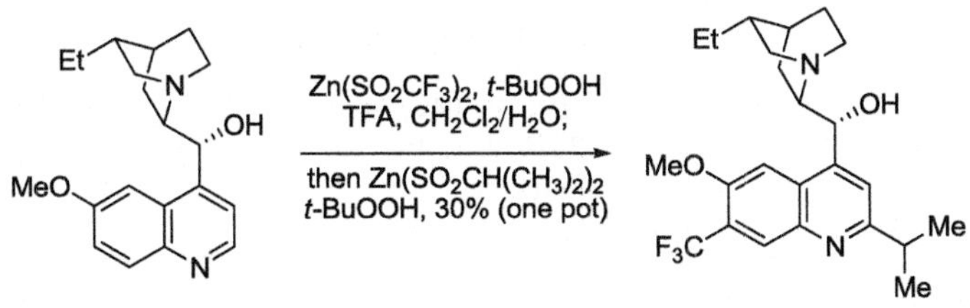

Cf. Example 3[1]

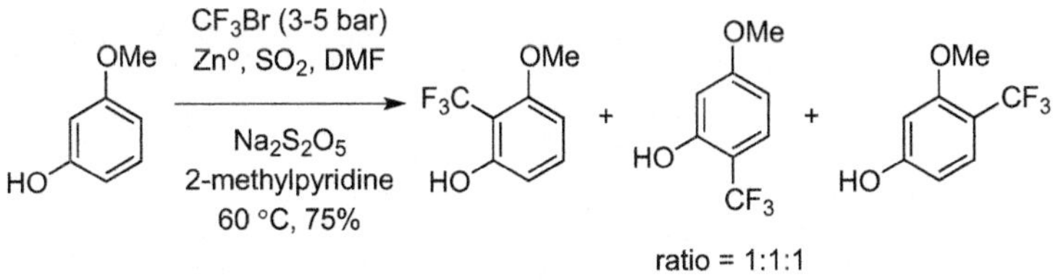

ratio = 1:1:1

Cf. Example 4, use of Langlois reagent (sodium trifluoromethanesulfinate)[2]

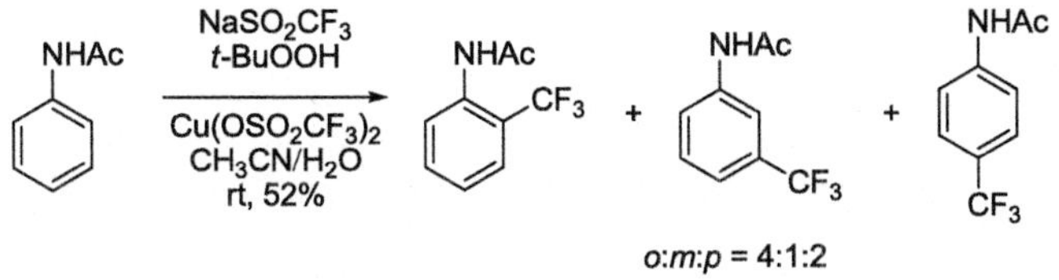

o:m:p = 4:1:2

Cf. Example 5[3]

ratio = 15:1

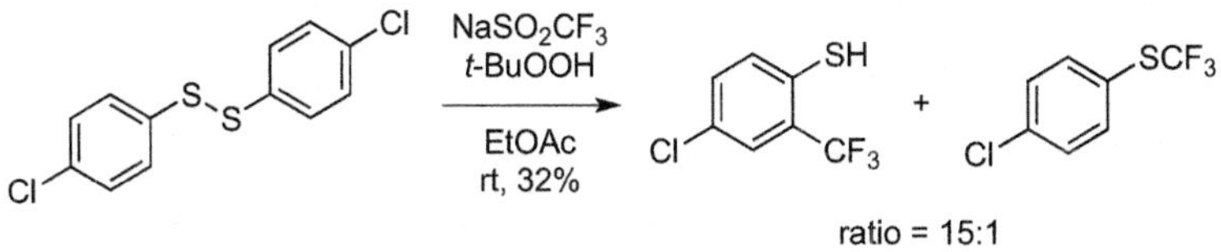

Cf. Example 6, Yamakawa's group[4]

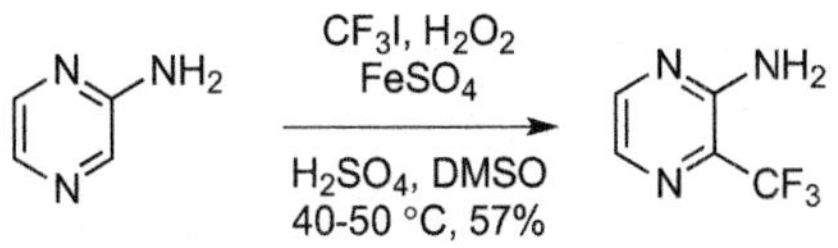

**References**

1.    Tordeux, M.; Langlois, B.; Wakselman, C. *J. Chem. Soc., Perkin Trans. 1* **1990**,

2293–2299.

2.   Langlois, B. R.; Laurent, E.; Roidot, N. *Tetrahedron Lett.* **1991**, *32*, 7525–7528.

3.   Clavel, J. L.; Langlois, B.; Laurent, E.; Roidot, N. *Phosphorus, Sulfur, and Silicon* **1991**, *58*, 463–466.

4.   Kino, T.; Nagase, Y.; Ohtsuka, Y.; Yamamoto, K.; Uraguchi, D.; Tokuhisa, K.; Yamakawa, T. *J. Fluorine Chem.* **2010**, *131*, 98–105.

5.   Ji, Y.; Brueckl, T.; Baxter, R. D.; Fujiwara, Y.; Seiple, I. B.; Su, S.; Blackmond, D. G.; Baran, P. S. *Proc. Natl. Acad. Sci. U. S. A.* **2011,** *108*, 14411–14415.   Phil S. Baran is currently a Professor in the Department of Chemistry at The Scripps Research Institute and Member of The Skaggs Institute for Chemical Biology. He was born in New Jersey in 1977, received his B.S. in chemistry from NYU in 1997, his Ph.D. at The Scripps Research Institute in 2001 (NSF fellow with K.C. Nicolaou), and from 2001–2003 he was an NIH postdoctoral fellow at Harvard in the laboratory of E. J. Corey. He has published over 120 scientific articles (H-Factor = 40) and has recently released a book in heterocyclic chemistry. His laboratory is focused on the invention of new reactions of broad utility and synthesizing complex natural products in a scalable, economic fashion.  He recently won the John D. and Catharine T. MacArthur award.

6.   Fujiwara, Y.; Dixon, J. A.; Rodriguez, R. A.; Baxter, R. D.; Dixon, D. D.; Collins, M. R.; Blackmond, D. G.; Baran, P. S. *J. Am. Chem. Soc.* **2012,** *134*, 1494–1497.

7.   Fujiwara, Y.; Dixon, J. A.; O'Hara, F.; Funder, E. D.; Dixon, D. D.; Rodriguez, R. A.; Baxter, R. D.; Herle, B.; Sach, N.; Collins, M. R.; Ishihara, Y.; Baran, P. S. *Nature* **2012,** *492*, 95–99.

8.   Zhou, Q.; Ruffoni, A.; Gianatassio, R.; Fujiwara, Y.; Sella, E.; Shabat, D.; Baran, P. S. *Angew. Chem. Int. Ed.* **2013,** *52,* 3949–3952

9.   O'Hara, F.; Baxter, R. D.; O'Brien, A. G.; Collins, M. R.; Dixon, J. A.; Fujiwara, Y.; Ishihara, Y.; Baran, P. S. *Nature Protocols* **2013,** *8*, 1042–1047.

10.  Baxter, R. D.; Blackmond, D. G. *Tetrahedron* **2013,** *69*, 5604–5608.

11.  O'Hara, F.; Blackmond, D. G.; Baran, P. S. *J. Am. Chem. Soc.* **2013,** *135*, 12122–12134.

# Barbier reaction

The Barbier reaction is an organic reaction between an alkyl halide and a carbonyl group as an electrophilic substrate in the presence of magnesium, aluminium, zinc, indium, tin or its salts. The reaction product is a primary, secondary or tertiary alcohol. *Cf.* Grignard reaction.

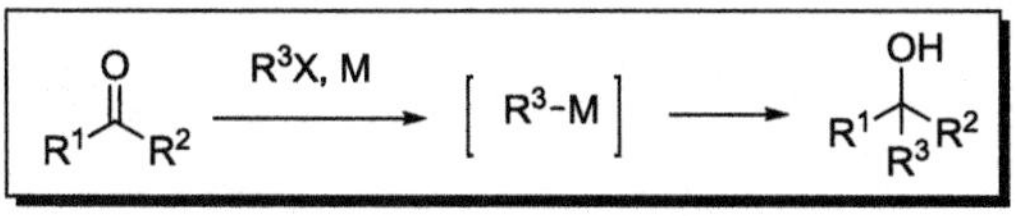

According to conventional wisdom,[3] the organometallic intermediate (M = Mg, Li, Sm, Zn, La, *etc.*) is generated *in situ*, which is intermediately trapped by the carbonyl compound. However, recent experimental and theoretical studies seem to suggest that the Barbier coupling reaction goes through a single electron transfer (SET) pathway.

Generation of the organometallic intermediate *in situ*:

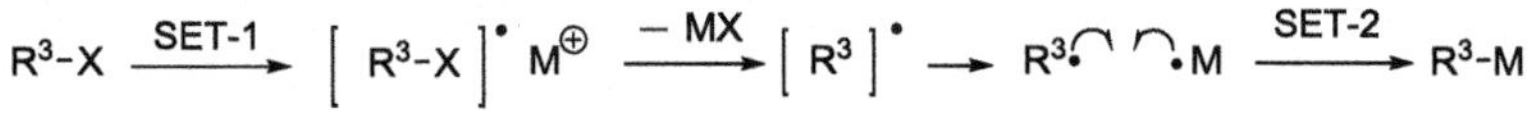

SET = single electron transfer

Ionic mechanism,

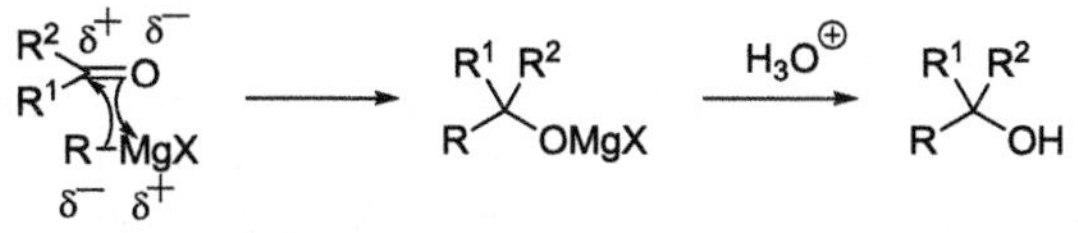

Single electron transfer (SET) mechanism:

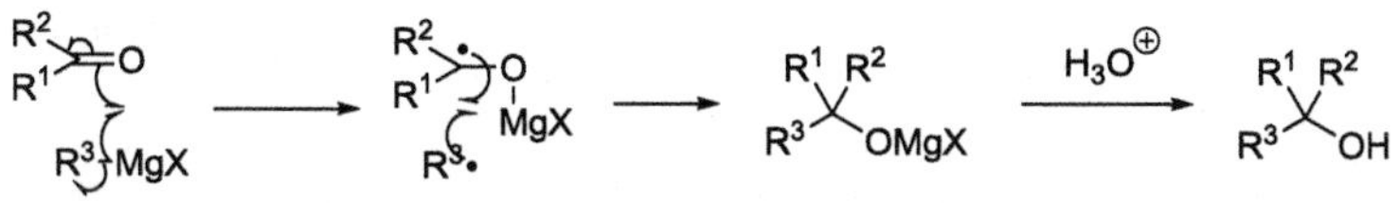

Example 1[6]

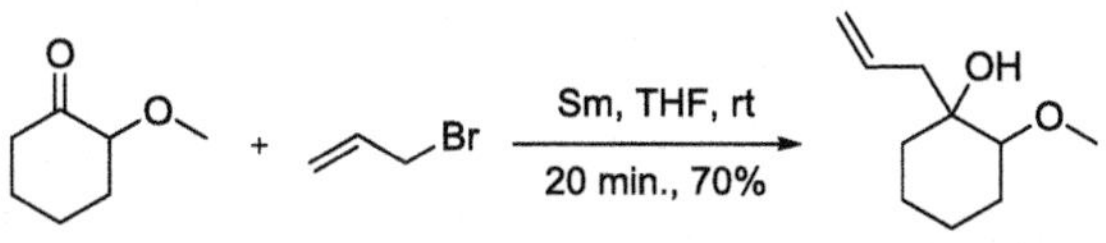

Example 2[9]

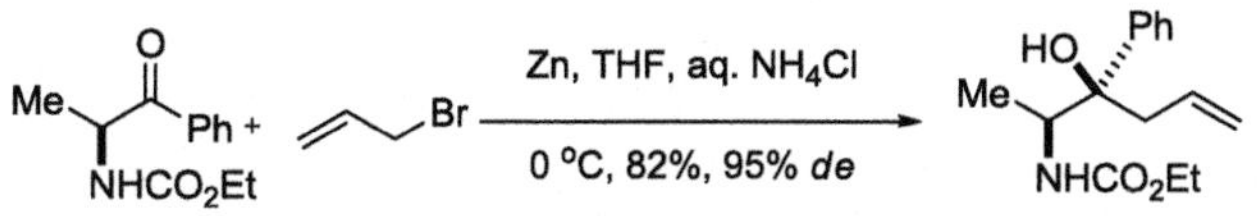

J.J. Li, *Name Reactions: A Collection of Detailed Mechanisms and Synthetic Applications*, DOI 10.1007/978-3-319-03979-4_10, © Springer International Publishing Switzerland 2014

Example 3[10]

Example 4[11]

Example 5, The following whole sequence of 5 steps can also be carried out in one-pot[12]

### References

1.  Barbier, P. *C. R. Hebd. Séances Acad. Sci.* **1899**, *128*, 110–111. Phillippe Barbier (1848–1922) was born in Luzy, Nièvre, France. He studied terpenoids using zinc and magnesium. Barbier suggested the use of magnesium to his student, Victor Grignard, who later discovered the Grignard reagent and won the Nobel Prize in 1912.

2.  Grignard, V. *C. R. Hebd. Séances Acad. Sci.* **1900**, *130*, 1322–1324.

3.  Moyano, A.; Pericás, M. A.; Riera, A.; Luche, J.-L. *Tetrahedron Lett.* **1990**, *31*, 7619–7622. (Theoretical study).

4.  Alonso, F.; Yus, M. *Rec. Res. Dev. Org. Chem.* **1997**, *1*, 397–436. (Review).

5.  Russo, D. A. *Chem. Ind.* **1996**, *64*, 405–409. (Review).

6.  Basu, M. K.; Banik, B. *Tetrahedron Lett.* **2001**, *42*, 187–189.

7.  Sinha, P.; Roy, S. *Chem. Commun.* **2001**, 1798–1799.

8.  Lombardo, M.; Gianotti, K.; Licciulli, S.; Trombini, C. *Tetrahedron* **2004**, *60*, 11725–11732.

9.  Resende, G. O.; Aguiar, L. C. S.; Antunes, O. A. C. *Synlett* **2005**, 119–120.

10. Erdik, E.; Kocoglu, M. *Tetrahedron Lett.* **2007**, *48*, 4211–4214.

11. Takeuchi, T.; Matsuhashi, M.; Nakata, T. *Tetrahedron Lett.* **2008**, *49*, 6462–6465.

12. Hirayama, L. C.; Haddad, T. D.; Oliver, A. G.; Singaram, B. *J. Org. Chem.* **2012**, *77*, 4342–4353.

13. Aslam, N. A.; Babu, S. A.; Sudha, A. J.; Yasuda, M.; Baba, A. *Tetrahedron* **2013**, *69*, 6598–6611.

# Bargellini reaction

Synthesis of hindered morpholinones or piperazinones from ketones (such as acetone) and 2-amino-2-methyl-1-propanol or 1,2-diaminopropanes.

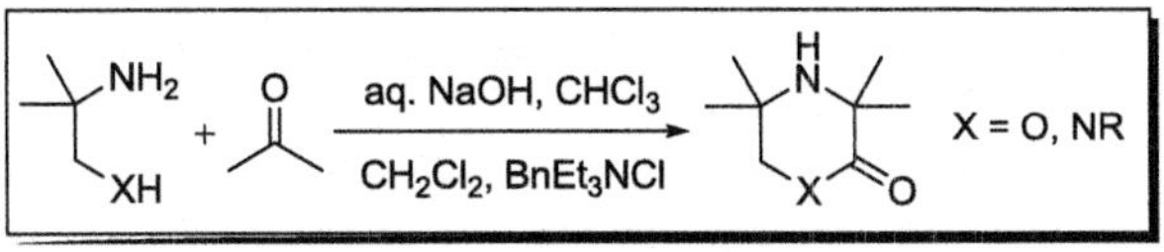

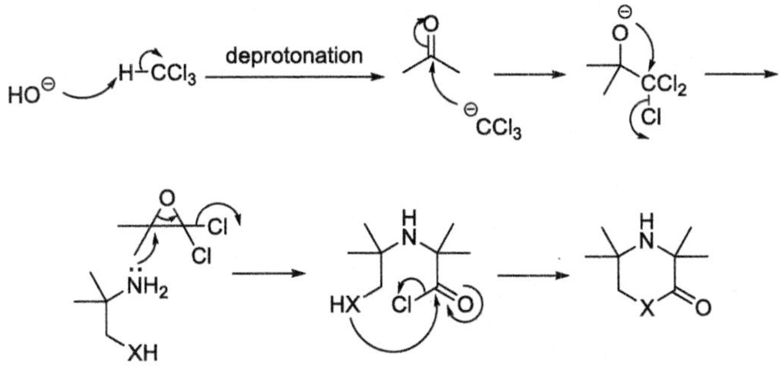

Example 1[2]

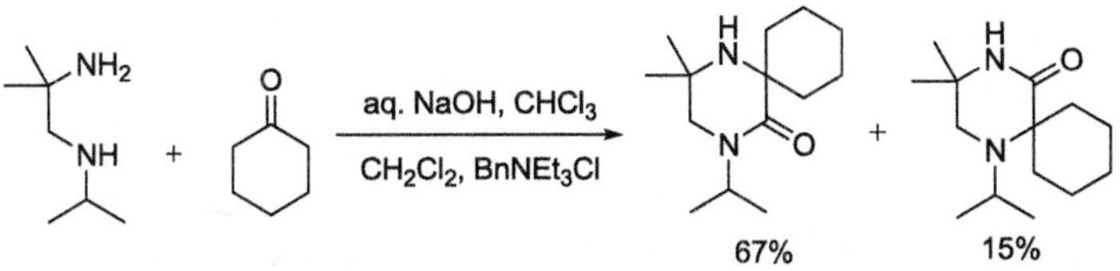

67%        15%

Example 2[4]

1. aq. NaOH, CHCl₃, acetone

2. CSA, toluene, 66%

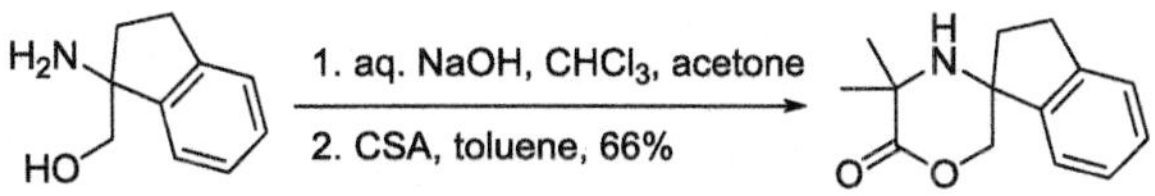

# References

1. Bargellini, G. *Gazz. Chim. Ital.* **1906**, *36*, 329–337.
2. Lai, J. T. *J. Org. Chem.* **1980**, *45*, 754.
3. Lai, J. T. *Synthesis* **1981**, 754; **1984**, 122; **1984**, 124.
4. Rychnovsky, S. D.; Beauchamp, T.; Vaidyanathan, R.; Kwan, T. *J. Org. Chem.* **1998**, *63*, 6363–6374.
5. Butcher, K. J.; Hurst, J. *Tetrahedron Lett.* **2009**, *50*, 2497–2500.
6. Rohman, M. R.; Myrboh, B. *Tetrahedron Lett.* **2010**, *50*, 4772–4775.
7. Snowden, T. S. *ARKIVOC* **2012**, *(ii)*, 24–40. (Review).

J.J. Li, *Name Reactions: A Collection of Detailed Mechanisms and Synthetic Applications*, DOI 10.1007/978-3-319-03979-4_11, © Springer International Publishing Switzerland 2014

## Bartoli indole synthesis

7-Substituted indoles from the reaction of *ortho*-substituted nitroarenes and vinyl Grignard reagents.

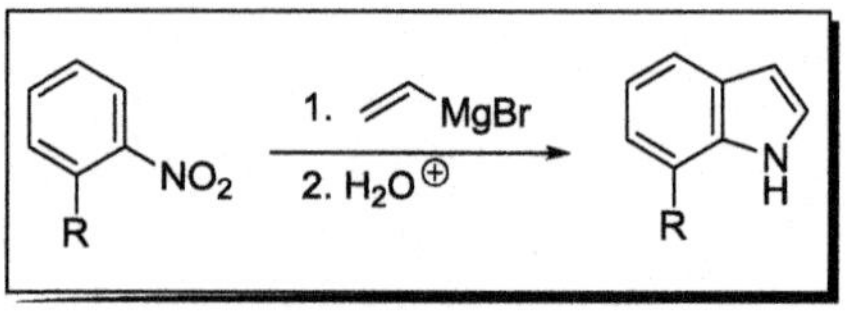

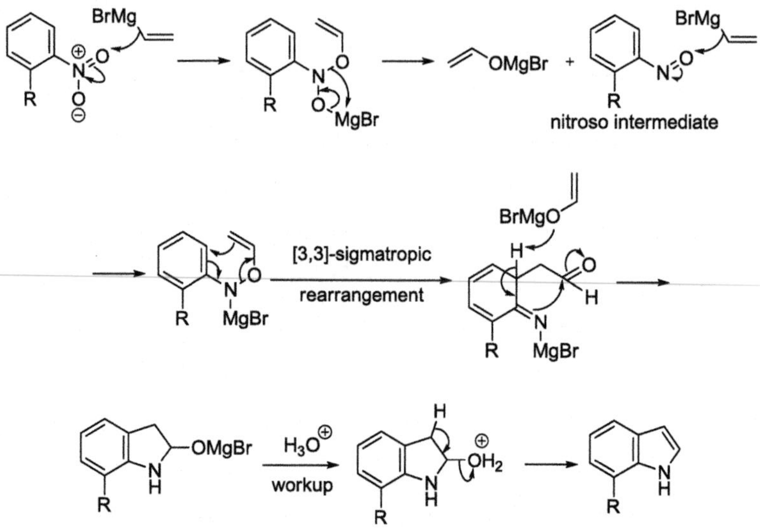

Example 1[3]

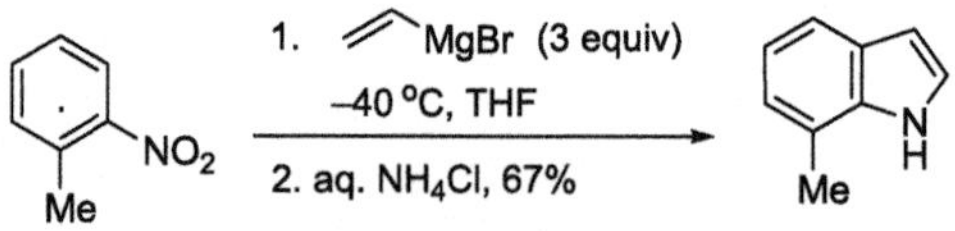

Example 2[6]

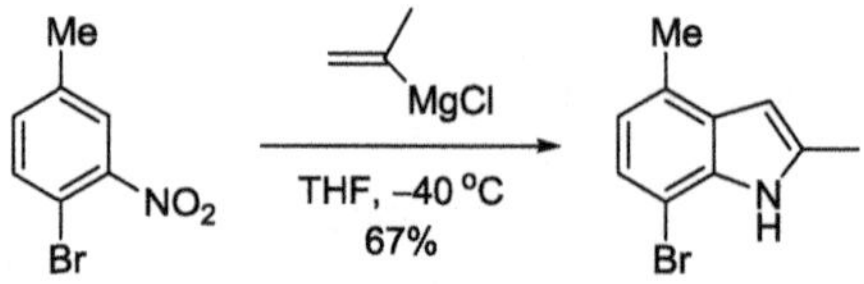

Example 3[10]

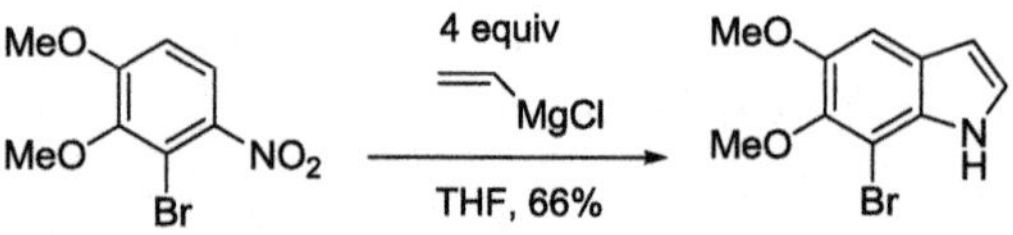

J.J. Li, *Name Reactions: A Collection of Detailed Mechanisms and Synthetic Applications*,
DOI 10.1007/978-3-319-03979-4_12, © Springer International Publishing Switzerland 2014

Example 4[11]

Example 5[12]

## References

1. Bartoli, G.; Leardini, R.; Medici, A.; Rosini, G. *J. Chem. Soc., Perkin Trans. 1* **1978,** 692–696. Giuseppe Bartoli is a professor at the Università di Bologna, Italy.
2. Bartoli, G.; Bosco, M.; Dalpozzo, R.; Todesco, P. E. *J. Chem. Soc., Chem. Commun.* **1988,** 807–805.
3. Bartoli, G.; Palmieri, G.; Bosco, M.; Dalpozzo, R. *Tetrahedron Lett.* **1989,** *30,* 2129–2132.
4. Bosco, M.; Dalpozzo, R.; Bartoli, G.; Palmieri, G.; Petrini, M. *J. Chem. Soc., Perkin Trans. 2* **1991,** 657–663. Mechanistic studies.
5. Bartoli, G.; Bosco, M.; Dalpozzo, R.; Palmieri, G.; Marcantoni, E. *J. Chem. Soc., Perkin Trans. 1* **1991,** 2757–2761.
6. Dobbs, A. *J. Org. Chem.* **2001,** *66,* 638–641.
7. Garg, N. K.; Sarpong, R.; Stoltz, B. M. *J. Am. Chem. Soc.* **2002,** *124,* 13179–13184.
8. Li, J.; Cook, J. M. *Bartoli Indole Synthesis.* In *Name Reactions in Heterocyclic Chemistry*; Li, J. J., Corey, E. J. Eds.; Wiley: Hoboken, NJ, **2005,** pp 100–103. (Review).
9. Dalpozzo, R.; Bartoli, G. *Current Org. Chem.* **2005,** *9,* 163–178. (Review).
10. Huleatt, P. B.; Choo, S. S.; Chua, S.; Chai, C. L. L. *Tetrahedron Lett.* **2008,** *49,* 5309–5311.
11. Buszek, K. R.; Brown, N.; Luo, D. *Org. Lett.* **2009,** *11,* 201–204.
12. Grant, S. W.; Gallagher, T. F.; Bobko, M. A.; Duquenne, C.; Axten, J. M. *Tetrahedron Lett.* **2011,** *52,* 3376–3378.
13. Chandrasoma, N.; Brown, N.; Brassfield, A.; Nerurkar, A.; Suarez, S.; Buszek, K. R. *Tetrahedron Lett.* **2013,** *54,* 913–917.

## Barton radical decarboxylation

Radical decarboxylation of the carboxylic acids.

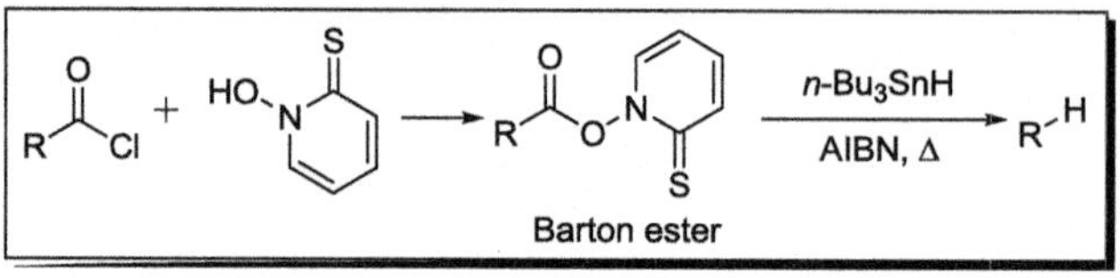

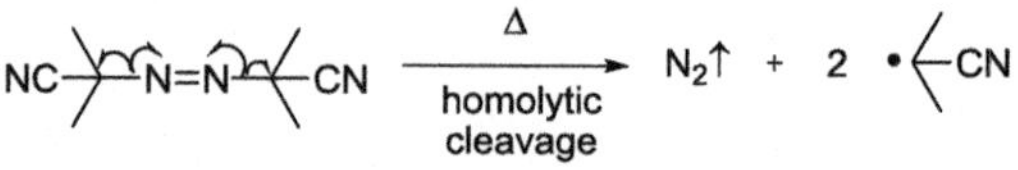

Barton ester

AIBN = 2,2′-azobisisobutyronitrile

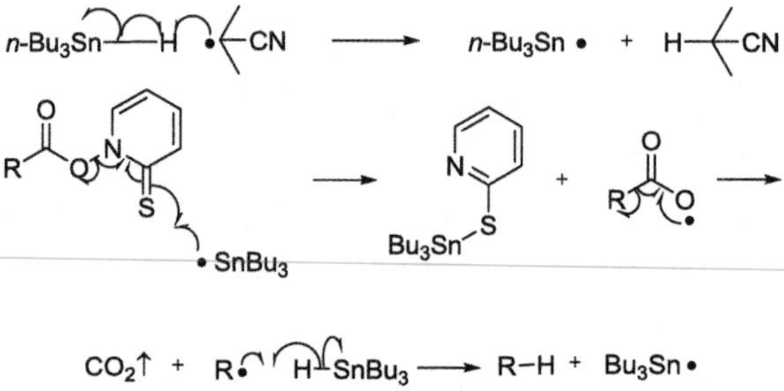

**Example 1**, Tin hydride not used and elimination occurs via thioether intermediate[3]

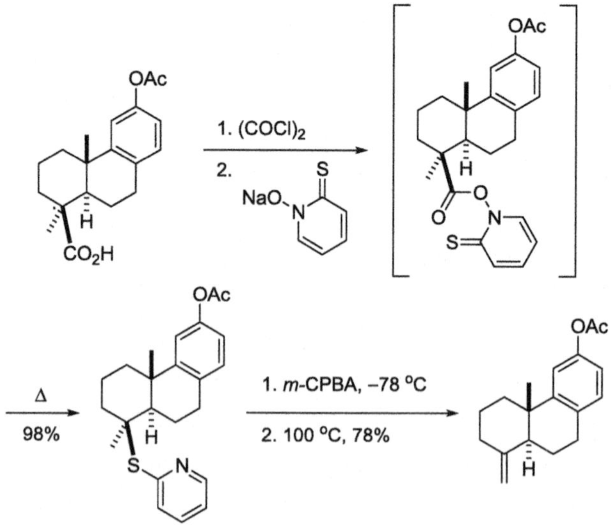

1. (COCl)$_2$

2. NaO–

$\Delta$
98%

1. m-CPBA, −78 °C
2. 100 °C, 78%

**Example 2**[6]

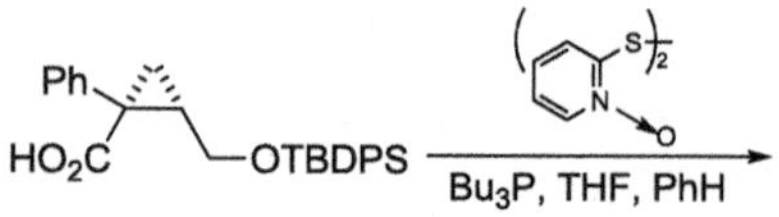

Bu$_3$P, THF, PhH

J.J. Li, *Name Reactions: A Collection of Detailed Mechanisms and Synthetic Applications*, DOI 10.1007/978-3-319-03979-4_13, © Springer International Publishing Switzerland 2014

Example 3[9]

Example 4[11]

## References

1.  Barton, D. H. R.; Crich, D.; Motherwell, W. B. *J. Chem. Soc., Chem. Commun.* **1983,** 939–941. Sir Derek Barton (United Kingdom, 1918–1998) studied under Ian Heilbron at Imperial College in his youth. He taught in England, France and the US. Barton won the Nobel Prize in Chemistry in 1969 for development of the concept of conformation. He passed away in his office at the University of Texas A&M in 1998.

2.  Barton, D. H. R.; Zard, S. Z. *Pure Appl. Chem.* **1986,** *58,* 675–684. (Review).

3.  Cochane, E. J.; Lazer, S. W.; Pinhey, J. T.; Whitby, J. D. *Tetrahedron Lett.* **1989,** *30,* 7111–7114.

4.  Barton, D. H. R. *Aldrichimica Acta* **1990,** *23,* 3. (Review).

5.  Crich, D.; Hwang, J.-T.; Yuan, H. *J. Org. Chem.* **1996,** *61,* 6189–6198.

6.  Yamaguchi, K.; Kazuta, Y.; Abe, H.; Matsuda, A.; Shuto, S. *J. Org. Chem.* **2003,** *68,* 9255–9262.

7.  Zard, S. Z. *Radical Reactions in Organic Synthesis* Oxford University Press: Oxford, UK, **2003.** (Book).

8.  Carry, J.-C.; Evers, M.; Barriere, J.-C.; Bashiardes, G.; Bensoussan, C.; Gueguen, J.-C.; Dereu, N.; Filoche, B.; Sable, S.; Vuilhorgne, M.; Mignani, S. *Synlett* **2004,** 316–320.

9.  Brault, L.; Denance, M.; Banaszak, E.; El Maadidi, S.; Battaglia, E.; Bagrel, D.; Samadi, M. *Eur. J. Med. Chem.* **2007,** *42,* 243–247.

10.  Guthrie, D. B.; Curran, D. P. *Org. Lett.* **2009,** *11,* 249–251.

11.  He, Z.; Trinchera, P.; Adachi, S.; St. Denis, J. D.; Yudin, A. K. *Angew. Chem. Int. Ed.* **2012,** *51,* 11092–11096.

# Barton–McCombie deoxygenation

Deoxygenation of alcohols by means of radical scission of their corresponding thiocarbonyl derivatives.

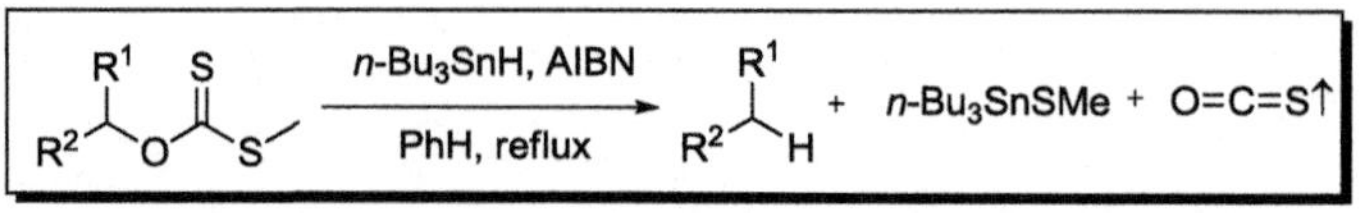

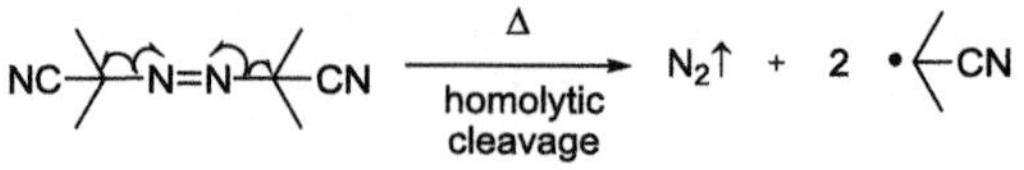

AIBN = 2,2'-azobisisobutyronitrile

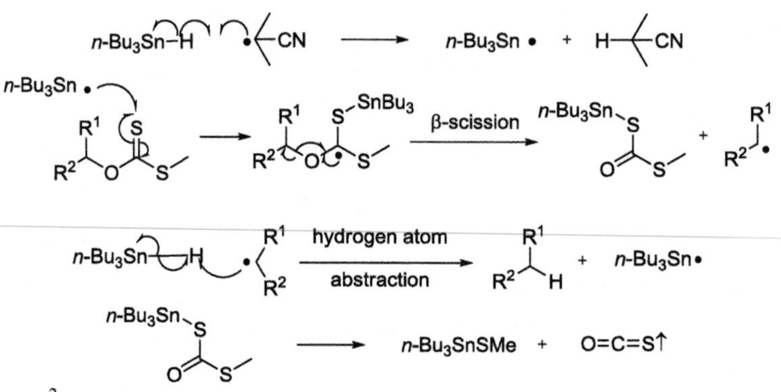

Example 1[2]

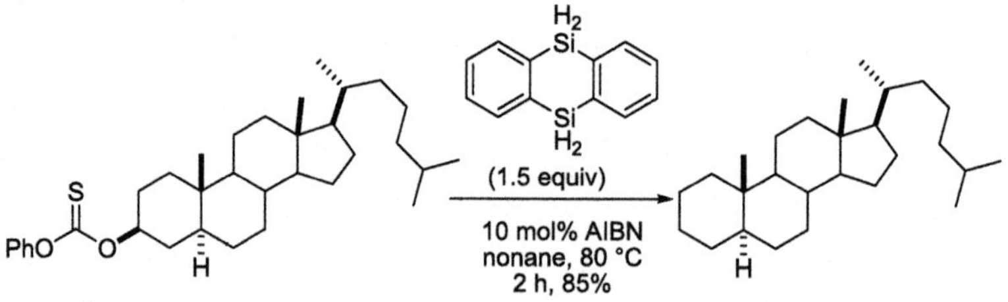

Example 2[6]

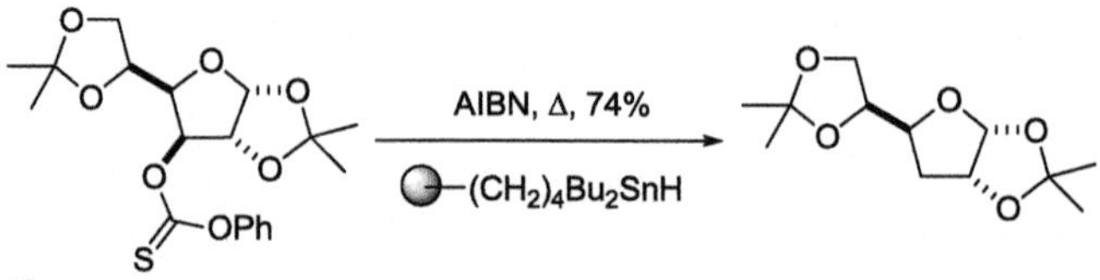

Example 3[10]

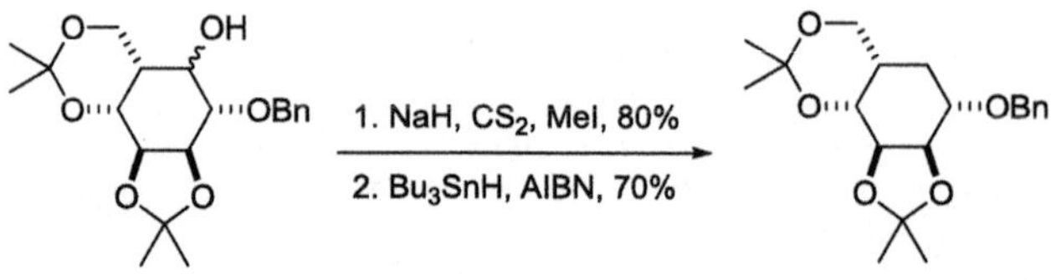

J.J. Li, *Name Reactions: A Collection of Detailed Mechanisms and Synthetic Applications*,
DOI 10.1007/978-3-319-03979-4_14, © Springer International Publishing Switzerland 2014

Example 4[11]

Ts-N / OH
MeO₂C''' / H
Boc

NaH, CS₂
MeI, THF
rt

Ts-N, O-C(=S)-S-Me
MeO₂C'''
Boc

Bu₃SnH, AIBN
toluene, reflux
72%, 2 steps

Ts-N / H
MeO₂C'''
Boc

Example 5[13]

1. NaH, imidazole
2. CS₂, MeI
12%, 2 steps

Bu₃SnH, AIBN
toluene, 43%

## References

1. Barton, D. H. R.; McCombie, S. W. *J. Chem. Soc., Perkin Trans. 1* **1975,** 1574–1585. Stuart McCombie, a Barton student, worked at Schering–Plough for many years, but is now retired after the company was bought by Merck.
2. Gimisis, T.; Ballestri, M.; Ferreri, C.; Chatgilialoglu, C.; Boukherroub, R.; Manuel, G. *Tetrahedron Lett.* **1995,** *36,* 3897–3900.
3. Zard, S. Z. *Angew. Chem. Int. Ed.* **1997,** *36,* 673–685.
4. Lopez, R. M.; Hays, D. S.; Fu, G. C. *J. Am. Chem. Soc.* **1997,** *119,* 6949–6950.
5. Hansen, H. I.; Kehler, J. *Synthesis* **1999,** 1925–1930.
6. Boussaguet, P.; Delmond, B.; Dumartin, G.; Pereyre, M. *Tetrahedron Lett.* **2000,** *41,* 3377–3380.
7. Cai, Y.; Roberts, B. P. *Tetrahedron Lett.* **2001,** *42,* 763–766.
8. Clive, D. L. J.; Wang, J. *J. Org. Chem.* **2002,** *67,* 1192–1198.
9. Rhee, J. U.; Bliss, B. I.; RajanBabu, T. V. *J. Am. Chem. Soc.* **2003,** *125,* 1492–1493.
10. Gómez, A. M.; Moreno, E.; Valverde, S.; López, J. C. *Eur. J. Org. Chem.* **2004,** 1830–1840.
11. Deng, H.; Yang, X.; Tong, Z.; Li, Z.; Zhai, H. *Org. Lett.* **2008,** *10,* 1791–1793.
12. Mancuso, J. *Barton–McCombie deoxygenation.* In *Name Reactions for Homologations-Part I*; Li, J. J., Ed.; Wiley: Hoboken, NJ, **2009,** pp 614–632. (Review).
13. McCombie, S. W.; Motherwell, W. B.; Tozer, M. J. *The Barton–McCombie Reaction,* In *Org. React.* **2012,** *77,* pp 161–591. (Review).
14. Jastrzebska, I.; Gorecki, M.; Frelek, J.; Santillan, R.; Siergiejczyk, L.; Morzycki, J. W. *J. Org. Chem.* **2012,** *77,* 11257–11269.

## Barton nitrite photolysis

Photolysis of a nitrite ester to a γ-oximino alcohol.

Example 1[2]

J.J. Li, *Name Reactions: A Collection of Detailed Mechanisms and Synthetic Applications*,
DOI 10.1007/978-3-319-03979-4_15, © Springer International Publishing Switzerland 2014

Example 2[6]

hν

PhH, 6 °C
46%

Example 3[7]

1. NOCl, CH₂Cl₂, –20 °C, 100%

2. Irradiation at 350 nM
5 h, PhH, 0 °C, 50%

## References

1. (a) Barton, D. H. R.; Beaton, J. M.; Geller, L. E.; Pechet, M. M. *J. Am. Chem. Soc.* **1960**, *82*, 2640–2641. In 1960, Derek Barton took a "vacation" in Cambridge, Massachusetts; he worked in a small research institute called the Research Institute for Medicine and Chemistry. In order to make the adrenocortical hormone aldosterol, Barton invented the Barton nitrite photolysis by simply writing down on a piece of paper what he thought would be an ideal process. His skilled collaborator, Dr. John Beaton, was able to reduce it to practice. They were able to make 40 to 50 g of aldosterol at a time when the total world supply was only about 10 mg. Barton considered it his most satisfying piece of work. (b) Barton, D. H. R.; Beaton, J. M. *J. Am. Chem. Soc.* **1960**, *82*, 2641–2641. (c) Barton, D. H. R.; Beaton, J. M. *J. Am. Chem. Soc.* **1961**, *83*, 4083–4089. (d) Barton, D. H. R.; Lier, E. F.; McGhie, J. M. *J. Chem. Soc., (C)* **1968**, 1031–1040.

2. Nickon, A; Iwadare, T.; McGuire, F. J.; Mahajan, J. R; Narang, S. A.; Umezawa, B. *J. Am. Chem. Soc.* **1970**, *92*, 1688–1696.

3. Barton, D. H. R.; Hesse, R. H.; Pechet, M. M.; Smith, L. C. *J. Chem. Soc., Perkin Trans. 1* **1979**, 1159–1165.

4. Barton, D. H. R. *Aldrichimica Acta* **1990**, *23*, 3–10. (Review).

5. Majetich, G.; Wheless, K. *Tetrahedron* **1995**, *51*, 7095–7129. (Review).

6. Sicinski, R. R.; Perlman, K. L.; Prahl, J.; Smith, C.; DeLuca, H. F. *J. Med. Chem.* **1996**, *22*, 4497–4506.

7. Anikin, A.; Maslov, M.; Sieler, J.; Blaurock, S.; Baldamus, J.; Hennig, L.; Findeisen, M.; Reinhardt, G.; Oehme, R.; Welzel, P. *Tetrahedron* **2003**, *59*, 5295–5305.

8. Suginome, H. *CRC Handbook of Organic Photochemistry and Photobiology* 2nd edn.; **2004**, 102/1–102/16. (Review).

9. Hagan, T. J. *Barton nitrite photolysis*. In *Name Reactions for Homologations-Part I*; Li, J. J., Ed.; Wiley: Hoboken, NJ, **2009**, pp 633–647. (Review).

# Barton–Zard reaction

## Barton–Zard reaction

Base-induced reaction of nitroalkenes with alkyl α-isocyanoacetates to afford pyrroles.

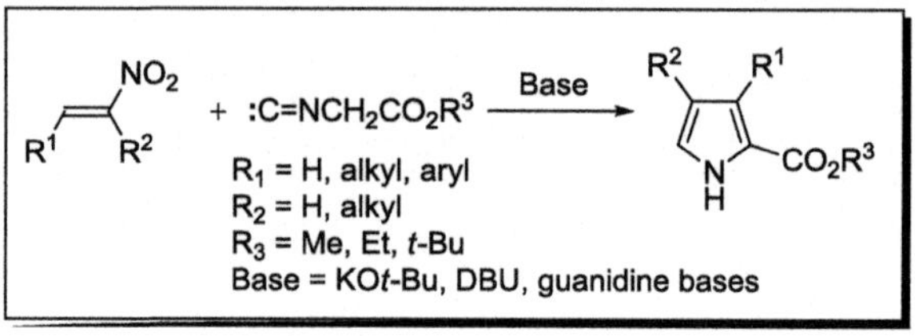

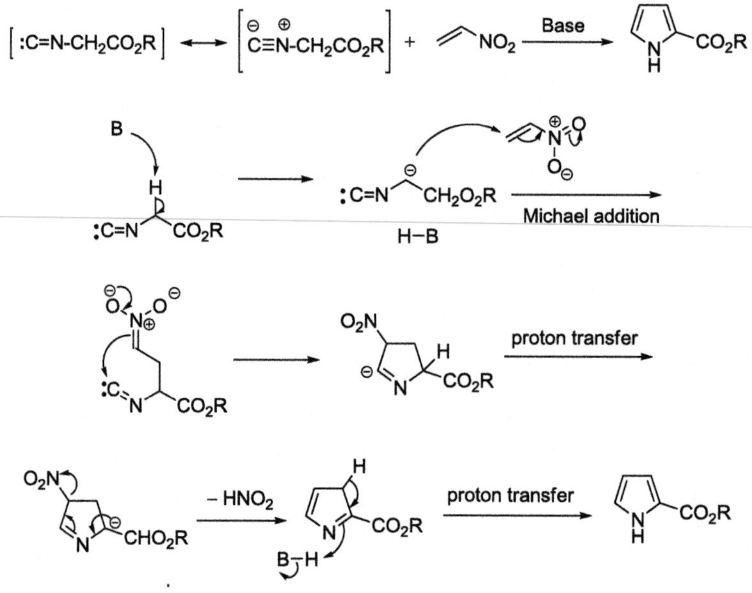

Example 1[5]

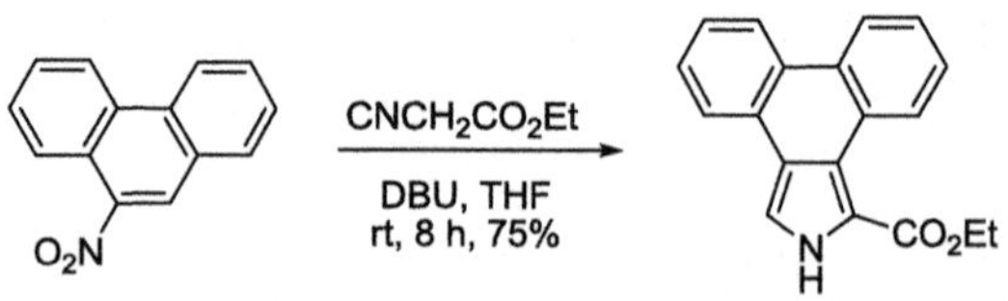

Example 2[7]

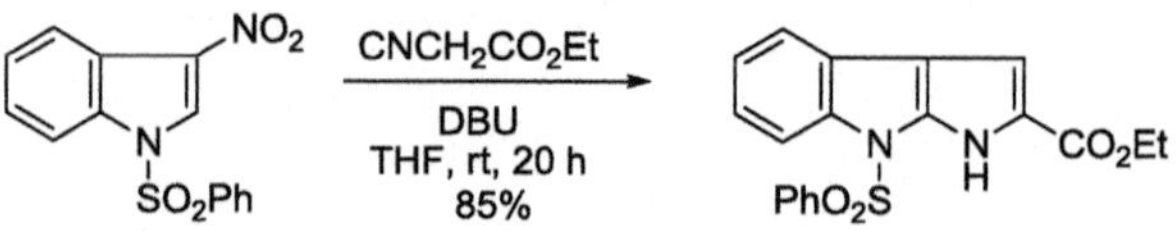

J.J. Li, *Name Reactions: A Collection of Detailed Mechanisms and Synthetic Applications*, DOI 10.1007/978-3-319-03979-4_16, © Springer International Publishing Switzerland 2014

Example 3, A Barton–Zard reaction on a ferrocene ring[12]

## References

1.  Barton, D. H. R.; Zard, S. Z. *J. Chem. Soc., Chem. Commun.* **1985**, 1098–1100. Samir Z. Zard, a Barton student, emigrated from Lebanon to the UK in 1975. He is a professor at CNRS and École Polytechnique in France.
2.  van Leusen, A. M.; Siderius, H.; Hoogenboom, B. E.; van Leusen, D. *Tetrahedron Lett.* **1972**, 5337–5340.
3.  Barton, D. H. R.; Kervagoret, J.; Zard, S. Z. *Tetrahedron* **1990**, *46*, 7587-5340.
4.  Sessler, J. L.; Mozaffari, A.; Johnson, M. R. *Org. Synth.* **1991**, *70*, 68-78.
5.  Ono, N.; Hironaga, H.; Ono, K.; Kaneko, S.; Murashima, T.; Ueda, T.; Tsukamura, C.; Ogawa, T. *J. Chem. Soc., Perkin Trans. 1* **1996**, 417–423.
6.  Murashima, T.; Fujita, K.; Ono, K.; Ogawa, T.; Uno, H.; Ono, N. *J. Chem. Soc., Perkin Trans. 1* **1996**, 1403–1407.
7.  Pelkey, E. T.; Chang, L.; Gribble, G. W. *Chem. Commun.* **1996**, 1909–1910.
8.  Fumoto, Y.; Uno, H.; Tanaka, K.; Tanaka, M.; Murashima, T.; Ono, N. *Synthesis* **2001**, 399–402.
9.  Lash, T. D.; Werner, T. M.; Thompson, M. L.; Manley, J. M. *J. Org. Chem.* **2001**, *66*, 3152–3159.
10. Ferreira, V. F.; de Souza, M. C. B. V.; Cunha, A. C.; Pereira, L. O. R.; Ferreira, M. L. G. *Org. Prep. Proc. Int.* **2001**, *33*, 411–454. (Review).
11. Gribble, G. W. *Barton–Zard Reaction* in *Name Reactions in Heterocyclic Chemistry*, Li, J. J., Ed.; Wiley: Hoboken, NJ, **2005**, 70–78. (Review).
12. Guillon, J.; Mouray, E.; Moreau, S.; Mullié, C.; Forfar, I.; Desplat, V.; Belisle-Fabre, S.; Pinaud, N.; Ravanello, F.; Le-Naour, A.; Léger, J. M.; Gosmann, G.; Jarry, C.; Déléris, G.; Sonnet, P.; Grellier, P. *E. J. Med. Chem.* **2011**, *46*, 2310–2326.

# Batcho–Leimgruber indole synthesis

Condensation of o-nitrotoluene derivatives with formamide acetals, followed by reduction of the trans-β-dimethylamino-2-nitrostyrene to furnish indole derivatives.

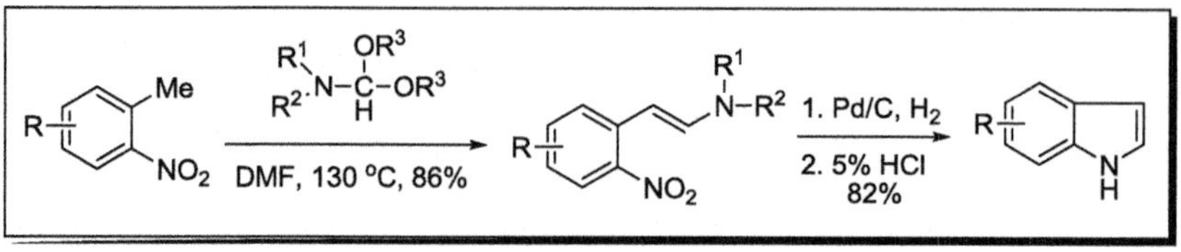

Example 1[4]

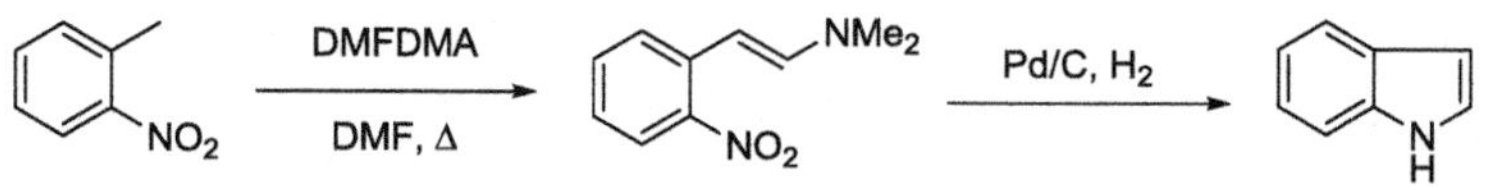

DMFDMA = N,N-dimethylformamide dimethyl acetal, $Me_2NCH(OMe)_2$

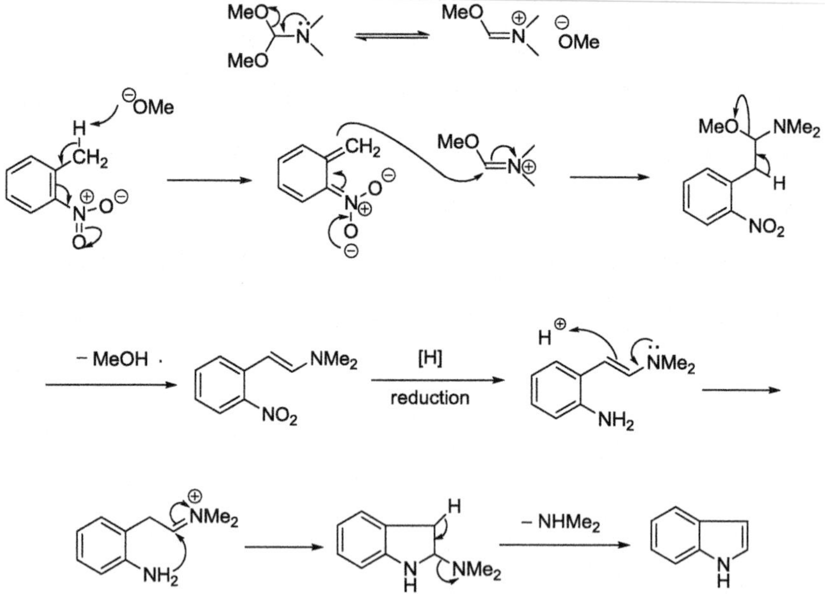

Example 2[4]

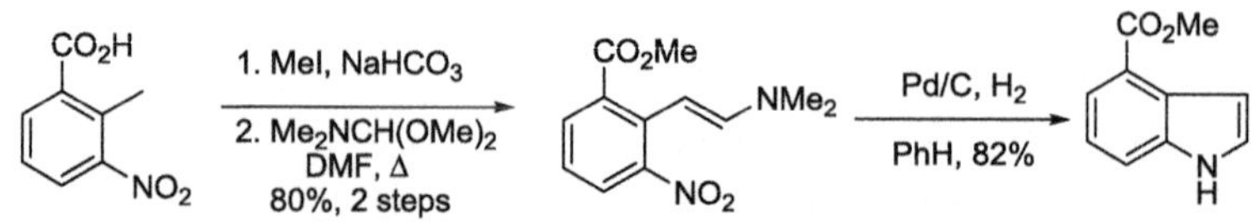

J.J. Li, Name Reactions: A Collection of Detailed Mechanisms and Synthetic Applications,
DOI 10.1007/978-3-319-03979-4_17, © Springer International Publishing Switzerland 2014

Example 3[5]

Example 4[10]

Example 5[12]

## References

1. Leimgruber, W.; Batcho, A. D. *Third International Congress of Heterocyclic Chemistry*: Japan, **1971**. Both Andrew D. Batcho and Willy Leimgruber were chemists at Hoffmann-La Roche in Nutley, NJ, a site that was closed in 2012.
2. Leimgruber, W.; Batcho, A. D. USP 3732245 (1973).
3. Sundberg, R. J. *The Chemistry of Indoles*; Academic Press: New York & London, **1970**. (Review).
4. Kozikowski, A. P.; Ishida, H.; Chen, Y.-Y. *J. Org. Chem.* **1980**, *45*, 3350–3352.
5. Batcho, A. D.; Leimgruber, W. *Org. Synth.* **1985**, *63*, 214–225.
6. Clark, R. D.; Repke, D. B. *Heterocycles* **1984**, *22*, 195–221. (Review).
7. Moyer, M. P.; Shiurba, J. F.; Rapoport, H. *J. Org. Chem.* **1986**, *51*, 5106–5110.
8. Siu, J.; Baxendale, I. R.; Ley, S. V. *Org. Biomol. Chem.* **2004**, *2*, 160–167.
9. Li, J.; Cook, J. M. *Batcho–Leimgruber Indole Synthesis*. In *Name Reactions in Heterocyclic Chemistry*; Li, J. J., Ed.; Wiley: Hoboken, NJ, **2005**, pp 104–109. (Review).
10. Braun, H. A.; Zall, A.; Brockhaus, M.; Schütz, M.; Meusinger, R.; Schmidt, B. *Tetrahedron Lett.* **2007**, *48*, 7990–7993.
11. Leze, M.-P.; Palusczak, A.; Hartmann, R. W.; Le Borgne, M. *Bioorg. Med. Chem. Lett.* **2008**, *18*, 4713–4715.
12. Gillmore, A. T.; Badland, M.; Crook, C. L.; Castro, N. M.; Critcher, D. J.; Fussell, S. J.; Jones, K. J.; Jones, M. C.; Kougoulos, E.; Mathew, J. S.; et al. *Org. Process Res. Dev.* **2012**, *16*, 1897–1904.

# Baylis–Hillman reaction

Also known as the Morita–Baylis–Hillman reaction. It is a carbon–carbon bond-forming transformation of an electron-poor alkene with a carbon electrophile. Electron-poor alkenes include acrylic esters, acrylonitriles, vinyl ketones, vinyl sulfones, and acroleins. On the other hand, carbon electrophiles may be aldehydes, α-alkoxycarbonyl ketones, aldimines, and Michael acceptors.
General scheme:

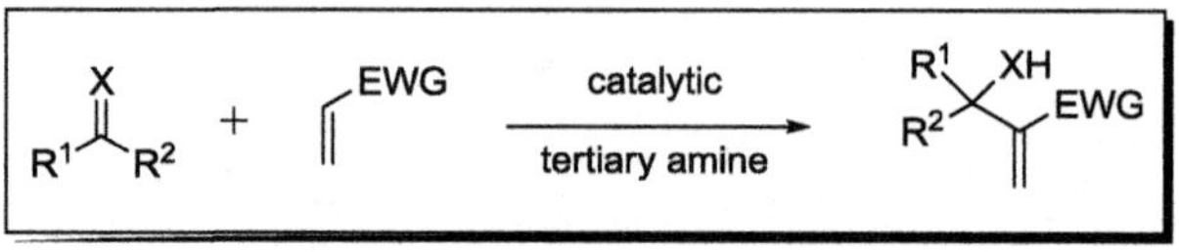
X
R1 R2
EWG
catalytic
tertiary amine
R1 XH
R2 EWG

X = O, $NR_2$, EWG = $CO_2R$, COR, CHO, CN, $SO_2R$, $SO_3R$, $PO(OEt)_2$, $CONR_2$, $CH_2{=}CHCO_2Me$

Catalytic tertiary amines:

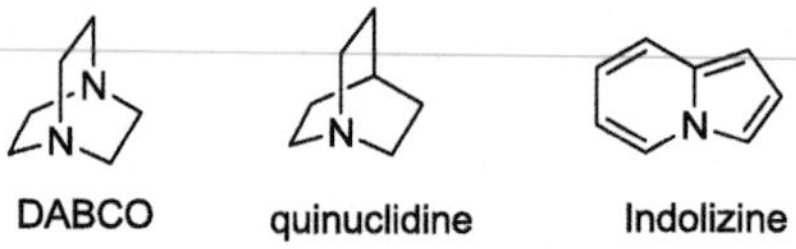
DABCO
quinuclidine
Indolizine

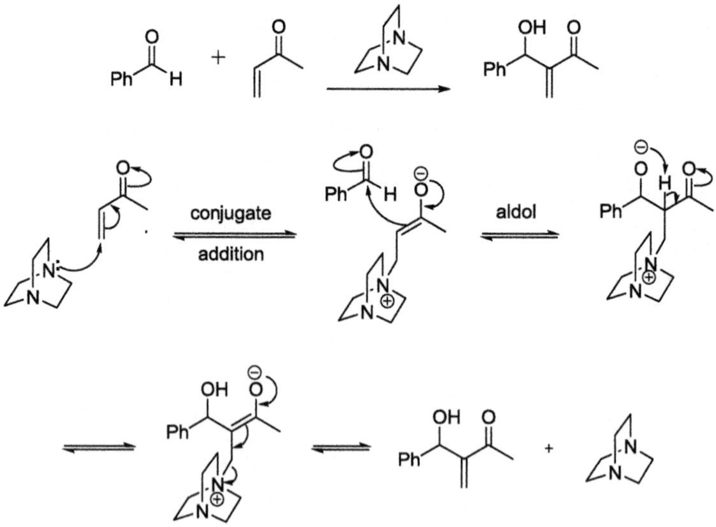
Ph H
OH O
Ph
conjugate
addition
aldol

J.J. Li, *Name Reactions: A Collection of Detailed Mechanisms and Synthetic Applications*,
DOI 10.1007/978-3-319-03979-4_18, © Springer International Publishing Switzerland 2014

E2 (bimolecular elimination) mechanism is also operative here:

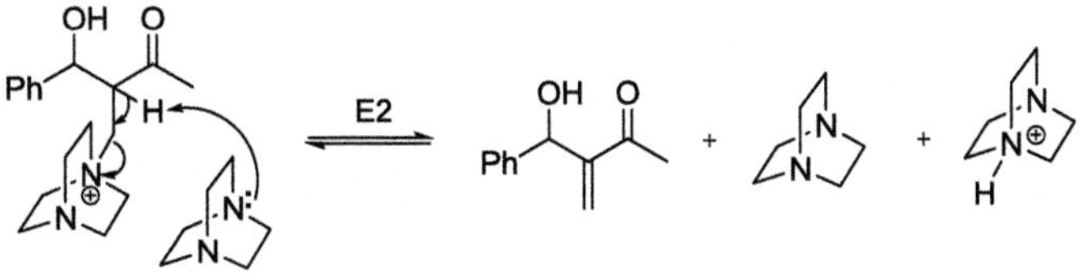

Example 1, Intramolecular Baylis–Hillman reaction[6]

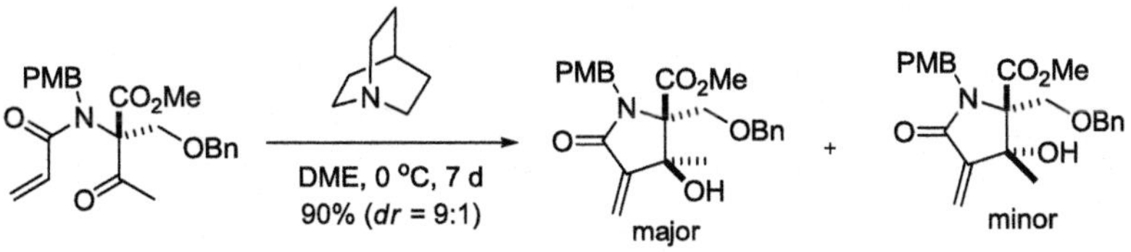

Example 2[7]

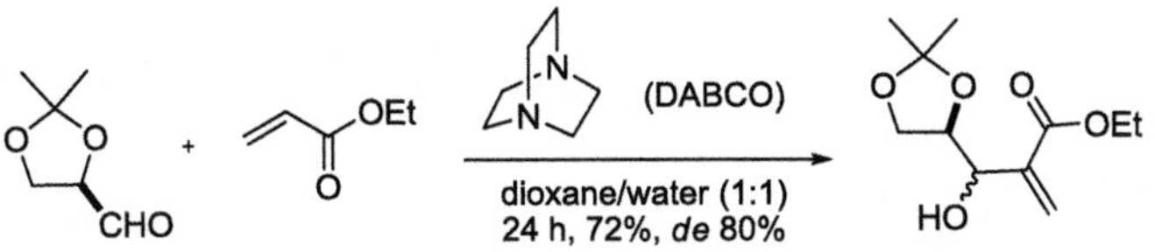

Example 3[8]

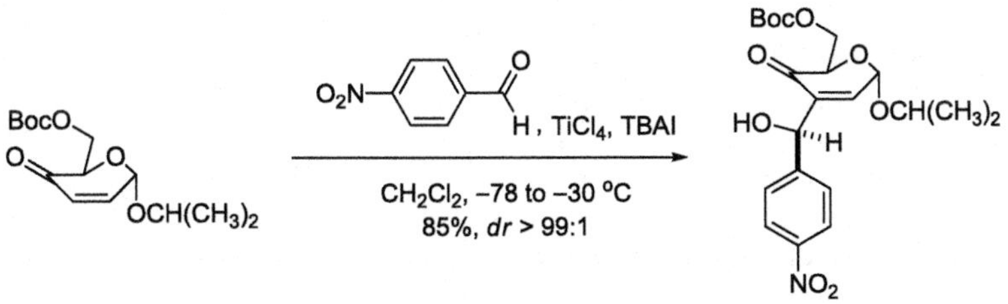

Example 4[9]

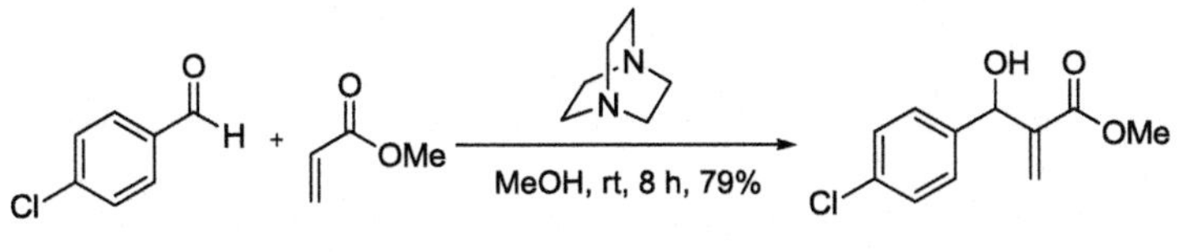

Example 5[10]

R = p-Cl-C$_6$H$_5$
R = p-OMe-C$_6$H$_5$
R = p-NO$_2$-C$_6$H$_5$
R = 2-furyl
R = 2-naphthyl

(10 mol%)

cyclopentyl methyl ether/toluene, −15 °C

87–100%, 88–95% ee

Example 6[13]

PhSeLi
THF

NH$_4$Cl
87%

**References**

1. Baylis, A. B.; Hillman, M. E. D. Ger. Pat. 2,155,113, (**1972**). Both Anthony B. Baylis and Melville E. D. Hillman were chemists at Celanese Corp. USA.
2. Basavaiah, D.; Rao, P. D.; Hyma, R. S. *Tetrahedron* **1996**, *52*, 8001–8062. (Review).
3. Ciganek, E. *Org. React.* **1997**, *51*, 201–350. (Review).
4. Wang, L.-C.; Luis, A. L.; Agapiou, K.; Jang, H.-Y.; Krische, M. J. *J. Am. Chem. Soc.* **2002**, *124*, 2402–2403.
5. Frank, S. A.; Mergott, D. J.; Roush, W. R. *J. Am. Chem. Soc.* **2002**, *124*, 2404–2405.
6. Reddy, L. R.; Saravanan, P.; Corey, E. J. *J. Am. Chem. Soc.* **2004**, *126*, 6230–6231.
7. Krishna, P. R.; Narsingam, M.; Kannan, V. *Tetrahedron Lett.* **2004**, *45*, 4773–4775.
8. Sagar, R,; Pant, C. S.; Pathak, R.; Shaw, A. K. *Tetrahedron* **2004**, *60*, 11399–11406.
9. Mi, X.; Luo, S.; Cheng, J.-P. *J. Org. Chem.* **2005**, *70*, 2338–2341.
10. Matsui, K.; Takizawa, S.; Sasai, H. *J. Am. Chem. Soc.* **2005**, *127*, 3680–3681.
11. Price, K. E.; Broadwater, S. J.; Jung, H. M.; McQuade, D. T. *Org. Lett.* **2005**, *7*, 147–150. A novel mechanism involving a hemiacetal intermediate is proposed.
12. Limberakis, C. *Morita–Baylis–Hillman Reaction*. In *Name Reactions for Homologations-Part I*; Li, J. J., Ed.; Wiley: Hoboken, NJ, **2009**, pp 350–380. (Review).
13. Cheng, P.; Clive, D. L. J. *J. Org. Chem.* **2012**, *77*, 3348–3364.
14. Chandrasoma, N.; Brown, N.; Brassfield, A.; Nerurkar, A.; Suarez, S.; Buszek, K. R. *Tetrahedron Lett.* **2013**, *54*, 913–917.

# Beckmann rearrangement

Acid-mediated isomerization of oximes to amides.
In protic acid:

the substituent *trans* to the leaving group migrates

With PCl$_5$:

Again, the substituent *trans* to the leaving group migrates

Example 1, Microwave (MW) reaction[3]

J.J. Li, *Name Reactions: A Collection of Detailed Mechanisms and Synthetic Applications*,
DOI 10.1007/978-3-319-03979-4_19, © Springer International Publishing Switzerland 2014

Example 2[4]

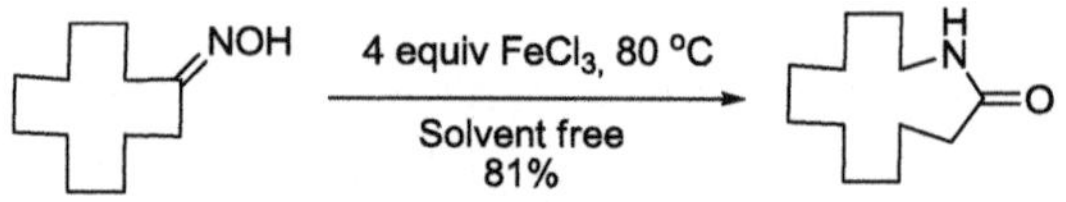

Example 3[6]

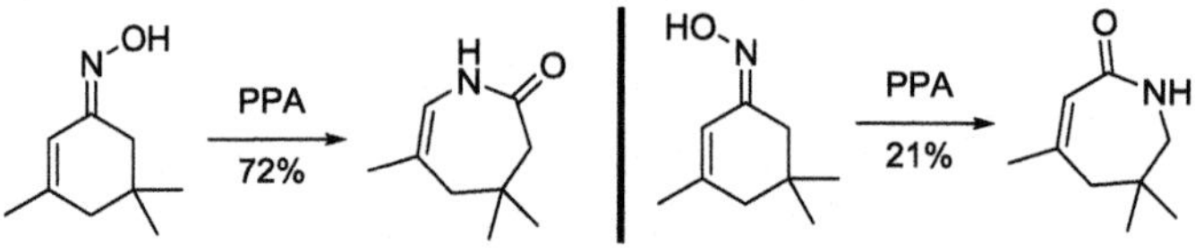

PPA = **p**oly**p**hosphoric **a**cid

Example 4[8]

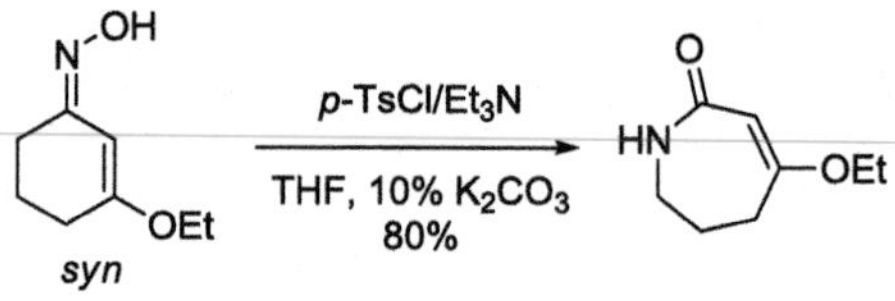

**Abnormal Beckmann rearrangement** is when the migrating fragment (e.g., R$^1$) departs from the intermediate, leaving a nitrile as a stable product.

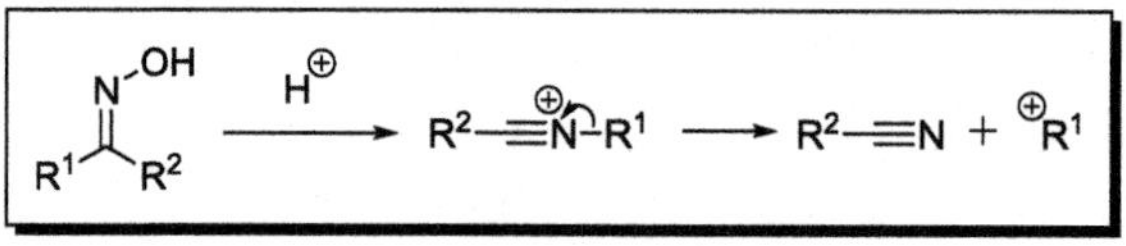

Example 1[9]

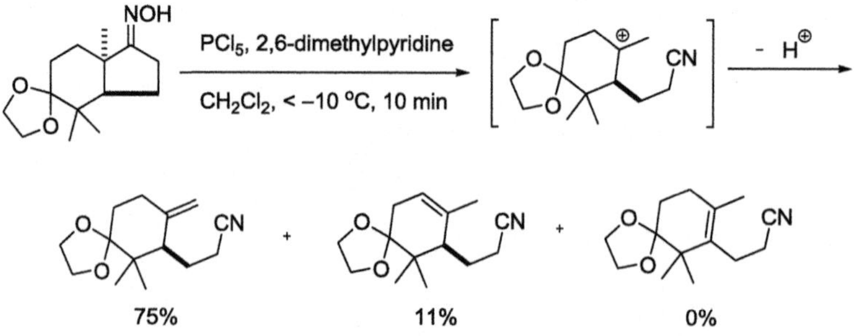

75%                          11%                          0%

Example 2[10]

**References**

1. Beckmann, E. *Chem. Ber.* **1886**, *89*, 988. Ernst Otto Beckmann (1853–1923) was born in Solingen, Germany. He studied chemistry and pharmacy at Leipzig. In addition to the Beckmann rearrangement of oximes to amides, his name is associated with the Beckmann thermometer, used to measure freezing and boiling point depressions to determine molecular weights.

2. Gawley, R. E. *Org. React.* **1988**, *35*, 1–420. (Review).

3. Thakur, A. J.; Boruah, A.; Prajapati, D.; Sandhu, J. S. *Synth. Commun.* **2000**, *30*, 2105–2011.

4. Khodaei, M. M.; Meybodi, F. A.; Rezai, N.; Salehi, P. *Synth. Commun.* **2001**, *31*, 2047–2050.

5. Torisawa, Y.; Nishi, T.; Minamikawa, J.-i. *Bioorg. Med. Chem. Lett.* **2002**, *12*, 387–390.

6. Hilmey, D. G.; Paquette, L. A. *Org. Lett.* **2005**, *7*, 2067–2069.

7. Fernández, A. B.; Boronat, M.; Blasco, T.; Corma, A. *Angew. Chem. Int. Ed.* **2005**, *44*, 2370–2373.

8. Collison, C. G.; Chen, J.; Walvoord, R. *Synthesis* **2006**, 2319–2322.

9. Cao, L.; Sun, J.; Wang, X.; Zhu, R.; Shi, H.; Hu, Y. *Tetrahedron* **2007**, *63*, 5036–5041.

10. Wang, C.; Rath, N. P.; Covey, D. F. *Tetrahedron* **2007**, *63*, 7977–7984.

11. Kumar, R. R.; Vanitha, K. A.; Balasubramanian, M. *Beckmann Rearrangement.* In *Name Reactions for Homologations-Part II*; Li, J. J., Ed.; Wiley: Hoboken, NJ, **2009**, pp 274–292. (Review).

12. Faraldos, J. A.; Kariuki, B. M.; Coates, R. M. *Org. Lett.* **2011**, *13*, 836–839.

13. Tian, B.-X.; An, N.; Deng, W.-P.; Eriksson, L. A. **2013**, *15*, 6782–6785.

# Beirut reaction

Synthesis of quinoxaline-1,4-dioxides from benzofurazan oxide.

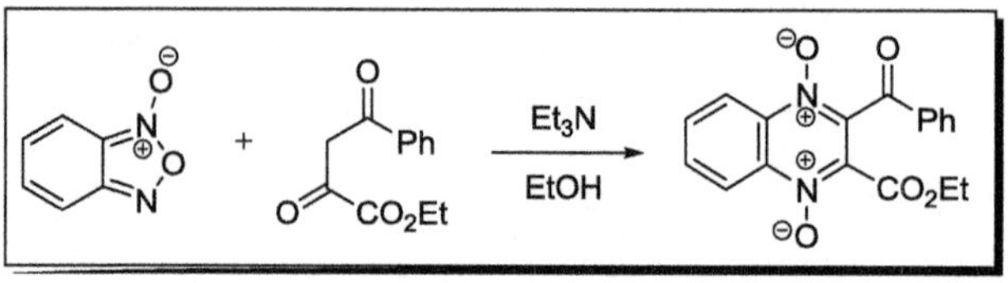

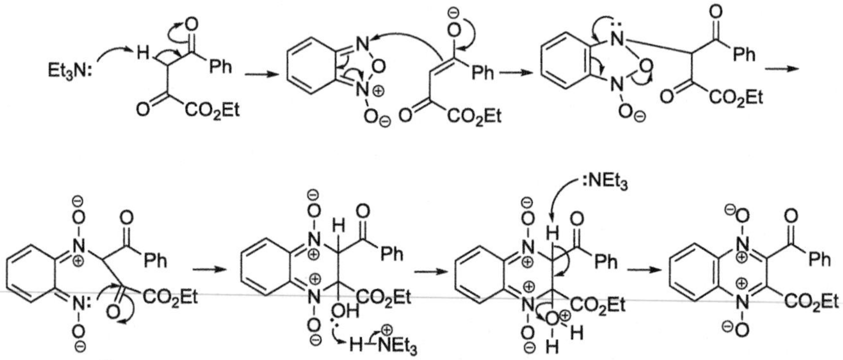

Example 1[3]

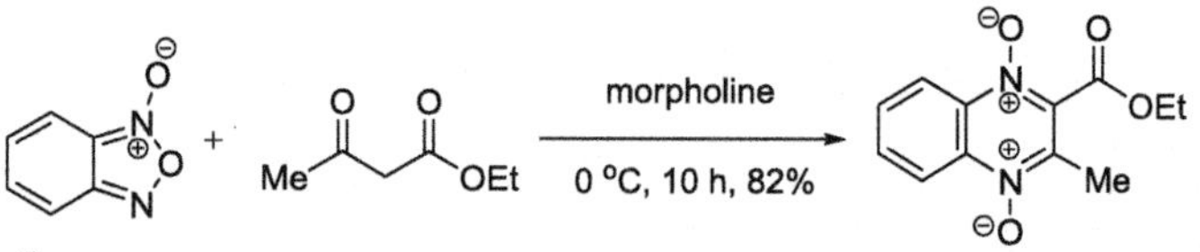

Example 2[7]

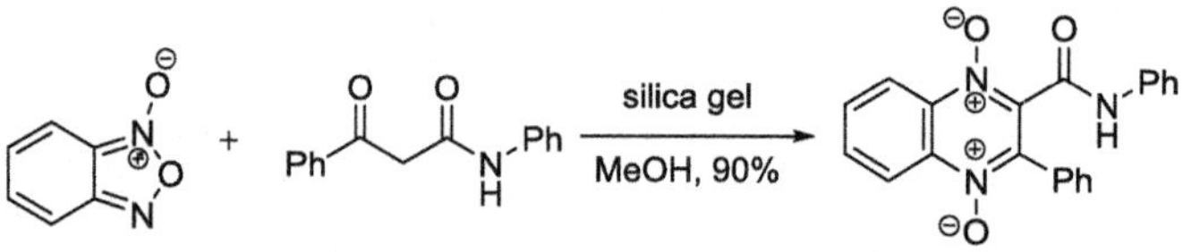

Example 4, Promoted by β-cyclodextrin[11]

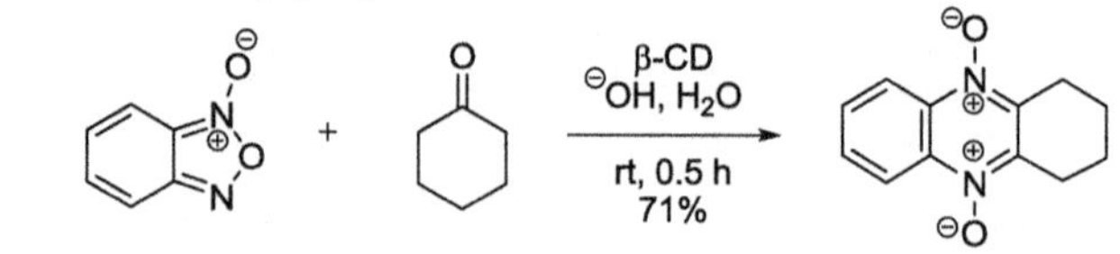

Example 5, An unusual Beirut reaction[12]

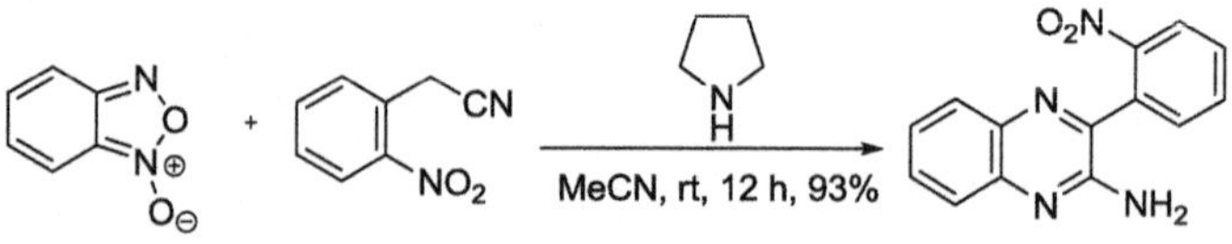

J.J. Li, *Name Reactions: A Collection of Detailed Mechanisms and Synthetic Applications*,
DOI 10.1007/978-3-319-03979-4_20, © Springer International Publishing Switzerland 2014

### References

1.  Haddadin, M. J.; Issidorides, C. H. *Heterocycles* **1976**, *4*, 767–816.  The authors named the reaction after the city where it was discovered, Beirut, the capital of Lebanon.
2.  Gaso, A.; Boulton, A. J. In *Advances in Heterocyclic Chem.;* Vol. 29, Katritzky, A. R.; Boulton, A. J., eds.; Academic Press Inc.: New York, **1981**, 251. (Review).
3.  Vega, A. M.; Gil, M. J.; Fernández-Alvarez, E. *J. Heterocycl. Chem.* **1984**, *21*, 1271.
4.  Atfah, A.; Hill, J. *J. Chem. Soc., Perkin Trans. 1* **1989**, 221–224.
5.  Haddadin, M. J.; Issidorides, C. H. *Heterocycles* **1993**, *35*, 1503–1525.
6.  El-Abadelah, M. M.; Nazer, M. Z.; El-Abadla, N. S.; Meier, H. *Heterocycles* **1995**, *41*, 2203–2219.
7.  Takabatake, T.; Miyazawa, T.; Kojo, M.; Hasegawa, M. *Heterocycles* **2000**, *53*, 2151–2162.
8.  Panasyuk, P. M.; Mel'nikova, S. F.; Tselinskii, I. V. *Russ. J. Org. Chem.* **2001**, *37*, 892.
9.  Turker, L.; Dura, E. *Theochem* **2002**, *593*, 143–147.
10. Tinsley, J. M. *Beirut Reaction* in *Name Reactions in Heterocyclic Chemistry*, Li, J. J., Ed.; Wiley: Hoboken, NJ, **2005**, 504–509. (Review).
11. Sun, T.; Zhao, W.-J.; Hao, A.-Y.; Sun, L.-Z. *Synthesis* **2011**, *41*, 3097–3105.
12. Haddadin, M. J.; El-Khatib, M.; Shoker, T. A.; Beavers, C. M.; Olmstead, M. M.; Fettinger, J. C.; Farber, K. M.; Kurth, M. J. *J. Org. Chem.* **2011**, *76*, 8421–8427.

## Benzilic acid rearrangement

Rearrangement of benzil to benzilic acid *via* aryl migration.

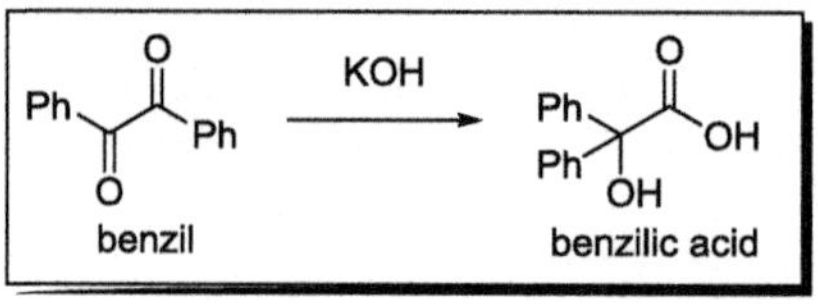
KOH
benzil
benzilic acid

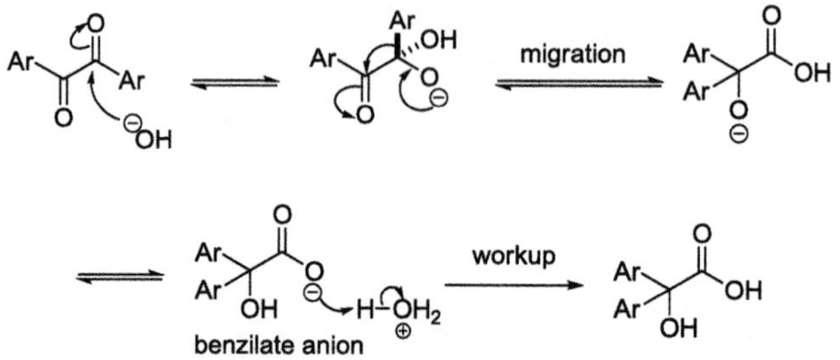
migration
workup
benzilate anion

Final deprotonation (before workup) of the carboxylate to afford the benzilate anion drives the reaction forward.

Example 1[3]

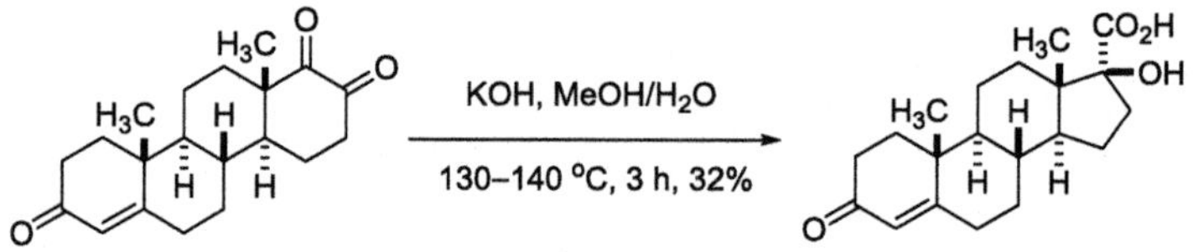
KOH, MeOH/H2O
130–140 °C, 3 h, 32%

Example 2[6]

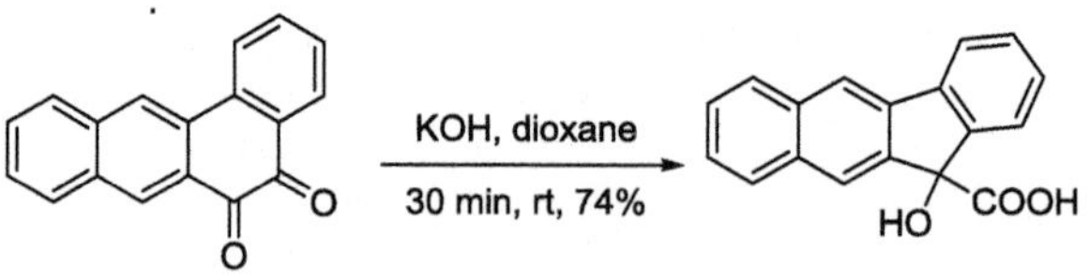
KOH, dioxane
30 min, rt, 74%

Example 3, Retro-benzilic acid rearrangement[7]

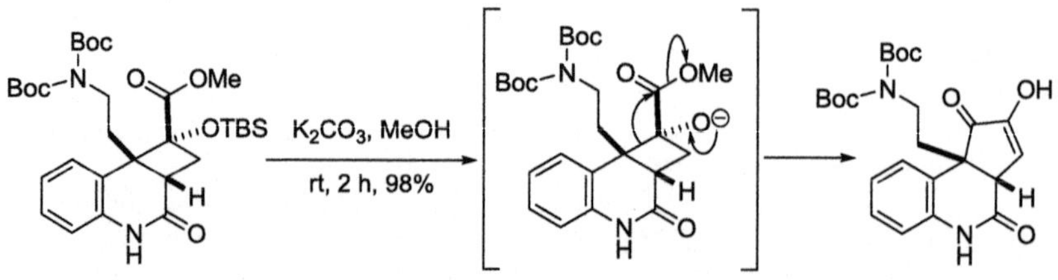
K2CO3, MeOH
rt, 2 h, 98%

J.J. Li, *Name Reactions: A Collection of Detailed Mechanisms and Synthetic Applications*, DOI 10.1007/978-3-319-03979-4_21, © Springer International Publishing Switzerland 2014

Example 4, Cyclobutane-1,2-diones (Computational Chemistry)[9]

**References**
1.  Liebig, J. *Justus Liebigs Ann. Chem.* **1838**, 27. Justus von Liebig (1803–1873) pursued his Ph.D. in organic chemistry in Paris under the tutelage of Joseph Louis Gay-Lussac (1778–1850). He was appointed the Chair of Chemistry at Giessen University, which incited a furious jealousy amongst several of the professors already working there because he was so young. Fortunately, time would prove the choice was a wise one for the department. Liebig would soon transform Giessen from a sleepy university to a mecca of organic chemistry in Europe. Liebig is now considered the father of organic chemistry. Many classic name reactions were published in the journal that still bears his name, *Justus Liebigs Annalen der Chemie.*[2]
2.  Zinin, N. *Justus Liebigs Ann. Chem.* **1839**, *31*, 329.
3.  Georgian, V.; Kundu, N. *Tetrahedron* **1963**, *19*, 1037–1049.
4.  Robinson, J. M.; Flynn, E. T.; McMahan, T. L.; Simpson, S. L.; Trisler, J. C.; Conn, K. B. *J. Org. Chem.* **1991**, *56*, 6709–6712.
5.  Fohlisch, B.; Radl, A.; Schwetzler-Raschke, R.; Henkel, S. *Eur. J. Org. Chem.* **2001**, 4357–4365.
6.  Patra, A.; Ghorai, S. K.; De, S. R.; Mal, D. *Synthesis* **2006**, *15*, 2556–2562.
7.  Selig, P.; Bach, T. *Angew. Chem. Int. Ed.* **2008**, *47*, 5082–5084.
8.  Kumar, R. R.; Balasubramanian, M. *Benzilic Acid Rearrangement.* In *Name Reactions for Homologations-Part II*; Li, J. J., Ed.; Wiley: Hoboken, NJ, **2009**, pp 395–405. (Review).
9.  Sultana, N.; Fabian, W. M. F. *Beilstein J. Org. Chem.* **2013**, *9*, 594–601.

## Benzoin condensation

Cyanide-catalyzed condensation of aryl aldehyde to benzoin.  Now cyanide is mostly replaced by thiazolium salts or *N*-heterocyclic carbenes.   *Cf.* Stetter reaction.

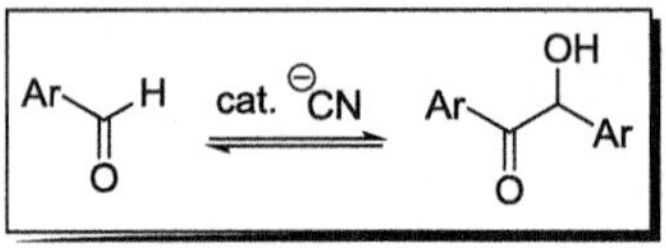

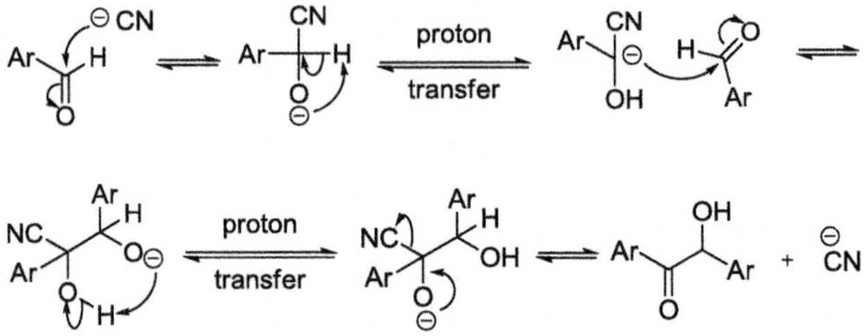

Example 1[2]

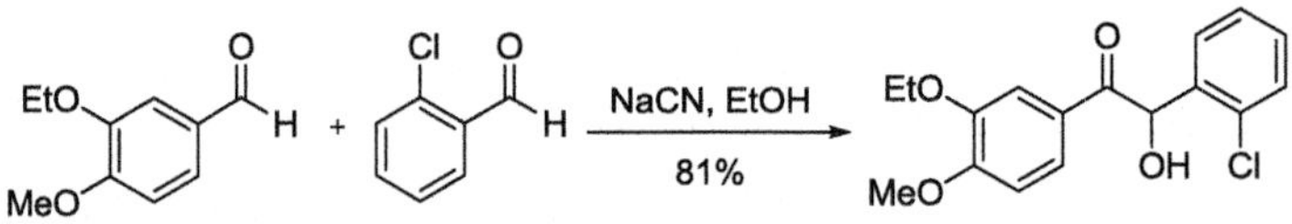

Example 2[7]

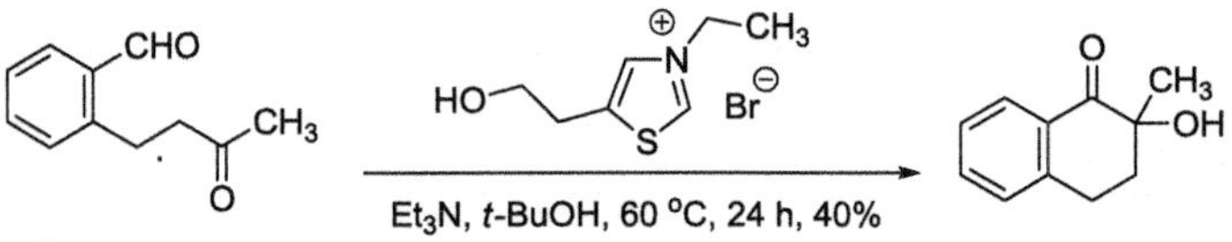

Example 3[7]

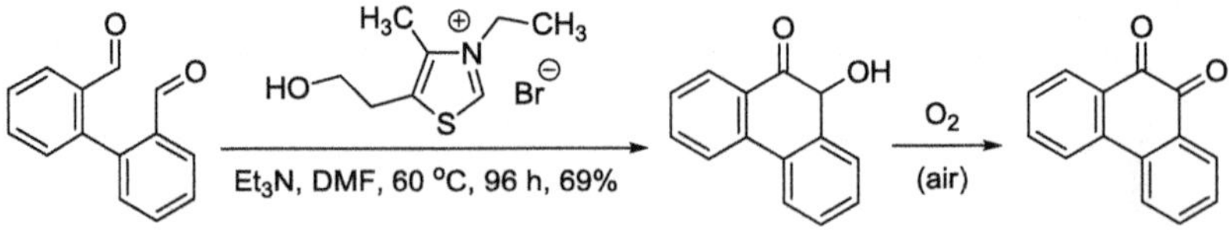

J.J. Li, *Name Reactions: A Collection of Detailed Mechanisms and Synthetic Applications*,
DOI 10.1007/978-3-319-03979-4_22, © Springer International Publishing Switzerland 2014

Example 4, With Brook rearrangement[9]

Example 5[10]

Example 6[12]

**References**

1.  Lapworth, A. J. *J. Chem. Soc.* **1903**, *83*, 995–1005. Arthur Lapworth (1872–1941) was born in Scotland. He was a figure in the development of the modern view of mechanisms of organic reactions. Lapworth investigated the benzoin condensation at the Chemical Department, The Goldsmiths' Institute, New Cross, UK.

2.  Buck, J. S.; Ide, W. S. *J. Am. Chem. Soc.* **1932**, *54*, 3302–3309.

3.  Ide, W. S.; Buck, J. S. *Org. React.* **1948**, *4*, 269–304. (Review).

4.  Stetter, H.; Kuhlmann, H. *Org. React.* **1991**, *40*, 407–496. (Review).

5.  White, M. J.; Leeper, F. J. *J. Org. Chem.* **2001**, *66*, 5124–5131.

6.  Hachisu, Y.; Bode, J. W.; Suzuki, K. *J. Am. Chem. Soc.* **2003**, *125*, 8432–8433.

7.  Enders, D.; Niemeier, O. *Synlett* **2004**, 2111–2114.

8.  Johnson, J. S. *Angew. Chem. Int. Ed.* **2004**, *43*, 1326–1328. (Review).

9.  Linghu, X.; Potnick, J. R.; Johnson, J. S. *J. Am. Chem. Soc.* **2004**, *126*, 3070–3071.

10. Enders, D.; Han, J. *Tetrahedron: Asymmetry* **2008**, *19*, 1367–1371.

11. Cee, V. J. *Benzoin Condensation.* In *Name Reactions for Homologations-Part I*; Li, J. J., Ed.; Wiley: Hoboken, NJ, **2009**, pp 381–392. (Review).

12. Kabro, A.; Escudero-Adan, E. C.; Grushin, V. V.; van Leeuwen, P. W. N. M. *Org. Lett.* **2012**, *14*, 4014–4017.

# Bergman cyclization

Formation of a substituted benzene through 1,4-benzenediyl diradical formation from enediyne *via* electrocyclization.

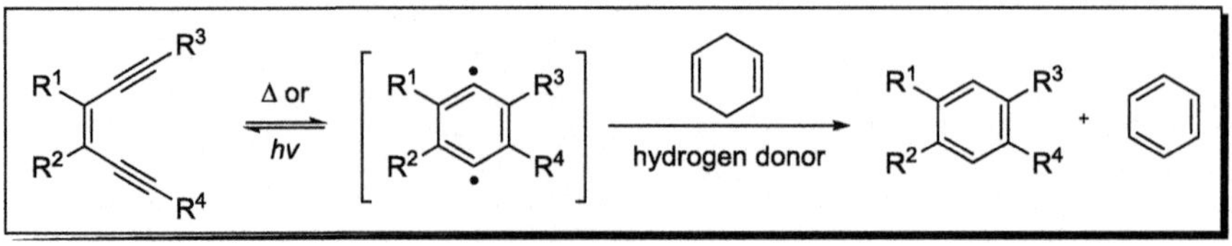

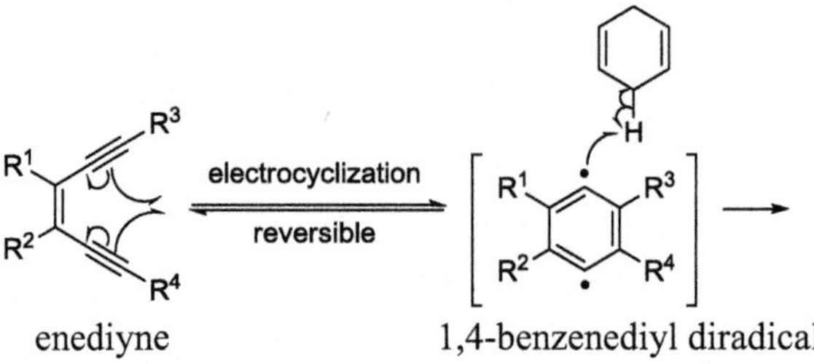

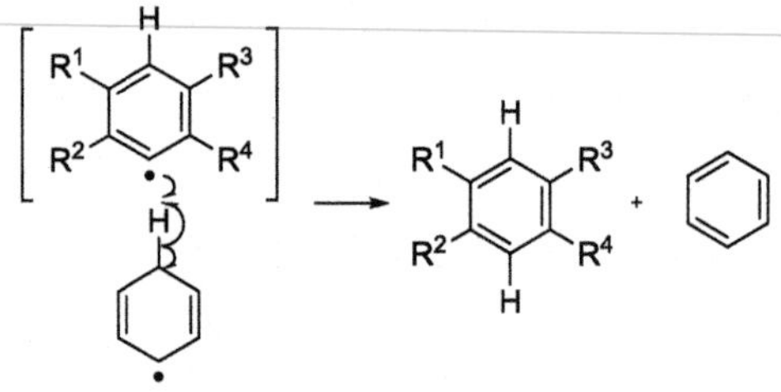

enediyne                                    1,4-benzenediyl diradical

Example 1[6]

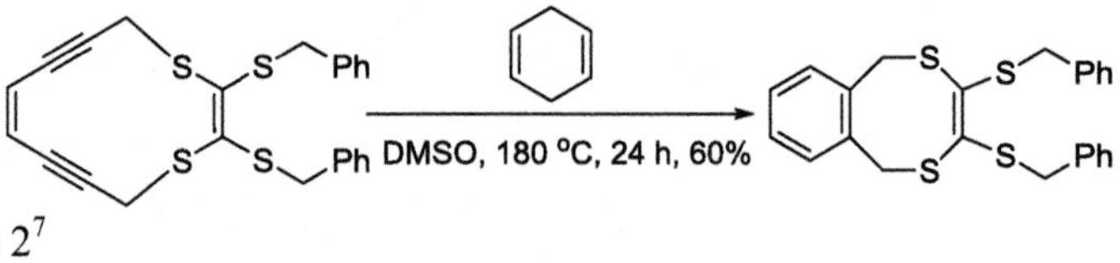

DMSO, 180 °C, 24 h, 60%

Example 2[7]

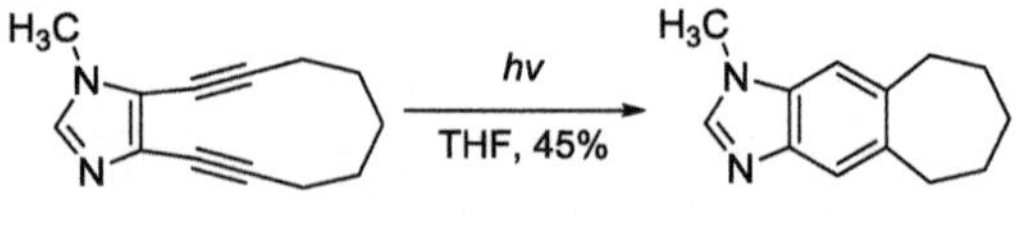

$hv$

THF, 45%

Example 3, Wolff rearrangement followed by Bergman cyclization[8]

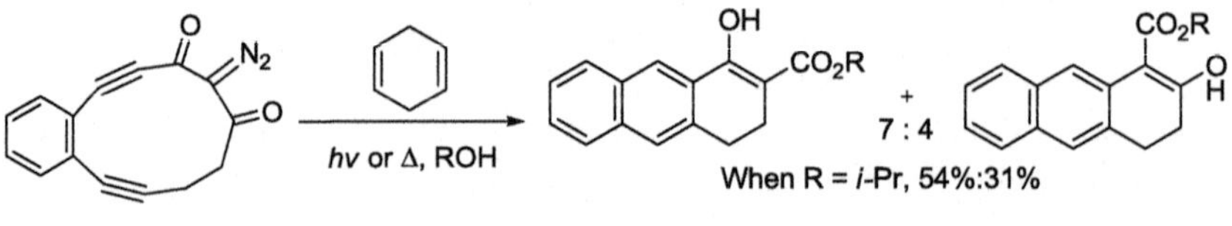

$hv$ or $\Delta$, ROH

7 : 4

When R = *i*-Pr, 54%:31%

J.J. Li, *Name Reactions: A Collection of Detailed Mechanisms and Synthetic Applications*,
DOI 10.1007/978-3-319-03979-4_23, © Springer International Publishing Switzerland 2014

Example 4[10]

142 °C, $t_{1/2}$ = 14.4 h

Example 5[12]

20% 1,4-cyclohexadiene

PhCl, 180 °C, 48 h, 65%

major product    minor product

major: minor

= 9:1

Example 5[13]

dry DMSO, 90 °C

8 h, 80%

## References

1. Jones, R. R.; Bergman, R. G. *J. Am. Chem. Soc.* **1972**, *94*, 660–661. Robert G. Bergman (1942–) is a professor at the University of California, Berkeley. His discovery of the Bergman cyclization was completed far in advance of the discovery of ene-diyne's anti-cancer properties.
2. Bergman, R. G. *Acc. Chem. Res.* **1973**, *6*, 25–31. (Review).
3. Myers, A. G.; Proteau, P. J.; Handel, T. M. *J. Am. Chem. Soc.* **1988**, *110*, 7212–7214.
4. Yus, M.; Foubelo, F. *Rec. Res. Dev. Org. Chem.* **2002**, *6*, 205–280. (Review).
5. Basak, A.; Mandal, S.; Bag, S. S. *Chem. Rev.* **2003**, *103*, 4077–4094. (Review).
6. Bhattacharyya, S.; Pink, M.; Baik, M.-H.; Zaleski, J. M. *Angew. Chem. Int. Ed.* **2005**, *44*, 592–595.
7. Zhao, Z.; Peacock, J. G.; Gubler, D. A.; Peterson, M. A. *Tetrahedron Lett.* **2005**, *46*, 1373–1375.
8. Karpov, G. V.; Popik, V. V. *J. Am. Chem. Soc.* **2007**, *129*, 3792–3793.
9. Kar, M.; Basak, A. *Chem. Rev.* **2007**, *107*, 2861–2890. (Review).
10. Lavy, S.; Pérez-Luna, A.; Kündig, E. P. *Synlett* **2008**, 2621–2624.
11. Pandithavidana, D. R.; Poloukhtine, A.; Popik, V. V. *J. Am. Chem. Soc.* **2009**, *131*, 351–356.
12. Spence, J. D.; Rios, A. C.; Frost, M. A.; etc. *J. Org. Chem.* **2012**, *77*, 10329–10339.
13. Roy, S.; Basak, A. *Tetrahedron* **2013**, *69*, 2184–2192.

## Biginelli reaction

Also known as Biginelli pyrimidone synthesis. One-pot condensation of an aromatic aldehyde, urea, and β-dicarbonyl compound in acidic ethanolic solution and expansion of such a condensation thereof. It belongs to a class of transformations called multicomponent reactions (MCRs).

Example 1[4]

J.J. Li, *Name Reactions: A Collection of Detailed Mechanisms and Synthetic Applications*,
DOI 10.1007/978-3-319-03979-4_24, © Springer International Publishing Switzerland 2014

Example 2[5]

Example 3, Microwave-induced Biginelli condensation[9]

Example 3[10]

**References**

1.  Biginelli, P. *Ber.* **1891**, *24*, 1317. Pietro Biginelli was at Lab. chim. della Sanita pubbl. in Roma, Italy when this paper was published.
2.  Kappe, C. O. *Tetrahedron* **1993**, *49*, 6937–6963. (Review).
3.  Kappe, C. O. *Acc. Chem. Res.* **2000**, *33*, 879–888. (Review).
4.  Kappe, C. O. *Eur. J. Med. Chem.* **2000**, *35*, 1043–1052. (Review).
5.  Ghorab, M. M.; Abdel-Gawad, S. M.; El-Gaby, M. S. A. *Farmaco* **2000**, *55*, 249–255.
6.  Bose, D. S.; Fatima, L.; Mereyala, H. B. *J. Org. Chem.* **2003**, *68*, 587–590.
7.  Kappe, C. O.; Stadler, A. *Org. React.* **2004**, *68*, 1–116. (Review).
8.  Limberakis, C. *Biginelli Pyrimidone Synthesis* In *Name Reactions in Heterocyclic Chemistry*; Li, J. J., Ed.; Wiley: Hoboken, NJ, **2005**, pp 509–520. (Review).
9.  Banik, B. K.; Reddy, A. T.; Datta, A.; Mukhopadhyay, C. *Tetrahedron Lett.* **2007**, *48*, 7392–7394.
10. Wang, R.; Liu, Z.-Q. *J. Org. Chem.* **2012**, *77*, 3952–3958.
11. Fuchs, D.; Nasr-Esfahani, M.; Diab, L.; Šmejkal, T.; Breit, B. *Synlett* **2013**, *24*, 1657–1662.
12. Liberto, N. A.; de Paiva Silva, S.; de Fátima, Â.; Fernandes, S. A. *Tetrahedron* **2013**, *69*, 8245–8249.

## Birch reduction

The Birch reduction is the 1,4-reduction of aromatics to their corresponding cyclohexadienes by alkali metals (Li, K, Na) dissolved in liquid ammonia in the presence of an alcohol.

Benzene ring bearing an electron-donating substituent:

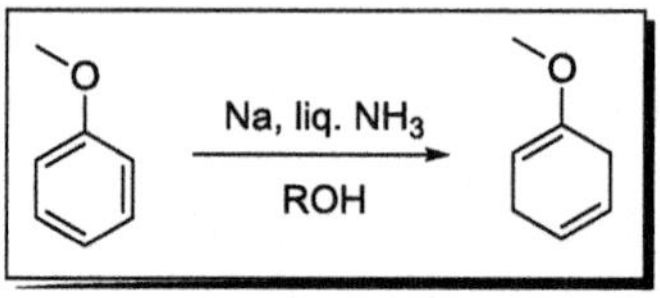

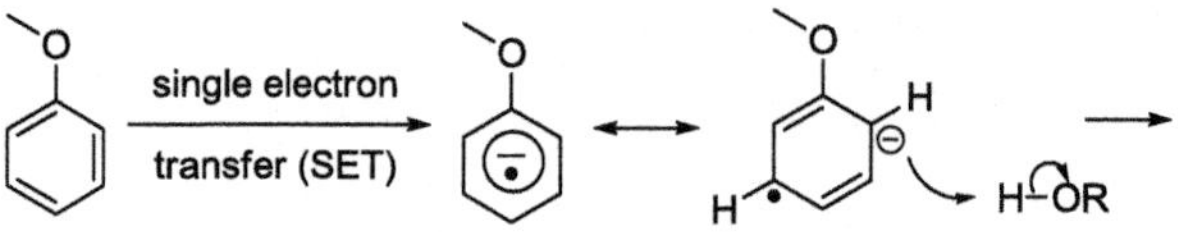

radical anion

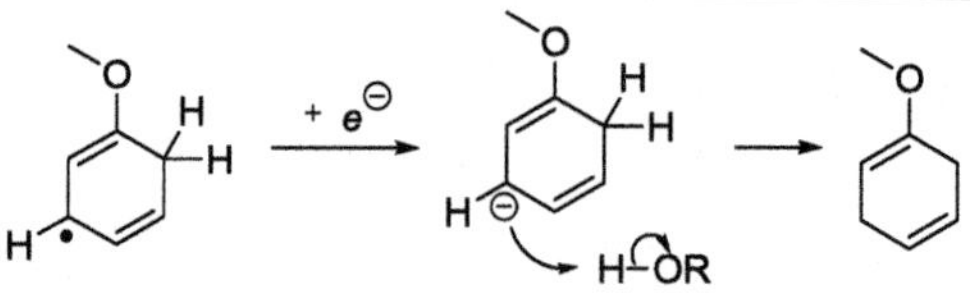

Benzene ring with an electron-withdrawing substituent:

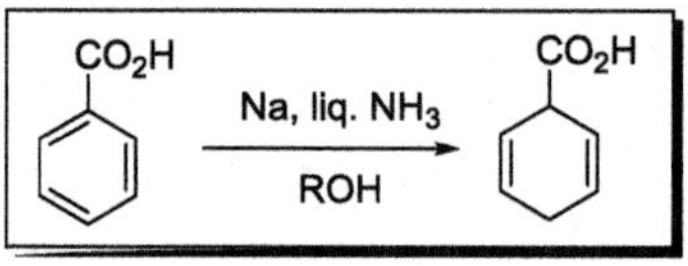

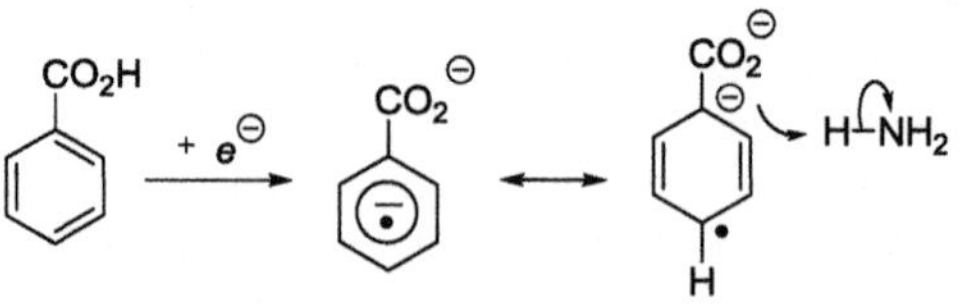

radical anion

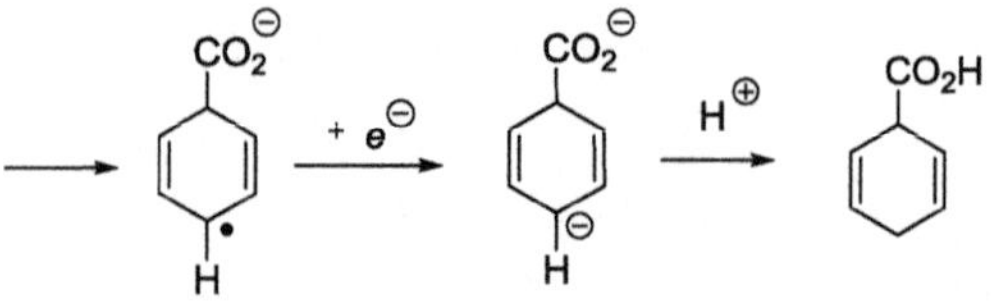

J.J. Li, *Name Reactions: A Collection of Detailed Mechanisms and Synthetic Applications*,
DOI 10.1007/978-3-319-03979-4_25, © Springer International Publishing Switzerland 2014

Example 1, Birch reductive alkylation[4]

1. Na, NH$_3$, THF, –78 °C

2. MeI, 98%, 30:1 *dr*

Example 2[7]

Na, NH$_3$, THF

–78 °C, quant.

Example 3, Fully reduced products[8]

Na, NH$_3$
THF, EtOH

–33 °C, 1 h
16.3%

Example 4, Birch reductive alkylation[9]

Li, THF, –78 °C

then

71%

## References

1. Birch, A. J. *J. Chem. Soc.* **1944**, 430–436. Arthur Birch (1915–1995), an Australian, developed the "Birch reduction" at Oxford University during WWII in Robert Robinson's laboratory. The Birch reduction was instrumental to the discovery of the birth control pills and many other drugs.
2. Rabideau, P. W.; Marcinow, Z. *Org. React.* **1992**, *42*, 1–334. (Review).
3. Birch, A. J. *Pure Appl. Chem.* **1996**, *68*, 553–556. (Review).
4. Donohoe, T. J.; Guillermin, J.-B.; Calabrese, A. A.; Walter, D. S. *Tetrahedron Lett.* **2001**, *42*, 5841–5844.
5. Pellissier, H.; Santelli, M. *Org. Prep. Proced. Int.* **2002**, *34*, 611–642. (Review).
6. Subba Rao, G. S. R. *Pure Appl. Chem.* **2003**, *75*, 1443–1451. (Review).
7. Kim, J. T.; Gevorgyan, V. *J. Org. Chem.* **2005**, *70*, 2054–2059.
8. Gealis, J. P.; Müller-Bunz, H.; Ortin, Y.; Condell, M.; Casey, M.; McGlinchey, M. J. *Chem. Eur. J.* **2008**, *14*, 1552–1560.
9. Fretz, S. J.; Hadad, C. M.; Hart, D. J.; Vyas, S.; Yang, D. *J. Org. Chem.* **2013**, *78*, 83–92.

## Bischler–Möhlau indole synthesis

The Bischler–Möhlau indole synthesis, also known as the Bischler indole synthesis, refers to the synthesis of 2-arylindoles from the cyclization of α-arylamino-ketones and excess anilines.

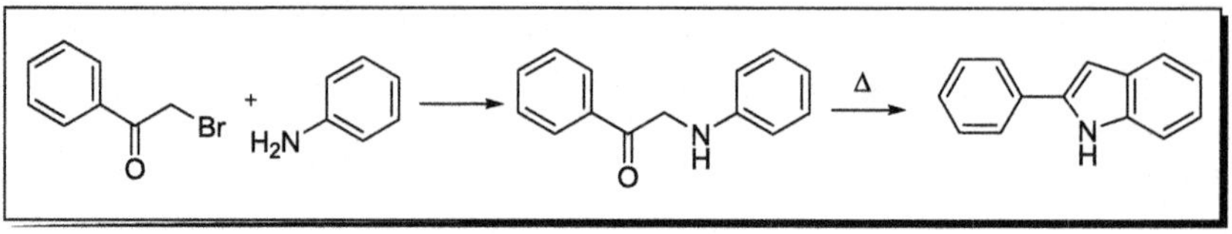

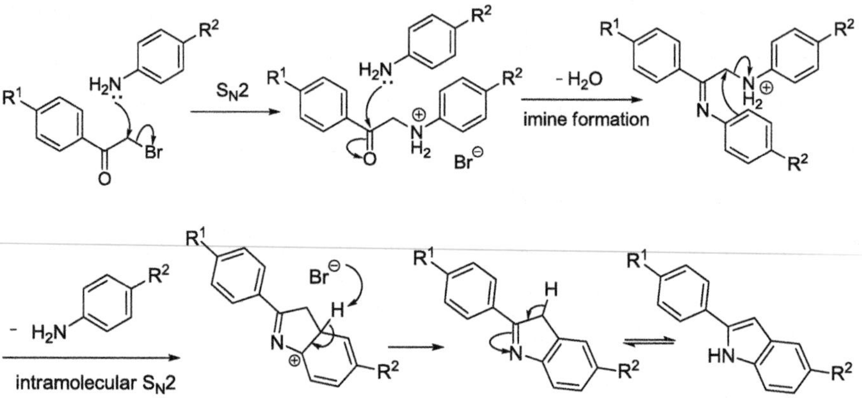

Example 1[5]

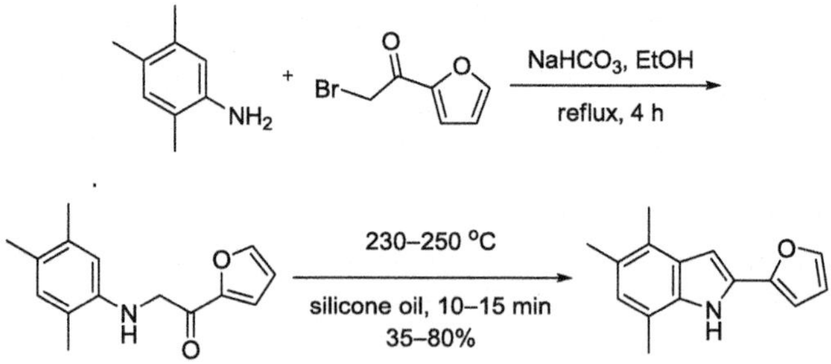

Example 3[9]

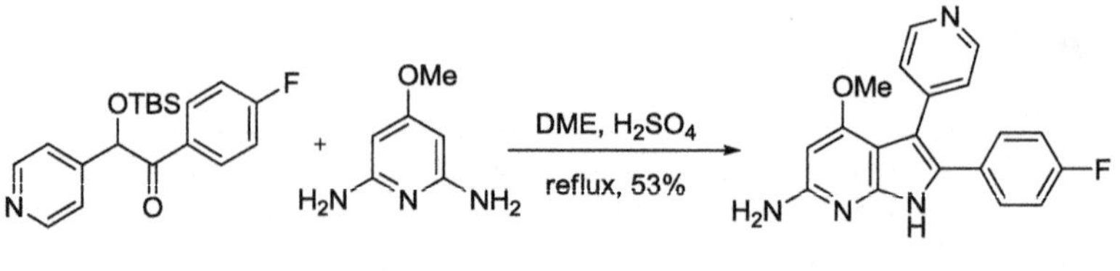

J.J. Li, *Name Reactions: A Collection of Detailed Mechanisms and Synthetic Applications*, DOI 10.1007/978-3-319-03979-4_26, © Springer International Publishing Switzerland 2014

Example 4[10]

Example 5, Microwave-assisted, solvent-free Bischler indole synthesis[11]

**References**
1. Möhlau, R. *Ber.* **1881,** *14*, 171–175. Möhlau, a German chemist, worked extensively in the dye industry.
2. Bischler, A.; Fireman, P. *Ber.* **1893,** *26*, 1346–1349. Augustus Bischler (1865–1957) was born in Southern Russia. He studied in Zurich with Arthur Hantzsch. He discovered the Bischler–Napieralski reaction while studying alkaloids at Basel Chemical Works, Switzerland with his coworker, B. Napieralski.
3. Sundberg, R. J. *The Chemistry of Indoles;* Academic Press: New York, **1970,** pp 164. (Book).
4. Buu-Hoï, N. P.; Saint-Ruf, G.; Deschamps, D.; Bigot, P. *J. Chem. Soc. (C)* **1971,** 2606–2609.
5. Houlihan, W. J., Ed.; *The Chemistry of Heterocyclic Compounds, Indoles (Part 1),* Wiley: New York, **1972.** (Book).
6. Bigot, P.; Saint-Ruf, G.; Buu-Hoï, N. P. *J. Chem. Soc., Perkin 1* **1972,** 2573–2576.
7. Bancroft, K. C. C.; Ward, T. J. *J. Chem. Soc., Perkin 1* **1974,** 1852–1858.
8. Coïc, J. P.; Saint-Ruf, G.; Brown, K. *J. Heterocycl. Chem.* **1978,** *15*, 1367–1371.
9. Henry, J. R.; Dodd, J. H. *Tetrahedron Lett.* **1998,** *39*, 8763–8764.
10. Pchalek, K.; Jones, A. W.; Wekking, M. M. T.; Black, D. S. C. *Tetrahedron* **2005,** *61*, 77–82.
11. Sridharan, V.; Perumal, S.; Avendaño, C.; Menéndez, J. C. *Synlett* **2006,** 91–95.
12. Zhang, J. *Bischler–Möhlau Indole Synthesis*, In *Name Reactions in Heterocyclic Chemistry II*; Li, J. J., Ed.; Wiley: Hoboken, NJ, **2011,** pp 84–90. (Review).

# Bischler–Napieralski reaction

Dihydroisoquinolines from β-phenethylamides in refluxing phosphorus oxychloride.

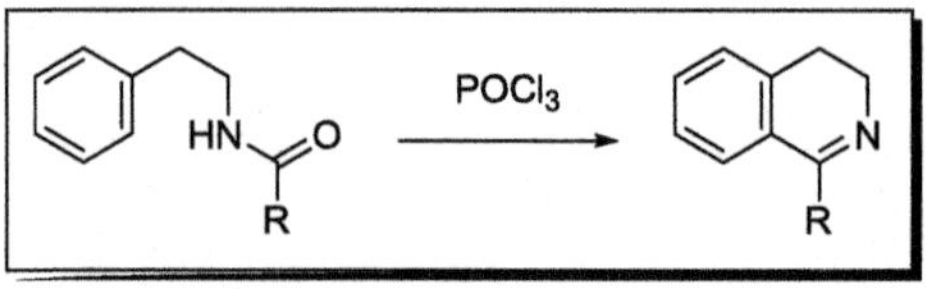

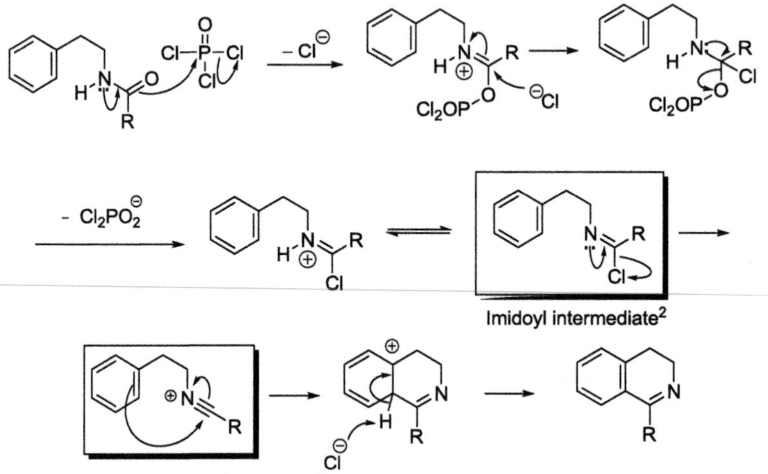

Imidoyl intermediate[2]

Nitrilium salt intermediate[2]

Example 1[3]

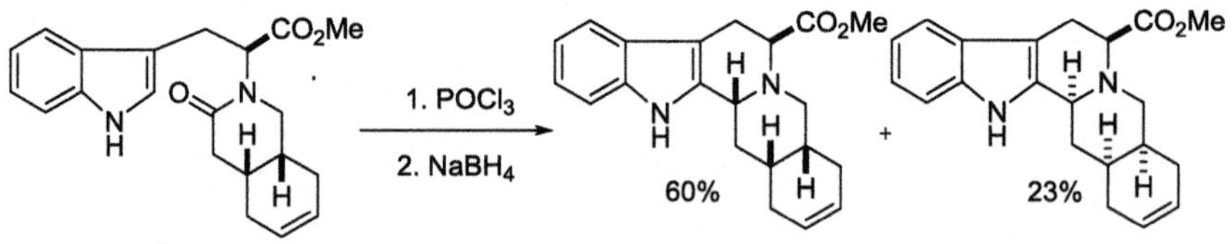

60%        +        23%

Example 2[5]

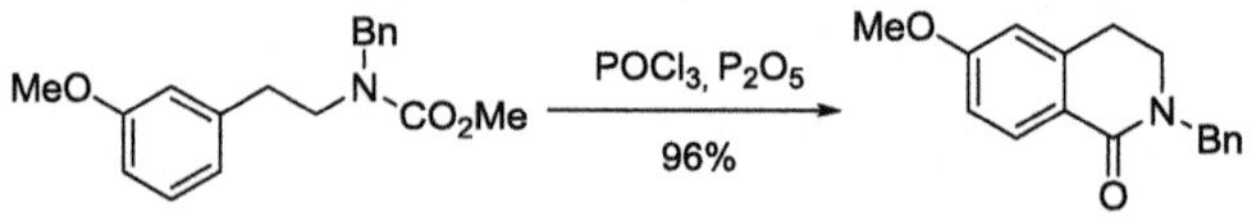

96%

Example 3[7]

80 °C, 81%

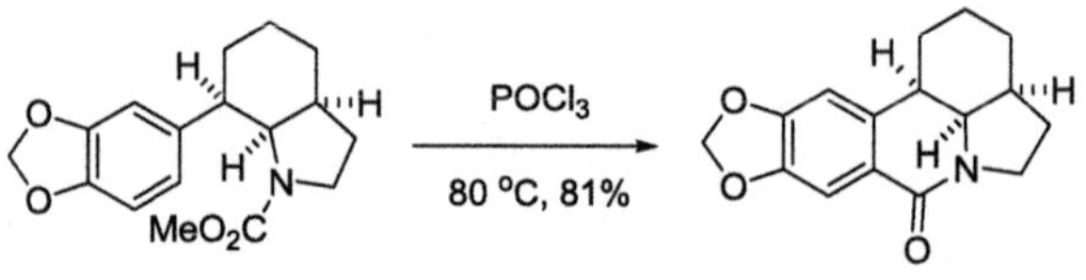

J.J. Li, *Name Reactions: A Collection of Detailed Mechanisms and Synthetic Applications*,
DOI 10.1007/978-3-319-03979-4_27, © Springer International Publishing Switzerland 2014

Example 4[8]

POCl₃

toluene, 95%

Example 5[10]

POCl₃, PhH, reflux, 3 h

then LiAlH₄, THF, 1.5 h
64%

Example 6, An unprecedented Bischler–Napieralski reaction[12]

POCl₃, sulfolane
80 °C, 51%

7-exo-trig-B-N reaction

## References

1. Bischler, A.; Napieralski, B. *Ber.* **1893**, *26*, 1903–1908. Augustus Bischler discovered the Bischler–Napieralski reaction while studying alkaloids at Basel Chemical Works, Switzerland with his coworker, B. Napieralski. Bernard Napieralski was affiliated with the University of Zurich.

2. Mechanistic studies: (a) Fodor, G.; Gal, J.; Phillips, B. A. *Angew. Chem. Int. Ed. Engl.* **1972**, *11*, 919–920. (b) Nagubandi, S.; Fodor, G. *J. Heterocycl. Chem.* **1980**, *17*, 1457–1463. (c) Fodor, G.; Nagubandi, S. *Tetrahedron* **1980**, *36*, 1279–1300.

3. Aubé, J.; Ghosh, S.; Tanol, M. *J. Am. Chem. Soc.* **1994**, *116*, 9009–9018.

4. Sotomayor, N.; Domínguez, E.; Lete, E. *J. Org. Chem.* **1996**, *61*, 4062–4072.

5. Wang, X.-j.; Tan, J.; Grozinger, K. *Tetrahedron Lett.* **1998**, *39*, 6609–6612.

6. Ishikawa, T.; Shimooka, K.; Narioka, T.; Noguchi, S.; Saito, T.; Ishikawa, A.; Yamazaki, E.; Harayama, T.; Seki, H.; Yamaguchi, K. *J. Org. Chem.* **2000**, *65*, 9143–9151.

7. Banwell, M. G.; Harvey, J. E.; Hockless, D. C. R., Wu, A. W. *J. Org. Chem.* **2000**, *65*, 4241–4250.

8. Capilla, A. S.; Romero, M.; Pujol, M. D.; Caignard, D. H.; Renard, P. *Tetrahedron* **2001**, *57*, 8297–8303.

9. Wolfe, J. P. *Bischler–Napieralski Reaction.* In *Name Reactions in Heterocyclic Chemistry*; Li, J. J., Ed.; Wiley: Hoboken, NJ, **2005**, pp 376–385. (Review).

10. Ho, T.-L.; Lin, Q.-x. *Tetrahedron* **2008**, *64*, 10401–10405.

11. Csomós, P.; Fodor, L.; Bernáth, G.; Csámpai, A.; Sohár, P. *Tetrahedron* **2009**, *65*, 1475–1480.

12. Buyck, T.; Wang, Q.; Zhu, J. *Org. Lett.* **2012**, *14*, 1338–1341.

## Blaise reaction

β-Ketoesters from nitriles, α-haloesters and Zn. *Cf.* Reformatsky reaction.

The Zn enolate itself is a *C*-enolate (in the crystal form), but for the reaction to occur, it equilibrates back into an *O*-enolate

Example 1, Preparation of the statin side chain[5]

Example 2[6]

J.J. Li, *Name Reactions: A Collection of Detailed Mechanisms and Synthetic Applications*,
DOI 10.1007/978-3-319-03979-4_28, © Springer International Publishing Switzerland 2014

Example 3[7]

Example 4, Chemoselective tandem acylation of a Blaise reaction intermediate[9]

Example 5, Chemoselective intramolecular alkylation of Blaise reaction intermediate[10]

Blaise reaction intermediate

**References**
1.  (a) Blaise, E. E. *C. R. Hebd. Seances Acad. Sci.* **1901**, *132*, 478–480. (b) Blaise, E. E. *C. R. Hebd. Seances Acad. Sci.* **1901**, *132*, 978–980. Blaise was at Institut Chimique de Nancy, France.
2.  Beard, R. L.; Meyers, A. I. *J. Org. Chem.* **1991**, *56*, 2091–2096.
3.  Deutsch, H. M.; Ye, X.; Shi, Q.; Liu, Z.; Schweri, M. M. *Eur. J. Med. Chem.* **2001**, *36*, 303–311.
4.  Creemers, A. F. L.; Lugtenburg, J. *J. Am. Chem. Soc.* **2002**, *124*, 6324–6334.
5.  Shin, H.; Choi, B. S.; Lee, K. K.; Choi, H.-w.; Chang, J. H.; Lee, K. W.; Nam, D. H.; Kim, N.-S. *Synthesis* **2004**, 2629–2632.
6.  Choi, B. S.; Chang, J. H.; Choi, H.-w.; Kim, Y. K.; Lee, K. K.; Lee, K. W.; Lee, J. H.; Heo, T.; Nam, D. H.; Shin, H. *Org. Proc. Res. Dev.* **2005**, *9*, 311–313.
7.  Pospíšil, J.; Markó, I. E. *J. Am. Chem. Soc.* **2007**, *129*, 3516–3517.
8.  Rao, H. S. P.; Rafi, S.; Padmavathy, K. *Tetrahedron* **2008**, *64*, 8037–8043. (Review).
9.  Chun, Y. S.; Lee, S.-g.; Ko, Y. O.; Shin, H. *Chem. Commun.* **2008**, 5098–5100.
10. Kim, J. H.; Shin, H.; Lee, S.-g. *J. Org. Chem.* **2012**, *77*, 1560–1565.
11. Chun, Y. S.; Xuan, Z.; Kim, J. H.; Lee, S.-g. *Org. Lett.* **2013**, *15*, 3162–3165.

## Blum–Ittah aziridine synthesis

Ring opening of oxiranes using azide followed by $PPh_3$ reduction of the intermediate azido alcohol to give the corresponding aziridines. *Cf.* Staudinger reduction.

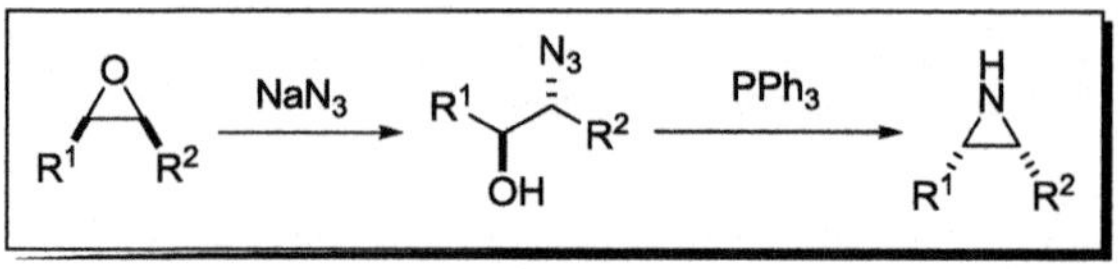

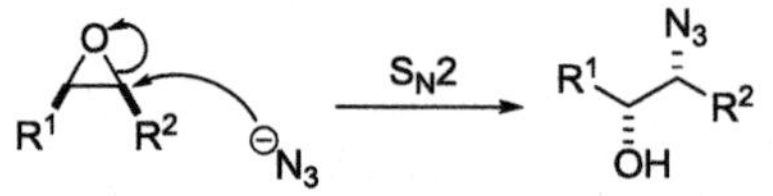

Regardless of the regioselectivity of the $S_N2$ reaction of the azide, the ultimate stereochemical outcome for the aziridine is the same.

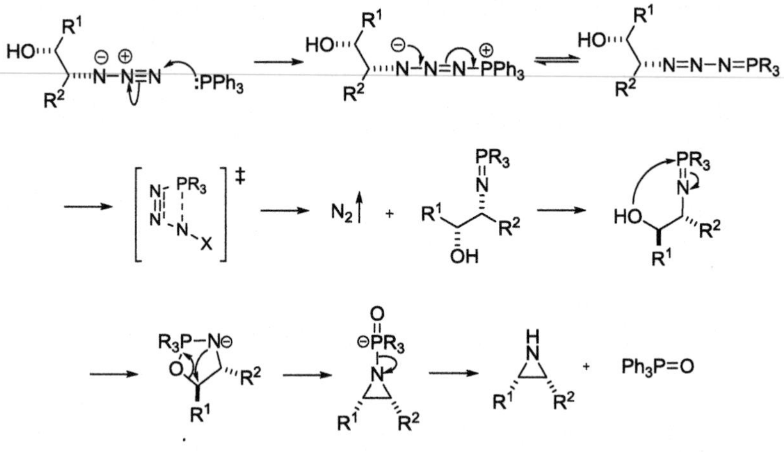

Example 1[3]

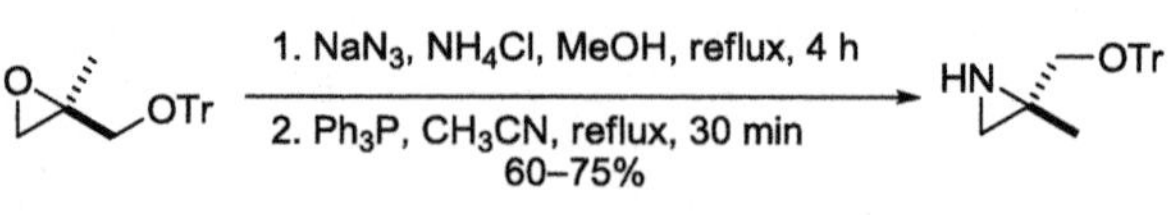

1. $NaN_3$, $NH_4Cl$, MeOH, reflux, 4 h
2. $Ph_3P$, $CH_3CN$, reflux, 30 min
60–75%

Example 2[4]

1. $NaN_3$, $NH_4Cl$, 55%
2. $PPh_3$, THF, 60 °C, 75%

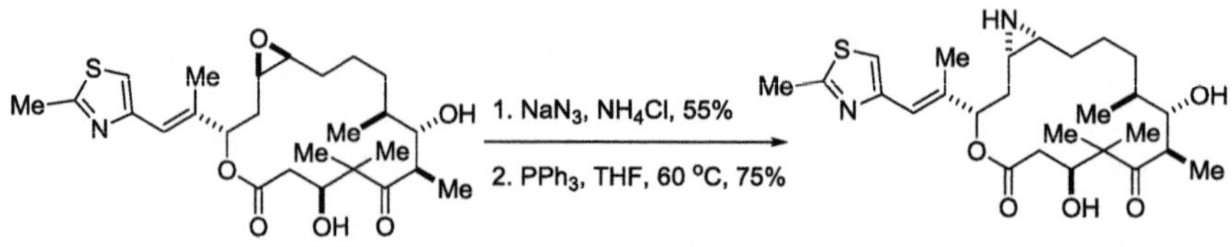

J.J. Li, *Name Reactions: A Collection of Detailed Mechanisms and Synthetic Applications*,
DOI 10.1007/978-3-319-03979-4_29, © Springer International Publishing Switzerland 2014

Example 3[6]

Example 4[8]

Example 5[9]

## References

1. Ittah, Y.; Sasson, Y.; Shahak, I.; Tsaroom, S.; Blum, J. *J. Org. Chem.* **1978,** *43*, 4271–4273. Jochanan Blum is a professor at The Hebrew University in Jerusalem, Israel.
2. Tanner, D.; Somfai, P. *Tetrahedron Lett.* **1987,** *28*, 1211–1214.
3. Wipf, P.; Venkatraman, S.; Miller, C. P. *Tetrahedron Lett.* **1995,** *36*, 3639–3642.
4. Regueiro-Ren, A.; Borzilleri, R. M.; Zheng, X.; Kim, S.-H.; Johnson, J. A.; Fairchild, C. R.; Lee, F. Y. F.; Long, B. H.; Vite, G. D. *Org. Lett.* **2001,** *3*, 2693–2696.
5. Fürmeier, S.; Metzger, J. O. *Eur. J. Org. Chem.* **2003,** 649–659.
6. Oh, K.; Parsons, P. J.; Cheshire, D. *Synlett* **2004,** 2771–2775.
7. Serafin, S. V.; Zhang, K.; Aurelio, L.; Hughes, A. B.; Morton, T. H. *Org. Lett.* **2004,** *6*, 1561–1564.
8. Torrado, A. *Tetrahedron Lett.* **2006,** *47*, 7097–7100.
9. Pulipaka, A. B.; Bergmeier, S. C. *J. Org. Chem.* **2008,** *73*, 1462–1467.
10. Richter, J. M. *Blum aziridine synthesis*, In *Name Reactions in Heterocyclic Chemistry II*; Li, J. J., Ed.; Wiley: Hoboken, NJ, **2011,** pp 2–10. (Review).

## Boekelheide reaction

Treatment of 2-methylpyridine $N$-oxide with trifluoroacetic anhydride, or acetic anhydride gives rise to 2-hydroxymethylpyridine.

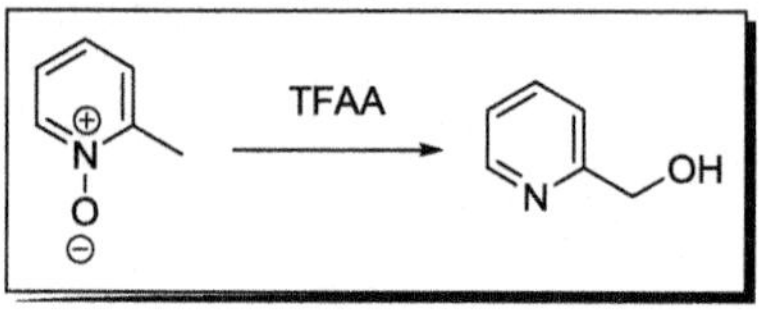

TFAA: trifluoroacetic anhydride

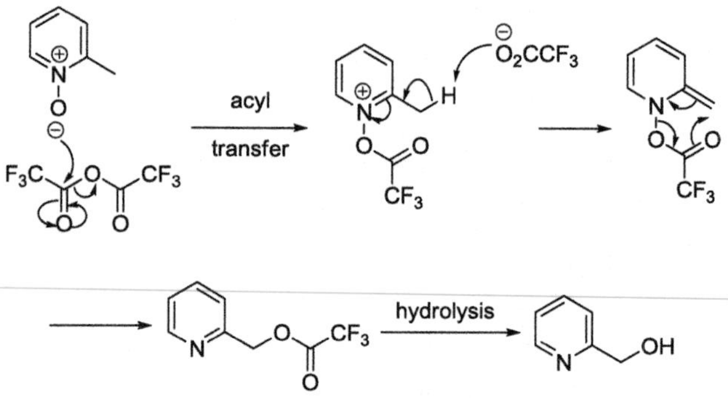

Example 1[4]

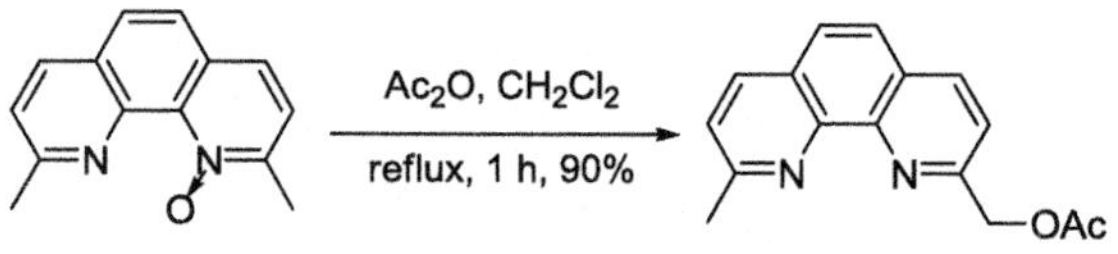

Example 2[6]

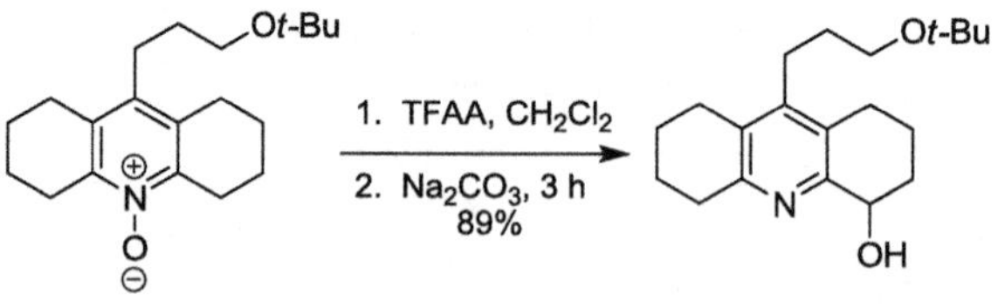

Example 3, Double Boekelheide reaction[8]

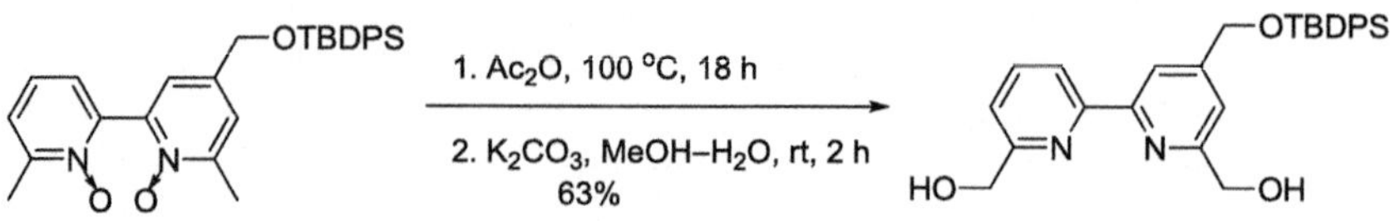

J.J. Li, *Name Reactions: A Collection of Detailed Mechanisms and Synthetic Applications*,
DOI 10.1007/978-3-319-03979-4_30, © Springer International Publishing Switzerland 2014

Example 4[9]

Example 5, Double Boekelheide reaction[10]

**References**

1. Boekelheide, V.; Linn, W. J. *J. Am. Chem. Soc.* **1954**, *76*, 1286–1291. Virgil Boekelheide (1919–2003) was a professor at the University of Oregon.
2. Boekelheide, V.; Harrington, D. L. *Chem. Ind.* **1955**, 1423–1424.
3. Katritzky, A. R.; Lagowski, J. M. *Chemistry of the Heterocylic N-Oxides,* Academic Press: NY, **1971**. (Review).
4. Newkome, G. R.; Theriot, K. J.; Gupta, V. K.; Fronczek, F. R.; Baker, G. R. *J. Org. Chem.* **1989**, *54*, 1766–1769.
5. Katritzky, A. R.; Lam, J. N. *Heterocycles* **1992,** *33*, 1011–1049. (Review).
6. Fontenas, C.; Bejan, E.; Haddou, H. A.; Balavoine, G. G. A. *Synth. Commun.* **1995,** *25*, 629–633.
7. Galatsis, P. *Boekelheide Reaction.* In *Name Reactions in Heterocyclic Chemistry*; Li, J. J., Ed.; Wiley: Hoboken, NJ, **2005**, pp 340–349. (Review).
8. Havas, F.; Danel, M.; Galaup, C.; Tisnès, P.; Picard, C. *Tetrahedron Lett.* **2007,** *48*, 999–1002.
9. Dai, L.; Fan, D.; Wang, X.; Chen, Y. *Synth. Commun.* **2008,** *38*, 576–582.
10. Das, S. K.; Frey, J. *Tetrahedron Lett.* **2012,** *53*, 3869–3872.

## Boger pyridine synthesis

Pyridine synthesis *via* hetero-Diels–Alder reaction of 1,2,4-triazines and dienophiles (e.g., enamine) followed by extrusion of $N_2$.

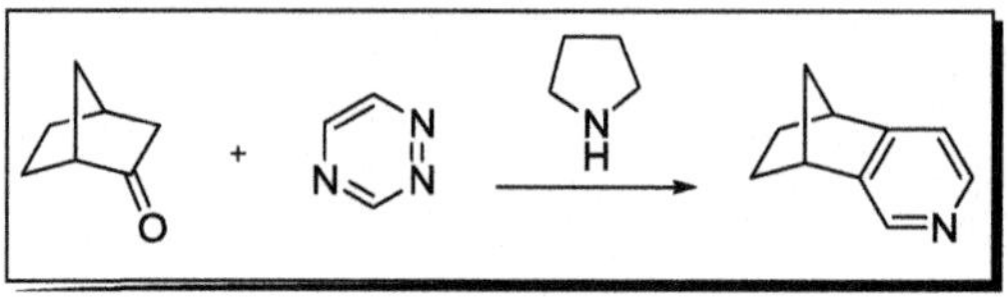

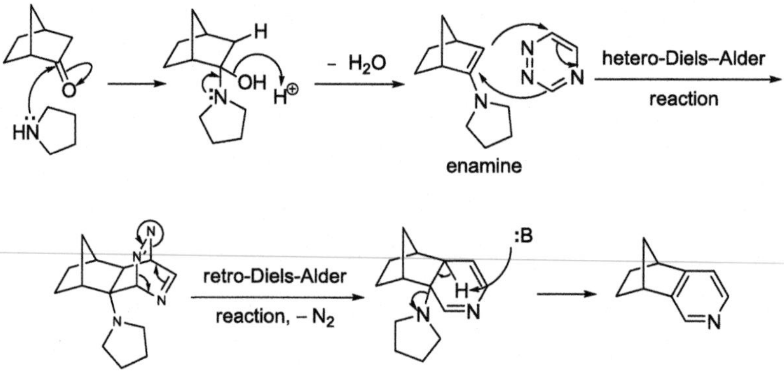

enamine

Example 1[3]

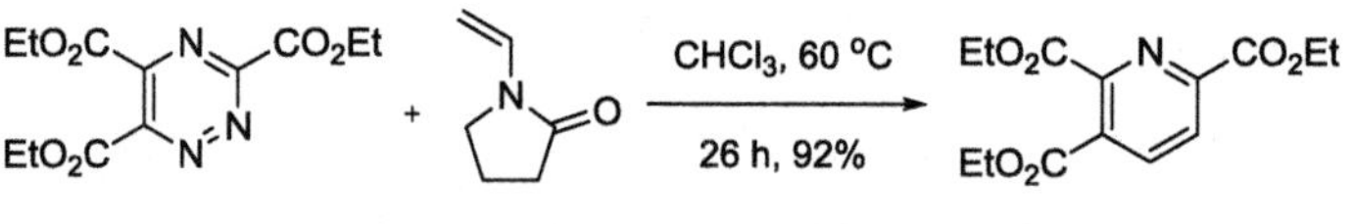

Example 2, Intramolecular Boger pyridine synthesis[8]

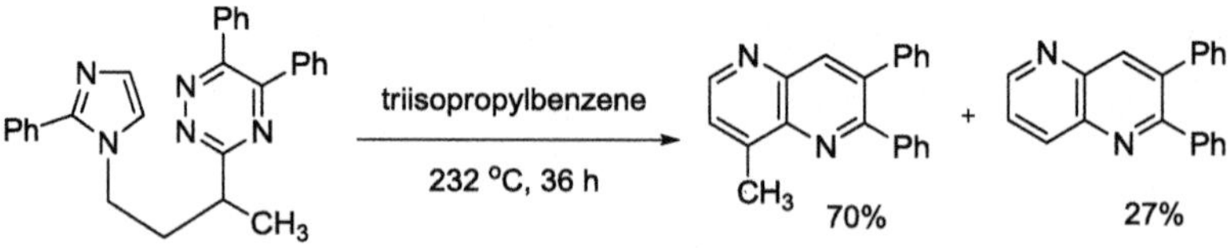

J.J. Li, *Name Reactions: A Collection of Detailed Mechanisms and Synthetic Applications*, DOI 10.1007/978-3-319-03979-4_31, © Springer International Publishing Switzerland 2014

Example 3[10]

1. pyrrolidine, xylene, Δ
   10 h, sealed tube
2. silica, 5 h, 67%

Example 4[11]

p-cumene, 150 °C, 30 h
67%

## References

1. Boger, D. L.; Panek, J. S. *J. Org. Chem.* **1981**, *46*, 2179–2182. Dale Boger obtained his Ph.D. under E.J. Corey at Harvard University in 1980. He started his independent career at the University of Kansas, moving onto Purdue University, and currently he is a professor at The Scripps Research Institute.
2. Boger, D. L. *Tetrahedron* **1983**, *39*, 2869–2939. (Review).
3. Boger, D. L.; Panek, J. S.; Yasuda, M. *Org. Synth.* **1988**, *66*, 142–150.
4. Boger, D. L. In *Comprehensive Organic Synthesis;* Trost, B. M.; Fleming, I., Eds.; Pergamon, **1991**, *Vol. 5*, 451–512. (Review).
5. Behforouz, M.; Ahmadian, M.*Tetrahedron* **2000**, *56*, 5259–5288. (Review).
6. Buonora, P.; Olsen, J.-C.; Oh, T. *Tetrahedron* **2001**, *57*, 6099–6138. (Review).
7. Jayakumar, S.; Ishar, M. P. S.; Mahajan, M. P. *Tetrahedron* **2002**, *58*, 379–471. (Review).
8. Lahue, B. R.; Lo, S.-M.; Wan, Z.-K.; Woo, G. H. C.; Snyder, J. K. *J. Org. Chem.* **2006**, *69*, 7171–7182.
9. Galatsis, P. *Boger Reaction*. In *Name Reactions in Heterocyclic Chemistry*; Li, J. J., Ed.; Wiley: Hoboken, NJ, **2005**, pp 323–339. (Review).
10. Catozzi, N.; Bromley, W. J.; Wasnaire, P.; Gibson, M.; Taylor, R. J. K. *Synlett* **2007**, 2217–2221.
11. Lawecka, J.; Bujnicki, B.; Drabowicz, J.; Rykowski, A. *Tetrahedron Lett.* **2008**, *49*, 719–722.

## Borch reductive amination

Reduction (often using NaCNBH$_3$) of an imine, formed by condensation of an amine and a carbonyl, to afford the corresponding amine.

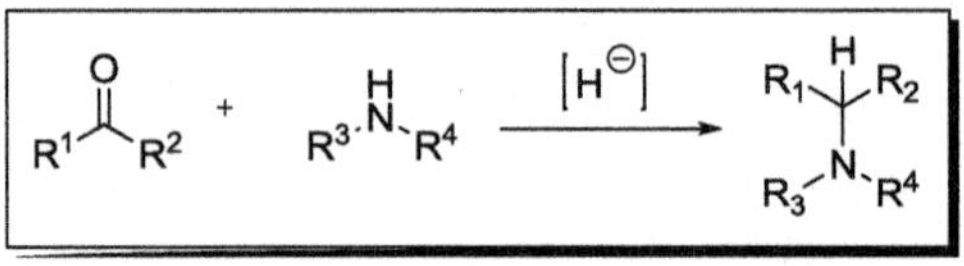

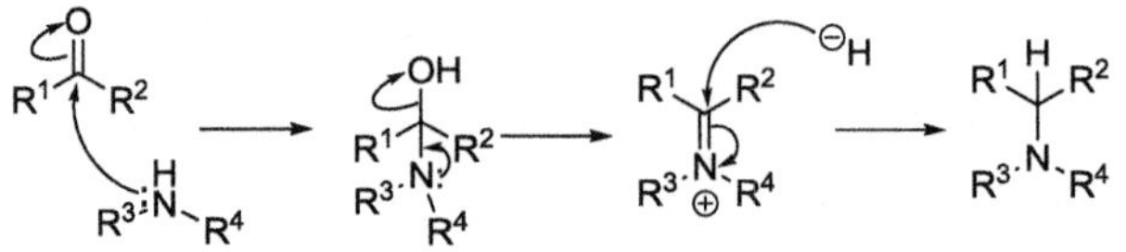

Example 1[4]

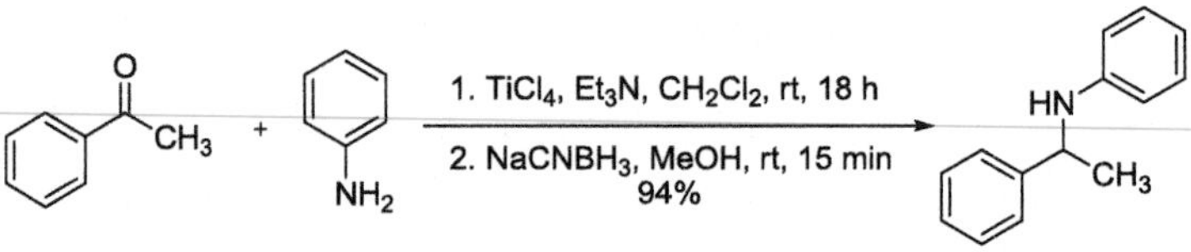

Example 2[5]

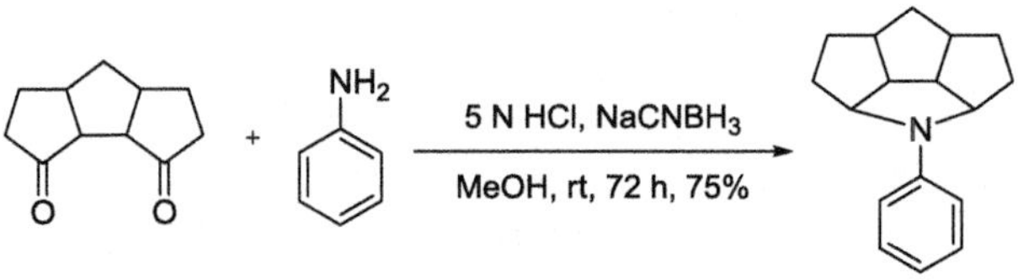

Example 3[8]

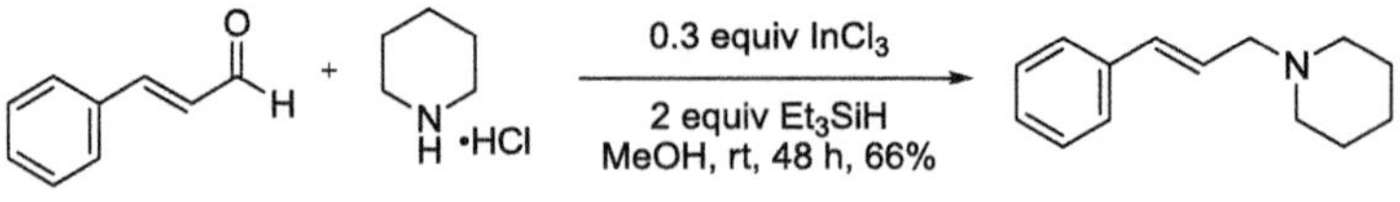

Example 4[9]

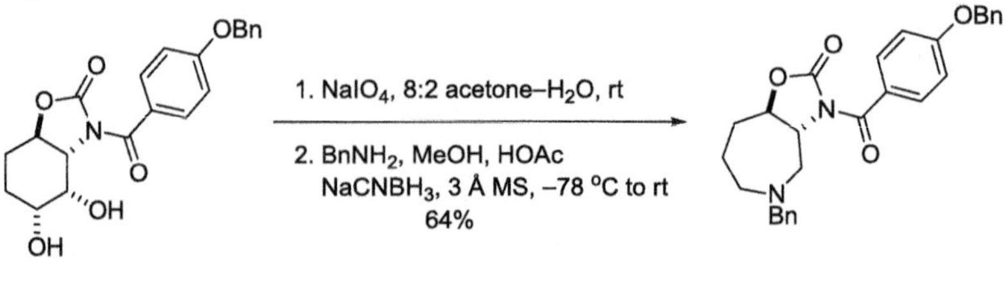

J.J. Li, *Name Reactions: A Collection of Detailed Mechanisms and Synthetic Applications*, DOI 10.1007/978-3-319-03979-4_32, © Springer International Publishing Switzerland 2014

**References**
1.  Borch, R. F., Durst, H. D. *J. Am. Chem. Soc.* **1969,** *91*, 3996–3997. Richard F. Borch, born in Cleveland, Ohio, was a professor at the University of Minnesota.
2.  Borch, R. F.; Bernstein, M. D.; Durst, H. D. *J. Am. Chem. Soc.* **1971,** *93*, 2897–2904.
3.  Borch, R. F.; Ho, B. C. *J. Org. Chem.* **1977,** *42*, 1225–1227.
4.  Barney, C. L.; Huber, E. W.; McCarthy, J. R. *Tetrahedron Lett.* **1990,** *31*, 5547–5550.
5.  Mehta, G.; Prabhakar, C. *J. Org. Chem.* **1995,** *60*, 4638–4640.
6.  Lewin, G.; Schaeffer, C. *Heterocycles* **1998,** *48*, 171–174.
7.  Lewin, G.; Schaeffer, C.; Hocquemiller, R.; Jacoby, E.; Léonce, S.; Pierré, A.; Atassi, G. *Heterocycles* **2000,** *53*, 2353–2356.
8.  Lee, O.-Y.; Law, K.-L.; Ho, C.-Y.; Yang, D. *J. Org. Chem.* **2008,** *73*, 8829–8837.
9.  Sullivan, B.; Hudlicky, T. *Tetrahedron Lett.* **2008,** *49*, 5211–5213.
10. Koszelewski, D.; Lavandera, I.; Clay, D.; Guebitz, G. M.; Rozzell, D.; Kroutil, W. *Angew. Chem. Int. Ed.* **2008,** *47*, 9337–9340.

## Borsche–Drechsel cyclization

Also known as Borsche carbazole synthesis, is the two-step conversion of phenyl
hydrazine and cyclohexanone derivatives to the corresponding carbazole. *Cf.*
Fischer indole synthesis.

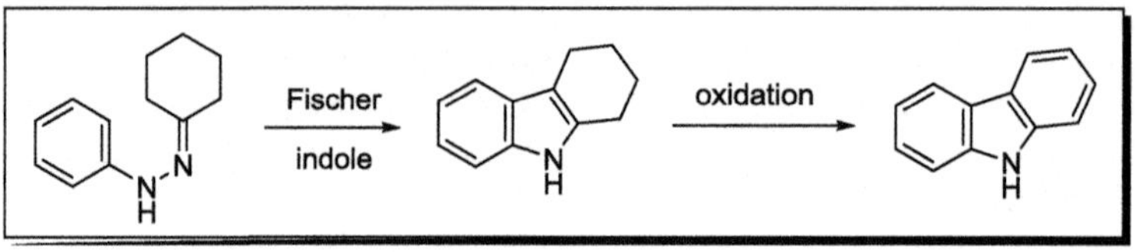

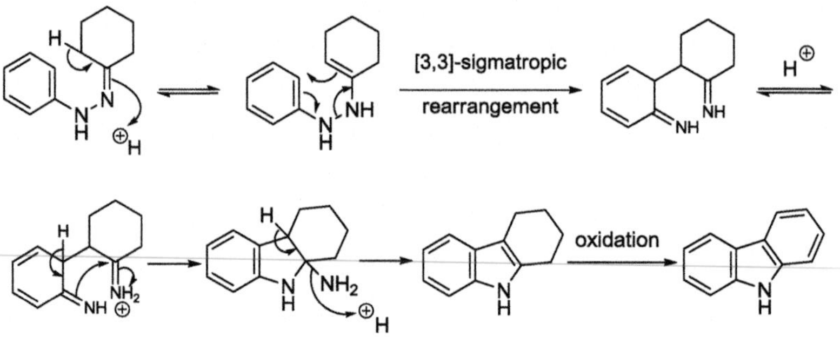

Example 1[6]

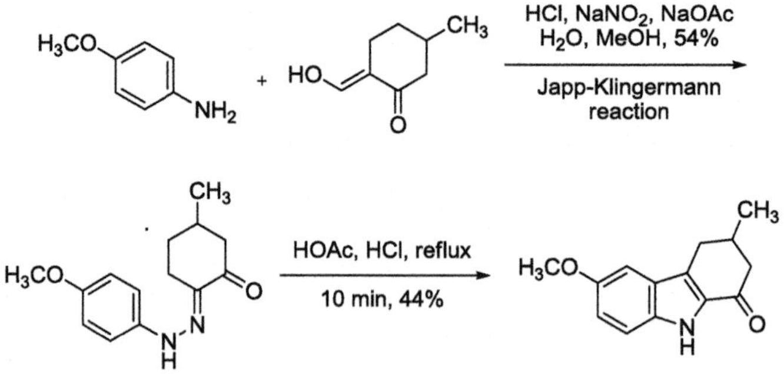

Example 2[9]

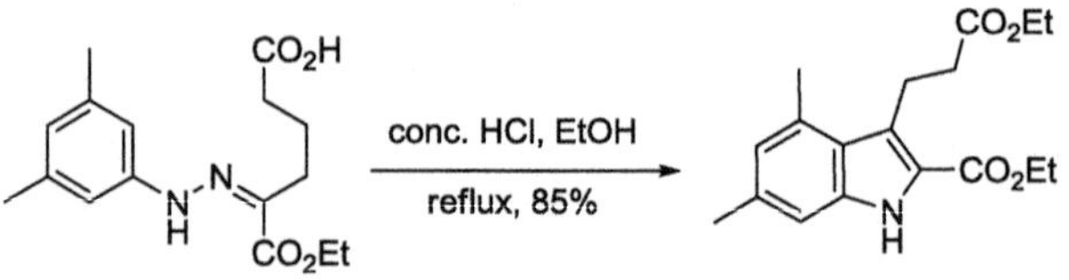

J.J. Li, *Name Reactions: A Collection of Detailed Mechanisms and Synthetic Applications*,
DOI 10.1007/978-3-319-03979-4_33, © Springer International Publishing Switzerland 2014

Example 3[10]

Japp-Klingemann hydrazone synthesis

**References**
1. Drechsel, E. *J. Prakt. Chem.* **1858**, *38*, 69.
2. Borsche, W.; Feise, M. *Ann.* **1908**, *359*, 49–80. Walther Borsche was a professor at Chemischen Institut, Universität Göttingen, Germany when this paper was published. Borsche was completely devoid of the arrogance shown by many of his contemporaries. Both Borsche and his colleague at Frankfurt, Julius von Braun, suffered under the Nazi regime for their independent minds.
3. Bruck, P. *J. Org. Chem.* **1970**, *35*, 2222–2227.
4. Gazengel, J.-M.; Lancelot, J.-C.; Rault, S.; Robba, M. *J. Heterocycl. Chem.* **1990**, *27*, 1947–1951.
5. Abramovitch, R. A.; Bulman, A. *Synlett* **1992**, 795–796.
6. Lin, G.; Zhang, A. *Tetrahedron* **2000**, *56*, 7163–7171.
7. Ergun, Y.; Bayraktar, N.; Patir, S.; Okay, G. *J. Heterocycl. Chem.* **2000**, *37*, 11–14.
8. Rebeiro, G. L.; Khadilkar, B. M. *Synthesis* **2001**, 370–372.
9. Takahashi, K.; Kasai, M.; Ohta, M.; Shoji, Y.; Kunishiro, K.; Kanda, M.; Kurahashi, K.; Shirahase, H. *J. Med. Chem.* **2008**, *51*, 4823–4833.
10. Pete, B. *Tetrahedron Lett.* **2008**, *49*, 2835–2838.
11. Fultz, M. W. *Borsche–Drechsel Cyclization*, In *Name Reactions in Heterocyclic Chemistry II*; Li, J. J., Ed.; Wiley: Hoboken, NJ, **2011**, pp 91–101. (Review).

# Boulton–Katritzky rearrangement

Rearrangement of one five-membered heterocycle into another under thermolysis.

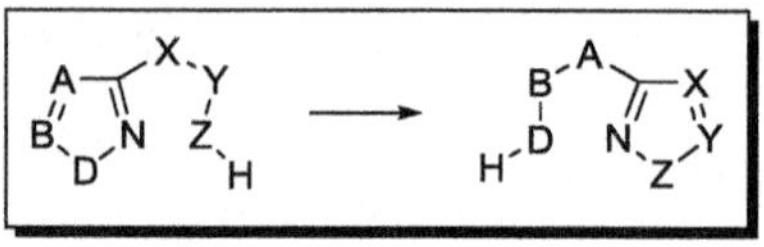

Example 1[4]

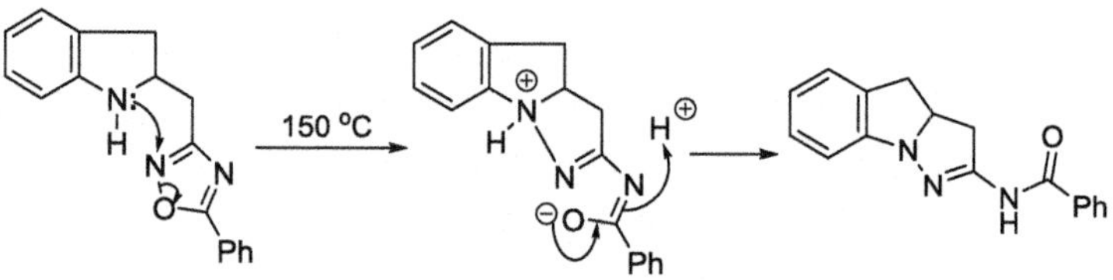

## Example 2, Hydrazinolysis[7]

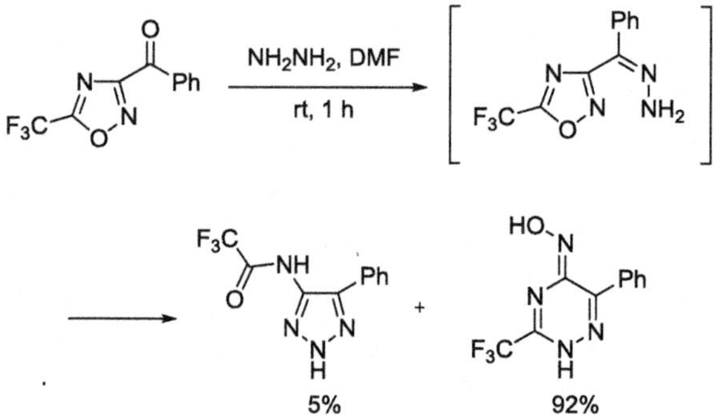

5%          92%

Example 4[3]

t-BuOK, DMF

reflux, 2 h, 75%

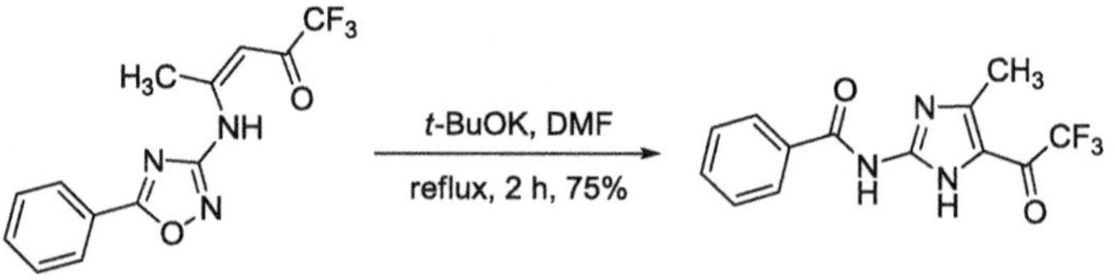

J.J. Li, *Name Reactions: A Collection of Detailed Mechanisms and Synthetic Applications*,
DOI 10.1007/978-3-319-03979-4_34, © Springer International Publishing Switzerland 2014

Example 5[12]

$$\xrightarrow{\Delta}$$
104 h, 62%

## References

1. Boulton, A. J.; Katritzky, A. R.; Majid Hamid, A. *J. Chem. Soc. (C)* **1967**, 2005–2007. Alan Katritzky, a professor at the University of Florida, is best known for his series *Advances of Heterocyclic Chemistry*, now in its 107th volume.
2. Ruccia, M.; Vivona, N.; Spinelli, D. *Adv. Heterocycl. Chem.* **1981**, *29*, 141–169. (Review).
3. Vivona, N.; Buscemi, S.; Frenna, V.; Gusmano, C. *Adv. Heterocyl. Chem.* **1993**, *56*, 49–154. (Review).
4. Katayama, H.; Takatsu, N.; Sakurada, M.; Kawada, Y. *Heterocycles* **1993**, *35*, 453–459.
5. Rauhut, G. *J. Org. Chem.* **2001**, *66*, 5444–5448.
6. Crampton, M. R.; Pearce, L. M.; Rabbitt, L. C. *J. Chem. Soc., Perkin Trans. 2* **2002**, 257–261.
7. Buscemi, S.; Pace, A.; Piccionello, A. P.; Macaluso, G.; Vivona, N.; Spinelli, D.; Giorgi, G. *J. Org. Chem.* **2005**, *70*, 3288–3291.
8. Pace, A.; Pibiri, I.; Piccionello, A. P.; Buscemi, S.; Vivona, N.; Barone, G. *J. Org. Chem.* **2007**, *64*, 7656–7666.
9. Piccionello, A. P.; Pace, A.; Buscemi, S.; Vivona, N.; Pani, M. *Tetrahedron* **2008**, *64*, 4004–4010.
10. Pace, A.; Pierro, P.; Buscemi, S.; Vivona, N.; Barone, G. *J. Org. Chem.* **2009**, *74*, 351–358.
11. Corbett M. T.; Mullins, R. J. *Boulton–Katritzky rearrangement*, In *Name Reactions in Heterocyclic Chemistry II*; Li, J. J., Ed.; Wiley: Hoboken, NJ, **2011**, pp 527–538. (Review).
12. Ott, G. R.; Anzalone, A. V. *Synlett* **2011**, 3018–3022.

## Bouveault aldehyde synthesis

Formylation of an alkyl or aryl halide to the homologous aldehyde by transformation to the corresponding organometallic reagent then addition of DMF (M = Li, Mg, Na, and K).

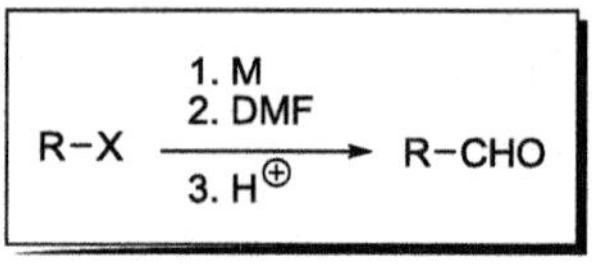

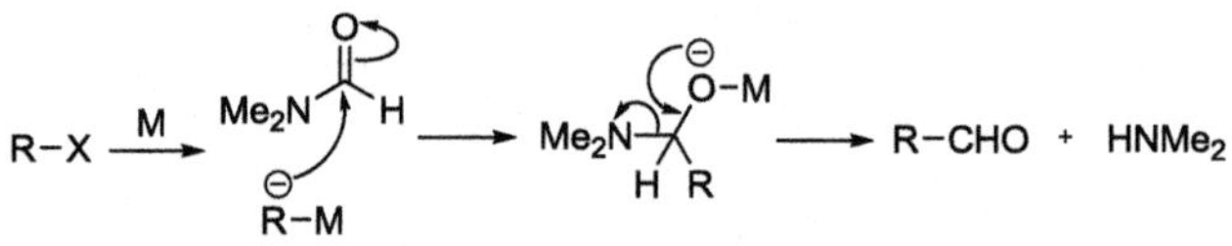

$$R-X \quad \xrightarrow[\substack{3.\ H^{\oplus}}]{\substack{1.\ M \\ 2.\ DMF}} \quad R-CHO$$

$$R-X \xrightarrow{M} \underset{\underset{R-M}{\ominus}}{Me_2N} \overset{O}{\underset{H}{\overset{\displaystyle\frown}{\parallel}}} H \longrightarrow Me_2N \overset{\ominus}{\underset{H\ R}{\overset{O-M}{\diagup}}} \longrightarrow R-CHO \ + \ HNMe_2$$

Comins modification:[4]

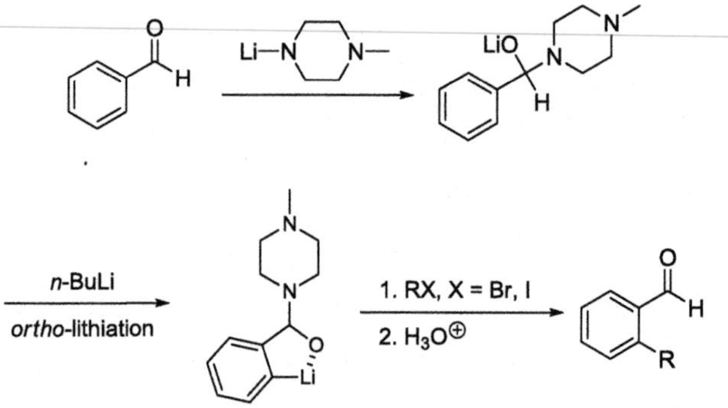

Example 1[3]

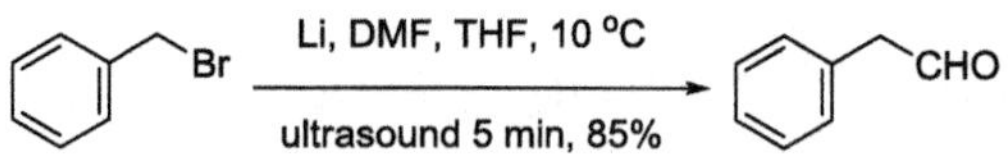

Li, DMF, THF, 10 °C

ultrasound 5 min, 85%

Example 2, A modified Bouveault reaction[7]

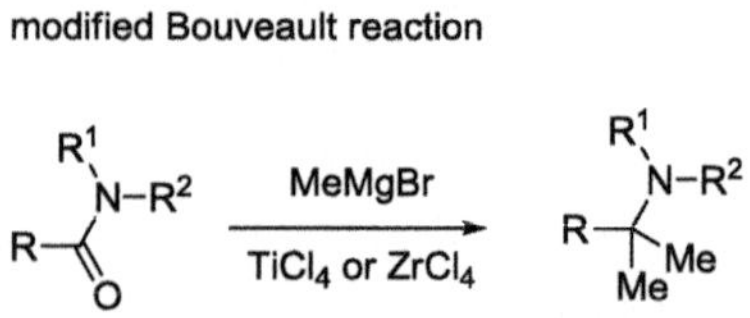

modified Bouveault reaction

MeMgBr

TiCl₄ or ZrCl₄

J.J. Li, *Name Reactions: A Collection of Detailed Mechanisms and Synthetic Applications*, DOI 10.1007/978-3-319-03979-4_35, © Springer International Publishing Switzerland 2014

## References

1. Bouveault, L. *Bull. Soc. Chim. Fr.* **1904,** *31,* 1306–1322, 1322–1327. Louis Bouveault (1864–1909) was born in Nevers, France. He devoted his short yet very productive life to teaching and to working in science.

2. Sicé, J. *J. Am. Chem. Soc.* **1953,** *75,* 3697–3700.

3. Pétrier, C.; Gemal, A. L.; Luche, J.-L. *Tetrahedron Lett.* **1982,** *23,* 3361–3364.

4. Comins, D. L.; Brown, J. D. *J. Org. Chem.* **1984,** *49,* 1078–1083.

5. Einhorn, J.; Luche, J. L. *Tetrahedron Lett.* **1986,** *27,* 1793–1796.

6. Meier, H.; Aust, H. *J. Prakt. Chem.* **1999,** *341,* 466–471.

7. Dillon, B R.; Roberts, D F.; Entwistle, D A.; Glossop, P A.; Knight, C. J.; Laity, D. A.; James, K.; Praquin, C. F.; Strang, R. S.; Watson, C. A. L. *Org. Process Res. Dev.* **2012,** *16,* 195–203.

## Bouveault–Blanc reduction

Also known as the Bouveault reaction. Reduction of esters to the corresponding alcohols using sodium in an alcoholic solvent.

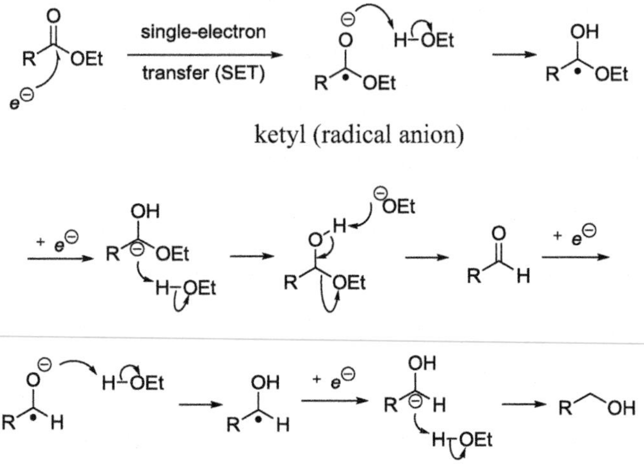

ketyl (radical anion)

Example[2]

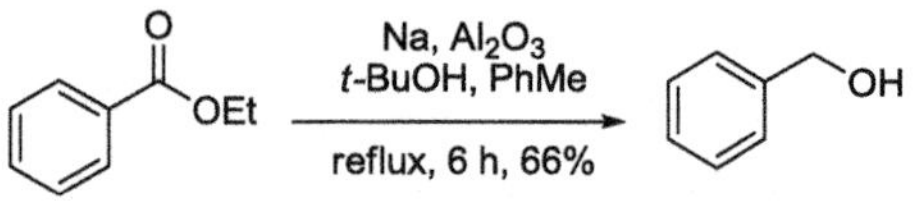

### References

1.  Bouveault, L.; Blanc, G. *Compt. Rend. Hebd. Seances Acad. Sci.* **1903**, *136*, 1676–1678.
2.  Bouveault, L.; Blanc, G. *Bull. Soc. Chim.* **1904**, *31*, 666–672.
3.  Rühlmann, K.; Seefluth, H.; Kiriakidis, T.; Michael, G.; Jancke, H.; Kriegsmann, H. *J. Organomet. Chem.* **1971**, *27*, 327–332.
4.  Seo, B.-I.; Wall, L. K.; Lee, H.; Buttrum, J. W.; Lewis, D. E. *Synth. Commun.* **1993**, *23*, 15–22.
5.  Singh, S.; Dev, S. *Tetrahedron* **1993**, *49*, 10959–10964.
6.  Schopohl, M. C.; Bergander, K.; Kataeva, O.; Föehlich, R.; Waldvogel, S. R. *Synthesis* **2003**, 2689–2694.

J.J. Li, *Name Reactions: A Collection of Detailed Mechanisms and Synthetic Applications*, DOI 10.1007/978-3-319-03979-4_36, © Springer International Publishing Switzerland 2014

# Boyland–Sims oxidation

Oxidation of anilines to phenols using alkaline persulfate.

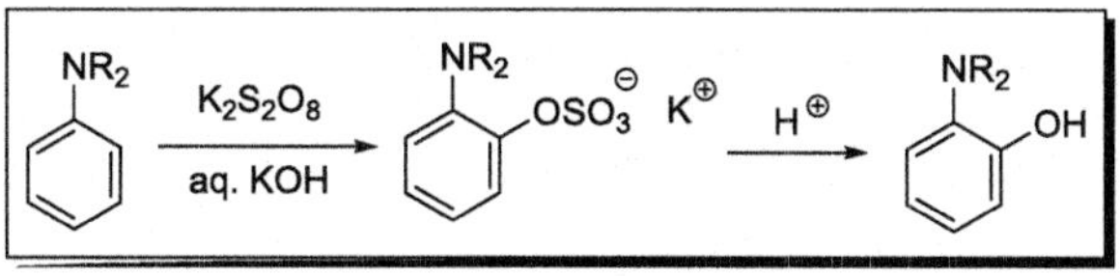

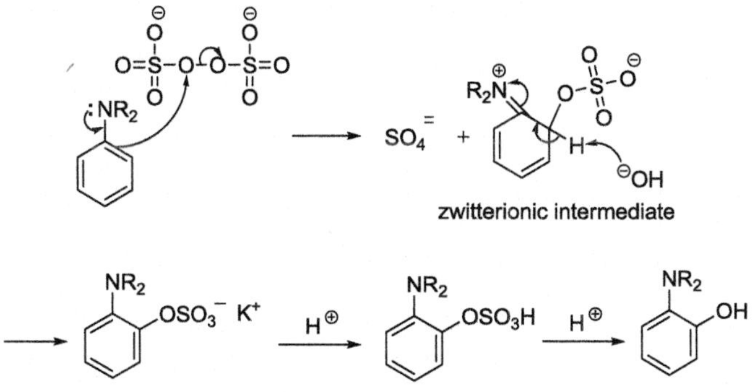

zwitterionic intermediate

Another pathway is also operative:[9–12]

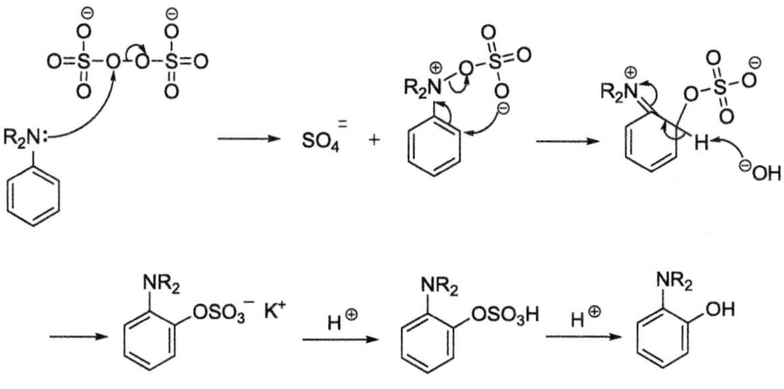

Example 1[3]

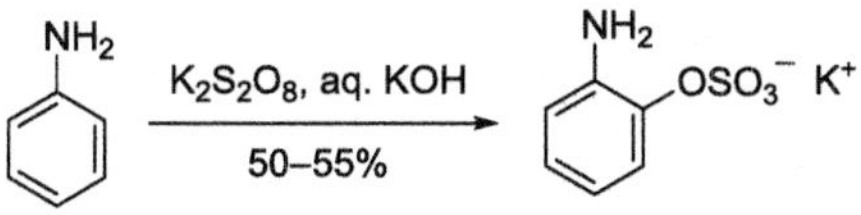

J.J. Li, *Name Reactions: A Collection of Detailed Mechanisms and Synthetic Applications*, DOI 10.1007/978-3-319-03979-4_37, © Springer International Publishing Switzerland 2014

## *Elbs oxidation*

Also known as the Elbs persulfate oxidation,[13-15] it is a variant of the Boyland–Sims oxidation except the substrate is phenol rather than aniline. Its mechanism is similar to that of the Boyland–Sims oxidation.

### References

1.  Boyland, E.; Manson, D.; Sims, P. *J. Chem. Soc.* **1953**, 3623–3628. Eric Boyland and Peter Sims were at the Royal Cancer Hospital in London, UK.
2.  Boyland, E.; Sims, P. *J. Chem. Soc.* **1954**, 980–985.
3.  Behrman, E. J. *J. Am. Chem. Soc.* **1967**, *89*, 2424–2428.
4.  Behrman, E. J.; Behrman, D. M. *J. Org. Chem.* **1978**, *43*, 4551–4552.
5.  Srinivasan, C.; Perumal, S.; Arumugam, N. *J. Chem. Soc., Perkin Trans. 2* **1985**, 1855–1858.
6.  Behrman, E. J. *Org. React.* **1988**, *35*, 421–511. (Review).
7.  Behrman, E. J. *J. Org. Chem.* **1992**, *57*, 2266–2270.
8.  Behrman, E. J. *Beilstein J. Org. Chem.* **2006**, *2*, 22.
9.  Behrman, E. J. *Chem. Educator* **2010**, *15*, 392–393.
10. Behrman, E. J. *J. Phys. Chem.* **2011**, *115*, 7863–7864.
11. Marjanović, B.; Juranić, I.; Ćiric-Marjanović, G. *J. Phys. Chem.* **2011**, *115*, 3536–3550.
12. Marjanović, B.; Juranić, I.; Ćiric-Marjanović, G. *J. Phys. Chem.* **2011**, *115*, 7865–7868.
13. Elbs, K. *J. Prakt. Chem.* **1893**, *48*, 179–185.
14. Watson, K. G.; Serban, A. *Aust. J. Chem.* **1995**, *48*, 1503–1509.
15. Dai, K.; Wen, Q.; Liu, X.; Chen, S. *Faming Zhuanli Shenqing* CN102408401 (**2012**).

# Bradsher reaction

The intramolecular Bradsher cyclization refers to the acid-catalyzed aromatic cyclodehydration of *ortho*-acyl diarylmethanes to form anthracenes. On the other hand, the intermolecular Bradsher cycloaddition often involves the Diels–Alder reaction of a pyridium with a vinyl ether or vinyl sulfide.

Example 1, Intramolecular Bradsher reaction[2]

Example 2, Intramolecular Bradsher reaction[5]

J.J. Li, *Name Reactions: A Collection of Detailed Mechanisms and Synthetic Applications*, DOI 10.1007/978-3-319-03979-4_38, © Springer International Publishing Switzerland 2014

Example 3, Intermolecular Bradsher cycloaddition[8]

DNP = dinitrophenyl

Example 4, Intermolecular Bradsher cycloaddition[11]

**References**
1.  (a) Bradsher, C. K. *J. Am. Chem. Soc.* **1940,** *62*, 486–488.  Charles K. Bradsher was born in Petersburg, VA in 1912.  After his Ph.D. under Louis F. Fieser at Harvard and postdoctoral training with R. C. Fuson, he became a professor at Duke University.  (b) Bradsher, C. K.; Smith, E. S. *J. Am. Chem. Soc.* **1943,** *65*, 451–452.  (c) Bradsher, C. K.; Vingiello, F. A. *J. Org. Chem.* **1948,** *13*, 786–789.  (d) Bradsher, C. K.; Sinclair, E. F. *J. Org. Chem.* **1957,** *22*, 79–81.
2.  Vingiello, F. A.; Spangler, M. O. L.; Bondurant, J. E. *J. Org. Chem.* **1960,** *25*, 2091–2094.
3.  Brice, L. K.; Katstra, R. D. *J. Am. Chem. Soc.* **1960,** *82*, 2669–2670.
4.  Saraf, S. D.; Vingiello, F. A. *Synthesis* **1970,** 655.
5.  Ahmed, M.; Ashby, J.; Meth-Cohn, O. *J. Chem. Soc., Chem. Commun.* **1970,** 1094–1095.
6.  Ashby, J.; Ayad, M.; Meth-Cohn, O. *J. Chem. Soc., Perkin Trans. 1* **1974,** 1744–1747.
7.  Bradsher, C. K. *Chem. Rev.* **1987,** *87*, 1277–1297.  (Review).
8.  Nicolas, T. E.; Franck, R. W. *J. Org. Chem.* **1995,** *60*, 6904–6911.
9.  Magnier, E.; Langlois, Y. *Tetrahedron Lett.* **1998,** *39*, 837–840.
10. Urban, D.; Duval, E.; Langlois, Y. *Tetrahedron Lett.* **2000,** *41*, 9251–9256.
11. Soll, C. E.; Franck, R. W. *Heterocycles* **2006,** *70*, 531–540.
13. Mondal, M.; Kerrigan, N. J. *Bradsher Reaction, In Name Reactions for Carbocyclic Ring Formations,* Li, J. J., Ed.; Wiley: Hoboken, NJ, **2010,** pp 251–266. (Review).

# Brook rearrangement

Rearrangement of α-silyl oxyanions to α-silyloxy carbanions *via* a reversible process involving a pentacoordinate silicon intermediate is known as the [1,2]-Brook rearrangement, or [1,2]-silyl migration.

[1,2]-Brook rearrangement

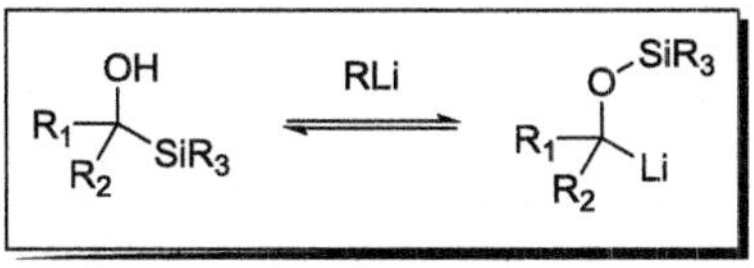

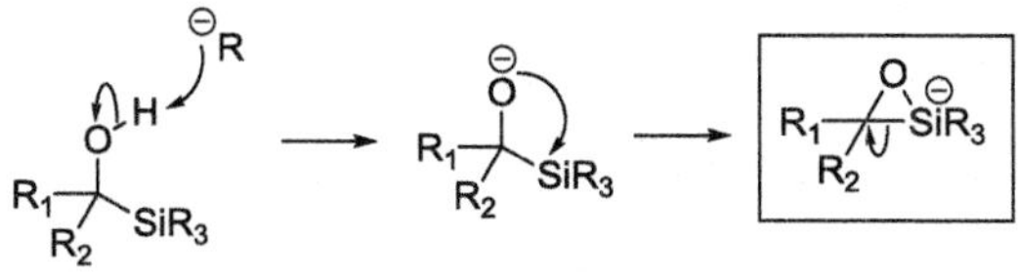

pentacoordinate silicon intermediate

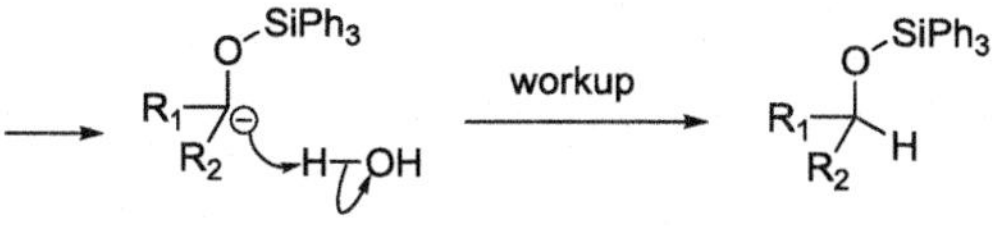

[1,3]-Brook rearrangement

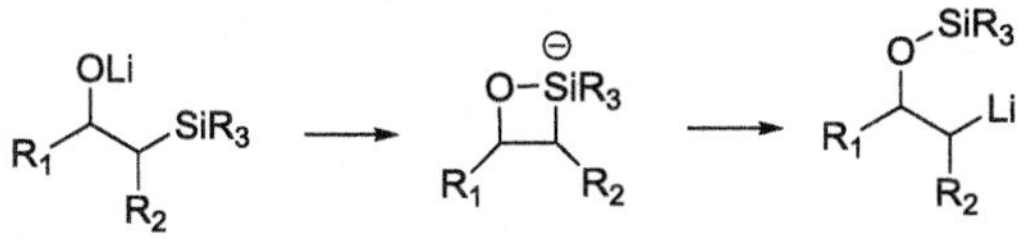

[1,4]-Brook rearrangement

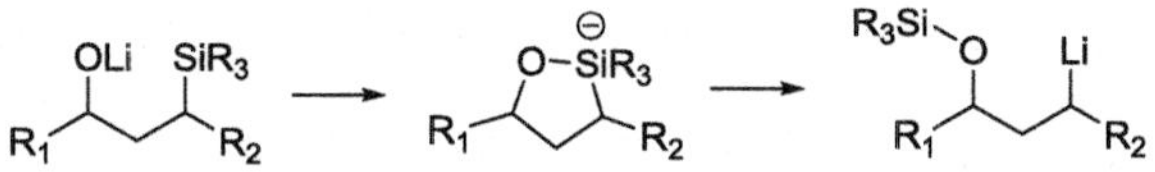

Example 1[6]

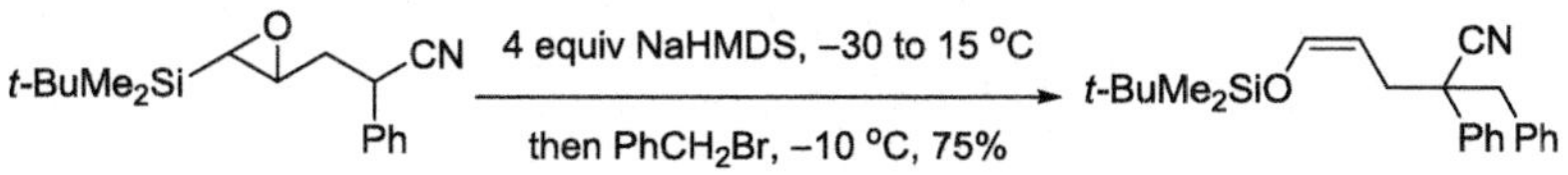

J.J. Li, *Name Reactions: A Collection of Detailed Mechanisms and Synthetic Applications*,
DOI 10.1007/978-3-319-03979-4_39, © Springer International Publishing Switzerland 2014

Example 2, [1,2]-Brook rearrangement followed by a retro-[1,5]-Brook rearrangement[8]

Example 3, [1,5]-Brook rearrangement[9]

Example 4, Retro-[1,4]-Brook rearrangement[10]

Example 5, Retro-Brook rearrangement[12]

**References**

1.  Brook, A. G. *J. Am. Chem. Soc.* **1958**, *80*, 1886–1889. Adrian G. Brook (1924–) was born in Toronto, Canada. He was a professor in Lash Miller Chemical Laboratories, University of Toronto, Canada.
2.  Brook, A. G. *Acc. Chem. Res.* **1974**, *7*, 77–84. (Review).
3.  Bulman Page, P. C.; Klair, S. S.; Rosenthal, S. *Chem. Soc. Rev.* **1990**, *19*, 147–195. (Review).
4.  Fleming, I.; Ghosh, U. *J. Chem. Soc., Perkin Trans. 1* **1994**, 257–262.
5.  Moser, W. H. *Tetrahedron* **2001**, *57*, 2065–2084. (Review).
6.  Okugawa, S.; Takeda, K. *Org. Lett.* **2004**, *6*, 2973–2975.
7.  Matsumoto, T.; Masu, H.; Yamaguchi, K.; Takeda, K. *Org. Lett.* **2004**, *6*, 4367–4369.
8.  Clayden, J.; Watson, D. W.; Chambers, M. *Tetrahedron* **2005**, *61*, 3195–3203.
9.  Smith, A. B., III; Xian, M.; Kim, W.-S.; Kim, D.-S. *J. Am. Chem. Soc.* **2006**, *128*, 12368–12369.
10. Mori, Y.; Futamura, Y.; Horisaki, K. *Angew. Chem. Int. Ed.* **2008**, *47*, 1091–1093.
11. Greszler, S. N.; Johnson, J. S. *Org. Lett.* **2009**, *11*, 827–830.
12. He, Y.; Hu, H.; Xie, X.; She, X. *Tetrahedron* **2013**, *69*, 559–563.

# Brown hydroboration

Addition of boranes to olefins followed by alkalinic oxidation of the organoborane adducts to afford alcohols. Regiochemistry is anti-Markovnikov's rule.

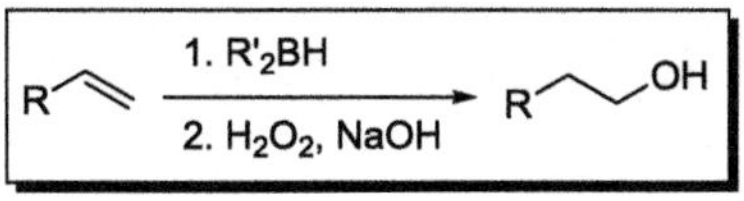

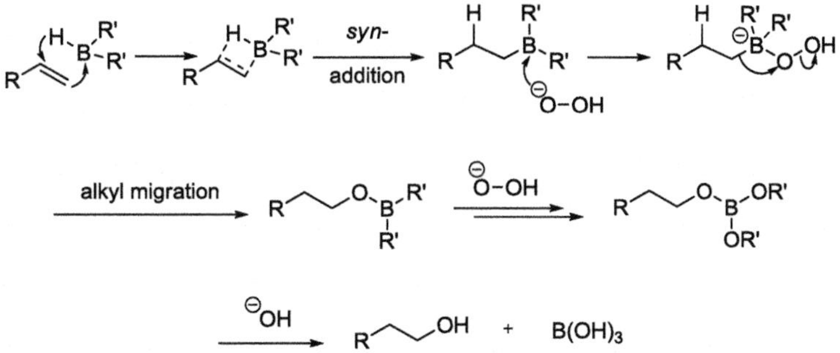

Example 1[2]

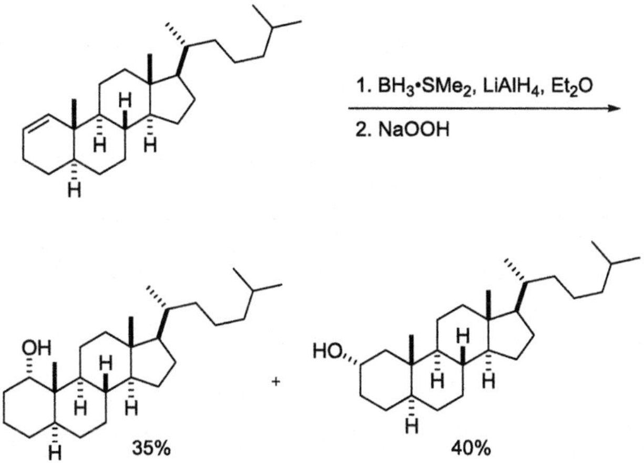

Example 2[7]

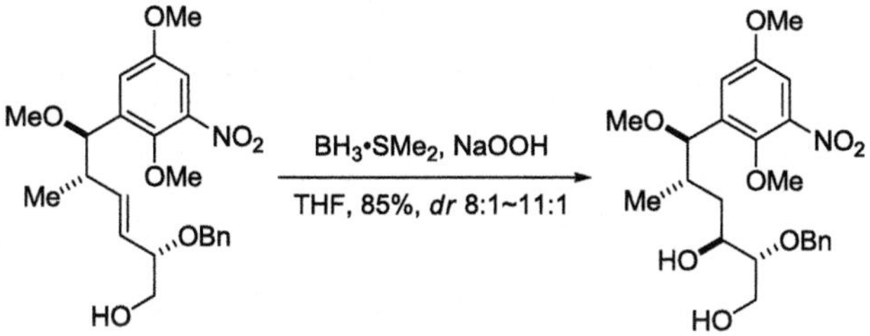

J.J. Li, *Name Reactions: A Collection of Detailed Mechanisms and Synthetic Applications*, DOI 10.1007/978-3-319-03979-4_40, © Springer International Publishing Switzerland 2014

Example 3[8]

Example 4, Asymmetric hydroboration[10]

Example 5[11]

## References

1. Brown, H. C.; Tierney, P. A. *J. Am. Chem. Soc.* **1958**, *80*, 1552–1558.  Herbert C. Brown (USA, 1912–2004) began his academic career at Wayne State University and moved on to Purdue University where he shared the Nobel Prize in Chemistry in 1981 with Georg Wittig (Germany, 1897–1987) for their development of organic boron and phosphorous compounds.
2. Nussim, M.; Mazur, Y.; Sondheimer, F. *J. Org. Chem.* **1964**, *29*, 1120–1131.
3. Pelter, A.; Smith, K.; Brown, H. C. *Borane Reagents,* Academic Press: New York, **1972**. (Book).
4. Brewster, J. H.; Negishi, E. *Science* **1980**, *207*, 44–46. (Review).
5. Fu, G. C.; Evans, D. A.; Muci, A. R. *Advances in Catalytic Processes* **1995**, *1*, 95–121. (Review).
6. Hayashi, T. *Comprehensive Asymmetric Catalysis I–III* **1995**, *1*, 351–364. (Review).
7. Carter K. D.; Panek J. S. *Org. Lett.* **2004**, *6*, 55–57.
8. Clay, J. M.; Vedejs, E. *J. Am. Chem. Soc.* **2005**, *127*, 5766–5767.
9. Clay, J. M. *Brown hydroboration reaction.* In *Name Reactions for Functional Group Transformations*; Li, J. J., Ed.; Wiley: Hoboken, NJ, **2007**, pp 183–188.  (Review).
10. Smith, S. M.; Thacker, N. C.; Takacs, J. M. *J. Am. Chem. Soc.* **2008**, *130*, 3734–3735.
11. Anderson, L. L.; Woerpel, K. A. *Org. Lett.* **2009**, *11*, 425–428.
12. Yadav, J. S.; Kavita, A.; Raghavendra Rao, K. V.; Mohapatra, D. K. *Tetrahedron Lett.* **2013**, *54*, 1710–1713.

# Bucherer carbazole synthesis

Carbazole formation from naphthols and aryl hydrazines promoted by sodium bisulfite. Another variant of the Fischer indole synthesis.

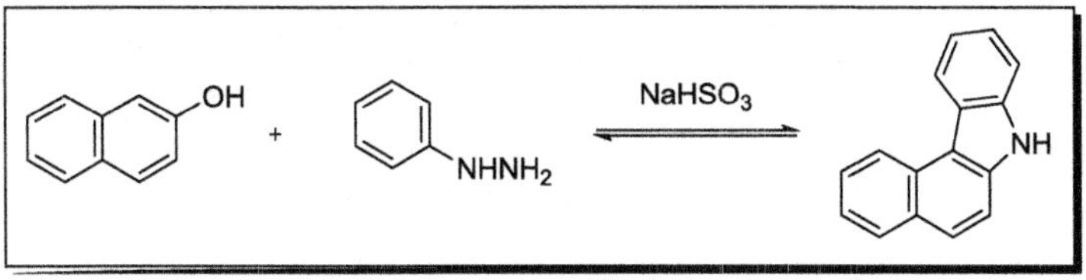

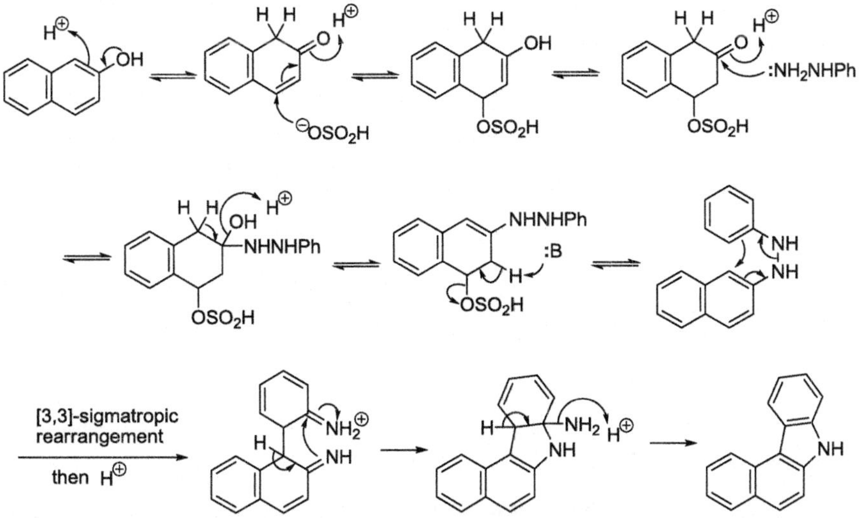

**Example 1[2]**

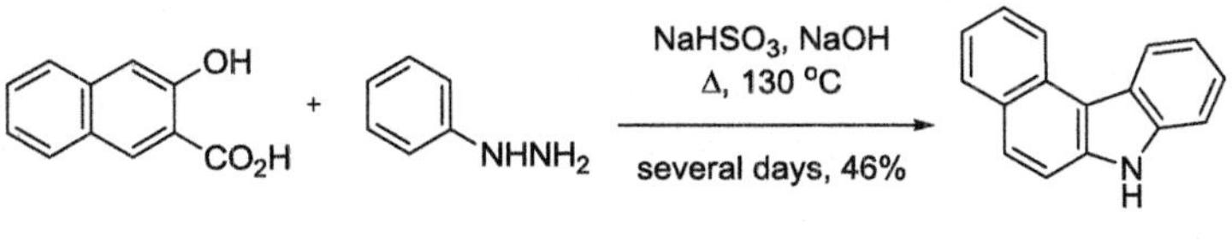

**Example 2[3]**

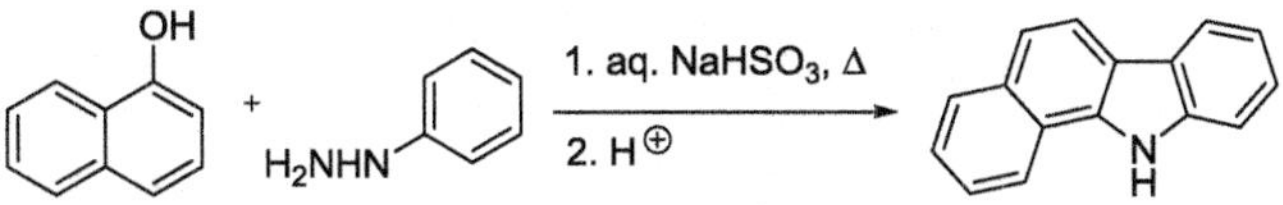

J.J. Li, *Name Reactions: A Collection of Detailed Mechanisms and Synthetic Applications*,
DOI 10.1007/978-3-319-03979-4_41, © Springer International Publishing Switzerland 2014

Example 3[4]

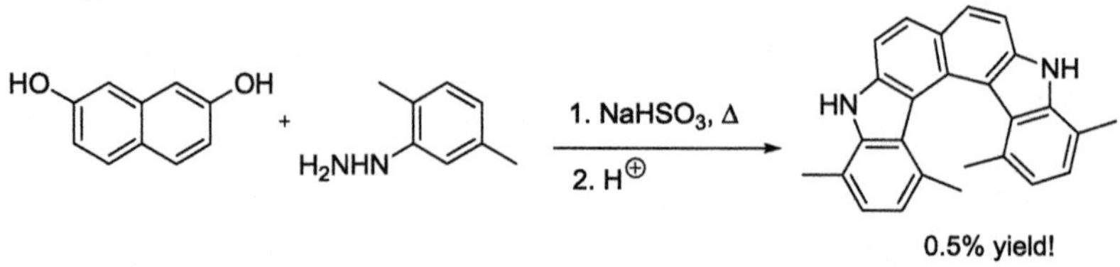

0.5% yield!

Example 4[7]

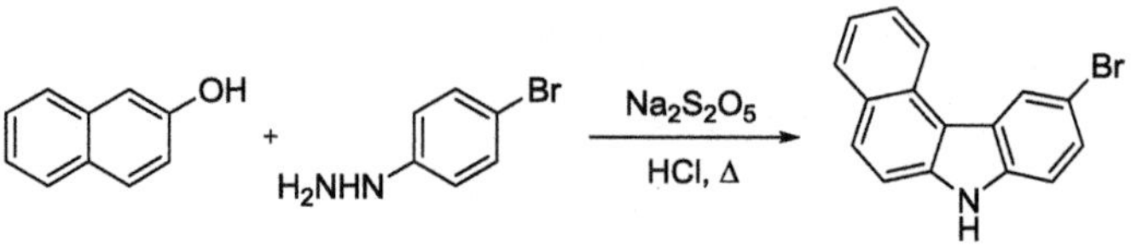

**References**

1.  Bucherer, H. T. *J. Prakt. Chem.* **1904,** *69*, 49–91.  Hans Th. Bucherer (1869–1949) was born in Ehrenfeld, Germany.  He shuttled between industry and academia all through his career.
2.  Bucherer, H. T.; Schmidt, M. *J. Prakt. Chem.* **1909,** *79*, 369–417.
3.  Bucherer, H. T.; Sonnenburg, E. F. *J. Prakt. Chem.* **1909,** *81*, 1–48.
4.  Drake, N. L. *Org. React.* **1942,** *1*, 105–128.  (Review).
5.  Seeboth, H. *Angew. Chem. Int. Ed.* **1967,** *6*, 307–317.  (Review).
6.  Robinson, B. *The Fischer Indole Synthesis*, Wiley-Interscience, New York, **1982.** (Book).
7.  Hill, J. A.; Eaddy, J. F. *J. Labelled Compd. Radiopharm.* **1994,** *34*, 697–706.
8.  Pischel, I.; Grimme, S.; Kotila, S.; Nieger, M.; Vögtle, F. *Tetrahedron: Asymmetry* **1996,** *7*, 109–116.
9.  Moore, A. J. *Bucherer Carbazole Synthesis*. In *Name Reactions in Heterocyclic Chemistry*; Li, J. J., Ed.; Wiley: Hoboken, NJ, **2005,** pp 110–115.  (Review).

# Bucherer reaction

Transformation of β-naphthols to β-naphthylamines using ammonium sulfite.

Example 1, Retro-Bucherer reaction[6]

Example 2, Although the classic Bucherer reaction requires high temperatures, it may be carried out at room temperature with the aid of microwave (150 Watts):[7]

J.J. Li, *Name Reactions: A Collection of Detailed Mechanisms and Synthetic Applications*,
DOI 10.1007/978-3-319-03979-4_42, © Springer International Publishing Switzerland 2014

Example 3[8]

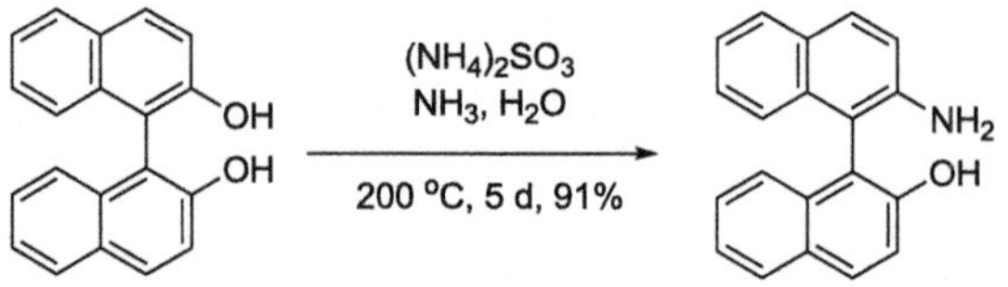

**References**

1.  Bucherer, H. T. *J. Prakt. Chem.* **1904**, *69*, 49–91.
2.  Drake, N. L. *Org. React.* **1942**, *1*, 105–128. (Review).
3.  Gilbert, E. E. *Sulfonation and Related Reactions* Wiley: New York, **1965**, p 166. (Review).
4.  Seeboth, H. *Angew. Chem. Int. Ed.* **1967**, *6*, 307–317.
5.  Gruszecka, E.; Shine, H. J. *J. Labelled Compd. Radiopharm.* **1983**, *20*, 1257–1264.
6.  Belica, P. S.; Manchand, P. S. *Synthesis* **1990**, 539–540.
7.  Canete, A.; Melendrez, M. X.; Saitz, C.; Zanocco, A. L. *Synth. Commun.* **2001**, *31*, 2143–2148.
8.  Körber, K.; Tang, W.; Hu, X.; Zhang, X. *Tetrahedron Lett.* **2002**, *43*, 7163–7165.
9.  Deady, L. W.; Devine, S. M. *Tetrahedron* **2006**, *62*, 2313–2320.
10. Budzikiewicz, H. *Mini-Reviews Org. Chem.* **2006**, *3*, 93–97. (Review).
11. Yu, J.; Zhang, P.i; Wu, J.; Shang, Z. *Tetrahedron Lett.* **2013**, *54*, 3167–3170.

# Bucherer–Bergs reaction

Formation of hydantoins from carbonyl compounds with potassium cyanide (KCN) and ammonium carbonate [$(NH_4)_2CO_3$] or from cyanohydrins and ammonium carbonate. It belongs to the category of multiple component reactions (MCRs).

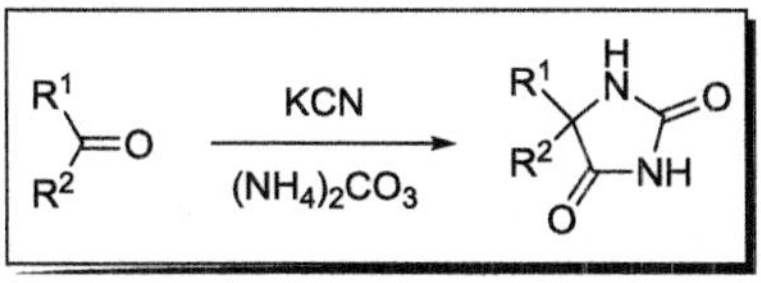

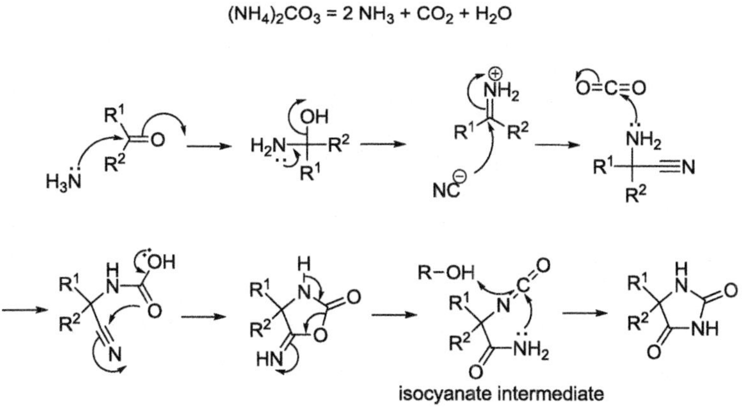

$$(NH_4)_2CO_3 = 2 NH_3 + CO_2 + H_2O$$

isocyanate intermediate

Example 1[5]

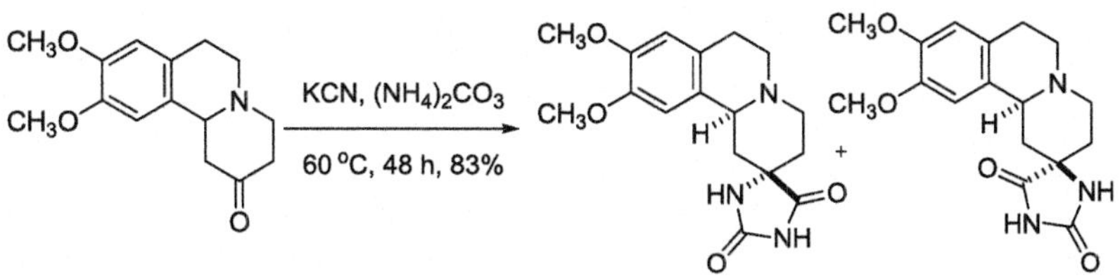

KCN, $(NH_4)_2CO_3$

60 °C, 48 h, 83%

Example 2[6]

KCN, $(NH_4)_2CO_3$

EtOH/$H_2O$, 70 °C, 50%

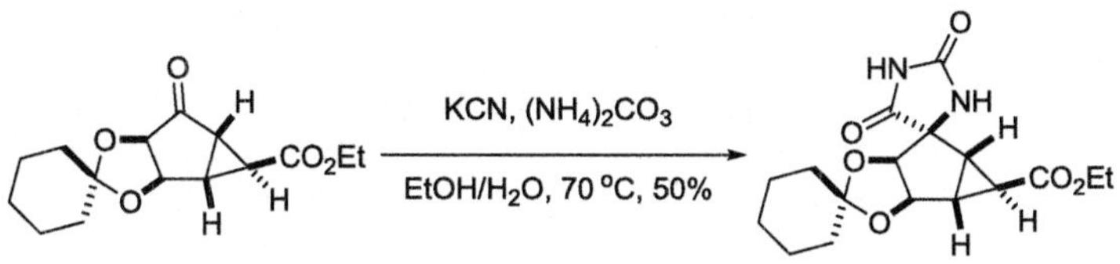

J.J. Li, *Name Reactions: A Collection of Detailed Mechanisms and Synthetic Applications*,
DOI 10.1007/978-3-319-03979-4_43, © Springer International Publishing Switzerland 2014

Example 3[7]

Example 4[9]

Example 5[11]

## References

1.  Bergs, H. Ger. Pat. 566, 094, **1929**. Hermann Bergs worked at I. G. Farben in Germany.
2.  Bucherer, H. T., Steiner, W. *J. Prakt. Chem.* **1934**, *140*, 291–316. (Mechanism).
3.  Ware, E. *Chem. Rev.* **1950**, *46*, 403–470. (Review).
4.  Wieland, H. In *Houben–Weyl's Methoden der organischen Chemie*, Vol. XI/2, **1958**, p 371. (Review).   .
5.  Menéndez, J. C.; Díaz, M. P.; Bellver, C.; Söllhuber, M. M. *Eur. J. Med. Chem.* **1992**, *27*, 61–66.
6.  Domínguez, C.; Ezquerra, A.; Prieto, L.; Espada, M.; Pedregal, C. *Tetrahedron: Asymmetry* **1997**, *8*, 511–514.
7.  Zaidlewicz, M.; Cytarska, J.; Dzielendziak, A.; Ziegler-Borowska, M. *ARKIVOC* **2004**, *iii*, 11–21.
8.  Li, J. J. *Bucherer–Bergs Reaction*. In *Name Reactions in Heterocyclic Chemistry*, Li, J. J., Ed.; Wiley: Hoboken, NJ, **2005**, pp 266–274. (Review).
9.  Sakagami, K.; Yasuhara, A.; Chaki, S.; Yoshikawa, R.; Kawakita, Y.; Saito, A.; Taguchi, T.; Nakazato, A. *Bioorg. Med. Chem.* **2008**, *16*, 4359–4366.
10. Wuts, P. G. M.; Ashford, S. W.; Conway, B.; Havens, J. L.; Taylor, B.; Hritzko, B.; Xiang, Y.; Zakarias, P. S. *Org. Proc. Res. Dev.* **2009**, *13*, 331–335.
11. Oba, M.; Shimabukuro, A.; Ono, M.; Doi, M.; Tanaka, M. *Tetrahedron: Asymmetry* **2013**, *24*, 464–467.

# Büchner ring expansion

Reaction of a phenyl ring with a diazoacetic ester to give a cyclohepta-2,4,6-trienecarboxylic acid ester.  Intramolecular Büchner reaction is more useful in synthesis.  *Cf.* Pfau–Platter azulene synthesis.

rhodium carbenoid

Example 1, Intramolecular Büchner reaction[7]

Example 2, Intramolecular Büchner reaction[8]

Example 3, An intramolecular Büchner reaction within the Grubbs' catalyst![9]

J.J. Li, *Name Reactions: A Collection of Detailed Mechanisms and Synthetic Applications*, DOI 10.1007/978-3-319-03979-4_44, © Springer International Publishing Switzerland 2014

Example 4[10]

Ar = Ph, or 3-thienyl

Example 5[12]

**References**

1.  Büchner, E. *Ber.* **1896,** *29,* 106–109. Eduard Büchner (1860–1917) won Nobel Prize in 1907 for his work on fermentation. His name is imortalized with the Büchner funnels that we still use daily in organic laboratories. 🍄

2.  von E. Doering, W.; Knox, L. H. *J. Am. Chem. Soc.* **1957,** *79,* 352–356.

3.  Marchard, A. P.; Brockway, N. M. *Chem. Rev.* **1974,** *74,* 431–469. (Review).

4.  Anciaux, A. J.; Demoncean, A.; Noels, A. F.; Hubert, A. J.; Warin, R.; Teyssié, P. *J. Org. Chem.* **1981,** *46,* 873–876.

5.  Duddeck, H.; Ferguson, G.; Kaitner, B.; Kennedy, M.; McKervey, M. A.; Maguire, A. R. *J. Chem. Soc., Perkin Trans. 1* **1990,** 1055–1063.

6.  Doyle, M. P.; Hu, W.; Timmons, D. *J. Org. Lett.* **2001,** *3,* 933–935.

7.  Manitto, P.; Monti, D.; Speranza, G. *J. Org. Chem.* **1995,** *60,* 484–485.

8.  Crombie, A. L; Kane, J. L., Jr.; Shea, K. M.; Danheiser, R. L. *J. Org. Chem.* **2004,** *69,* 8652–8667.

9.  Galan, B. R.; Gembicky, M.; Dominiak, P. M.; Keister, J. B.; Diver, S. T. *J. Am. Chem. Soc.* **2005,** *127,* 15702–15703.

10. Panne, P.; Fox, J. M. *J. Am. Chem. Soc.* **2007,** *129,* 22–23.

11. Gomes, A. T. P. C.; Leão, R. A. C.; Alonso, C. M. A.; Neves, M. G. P. M. S.; Faustino, M. A. F.; Tomé, A. C.; Silva, A.M. S.; Pinheiro, S.; de Souza, M. C. B. V.; Ferreira, V. F.; Cavaleiro, J. A. S. *Helv. Chim. Acta* **2008,** *91,* 2270–2283.

12. Foley, D. A.; O'Leary, P.; Buckley, N. R.; Lawrence, S. E.; Maguire, A. R. *Tetrahedron* **2013,** *69,* 1778–1794.

# Buchwald–Hartwig amination

The Buchwald–Hartwig amination is an exceedingly general method for generating an aromatic amine from an aryl halide or an aryl sulfonates. The key feature of this methodology is the use of catalytic palladium modulated by various electron-rich ligands. Strong bases, such as sodium *tert*-butoxide, are essential for catalyst turnover.

Mechanism:

The catalytic cycle is shown on the next page.

Example 1[3]

$R^1$ = EWG or EDG
amine = 2° cyclic or acyclic
amine = 1° aliphatic: low yield, unless $R^1$ *ortho*

J.J. Li, *Name Reactions: A Collection of Detailed Mechanisms and Synthetic Applications*,
DOI 10.1007/978-3-319-03979-4_45, © Springer International Publishing Switzerland 2014

Example 2[4]

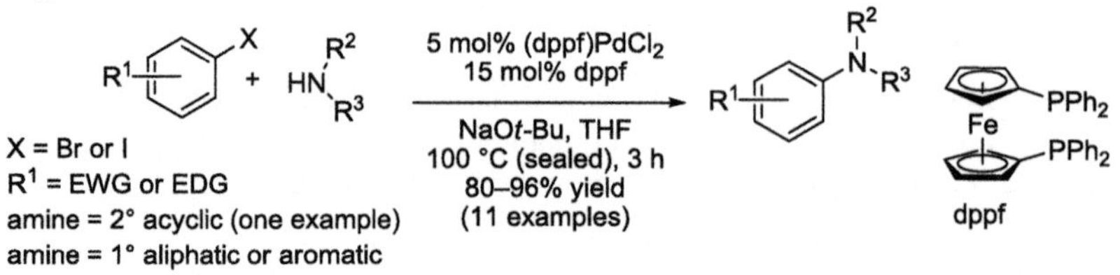
X
R2
HN
R3
5 mol% (dppf)PdCl2
15 mol% dppf
NaOt-Bu, THF
100 °C (sealed), 3 h
80–96% yield
(11 examples)
X = Br or I
R1 = EWG or EDG
amine = 2° acyclic (one example)
amine = 1° aliphatic or aromatic
PPh2
Fe
PPh2
dppf

Catalytic cycle:

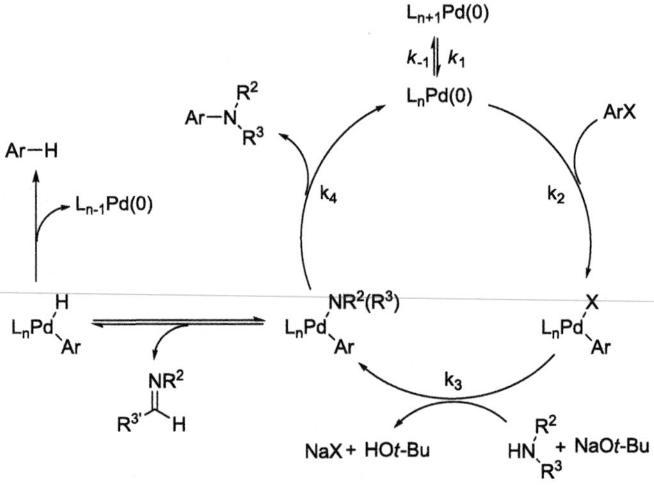
Ln+1Pd(0)
k-1 k1
LnPd(0)
ArX
k2
k4
k3
Ar—H
Ln-1Pd(0)
NR2(R3)
LnPd
Ar
X
NR2
R3' H
NaX + HOt-Bu
HN + NaOt-Bu

Pd(BINAP)$_2$ catalyzed

$$\frac{-d[ArX]}{dt} = \frac{k_1 k_2}{k_{-1}[L]}[ArX][Pd]$$

Example 3, Room temperature Buchwald–Hartwig amination[9]

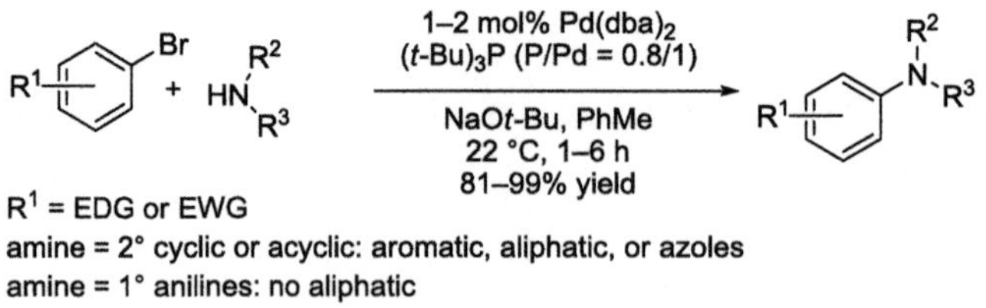
Br
1–2 mol% Pd(dba)2
(t-Bu)3P (P/Pd = 0.8/1)
NaOt-Bu, PhMe
22 °C, 1–6 h
81–99% yield
R1 = EDG or EWG
amine = 2° cyclic or acyclic: aromatic, aliphatic, or azoles
amine = 1° anilines: no aliphatic

Example 4[10]

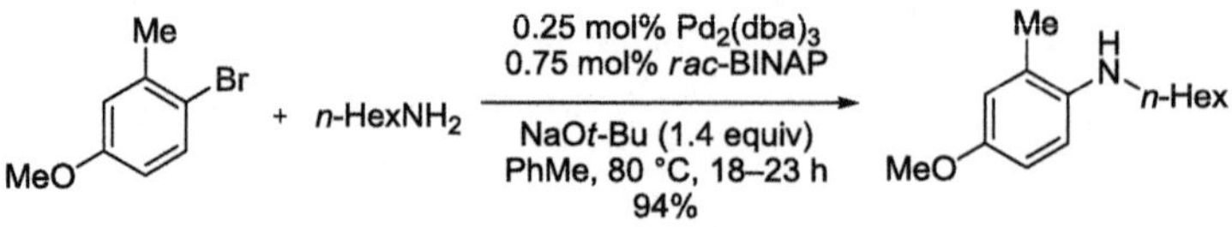
Me
Br
MeO
n-HexNH2
0.25 mol% Pd2(dba)3
0.75 mol% rac-BINAP
NaOt-Bu (1.4 equiv)
PhMe, 80 °C, 18–23 h
94%
n-Hex

Example 5[11]

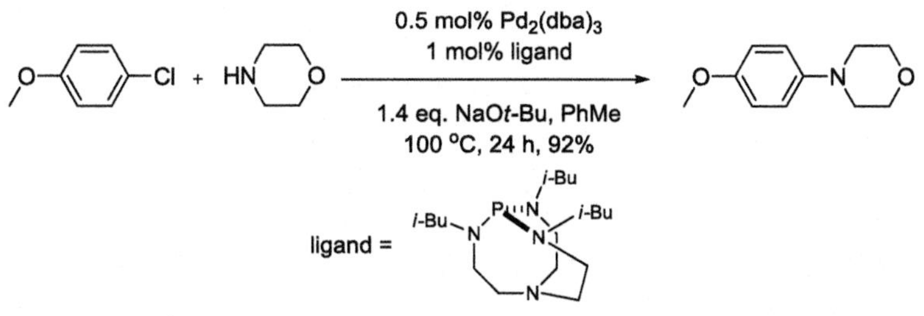

ligand =

Example 6[12]

DPE-Phos =

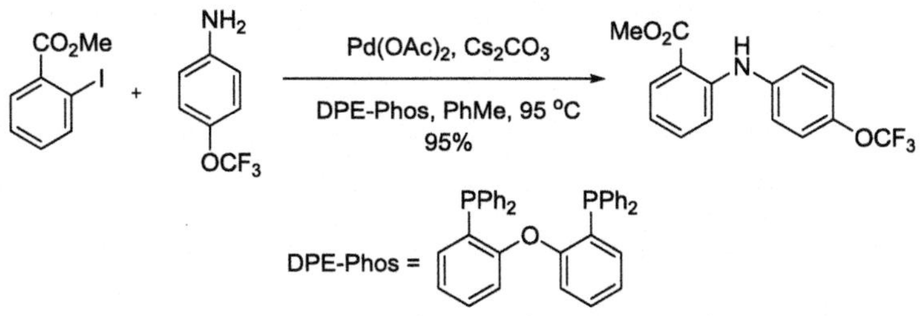

Example 7, Amination of volatile amines[14]

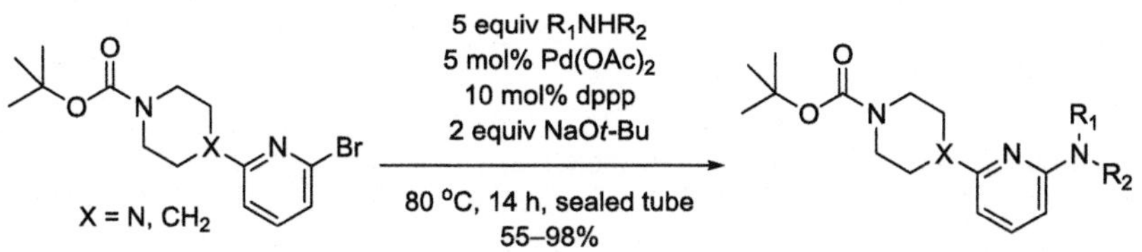

X = N, CH₂

Example 8[15]

XPhos =

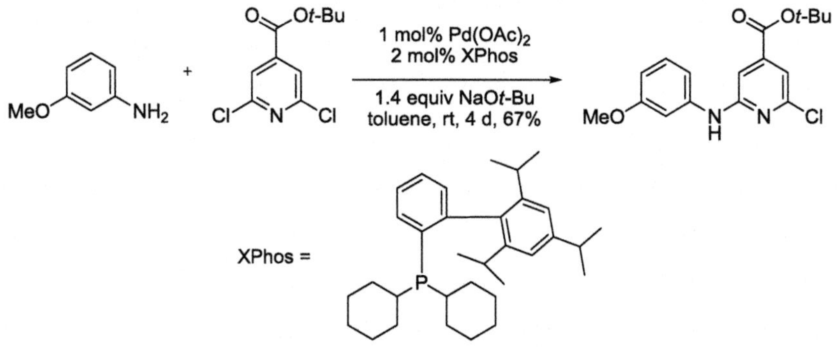

**References**
1.   (a) Paul, F.; Patt, J.; Hartwig, J. F. *J. Am. Chem. Soc.* **1994,** *116*, 5969–5970. John Hartwig earned his Ph.D. at the University of California-Berkeley in 1990 under the guidance of Robert Bergman and Richard Anderson. He moved from Yale University to the University of Illinois at Urbana-Champaign in 2006 and moved from UI-UC to UC Berkeley in 2011. Hartwig and Buchwald independently discovered this chemistry. (b) Mann, G.; Hartwig, J. F. *J. Org. Chem.* **1997,** *62*, 5413–5418. (c) Mann, G.; Hartwig, J. F. *Tetrahedron Lett.* **1997,** *38*, 8005–8008.
2.   (a) Guram, A. S.; Buchwald, S. L. *J. Am. Chem. Soc.* **1994,** *116*, 7901–7902. Stephen Buchwald received his Ph.D. in 1982 under Jeremy Knowles at Harvard University. He is currently a professor at MIT. (b) Palucki, M.; Wolfe, J. P.; Buchwald, S. L. *J. Am. Chem. Soc.* **1996,** *118*, 10333–10334.
3.   Wolfe, J. P.; Buchwald, S. L. *J. Org. Chem.* **1996,** *61*, 1133–1135.
4.   Driver, M. S.; Hartwig, J. F. *J. Am. Chem. Soc.* **1996,** *118*, 7217–7218.
5.   Wolfe, J. P.; Wagaw, S.; Marcoux, J.-F.; Buchwald, S. L. *Acc. Chem. Res.* **1998,** *31*, 805–818. (Review).
6.   Hartwig, J. F. *Acc. Chem. Res.* **1998,** *31*, 852–860. (Review).
7.   Frost, C. G.; Mendonça, P. *J. Chem. Soc., Perkin Trans. 1* **1998,** 2615–2624. (Review).
8.   Yang, B. H.; Buchwald, S. L. *J. Organomet. Chem.* **1999,** *576*, 125–146. (Review).
9.   Hartwig, J. F.; Kawatsura, M.; Hauck, S. I.; Shaughnessy, K. H.; Alcazar-Roman, L. M. *J. Org. Chem.* **1999,** *64*, 5575–5580.
10.  Wolfe, J. P.; Buchwald, S. L. *Org. Syn.* **2002,** *78*, 23–30.
11.  Urgaonkar, S.; Verkade, J. G. *J. Org. Chem.* **2004,** *69*, 9135–9142.
12.  Csuk, R.; Barthel, A.; Raschke, C. *Tetrahedron* **2004,** *60*, 5737–5750.
13.  Janey, J. M. *Buchwald–Hartwig amination,* In *Name Reactions for Functional Group Transformations*; Li, J. J., Corey, E. J. Eds.; Wiley: Hoboken, NJ, **2007,** pp 564–609. (Review).
14.  Li, J. J.; Wang, Z.; Mitchell, L. H. *J. Org. Chem.* **2007,** *72*, 3606–3607.
15.  Lorimer, A. V.; O'Connor, P. D.; Brimble, M. A. *Synthesis* **2008,** 2764–2770.
16.  Nodwell, M.; Pereira, A.; Riffell, J. L.; Zimmerman, C.; Patrick, B. O.; Roberge, M.; Andersen, R. J. *J. Org. Chem.* **2009,** *74*, 995–1006.
17.  Witt, A.; Teodorovic, P.; Linderberg, M.; Johansson, P.; Minidis, A. *Org. Process Res. Dev.* **2013,** *17*, 672–678.
18.  Raders, S. M.; Moore, J. N.; et al. *Org. Chem.* **2013,** *78*, 4649–4664.

# Burgess reagent

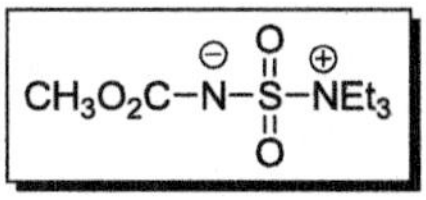

$$CH_3O_2C-\overset{\ominus}{N}-\overset{\overset{O}{\|}}{\underset{\underset{O}{\|}}{S}}-\overset{\oplus}{N}Et_3$$

The Burgess reagent [methyl N-(triethylammoniumsulfonyl)carbamate], a neutral, white crystalline solid, is efficient at generating olefins from secondary and tertiary alcohols where the first-order thermolytic Ei (during the elimination—the two groups leave at about the same time and bond to each other concurrently) mechanism prevails.

Preparation[2]

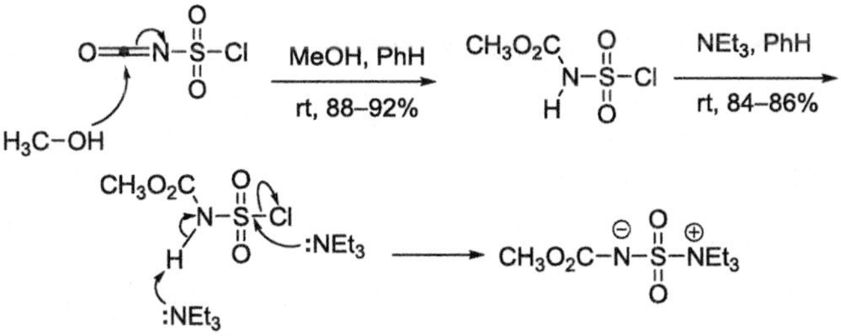

Mechanism of dehydration[5]

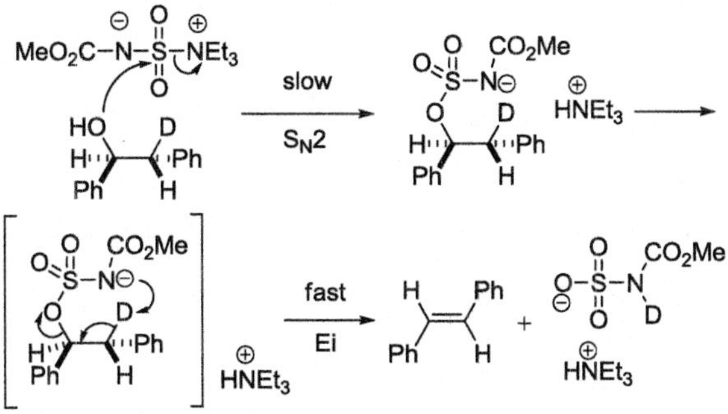

Example 1, On primary alcohols, the hydroxyl group does not eliminate but rather undergoes substitution[3]

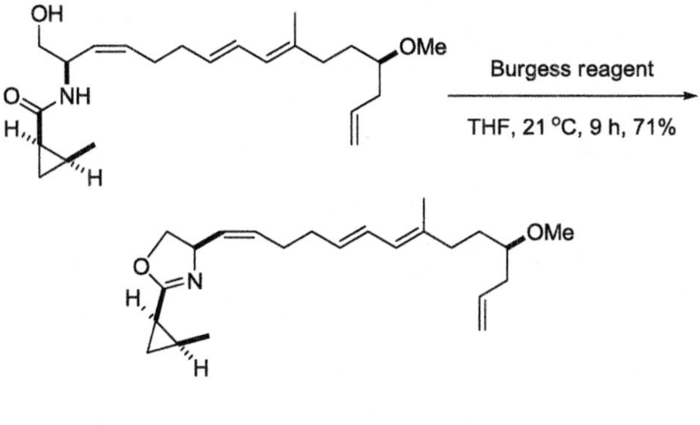

J.J. Li, *Name Reactions: A Collection of Detailed Mechanisms and Synthetic Applications*, DOI 10.1007/978-3-319-03979-4_46, © Springer International Publishing Switzerland 2014

Example 2[6]

Example 3[7]

Example 4[8]

Example 5, Cyclodehydration followed by a novel carbamoylsulfonylation[10]

## References

1    (a) Atkins, G. M., Jr.; Burgess, E. M. *J. Am. Chem. Soc.* **1968**, *90*, 4744–4745. (b) Burgess, E. M.; Penton, H. R., Jr.; Taylor, E. A., Jr. *J. Am. Chem. Soc.* **1970**, *92*, 5224–5226. (c) Atkins, G. M., Jr.; Burgess, E. M. *J. Am. Chem. Soc.* **1972**, *94*, 6135–6141. (d) Burgess, E. M.; Penton, H. R., Jr.; Taylor, E. A. *J. Org. Chem.* **1973**, *38*, 26–31.

2    (a) Burgess, E. M.; Penton, H. R., Jr.; Taylor, E. A.; Williams, W. M. *Org. Synth. Coll. Edn.* **1987**, *6*, 788–791. (b) Duncan, J. A.; Hendricks, R. T.; Kwong, K. S. *J. Am. Chem. Soc.* **1990**, *112*, 8433–8442.

3    Wipf, P.; Xu, W. *J. Org. Chem.* **1996**, *61*, 6556–6562.

4    Lamberth, C. *J. Prakt. Chem.* **2000**, *342*, 518–522. (Review).

5    Khapli, S.; Dey, S.; Mal, D. *J. Indian Inst. Sci.* **2001**, *81*, 461–476. (Review).

6    Miller, C. P.; Kaufman, D. H. *Synlett* **2000**, *8*, 1169–1171.

7    Keller, L.; Dumas, F.; D'Angelo, J. *Eur. J. Org. Chem.* **2003**, 2488–2497.

8    Nicolaou, K. C.; Snyder, S. A.; Longbottom, D. A.; Nalbandian, A. Z.; Huang, X. *Chem. Eur. J.* **2004**, *10*, 5581–5606.

9    Holsworth, D. D. *The Burgess Dehydrating Reagent.* In *Name Reactions for Functional Group Transformations*; Li, J. J., Ed.; Wiley: Hoboken, NJ, **2007**, pp 189–206. (Review).

10   Li, J. J.; Li, J. J.; Li, J.; et al. *Org. Lett.* **2008**, *10*, 2897–2900.

11   Werner, L.; Wernerova, M.; Hudlicky, T. et al. *Adv. Synth. Catal.* **2012**, *354*, 2706–2712.

# Burke boronates

Burke boronates can serve as B-protected haloboronic acids for a wide variety of applications in iterative cross-coupling.[1-6] The corresponding boronic acids can be liberated using mild aqueous bases such as NaOH or NaHCO$_3$.[1-4] Burke boronates are also compatible with many synthetic reagents, enabling the synthesis of complex boronic acids from simple B-containing starting materials.[3,6] They can also serve as stable building blocks for cross-coupling, i.e., under aqueous basic conditions, the corresponding boronic acid is released and coupled in situ.[2,3,7] Moreover, Burke boronates are highly crystalline, monomeric, free-flowing solids that are indefinitely stable to benchtop storage under air and compatible with silica gel chromatography.[1-3,6]

Preparation:[1,2,4,6,]

Alternatively, many of these building blocks are now commercially available.

Example 1[2]
A wide range of selective couplings can be performed at the halide terminus of a B-protected haloboronic acid.

J.J. Li, *Name Reactions: A Collection of Detailed Mechanisms and Synthetic Applications*, DOI 10.1007/978-3-319-03979-4_47, © Springer International Publishing Switzerland 2014

Suzuki coupling

Ph—\_\_B(OH)$_2$
Pd(OAc)$_2$
SPhos, KF
PhMe, 23°C
92 %

TMS——≡
Pd(PPh$_3$)$_4$, CuI
piperidine
THF
23 °C, 73%

Sonogashira coupling

Stille coupling

SnBu$_3$
Pd$_2$dba$_3$, Fur$_3$P
DMF, 45 °C, 91%

PdCl$_2$(CH$_3$CN)$_2$
SPhos, KOAc

PhMe
45 °C, 71%

Miyaura coupling

Heck reaction

MeO
Pd(OAc)$_2$, PPh$_3$
Et$_3$N, DMF, 45 °C
90%

Bu$_3$Sn—\_\_ZnCl
Pd(OAc)$_2$, SPhos
THF, 0 °C
66%

Negishi coupling

Example 2[9,10]

Small molecule natural products and their derivatives can be prepared via iterative cross-coupling with B-protected haloboronic acids.

NaOH
THF:H$_2$O
95%

XPhosPd, K$_3$PO$_4$
PhMe:THF, 79%

1. pinacol, NaHCO$_3$, MeOH, 99%
2.

PdCl$_2$(PPh$_3$)$_2$, Ag$_2$O ·
DMSO, 45%

1.

XPhosPd, NaOH
THF:H$_2$O, 60%

2. HF pyr, THF, 65%

peridinin

Example 3[3]

Burke boronates are stable to a wide range of synthetic reagents, including acids, non-aqueous bases, oxidants, reductants, electrophiles, and soft nucleophiles. This reagent compatibility enables multistep synthesis of complex boranes from simple boron-containing starting materials.

Example 4[7,8]

Burke boronates can be used directly in cross-couplings as shelf-stable surrogates for unstable boronic acids via slow-release cross-coupling.

## References

1.  Gillis, E. P.; Burke, M. D. *J. Am. Chem. Soc.* **2007**, *129*, 6716–6717.
2.  Lee, S. J., Gray, K. C., Paek, J. S., Burke, M. D. *J. Am. Chem. Soc.* **2008**, *130*, 466–468.
3.  Gillis, E. P.; Burke, M. D. *J. Am. Chem. Soc.* **2008**, *130*, 14084–14085.
4.  Ballmer, S. G.; Gillis, E. P.; Burke, M. D. *Org. Synth.* **2009**, 86, 344–359.
5.  Gillis, E. P.; Burke, M. D. *Aldrichimica Acta* **2009**, *42*, 17–27.
6.  Uno, B. E.; Gillis, E. P.; Burke, M. D. *Tetrahedron* **2009**, *65*, 3130–3138.
7.  Knapp, D. M.; Gillis, E. P., Burke, M. D. *J. Am. Chem. Soc.* **2009**, *131*, 6961–6963.
8.  Dick, G. R.; Woerly, E. M.; Burke, M. D. *Angew. Chem. Int. Ed.* **2012**, *51*, 2667–2672.
9.  Woerly, E. M.; Cherney, A. H.; Davis, E. K.; Burke, M. D. *J. Am. Chem. Soc.* **2010**, *132*, 6941–6943.
10. Gray, K. C.; Palacios, D. S.; Dailey, I.; Endo, M. M.; Uno, B. E.; Wilcock, B. C.; Burke, M. D. *Proc. Natl. Acad. Sci. U.S.A.* **2012**, *109*, 2234–2239.

## Cadiot–Chodkiewicz coupling

Bis-acetylene synthesis from alkynyl halides and alkynyl copper reagents.
*Cf.* Castro–Stephens reaction.

$$R^1\!\!-\!\!\equiv\!\!-X + Cu\!\!-\!\!\equiv\!\!-R^2 \longrightarrow R^1\!\!-\!\!\equiv\!\!-\!\!\equiv\!\!-R^2$$

$$R^1\!\!-\!\!\equiv\!\!-X + Cu\!\!-\!\!\equiv\!\!-R^2 \xrightarrow[\text{addition}]{\text{oxidative}} R^1\!\!-\!\!\equiv\!\!-\overset{X}{\underset{|}{Cu}}\!\!-\!\!\equiv\!\!-R^2$$

Cu(III) intermediate

$$\xrightarrow[\text{elimination}]{\text{reductive}} CuX + R^1\!\!-\!\!\equiv\!\!-\!\!\equiv\!\!-R^2$$

Example 1[3]

Example 2[7]

Example 3[9]

| | n = 1 to 7 |
| --- | --- |
| | n = 1, 8% |
| | n = 2, 11% |
| | n = 3, 32% |
| | n = 4, 8% |
| | n = 5, 13% |
| | n = 6, 3% |
| | n = 7, 8% |

Example 4, Cadiot–Chodkiewicz active template synthesis of rotaxanes and switchable molecular shuttles with weak intercomponent interactions[10]

1. 1 equiv *n*-BuLi
   THF, −78 °C

2. 1 equiv CuI, 0 °C
3. 1 equiv **3**
   1 equiv **2**

J.J. Li, *Name Reactions: A Collection of Detailed Mechanisms and Synthetic Applications*,
DOI 10.1007/978-3-319-03979-4_48, © Springer International Publishing Switzerland 2014

trapping molecule **3** =

**2**

**References**

1. Chodkiewicz, W.; Cadiot, P. *C. R. Hebd. Seances Acad. Sci.* **1955**, *241*, 1055–1057. Both Paul Cadiot (1923–) and Wladyslav Chodkiewicz (1921–) are French chemists.
2. Cadiot, P.; Chodkiewicz, W. In *Chemistry of Acetylenes;* Viehe, H. G., ed.; Dekker: New York, **1969**, 597–647. (Review).
3. Gotteland, J.-P.; Brunel, I.; Gendre, F.; Désiré, J.; Delhon, A.; Junquéro, A.; Oms, P.; Halazy, S. *J. Med. Chem.* **1995**, *38*, 3207–3216.
4. Bartik, B.; Dembinski, R.; Bartik, T.; Arif, A. M.; Gladysz, J. A. *New J. Chem.* **1997**, *21*, 739–750.
5. Montierth, J. M.; DeMario, D. R.; Kurth, M. J.; Schore, N. E. *Tetrahedron* **1998**, *54*, 11741–11748.
6. Negishi, E.-i.; Hata, M.; Xu, C. *Org. Lett.* **2000**, *2*, 3687–3689.
7. Marino, J. P.; Nguyen, H. N. *J. Org. Chem.* **2002**, *67*, 6841–6844.
8. Utesch, N. F.; Diederich, F.; Boudon, C.; Gisselbrecht, J.-P.; Gross, M. *Helv. Chim. Acta* **2004**, *87*, 698–718.
9. Bandyopadhyay, A.; Varghese, B.; Sankararaman, S. *J. Org. Chem.* **2006**, *71*, 4544–4548–4548.
10. Berna, J.; Goldup, S. M.; Lee, A.-L.; Leigh, D. A.; Symes, M. D.; Teobaldi, G.; Zerbetto, F. *Angew. Chem. Int. Ed.* **2008**, *47*, 4392–4396.
11. Glen, P. E.; O'Neill, J. A. T.; Lee, A.-L. *Tetrahedron* **2013**, *69*, 57–68.

## Cadogan–Sundberg indole synthesis

The Cadogan reaction refers to the deoxygenation of $o$-nitrostyrenes **1** or $o$-nitrostilbenes with trialkyl phosphite or trialkylphosphine and subsequent cyclization of the resulting intermediate nitrene **2** to form indoles **3**. The Sundberg indole synthesis refers to the synthesis of indoles **3** via either thermolysis or irradiation of $o$-azidostyrene **4** via the intermediacy of nitrene **2**.

Example 1, Söderberg's modified conditions to prepare indole[3]

J.J. Li, *Name Reactions: A Collection of Detailed Mechanisms and Synthetic Applications*, DOI 10.1007/978-3-319-03979-4_49, © Springer International Publishing Switzerland 2014

Example 2[4]

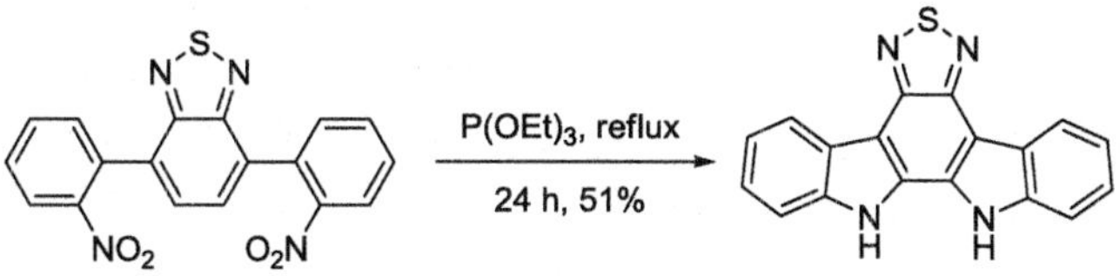
N S N
$NO_2$
$O_2N$
$P(OEt)_3$, reflux
24 h, 51%
N S N
N H
N H

Example 3[5]

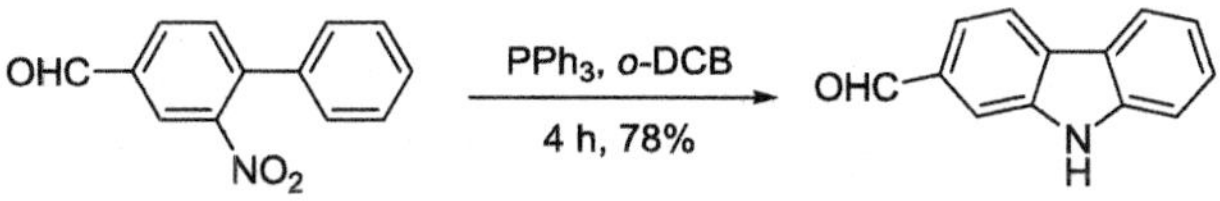
OHC
$NO_2$
$PPh_3$, o-DCB
4 h, 78%
OHC
N H

Example 4[6]

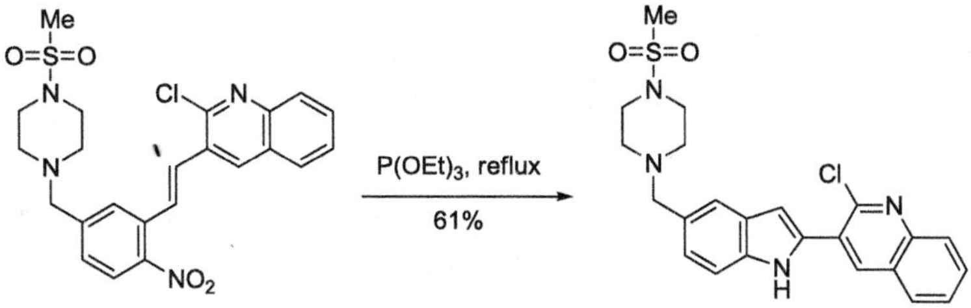
Me
O=S=O
N
N
Cl
N
$NO_2$
$P(OEt)_3$, reflux
61%
Me
O=S=O
N
N
Cl
N
N H

Example 5[7]

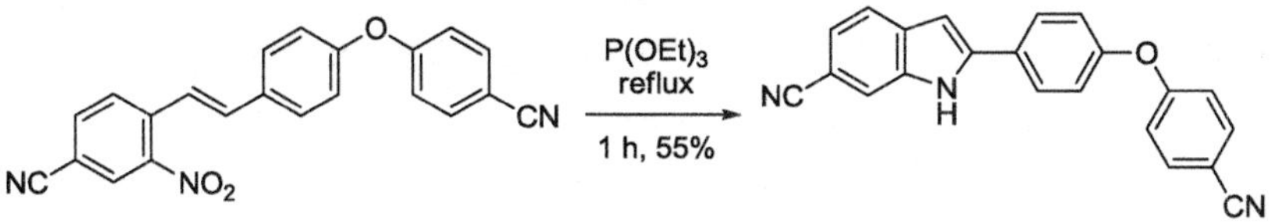
O
CN
NC
$NO_2$
$P(OEt)_3$
reflux
1 h, 55%
NC
N H
O
CN

**References**

1 Cadogan, J. I. G.; Cameron-Wood, M. *Proc. Chem. Soc.* **1962,** 361.
2 Sundberg, R. J. *J. Org. Chem.* **1965,** *30*, 3604–3610.
3 Scott, T. L.; Söderberg, B. C. G. *Tetrahedron Lett.* **2002,** *43*, 1621–1624.
4 Kuethe, J. T.; Wong, A.; Qu, C.; Smitrovich, J.; Davies, I. W.; Hughes, D. L. *J. Org. Chem.* **2005,** *70,* 2555–2567.
5 Freeman, A. W.; Urvoy, M.; Criswell, M. E. *J. Org. Chem.* **2005,** *70,* 5014–5019.
6 Balaji, G.; Shim, W. L.; Parameswaran, M.; Valiyaveettil, S. *Org. Lett.* **2009,** *11,* 4450–4453.
7 Li, B.; Pai, R.; Cardinale, S. C.; Butler, M. M.; Peet, N. P.; Moir, D. T.; Bavari, S.; Bowlin, T. L. *J. Med. Chem.* **2010,** *53,* 2264–2276.

## Camps quinoline synthesis

Base-catalyzed intramolecular condensation of a 2-acetamido acetophenone (**1**) to a 2-(and possibly 3)-substituted-quinolin-4-ol (**2**), a 4-(and possibly 3)-substituted-quinolin-2-ol (**3**), or a mixture.

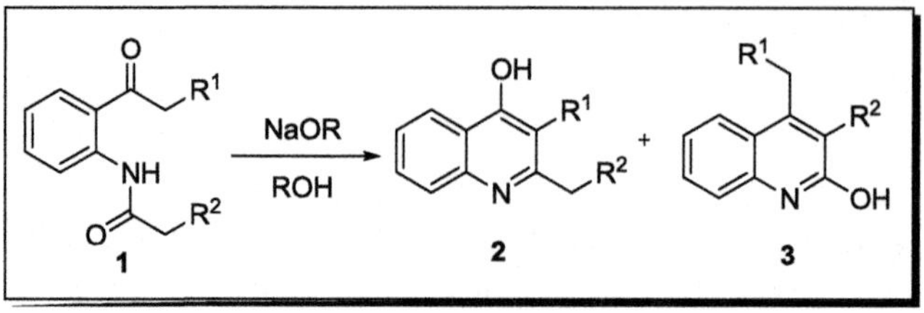

Pathway A:

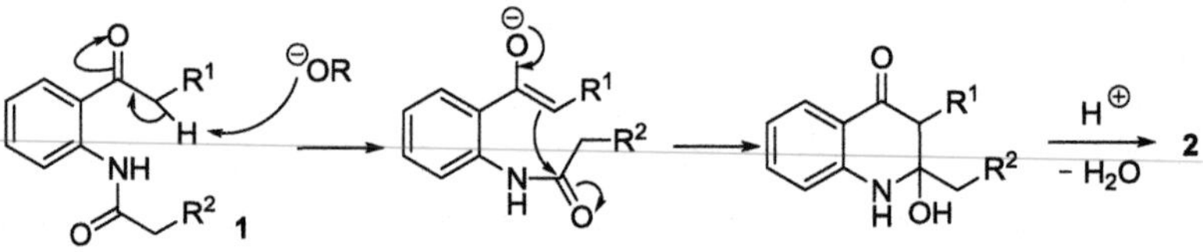

Pathway B:

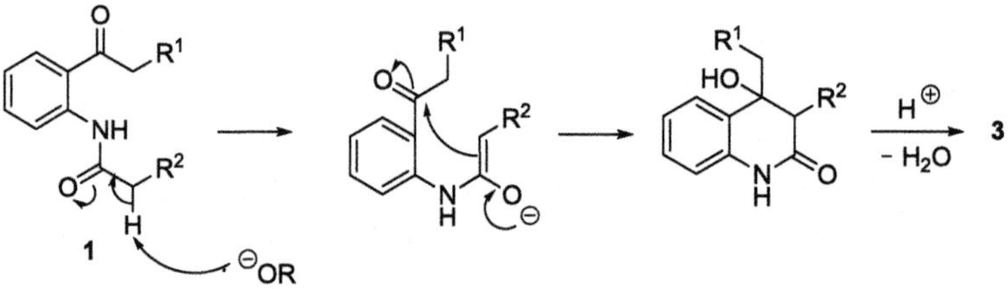

Example 1[1]

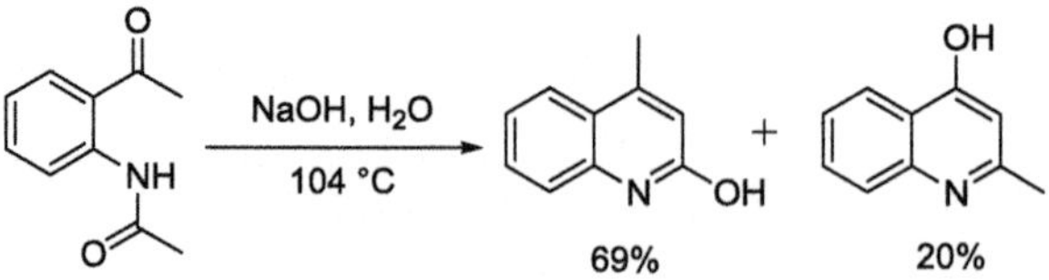

69%          20%

Example 2[6]

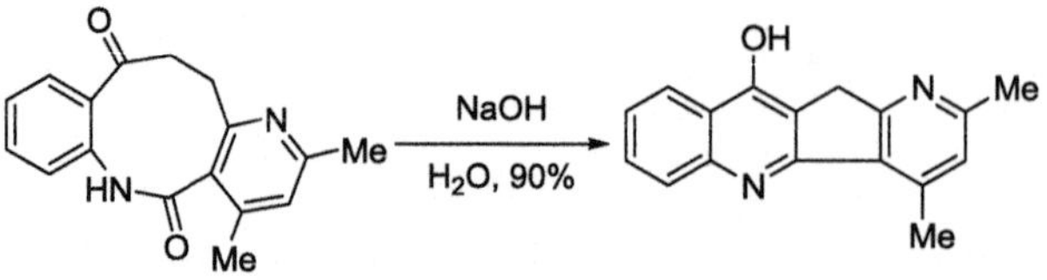

J.J. Li, *Name Reactions: A Collection of Detailed Mechanisms and Synthetic Applications*, DOI 10.1007/978-3-319-03979-4_50, © Springer International Publishing Switzerland 2014

## References

1.  (a) Camps, R. *Chem. Ber.* **1899,** *32*, 3228–3234. Rudolf Camps worked under Professor Engler from 1899 to 1902 at the Technische Hochschule in Karlsruhe, Germany. (b) Camps, R. *Arch. Pharm.* **1899,** *237*, 659–691.
2.  Elderfield, R. C.; Todd, W. H.; Gerber, S. *Heterocyclic Compounds* Vol. 6, Elderfield, R. C., ed.; Wiley and Sons, New York, **1957,** 576. (Review).
3.  Clemence, F.; LeMartret, O.; Collard, J. *J. Heterocycl. Chem.* **1984,** *21*, 1345–1353.
4.  Hino, K.; Kawashima, K.; Oka, M.; Nagai, Y.; Uno, H.; Matsumoto, J. *Chem. Pharm. Bull.* **1989,** *37*, 110–115.
5.  Witkop, B.; Patrick, J. B.; Rosenblum, M. *J. Am. Chem. Soc.* **1951,** *73*, 2641–2647.
6.  Barret, R.; Ortillon, S.; Mulamba, M.; Laronze, J. Y.; Trentesaux, C.; Lévy, J. *J. Heterocycl. Chem.* **2000,** *37*, 241–244.
7.  Pflum, D. A. *Camps Quinolinol Synthesis.* In *Name Reactions in Heterocyclic Chemistry*; Li, J. J., Ed.; Wiley: Hoboken, NJ, **2005,** pp 386–389. (Review).

## Cannizzaro reaction

Redox reaction between aromatic aldehydes, formaldehyde or other aliphatic aldehydes without α-hydrogen. Base is used to afford the corresponding alcohols and carboxylic acids.

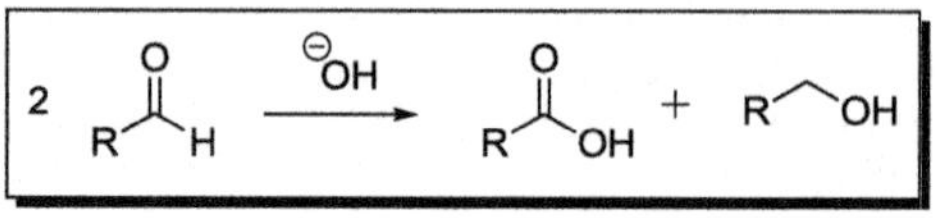

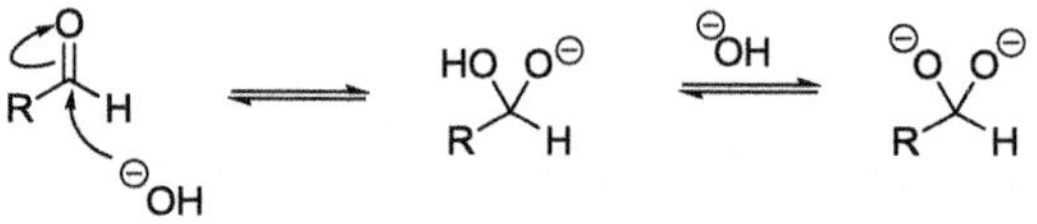

Pathway A:

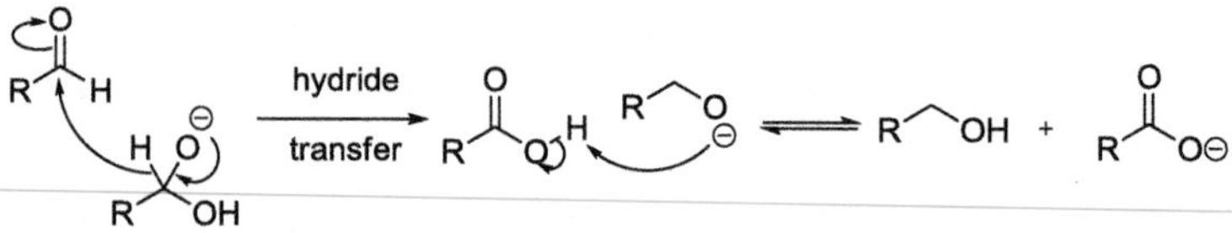

Final deprotonation of the carboxylic acid drives the reaction forward.

Pathway B:

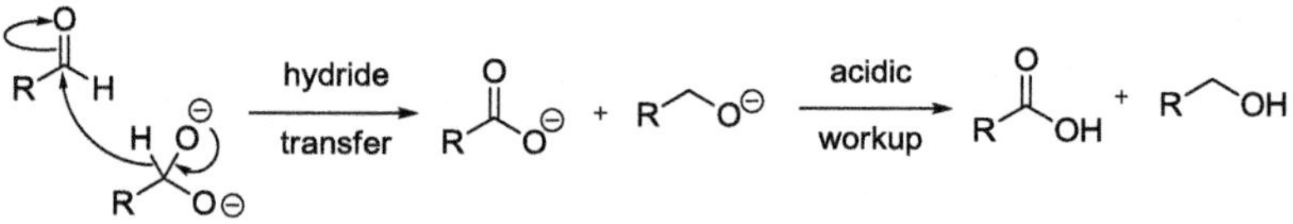

Example 1[4]

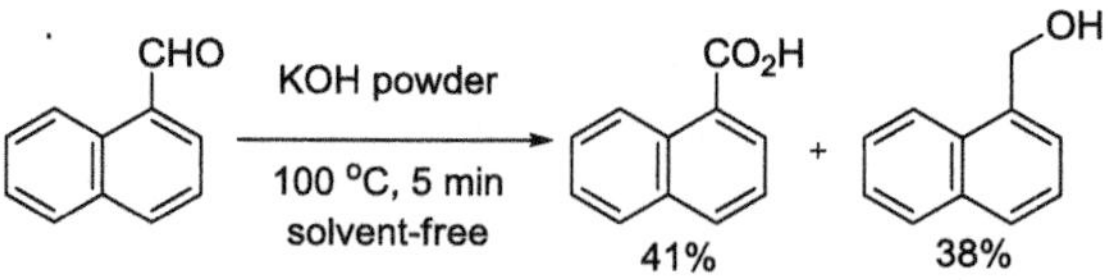

KOH powder

100 °C, 5 min
solvent-free

41%        38%

Example 2[6]

NaN

THF, 0 °C to rt, 5 h

79%

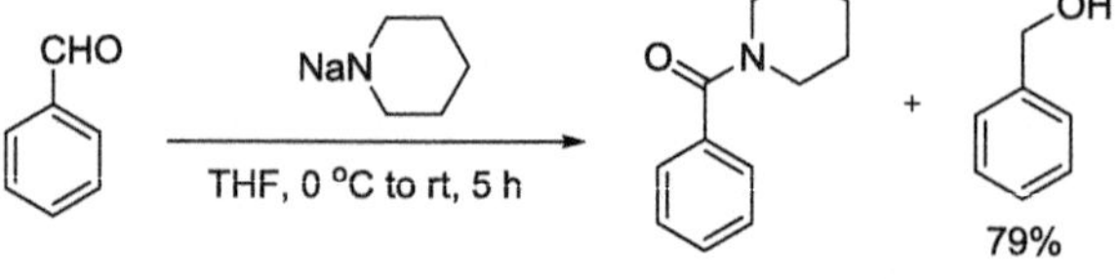

J.J. Li, *Name Reactions: A Collection of Detailed Mechanisms and Synthetic Applications*,
DOI 10.1007/978-3-319-03979-4_51, © Springer International Publishing Switzerland 2014

Example 3[8]

42%    43%

TMG = 1,1,3,3-tetramethylguanidine

Example 4, Desymmetrization by intramolecular Cannizzaro reaction[9]

**References**

1. Cannizzaro, S. *Ann.* **1853**, *88*, 129–130. Stanislao Cannizzaro (1826–1910) was born in Palermo, Sicily, Italy. In 1847, he had to escape to Paris for participating in the Sicilian Rebellion. Upon his return to Italy, he discovered benzyl alcohol synthesis by the action of potassium hydroxide on benzaldehyde. Political interests brought Cannizzaro to the Italian Senate and he later became its vice president.
2. Geissman, T. A. *Org. React.* **1944**, *1*, 94–113. (Review).
3. Russell, A. E.; Miller, S. P.; Morken, J. P. *J. Org. Chem.* **2000**, *65*, 8381–8383.
4. Yoshizawa, K.; Toyota, S.; Toda, F. *Tetrahedron Lett.* **2001**, *42*, 7983–7985.
5. Reddy, B. V. S.; Srinvas, R.; Yadav, J. S.; Ramalingam, T. *Synth. Commun.* **2002**, *32*, 219–223.
6. Ishihara, K.; Yano, T. *Org. Lett.* **2004**, *6*, 1983–1986.
7. Curini, M.; Epifano, F.; Genovese, S.; Marcotullio, M. C.; Rosati, O. *Org. Lett.* **2005**, *7*, 1331–1333.
8. Basavaiah, D.; Sharada, D. S.; Veerendhar, A. *Tetrahedron Lett.* **2006**, *47*, 5771–5774.
9. Ruiz-Sanchez, A. J.; Vida, Y.; Suau, R.; Perez-Inestrosa, E. *Tetrahedron* **2008**, *64*, 11661–11665.
10. Yamabe, S.; Yamazaki, S. *Org. Biomol. Chem.* **2009**, *7*, 951–961.
11. Shen, M.-G.; Shang, S.-B.; Song, Z.-Q.; Wang, D.; Rao, X.-P.; Gao, H.; Liu, H. *J. Chem. Res.* **2013**, *37*, 51–52.

# Carroll rearrangement

Thermal rearrangement of β-ketoesters followed by decarboxylation to yield γ-unsaturated ketones *via* anion-assisted Claisen rearrangement. It is a variant of the Claisen rearrangement.

Example 1, Asymmetric Carroll rearrangement[4,5]

Example 2, Hetero-Carroll rearrangement[6]

Example 3[7]

J.J. Li, *Name Reactions: A Collection of Detailed Mechanisms and Synthetic Applications*,
DOI 10.1007/978-3-319-03979-4_52, © Springer International Publishing Switzerland 2014

Example 4, Similar to Example 3[7]

Example 5[8]

**References**

1. (a) Carroll, M. F. *J. Chem. Soc.* **1940**, 704–706. Michael F. Carroll worked at A. Boake, Roberts and Co. Ltd., in London, UK. (b) Carroll, M. F. *J. Chem. Soc.* **1941**, 507–511.
2. Ziegler, F. E. *Chem. Rev.* **1988**, *88*, 1423–1452. (Review).
3. Echavarren, A. M.; Mendosa, J.; Prados, P.; Zapata, A. *Tetrahedron Lett.* **1991**, *32*, 6421–6424.
4. Enders, D.; Knopp, M.; Runsink, J.; Raabe, G. *Angew. Chem. Int. Ed.* **1995**, *34*, 2278–2280.
5. Enders, D.; Knopp, M. *Tetrahedron* **1996**, *52*, 5805–5818.
6. Coates, R. M.; Said, I. M. *J. Am. Chem. Soc.* **1977**, *99*, 2355–2357.
7. Hatcher, M. A.; Posner, G. H. *Tetrahedron Lett.* **2002**, *43*, 5009–5012.
8. Jung, M. E.; Duclos, B. A. *Tetrahedron Lett.* **2004**, *45*, 107–109.
9. Defosseux, M.; Blanchard, N.; Meyer, C.; Cossy, J. *J. Org. Chem.* **2004**, *69*, 4626–4647.
10. Williams, D. R.; Nag, P. P. *Claisen and Related Rearrangements*. In *Name Reactions for Homologations-Part II*; Li, J. J., Ed.; Wiley: Hoboken, NJ, **2009**, pp 33–87. (Review).
11. Naruse, Y.; Todo, Y.; Shiomi, M. *Tetrahedron Lett.* **2011**, *52*, 4456–4460.
12. Abe, H.; Sato, A.; Kobayashi, T.; Ito, H. *Org. Lett.* **2013**, *15*, 1298–1301.

## Castro–Stephens coupling

Aryl–acetylene synthesis, *Cf.* Cadiot–Chodkiewicz coupling and Sonogashira coupling. The Castro–Stephens coupling uses stoichiometric copper, whereas the Sonogashira variant uses catalytic palladium and copper.

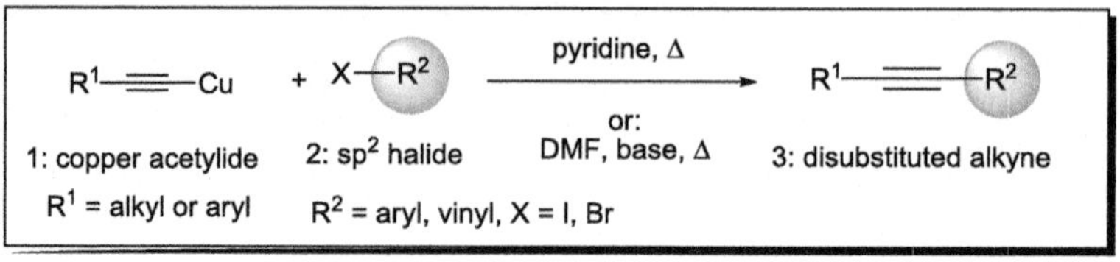

$R^1$—≡—Cu   +   X—$R^2$   $\xrightarrow{\text{pyridine, } \Delta}$   $R^1$—≡—$R^2$

or:

DMF, base, Δ

1: copper acetylide    2: sp² halide           3: disubstituted alkyne

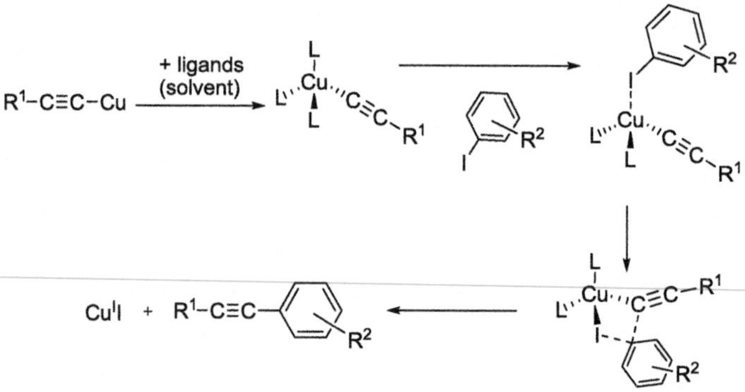

$R^1$ = alkyl or aryl    $R^2$ = aryl, vinyl, X = I, Br

$R^1$-C≡C-Cu $\xrightarrow[\text{(solvent)}]{\text{+ ligands}}$

An alternative mechanism similar to that of the Cadiot–Chodkiewicz coupling:

$Ar-X + Cu$—≡—$R$ $\xrightarrow[\text{addition}]{\text{oxidative}}$ $Ar-Cu$—≡—$R$ $\xrightarrow[\text{elimination}]{\text{reductive}}$ $CuX$ + $Ar$—≡—$R$

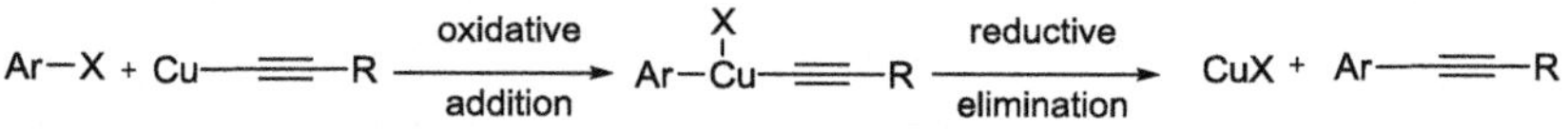

Cu(III) intermediate

Example 1, A variant, also known as the Rosenmund–von Braun synthesis of aryl nitriles[2]

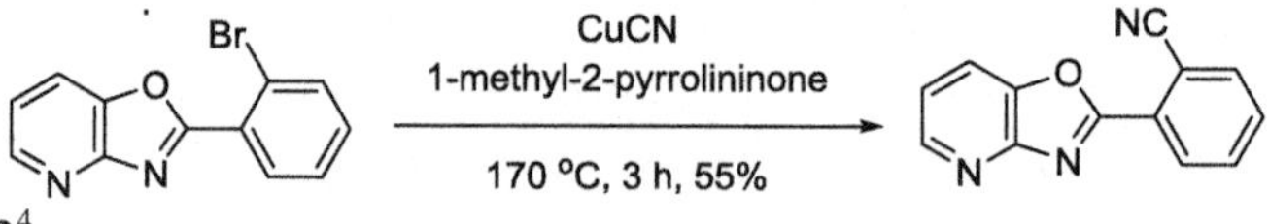

$\xrightarrow[\text{170 °C, 3 h, 55\%}]{\substack{\text{CuCN} \\ \text{1-methyl-2-pyrrolininone}}}$

Example 2[4]

—≡—H $\xrightarrow[\text{ca. 95\%}]{\substack{\text{CuSO}_4\text{, NH}_2\text{OH•HCl} \\ \text{aq. NH}_3\text{, EtOH}}}$ —≡—Cu

$\xrightarrow[\substack{\text{pyridine, } \Delta \\ 75\%}]{}$

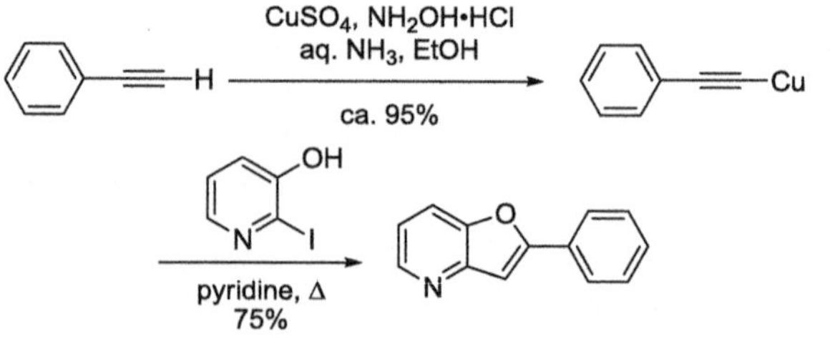

J.J. Li, *Name Reactions: A Collection of Detailed Mechanisms and Synthetic Applications*, DOI 10.1007/978-3-319-03979-4_53, © Springer International Publishing Switzerland 2014

Example 3[5]

Example 4[8]

Example 5, *In situ* Castro–Stephens reaction[10]

Example 6[13]

**References**

1.  (a) Castro, C. E.; Stephens, R. D. *J. Org. Chem.* **1963**, *28*, 2163. Castro and Stephens worked in the Department of Nematology and Chemistry at University of California, Riverside. (b) Stephens, R. D.; Castro, C. E. *J. Org. Chem.* **1963**, *28*, 3313–3315.
2.  Clark, R. L.; Pessolano, A. A.; Witzel, B.; Lanza, T.; Shen, T. Y.; Van Arman, C. G.; Risley, E. A. *J. Med. Chem.* **1978**, *21*, 1158–1162.
3.  Staab, H. A.; Neunhoeffer, K. *Synthesis* **1974**, 424.
4.  Owsley, D.; Castro, C. *Org. Synth.* **1988**, *52*, 128–131.
5.  Kundu, N. G.; Chaudhuri, L. N. *J. Chem. Soc., Perkin Trans 1* **1991**, 1677–1682.
6.  Kabbara, J.; Hoffmann, C.; Schinzer, D. *Synthesis* **1995**, 299–302.
7.  White, J. D.; Carter, R. G.; Sundermann, K. F.; Wartmann, M. *J. Am. Chem. Soc.* **2001**, *123*, 5407–5413.
8.  Coleman, R. S.; Garg, R. *Org. Lett.* **2001**, *3*, 3487–3490.
9.  Rawat, D. S.; Zaleski, J. M. *Synth. Commun.* **2002**, *32*, 1489–1494.
10. Bakunova, A.; Bakunov, S.; Wenzler, T.; Barszcz, T.; Werbovetz, K.; Brun, R.; Hall, J.; Tidwell, R. *J. Med. Chem.* **2007**, *50*, 5807–5823.
11. Gray, D. L. *Castro–Stephens coupling.* In *Name Reactions for Homologations-Part I*; Li, J. J., Ed.; Wiley: Hoboken, NJ, **2009**, pp 212–235. (Review).
12. Wang, Z.-L.; Zhao, L.; Wang, M.-X. *Org. Lett.* **2012**, *14*, 1472–1475.

# C−H activation

The C−H activation reaction is a reaction that cleaves a carbon–hydrogen bond. Here the carbon–hydrogen bond is mostly referred to unactivated carbon–hydrogen bonds.

## *Catellani reaction*

Selective *ortho*-alkylation and -arylation of aryl iodides can be achieved by the cooperative catalytic action of palladium and norbornene.[1] The first reported case was the *ortho*-dialkylation of aryl iodides, followed by Heck reaction.[2] Here an aryl iodide with free *o*-positions reacts with an aliphatic iodide and a terminal olefin in the presence of palladium/norbornene as catalyst and a base, to give a 2,6-substituted vinylarene. Analogously, an aryl iodide with one substituted *o*-position leads to a vinylarene containing two different *ortho* groups.[3]

Example 1, A three-component reaction allowing the construction of three adjacent C–C bonds through C–I and C–H activation.[2]

Mechanism for the reaction of an *o*-substitued aryl iodide: Pd(0), Pd(II) and Pd(IV) intermediates and catalytic role of palladium and norbornene.[1–3]

J.J. Li, *Name Reactions: A Collection of Detailed Mechanisms and Synthetic Applications*, DOI 10.1007/978-3-319-03979-4_54, © Springer International Publishing Switzerland 2014

The mechanism involves initial oxidative addition of an $o$-substituted aryl iodide to Pd(0) followed by a stereoselective norbornene insertion leading to the *cis,exo* complex **2**. β-Hydrogen elimination is prevented by geometric constraints, and a five-membered palladacycle (**3**) readily forms through intramolecular C–H activation. Oxidative addition of an alkyl iodide to **3** affords a Pd(IV) intermediate (**4**) which undergoes reductive elimination by selective migration of the alkyl moiety onto the aromatic ring to form **5**. Norbornene deinsertion occurs spontaneously at this point, likely due to steric hindrance, giving 2,6-disubstituted phenylpalladium(II) species (**6**) which finally react with the terminal olefin to liberate the organic product and Pd(0). Alternatively the sequence can be terminated by other well-known reactions of the aryl-Pd bond such as the Suzuki or Sonogashira couplings, hydrogenolysis, amination, or cyanation. The described methodology can also be extended to ring-forming reactions.[1e] Thus, the reaction is very versatile and offers countless possibilities for building up many types of functionalized aromatic compounds.

Example 2, The synthesis of fused aromatic compounds through final intramolecular Heck reaction was first reported by the Lautens group.[4,1e]

Example 3, The high tolerance to functional groups enabled a key step to the synthesis of a precursor of (+)-linoxepin by Lautens.[5]

92%

*ortho*-Arylation of an aryl iodide leading to the construction of a biaryl moiety is also possible, provided that the starting aryl iodide bears an *ortho* substituent. The $o$-substituent in palladacycles of type **3** is essential for selectively directing the attack of an aryl halide onto the aromatic site (*ortho effect*).[1,6]

Example 4, Aryl-aryl coupling combined with Heck reaction.[7]

Example 5. The non symmetrical coupling of an aryl iodide bearing an *o*-electron-donating group, an aryl bromide containing an electron-withdrawing substituent, and a terminal olefin illustrates the importance of correctly tuning the electronic properties of the two aryl halides for selectivity control.[8]

Example 6 shows that internal chelation to Pd(IV)[9] can cancel the *ortho* effect.[10]

**References**
1. (a) Tsuji, J. Palladium Reagents and Catalysts – New Perspective for the 21st Century, 2004, John Wiley & Sons, pp. 409–416. (b) Catellani, M. *Synlett* **2003**, 298–313. (c) Catellani, M. *Top. Organomet. Chem.* **2005**, *14*, 21–53. (d) Catellani M.; Motti E.; Della Ca' N. *Acc. Chem. Res.* **2008**, *41*, 1512–1522. (e) Martins, A.; Mariampillai, B.; Lautens, M. *Top Curr Chem* **2010**, *292*, 1–33. (f) Chiusoli, G. P.; Catellani, M.; Costa, M.; Motti, E.; Della Ca', N.; Maestri, G. *Coord. Chem. Rev.* **2010**, *254*, 456–469.
2. (a) Catellani, M.; Frignani, F.; Rangoni, A. *Angew. Chem. Int. Ed. Engl.* **1997**, *36*, 119–122. (b) Catellani, M.; Fagnola, M. C. *Angew. Chem. Int. Ed. Engl.* **1994**, *33*, 2421–2422.
3. Catellani, M;..Cugini, F. *Tetrahedron*, **1999**, *55*, 6595–6602.
4. (a) Lautens, M.; Piguel, S.; Dahlmann, M. *Angew. Chem. Int. Ed. Engl.* **2000**, *39*, 1045–1046. (b) Lautens, M.; Paquin, J.-F.; Piguel, S. *J. Org. Chem.* **2001**, *66*, 8127–8134. (c) Lautens, M.; Paquin, J.-F.; Piguel, S. *J. Org. Chem.* **2002**, *67*, 3972–3974.
5. Weinstabl, H.; Suhartono, M.; Qureshi, Z.; Lautens, M. *Angew. Chem. Int. Ed.* **2013**, *125*, 5413–5416.
6. Maestri, G.; Motti, E.; Della Ca', N.; Malacria, M.; Derat, E.; Catellani, M. *J. Am. Chem. Soc.* **2011**, *133*, 8574–8585.
7. Motti, E.; Ippomei, G.; Deledda, S.; Catellani, M. *Synthesis* **2003**, 2671–2678.
8. Faccini, F.; Motti, E.; Catellani, M. *J. Am. Chem. Soc.* **2004**, *126*, 78–79.
9. Vicente, J.; Arcas, A.; Juliá-Hernández, F.; Bautista, D. *Angew. Chem. Int. Ed.* **2011**, *50*, 6896–6899.
10. Della Ca', N.; Maestri, G.; Malacria, M.; Derat, E.; Catellani, M. *Angew. Chem. Int. Ed.* **2011**, *50*, 12257–12261.

## *Sanford reaction*

C–H acetoxylations using directing groups L, such as pyridine and pyrimidine.[1,6]

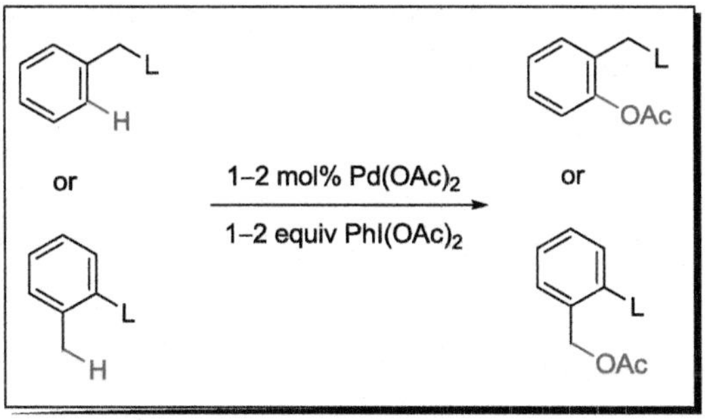

Catalytic Cycle for Ligand-Directed C–H Acetoxylation:[2]

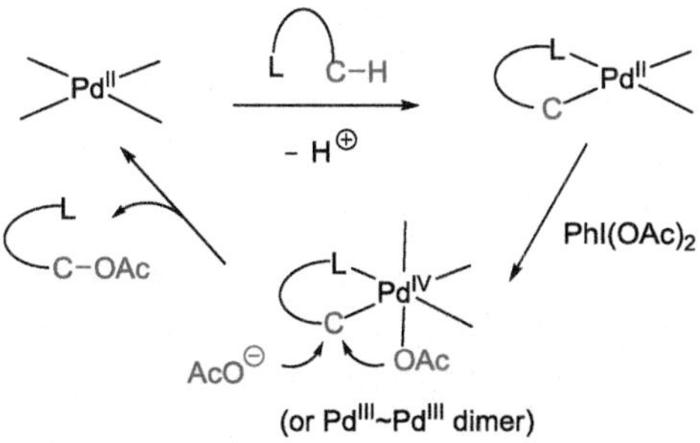

(or $Pd^{III}$~$Pd^{III}$ dimer)

Example 1.[3]

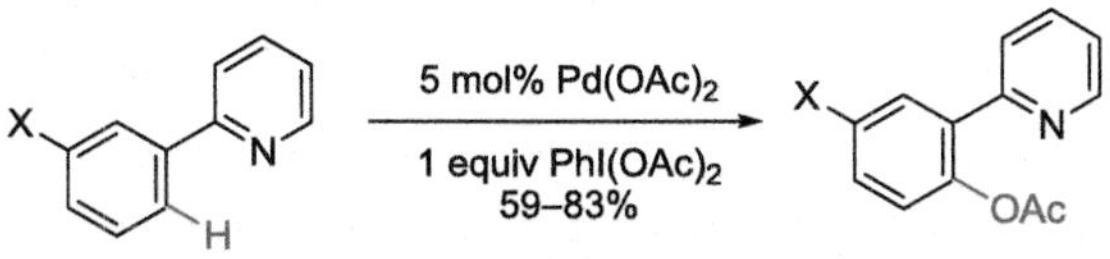

X = OMe, OMOM, Me, Br, F, $CF_3$, Ac, oxime, $NO_2$,
regioselectivity: 6:1 to > 20:1

Example 2.[5]

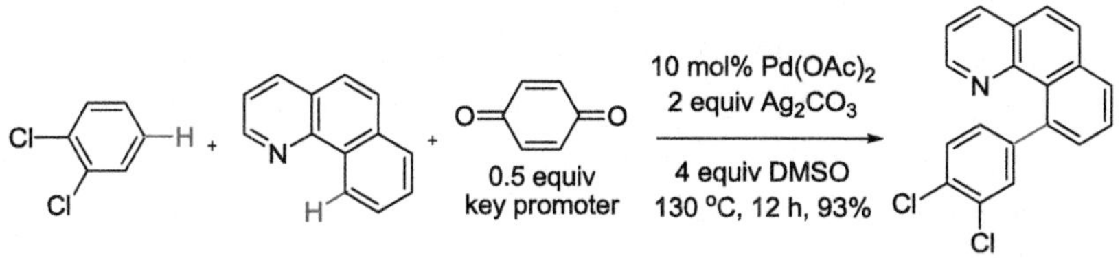

Single regioisomer

Example 3, Directing Group Ability in Palladium-Catalyzed C–H Bond Functionalization.[6]

0.5 equiv                    1 mol% Pd(OAc)₂                    > 99%

1.02 equiv PhI(OAc)₂
Ac₂O/AcOH, 100 °C

0.5 equiv                                                      < 1%

Example 4.[10]

2 mol% Pd(OAc)₂
2 mol% pyridine

ArI(OAc)₂, 68%

## References

1.  Dick, A. R.; Hull, K. L.; Sanford, M. S. *J. Am. Chem. Soc.* **2004**, *126*, 2300–2301. Melanie S. Sanford was born in New Bedford, MA, in 1975. She received a B.S. and M.S. in chemistry from Yale University in 1996 before obtaining her Ph.D. at California Institute of Technology with Professor Robert Grubbs in 2001. She then completed postdoctoral studies at Princeton University working with Professor John Groves. In 2003, she joined the faculty of the University of Michigan where she is currently the Moses Gomberg Collegiate Professor of Chemistry as well as Arthur F. Thurnau Professor of Chemistry. Her research program focuses on the development of new catalysts and catalytic transformations for applications in the synthesis of useful organic molecules.
2.  Kalyani, D.; Deprez, N. R.; Desai, L. V.; Sanford, M. S. *J. Am. Chem. Soc.* **2005**, *127*, 7330–7331.
3.  Kalyani, D.; Sanford, M. S. *Org. Lett.* **2005**, *7*, 4149–4152.
4.  Hull, K. L.; Lanni, E. L.; Sanford, M. S. *J. Am. Chem. Soc.* **2006**, *128*, 14047–14047. (Mechanistic insight).
5.  Hull, K. L.; Sanford, M. S. *J. Am. Chem. Soc.* **2007**, *129*, 11904–11905.
6.  Desai, L. V.; Stowers, K. J.; Sanford, M. S. *J. Am. Chem. Soc.* **2008**, *130*, 13285–13293.
7.  Stowers, K. J.; Sanford, M. S. *Org. Lett.* **2009**, *11*, 4584–4587. (Mechanistic insight).
8.  Lyons, T. W.; Sanford, M. S. *Chem. Rev.* **2010**, *110*, 1147–1169. (Review).
9.  Lyons, T. W.; Hull, K. L.; Sanford M. S. *J. Am. Chem. Soc.* **2011**, *133*, 4455–4464. (Regioselectivity).
10. Emmert, M. H.; Cook, A. K.; Xie, Y. J.; Sanford, M. S. *Angew. Chem. Int. Ed.* **2011**, *50*, 9409–9412.
11. Neufeldt, S. R.; Sanford, M. S. *Acc. Chem. Res.* **2012**, *45*, 936–946. (Review).

## White catalyst

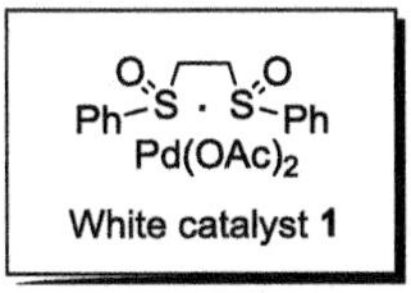

White catalyst 1

The White catalyst **1** is a highly versatile, commercially-available catalyst for allylic C–H oxidation which allows for the construction of useful C–O, C–N, and C–C bonds directly from relatively inert allylic C–H bonds (Figure 1).[1–11] The White catalyst enables novel and predictable disconnections for the synthesis of complex molecules which can streamline their synthesis.[2,4,7,8] Widely available α-olefins undergo intra- and intermolecular C–H oxidation with remarkably high levels of chemo-, regio-, and stereoselectivity. Mechanistic studies provide evidence that the White catalyst promotes allylic C–H cleavage to generate π-allylpalladium intermediate **2** which can then be functionalized with an oxygen, nitrogen or carbon nucleophile (Figure 1).[3]

Figure 1

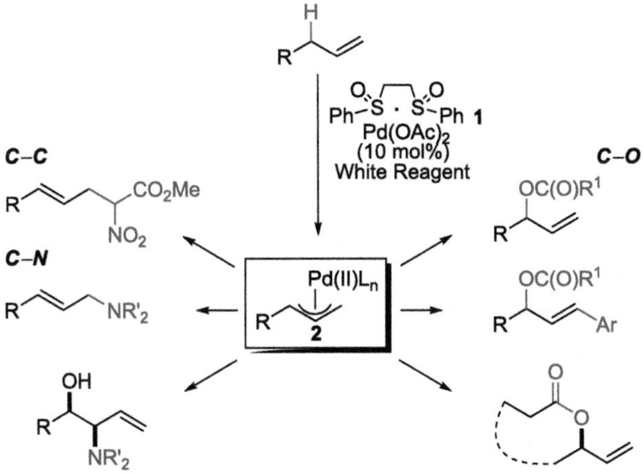

Common organic functionality such as Lewis basic phenol **3**,[3] acid-labile acetal **4**,[8] highly reactive aryl triflate **6**,[11] and depsipeptide **5**[5] are well-tolerated under the mild reaction conditions (Figure 2). In all cases the products are isolated as one regioisomer and olefin isomer after column purification.

Current state-of-the-art methods for constructing C–N bonds rely on functional group interconversions or C–C bond forming reactions using preoxidized materials. Allylic amination using the White catalyst can streamline the synthesis of nitrogen-containing molecules by reducing the functional group manipulations necessary for working with oxygenated intermediates. Allylic C–H amination was used to synthesize (–)-**8**, an intermediate in the synthesis of *L*-acosamine derivative **9** (Figure 3A).[7] The C–H amination route to (–)-**8** proceeded in half the total number of steps, no functional group manipulations, and comparable overall yield to the alternative C–O to C–N bond-forming route. Intermolecular C–H amination has also led to the construction of (+)-deoxynegamycin analogue **12** in five less steps and improved overall yield compared to the alternative route relying on C–O substitution (Figure 3B).[8]

Figure 2

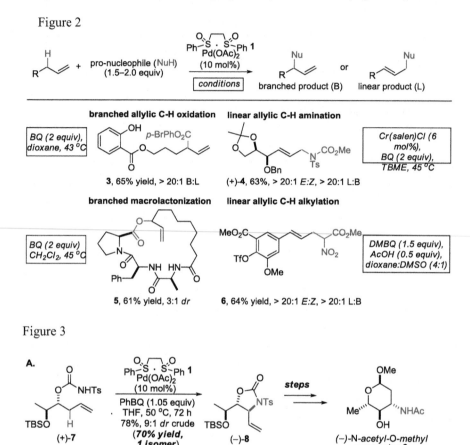

**branched allylic C-H oxidation**

**3**, 65% yield, > 20:1 B:L

BQ (2 equiv), dioxane, 43 °C

**linear allylic C-H amination**

(+)-**4**, 63%, > 20:1 E:Z, > 20:1 L:B

Cr(salen)Cl (6 mol%), BQ (2 equiv), TBME, 45 °C

**branched macrolactonization**

**5**, 61% yield, 3:1 dr

BQ (2 equiv) CH₂Cl₂, 45 °C

**linear allylic C-H alkylation**

**6**, 64% yield, > 20:1 E:Z, > 20:1 L:B

DMBQ (1.5 equiv), AcOH (0.5 equiv), dioxane:DMSO (4:1)

Figure 3

**A.**

(+)-**7** → PhBQ (1.05 equiv), THF, 50 °C, 72 h, 78%, 9:1 dr crude *(70% yield, 1 isomer)* → (–)-**8** → steps → (–)-N-acetyl-O-methyl acosamine **9**

**B.**

(–)-**10** → Cr(salen)Cl (6 mol%), CbzNHTs (2 equiv), BQ (2 equiv), TBME, 54%, 1 isomer → (+)-**11** → steps → (+)-deoxynegamycin analogue **12**

Similarly, allylic C–H oxidation can streamline the construction of oxygenated compounds by reducing functional group manipulations necessary for working with bisoxygenated intermediates. For example, a chiral allylic C–H oxidation/enzymatic resolution sequence furnished bisoxygenated compound **14** in 97% ee and in 42% overall yield in just 3 steps from a commercially available monooxygenated precursor, 11-undecenoic acid (Figure 4).[10] Alternative routes to similar molecules require protection/deprotection sequences and use a kinetic resolution giving a maximum of 50% yield.

Figure 4

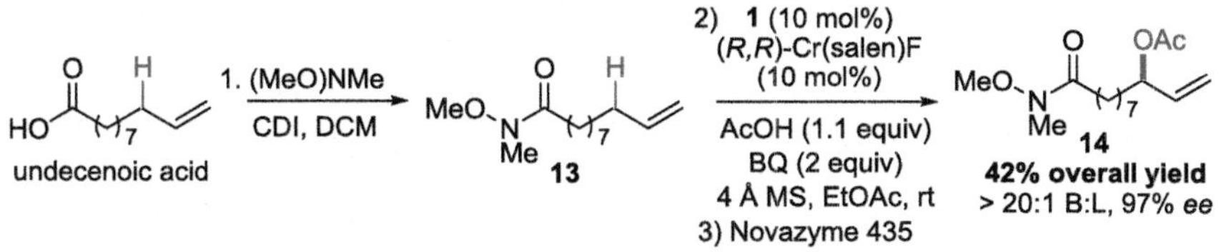

In addition to allylic C–H oxidation, the White catalyst also catalyzes intermolecular Heck arylations.[6]   Notably, the arylation uses electronically *unbiased* α-olefins and aryl boronic acids and occurs under acidic, oxidative conditions.   A one-pot allylic C–H oxidation/vinylic C–H arylation reaction furnishes *E*-arylated allylic esters with high regio- and stereoselectivities (Figure 5).   This three-component coupling can be used to rapidly synthesize densely functionalized products from inexpensive hydrocarbon feedstocks.   *N*-Boc glylcine allylic ester **9** was synthesized in one step using commercially available olefin, amino acid, and boronic acid reagents. Compounds similar to **15** have been transformed into medicinally relevant dipeptidyl peptidase IV inhibitors.[6]

Figure 5

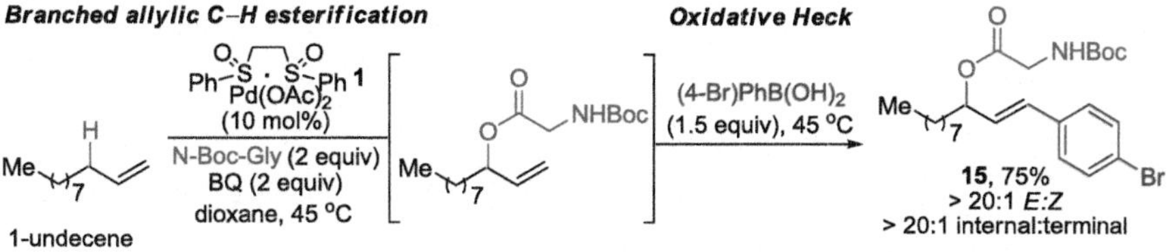

Besides the one-pot process described above, the White catalyst catalyzes a chelate-controlled oxidative Heck arylation between a wide range of α-olefins and organoborane compounds in good yields and with excellent regio- and stereoselectivities (Figure 6).[9]   Unlike other Heck arylation methods, no Pd–H isomerization is observed under the mild reaction conditions. Aryl boronic acids, styrenylpinacol boronic esters, and aryl potassium trifluoroborates (activated with boric acid) are all compatible with the general reaction conditions.

Figure 6

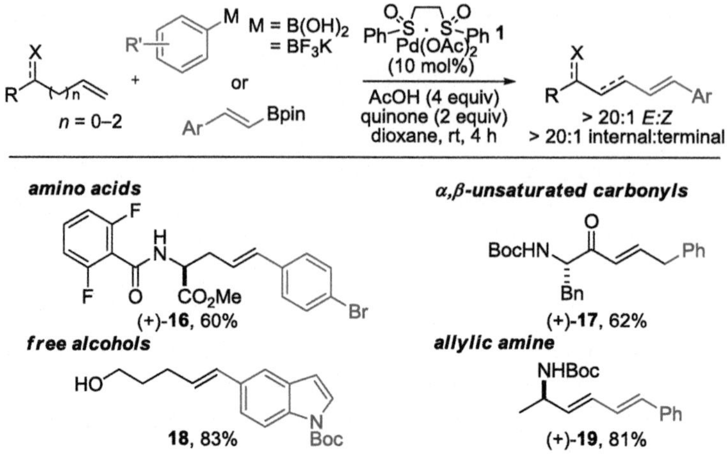

**amino acids**

(+)-**16**, 60%

**free alcohols**

**18**, 83%

**α,β-unsaturated carbonyls**

(+)-**17**, 62%

**allylic amine**

(+)-**19**, 81%

## References

1. Chen, M. S.; White, M. C. *J. Am. Chem. Soc.* **2004,** *126*, 1346–1347. Christina M. White is a professor at the University of Illinois at Urbana-Champagne.
2. Fraunhoffer, K. J.; Bachovchin, D. A.; White, M. C. *Org. Lett.* **2005,** *7*, 223–226.
3. Chen. M. S.; Prabagaran, N.; Labenz, N. A.; White, M. C. *J. Am. Chem. Soc.* **2005,** *127*, 6970–6971.
4. Covell, D. J.; Vermeulen, N. A.; White, M. C. *Angew. Chem. Int. Ed.* **2006,** *45*, 8217–8220.
5. Fraunhoffer, K. J.; Prabagaran, N.; Sirois, L. E.; White, M. C. *J. Am. Chem. Soc.* **2006,** *128*, 9032–9033.
6. Delcamp, J. H.; White, M. C. *J. Am. Chem. Soc.* **2006,** *128*, 15076–15077.
7. Fraunhoffer, K. J.; White, M. C. *J. Am. Chem. Soc.* **2007,** *129*, 7274–7276.
8. Reed, S. A.; White, M. C. *J. Am. Chem. Soc.* **2008,** *129*, 3316–3318.
9. Delcamp, J. H.; Brucks, A. P.; White, M. C. *J. Am. Chem. Soc.* **2008,** *129*, 11270–11271.
10. Covell, D. J.; White, M. C. *Angew. Chem. Int. Ed.* **2008,** *47*, 6448–6451.
11. Young, A. J.; White, M. C. *J. Am. Chem. Soc.* **2008,** *129*, 14090–14091.

## *Yu C–H activation*

A variety of position-selective or stereoselective C–H activation reactions have been developed by Yu and co-workers.[1,7] These transformations are characterized by the use of a Pd catalyst, an oxidant, often with built-in directing groups and/or optimized ligands that enhance selectivity as well as reaction rate.

Canonical example of sp$^2$ C–H activation:

Canonical example of sp$^3$ C–H activation:

Example 1, Hydroxyl-directed C–H activation/C–O cyclization[2]

Example 2, Amide-directed sp$^3$ C–H carbonylation[3]

Example 3, Sulfonamide-directed C–H methylation[4]

Example 4, *ortho*-Selective C–H arylation of arene[5]

Example 5, C$_3$-Selective C–H arylation of pyridine[6]

Example 6, *meta*-Selective C–H vinylation of arene[8]

**References**
1.      Wang, D. H.; Engle, K. M.; Shi, B. F.; Yu, J. Q. *Science* **2010**, *327*, 315−319.
2.      Wang, X.; Lu, Y.; Dai, H. X.; Yu, J. Q. *J. Am. Chem. Soc.* **2010**, *132*,
        12203−12205.
3.      Yoo, E. J.; Wasa, M.; Yu, J. Q. *J. Am. Chem. Soc.* **2010**, *132*, 17378−17380.
4.      Dai, H. X.; Stepan, A. F.; Plummer, M. S.; Zhang, Y. H.; Yu, J. Q. *J. Am. Chem.
        Soc.* **2011**, *133*, 7222−7228.
5.      Engle, K, M.; Thuy-Boun, P. S.; Dang, M.; Yu, J. Q. *J. Am. Chem. Soc.* **2011**,
        *133*, 18183−18193.
6.      Ye, M.; Gao, G. L.; Edmunds, A. J.; Worthington, P. A.; Morris, J. A.; Yu, J. Q.
        *J. Am. Chem. Soc.* **2011**, *133*, 19090−19093.
7.      Engle, K. M.; Mei, T. S.; Wasa, M.; Yu, J. Q. *Acc. Chem. Res.* **2012**, *45*, 788−802
        (Review).
8.      Leow, D.; Li, G.; Mei, T. S.; Yu, J. Q. *Nature* **2012**, *486*, 518−522.

/

# Chan alkyne reduction

Stereoselective reduction of acetylenic alcohols to $E$-allylic alcohols using sodium bis(2-methoxyethoxy)aluminum hydride (SMEAH, also known as Red-Al) or LiAlH$_4$.

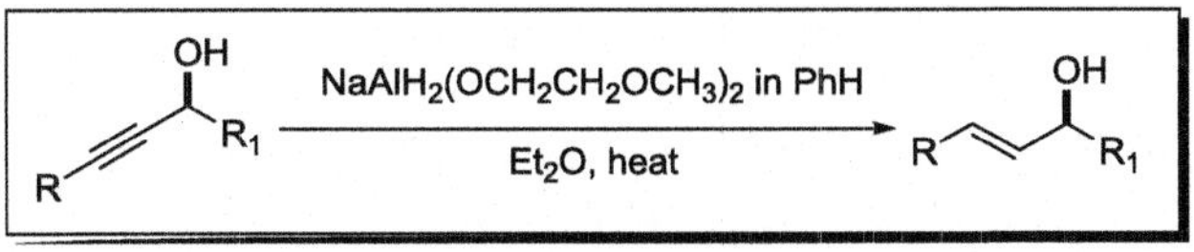

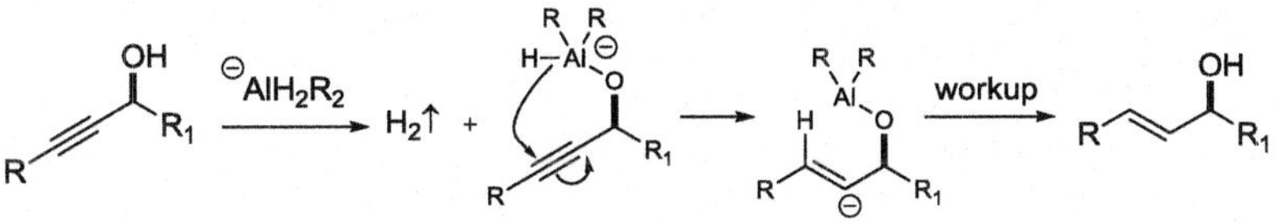

Example 1[3]

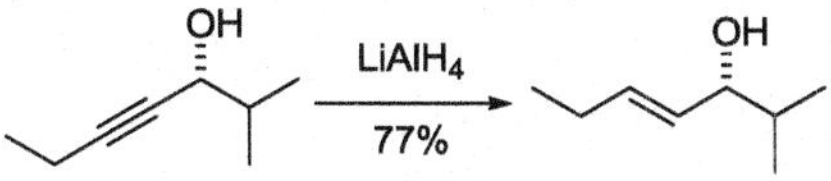

Example 2[4]

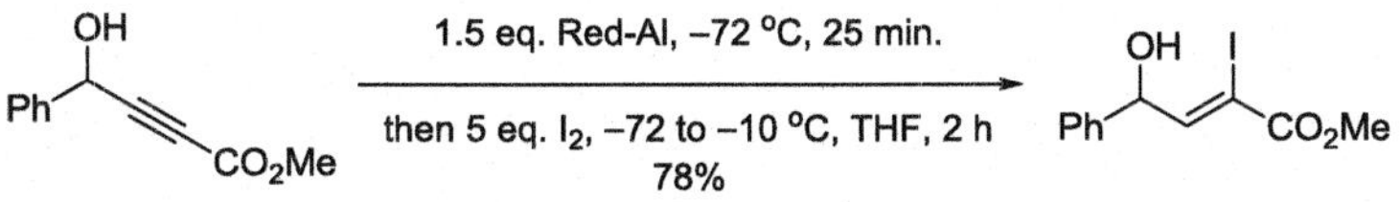

Example 3[6]

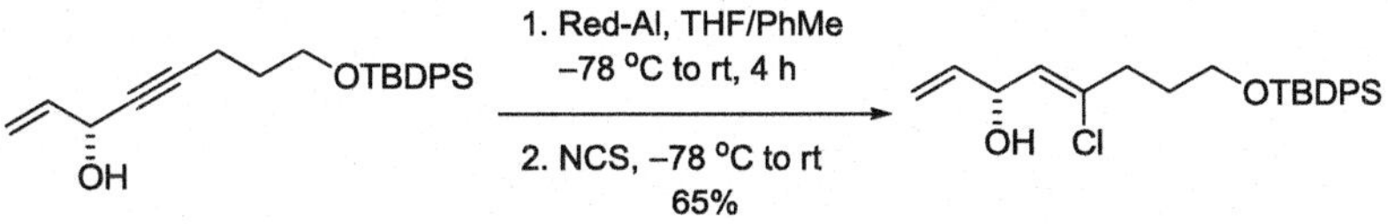

Example 4[7]

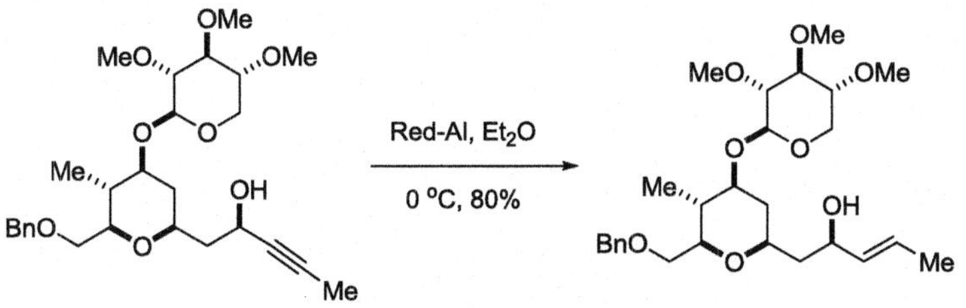

J.J. Li, *Name Reactions: A Collection of Detailed Mechanisms and Synthetic Applications*, DOI 10.1007/978-3-319-03979-4_55, © Springer International Publishing Switzerland 2014

Example 5[8]

## References

1. Chan, K.-K.; Cohen, N.; De Noble, J. P.; Specian, A. C., Jr.; Saucy, G. *J. Org. Chem.* **1976,** *41,* 3497–3505. Ka-Kong Chan was a chemist at Hoffmann–La Roche, Inc. in Nutley, NJ, USA.
2. Blunt, J. W.; Hartshorn, M. P.; Munro, M. H. G.; Soong, L. T.; Thompson, R. S.; Vaughan, J. *J. Chem. Soc., Chem. Commun.* **1980,** 820–821.
3. Midland, M. M.; Gabriel, J. *J. Org. Chem.* **1985,** *50,* 1143–1144.
4. Meta, C. T.; Koide, K. *Org. Lett.* **2004,** *6,* 1785–1787.
5. Yamazaki, T.; Ichige, T.; Kitazume, T. *Org. Lett.* **2004,** *6,* 4073–4076.
6. Xu, S.; Arimoto, H.; Uemura, D. *Angew. Chem. Int. Ed.* **2007,** *46,* 5746–5749.
7. Chakraborty, T. K.; Reddy, V. R.; Gajula, P. K. *Tetrahedron* **2008,** *64,* 5162–5167.
8. Krishna, P. R.; Krishnarao, L.; Reddy, K. L. N. *Beilstein J. Org. Chem.* **2009,** *5,* No. 14.
9. Yadav, J. S.; Krishna, V. H.; Srilatha, A.; Somaiah, R.; Reddy, B. V. S. *Synthesis* **2011,** 3004–3012.
10. Krishna, P. R.; Alivelu, M. *Helv. Chim. Acta* **2011,** *94,* 1102–1107.

# Chan–Lam C–X coupling reaction

Arylation, vinylation and alkylation of a wide range of NH/OH/SH substrates by oxidative cross-coupling with boronic acids in the presence of catalytic cupric acetate, weak base and in air (open-flask chemistry). The reaction works for amides, amines, amidines, anilines, azides, azoles, hydantoins, hydrazines, imides, imines, nitroso, pyrazinones,  yridines, purines, pyrimidines, sulfonamides, sulfinates, sulfoximines, ureas, alcohols, phenols, thiols, *etc.* The boronic acids can be replaced with siloxanes, stannanes or other organometalloids. The mild condition of this reaction is an advantage over Buchwald–Hartwig's Pd-catalyzed cross-coupling using halides, though boronic acids are more expensive than halides. The Chan–Lam C–X bond cross-coupling reaction has emerged as a powerful and popular methodogy similar to Suzuki–Miyaura's C–C bond cross-coupling reaction.

$$Ar{-}M \;+\; H{-}XR \xrightarrow[\text{weak base, MC, air}]{\text{cat. Cu(AcO)}_2} Ar{-}XR$$

M = B(OH)$_2$, B(OR)$_2$, B(OR)$_3^-$, BF$_3^-$, SnMe$_3$, Si(OR)$_3$.
X = N, O, S, Se, Te, F, Cl, Br, I.

Example 1 [1a,d]

Proposed Mechanism:[4]

J.J. Li, *Name Reactions: A Collection of Detailed Mechanisms and Synthetic Applications*,
DOI 10.1007/978-3-319-03979-4_56, © Springer International Publishing Switzerland 2014

Example 2[5]

1.1 eq Cu(OAc)₂, TEA
rt, 24 h, air
52%

Example 3[6]

Cu(OAc)₂
TEA, air

52%

Example 4[14]

(HO)₂B—⟨  ⟩—OMe

Cu(OAc)₂/Et₃N/Py
1eq/3 eq/ 3 eq
4 Å MS, air

93% (α-ester assistance,
acetal lower yield)

Example 5[15]

NaHMDS, Cu(AcO)₂
DMAP, air, 95 °C
93%

**References**

1.        (a) Chan, D. M. T.; Monaco, K. L.; Wang, R.-P.; Winters, M. P. *Tetrahe
dron Lett.* **1998**, *39*, 2933–2936. (b) Lam, P. Y. S.; Clark, C. G.;      Saubern, S.;
Adams, J.; Winters, M. P.; Chan, D. M. T.; Combs, A. *Tet rahedron   Lett.*   **1998,**
*39*, 2941–2949. Dominic Chan is a chemist at DuPont Crop Protection, Wilming
ton, DE, USA. He did his PhD research with Prof. Barry Trost at the University
of Wisconson, Madison. Patrick Lam is a research director at Bristol–Myers
Squibb, Princeton, NJ, USA. He was formerly with DuPont Pharmaceuticals
Company. He did his PhD research with Prof. Louis Friedrich in the Univeristy of
Rochester and Post-doc research with Prof. Michael Jung and the late Prof. Don
ald Cram in UCLA. (c) Evans, D. A.; Katz, J. L.; West, T. R. *Tetrahedron Lett.*
**1998**, *39*, 2937–2940. Prof. Evans' group found out about the discovery of this
reaction on a National Organic Symposium poster and be  came interested in the
*O*-arylation because of his long interest in vancomycin total synthesis. (d) Lam, P.
Y. S.; Clark, C. G.; Saubern, S.; Adams, J.; Averill, K. M.; Chan, D. M. T.;
Combs, A. *Synlett* **2000**, 674–676. (e) Lam, P. Y. S.; Bonne, D.; Vincent, G.;
Clark, C. G.; Combs, A. P. *Tetrahedron Lett.* **2003**, *44*, 1691–1694.

2. Reviews: (a) Qiao, J. X.; Lam, P. Y. S. *Syn.* **2011**, 829–856; (b) Chan, D. M. T.; Lam, P. Y. S., Book chapter in *Boronic Acids* Hall, ed. **2005**, Wiley–VCH, 205–240. (c) Ley, S. V.; Thomas, A. W. *Angew. Chem., Int. Ed. Engl.* **2003**, *42*, 5400–5449.

3. Catalytic copper: (a) Lam, P. Y. S.; Vincent, G.; Clark, C. G.; Deudon, S.; Jadhav, P. K. *Tetrahedron Lett.* **2001**, *42*, 3415–3418. (b) Antilla, J. C.; Buch wald, S. L. *Org. Lett.* **2001**, *3*, 2077–2079. (c) Quach, T. D.; Batey, R. A. *Org. Lett.* **2003**, *5*, 4397–4400. (d) Collman, J. P.; Zhong, M. *Org. Lett.* **2000**, *2*, 1233–1236. (e) Lan, J.-B.; Zhang, G.-L.; Yu, X.-Q.; You, J.-S.; Chen, L.; Yan, M.; Xie, R.-G. *Synlett* **2004**, 1095–1097.

4. Mechanism (Part of the mechanistic work from Shannon's lab was funded and in collaboration with BMS: (a) Huffman, L. M.; Stahl, S. S. *J. Am. Chem. Soc.* **2008**, *130*, 9196–9197. (b) King, A. E.; Brunold, T. C.; Stahl, S. S. *J. Am. Chem. Soc.* **2009**, *131*, 5044. (c) King, A. E.; Huffman, L. M.; Casitas, A.; Costas, M.; Ribas, X.; Stahl, S. S. *J. Am. Chem. Soc.* **2010**, *132*, 12068–12073. (d) Casita, A.; King, A. E.; Prella, T.; Costas, M.; Stahl, S. S.; Ribas, X. *J. Chem. Sci.* **2010**, *1*, 326–330.

5. Vinyl boronic acids: Lam, P. Y. S.; Vincent, G.; Bonne, D.; Clark, C. G. *Tetrahe dron Lett.* **2003**, *44*, 4927–4931.

6. Intramolecular: Decicco, C. P.; Song, Y.; Evans, D.A. *Org. Lett.* **2001**, *3*, 1029–1032.

7. Solid phase: (a) Combs, A. P.; Saubern, S.; Rafalski, M.; Lam, P. Y. S. *Tetrahe dron Lett.* **1999**, *40*, 1623–1626. (b) Combs, A. P.; Tadesse, S.; Rafalski, M.; Haque, T. S.; Lam, P. Y. S. *J. Comb. Chem.* **2002**, *4*, 179–182.

8. Boronates/borates: (a) Chan, D. M. T.; Monaco, K. L.; Li, R.; Bonne, D.; Clark, C. G.; Lam, P. Y. S. *Tetrahedron Lett.* **2003**, *44*, 3863–3865. (b) Yu, X. Q.; Yamamoto, Y.; Miyuara, N. *Chem. Asian J.* **2008**, *3*, 1517–1522.

9. Siloxanes: (a) Lam, P. Y. S.; Deudon, S.; Averill, K. M.; Li, R.; He, M. Y.; DeShong, P.; Clark, C. G. *J. Am. Chem. Soc.* **2000**, *122*, 7600–7601. (b) Lam, P. Y. S.; Deudon, S.; Hauptman, E.; Clark, C. G. *Tetrahedron Lett.* **2001**, *42*, 2427–2429.

10. Stannanes: Lam, P. Y. S.; Vincent, G.; Bonne, D.; Clark, C. G. *Tetrahedron Lett.* **2002**, *43*, 3091–3094.

11. Thiols: (a) Herradura, P. S.; Pendora, K. A.; Guy, R. K. *Org. Lett.* **2000**, *2*, 2019–2022. (b) Savarin, C.; Srogl, J.; Liebeskind, L. S. . *Org. Lett.* **2002**, *4*, 4309–4312. (c) Xu, H.-J.; Zhao, Y.-Q.; Feng, T.; Feng, Y.-S. *J. Org. Chem.* **2012**, *77*, 2878–2884.

12. Sulfinates: (a) Beaulieu, C.; Guay. D.; Wang, C.; Evans, D. A. *Tetrahedron Lett.* **2004**, *45*, 3233–3236. (b) Huang, H.; Batey, R. A. *Tetrahedron.* **2007**, *63*, 7667–7672. (c) Kar, A.; Sayyed, L.A.; Lo, W.F.; Kaiser, H.M.; Beller, M.; Tse, M. K. *Org. Lett.* **2007**, *9*, 3405–3408.

13. Sulfoximines: Moessner, C.; Bolm, C. *Org. Lett.* **2005**, *7*, 2667–2669.

14. β-Lactam: Wang, W.; *et al. Bio. Med. Chem. Lett.* **2008**, *18*, 1939–1944.

15. Cyclopropyl boronic acid: Tsuritani, T.; Strotman, N. A.; Yamamoto, Y.; Kawa saki, M.; Yasuda, N.; Mase, T. *Org. Lett.* **2008**, *10*, 1653–1655.

16. Alcohols: Quach, T. D.; Batey, R. A. *Org. Lett.* **2003**, *5*, 1381–1384.

17. Fluorides: (a) Ye, Y.; Sanford, M. S. *J. Am. Chem. Soc.* **2013**, *135*, 4648–4651. (b) Fier, P. S.; Luo, J.; Hartwig, J. F. *J. Am. Chem. Soc.* **2013**, *135*, 2552–2559.

# Chapman rearrangement

Thermal aryl rearrangement of $O$-aryliminoethers to amides.

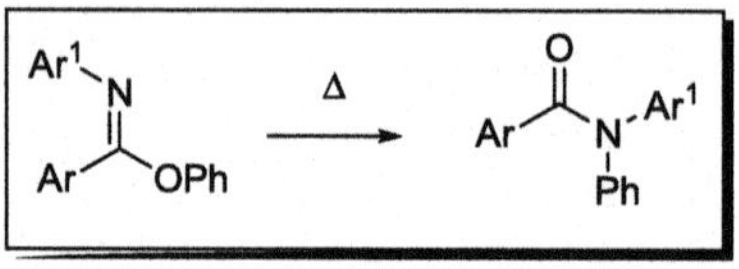

Mechanism:

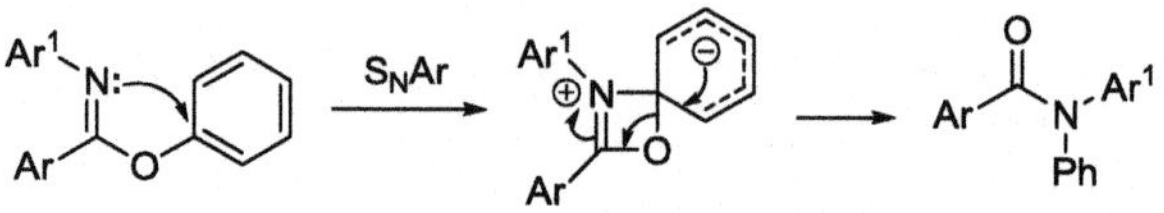

oxazete intermediate

Example 1[2]

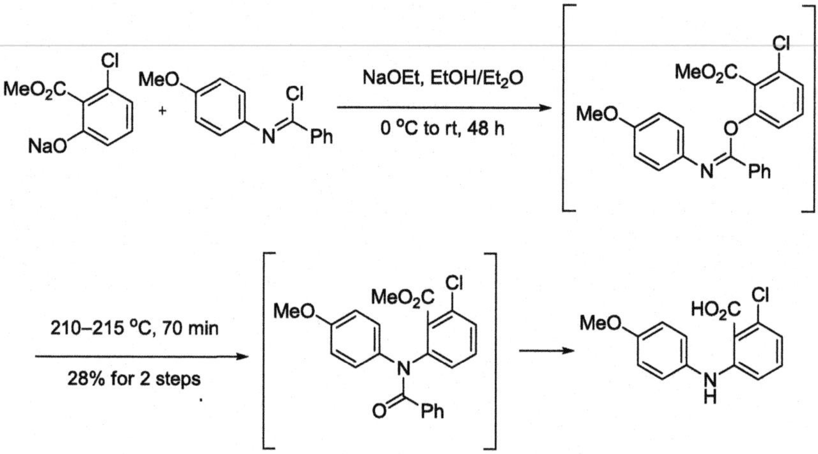

Example 2[4]

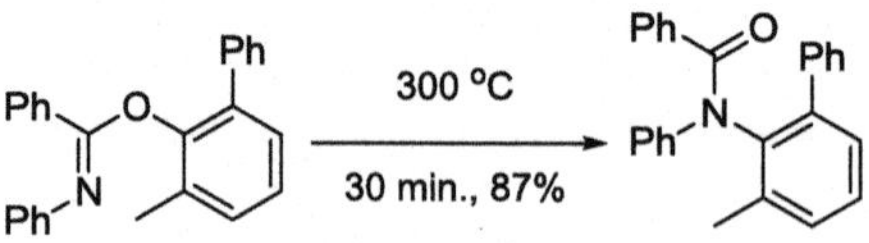

J.J. Li, *Name Reactions: A Collection of Detailed Mechanisms and Synthetic Applications*,
DOI 10.1007/978-3-319-03979-4_57, © Springer International Publishing Switzerland 2014

Example 3, Double Chapman rearrangement[9]

Example 4, Chapman-like thermal rearrangement[11]

**References**
1.  Chapman, A. W. *J. Chem. Soc.* **1925**, *127*, 1992–1998. Arthur William Chapman was born in 1898 in London, England. He was a Lecturer in Organic Chemistry and later became Registrar of the University of Sheffield from 1944 to 1963.
2.  Dauben, W. G.; Hodgson, R. L. *J. Am. Chem. Soc.* **1950**, *72*, 3479–3480.
3.  Schulenberg, J. W.; Archer, S. *Org. React.* **1965**, *14*, 1–51. (Review).
4.  Relles, H. M. *J. Org. Chem.* **1968**, *33*, 2245–2253.
5.  Shawali, A. S.; Hassaneen, H. M. *Tetrahedron* **1972**, *28*, 5903–5909.
6.  Kimura, M.; Okabayashi, I.; Isogai, K. *J. Heterocycl. Chem.* **1988**, *25*, 315–320.
7.  Farouz, F.; Miller, M. J. *Tetrahedron Lett.* **1991**, *32*, 3305–3308.
8.  Dessolin, M.; Eisenstein, O.; Golfier, M.; Prange, T.; Sautet, P. *J. Chem. Soc., Chem. Commun.* **1992**, 132–134.
9.  Marsh, A.; Nolen, E. G.; Gardinier, K. M.; Lehn, J. M. *Tetrahedron Lett.* **1994**, *35*, 397–400.
10. Almeida, R.; Gomez-Zavaglia, A.; Kaczor, A.; Cristiano, M. L. S.; Eusebio, M. E. S.; Maria, T. M. R.; Fausto, R. *Tetrahedron* **2008**, *64*, 3296–3305.
11. Noorizadeh, S.; Ozhand, A. *Chin. J. Chem.* **2010**, *28*, 1876–1884.

# Chichibabin pyridine synthesis

Also known as the Chichibabin reaction. Condensation of aldehydes with ammonia to afford pyridines.

## Example 1[4]

## Example 2[8]

J.J. Li, *Name Reactions: A Collection of Detailed Mechanisms and Synthetic Applications*,
DOI 10.1007/978-3-319-03979-4_58, © Springer International Publishing Switzerland 2014

Example 3[9]

Example 4, An abnormal Chichibabin reaction[10]

## References

1. Chichibabin, A. E. *J. Russ. Phys. Chem. Soc.* **1906,** *37,* 1229. Alexei E. Chichibabin (1871–1945) was born in Kuzemino, Russia. He was Markovnikov's favorite student. Markovnikov's successor, Zelinsky (of Hell–Volhard–Zelinsky reaction fame) did not want to cooperate with the pupil and gave Chichibabin a negative judgment on his Ph.D. work, earning Chichibabin the nickname "the self-educated man."
2. Sprung, M. M. *Chem. Rev.* **1940,** *40,* 297–338. (Review).
3. Frank, R. L.; Riener, E. F. *J. Am. Chem. Soc.* **1950,** *72,* 4182–4183.
4. Weiss, M. *J. Am. Chem. Soc.* **1952,** *74,* 200–202.
5. Kessar, S. V.; Nadir, U. K.; Singh, M. *Indian J. Chem.* **1973,** *11,* 825–826.

6. Shimizu, S.; Abe, N.; Iguchi, A.; Dohba, M.; Sato, H.; Hirose, K.-I. *Microporous Mesoporous Materials* **1998,** *21*, 447–451.

7. Galatasis, P. *Chichibabin (Tschitschibabin) Pyridine Synthesis*. In *Name Reactions in Heterocyclic Chemistry*; Li, J. J., Ed.; Wiley: Hoboken, NJ, **2005,** pp 308–309. (Review).

8. Snider, B. B.; Neubert, B. J. *Org. Lett.* **2005,** *7*, 2715–2718.

9. Wang, X.-L.; Li, Y.-F.; Gong, C.-L.; Ma, T.; Yang, F.-C. *J. Fluorine Chem.* **2008,** *129*, 56–63.

10. Burns, N. Z.; Baran, P. S. *Angew. Chem. Int. Ed.* **2008,** *47*, 205–208.

11. Huang, Y.-C.; Wang, K.-L.; Chang, C.-H.; Liao, Y.-A.; Liaw, D.-J.; Lee, K.-R.; Lai, J.-Y. *Macromolecules* **2008,** *46*, 7443–7450.

# Chugaev elimination

Thermal elimination of xanthates to olefins.

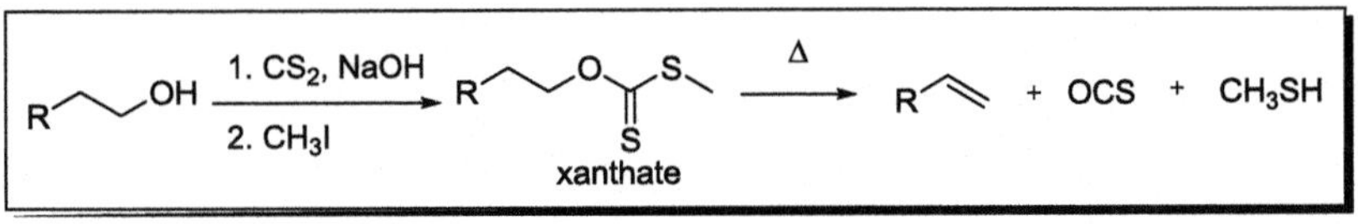

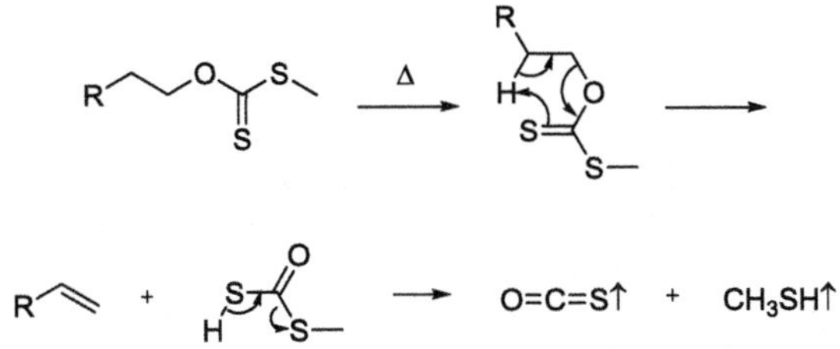

Example 1[4]

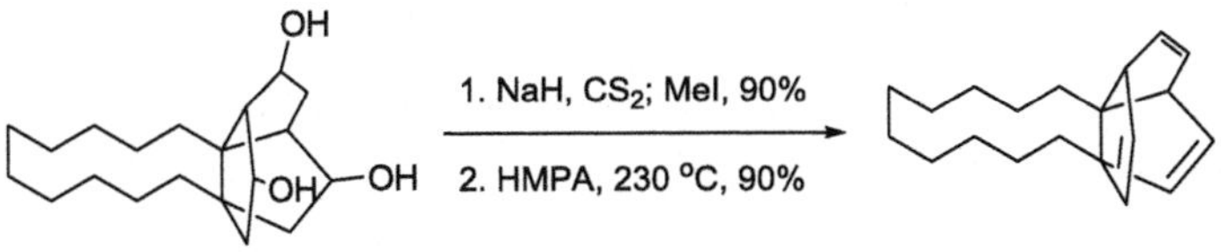

Example 2[5]

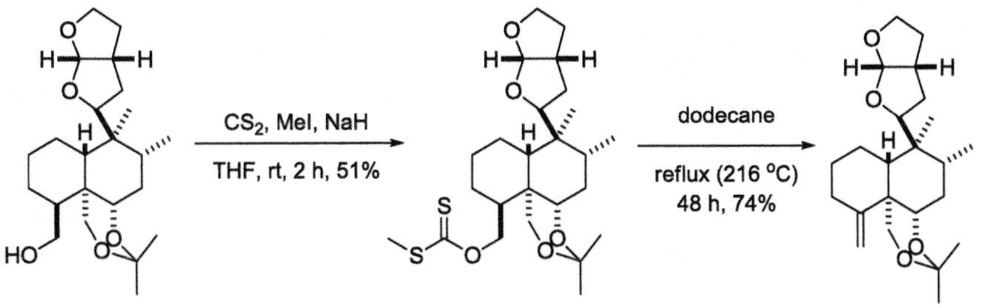

Example 3, Chugaev *syn*-elimination is followed by an intramolecular ene reaction[6]

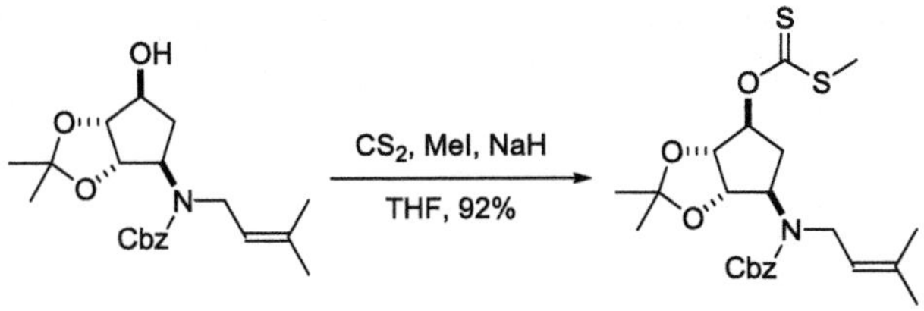

J.J. Li, *Name Reactions: A Collection of Detailed Mechanisms and Synthetic Applications*,
DOI 10.1007/978-3-319-03979-4_59, © Springer International Publishing Switzerland 2014

## References

1. Chugaev, L. *Ber.* **1899,** *32*, 3332. Lev A. Chugaev (1873–1922) was born in Moscow, Russia. He was a Professor of Chemistry at Petrograd, a position once held by Dimitri Mendeleyev and Paul Walden. In addition to terpenoids, Chugaev also investigated nickel and platinum chemistry. He completely devoted his life to science. The light in Chugaev's study would invariably burn until 4 or 5 a.m.
2. Harano, K.; Taguchi, T. *Chem. Pharm. Bull.* **1975,** *23*, 467–472.
3. Ho, T.-L.; Liu, S.-H. *J. Chem. Soc., Perkin Trans. 1* **1984,** 615–617.
4. Fu, X.; Cook, J. M. *Tetrahedron Lett.* **1990,** *31*, 3409–3412.
5. Meulemans, T. M.; Stork, G. A.; Macaev, F. Z.; Jansen, B. J. M.; de Groot, A. *J. Org. Chem.* **1999,** *64*, 9178–9188.
6. Nakagawa, H.; Sugahara, T.; Ogasawara, K. *Org. Lett.* **2000,** *2*, 3181–3183.
7. Nakagawa, H.; Sugahara, T.; Ogasawara, K. *Tetrahedron Lett.* **2001,** *42*, 4523–4526.
8. Fuchter, M. J. *Chugaev elimination.* In *Name Reactions for Functional Group Transformations*; Li, J. J., Ed.; Wiley: Hoboken, NJ, **2007,** pp 334–342. (Review).
9. Ahmed, S.; Baker, L. A.; Grainger, R. S.; Innocenti, P.; Quevedo, C. E. *J. Org. Chem.* **2008,** *73*, 8116–8119.
10. Tang, P.; Wang, L.; Chen, Q.-F.; Chen, Q.-H.; Jian, X.-X.; Wang, F.-P. *Tetrahedron* **2012,** *68*, 5031–5036.

# Ciamician–Dennsted rearrangement

Cyclopropanation of a pyrrole with dichlorocarbene generated from $CHCl_3$ and NaOH. Subsequent rearrangement takes place to give 3-chloropyridine.

carbene

Example 1[4]

Example 2[5]

## References

1. Ciamician, G. L.; Dennsted, M. *Ber.* **1881**, *14*, 1153. Giacomo Luigi Ciamician (1857–1922) was born in Trieste, Italy. Ciamician is considered the father of modern organic photochemistry.
2. Wynberg, H. *Chem. Rev.* **1960**, *60*, 169–184. (Review).
3. Wynberg, H. and Meijer, E. W. *Org. React.* **1982**, *28*, 1–36. (Review).
4. Parham, W. E.; Davenport, R. W.; Biasotti, J. B. *J. Org. Chem.* **1970**, *35*, 3775–3779.
5. Král, V.; Gale, P. A.; Anzenbacher, P. Jr.; K. Jursíková; Lynch, V.; Sessler, J. L. *Chem. Comm.* **1998**, 9–10.
6. Pflum, D. A. *Ciamician–Dennsted Rearrangement.* In *Name Reactions in Heterocyclic Chemistry*; Li, J. J., Ed.; Wiley: Hoboken, NJ, **2005**, pp 350–354. (Review).

J.J. Li, *Name Reactions: A Collection of Detailed Mechanisms and Synthetic Applications*,
DOI 10.1007/978-3-319-03979-4_60, © Springer International Publishing Switzerland 2014

# Claisen condensation

Base-catalyzed condensation of esters to afford β-keto esters.

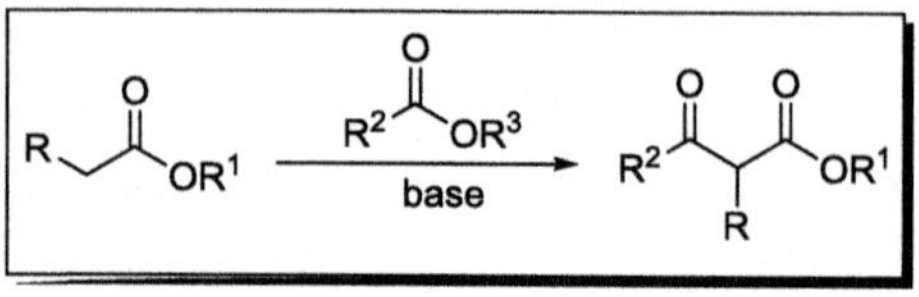

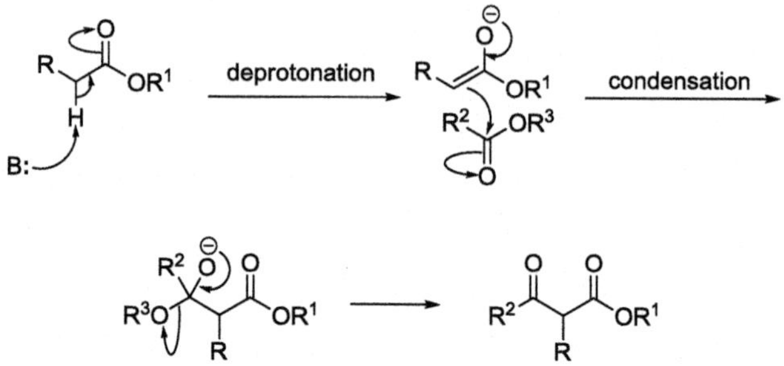

Example 1[4]

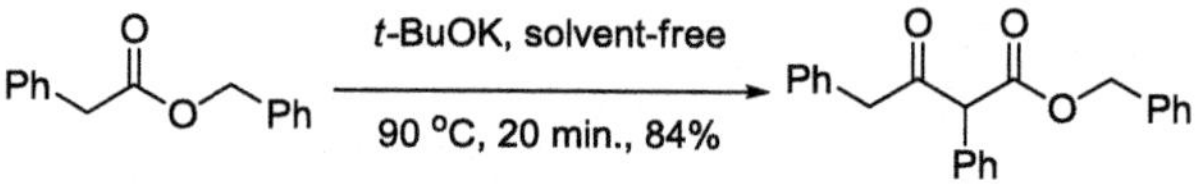

Example 1[4]

Example 2[6]

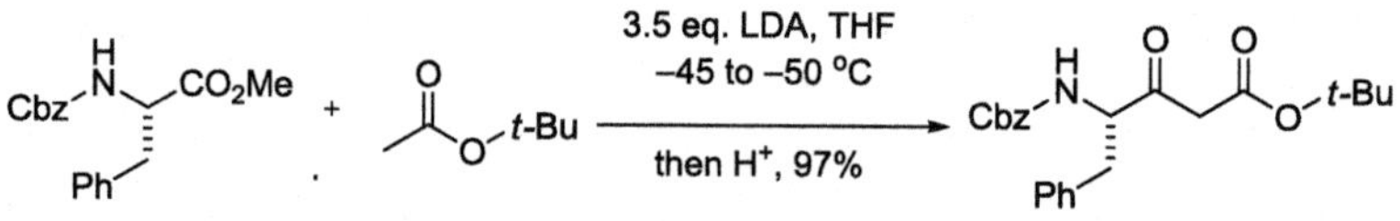

Example 3, Retro-Claisen condensation[9]

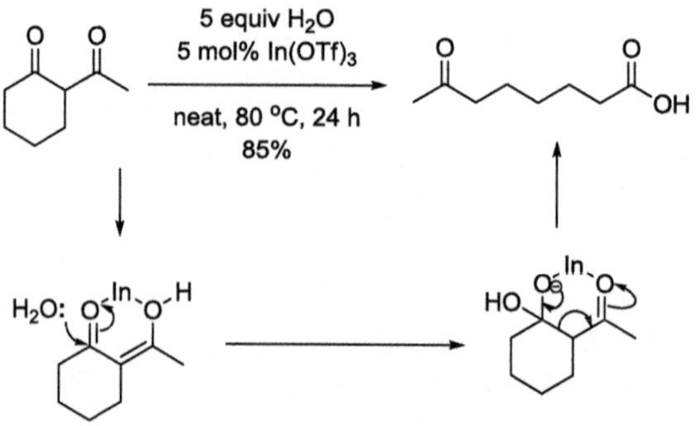

J.J. Li, *Name Reactions: A Collection of Detailed Mechanisms and Synthetic Applications*,
DOI 10.1007/978-3-319-03979-4_61, © Springer International Publishing Switzerland 2014

Example 4, Solvent-free Claisen condensation[10]

Example 5, Intramolecualr Claisen condensation (Dieckmann condensation)[11]

## References

1   Claisen, R. L.; Lowman, O. *Ber.* **1887**, *20*, 651. Rainer Ludwig Claisen (1851–1930), born in Cologne, Germany, probably had the best pedigree in the history of organic chemistry. He apprenticed under Kekulé, Wöhler, von Baeyer, and Fischer before embarking on his own independent research.

2   Hauser, C. R.; Hudson, B. E. *Org. React.* **1942**, *1*, 266–302. (Review).

3   Schäfer, J. P.; Bloomfield, J. J. *Org. React.* **1967**, *15*, 1–203. (Review).

4   Yoshizawa, K.; Toyota, S.; Toda, F. *Tetrahedron Lett.* **2001**, *42*, 7983–7985.

5   Heath, R. J.; Rock, C. O. *Nat. Prod. Rep.* **2002**, *19*, 581–596. (Review).

6   Honda, Y.; Katayama, S.; Kojima, M.; Suzuki, T.; Izawa, K. *Org. Lett.* **2002**, *4*, 447–449.

7   Mogilaiah, K.; Reddy, N. V. *Synth. Commun.* **2003**, *33*, 73–78.

8   Linderberg, M. T.; Moge, M.; Sivadasan, S. *Org. Process Res. Dev.* **2004**, *8*, 838–845.

9   Kawata, A.; Takata, K.; Kuninobu, Y.; Takai, K. *Angew. Chem. Int. Ed.* **2007**, *46*, 7793–7795.

10  Iida, K.; Ohtaka, K.; Komatsu, T.; Makino, T.; Kajiwara, M. *J. Labelled Compd. Radiopharm.* **2008**, *51*, 167–169.

11  Song, Y. Y.; He, H. G.; Li, Y.; Deng, Y. *Tetrahedron Lett.* **2013**, *54*, 2658–2660.

## Claisen isoxazole synthesis

Cyclization of β-keto esters with hydroxylamine to provide 3-hydroxy-isoxazoles (3-isoxazolols).

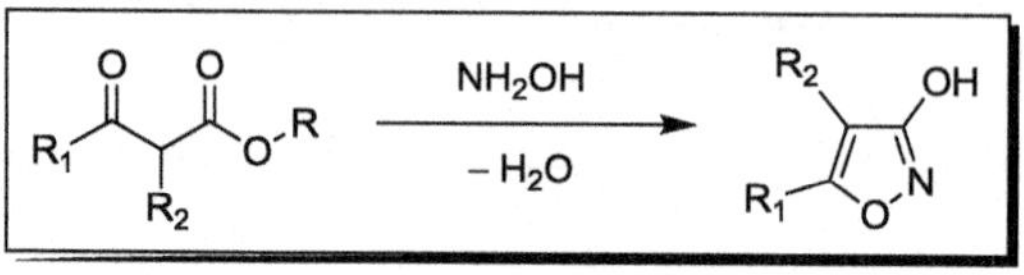

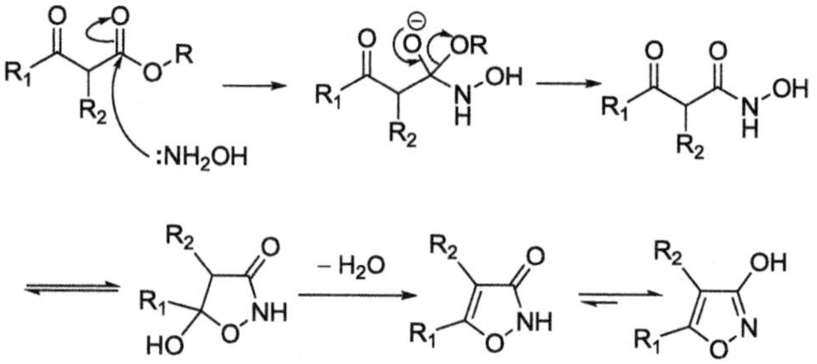

3-isoxazolol

A side reaction:

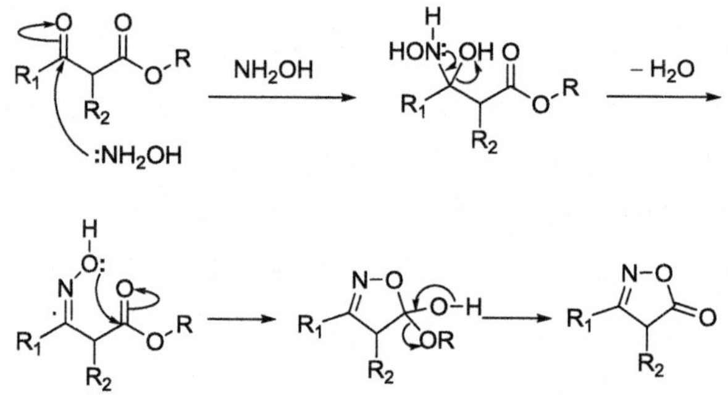

5-isoxazolone

Example 1, A thio-analog[6]

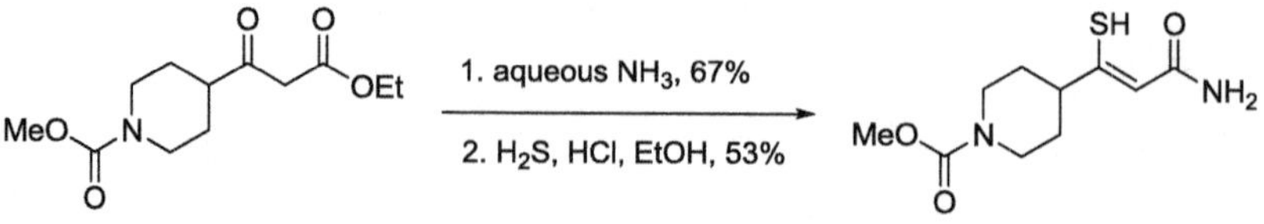

J.J. Li, *Name Reactions: A Collection of Detailed Mechanisms and Synthetic Applications*,
DOI 10.1007/978-3-319-03979-4_62, © Springer International Publishing Switzerland 2014

I$_2$, K$_2$CO$_3$, 59%

**Example 2[7]**

Meldrum's acid

R = Me, Et, *i*-Pr, cyclopropyl, cyclohexyl, Ph, Bn, neopentyl

**Example 3[8]**

NH$_2$OH·HCl, NaOH

MeOH, −50 °C, 2 h
then 85 °C, 1 h, 65%

## References

1.  (a) Claisen, L; Lowman, O. E. *Ber.* **1888**, *21*, 784. (b) Claisen, L.; Zedel, W. *Ber.* **1891**, *24*, 140. (c) Hantzsch, A. *Ber.* **1891**, *24*, 495–506.
2.  Barnes, R. A. In *Heterocyclic Compounds*; Elderfield, R. C., Ed.; Wiley: New York, **1957**; *Vol. 5*, p 474ff. (Review).
3.  Loudon, J. D. In *Chemistry of Carbon Compounds*; Rodd, E. H., Ed.; Elsevier: Amsterdam, **1957**; Vol. 4a, p. 345ff. (Review).
4.  McNab, H. *Chem. Soc. Rev.* **1978**, *7*, 345–358. (Review).
5.  Chen, B.-C. *Heterocycles* **1991**, *32*, 529–597. (Review).
6.  Frølund, B.; Kristiansen, U.; Brehm, L.; Hansen, A. B.; Krogsgaard-Larsen, K.; Falch, E. *J. Med. Chem.* **1995**, *38*, 3287–3296.
7.  Sorensen, U. S.; Falch, E.; Krogsgaard-Larsen, K. *J. Org. Chem.* **2000**, *65*, 1003–1007.
8.  Madsen, U.; Bräuner-Osborne, H.; Frydenvang, K.; Hvene, L.; Johansen, T.N.; Nielsen, B.; Sánchez, C.; Stensbøl, T.B.; Bischoff, F.; Krogsgaard-Larsen, K. *J. Med. Chem.* **2001**, *44*, 1051–1059.
9.  Brooks, D. A. *Claisen Isoxazole Synthesis*. In *Name Reactions in Heterocyclic Chemistry*; Li, J. J., Ed.; Wiley: Hoboken, NJ, **2005**, pp 220–224. (Review).
10. El Shehry, M. F.; Swellem, R. H.; Abu-Bakr, Sh. M.; El-Telbani, E. M. *Eur. J. Med. Chem.* **2010**, *45*, 4783–4787.

## Claisen rearrangements

The Claisen, *para*-Claisen rearrangements, Belluš–Claisen rearrangement; Corey–Claisen, Eschenmoser–Claisen rearrangement, Ireland–Claisen, Kazmaier–Claisen, Saucy–Claisen; orthoester Johnson–Claisen, along with the Carroll rearrangement, belong to the category of *[3,3]-sigmatropic rearrangements*. The Claisen rearrangement is a concerted process and the arrow pushing here is merely illustrative.

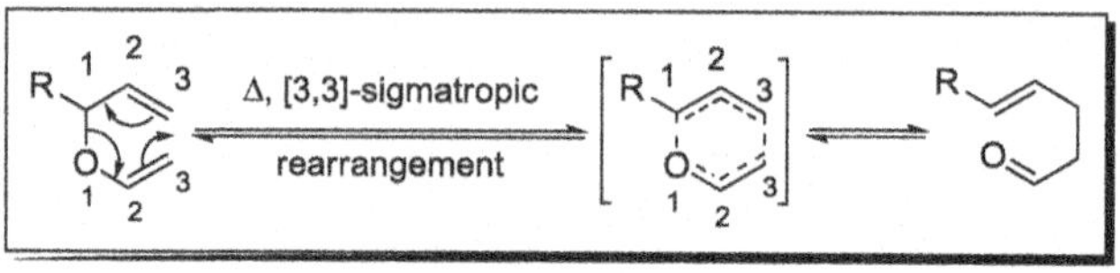

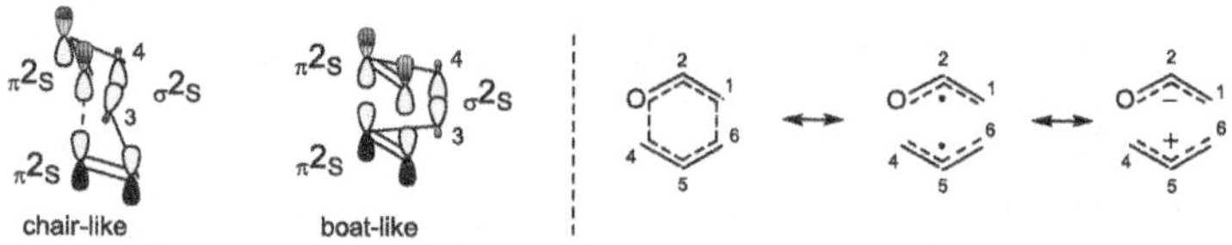

Example 1[7]

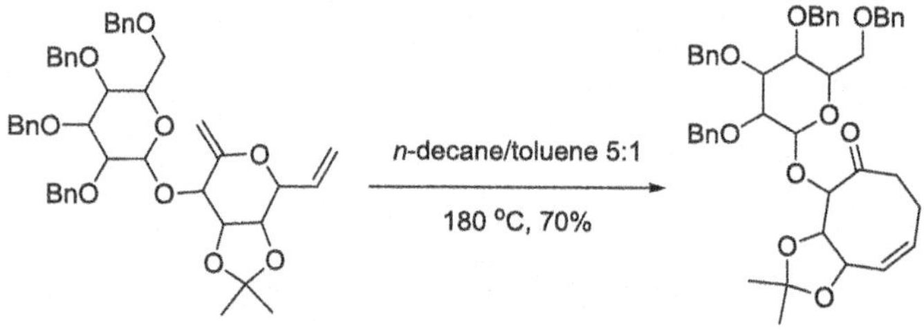

Example 2[8]

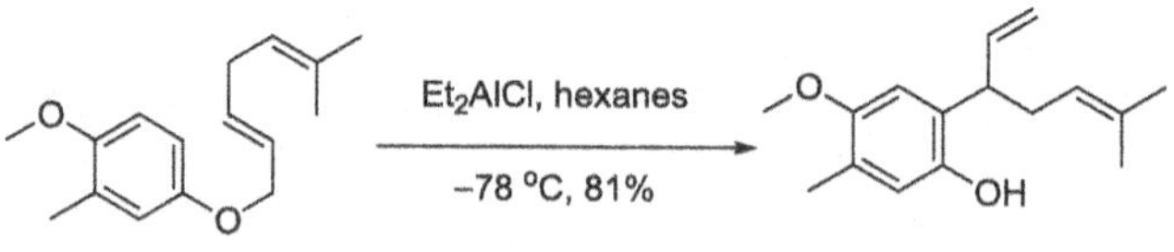

Example 3[9]

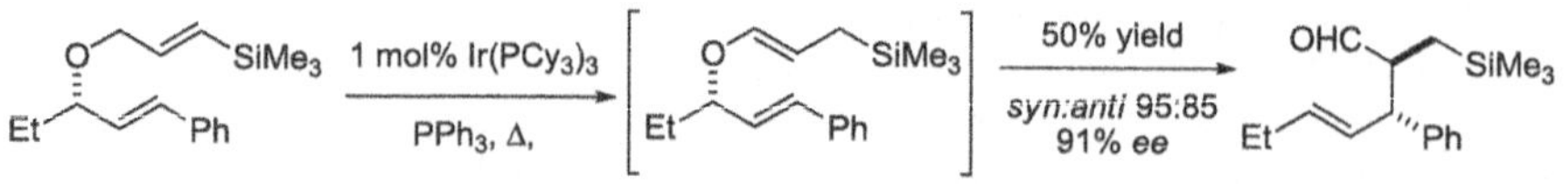

J.J. Li, *Name Reactions: A Collection of Detailed Mechanisms and Synthetic Applications*, DOI 10.1007/978-3-319-03979-4_63, © Springer International Publishing Switzerland 2014

## Example 4, Asymmetric Claisen rearrangement[10]

98%, > 90% *de*, 99% *ee*

## Example 5, Asymmetric Claisen rearrangement[11]

73%, 96% *ee*

## Example 6[13]

ArCHO, DBU, MeOH

rt to 65–85 °C, 55–80%

## References

1. Claisen, L. *Ber.* **1912**, *45*, 3157–3166.
2. Rhoads, S. J.; Raulins, N. R. *Org. React.* **1975**, *22*, 1–252. (Review).
3. Wipf, P. In *Comprehensive Organic Synthesis;* Trost, B. M.; Fleming, I., Eds.; Pergamon, **1991**, *Vol. 5*, 827–873. (Review).
4. Ganem, B. *Angew. Chem. Int. Ed.* **1996**, *35*, 937–945. (Review).
5. Ito, H.; Taguchi, T. *Chem. Soc. Rev.* **1999**, *28*, 43–50. (Review).
6. Castro, A. M. M. *Chem. Rev.* **2004**, *104*, 2939–3002. (Review).
7. Jürs, S.; Thiem, J. *Tetrahedron: Asymmetry* **2005**, *16*, 1631–1638.
8. Vyvyan, J. R.; Oaksmith, J. M.; Parks, B. W.; Peterson, E. M. *Tetrahedron Lett.* **2005**, *46*, 2457–2460.
9. Nelson, S. G.; Wang, K. *J. Am. Chem. Soc.* **2006**, *128*, 4232–4233.
10. Körner, M.; Hiersemann, M. *Org. Lett.* **2007**, *9*, 4979–4982.
11. Uyeda, C.; Jacobsen, E. N. *J. Am. Chem. Soc.* **2008**, *130*, 9228–9229.
12. Williams, D. R.; Nag, P. P. *Claisen and Related Rearrangements*. In *Name Reactions for Homologations-Part II*; Li, J. J., Ed.; Wiley: Hoboken, NJ, **2009**, pp 33–43. (Review).
13. Alwarsh, S.; Ayinuola, K.; Dormi, S. S.; McIntosh, M. C. *Org. Lett.* **2013**, *15*, 3–5.

## para-*Claisen rearrangement*

Further rearrangement of the normal *ortho*-Claisen rearrangement product gives
the *para*-Claisen rearrangement product.

Mechanism 1:

Mechanism 2:

Mechanism 3:

Example 1[6]

Example 2[7]

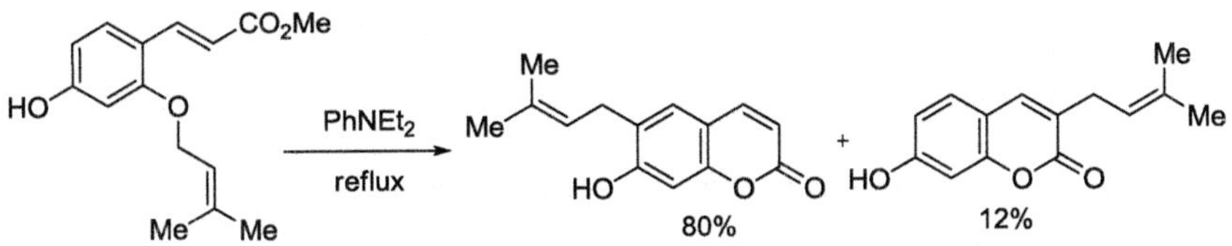

Example 3[8]

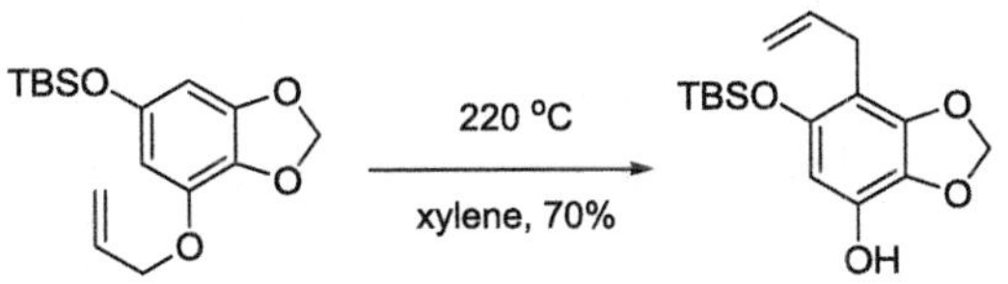

Example 4[10]

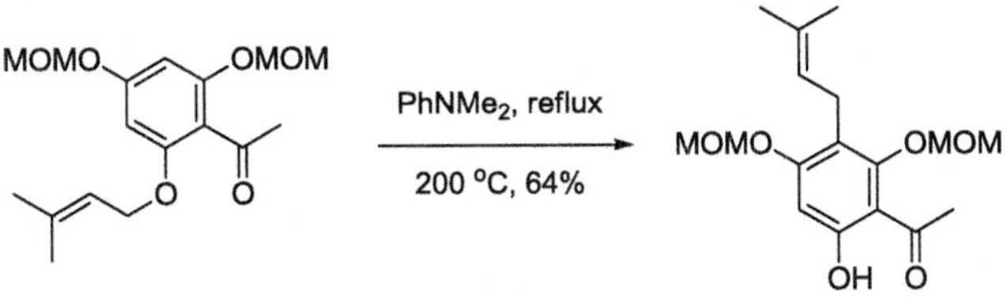

Example 5[11]

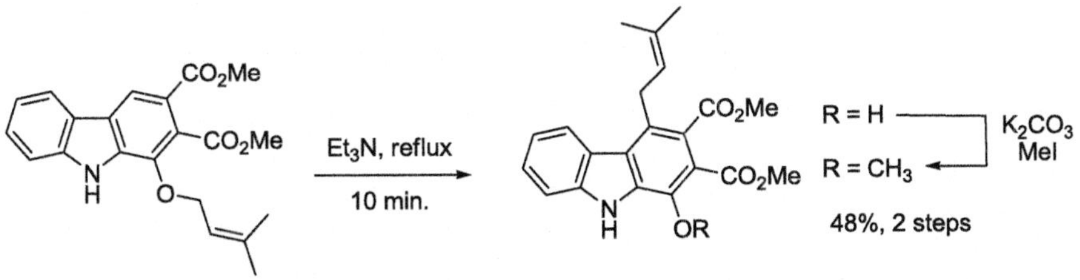

**References**
1.  Alexander, E. R.; Kluiber, R. W. *J. Am. Chem. Soc.* **1951,** *73*, 4304–4306.
2.  Rhoads, S. J.; Raulins, R.; Reynolds, R. D. *J. Am. Chem. Soc.* **1953,** *75*, 2531–2532.
3.  Dyer, A.; Jefferson, A.; Scheinmann, F. *J. Org. Chem.* **1968,** *33*, 1259–1261.
4.  Murray, R. D. H.; Lawrie, K. W. M. *Tetrahedron* **1979,** *35*, 697–699.
5.  Cairns, N.; Harwood, L. M.; Astles, D. P. *J. Chem. Soc., Chem. Commun.* **1986,** 1264–1266.
6.  Kilényi, S. N.; Mahaux, J.-M.; van Durme, E. *J. Org. Chem.* **1991,** *56*, 2591–2594.
7.  Cairns, N.; Harwood, L. M.; Astles, D. P. *J. Chem. Soc., Perkin Trans. 1* **1994,** 3101–3107.

8.   Pettus, T. R. R.; Inoue, M.; Chen, X.-T.; Danishefsky, S. J. *J. Am. Chem. Soc.* **2000**, *122*, 6160–6168.
9.   Al-Maharik, N.; Botting, N. P. *Tetrahedron* **2003**, *59*, 4177–4181.
10.  Khupse, R. S.; Erhardt, P. W. *J. Nat. Prod.* **2007**, *70*, 1507–1509.
11.  Jana, A. K.; Mal, D. *Chem. Commun.* **2010**, *46*, 4411–4413.

## *Abnormal Claisen rearrangement*

Further rearrangement of the normal Claisen rearrangement product with the β-carbon becoming attached to the ring.

Example 1[3]

Example 2, Enantioselective aromatic Claisen rearrangement[4]

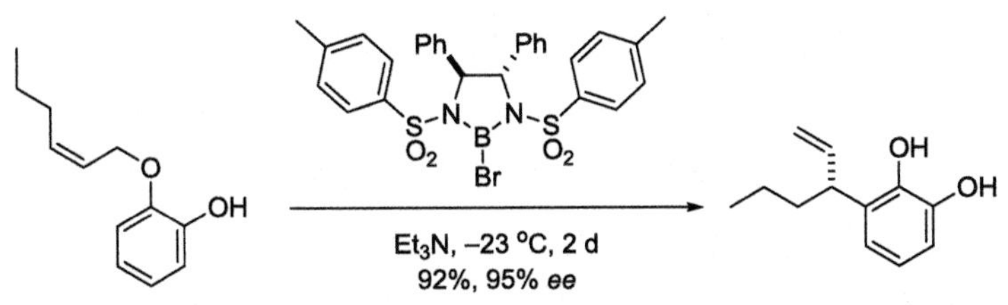

Example 3[5]

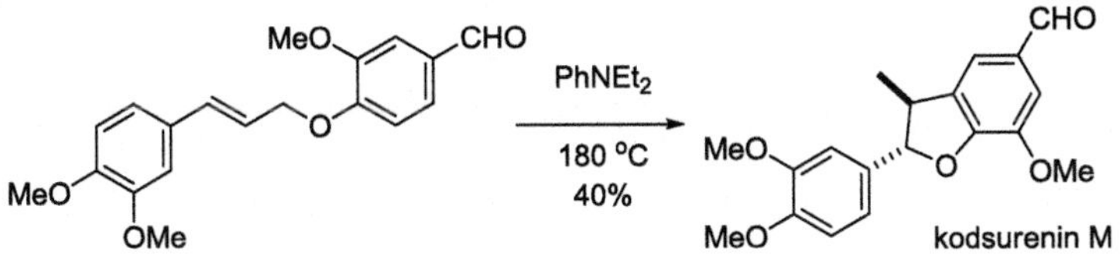

kodsurenin M

Example 4[6]

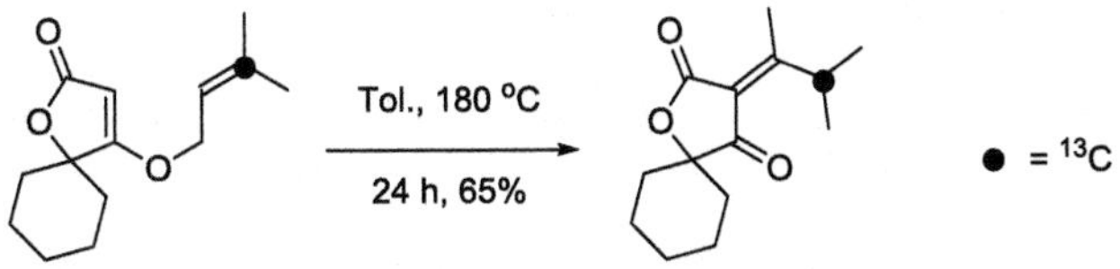

● = $^{13}$C

Example 5[7]

+

2 : 1

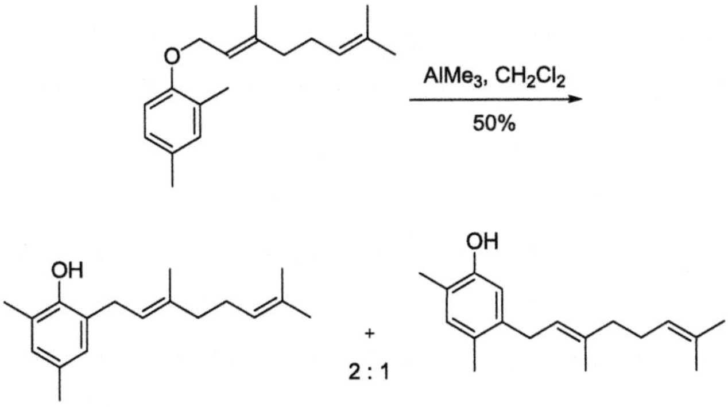

Example 6[10]

Microwave irradiation

180 °C, 20 h, 73%

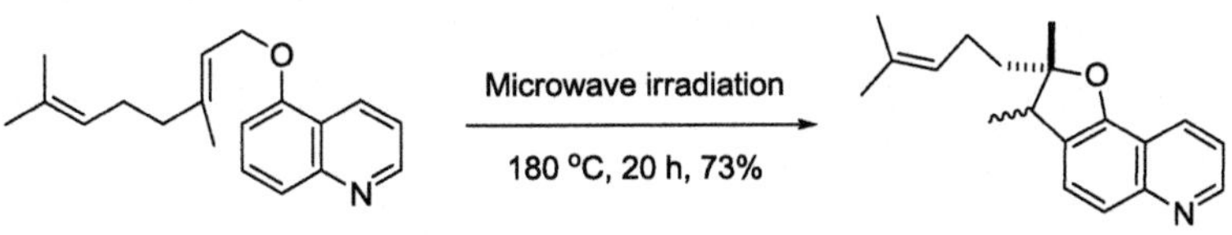

**References**

1.    Hansen, H.-J. In *Mechanisms of Molecular Migrations;* vol. 3, Thyagarajan, B. S., ed.; Wiley-Interscience: New York, **1971,** pp 177–236. (Review).

2.    Kilényi, S. N.; Mahaux, J.-M.; van Durme, E. *J. Org. Chem.* **1991,** *56,* 2591–2594.

3.    Fukuyama, T.; Li, T.; Peng, G. *Tetrahedron Lett.* **1994,** *35,* 2145–2148.

4.    Ito, H.; Sato, A.; Taguchi, T. *Tetrahedron Lett.* **1997,** *38,* 4815–4818.

5.    Yi, W. M.; Xin, W. A.; Fu, P. X. *J. Chem. Soc., (S),* **1998,** 168.

6.    Schobert, R.; Siegfried, S.; Gordon, G.; Mulholland, D.; Nieuwenhuyzen, M. *Tetrahedron Lett.* **2001,** *42,* 4561–4564.

7.    Wipf, P.; Rodriguez, S. *Ad. Synth. Catal.* **2002,** *344,* 434–440.

8.    Puranik, R.; Rao, Y. J.; Krupadanam, G. L. D. *Indian J. Chem., Sect. B* **2002,** *41B,* 868–870.

9.    Williams, D. R.; Nag, P. P. *Claisen and Related Rearrangements.* In *Name Reactions for Homologations-Part II*; Li, J. J., Ed.; Wiley: Hoboken, NJ, **2009,** pp 33–87. (Review).

10.   Torincsi, M.; Kolonits, P.; Fekete, J.; Novak, L. *Synth.Commun.* **2012,** *42,* 3187–3199.

## *Eschenmoser–Claisen amide acetal rearrangement*

[3,3]-Sigmatropic rearrangement of *N,O*-ketene acetals to yield γ,δ-unsaturated amides. Since Eschenmoser was inspired by Meerwein's observations on the interchange of amide, the Eschenmoser–Claisen rearrangement is sometimes known as the Meerwein–Eschenmoser–Claisen rearrangement.

Example 1[4]

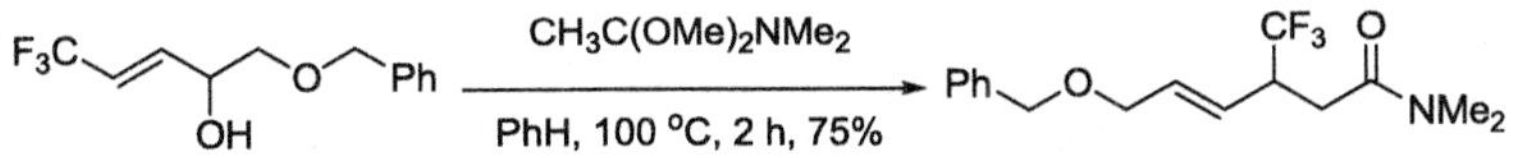

Example 2[5]

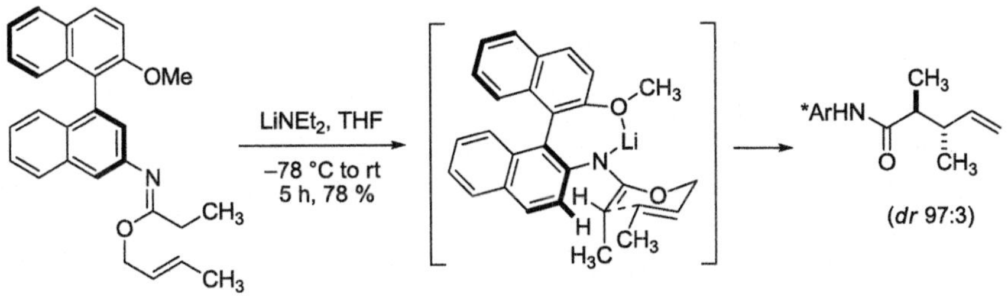

(*dr* 97:3)

Example 3[6]

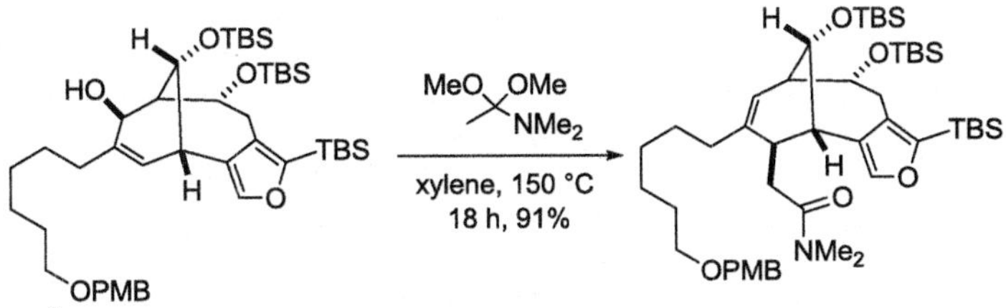

Example 4[8]

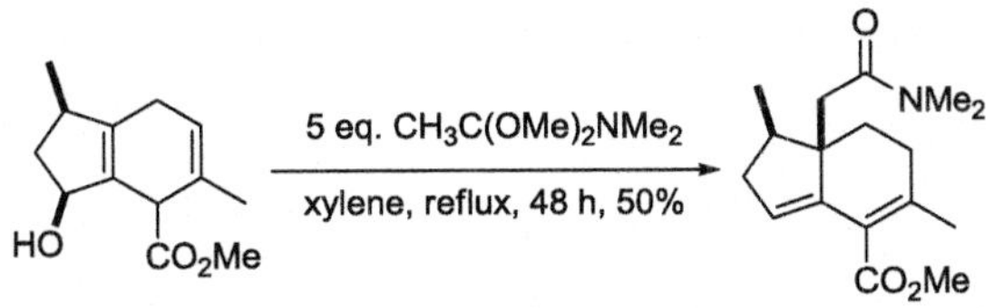

Example 5[9]

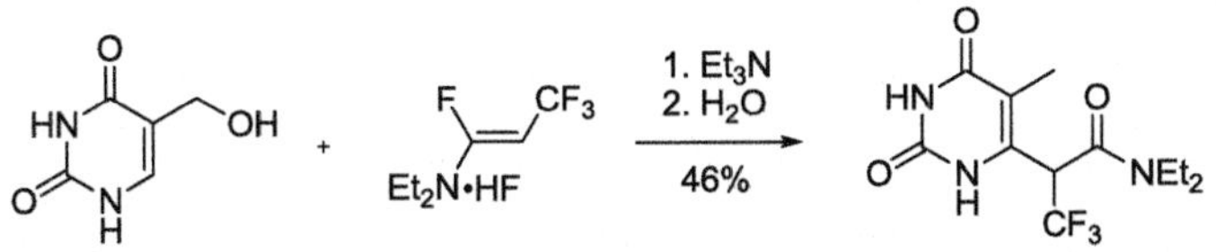

**References**
1.  Meerwein, H.; Florian, W.; Schön, N.; Stopp, G. *Ann.* **1961**, *641*, 1–39.
2.  Wick, A. E.; Felix, D.; Steen, K.; Eschenmoser, A. *Helv. Chim. Acta* **1964**, *47*, 2425–2429.  Albert Eschenmoser (Switzerland, 1925–) is known for his work on,

among many others, the monumental total synthesis of Vitamin B$_{12}$ with R. B. Woodward in 1973. He now holds dual appointments at both ETH Zürich and the Scripps Research Institute in La Jolla, CA.

3.  Wipf, P. In *Comprehensive Organic Synthesis;* Trost, B. M.; Fleming, I., Eds.; Pergamon, **1991,** *Vol. 5,* 827–873. (Review).
4.  Konno, T.; Nakano, H.; Kitazume, T. *J. Fluorine Chem.* **1997,** *86,* 81–87.
5.  Metz, P.; Hungerhoff, B. *J. Org. Chem.***1997,** *62,* 4442–4448.
6.  Kwon, O. Y.; Su, D. S.; Meng, D. F.; Deng, W.; D'Amico, D. C.; Danishefsky, S. J. *Angew. Chem. Int. Ed.* **1998,** *37,* 1877–1880.
7.  Ito, H.; Taguchi, T. *Chem. Soc. Rev.* **1999,** *28,* 43–50. (Review).
8.  Loh, T.-P.; Hu, Q.-Y. *Org. Lett.* **2001,** *3,* 279–281.
9.  Castro, A. M. M. *Chem. Rev.* **2004,** *104,* 2939–3002. (Review).
10. Williams, D. R.; Nag, P. P. *Claisen and Related Rearrangements.* In *Name Reactions for Homologations-Part II*; Li, J. J., Ed.; Wiley: Hoboken, NJ, **2009,** pp 60–68. (Review).
11. Walkowiak, J.; Tomas-Szwaczyk, M.; Haufe, G.; Koroniak, H. *J. Fluorine Chem.* **2012,** *143,* 189–197.

## *Ireland–Claisen (silyl ketene acetal) rearrangement*

Rearrangement of allyl trimethylsilyl ketene acetal, prepared by reaction of allylic ester enolates with trimethylsilyl chloride, to yield γ,δ-unsaturated carboxylic acids. The Ireland–Claisen rearrangement seems to be advantageous to the other variants of the Claisen rearrangement in terms of *E/Z* geometry control and mild conditions.

Example 1[2]

Example 2[3]

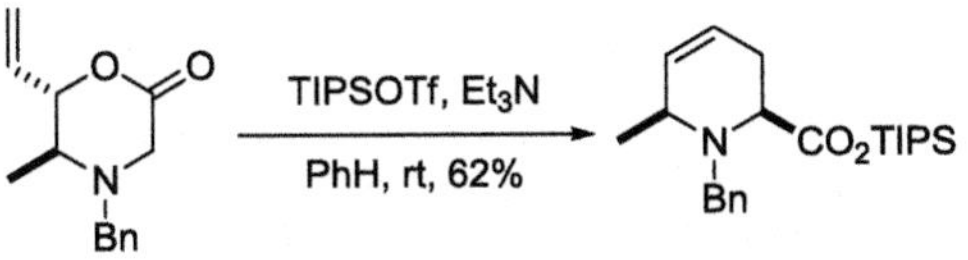

Example 3, Enantioselective ester enolate-Claisen Rearrangement[6]

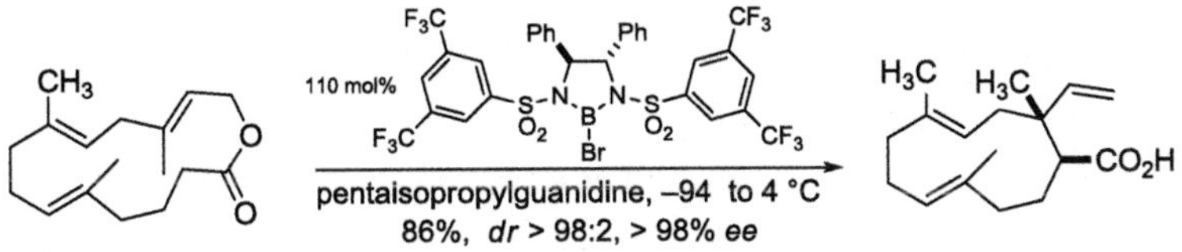

Example 4, A modified Ireland–Claisen rearrangement[8]

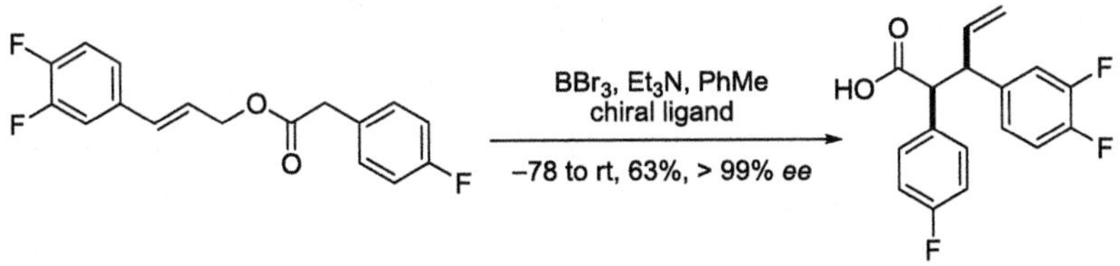

Example 5[9]

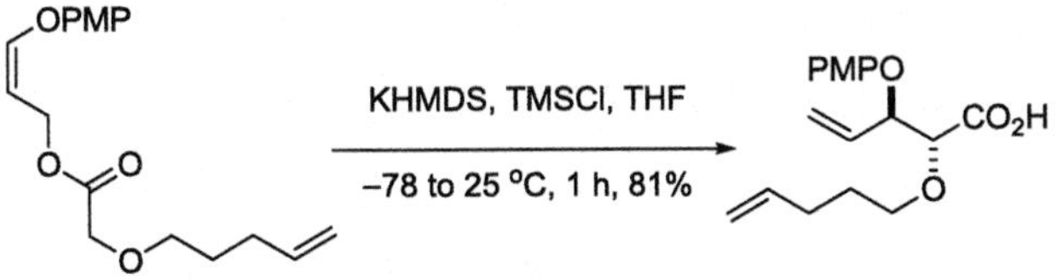

Example 6, chirality-transferring Ireland-Claisen rearrangement[11]

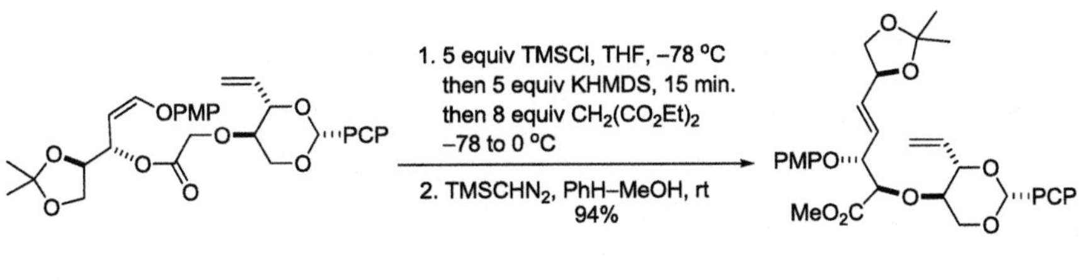

**References**

1.   Ireland, R. E.; Mueller, R. H. *J. Am. Chem. Soc.* **1972,** *94,* 5897–5898.  Also *J. Am. Chem. Soc.* **1976,** *98,* 2868–2877.  Robert E. Ireland obtained his Ph.D. from William

S. Johnson before becoming a professor at the University of Virginia and later at the
California Institute of Technology.  He is now retired.

2.   Begley, M. J.; Cameron, A. G.; Knight, D. W. *J. Chem. Soc., Perkin Trans. 1* **1986,**
     1933–1938.
3.   Angle, S. R.; Breitenbucher, J. G. *Tetrahedron Lett.* **1993,** *34,* 3985–3988.
4.   Pereira, S.; Srebnik, M. *Aldrichimica Acta* **1993,** *26,* 17–29. (Review).
5.   Ganem, B. *Angew. Chem. Int. Ed.* **1996,** *35,* 936–945. (Review).
6.   Corey, E.; Kania, R. S. *J. Am. Chem. Soc.* **1996,** *118,* 1229–1230.
7.   Chai, Y.; Hong, S.-p.; Lindsay, H. A.; McFarland, C.; McIntosh, M. C. *Tetrahedron*
     **2002,** *58,* 2905–2928. (Review).
8.   Churcher, I.; Williams, S.; Kerrad, S.; Harrison, T.; Castro, J. L.; Shearman, M. S.;
     Lewis, H. D.; Clarke, E. E.; Wrigley, J. D. J.; Beher, D.; Tang, Y. S.; Liu, W. *J. Med.*
     *Chem.* **2003,** *46,* 2275–2278.
9.   Fujiwara, K.; Goto, A.; Sato, D.; Kawai, H.; Suzuki, T. *Tetrahedron Lett.* **2005,** *46,*
     3465–3468.
10.  Williams, D. R.; Nag, P. P. *Claisen and Related Rearrangements.* In *Name Reactions*
     *for Homologations-Part II*; Li, J. J., Ed.; Wiley: Hoboken, NJ, **2009,** pp 45–51. (Re-
     view).
11.  Nogoshi, K.; Domon, D.; Fujiwara, K.; Kawamura, N.; Katoono, R.; Kawai, H.; Suzu-
     ki, T. *Tetrahedron Lett.* **2013,** *54,* 676–680.

## *Johnson–Claisen orthoester rearrangement*

Heating of an allylic alcohol with an excess of trialkyl orthoacetate in the presence
of trace amounts of a weak acid gives a mixed orthoester.  Mechanistically, the
orthoester loses alcohol to generate the ketene acetal, which undergoes [3,3]-
sigmatropic rearrangement to give a γ,δ-unsaturated ester.

Example 1[2]

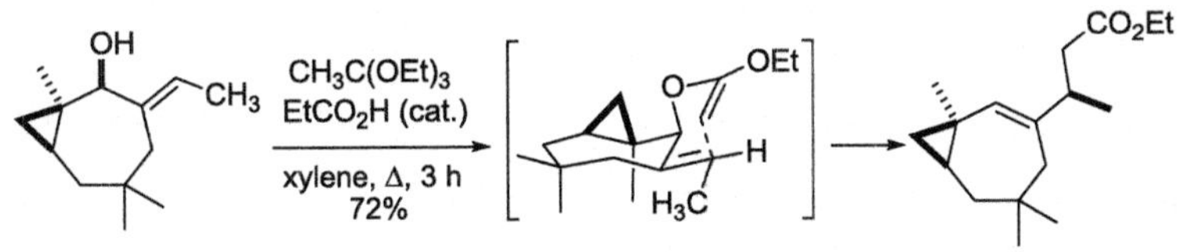

CH$_3$C(OEt)$_3$
EtCO$_2$H (cat.)

xylene, Δ, 3 h
72%

Example 2[3]

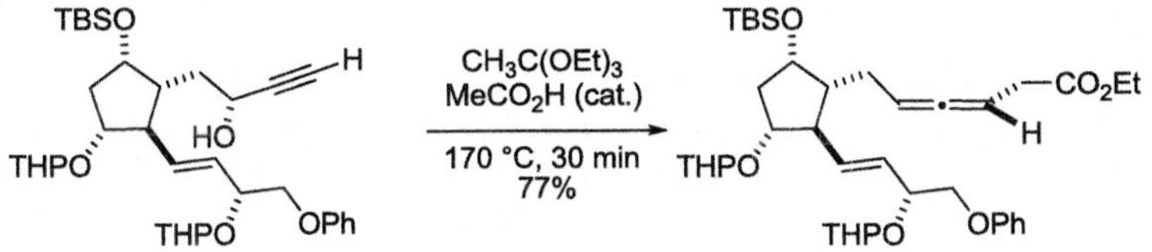

CH$_3$C(OEt)$_3$
MeCO$_2$H (cat.)

170 °C, 30 min
77%

Example 3[4]

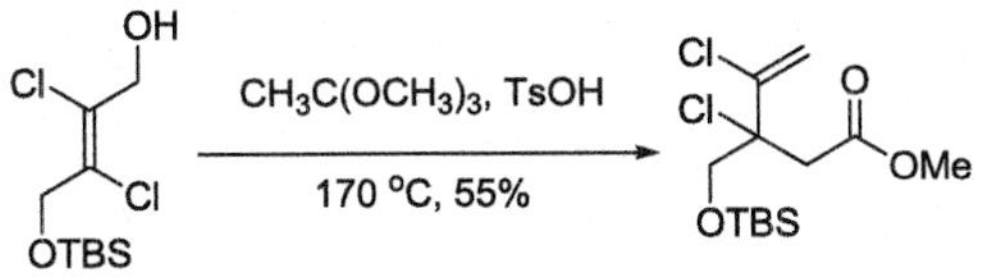

CH$_3$C(OCH$_3$)$_3$, TsOH

170 °C, 55%

Example 4[9]

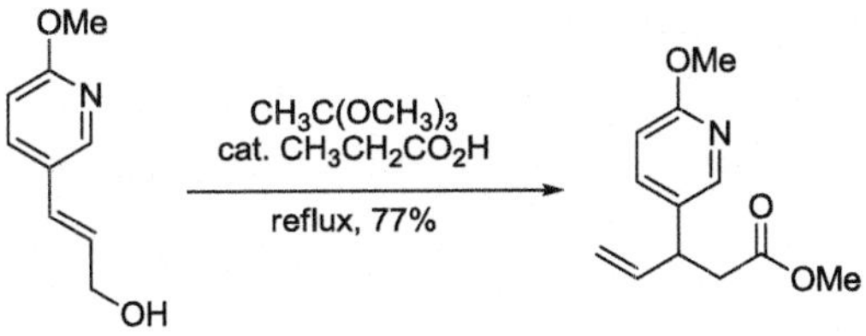

CH$_3$C(OCH$_3$)$_3$
cat. CH$_3$CH$_2$CO$_2$H

reflux, 77%

Example 5[10]

(EtO)$_3$CCH$_3$
pivalic acid
xylene, 140 °C
54%, E:Z > 95:5

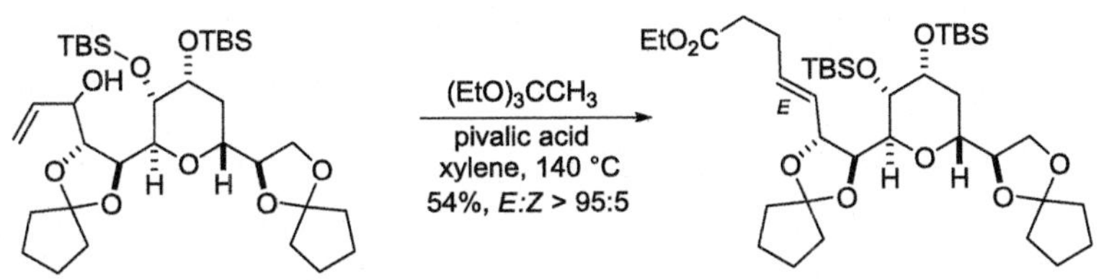

**References**

1.  Johnson, W. S.; Werthemann, L.; Bartlett, W. R.; Brocksom, T. J.; Li, T.-t.; Faulkner, D. J.; Peterson, M. R. *J. Am. Chem. Soc.* **1970**, *92*, 741–743.  William S. Johnson (1913–1995) was born in New Rochelle, New York.  He earned his Ph.D. in only two

years at Harvard under Louis Fieser. He was a professor at the University of Wisconsin for 20 years before moving to Stanford University, where he was credited with building the modern-day Stanford Chemistry Department.

2.    Paquette, L.; Ham, W. H. *J. Am. Chem. Soc.* **1987**, *109*, 3025–3036.
3.    Cooper, G. F.; Wren, D. L.; Jackson, D. Y.; Beard, C. C.; Galeazzi, E.; Van Horn, A. R.; Li, T. T. *J. Org. Chem.* **1993**, *58*, 4280–4286.
4.    Schlama, T.; Baati, R.; Gouverneur, V.; Valleix, A.; Falck, J. R.; Mioskowski, C. *Angew. Chem. Int. Ed.* **1998**, *37*, 2085–2087.
5.    Giardiná, A.; Marcantoni, E.; Mecozzi, T.; Petrini, M. *Eur. J. Org. Chem.* **2001**, 713–718.
6.    Funabiki, K.; Hara, N.; Nagamori, M.; Shibata, K.; Matsui, M. *J. Fluorine Chem.* **2003**, *122*, 237–242.
7.    Montero, A.; Mann, E.; Herradón, B. *Eur. J. Org. Chem.* **2004**, 3063–3073.
8.    Scaglione, J. B.; Rath, N. P.; Covey, D. F. *J. Org. Chem.* **2005**, *70*, 1089–1092.
9.    Zartman, A. E.; Duong, L. T.; Fernandez-Metzler, C.; Hartman, G. D.; Leu, C.-T.; Prueksaritanont, T.; Rodan, G. A.; Rodan, S. B.; Duggan, M. E.; Meissner, R. S. *Bioorg. Med. Chem. Lett.* **2005**, *15*, 1647–1650.
10.   Hicks, J. D.; Roush, W. R. *Org. Lett.* **2008**, *10*, 681–684.
11.   Williams, D. R.; Nag, P. P. *Claisen and Related Rearrangements.* In *Name Reactions for Homologations-Part II*; Li, J. J., Ed.; Wiley: Hoboken, NJ, **2009**, pp 68–72. (Review).
12.   Sydlik, S. A.; Swager, T. M. *Adv. Funct. Mater.* **2013**, *23*, 1873–1882.

## Clemmensen reduction

Reduction of aldehydes or ketones to the corresponding methylene compounds using amalgamated zinc in hydrochloric acid.

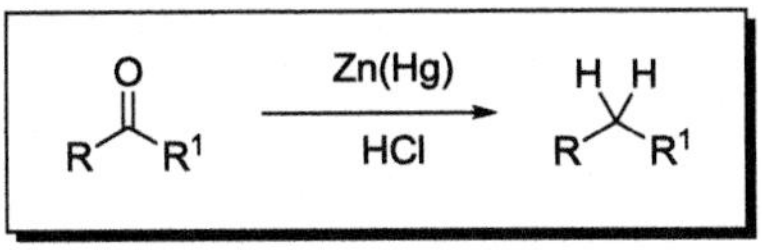

The zinc-carbenoid mechanism:[3]

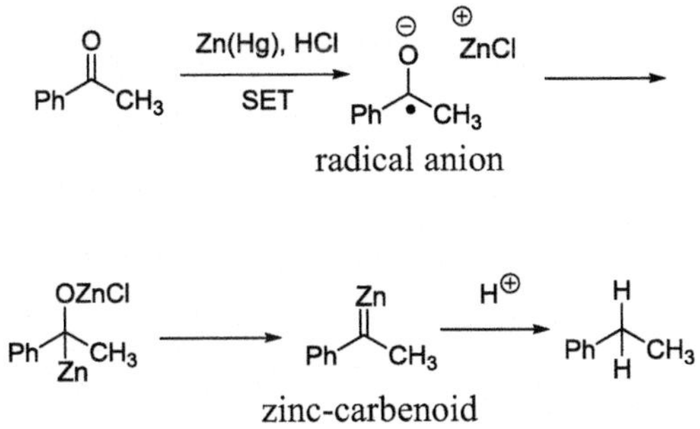

radical anion

zinc-carbenoid

The radical anion mechanism:

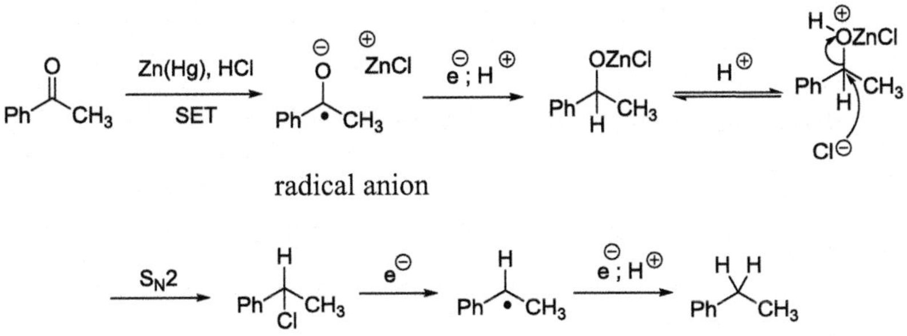

radical anion

Example 1[5]

18%     67%

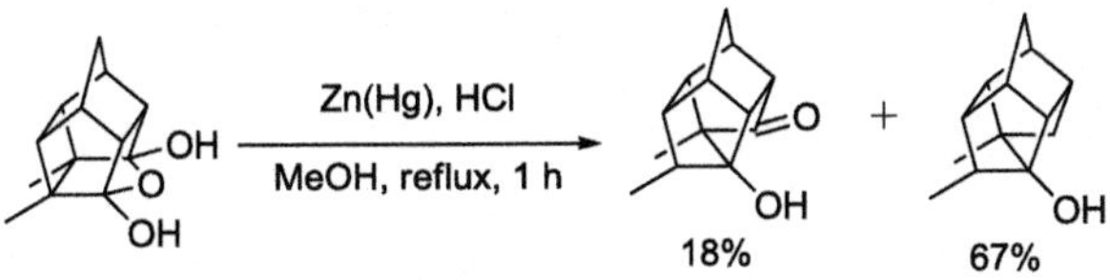

J.J. Li, *Name Reactions: A Collection of Detailed Mechanisms and Synthetic Applications*,
DOI 10.1007/978-3-319-03979-4_64, © Springer International Publishing Switzerland 2014

Example 2[6]

Zn(Hg), HCl
Et₂O, –5 °C, 57%

Example 3[7]

diosgenin

Zn dust
37% HCl, EtOH

reflux, 15 min.
50%

Example 4[9]

O₃, –78 °C

i-PrOH/CH₂Cl₂ (3:1)

Zn, TMSCl

–78 to 0 °C
0.5 h, 87%

## References

1.  Clemmensen, E. *Ber.* **1913**, *46*, 1837–1843. Erik C. Clemmensen (1876–1941) was born in Odense, Denmark. He received the M.S. degree from the Royal Polytechnic Institute in Copenhagen. In 1900, Clemmensen immigrated to the United States, and worked at Parke, Davis and Company in Detroit (coincidently, this author's first employer!) as a research chemist for 14 years, where he discovered the reduction of carbonyl compounds with amalgamated zinc. Clemmensen later founded a few chemical companies and was the president of one of them, the Clemmensen Chemical Corporation in Newark, New Jersey.
2.  Martin, E. L. *Org. React.* **1942**, *1*, 155–209. (Review).
3.  Vedejs, E. *Org. React.* **1975**, *22*, 401–422. (Review).
4.  Talpatra, S. K.; Chakrabarti, S.; Mallik, A. K.; Talapatra, B. *Tetrahedron* **1990**, *46*, 6047–6052.
5.  Martins, F. J. C.; Viljoen, A. M.; Coetzee, M.; Fourie, L.; Wessels, P. L. *Tetrahedron* **1991**, *47*, 9215–9224.
6.  Naruse, M.; Aoyagi, S.; Kibayashi, C. *J. Chem. Soc., Perkin Trans. 1* **1996**, 1113–1124.
7.  Alessandrini, L.; et al. *Steroids* **2004**, *69*, 789–794.
8.  Dey, S. P.; et al. *J. Indian Chem. Soc.* **2008**, *85*, 717–720.
9.  Xu, S.; Toyama, T.; Nakamura, J.; Arimoto, H. *Tetrahedron Let.* **2010**, *51*, 4534–4537.

# Combes quinoline synthesis

Acid-catalyzed condensation of anilines and β-diketones to assemble quinolines. *Cf.* Conrad–Limpach reaction.

An electrocyclization mechanism is also possible:

J.J. Li, *Name Reactions: A Collection of Detailed Mechanisms and Synthetic Applications*, DOI 10.1007/978-3-319-03979-4_65, © Springer International Publishing Switzerland 2014

Example 1[6]

Example 2[7]

## References

1.  Combes, A. *Bull. Soc. Chim. Fr.* **1888,** *49,* 89.    Alphonse-Edmond Combes
    (1858–1896) was born in St. Hippolyte-du-Fort, France. He apprenticed with Wurtz at
    Paris. He also collaborated with Charles Friedel of the Friedel–Crafts reaction fame.
    He became the president of the French Chemical Society in 1893 at the age of 35. His
    sudden death shortly after his 38th birthday was a great loss to organic chemistry.
2.  Roberts, E. and Turner, E. *J. Chem Soc.* **1927,** 1832–1857. (Review).
3.  Elderfield, R. C. In *Heterocyclic Compounds,* Elderfield, R. C., ed.; Wiley: New York,
    **1952,** vol. 4, 36–38. (Review).
4.  Popp, F. D. and McEwen, W. E. *Chem. Rev.* **1958,** *58,* 321–401. (Review).
5.  Jones, G. In *Chemistry of Heterocyclic Compounds,* Jones, G., ed.; Wiley & Sons,
    New York, **1977,** Quinolines *Vol. 32,* pp 119–125. (Review).
6.  Alunni-Bistocchi, G.; Orvietani, P., Bittoun, P., Ricci, A.; Lescot, E. *Pharmazie* **1993,**
    *48,* 817–820.
7.  El Ouar, M.; Knouzi, N.; Hamelin, J. *J. Chem. Res. (S)* **1998,** 92–93.
8.  Curran, T. T. *Combes Quinoline Synthesis.* In *Name Reactions in Heterocyclic Chem-
    istry*; Li, J. J., Ed.; Wiley: Hoboken, NJ, **2005,** pp 390–397. (Review).

# Conrad–Limpach reaction

Thermal or acid-catalyzed condensation of anilines with β-ketoesters leads to quinolin-4-ones. *Cf.* Combes quinoline synthesis.

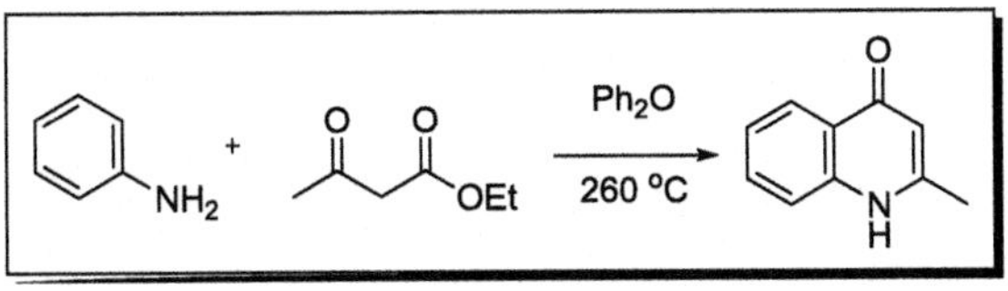

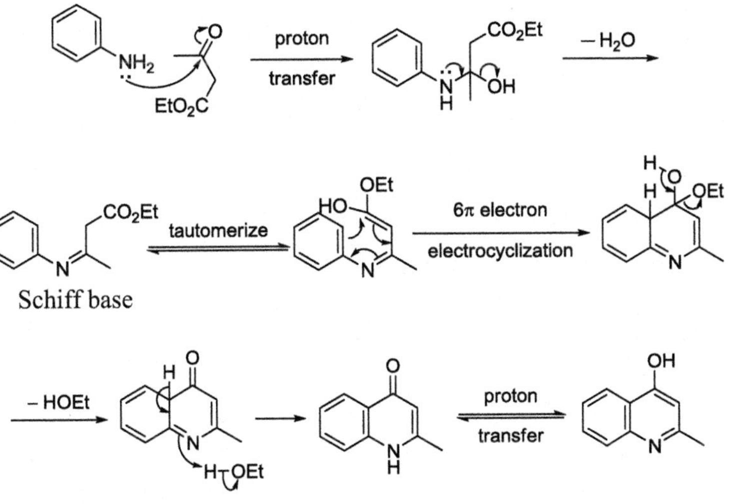

Example 1[3]

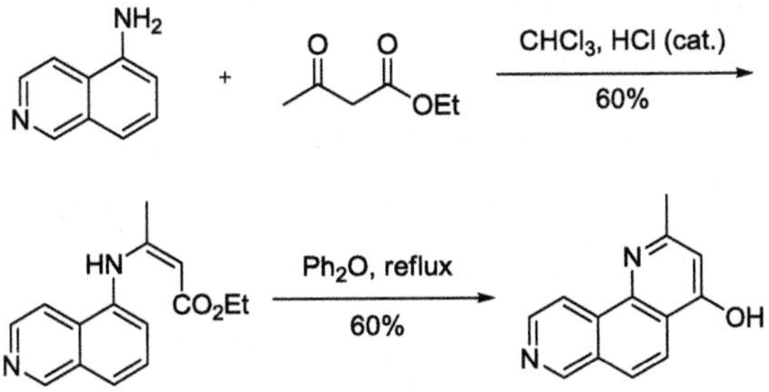

Example 2[7]

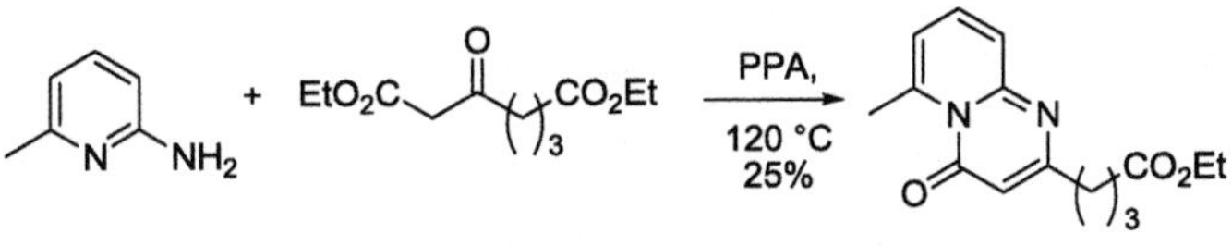

J.J. Li, *Name Reactions: A Collection of Detailed Mechanisms and Synthetic Applications*, DOI 10.1007/978-3-319-03979-4_66, © Springer International Publishing Switzerland 2014

Example 3[8]

Example 4, Thermal Conrad–Limpach cyclization[11]

## References

1. Conrad, M.; Limpach, L. *Ber.* **1887**, *20*, 944. Max Conrad (1848–1920), born in Munich, Germany, was a professor of the University of Würzburg, where he collaborated with Leonhard Limpach (1852–1933) on the synthesis of quinoline derivatives.
2. Manske, R. F. *Chem Rev.* **1942**, *30*, 113–114. (Review).
3. Misani, F.; Bogert, M. T. *J. Org. Chem.* **1945**, *10*, 347–365
4. Reitsema, R. H. *Chem. Rev.* **1948**, *43*, 43–68. (Review).
5. Elderfield, R. C. In *Chemistry of Heterocyclic Compounds*, Elderfield, R. C., Wiley & Sons, New York, **1952**, *vol. 4*, 31–36. (Review).
6. Jones, G. In *Heterocyclic Compounds*, Jones, G., ed.; Wiley & Sons, New York, **1977**, Quinolines, Vol 32, 137–151. (Review).
7. Deady, L. W.; Werden, D. M. *Synth. Commun.* **1987**, *17*, 319–328.
8. Kemp, D. S.; Bowen, B. R. *Tetrahedron Lett.* **1988**, *29*, 5077–5080.
9. Curran, T. T. *Conrad–Limpach Reaction.* In *Name Reactions in Heterocyclic Chemistry*; Li, J. J., Ed.; Wiley: Hoboken, NJ, **2005**, pp 398–406. (Review).
10. Chan, B. K.; Ciufolini, M. A. *J. Org. Chem.* **2008**, *72*, 8489–8495.
11. Lengyel, L.; Nagy, T. Z.; Sipos, G.; Jones, R.; Dormán, G.; Üerge, L.; Darvas, F. *Tetrahedron Lett.* **2012**, *53*, 738–743.

# Cope elimination reaction

Thermal elimination of *N*-oxides to olefins and *N*-hydroxyl amines.

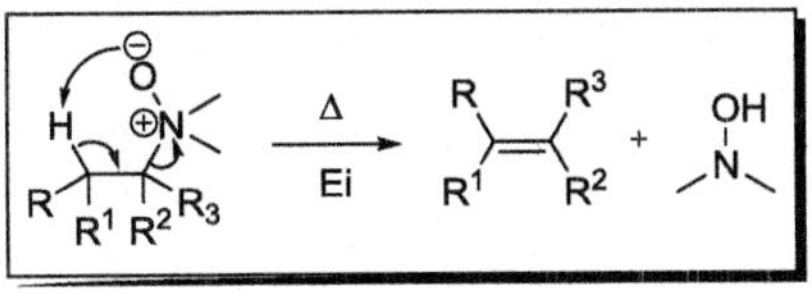

Example 1, Solid-phase Cope elimination[5]

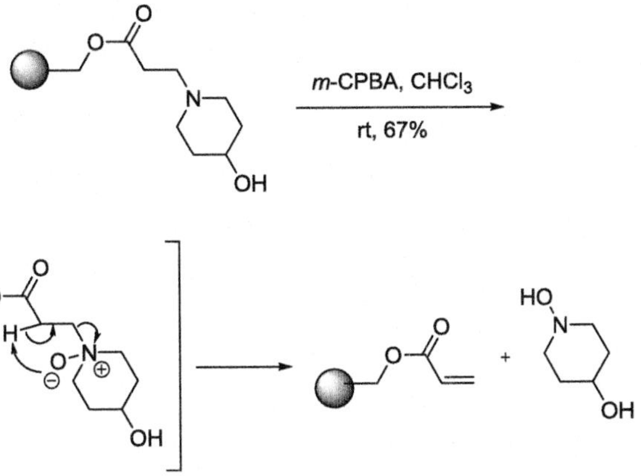

Example 2[6]

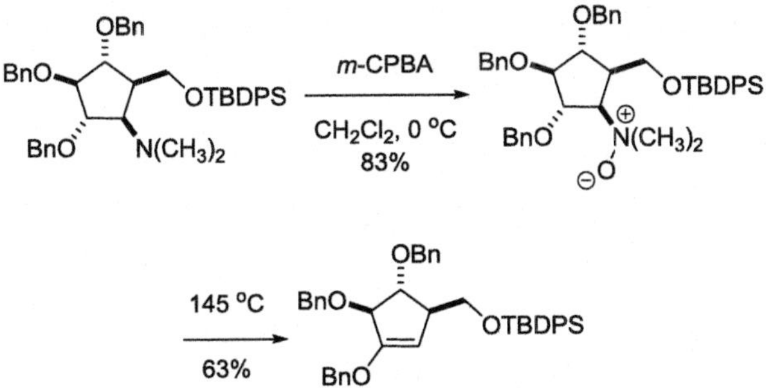

Example 3[8]

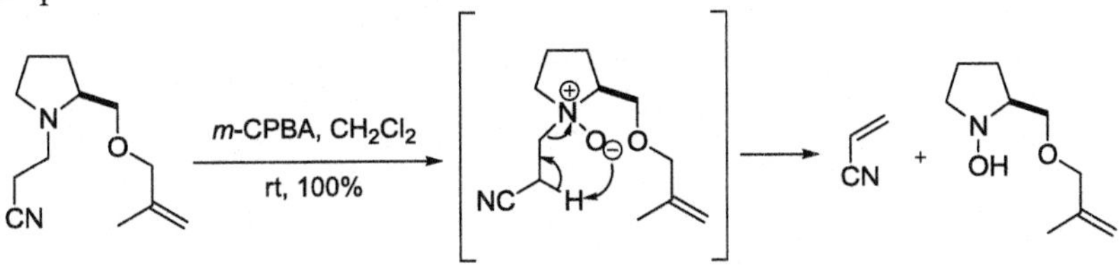

J.J. Li, *Name Reactions: A Collection of Detailed Mechanisms and Synthetic Applications*,
DOI 10.1007/978-3-319-03979-4_67, © Springer International Publishing Switzerland 2014

Example 4, Retro-Cope elimination[9]

Example 5[12]

**References**

1.  Cope, A. C.; Foster, T. T.; Towle, P. H. *J. Am. Chem. Soc.* **1949**, *71*, 3929–3934. Arthur Clay Cope (1909–1966) was born in Dunreith, Indiana. He was a professor and head at MIT where he discovered the Cope elimination reaction after he taught at Bryn Mawr and Columbia where he discovered the Cope rearrangement. The Arthur Cope Award is a prestigious award in organic chemistry administered by the American Chemical Society.
2.  Cope, A. C.; Trumbull, E. R. *Org. React.* **1960**, *11*, 317–493. (Review).
3.  DePuy, C. H.; King, R. W. *Chem. Rev.* **1960**, *60*, 431–457. (Review).
4.  Gallagher, B. M.; Pearson, W. H. *Chemtracts: Org. Chem.* **1996**, *9*, 126–130. (Review).
5.  Sammelson, R. E.; Kurth, M. J. *Tetrahedron Lett.* **2001**, *42*, 3419–3422.
6.  Vasella, A.; Remen, L. *Helv. Chim. Acta.* **2002**, *85*, 1118–1127.
7.  Garcia Martinez, A.; Teso Vilar, E.; Garcia Fraile, A.; de la Moya Cerero, S.; Lora Maroto, B. *Tetrahedron: Asymmetry* **2002**, *13*, 17–19.
8.  O'Neil, I. A.; Ramos, V. E.; Ellis, G. L.; Cleator, E.; Chorlton, A. P.; Tapolczay, D. J.; Kalindjian, S. B. *Tetrahedron Lett.* **2004**, *45*, 3659–3661.
9.  Henry, N.; O'Meil, I. A. *Tetrahedron Lett.* **2007**, *48*, 1691–1694.
10. Fuchter, M. J. *Cope Elimination Reaction.* In *Name Reactions for Functional Group Transformations*; Li, J. J., Ed.; Wiley: Hoboken, NJ, **2007**, pp 342–353. (Review).
11. Bourgeois, J.; Dion, I.; Cebrowski, P. H.; Loiseau, F.; Bedard, A.-C.; Beauchemin, A. M. *J. Am. Chem. Soc.* **2009**, *131*, 874–875.
12. Miyatake-Ondozabal, H.; Bannwart, L. M.; Gademann, K. *Chem. Commun.* **2013**, *49*, 1921–1923.

# Cope rearrangement

The Cope, aza-Cope, anionic oxy-Cope, and oxy-Cope rearrangements belong to the category of *[3,3]-sigmatropic rearrangements*. Since it is a concerted process, the arrow pushing here is only illustrative. This reaction is an equilibrium process. *Cf.* Claisen rearrangement.

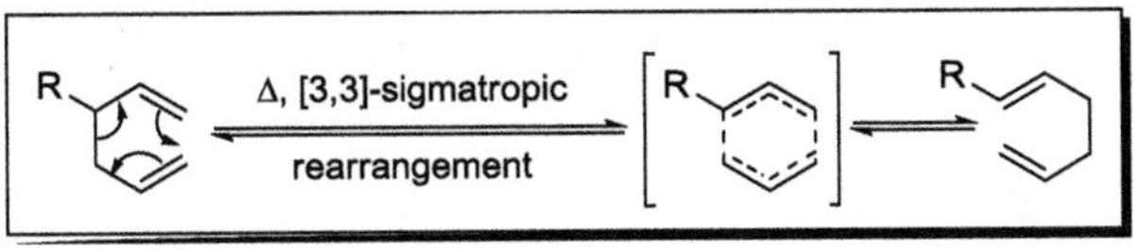

Example 1[4]

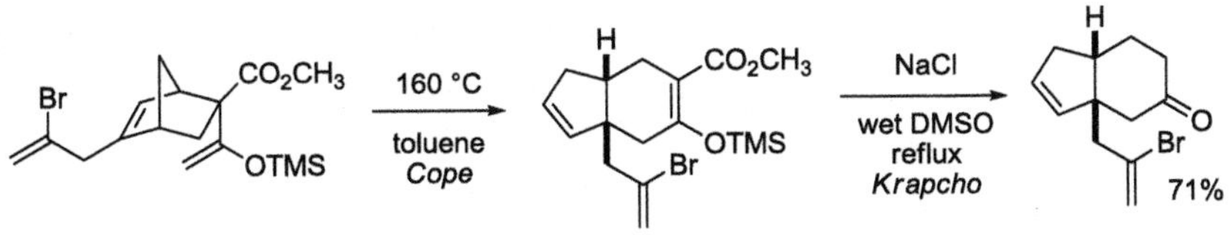

Example 2[6]

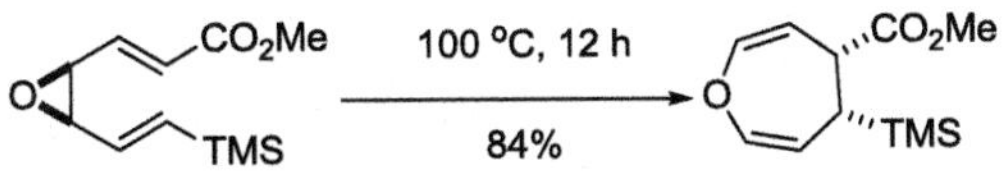

Example 3[9]

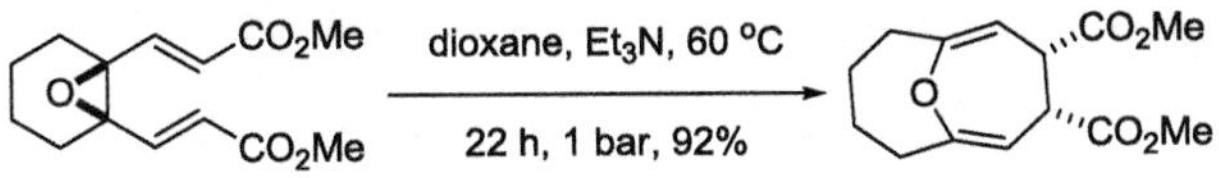

Example 4[10]

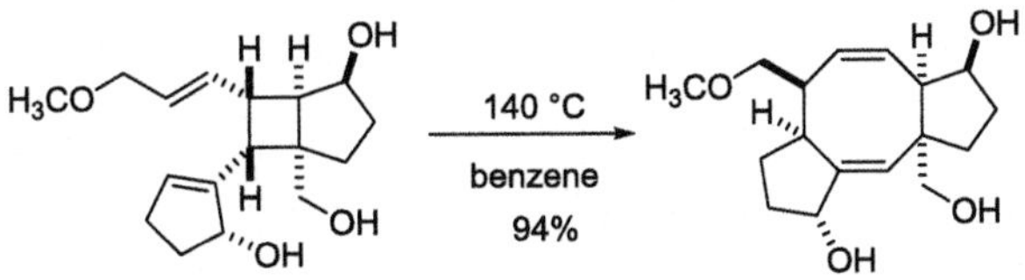

Example 5[11]

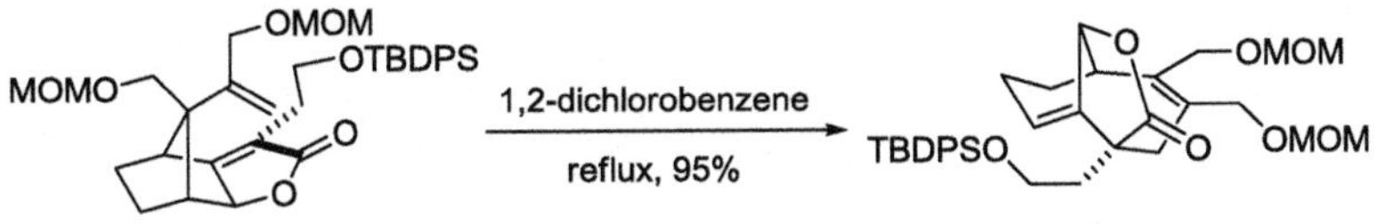

J.J. Li, *Name Reactions: A Collection of Detailed Mechanisms and Synthetic Applications*, DOI 10.1007/978-3-319-03979-4_68, © Springer International Publishing Switzerland 2014

Example 6[12]

Example 7, Cope rearrangement[14]

Example 8[15]

**References**
1.  Cope, A. C.; Hardy, E. M. *J. Am. Chem. Soc.* **1940**, *62*, 441–444.
2.  Frey, H. M.; Walsh, R. *Chem. Rev.* **1969**, *69*, 103–124. (Review).
3.  Rhoads, S. J.; Raulins, N. R. *Org. React.* **1975**, *22*, 1–252. (Review).
4.  Wender, P. A.; Schaus, J. M. White, A. W. *J. Am. Chem. Soc.* **1980**, *102*, 6159–6161.
5.  Hill, R. K. In *Comprehensive Organic Synthesis* Trost, B. M.; Fleming, I., Eds.; Pergamon, **1991**, *Vol. 5*, 785–826. (Review).
6.  Chou, W.-N.; White, J. B.; Smith, W. B. *J. Am. Chem. Soc.* **1992**, *114*, 4658–4667.
7.  Davies, H. M. L. *Tetrahedron* **1993**, *49*, 5203–5223. (Review).
8.  Miyashi, T.; Ikeda, H.; Takahashi, Y. *Acc. Chem. Res.* **1999**, *32*, 815–824. (Review).
9.  Von Zezschwitz, P.; Voigt, K.; Lansky, A.; Noltemeyer, M.; De Meijere, A. *J. Org. Chem.* **1999**, *64*, 3806–3812.
10. Lo, P. C.-K.; Snapper, M. L. *Org. Lett.* **2001**, *3*, 2819–2821.
11. Clive, D. L. J.; Ou, L. *Tetrahedron Lett.* **2002**, *43*, 4559–4563.
12. Malachowski, W. P.; Paul, T.; Phounsavath, S. *J. Org. Chem.* **2007**, *72*, 6792–6796.
13. Mullins, R. J.; McCracken, K. W. *Cope and Related Rearrangements*. In *Name Reactions for Homologations-Part II*; Li, J. J., Ed.; Wiley: Hoboken, NJ, **2009**, pp 88–135. (Review).
14. Ren, H.; Wulff, W. D. *Org. Lett.* **2013**, *15*, 242–245.
15. Yamada, T.; Yoshimura, F.; Tanino, K. *Tetrahedron Lett.* **2013**, *54*, 522–525.

## Anionic oxy-Cope rearrangement

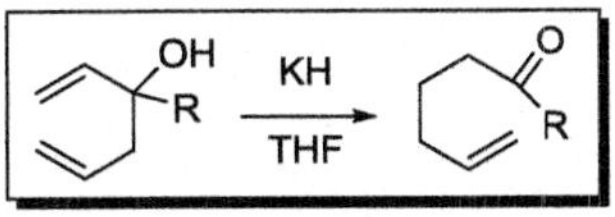

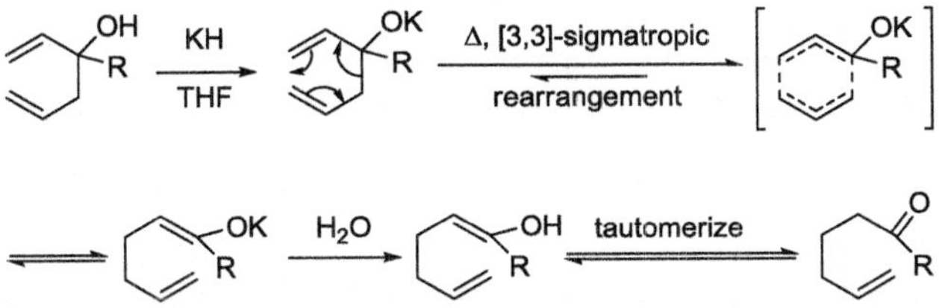

Example 1[1]

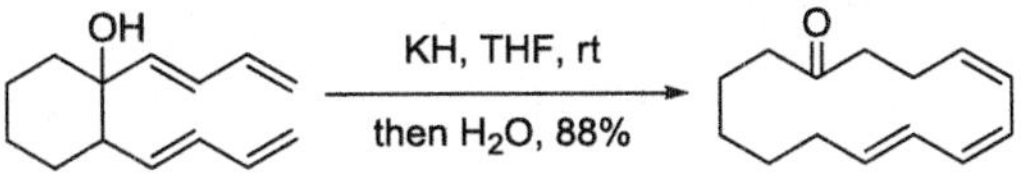

KH, THF, rt

then H₂O, 88%

Example 2[4]

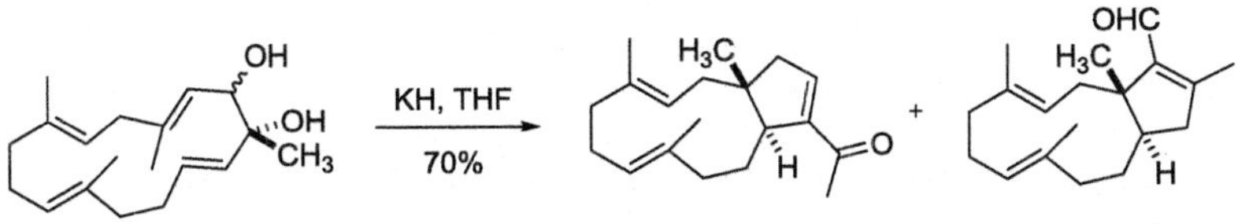

KH, THF

70%

Example 3[5]

KHMDS
18-crown-6, THF
temperature;

then NH₄Cl, H₂O

X = OCH₂CH₂TMS      0 °C; 71%
X = SPh             −78 °C; 85%

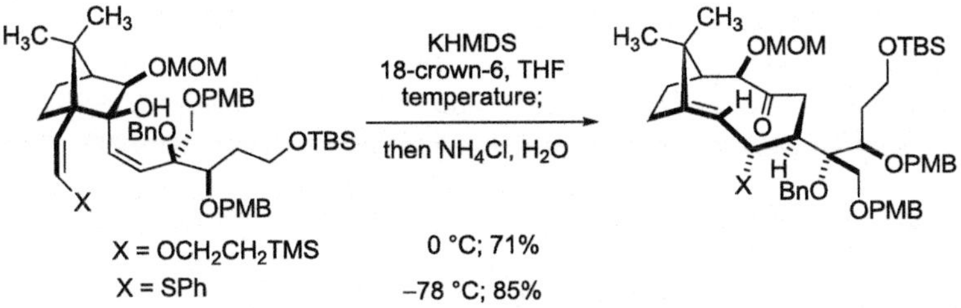

Example 4[8]

NaH, THF, reflux

22 h, 88%

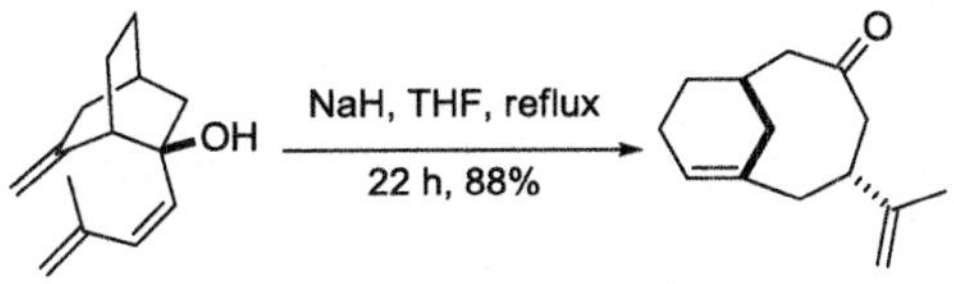

Example 5[9]

Example 6[11]

## References

1. Wender, P. A.; Sieburth, S. M.; Petraitis, J. J.; Singh, S. K. *Tetrahedron* **1981**, *37*, 3967–3975.
2. Wender, P. A.; Ternansky, R. J.; Sieburth, S. M. *Tetrahedron Lett.* **1985**, *26*, 4319–4322.
3. Paquette, L. A. *Tetrahedron* **1997**, *53*, 13971–14020. (Review).
4. Corey, E. J.; Kania, R. S. *Tetrahedron Lett.* **1998**, *39*, 741–744.
5. Paquette, L. A.; Reddy, Y. R.; Haeffner, F.; Houk, K. N. *J. Am. Chem. Soc.* **2000**, *122*, 740–741.
6. Voigt, B.; Wartchow, R.; Butenschon, H. *Eur. J. Org. Chem.* **2001**, 2519–2527.
7. Hashimoto, H.; Jin, T.; Karikomi, M.; Seki, K.; Haga, K.; Uyehara, T. *Tetrahedron Lett.* **2002**, *43*, 3633–3636.
8. Gentric, L.; Hanna, I.; Huboux, A.; Zaghdoudi, R. *Org. Lett.* **2003**, *5*, 3631–3634.
9. Jones, S. B.; He, L.; Castle, S. L. *Org. Lett.* **2006**, *8*, 3757–3760.
10. Mullins, R. J.; McCracken, K. W. *Cope and Related Rearrangements*. In *Name Reactions for Homologations-Part II*; Li, J. J., Ed.; Wiley: Hoboken, NJ, **2009**, pp 88–135. (Review).
11. Taber, D. F.; Gerstenhaber, D. A.; Berry, J. F. *J. Org. Chem.* **2013**, *76*, 7614–7617.

## *Oxy-Cope rearrangement*

While the anionic oxy-Cope rearrangements work at low temperature, the oxy-Cope rearrangements require high temperature but provide a thermodynamic sink.

Example 1[2]

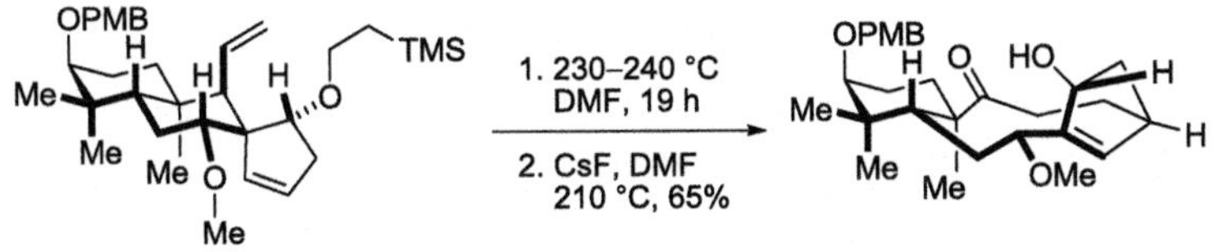

1. 230–240 °C
DMF, 19 h

2. CsF, DMF
210 °C, 65%

Example 2[3]

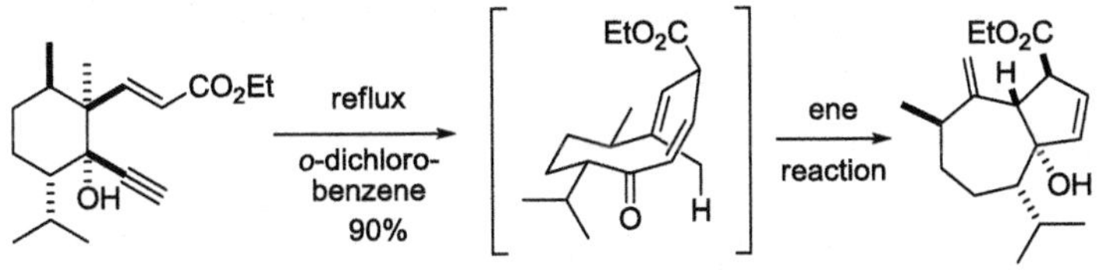

reflux

o-dichloro-
benzene
90%

ene
reaction

Example 3[4]

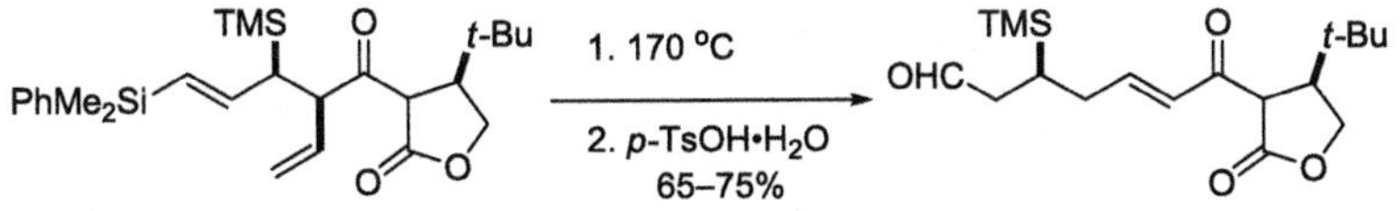

1. 170 °C

2. p-TsOH•H$_2$O
65–75%

Example 4[6]

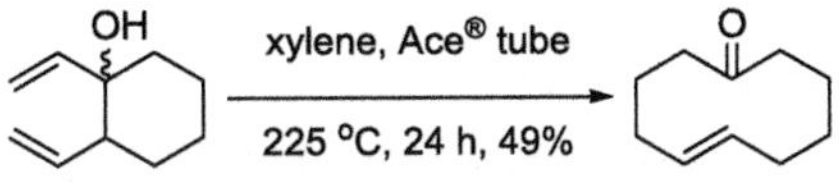

xylene, Ace® tube

225 °C, 24 h, 49%

Example 5[8]

Toluene, 120 °C

7 h, 42%

furanogermenone

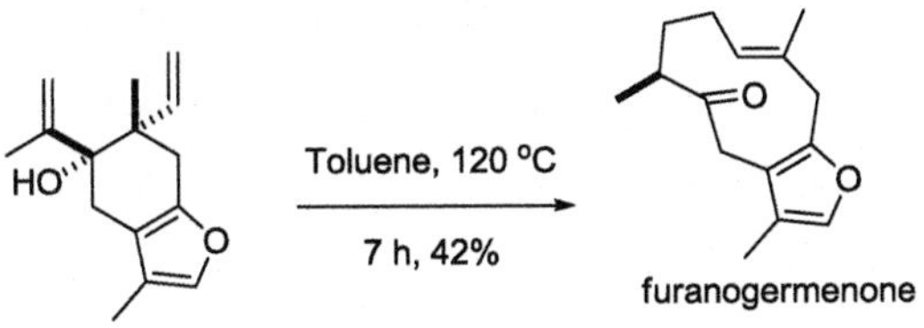

## References

1. Paquette, L. A. *Angew. Chem. Int. Ed.* **1990**, *29*, 609–626. (Review).
2. Paquette, L. A.; Backhaus, D.; Braun, R. *J. Am. Chem. Soc.* **1996**, *118*, 11990–11991.
3. Srinivasan, R.; Rajagopalan, K. *Tetrahedron Lett.* **1998**, *39*, 4133–4136.
4. Schneider, C.; Rehfeuter, M. *Chem. Eur. J.* **1999**, *5*, 2850–2858.
5. Schneider, C. *Synlett* **2001**, 1079–1091. (Review on siloxy-Cope rearrangement).
6. DiMartino, G.; Hursthouse, M. B.; Light, M. E.; Percy, J. M.; Spencer, N. S.; Tolley, M. *Org. Biomol. Chem.* **2003**, *1*, 4423–4434.

7.   Mullins, R. J.; McCracken, K. W. *Cope and Related Rearrangements*. In *Name Reactions for Homologations-Part II*; Li, J. J., Ed.; Wiley: Hoboken, NJ, **2009**, pp 88–135. (Review).
8.   Anagnostaki, E. E.; Zografos, A. L. *Org. Lett.* **2013**, *15*, 152–155.

## *Siloxy-Cope rearrangement*

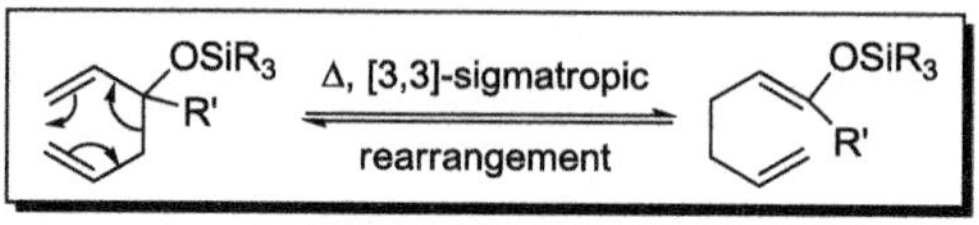

Example 1[1]

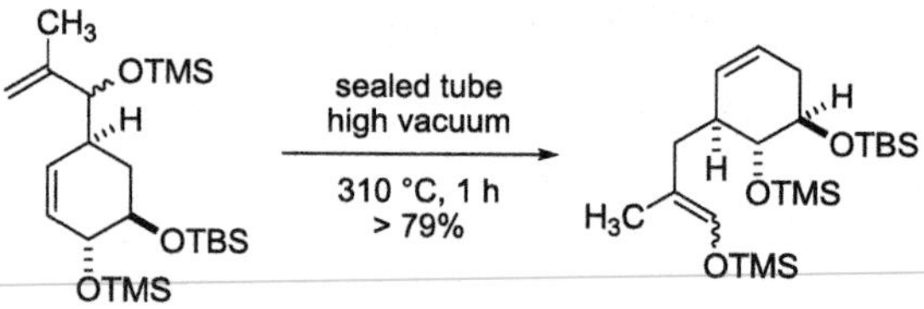

Example 2[2]

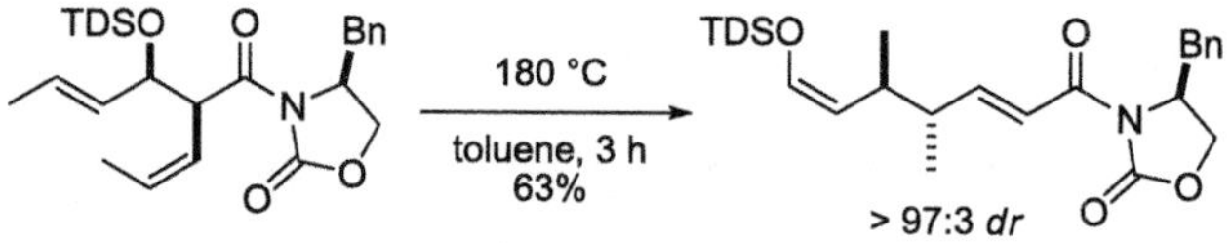

TDS = thexyldimethylsilyl

Example 3[3]

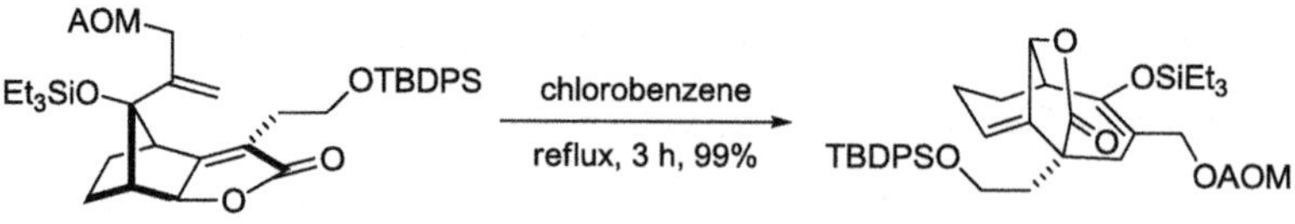

AOM = *p*-Anisyloxymethyl = *p*-MeOC$_6$H$_4$OCH$_2$-

Example 4[4]

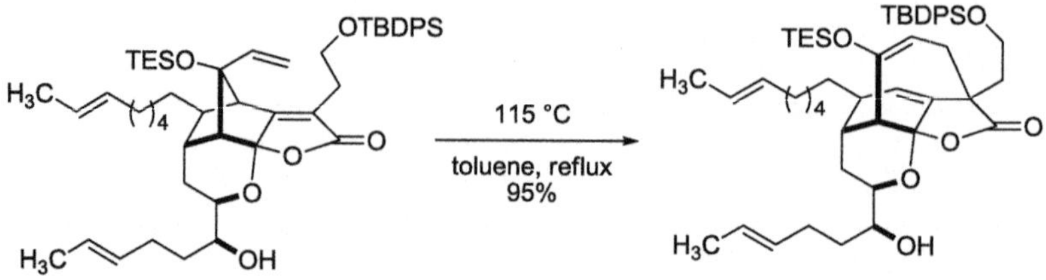

Example 5, Tandem aldol reaction/siloxy-Cope rearrangement[6]

**References**

1.  Askin, D.; Angst, C.; Danishefsky, D. J. *J. Org. Chem.* **1987,** *52*, 622–635.
2.  Schneider, C. *Eur. J. Org. Chem.* **1998,** 1661–1663.
3.  Clive, D. L. J.; Sun, S.; Gagliardini, V.; Sano, M. K. *Tetrahedron Lett.* **2000,** *41*, 6259–6263.
4.  Bio, M. M.; Leighton, J. L. *J. Org. Chem.* **2003,** *68*, 1693–1700.
5.  Mullins, R. J.; McCracken, K. W. *Cope and Related Rearrangements*. In *Name Reactions for Homologations-Part II*; Li, J. J., Ed.; Wiley: Hoboken, NJ, **2009,** pp 88–135. (Review).
6.  Davies, H. M. L.; Lian, Y. *Acc. Chem. Res.* **2013,** *45*, 923–935. (Review).

## Corey–Bakshi–Shibata (CBS) reagent

The CBS (Corey–Bakshi–Shibata) reagent is a chiral catalyst derived from proline. Also known as Corey's oxazaborolidine, it is used in enantioselective borane reduction of ketones, asymmetric Diels–Alder reactions and [3 + 2] cycloadditions.

Preparation[1,3]

The mechanism and catalytic cycle:[1,3]

J.J. Li, *Name Reactions: A Collection of Detailed Mechanisms and Synthetic Applications*, DOI 10.1007/978-3-319-03979-4_69, © Springer International Publishing Switzerland 2014

Example 1[6]

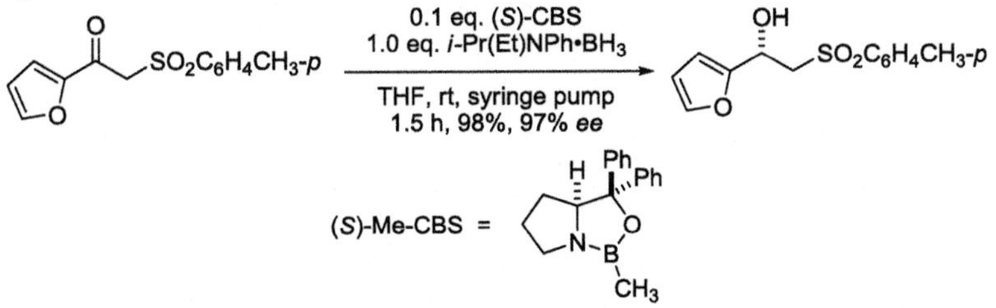

$(S)$-Me-CBS =

Example 2[9]

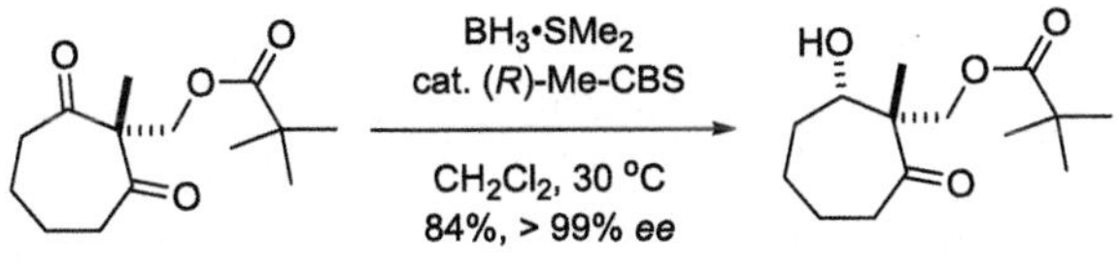

Example 3[11]

oseltamivir (Tamiflu®)

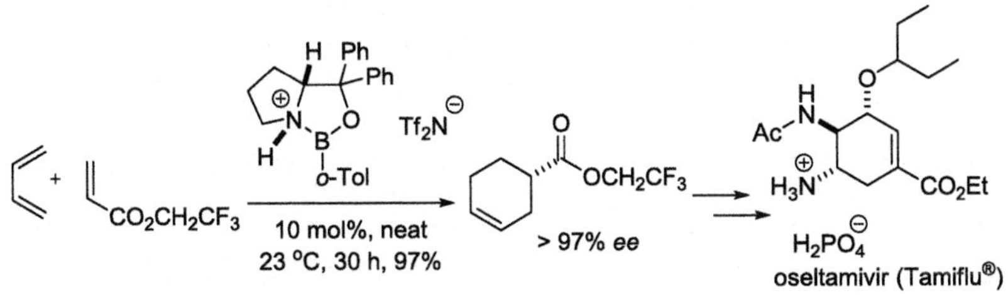

Example 4, Asymmetric [3 + 2]-cycloaddition[10]

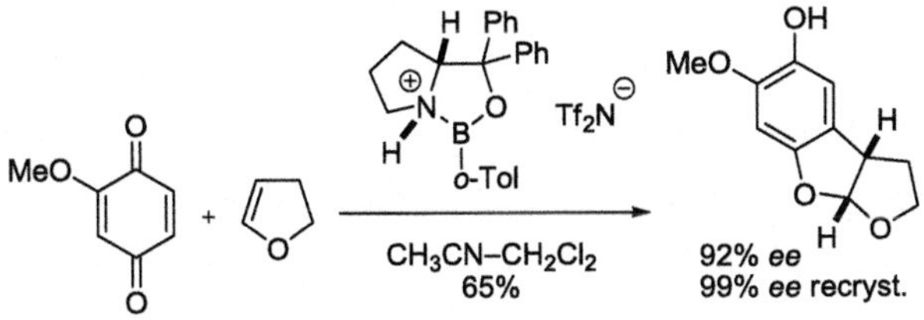

Example 5[13]

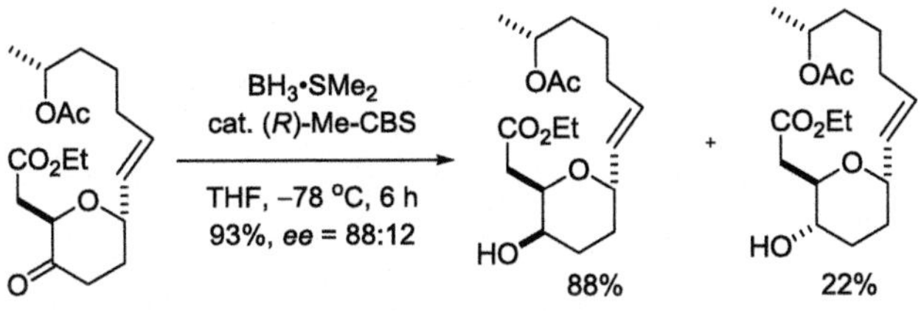

**References**
1.  (a) Corey, E. J.; Bakshi, R. K.; Shibata, S. *J. Am. Chem. Soc.* **1987,** *109*, 5551–5553. (b) Corey, E. J.; Bakshi, R. K.; Shibata, S.; Chen, C.-P.; Singh, V. K. *J. Am. Chem. Soc.* **1987,** *109*, 7925–7926. (c) Corey, E. J.; Shibata, S.; Bakshi, R. K. *J. Org. Chem.* **1988,** *53*, 2861–2863.
2.  Reviews: (a) Corey, E. J. *Pure Appl. Chem.* **1990,** *62*, 1209–1216. (b) Wallbaum, S.; Martens, J. *Tetrahedron: Asymm.* **1992,** *3*, 1475–1504. (c) Singh, V. K. *Synthesis* **1992,** 605–617. (d) Deloux, L.; Srebnik, M. *Chem. Rev.* **1993,** *93*, 763–784. (e) Taraba, M.; Palecek, J. *Chem. Listy* **1997,** *91*, 9–22. (f) Corey, E. J.; Helal, C. J. *Angew. Chem. Int. Ed.* **1998,** *37*, 1986–2012. g) Corey, E. J. *Angew. Chem. Int. Ed.* **2002,** *41*, 1650–1667. (h) Itsuno, S. *Org. React.* **1998,** *52*, 395–576. (i) Cho, B. T. *Aldrichimica Acta* **2002,** *35*, 3–16. (j) Glushkov, V. A.; Tolstikov, A. G. *Russ. Chem. Rev.* **2004,** *73*, 581–608. (k) Cho, B .T. *Tetrahedron* **2006,** *62*, 7621–7643.
3.  (a) Mathre, D. J.; Thompson, A. S.; Douglas, A. W.; Hoogsteen, K.; Carroll, J. D.; Corley, E. G.; Grabowski, E. J. J. *J. Org. Chem.* **1993,** *58*, 2880–2888. (b) Xavier, L. C.; Mohan, J. J.; Mathre, D. J.; Thompson, A. S.; Carroll, J. D.; Corley, E. G.; Desmond, R. *Org. Synth.* **1997,** *74*, 50–71.
4.  Corey, E. J.; Helal, C. J. *Tetrahedron Lett.* **1996,** *37*, 4837–4840.
5.  Clark, W. M.; Tickner-Eldridge, A. M.; Huang, G. K.; Pridgen, L. N.; Olsen, M. A.; Mills, R. J.; Lantos, I.; Baine, N. H. *J. Am. Chem. Soc.* **1998,** *120*, 4550–4551.
6.  Cho, B. T.; Kim, D. J. *Tetrahedron: Asymmetry* **2001,** *12*, 2043–2047.
7.  Price, M. D.; Sui, J. K.; Kurth, M. J.; Schore, N. E. *J. Org. Chem.* **2002,** *67*, 8086–8089.
8.  Degni, S.; Wilen, C.-E.; Rosling, A. *Tetrahedron: Asymmetry* **2004,** *15*, 1495–1499.
9.  Watanabe, H.; Iwamoto, M.; Nakada, M. *J. Org. Chem.* **2005,** *70*, 4652–4658.
10. Zhou, G.; Corey, E. J. *J. Am. Chem. Soc.* **2005,** *127*, 11958–11959.
11. Yeung, Y.-Y.; Hong, S.; Corey, E. J. *J. Am. Chem. Soc.* **2006,** *128*, 6310–6311.
12. Patti, A.; Pedotti, S. *Tetrahedron: Asymmetry* **2008,** *19*, 1891–1897.
13. Sridhar, Y.; Srihari, P. *Eur. J. Org. Chem.* **2013,** 578–587.

# Corey–Chaykovsky reaction

The Corey–Chaykovsky reaction entails the reaction of a sulfur ylide, either dimethylsulfoxonium methylide **1** (Corey's ylide) or dimethylsulfonium methylide **2**, with electrophile **3** such as carbonyl, olefin, imine, or thiocarbonyl, to offer **4** as the corresponding epoxide, cyclopropane, aziridine, or thiirane.

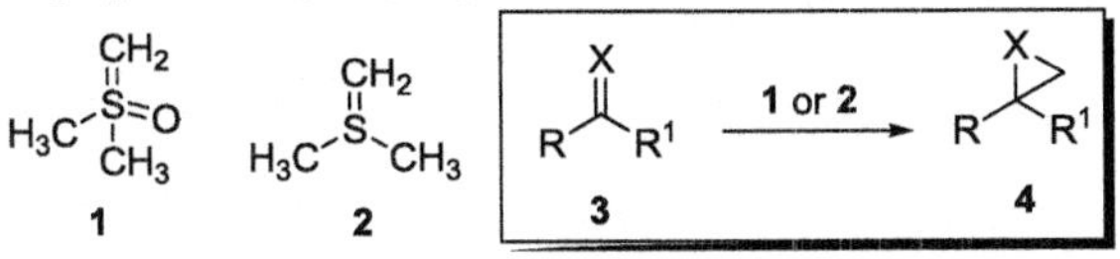

$X = O, CH_2, NR^2, S, CHCOR^3,$
$CHCO_2R^3, CHCONR_2, CHCN$

Preparation[1]

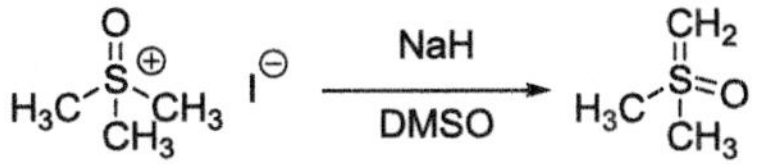

Mechanism[1]

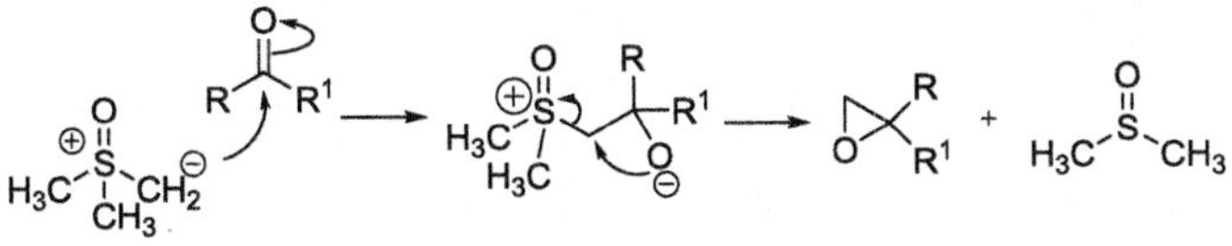

Example 1[11]

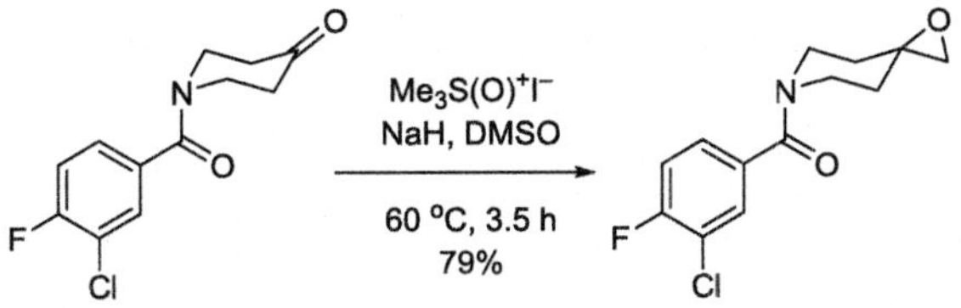

Example 2[9]

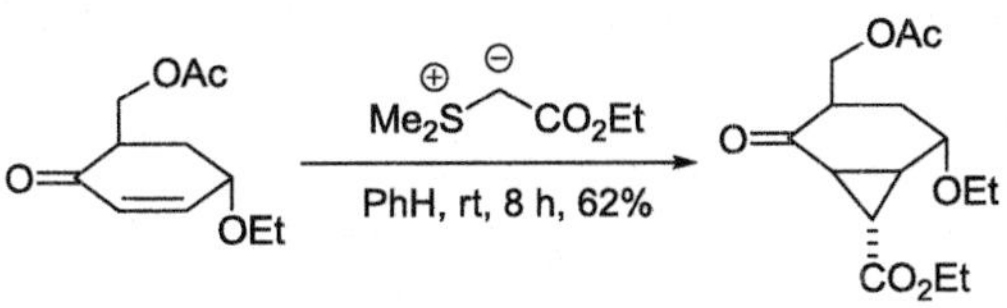

Example 3[10]

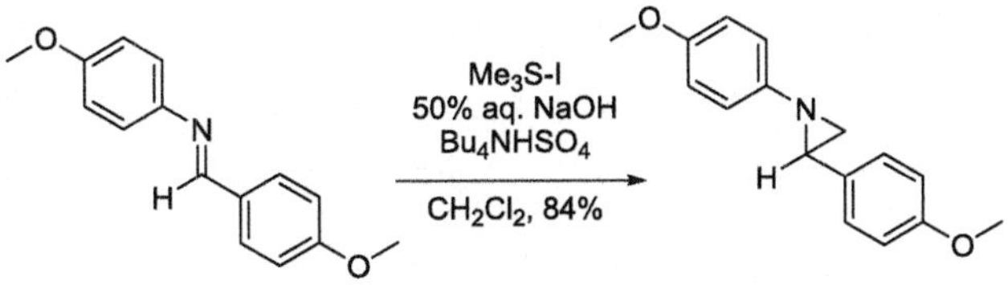

J.J. Li, *Name Reactions: A Collection of Detailed Mechanisms and Synthetic Applications*, DOI 10.1007/978-3-319-03979-4_70, © Springer International Publishing Switzerland 2014

Example 4[14]

Example 5[15]

Example 6[16]

## References

1    (a) Corey, E. J.; Chaykovsky, M. *J. Am. Chem. Soc.* **1962**, *84*, 867–868. (b) Corey, E.
     J.; Chaykovsky, M. *J. Am. Chem. Soc.* **1962**, *84*, 3782. (c) Corey, E. J.; Chaykovsky,
     M. *Tetrahedron Lett.* **1963**, 169–171. (d) Corey, E. J.; Chaykovsky, M. *J. Am. Chem.
     Soc.* **1964**, *86*, 1639–1640. (e) Corey, E. J.; Chaykovsky, M. *J. Am. Chem. Soc.* **1965**,
     *87*, 1353–1364.
2    Okazaki, R.; Tokitoh, N. In *Encyclopedia of Reagents in Organic Synthesis;* Paquette,
     L. A., Ed.; Wiley: New York, **1995**, pp 2139–2141. (Review).
3    Ng, J. S.; Liu, C. In *Encyclopedia of Reagents in Organic Synthesis;* Paquette, L. A.,
     Ed.; Wiley: New York, **1995**, pp 2159–2165. (Review).
4    Trost, B. M.; Melvin, L. S., Jr. *Sulfur Ylides;* Academic Press: New York, **1975**. (Re-
     view).
5    Block, E. *Reactions of Organosulfur Compounds* Academic Press: New York, **1978**.
     (Review).
6    Gololobov, Y. G.; Nesmeyanov, A. N. *Tetrahedron* **1987**, *43*, 2609–2651. (Review).
7    Aubé, J. In *Comprehensive Organic Synthesis;* Trost, B. M.; Fleming, I., Ed.;
     Pergamon: Oxford, **1991**, *Vol. 1*, pp 820–825. (Review).
8    Li, A.-H.; Dai, L.-X.; Aggarwal, V. K. *Chem. Rev.* **1997**, *97*, 2341–2372. (Review).

9    Rosenberger, M.; Jackson, W.; Saucy, G. *Helv. Chim. Acta* **1980,** *63*, 1665–1674.

10   Tewari, R. S.; Awatsthi, A. K.; Awasthi, A. *Synthesis* **1983,** 330–331.

11   Vacher, B.; Bonnaud, B. Funes, P.; Jubault, N.; Koek, W.; Assie, M.-B.; Cosi, C.; Kleven, M. *J. Med. Chem.* **1999,** *42*, 1648–1660.

12   Chandrasekhar, S.; Narasihmulu, Ch.; Jagadeshwar, V.; Reddy, K. V. *Tetrahedron Lett.* **2003,** *44*, 3629–3630.

13   Li, J. J. *Corey–Chaykovsky Reaction.* In *Name Reactions in Heterocyclic Chemistry*; Li, J. J., Ed.; Wiley: Hoboken, NJ, **2005,** pp 1–14. (Review).

14   Nishimura, Y.; Shiraishi, T.; Yamaguchi, M. *Tetrahedron Lett.* **2008,** *49*, 3492–3495.

15   Chittimalla, S. K.; Chang, T.-C.; Liu, T.-C.; Hsieh, H.-P.; Liao, C.-C. *Tetrahedron* **2008,** *64*, 2586–2595.

16   Palko, J. W.; Buist, P. H.; Manthorpe, J. M. *Tetrahedron: Asymmetry* **2013,** *24*, 165–168.

## Corey–Fuchs reaction

One-carbon homologation of an aldehyde to dibromoolefin, which is then treated with *n*-BuLi to produce a terminal alkyne.

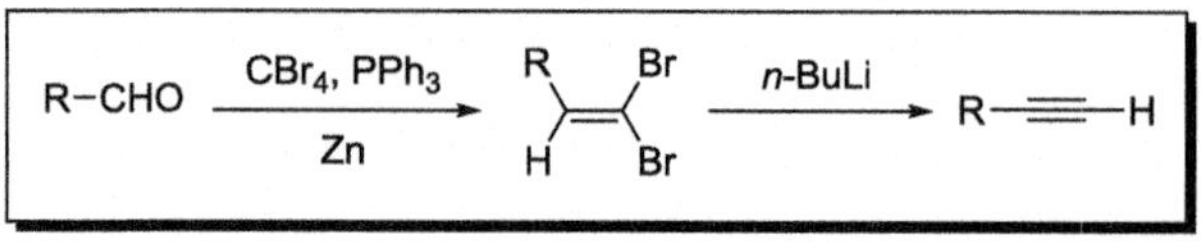

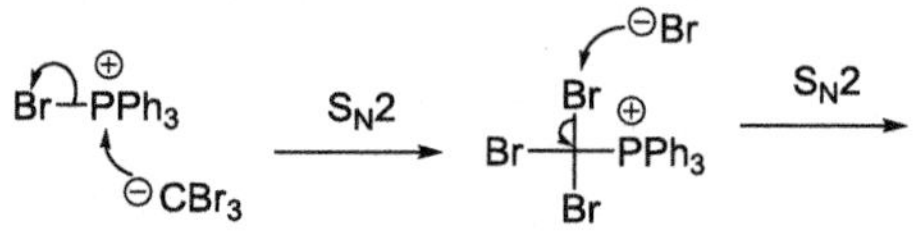

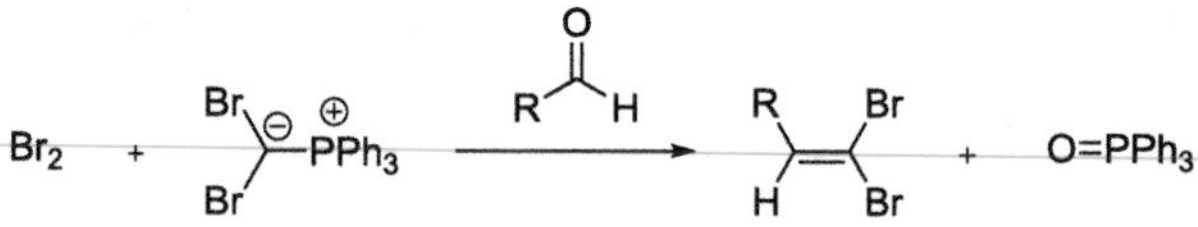

Wittig reaction

$$Br_2 \ + \ Zn \ \longrightarrow \ ZnBr_2$$

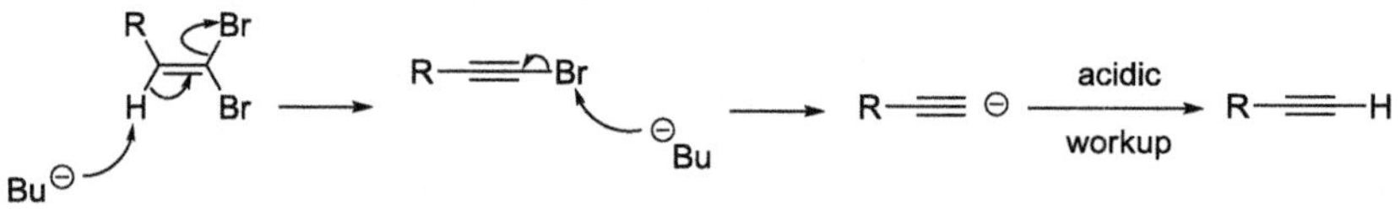

Example 1[3]

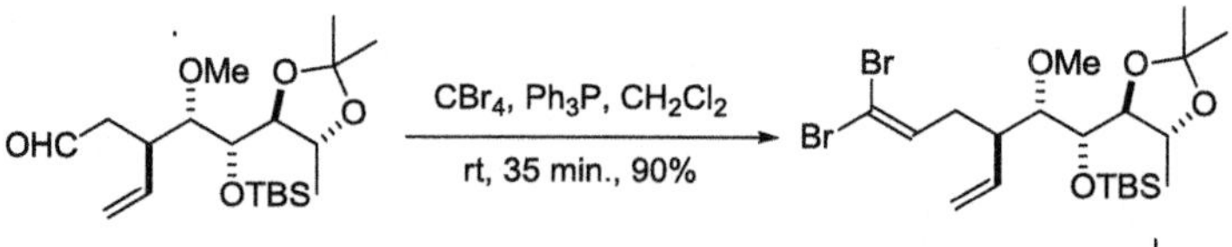

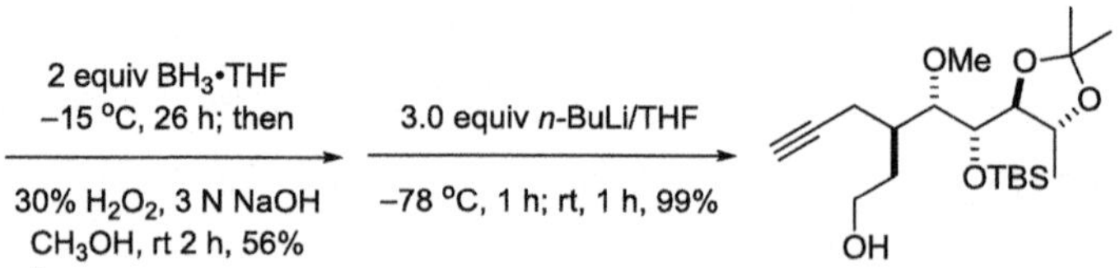

Example 2[7]

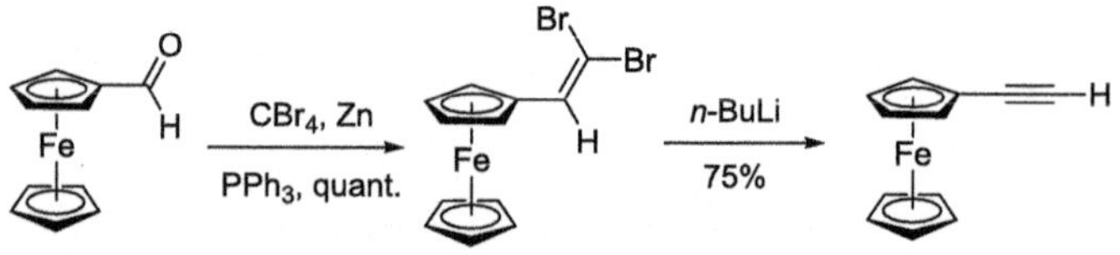

J.J. Li, *Name Reactions: A Collection of Detailed Mechanisms and Synthetic Applications*, DOI 10.1007/978-3-319-03979-4_71, © Springer International Publishing Switzerland 2014

## Example 3[8]

Ph ⋯⋯ =〈 OTBS 〉= CHO  →[1. CBr₄, Ph₃P, Zn, CH₂Cl₂][2. n-BuLi, then NH₄Cl, 90%]  Ph ⋯⋯ OTBS ⋯⋯ ≡

## Example 4[10]

→[4 equiv CBr₄][8 equiv PPh₃][8 equiv Zn][CH₂Cl₂, 0 °C to rt][3 h, 94%]

→[n-BuLi, n-hexane][−78 °C to rt, 1 h, 96%]

## Example 5[12]

→[a. (COCl)₂, DMSO, Et₃N, CH₂Cl₂, −78 °C, 2 h;][b. CBr₄, TPP, Et₃N, CH₂Cl₂, 0 °C, 2 h; 70%]

## References

1  Corey, E. J.; Fuchs, P. L. *Tetrahedron Lett.* **1972**, *13,* 3769–3772. Phil Fuchs is a professor at Purdue University.

2  For the synthesis of 1-bromalkynes see Grandjean, D.; Pale, P.; Chuche, J. *Tetrahedron Lett.* **1994**, *35*, 3529–3530.

3  Gilbert, A. M.; Miller, R.; Wulff, W. D. *Tetrahedron* **1999**, *55*, 1607–1630.

4  Muller, T. J. J. *Tetrahedron Lett.* **1999**, *40*, 6563–6566.

5  Serrat, X.; Cabarrocas, G.; Rafel, S.; Ventura, M.; Linden, A.; Villalgordo, J. M. *Tetrahedron: Asymmetry* **1999**, *10*, 3417–3430.

6  Okamura, W. H.; Zhu, G.-D.; Hill, D. K.; Thomas, R. J.; Ringe, K.; Borchardt, D. B.; Norman, A. W.; Mueller, L. J. *J. Org. Chem.* **2002**, *67*, 1637–1650.

7  Tsuboya, N.; Hamasaki, R.; Ito, M.; Mitsuishi, M.; Miyashita, T. Yamamoto, Y. *J. Mater. Chem.* **2003**, *13*, 511–513

8  Zeng, X.; Zeng, F.; Negishi, E.-i. *Org. Lett.* **2004**, *6*, 3245–3248.

9  Quéron, E.; Lett, R. *Tetrahedron Lett.* **2004**, *45*, 4527–4531.

10  Sahu, B.; Muruganantham, R.; Namboothiri, I. N. N. *Eur. J. Org. Chem.* **2007**, 2477–2489.

11  Han, X. *Corey–Fuchs reaction*. In *Name Reactions for Homologations-Part I*; Li, J. J., Ed.; Wiley: Hoboken, NJ, **2009**, pp 393–403. (Review).

12  Pradhan, T. K.; Lin, C. C.; Mong, K. K. T. *Synlett* **2013**, *24,* 219–222.

## Corey–Kim oxidation

Oxidation of alcohol to the corresponding aldehyde or ketone using NCS/DMS, followed by treatment with a base. *Cf.* Swern oxidation.

NCS = *N*-Chlorosuccinimide;  DMS = **D**imethylsulfide.

Example 1, Fluorous Corey–Kim reaction[5]

Example 2, Odorless Corey–Kim reaction[7]

J.J. Li, *Name Reactions: A Collection of Detailed Mechanisms and Synthetic Applications*, DOI 10.1007/978-3-319-03979-4_72, © Springer International Publishing Switzerland 2014

Example 3[9]

Example 4[10]

**References**

1   Corey, E. J.; Kim, C. U. *J. Am. Chem. Soc.* **1972**, *94*, 7586–7587. Choung U. Kim worked at Gilead Sciences, a company specialized in antiviral drugs in Foster City, California, where he co-discovered oseltamivir (Tamiflu).

2   Katayama, S.; Fukuda, K.; Watanabe, T.; Yamauchi, M. *Synthesis* **1988**, 178–183.

3   Shapiro, G.; Lavi, Y. *Heterocycles* **1990**, *31*, 2099–2102.

4   Pulkkinen, J. T.; Vepsäläinen, J. J. *J. Org. Chem.* **1996**, *61*, 8604–8609.

5   Crich, D.; Neelamkavil, S. *Tetrahedron* **2002**, *58*, 3865–3870.

6   Ohsugi, S.-I.; Nishide, K.; Oono, K.; Okuyama, K.; Fudesaka, M.; Kodama, S.; Node, M. *Tetrahedron* **2003**, *59*, 8393–8398.

7   Nishide, K.; Patra, P. K.; Matoba, M.; Shanmugasundaram, K.; Node, M. *Green Chem.* **2004**, *6*, 142–146.

8   Iula, D. M. *Corey–Kim Oxidation*. In *Name Reactions for Functional Group Transformations*; Li, J. J., Corey, E. J. (eds), Wiley: Hoboken, NJ, **2007**, pp 207–217. (Review).

9   Yin, W.; Ma, J.; Rivas, F. M.; Cook, M. *Org. Lett.* **2007**, *9*, 295–298.

10  Cink, R. D.; Chambournier, G.; Surjono, H.; Xiao, Z.; Richter, S.; Naris, M.; Bhatia, A. V. *Org. Process Res. Dev.* **2007**, *11*, 270–274.

11  Berger, O.; Gavara, L.; Montchamp, J.-L. *Org. Lett.* **2013**, *14*, 3404–3407.

## Corey–Nicolaou macrolactonization

Macrolactonization of ω-hydroxyl-acid using 2,2′-dipyridyl disulfide.   Also known as the Corey–Nicolaou double activation method.

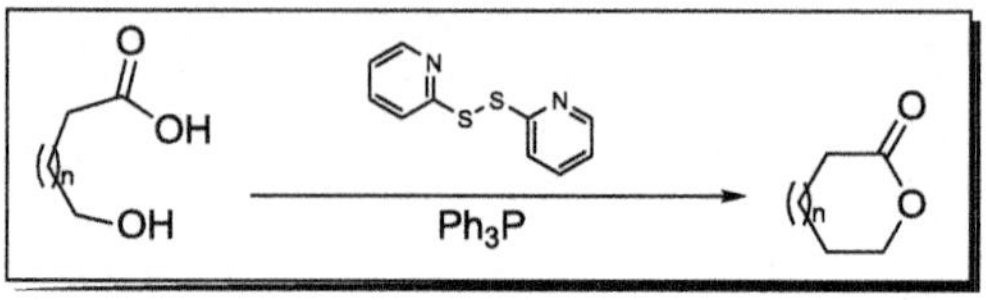

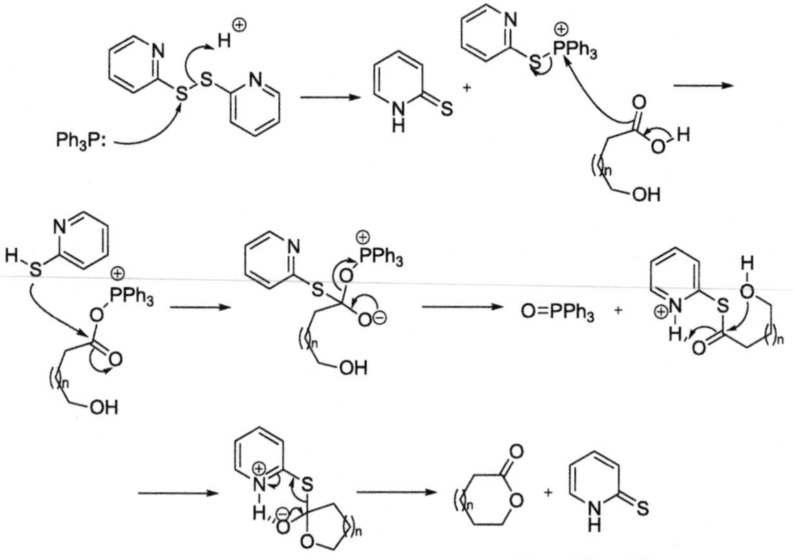

2-pyridinethione

Example 1[3]

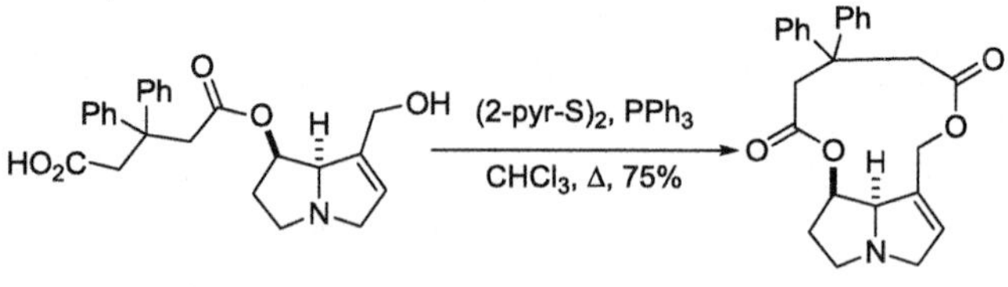

Example 2[6]

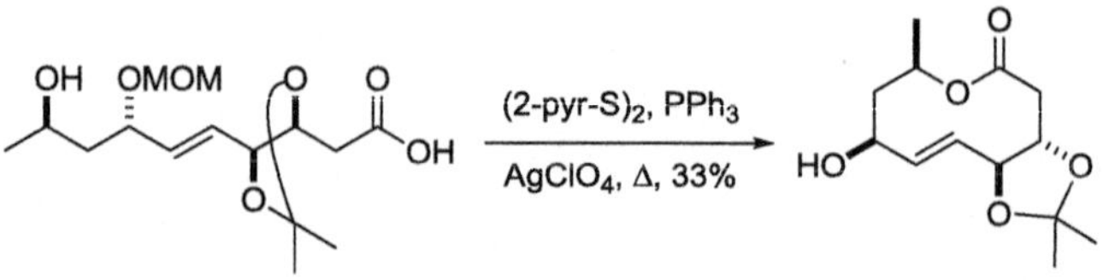

J.J. Li, *Name Reactions: A Collection of Detailed Mechanisms and Synthetic Applications*, DOI 10.1007/978-3-319-03979-4_73, © Springer International Publishing Switzerland 2014

Example 3[9]

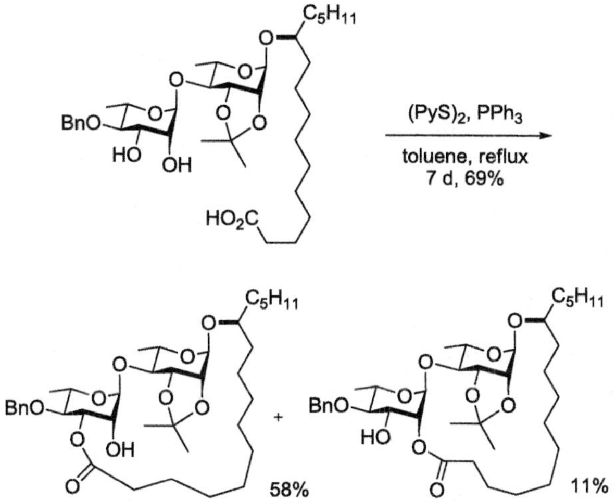

**References**

1.  Corey, E. J.; Nicolaou, K. C. *J. Am. Chem. Soc.* **1974,** *96,* 5614–5616.
2.  Nicolaou, K. C. *Tetrahedron* **1977,** *33,* 683–710. (Review).
3.  Devlin, J. A.; Robins, D. J.; Sakdarat, S. *J. Chem. Soc., Perkin Trans. 1* **1982,** 1117–1121.
4.  Barbour, R. H.; Robins, D. J. *J. Chem. Soc., Perkin Trans. 1* **1985,** 2475–2478.
5.  Barbour, R. H.; Robins, D. J. *J. Chem. Soc., Perkin Trans. 1* **1988,** 1169–1172.
6.  Andrus, M. B.; Shih, T.-L. *J. Org. Chem.* **1996,** *61,* 8780–8785.
7.  Lu, S.-F.; O'yang, Q. Q.; Guo, Z.-W.; Yu, B.; Hui, Y.-Z. *J. Org. Chem.* **1997,** *62,* 8400–8405.
8.  Sasaki, T.; Inoue, M.; Hirama, M. *Tetrahedron Lett.* **2001,** *42,* 5299–5303.
9.  Zhu, X.-M.; He, L.-L.; Yang, G.-L.; Lei, M.; Chen, S.-S.; Yang, J.-S. *Synlett* **2006,** 3510–3512.
10. Cochrane, J. R.; Yoon, D. H.; McErlean, C. S. P.; Jolliffe, K. A. *Beilstein J. Org. Chem.* **2012,** *8,* 1344–1351.

# Corey–Seebach reaction

Dithiane as a nucleophile, serving as a masked carbonyl equivalent.  This is an example of umpolung.

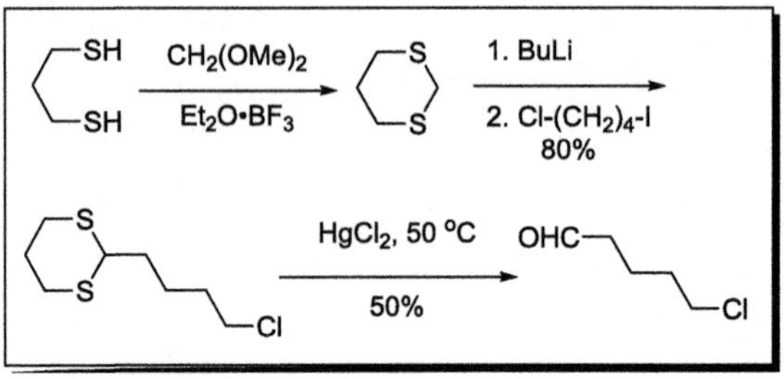

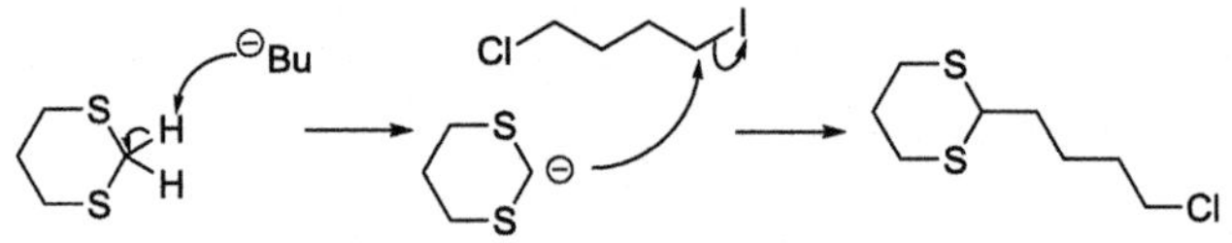

Example 1[2]

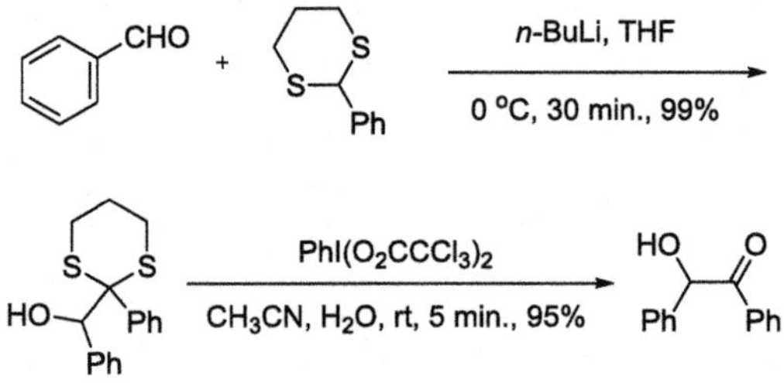

Example 2[4]

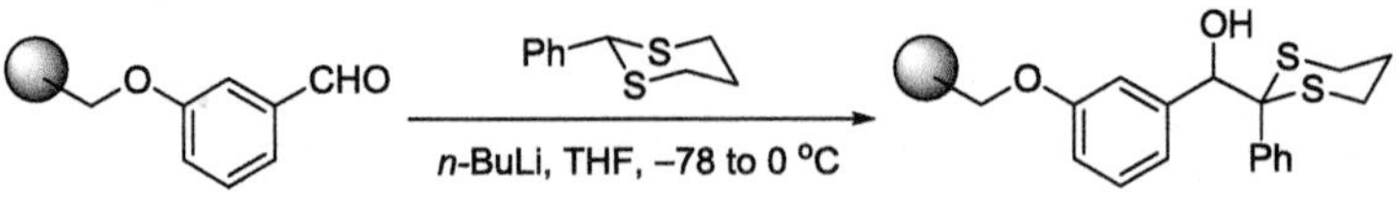

Example 3, Ethyl is infinitely different from methyl[6]

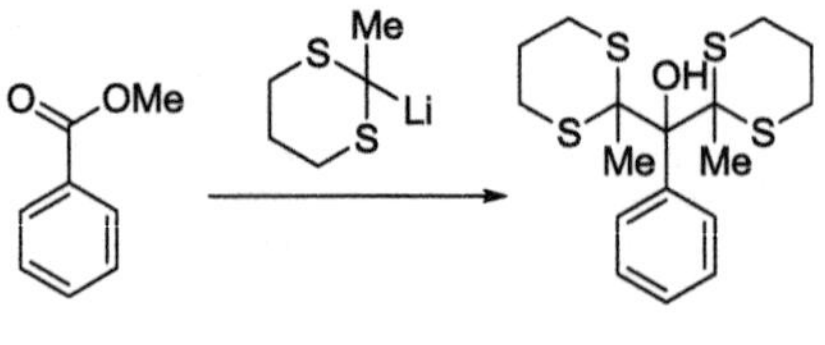

J.J. Li, *Name Reactions: A Collection of Detailed Mechanisms and Synthetic Applications*,
DOI 10.1007/978-3-319-03979-4_74, © Springer International Publishing Switzerland 2014

Example 4[8]

TFA = Trifluoroacetyl

**References**

1.  (a) Corey, E. J.; Seebach, D. *Angew. Chem. Int. Ed.* **1965**, *4*, 1075–1077. Dieter Seebach is a professor at ETH in Zürich, Switzerland. (b) Corey, E. J.; Seebach, D. *J. Org. Chem.* **1966**, *31*, 4097–4099. (c) Seebach, D.; Jones, N. R.; Corey, E. J. *J. Org. Chem.* **1968**, *33*, 300–305. (d) Seebach, D.; Corey, E. J. *Org. Synth.* **1968**, *50*, 72. (e) Seebach, D.; Corey, E. J. *J. Org. Chem.* **1975**, *40*, 231–237.

2.  Stowell, M. H. B.; Rock, R. S.; Rees, D. C.; Chan, S. I. *Tetrahedron Lett.* **1996**, *37*, 307–310.

3.  Hassan, H. H. A. M.; Tamm, C. *Helv. Chim. Acta* **1996**, *79*, 518–526.

4.  Lee, H. B.; Balasubramanian, S. *J. Org. Chem.* **1999**, *64*, 3454–3460.

5.  Bräuer, M.; Weston, J.; Anders, E. *J. Org. Chem.* **2000**, *65*, 1193–1199.

6.  Valiulin, R. A.; Kottani, R.; Kutateladze, A. G. *J. Org. Chem.* **2006**, *71*, 5047–5049.

7.  Chen, Y.-L.; Leguijt, R.; Redlich, H. *J. Carbohydrate Chem.* **2007**, *26*, 279–303.

8.  Chen, Y.-L.; Redlich, H.; Bergander, K.; Froehlich, R. *Org. Biomol. Chem.* **2007**, *5*, 3330–3339.

9.  Wright, P. M.; Myers, A. G. *Tetrahedron* **2011**, *67*, 9853–9869.

# Corey–Winter olefin synthesis

Transformation of diols to the corresponding olefins by sequential treatment with 1,1'-thiocarbonyldiimidazole (TCDI) and trimethylphosphite.  Also known as Corey–Winter reductive elimination, or Corey–Winter reductive olefination.

TCDI = Thiocarbonyldiimidazole

1,3-dioxolane-2-thione (cyclic thiocarbonate)

A mechanism involving a carbene intermediate can also be drawn and is supported by pyrolysis studies:

J.J. Li, *Name Reactions: A Collection of Detailed Mechanisms and Synthetic Applications*, DOI 10.1007/978-3-319-03979-4_75, © Springer International Publishing Switzerland 2014

Example 1[2]

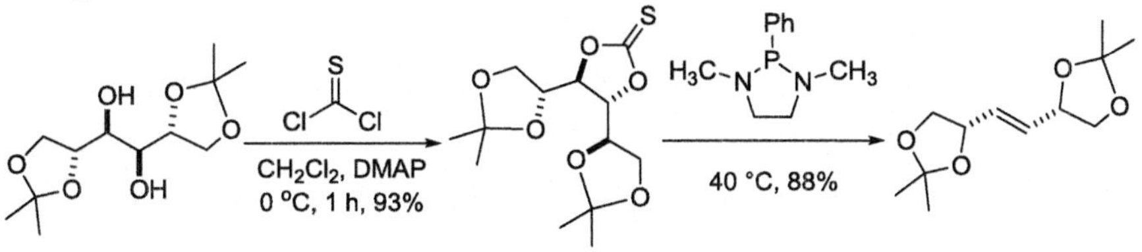

Example 2[4]

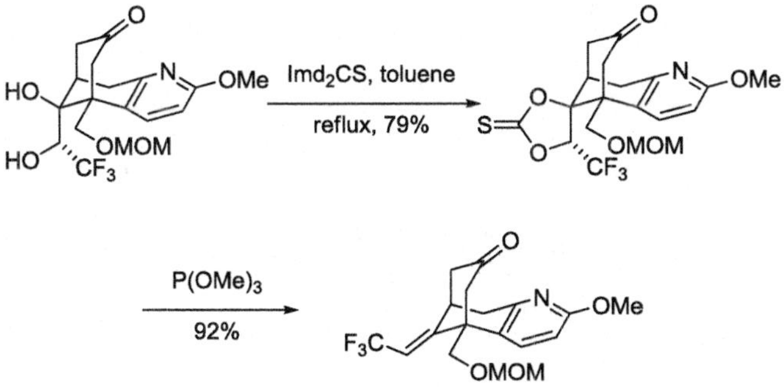

Single olefin isomer

Example 3[8]

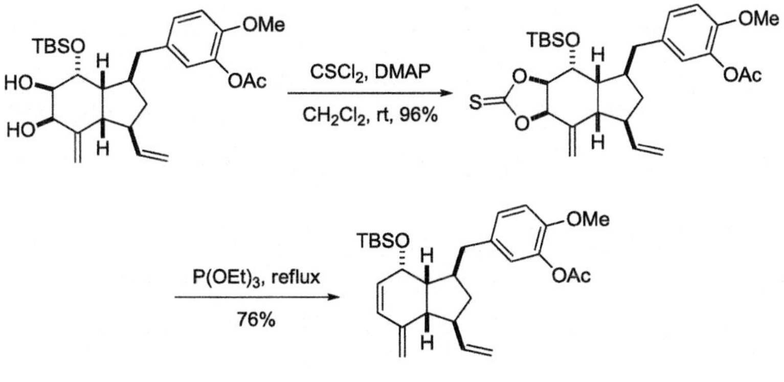

Example 4[9]

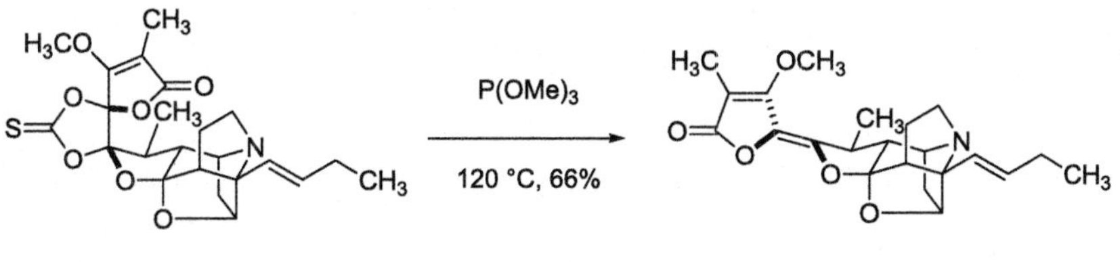

Example 5[10]

Example 6[11]

### References

1. Corey, E. J.; Winter, R. A. E. *J. Am. Chem. Soc.* **1963**, *85*, 2677–2678. Roland A. E. Winter works at Ciba Specialty Chemicals Corporation, USA.
2. Corey, E. J.; Carey, F. A.; Winter, R. A. E. *J. Am. Chem. Soc.* **1965,** *87,* 934–935.
3. Block, E. *Org. React.* **1984**, *30*, 457–566. (Review).
4. Kaneko, S.; Nakajima, N.; Shikano, M.; Katoh, T.; Terashima, S. *Tetrahedron* **1998,** *54*, 5485–5506.
5. Crich, D.; Pavlovic, A. B.; Wink, D. J. *Synth. Commun.* **1999**, *29*, 359–377.
6. Palomo, C.; Oiarbide, M.; Landa, A.; Esnal, A.; Linden, A. *J. Org. Chem.* **2001**, *66*, 4180–4186.
7. Saito, Y.; Zevaco, T. A.; Agrofoglio, L. A. *Tetrahedron* **2002**, *58*, 9593–9603.
8. Araki, H.; Inoue, M.; Katoh, T. *Synlett* **2003**, 2401–2403.
9. Brüggermann, M.; McDonald, A. I.; Overman, L. E.; Rosen, M. D.; Schwink, L.; Scott, J. P. *J. Am. Chem. Soc.* **2003**, *125*, 15284–15285.
10. Freiría, M.; Whitehead, A. J.; Motherwell, W. B. *Synthesis* **2005**, 3079–3084.
11. Mergott, D. J. *Corey–Winter olefin synthesis.* In *Name Reactions for Functional Group Transformations*; Li, J. J., Ed.; Wiley: Hoboken, NJ, **2007,** pp 354–362. (Review).
12. Xu, L.; Desai, M. C.; Liu, H. *Tetrahedron Lett.* **2009**, *50*, 552–554.
13. Iyoda, M.; Kuwatani, Y.; Nishinaga, T.; Takase, M.; Nishiuchi, T. In *Fragments of Fullerenes and Carbon Nanotube,* Petrukhina, M. A.; Scott, L. T. eds.; Wiley: Hoboken, NJ, 2012; pp 311–342. (Review).

# Criegee glycol cleavage

Vicinal diol is oxidized to the two corresponding carbonyl compounds using Pb(OAc)$_4$, (**lead tetraacetate, LTA**).

An acyclic mechanism is possible as well. It is much slower than the cyclic mechanism, but is operative when the cyclic intermediate can not form:[3]

Example 1[7]

J.J. Li, *Name Reactions: A Collection of Detailed Mechanisms and Synthetic Applications*, DOI 10.1007/978-3-319-03979-4_76, © Springer International Publishing Switzerland 2014

Example 2[9]

Pb(OAc)₄, CH₂Cl₂, K₂CO₃
−15 °C to rt, 1 h, then

Ph₃P=CHCO₂Me, CH₃CN, 80%

$E:Z$ = 10:1

Example 3[10]

Pb(OAc)₄, PhH

rt, 1.5 h, quant.

Example 4[11]

Pb(OAc)₄, EtOAc

0 °C, 5 min., 99%

## References

1   Criegee, R. *Ber.* **1931**, *64*, 260–266. Rudolf Criegee (1902–1975) was born in Düsseldorf, Germany. He earned his Ph.D. at age 23 under K. Dimroth at Würzburg. Criegee became a professor at Technical Institute at Karlsruhe in 1937, a chair in 1947. He was known for his modesty, mater-of-factness, and his breadth of interests.
2   Mihailovici, M. L.; Cekovik, Z. *Synthesis* **1970**, 209–224. (Review).
3   March, J. *Advanced Organic Chemistry,* 5th ed., Wiley: Hoboken, NJ, **2003**. (Review).
4   Danielmeier, K.; Steckhan, E. *Tetrahedron: Asymmetry* **1995**, *6*, 1181–1190.
5   Masuda, T.; Osako, K.; Shimizu, T.; Nakata, T. *Org. Lett.* **1999**, *1*, 941–944.
6   Lautens, M.; Stammers, T. A. *Synthesis* **2002**, 1993–2012.
7   Hartung, I. V.; Eggert, U.; Haustedt, L. O.; Niess, B.; Schäfer, P. M.; Hoffmann, H. M. R. *Synthesis* **2003**, 1844–1850.
8   Gaul, C.; Njardarson, J. T.; Danishefsky, S. J. *J. Am. Chem. Soc.* **2003**, *125*, 6042–6043.
9   Gorobets, E.; Stepanenko, V.; Wicha, J. *Eur. J. Org. Chem.* **2004**, 783–799.
10  Prasad, K. R.; Anbarasan, P. *Tetrahedron* **2006**, *63*, 1089–1092.
11  Prasad, K. R.; Anbarasan, P. *J. Org. Chem.* **2007**, *72*, 3155–3157.
12  Perez, L. J.; Micalizio, G. C. *Synthesis* **2008**, 627–648.

# Criegee mechanism of ozonolysis

primary ozonide (1,2,3-trioxolane)

zwitterionic peroxide        secondary ozonide (1,2,4-trioxolane)
(Criegee zwitterion)
also known as "carbonyl oxide"

Example 1[7]

Example 2[8]

## References

1.   (a) Criegee, R.; Wenner, G. *Ann.* **1949**, *564*, 9–15. (b) Criegee, R. *Rec. Chem. Prog.* **1957**, *18*, 111–120. (c) Criegee, R. *Angew. Chem.* **1975**, *87*, 765–771.
2.   Bunnelle, W. H. *Chem. Rev.* **1991**, *91*, 335–362. (Review).
3.   Kuczkowski, R. L. *Chem. Soc. Rev.* **1992**, *21*, 79–83. (Review).
4.   Marshall, J. A.; Garofalo, A. W. *J. Org. Chem.* **1993**, *58*, 3675–3680.
5.   Ponec, R.; Yuzhakov, G.; Haas, Y.; Samuni, U. *J. Org. Chem.* **1997**, *62*, 2757–2762.
6.   Dussault, P. H.; Raible, J. M. *Org. Lett.* **2000**, *2*, 3377–3379.
7.   Jiang, L.; Martinelli, J. R.; Burke, S. D. *J. Org. Chem.* **2003**, *68*, 1150–1153.
8.   Schank, K.; Beck, H.; Pistorius, S. *Helv. Chim. Acta* **2004**, *87*, 2025–2049.
9.   Coleman, B. E.; Ault, B. S. *J. Mol. Struct.* **2013**, *1031*, 138–143.

J.J. Li, *Name Reactions: A Collection of Detailed Mechanisms and Synthetic Applications*, DOI 10.1007/978-3-319-03979-4_77, © Springer International Publishing Switzerland 2014

# Curtius rearrangement

Alkyl-, vinyl-, and aryl-substituted acyl azides undergo thermal 1,2-carbon-to-nitrogen migration with extrusion of dinitrogen — the Curtius rearrangement — producing isocyantes. Reaction of the isocyanate products with nucleophiles, often *in situ*, provides carbamates, ureas, and other *N*-acyl derivatives. Alternatively, hydrolysis of the isocyanates leads to primary amines.

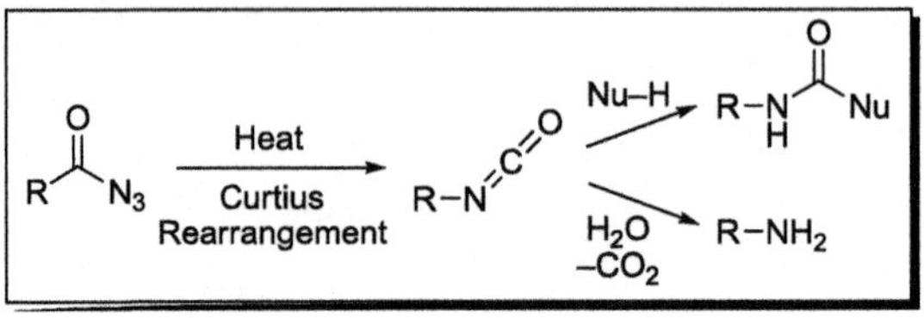

The thermal rearrangement:

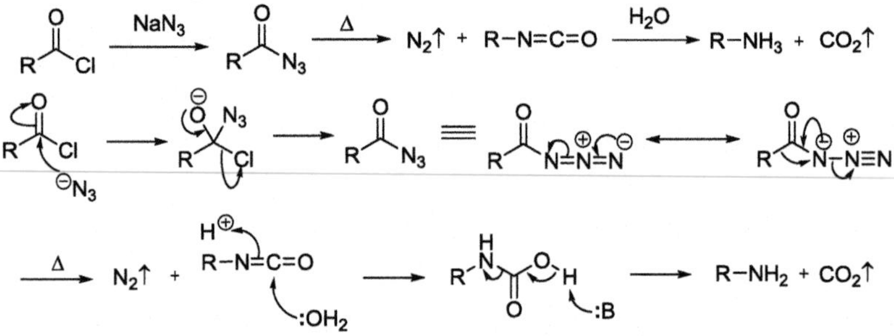

isocyanate intermediate

The photochemical rearrangement:

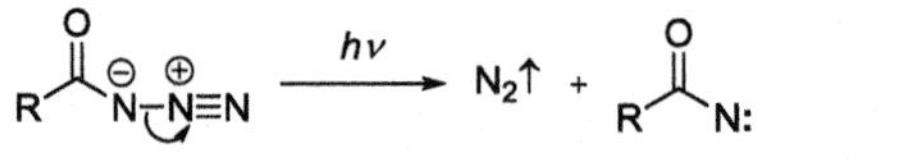

Nitrene

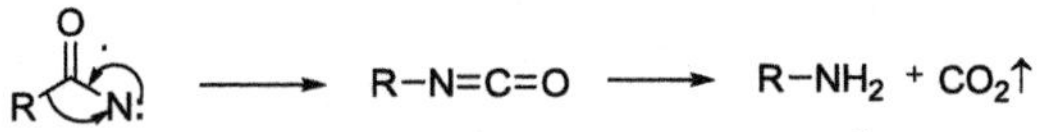

Example 1, The Shioiri–Ninomiya–Yamada modification[2]

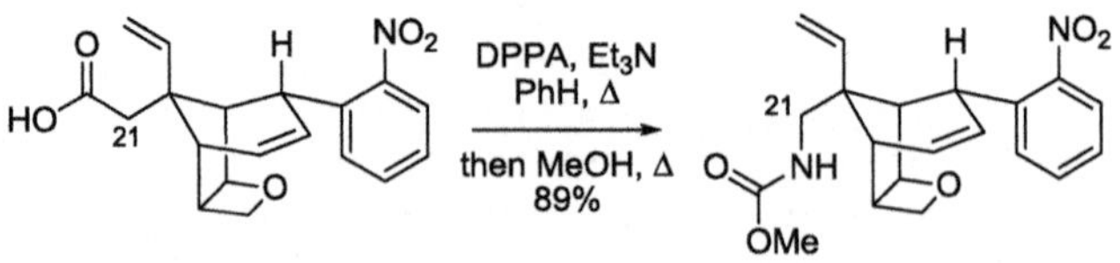

DPPA = diphenylphosphoryl azide

Example 2[3]

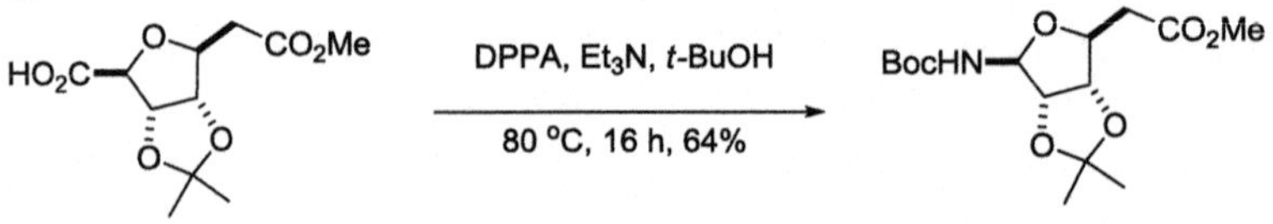

J.J. Li, *Name Reactions: A Collection of Detailed Mechanisms and Synthetic Applications*,
DOI 10.1007/978-3-319-03979-4_78, © Springer International Publishing Switzerland 2014

Example 3[4]

$$\text{(reagents)} \quad \xrightarrow[\text{EtOH, PhH, reflux, 55\%}]{\text{EtO(CO)Cl, then NaN}_3}$$

Example 4, The Weinstock variant of the Curtius rearrangement[6]

$i\text{-PrNEt}_2$, acetone, 0 °C

then NaN$_3$, rt, 12 h
75%

Example 5[7]

1. $n\text{-Bu}_3\text{SnN}_3$
PhBr, 0 °C to RT
30 min., 97%

2. $t\text{-BuOH}/o\text{-xylene}$
Δ, 6 h, 77%

Example 6, The Lebel modification[8]

2 equiv DPPA
0.1 equiv Ag$_2$CO$_3$
2 equiv K$_2$CO$_3$

PhH, Δ, 16 h, 81%

(2 equiv)

**References**

1. Curtius, T. *Ber.* **1890**, *23*, 3033–3041. Theodor Curtius (1857–1928) was born in Duisburg, Germany. He studied music before switching to chemistry under Bunsen, Kolbe, and von Baeyer before succeeding Victor Meyer as a Professor of Chemistry at Heidelberg. He discovered diazoacetic ester, hydrazine, pyrazoline derivatives, and many nitrogen-heterocycles. Curtius also sang in concerts and composed music.

2. Ng, F. W.; Lin, H.; Danishefsky, S. J. *J. Am. Chem. Soc.* **2002**, *124*, 9812–9824.

3. van Well, R. M.; Overkleeft, H. S.; van Boom, J. H.; Coop, A.; Wang, J. B.; Wang, H.; van der Marel, G. A.; Overhand, M. *Eur. J. Org. Chem.* **2003**, 1704–1710.

4. Dussault, P. H.; Xu, C. *Tetrahedron Lett.* **2004**, *45*, 7455–7457.

5. Holt, J.; Andreassen, T.; Bakke, J. M.; Fiksdahl, A. *J. Heterocycl. Chem.* **2005**, *42*, 259–264.

6. Crawley, S. L.; Funk, R. L. *Org. Lett.* **2006**, *8*, 3995–3998.

7. Tada, T.; Ishida, Y.; Saigo, K. *Synlett* **2007**, 235–238.

8. Sawada, D.; Sasayama, S.; Takahashi, H.; Ikegami, S. *Eur. J. Org. Chem.* **2007**, 1064–1068.

9. Rojas, C. M. *Curtius Rearrangements*. In *Name Reactions for Homologations-Part II*; Li, J. J., Ed.; Wiley: Hoboken, NJ, **2009**, pp 136–163. (Review).

10. Koza, G.; Keskin, S.; Özer, M. S.; Cengiz, B.; Şahin, E.; Balci, M. *Tetrahedron* **2013**, *69*, 395–409.

# Dakin oxidation

Oxidation of aryl aldehydes or aryl ketones to phenols using basic hydrogen peroxide conditions. *Cf.* A variant of the Baeyer–Villiger oxidation.

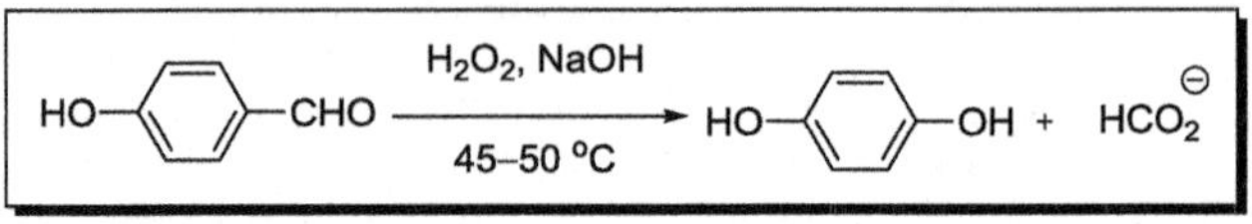

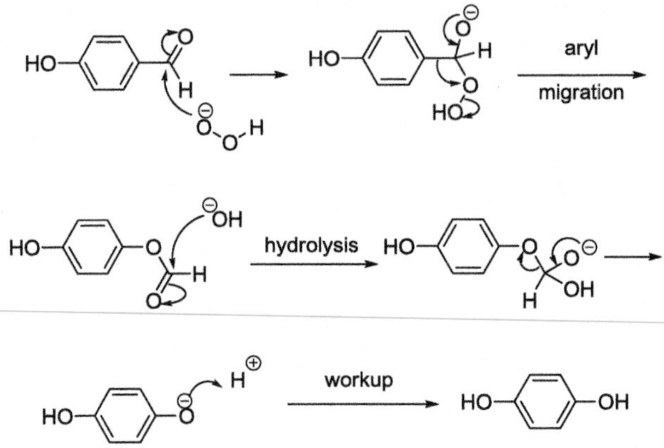

Example 1[6]

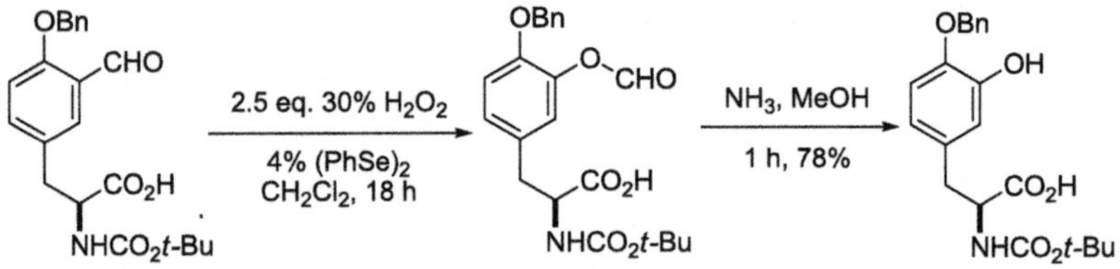

Example 2[7]

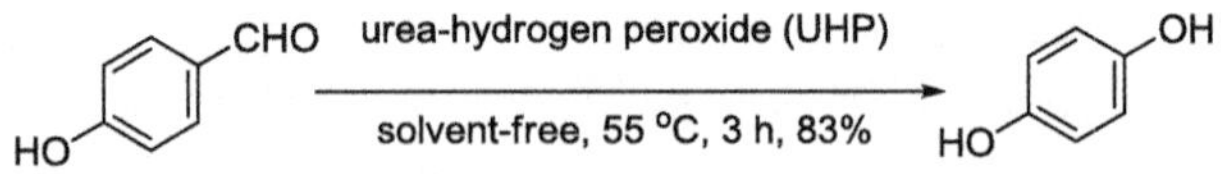

Example 3, Improved solvent-free Dakin oxidation protocol[9]

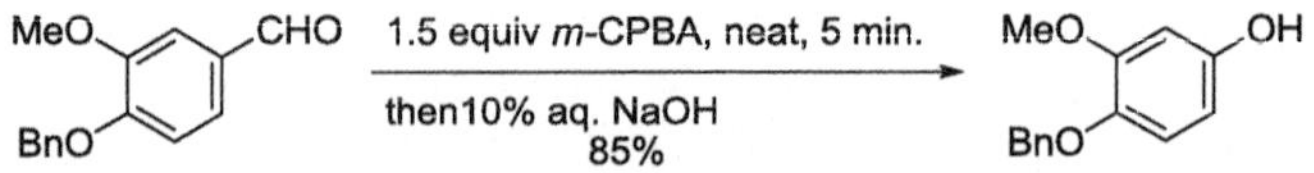

J.J. Li, *Name Reactions: A Collection of Detailed Mechanisms and Synthetic Applications*,
DOI 10.1007/978-3-319-03979-4_79, © Springer International Publishing Switzerland 2014

Example 4[10]

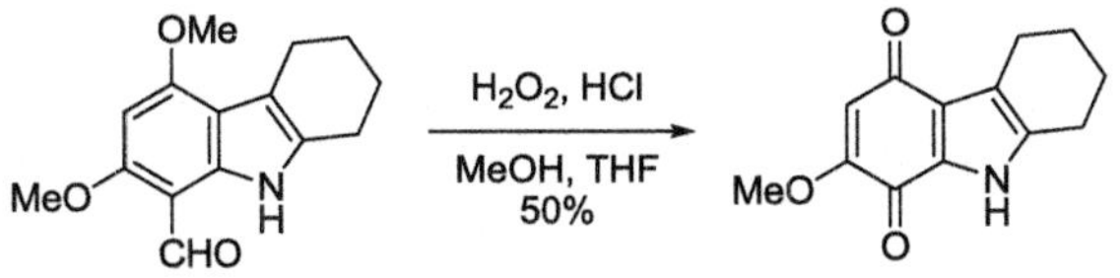

Example 5[11]

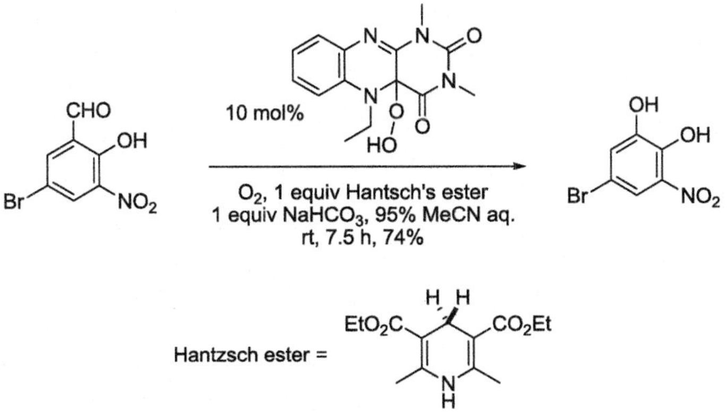

**References:**

1.  Dakin, H. D. *Am. Chem. J.* **1909**, *42*, 477–498. Henry D. Dakin (1880–1952) was born in London, England. During WWI, he invented his hypochlorite solution (Dakin's solution), which became a popular antiseptic for the treatment of wounds. After the Great War, he emmigrated to New York, where he investigated the B vitamins.

2.  Hocking, M. B.; Bhandari, K.; Shell, B.; Smyth, T. A. *J. Org. Chem.* **1982**, *47*, 4208–4215.

3.  Matsumoto, M.; Kobayashi, H.; Hotta, Y. *J. Org. Chem.* **1984**, *49*, 4740–4741.

4.  Zhu, J.; Beugelmans, R.; Bigot, A.; Singh, G. P.; Bois-Choussy, M. *Tetrahedron Lett.* **1993**, *34*, 7401–7404.

5.  Guzmán, J. A.; Mendoza, V.; García, E.; Garibay, C. F.; Olivares, L. Z.; Maldonado, L. A. *Synth. Commun.* **1995**, *25*, 2121–2133.

6.  Jung, M. E.; Lazarova, T. I. *J. Org. Chem.* **1997**, *62*, 1553–1555.

7.  Varma, R. S.; Naicker, K. P. *Org. Lett.* **1999**, *1*, 189–191.

8.  Lawrence, N. J.; Rennison, D.; Woo, M.; McGown, A. T.; Hadfield, J. A. *Bioorg. Med. Chem. Lett.* **2001**, *11*, 51–54.

9.  Teixeira da Silva, E.; Camara, C. A.; Antunes, O. A. C.; Barreiro, E. J.; Fraga, C. A. M. *Synth. Commun.* **2008**, *38*, 784–788.

10. Alamgir, M.; Mitchell, P. S. R.; Bowyer, P. K.; Kumar, N.; Black, D. St. C. *Tetrahedron* **2008**, *64*, 7136–7142.

11. Chen, S.; Foss, F. W. *Org. Lett.* **2012**, *14*, 5150–5153.

# Dakin–West reaction

The direct conversion of an α-amino acid into the corresponding α-acetylamino-alkyl methyl ketone, *via* oxazoline (azalactone) intermediates. The reaction proceeds in the presence of acetic anhydride and a base, such as pyridine, with the evolution of $CO_2$.

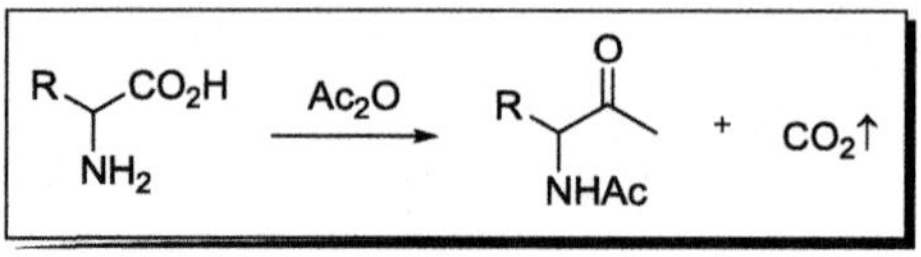

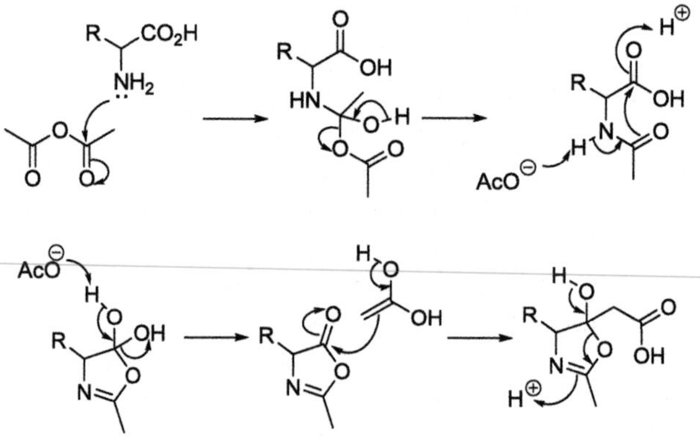

oxazolone (azalactone) intermediate

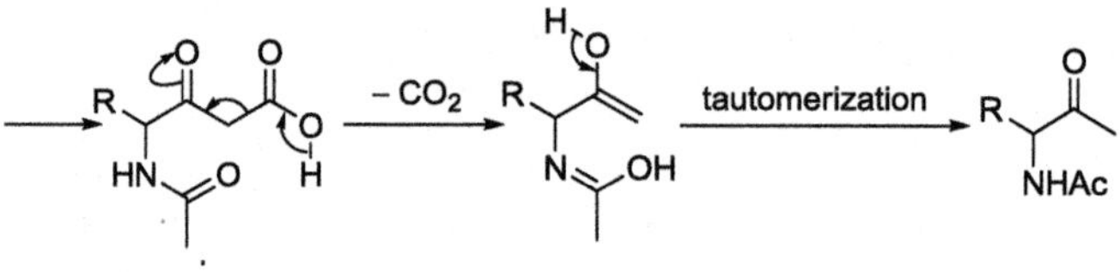

Example 1[6]

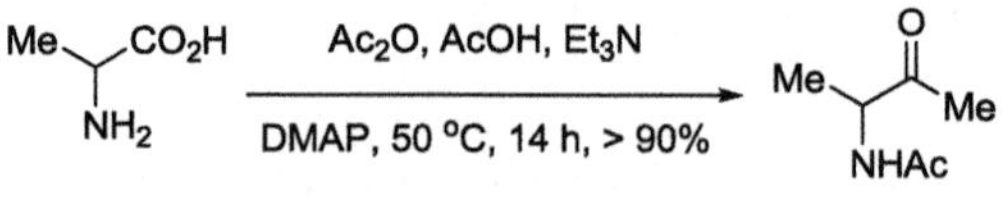

Example 2[7]

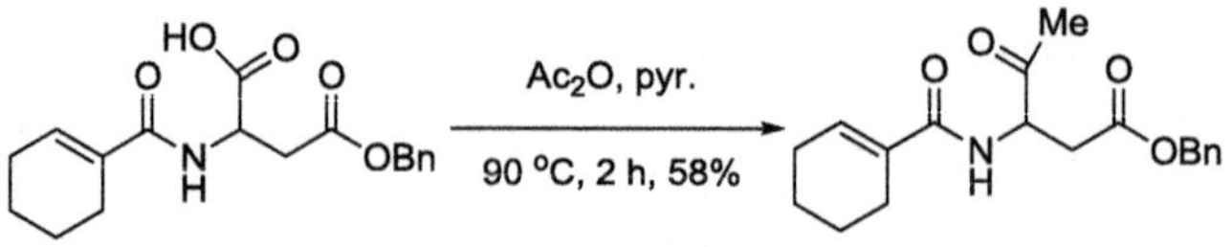

J.J. Li, *Name Reactions: A Collection of Detailed Mechanisms and Synthetic Applications*,
DOI 10.1007/978-3-319-03979-4_80, © Springer International Publishing Switzerland 2014

Example 3, A green Dakin–West reaction using the heteropoly acid catalyst, acetonitrile is a reactant[9]

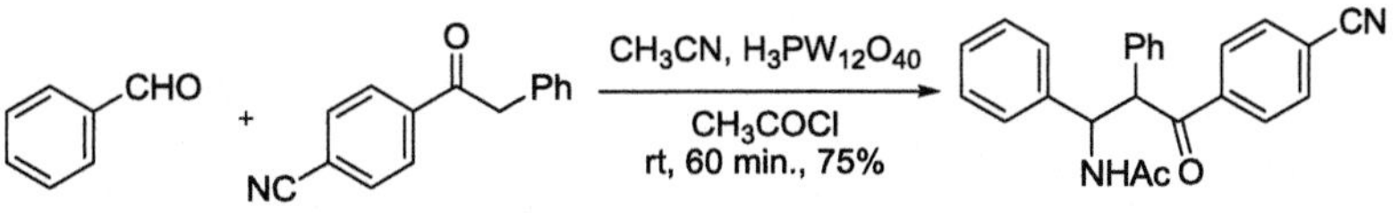

**References:**

1.  Dakin, H. D.; West, R. *J. Biol. Chem.* **1928,** *78,* 91, 745, and 757. In 1928, Henry Dakin and Rudolf West, a clinician, reported on the reaction of α-amino acids with acetic anhydride to give α-acetamido ketones *via* azalactone intermediates. Interestingly, one year before this paper by Dakin and West, Levene and Steiger had observed both tyrosine and α-phenylananine gave "abnormal" products when acetylated under these conditions.[2,3] Unfortunately, they were slow to identify the products and lost an opportunity to be immortalized by a name reaction.
2.  Buchanan, G. L. *Chem. Soc. Rev.* **1988,** *17,* 91–109. (Review).
3.  Jung, M. E.; Lazarova, T. I. *J. Org. Chem.* **1997,** *62,* 1553–1555.
4.  Kawase, M.; Hirabayashi, M.; Koiwai, H.; Yamamoto, K.; Miyamae, H. *Chem. Commun.* **1998,** 641–642.
5.  Kawase, M.; Hirabayashi, M.; Saito, S. *Recent Res. Dev. Org. Chem.* **2001,** *4,* 283–293. (Review).
6.  Fischer, R. W.; Misun, M. *Org. Proc. Res. Dev.* **2001,** *5,* 581–588.
7.  Godfrey, A. G.; Brooks, D. A.; Hay, L. A.; Peters, M.; McCarthy, J. R.; Mitchell, D. *J. Org. Chem.* **2003,** *68,* 2623–2632.
8.  Khodaei, M. M.; Khosropour, A. R.; Fattahpour, P. *Tetrahedron Lett.* **2005,** *46,* 2105–2108.
9.  Rafiee, E.; Tork, F.; Joshaghani, M. *Bioorg. Med. Chem. Lett.* **2006,** *16,* 1221–1226.
10. Tiwari, A. K.; Kumbhare, R. M.; Agawane, S. B.; Ali, A. Z.; Kumar, K. V. *Bioorg. Med. Chem. Lett.* **2008,** *18,* 4130–4132.
11. Dalla-Vechia, L.; Santos, V. G.; Godoi, M. N.; Cantillo, D.; Kappe, C. O.; Eberlin, M. N.; de Souza, R. O. M. A.; Miranda, L. S. M. *Org. Biomol. Chem.* **2012,** *10,* 9013–9020. (Mechanism).

# Danheiser annulation

Trimethylsilylcyclopentene annulation from an α,β-unsaturated ketone and trimethylsilylallene in the presence of a Lewis acid.

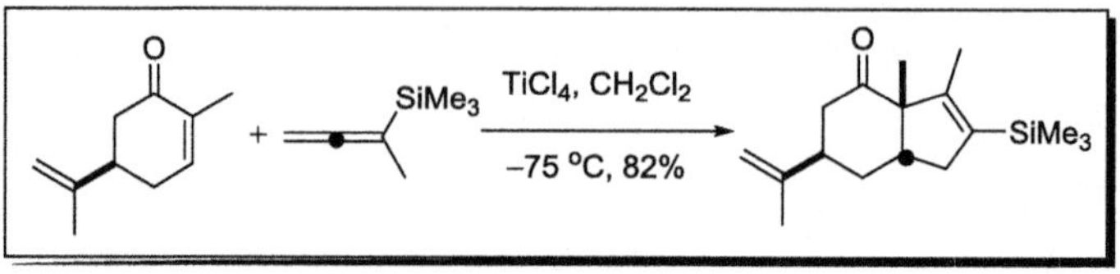

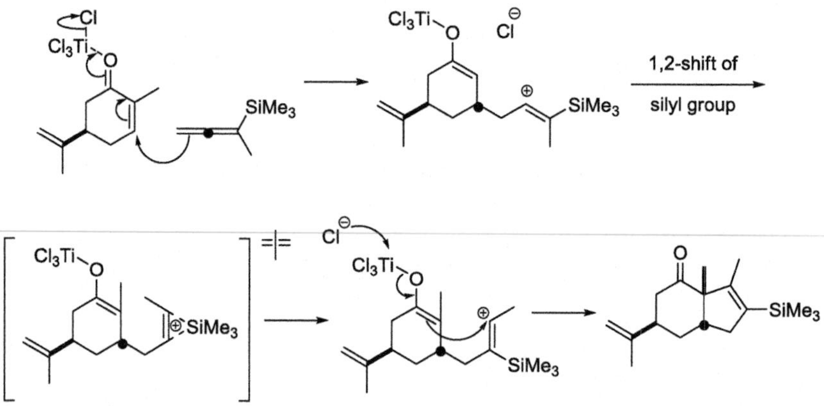

Transition State

Example 1[7]

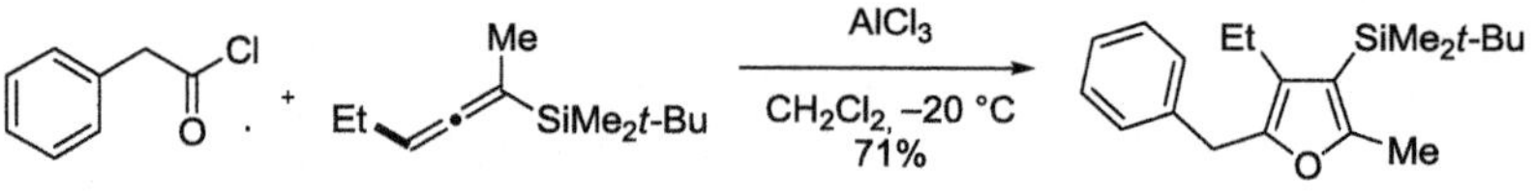

Example 2[8]

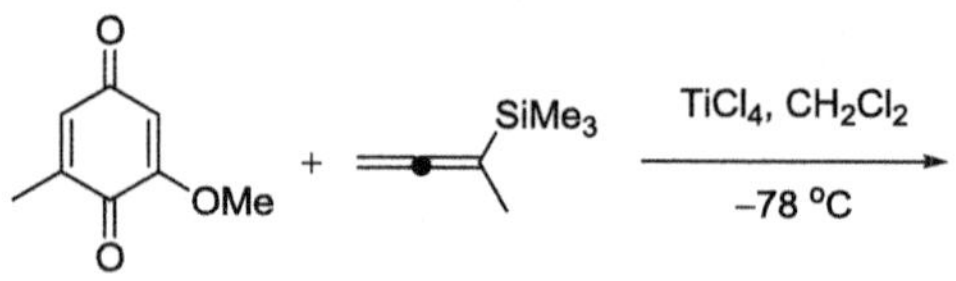

J.J. Li, *Name Reactions: A Collection of Detailed Mechanisms and Synthetic Applications*, DOI 10.1007/978-3-319-03979-4_81, © Springer International Publishing Switzerland 2014

Example 3[9]

## References

1.  Danheiser, R. L.; Carini, D. J.; Basak, A. *J. Am. Chem. Soc.* **1981**, *103*, 1604. Born in 1951, Rick L. Danheiser worked as an undergraduate under the direction of Professor Gilbert Stork and developed a method for the regiospecific alkylation of β-diketone enol ethers (the "Stork-Danheiser Alkylation"). He earned his Ph.D. in 1978 at Harvard under E. J. Corey. He began his independent academic career at MIT and now he is the Arthur C. Cope Professor of Chemistry. In addition to the Danheiser annulation, his another popular methodology is for the synthesis of aromatic and dihydroaromatic compounds based on cycloadditions of highly unsaturated conjugated molecules such as conjugated 1,3-enynes; and formal [2+2+2] cycloadditions based on propargylic ene reaction/Diels-Alder cycloaddition cascades. His group also applied these methodologies and completed the total synthesis of natural products. He is also a superb teacher, a maestro in the classroom, winning many teaching awards at MIT.
2.  Danheiser, R. L.; Carini, D. J.; Fink, D. M.; Basak, A. *Tetrahedron* **1983**, *39*, 935.
3.  Danheiser, R. L.; Kwasigroch, C. A.; Tsai, Y.-M. *J. Am. Chem. Soc.* **1985**, *107*, 7233.
4.  Danheiser, R. L.; Carini, D. J.; Kwasigroch, C. A. *J. Org. Chem.* **1986**, *51*, 3870.
5.  Danheiser, R. L.; Tsai, Y.-M.; Fink, D. M. *Org. Synth.* **1988**, *66*, 1.
6.  Danheiser, R. L.; Dixon, B. R.; Gleason, R. W. *J. Org. Chem.* **1992**, *57*, 6094.
7.  Sibi, M. P.; Christensen, J. W.; Kim, S.; Eggen, F.; Stessman, C.; Oien, L. *Tetrahedron Lett.* **1995**, *36*, 6209–6212.
8.  Engler, T. A.; Agrios, K.; Reddy, J. P.; Iyengar, R. *Tetrahedron Lett.* **1996**, *37*, 327.
9.  Friese, J. C.; Krause, S.; Schäfer, H. J. *Tetrahedron Lett.* **2002**, *43*, 2683.
10. Peese, K. M. *Danheiser Annulation*, In *Name Reactions in Carbocyclic Ring Formations*, Li, J. J., Ed., Wiley: Hoboken, NJ, **2010**, pp 72–92.

## Darzens condensation

α,β-Epoxy esters (glycidic esters) from base-catalyzed condensation of α-haloesters with carbonyl compounds.

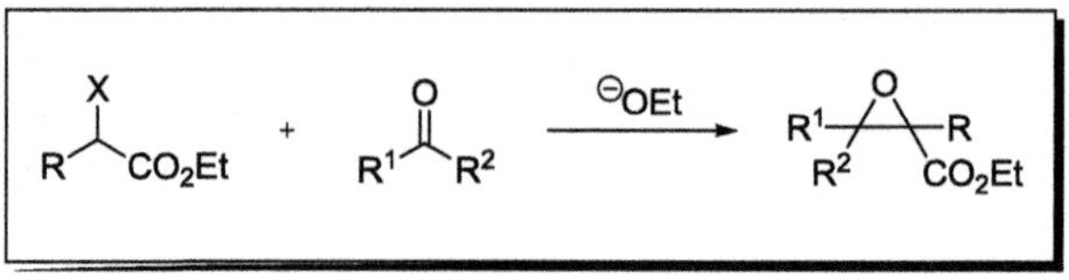

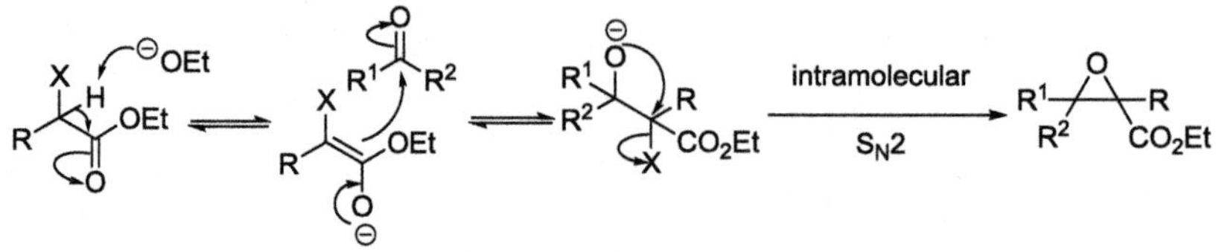

Example 1[4]

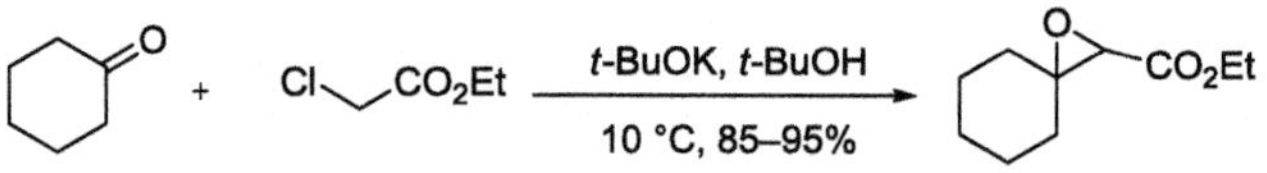

Example 2[6]

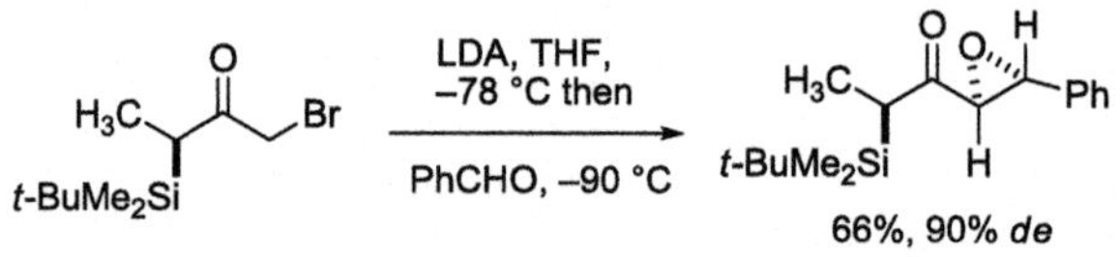

Example 3, the phenyl ring substituting for the carbonyl to acidify the protons[10]

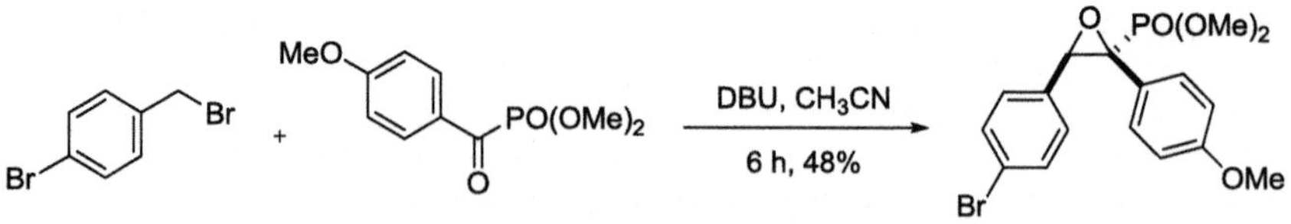

Example 4[11]

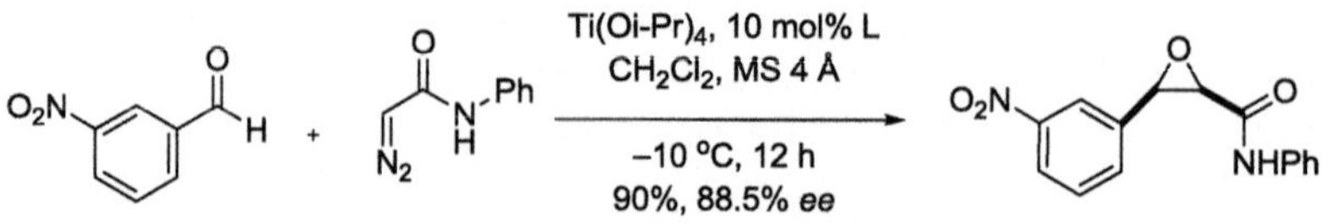

J.J. Li, *Name Reactions: A Collection of Detailed Mechanisms and Synthetic Applications*,
DOI 10.1007/978-3-319-03979-4_82, © Springer International Publishing Switzerland 2014

L =

## References

1   Darzens, G. A. *Compt. Rend. Acad. Sci.* **1904,** *139,* 1214–1217.  George Auguste Darzens (1867–1954), born in Moscow, Russia, studied at École Polytechnique in Paris and stayed there as a professor.

2   Newman, M. S.; Magerlein, B. J. *Org. React.* **1949,** *5*, 413–441. (Review).

3   Ballester, M. *Chem. Rev.* **1955,** *55*, 283–300. (Review).

4   Hunt, R. H.; Chinn, L. J.; Johnson, W. S. *Org. Syn. Coll. IV,* **1963,** 459.

5   Rosen, T. *Darzens Glycidic Ester Condensation* In *Comprehensive Organic Synthesis;* Trost, B. M.; Fleming, I., Eds.; Pergamon: Oxford, **1991,** *Vol. 2*, pp 409–439. (Review).

6   Enders, D.; Hett, R. *Synlett* **1998,** 961–962.

7   Davis, F. A.; Wu, Y.; Yan, H.; McCoull, W.; Prasad, K. R. *J. Org. Chem.* **2003,** *68*, 2410–2419.

8   Myers, B. J. *Darzens Glycidic Ester Condensation*. In *Name Reactions in Heterocyclic Chemistry*; Li, J. J., Ed.; Wiley: Hoboken, NJ, **2005,** pp 15–21. (Review).

9   Achard, T. J. R.; Belokon, Y. N.; Ilyin, M.; Moskalenko, M.; North, M.; Pizzato, F. *Tetrahedron Lett.* **2007,** *48*, 2965–2969.

10  Demir, A. S.; Emrullahoglu, M.; Pirkin, E.; Akca, N. *J. Org. Chem.* **2008,** *73*, 8992–8997.

11  Liu, G.; Zhang, D.; Li, J.; Xu, G.; Sun, J. *Org. Biomol. Chem.* **2013,** *11*, 900–904.

## Delépine amine synthesis

The reaction between alkyl halides and hexamethylenetetramine, followed by cleavage of the resulting salt with ethanolic HCl to yield primary amines.

*Cf.* Gabriel synthesis, where the product is also an amine and Sommelet reaction, where the product is an aldehyde. The Delépine works well for active halides such as benzyl, allyl halides, and α-halo-ketones.

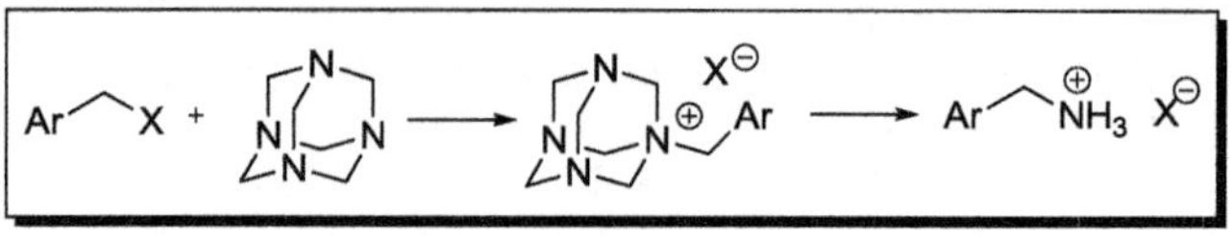

Hexamethylenetetramine

$$[ArCH_2C_6H_{12}N_4]^+X^- \quad + \quad 3\,HCl \quad + 6\,H_2O$$

$$\longrightarrow \quad ArCH_2NH_2 \cdot HX \quad + \quad 6\,CH_2O \quad + \quad 3\,NH_4Cl$$

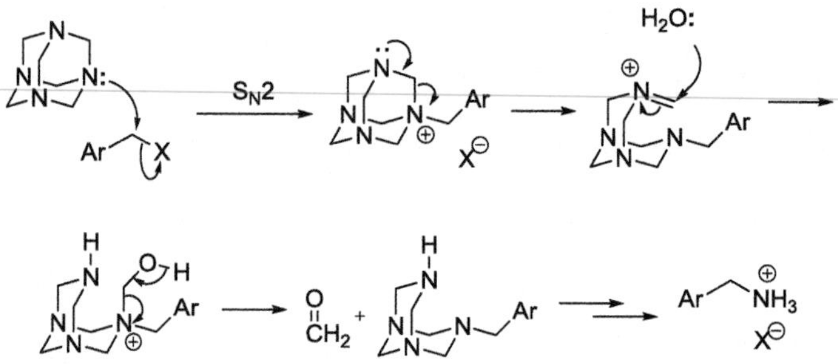

Example 1[3]

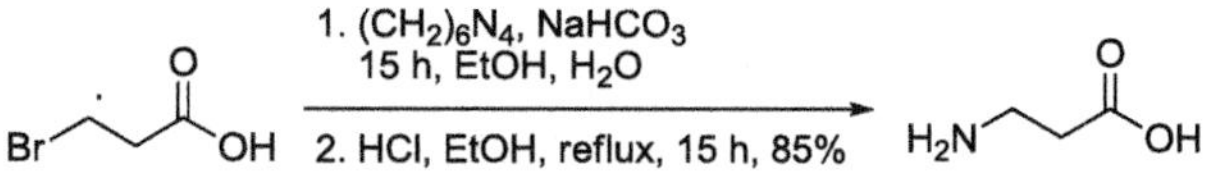

1. (CH₂)₆N₄, NaHCO₃
   15 h, EtOH, H₂O
2. HCl, EtOH, reflux, 15 h, 85%

Example 2[7]

hexamethylenetetramine

CH₂Cl₂, reflux, 5 h, then
–30 °C

conc. HCl, EtOH

reflux, 1 d
78%, 2 steps

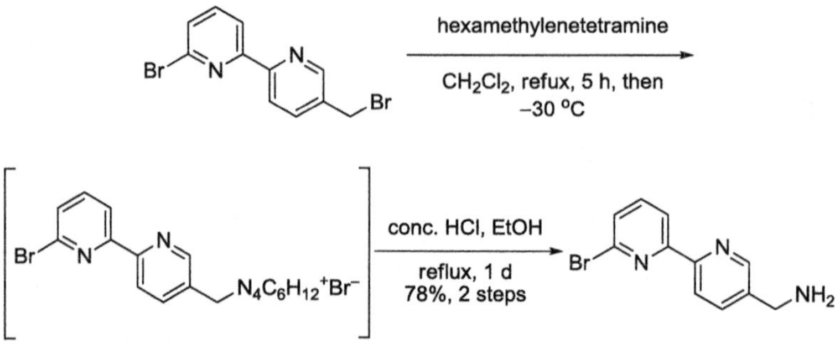

J.J. Li, *Name Reactions: A Collection of Detailed Mechanisms and Synthetic Applications*, DOI 10.1007/978-3-319-03979-4_83, © Springer International Publishing Switzerland 2014

## Example 3[8]

## Example 4[9]

**References**
1. (a) Delépine, M. *Bull. Soc. Chim. Paris* **1895,** *13*, 352–355; (b) Delépine, M. *Bull. Soc. Chim. Paris* **1897,** *17*, 292–295. Stephe Marcel Delépine (1871–1965) was born in St. Martin le Gaillard, France. He was a professor at the Collège de France after working for M. Bertholet at that institute. Delépine's long and fruitful career in science encompassed organic chemistry, inorganic chemistry, and pharmacy.
2. Galat, A.; Elion, G. *J. Am. Chem. Soc.* **1939,** *61*, 3585–3586.
3. Wendler, N. L. *J. Am. Chem. Soc.* **1949,** *71*, 375–384.
4. Quessy, S. N.; Williams, L. R.; Baddeley, V. G. *J. Chem. Soc., Perkin Trans. 1* **1979,** 512–516.
5. Blažzević, N.; Kolnah, D.; Belin, B.; Šunjić, V.; Kafjež, F. *Synthesis* **1979,** 161–176. (Review).
6. Henry, R. A.; Hollins, R. A.; Lowe-Ma, C.; Moore, D. W.; Nissan, R. A. *J. Org. Chem.* **1990,** *55*, 1796–1801.
7. Charbonnière, L. J.; Weibel, N.; Ziessel, R. *Synthesis* **2002,** 1101–1109.
8. Xie, L.; Yu, D.; Wild, C.; Allaway, G.; Turpin, J.; Smith, P. C.; Lee, K.-H. *J. Med. Chem.* **2004,** *47*, 756–760.
9. Loughlin, W. A.; Henderson, L. C.; Elson, K. E.; Murphy, M. E. *Synthesis* **2006,** 1975–1980.

# de Mayo reaction

[2 + 2]-Photochemical cyclization of enones with olefins is followed by a retro-aldol reaction to give 1,5-diketones.

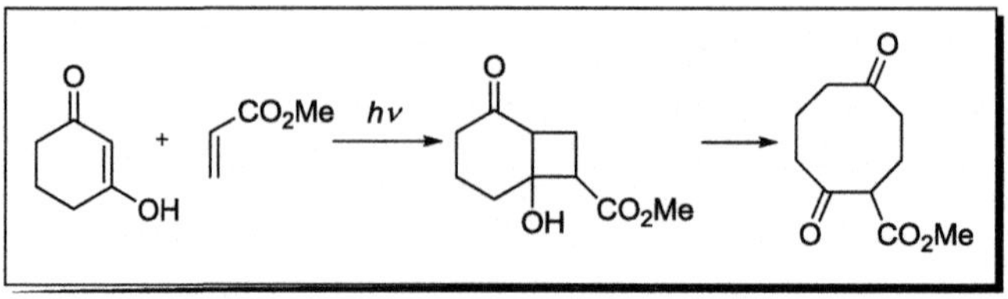

Head-to-tail alignment gives the major product:[1b]

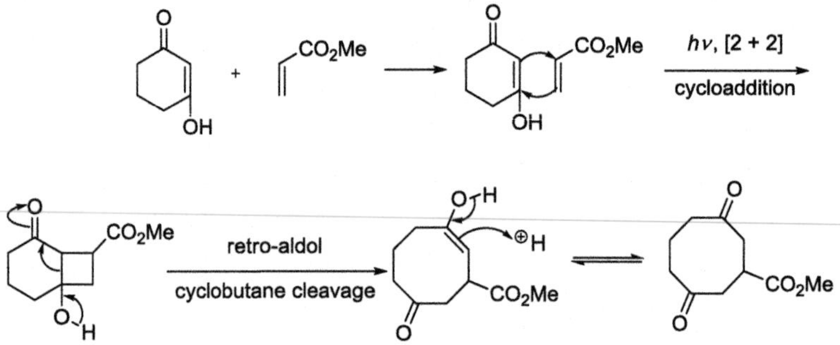

Head-to-head alignment gives the minor regioisomer:

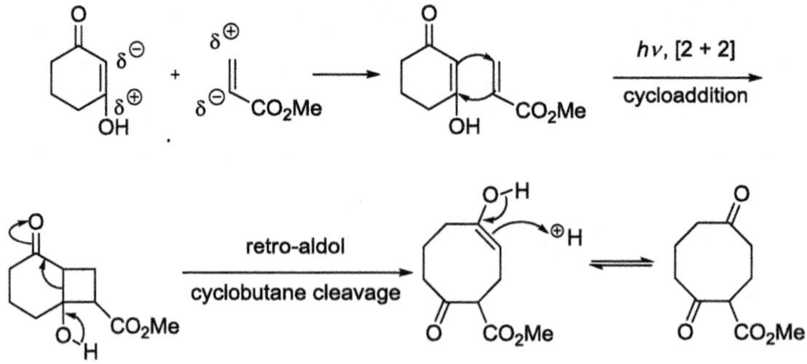

Example 1[3]

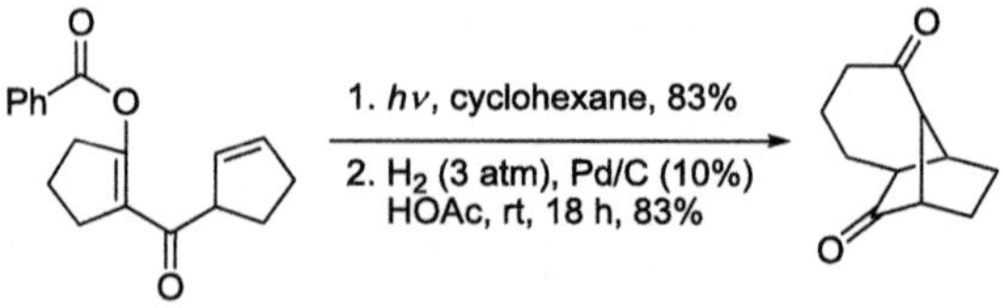

1. $hv$, cyclohexane, 83%

2. $H_2$ (3 atm), Pd/C (10%)
   HOAc, rt, 18 h, 83%

J.J. Li, *Name Reactions: A Collection of Detailed Mechanisms and Synthetic Applications*,
DOI 10.1007/978-3-319-03979-4_84, © Springer International Publishing Switzerland 2014

Example 2[6]

Example 3[9]

Example 4[10]

| R = H | 70% | 100 | : | 0 |
| R = Me | 58% | 50 | : | 50 |
| R = t-Bu | 72% | 0 | : | 100 |

**References**
1.  (a) de Mayo, P.; Takeshita, H.; Sattar, A. B. M. A. *Proc. Chem. Soc., London* **1962,** 119.  Paul de Mayo received his doctorate from Sir Derek Barton at Birkbeck College, University of London.  He later became a professor at the University of Western Ontario in London, Ontario, Canada, where he discovered the de Mayo reaction.  (b) Challand, B. D.; Hikino, H.; Kornis, G.; Lange, G.; de Mayo, P. *J. Org. Chem.* **1969,** *34,* 794–806.
2.  de Mayo, P. *Acc. Chem. Res.* **1971,** *4,* 41–48. (Review).
3.  Oppolzer, W.; Godel, T. *J. Am. Chem. Soc.* **1978,** *100,* 2583–2584.
4.  Oppolzer, W. *Pure Appl. Chem.* **1981,** *53,* 1181–1201. (Review).
5.  Kaczmarek, R.; Blechert, S. *Tetrahedron Lett.* **1986,** *27,* 2845–2848.
6.  Disanayaka, B. W.; Weedon, A. C. *J. Org. Chem.* **1987,** *52,* 2905–2910.
7.  Crimmins, M. T.; Reinhold, T. L. *Org. React.* **1993,** *44,* 297–588. (Review).
8.  Quevillon, T. M.; Weedon, A. C. *Tetrahedron Lett.* **1996,** *37,* 3939–3942.
9.  Minter, D. E.; Winslow, C. D. *J. Org. Chem.* **2004,** *69,* 1603–1606.
10. Kemmler, M.; Herdtweck, E.; Bach, T. *Eur. J. Org. Chem.* **2004,** 4582–4595.
11. Wu, Y.-J. *Name Reactions in Carbocyclic Ring Formations,* Li, J. J., Ed., Wiley: Hoboken, NJ, 2010; pp 451–488.

## Demjanov rearrangement

Carbocation rearrangement of primary amines *via* diazotization to give alcohols through C–C bond migration.

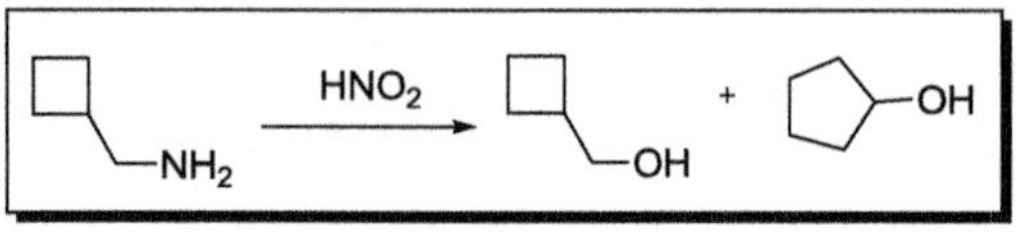

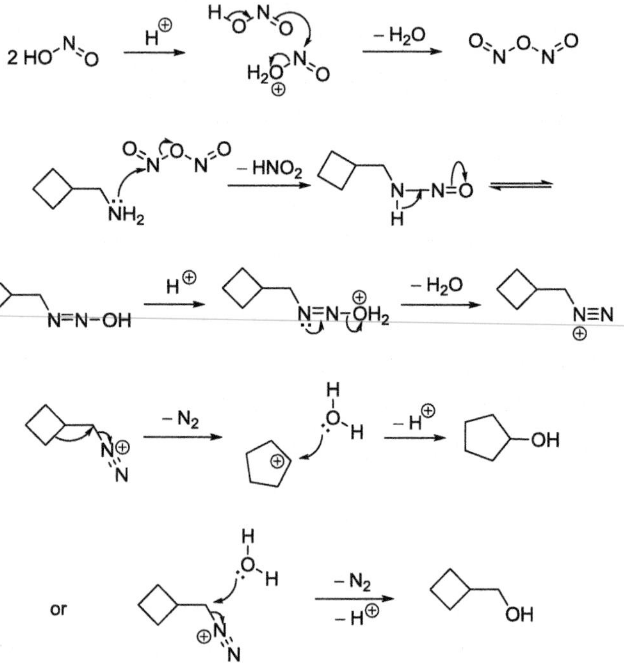

Example 1[3]

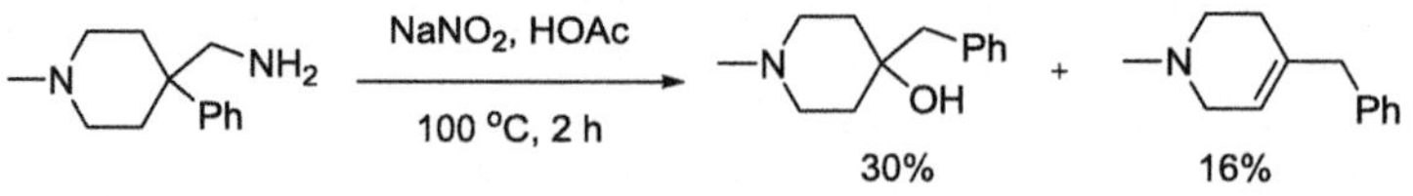

Example 2[6]

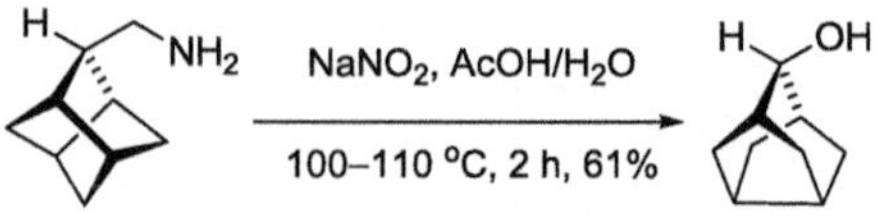

J.J. Li, *Name Reactions: A Collection of Detailed Mechanisms and Synthetic Applications*, DOI 10.1007/978-3-319-03979-4_85, © Springer International Publishing Switzerland 2014

Example 3[7]

Example 4[8]

**References**

1. Demjanov, N. J.; Lushnikov, M. *J. Russ. Phys. Chem. Soc.* **1903**, *35*, 26–42. Nikolai J. Demjanov (1861–1938) was a Russian chemist.
2. Smith, P. A. S.; Baer, D. R. *Org. React.* **1960**, *11*, 157–188. (Review).
3. Diamond, J.; Bruce, W. F.; Tyson, F. T. *J. Org. Chem.* **1965**, *30*, 1840–184.
4. Kotani, R. *J. Org. Chem.* **1965**, *30*, 350–354.
5. Diamond, J.; Bruce, W. F.; Tyson, F. T. *J. Org. Chem.* **1965**, *30*, 1840–1844.
6. Nakazaki, M.; Naemura, K.; Hashimoto, M. *J. Org. Chem.* **1983**, *48*, 2289–2291.
7. Fattori, D.; Henry, S.; Vogel, P. *Tetrahedron* **1993**, *49*, 1649–1664.
8. Kürti, L.; Czakó, B.; Corey, E. J. *Org. Lett.* **2008**, *10*, 5247–5250.
9. Curran, T. T. *Demjanov and Tiffeneau–Demjanov Rearrangement*. In *Name Reactions for Homologations-Part II*; Li, J. J., Ed.; Wiley: Hoboken, NJ, **2009**, pp 2–32. (Review).

## *Tiffeneau–Demjanov rearrangement*

Carbocation rearrangement of β-aminoalcohols *via* diazotization to afford carbonyl compounds through C–C bond migration.

Step 1, Generation of $N_2O_3$

*N*-nitrosonium ion

Step 2, Transformation of amine to diazonium salt

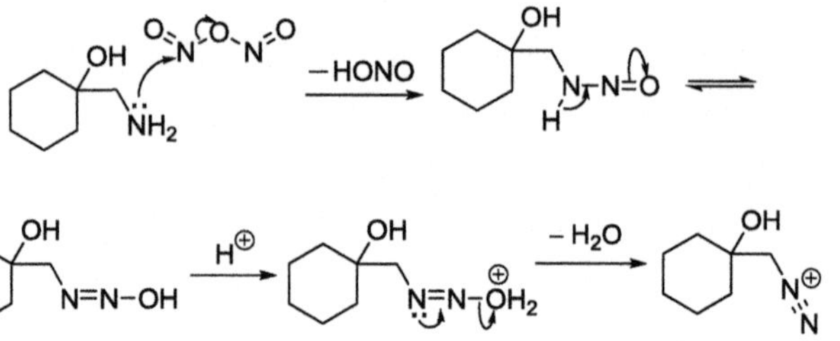

Step 3, Ring-expansion *via* rearrangement

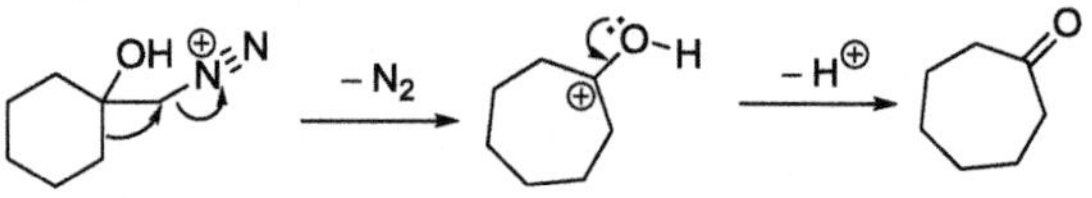

Example 1[5]

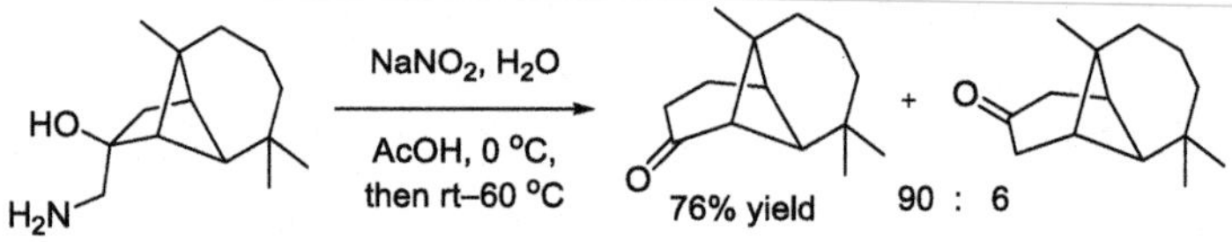

Example 2[6]

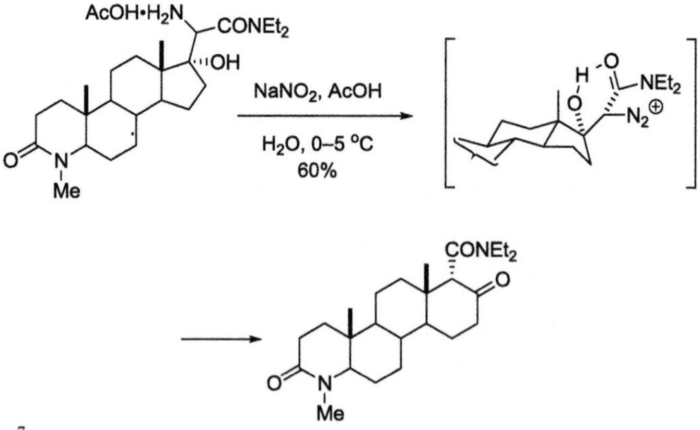

Example 3[7]

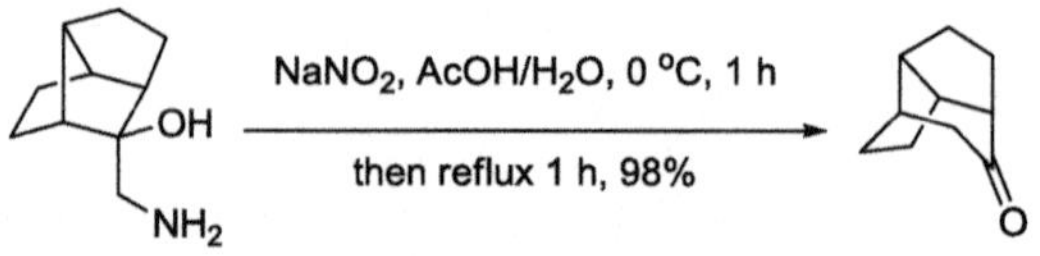

Example 4[9]

**References**
1.  Tiffeneau, M.; Weill, P.; Tehoubar, B. *Compt. Rend.* **1937**, *205*, 54–56.
2.  Smith, P. A. S.; Baer, D. R. *Org. React.* **1960**, *11*, 157–188. (Review).
3.  Parham, W. E.; Roosevelt, C. S. *J. Org. Chem.* **1972**, *37*, 1975–1979.
4.  Jones, J. B.; Price, P. *Tetrahedron* **1973**, *29*, 1941–1947.
5.  Miyashita, M.; Yoshikoshi, A. *J. Am. Chem Soc.* **1974**, *96*, 1917–1925.
6.  Steinberg, N. G.; Rasmusson, G. H.; Reynolds, G. F.; Hirshfield, J. H.; Arison, B. H. *J. Org. Chem.* **1984**, *49*, 4731–4733.
7.  Stern, A. G.; Nickon, A. *J. Org. Chem.* **1992**, *57*, 5342–5352.
8.  Fattori, D.; Henry, S.; Vogel, P. *Tetrahedron* **1993**, *49*, 1649–1664.
9.  Chow, L.; McClure, M.; White, J. *Org. Biomol. Chem.* **2004**, *2*, 648–650.
10. Curran, T. T. *Demjanov and Tiffeneau–Demjanov Rearrangement*. In *Name Reactions for Homologations-Part II*; Li, J. J., Ed.; Wiley: Hoboken, NJ, **2009**, pp 293–304. (Review).
11. Shi, L.; Meyer, K.; Greaney, M. F. *Angew. Chem. Int. Ed.* **2010**, *49*, 9250–9253,

## Dess–Martin periodinane oxidation

Oxidation of alcohols to the corresponding carbonyl compounds using triacetoxyperiodinane. The Dess–Martin periodinane, 1,1,1-triacetoxy-1,1-dihydro-1,2-benziodoxol-3(1$H$)-one, is one of the most useful oxidant for the conversion of primary and secondary alcohols to their corresponding aldehyde or ketone products, respectively.

Preparation,[1,2] the oxone preparation is much safer and easier than $KBrO_3$. The IBX intermediate that comes out of it has proven to be far less explosive[12]

However, The Dess–Martin periodinane is hydrolyzed by moisture to o-iodoxybenzoic acid (IBX), which is a more powerful oxidizing agent[3]

Mechanism[1]

J.J. Li, *Name Reactions: A Collection of Detailed Mechanisms and Synthetic Applications*,
DOI 10.1007/978-3-319-03979-4_86, © Springer International Publishing Switzerland 2014

Example 1[6]

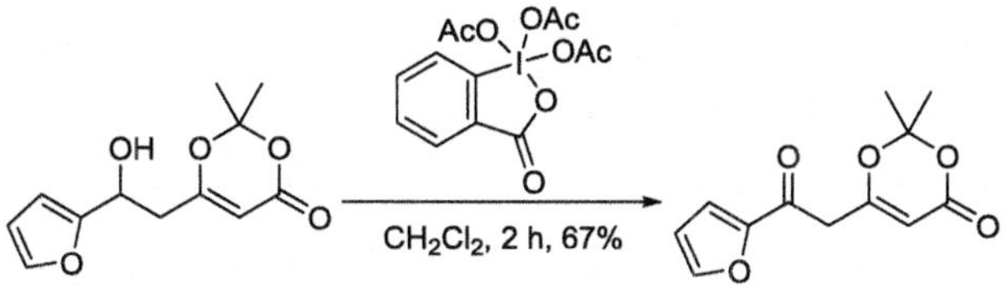

Example 2, An atypical Dess–Martin periodinane reactivity[7]

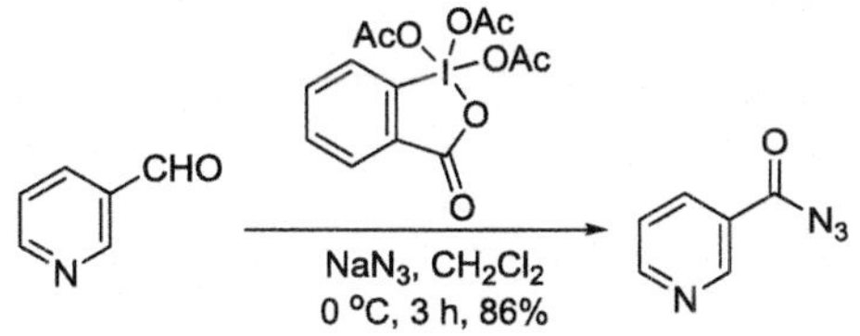

Example 3[10]

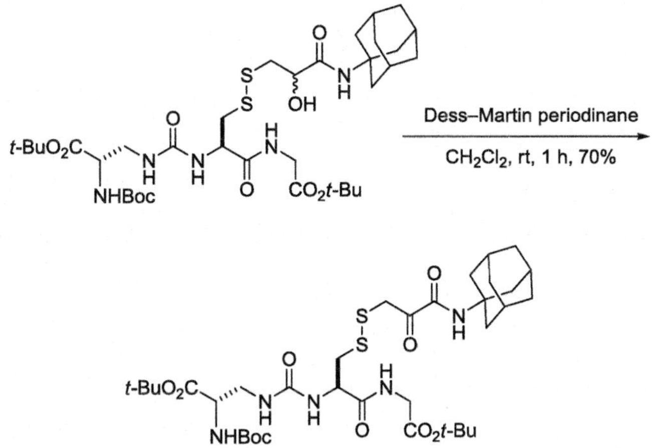

Example 4[11]

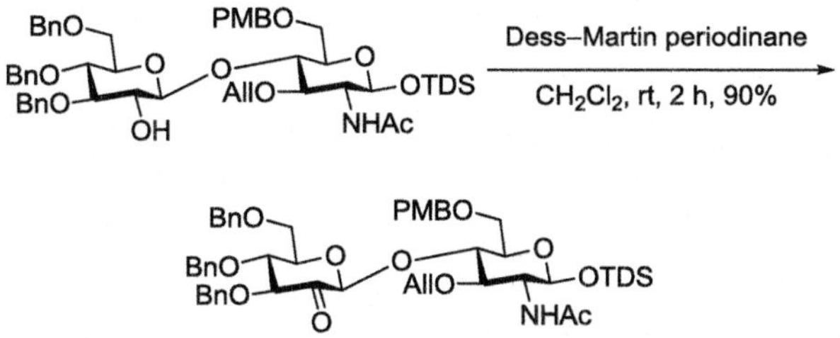

Example 5[12]

### References
1.  (a) Dess, D. B.; Martin, J. C. *J. Org. Chem.* **1983**, *48*, 4155–4156. James Cullen (J. C.) Martin (1928–1999) had a distinguished career spanning 36 years both at the University of Illinois at Urbana-Champaign and Vanderbilt University. J. C.'s formal training in physical organic chemistry with Don Pearson at Vanderbilt and P. D. Bartlett at Harvard prepared him well for his early studies on carbocations and radicals. However, it was his interest in understanding the limits of chemical bonding that led to his landmark investigations into hypervalent compounds of the main group elements. Over a 20-year period the Martin laboratories successfully prepared unprecedented chemical structures from sulfur, phosphorus, silicon and bromine while the ultimate "Holy Grail" of stable pentacoordinate carbon remained elusive. Although most of these studies were driven by J. C.'s fascination with unusual bonding schemes, they were not without practical value. Two hypervalent compounds, Martin's sulfurane (for dehydration) and the Dess–Martin periodinane have found widespread application in synthetic organic chemistry. J. C. Martin and his student Daniel Dess developed this methodology at the University of Illinois at Urbana. (Martin's biography was kindly supplied by Prof. Scott E. Denmark). (b) Dess, D. B.; Martin, J. C. *J. Am. Chem. Soc.* **1991**, *113*, 7277–7287.
2.  Ireland, R. E.; Liu, L. *J. Org. Chem.* **1993**, *58*, 2899.
3.  Meyer, S. D.; Schreiber, S. L. *J. Org. Chem.* **1994**, *59*, 7549–7552.
4.  Frigerio, M.; Santagostino, M.; Sputore, S. *J. Org. Chem.* **1999**, *64*, 4537–4538.
5.  Nicolaou, K. C.; Zhong, Y.-L.; Baran, P. S. *Angew. Chem. Int. Ed.* **2000**, *39*, 622–625.
6.  Bach, T.; Kirsch, S. *Synlett* **2001**, 1974–1976.
7.  Bose, D. S.; Reddy, A. V. N. *Tetrahedron* **2003**, *44*, 3543–3545.
8.  Tohma, H.; Kita, Y. *Adv. Synth. Cat.* **2004**, *346*, 111–124. (Review).
9.  Holsworth, D. D. *Dess–Martin oxidation.* In *Name Reactions for Functional Group Transformations*; Li, J. J., Ed.; Wiley: Hoboken, NJ; **2007**, pp 218–236. (Review).
10. More, S. S.; Vince, R. *J. Med. Chem.* **2008**, *51*, 4581–4588.
11. Crich, D.; Li, M.; Jayalath, P. *Carbohydrate Res.* **2009**, *344*, 140–144.
12. Howard, J. K.; Hyland, C. J. T.; Just, J.; J. A. *Org. Lett.* **2013**, *15*, 1714–1717.

# Dieckmann condensation

The Dieckmann condensation is the intramolecular version of the Claisen condensation.

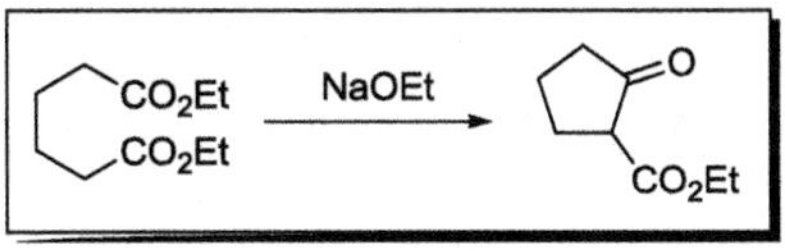

Example 1[4]

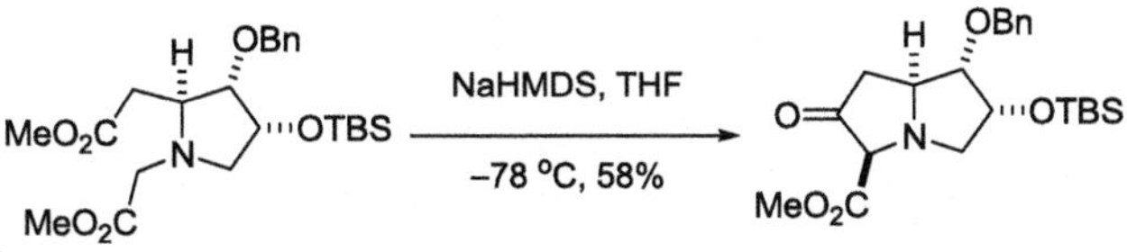

Example 2[6]

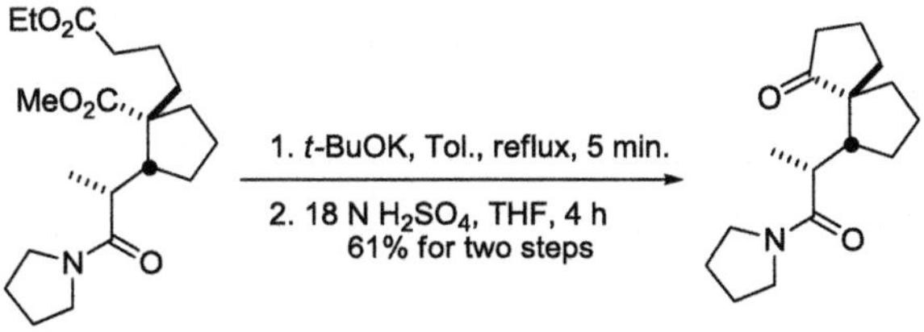

Example 3[7]

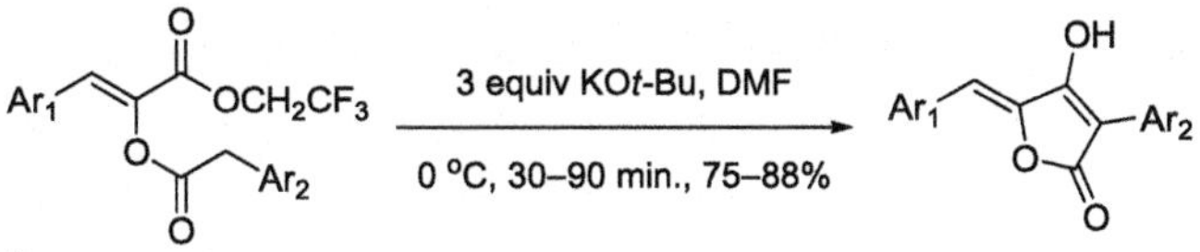

Example 4[8]

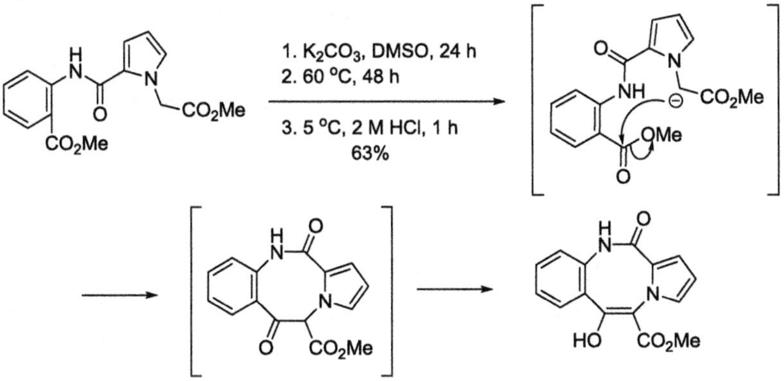

J.J. Li, *Name Reactions: A Collection of Detailed Mechanisms and Synthetic Applications*, DOI 10.1007/978-3-319-03979-4_87, © Springer International Publishing Switzerland 2014

Example 5, Michael–Dieckmann condensation[10]

Example 6, Michael–Dieckmann condensation[10]

oseltamivir (Tamiflu)

**References**

1. Dieckmann, W. *Ber.* **1894**, *27*, 102. Walter Dieckman (1869–1925), born in Hamburg, Germany, studied with E. Bamberger at Munich. After serving as an assistant to von Baeyer in his private laboratory, he became a professor at Munich. At age 56, he died while working in his chemical laboratory at the Barvarian Academy of Science.
2. Davis, B. R.; Garratt, P. J. *Comp. Org. Synth.* **1991**, *2*, 795–863. (Review).
3. Shindo, M.; Sato, Y.; Shishido, K. *J. Am. Chem. Soc.* **1999**, *121*, 6507–6508.
4. Rabiczko, J.;. Urbańczyk-Lipkowska, Z.; Chmielewski, M. *Tetrahedron* **2002**, *58*, 1433–1441.
5. Ho, J. Z.; Mohareb, R. M.; Ahn, J. H.; Sim, T. B.; Rapoport, H. *J. Org. Chem.* **2003**, *68*, 109–114.
6. de Sousa, A. L.; Pilli, R. A. *Org. Lett.* **2005**, *7*, 1617–1617.
7. Bernier, D.; Brueckner, R. *Synthesis* **2007**, 2249–2272.
8. Koriatopoulou, K.; Karousis, N.; Varvounis, G. *Tetrahedron* **2008**, *64*, 10009–10013.
9. Takao, K.-i.; Kojima, Y.; Miyashita, T.; Yashiro, K.; Yamada, T.; Tadano, K.-i. *Heterocycles* **2009**, *77*, 167–172.
10. Garrido, N. M.; Nieto, C. T.; Diez, D. *Synlett* **2013**, *24*, 169–172.
11. Kaliyamoorthy; A.; Makoto; F.; Kenzo; Y.; Naoya; K.; Takumi; W.; Masakatsu, S. *J. Org. Chem.* **2013**, *78*, 4019–4026.

# Diels–Alder reaction

The Diels–Alder reaction, inverse electronic demand Diels–Alder reaction, as well as the hetero-Diels–Alder reaction, belong to the category of *[4+2]-cycloaddition reactions*, which are concerted processes. The arrow pushing here is merely illustrative.

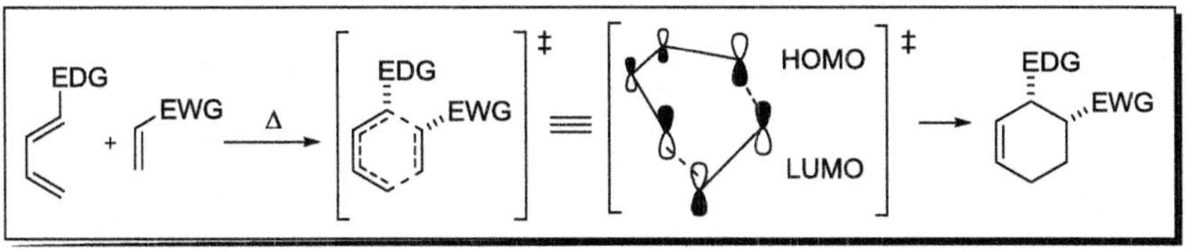

diene    dienophile                                          adduct

EDG = electron-donating group; EWG = electron-withdrawing group

Example 1[6]

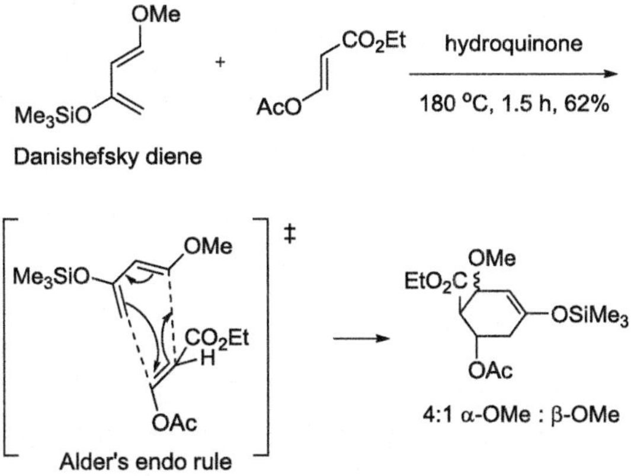

Danishefsky diene

Alder's endo rule

4:1 α-OMe : β-OMe

Example 2, Intramolecular Diels–Alder reaction[7]

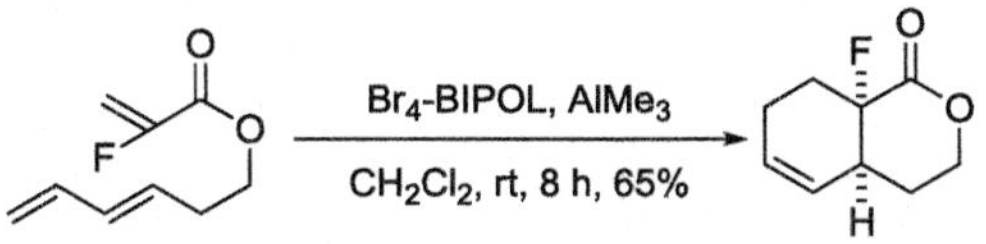

Example 3, Asymmetric Diels–Alder reaction[5,8]

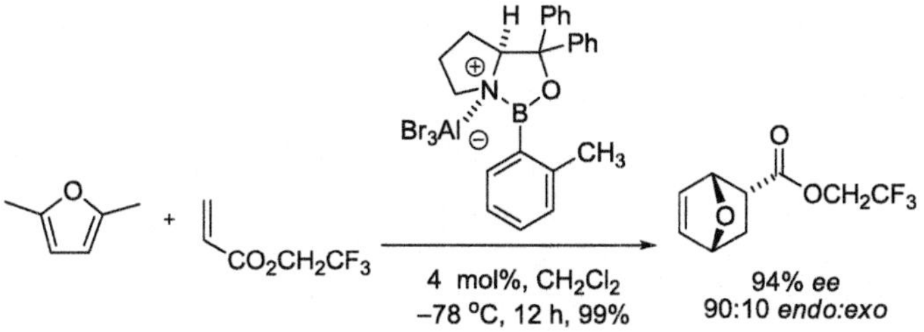

4 mol%, CH₂Cl₂
–78 °C, 12 h, 99%

94% ee
90:10 endo:exo

J.J. Li, *Name Reactions: A Collection of Detailed Mechanisms and Synthetic Applications*,
DOI 10.1007/978-3-319-03979-4_88, © Springer International Publishing Switzerland 2014

Example 4, Retro-Diels–Alder reaction[4,9]

Example 5, Intramolecular Diels–Alder reaction[11]

Example 6[11]

**References**
1.  Diels, O.; Alder, K. *Ann.* **1928**, *460*, 98–122. Otto Diels (Germany, 1876–1954) and his student, Kurt Alder (Germany, 1902–1958), shared the Nobel Prize in Chemistry in 1950 for development of the diene synthesis. In this article they claimed their territory in applying the Diels–Alder reaction in total synthesis: "We explicitly reserve for ourselves the application of the reaction developed by us to the solution of such problems."
2.  Oppolzer, W. In *Comprehensive Organic Synthesis;* Trost, B. M.; Fleming, I., Eds.; Pergamon, **1991**, *Vol. 5*, 315–399. (Review).
3.  Weinreb, S. M. In.*Comprehensive Organic Synthesis;* Trost, B. M.; Fleming, I., Eds.; Pergamon, **1991**, *Vol. 5*, 401–449. (Review).
4.  (a) Rickborn, B. The *retro-Diels–Alder reaction. Part I. C–C dienophiles* in *Org. React.* Wiley: Hoboken, NJ, **1998**, *52*. (b) Rickborn, B. *The retro-Diels–Alder reaction. Part II. Dienophiles with one or more heteroatom* in *Org. React.* Wiley: Hoboken, NJ, **1998**, *53*.
5.  Corey, E. J. *Angew. Chem. Int. Ed.* **2002**, *41*, 1650–1667. (Review).
6.  Wang, J.; Morral, J.; Hendrix, C.; Herdewijn, P. *J. Org. Chem.* **2001**, *66*, 8478–8482.
7.  Saito, A.; Yanai, H.; Sakamoto, W.; Takahashi, K.; Taguchi, T. *J. Fluorine Chem.* **2005**, *126*, 709–714.
8.  Liu, D.; Canales, E.; Corey, E. J. *J. Am. Chem. Soc.* **2007**, *129*, 1498–1499.
9.  Iqbal, M.; Duffy, P.; Evans, P.; Cloughley, G.; Allan, B.; Lledo, A.; Verdaguer, X.; Riera, A. *Org. Biomol. Chem.* **2008**, *6*, 4649–4661.
10. Ibrahim-Ouali, M. *Steroids* **2009**, *74*, 133–162.
11. Gao, S.; Wang, Q.; Chen, C. *J. Am. Chem. Soc.* **2009**, *131*, 1410–1412.
12. Martin, R. M.; Bergman, R. G.; Ellman, J. A. *Org. Lett.* **2013**, *15*, 444–447.

# Inverse electronic demand Diels–Alder reaction

Example 1, Catalytic asymmetric inverse electronic demand Diels–Alder reaction[2]

Example 2[3]

Example 3, Catalytic asymmetric inverse-electron-demand Diels–Alder reaction[4]

DBFOX-Ph =

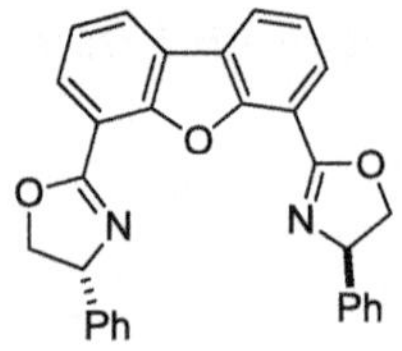

Example 4[5]

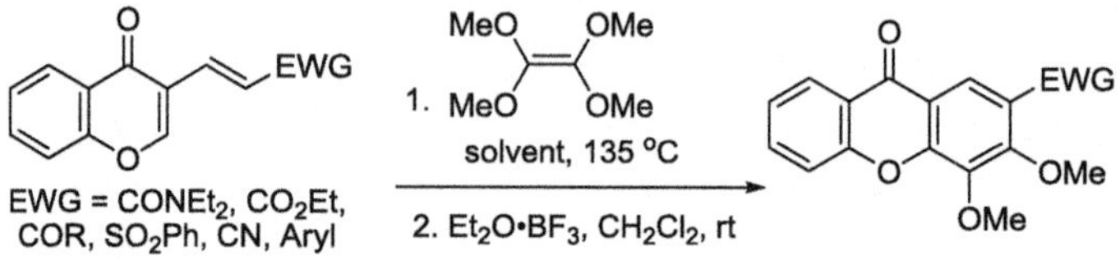

EWG = CONEt$_2$, CO$_2$Et,
COR, SO$_2$Ph, CN, Aryl

1. MeO, OMe, MeO, OMe
solvent, 135 °C

2. Et$_2$O·BF$_3$, CH$_2$Cl$_2$, rt

Example 5[6]

DMF, 50 °C

13 h, 85%

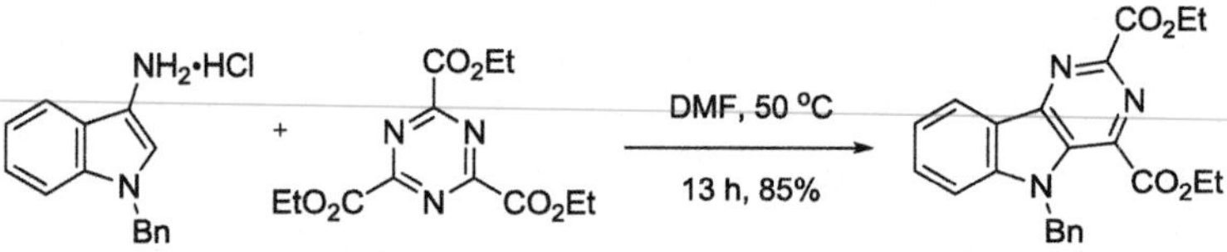

**References**
1. Boger, D. L.; Patel, M. *Prog. Heterocycl. Chem.* **1989**, *1*, 30–64. (Review).
2. Gao, X.; Hall, D. G. *J. Am. Chem. Soc.* **2005**, *127*, 1628–1629.
3. He, M.; Uc, G. J.; Bode, J. W. *J. Am. Chem. Soc.* **2006**, *128*, 15088–15089.
4. Esquivias, J.; Gomez Arrayas, R.; Carretero, J. C. *J. Am. Chem. Soc.* **2007**, *129*, 1480–1481.
5. Dang, A.-T.; Miller, D. O.; Dawe, L. N.; Bodwell, G. J. *Org. Lett.* **2008**, *10*, 233–236.
6. Xu, G.; Zheng, L.; Dang, Q.; Bai, X. *Synthesis* **2013**, *45*, 743–752.

## Hetero–Diels–Alder reaction

Heterodiene addition to dienophile or heterodienophile addition to diene. Typical hetero-Diels–Alder reactions are aza-Diels–Alder reaction and oxo-Diels–Alder reaction.

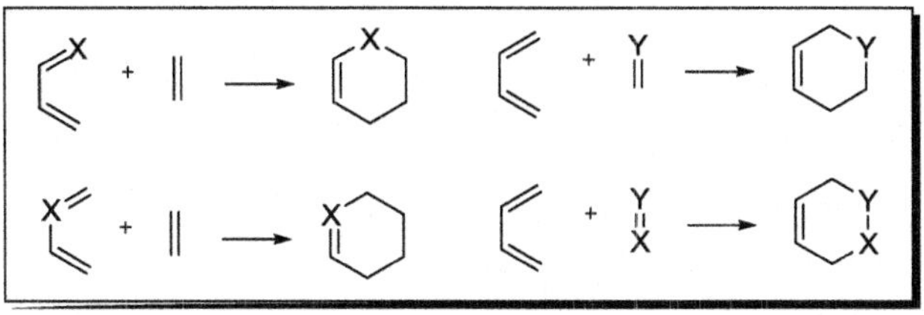

Example 1,

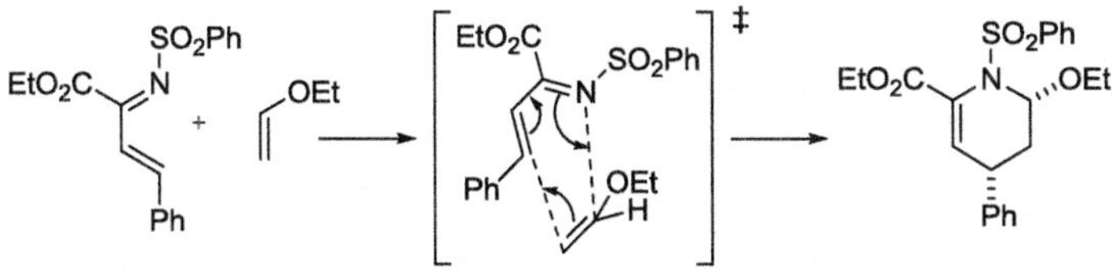

Example 2, Heterodienophile addition to diene[1]

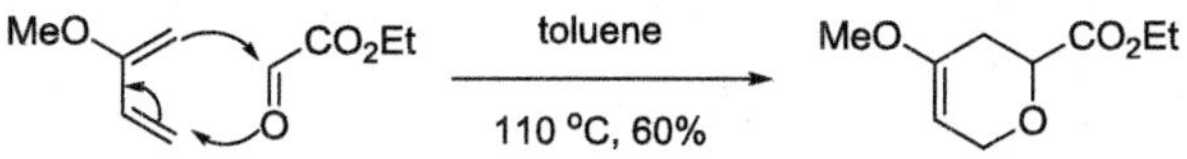

Example 3, Similar to the **Boger pyridine synthesis** (see page 59)[2]

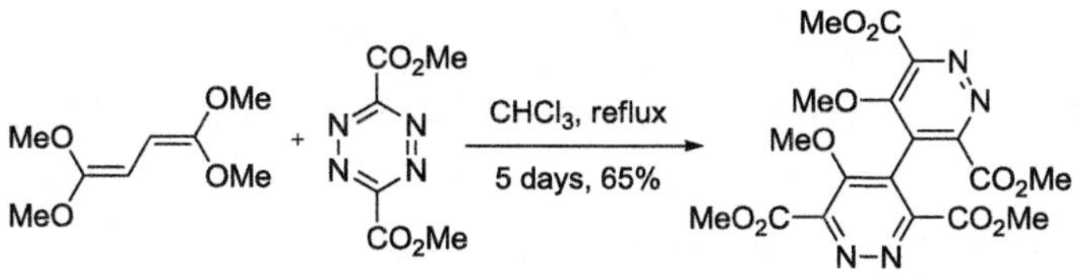

Example 4, Using **the Rawal diene**[4]

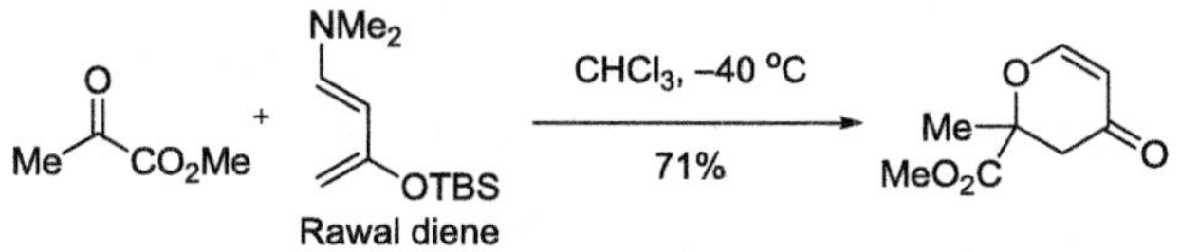

Rawal diene

Example 5, Also similar to the Boger pyridine synthesis[6]

| | |
|---|---|
| n = 1, 75% | |
| n = 2, 65% | |
| n = 3, 54% | |
| n = 4, 30% | |

Example 6, Asymmetric hetero-Diels–Alder reaction[7]

24:1 *endo/exo*
97% *ee*

Example 7, Asymmetric hetero-Diels–Alder reaction[8]

97% yield, 20:1 *rr*, 96% *ee*

## References

1.  Wender, P. A.; Keenan, R. M.; Lee, H. Y. *J. Am. Chem. Soc.* **1987**, *109*, 4390–4392.
2.  Boger, D. L. In *Comprehensive Organic Synthesis;* Trost, B. M.; Fleming, I., Eds.; Pergamon, **1991**, *Vol. 5*, 451–512. (Review).
3.  Boger, D. L.; Baldino, C. M. *J. Am. Chem. Soc.* **1993**, *115*, 11418–11425.
4.  Huang, Y.; Rawal, V. H. *Org. Lett.* **2000**, *2*, 3321–3323.
5.  Jørgensen, K. A. *Eur. J. Org. Chem.* **2004**, 2093–2102. (Review).
6.  Lipińska, T. M. *Tetrahedron* **2006**, *62*, 5736–5747.
7.  Evans, D. A.; Kvaerno, L.; Dunn, T. B.; Beauchemin, A.; Raymer, B.; Mulder, J. A.; Olhava, E. J.; Juhl, M.; Kagechika, K.; Favor, D. A. *J. Am. Chem. Soc.* **2008**, *130*, 16295–16309.
8.  Liu, B.; Li, K.-N.; Luo, S.-W.; Huang, J.-Z.; Pang, H.; Gong, L.-Z. *J. Am. Chem. Soc.* **2013**, *135*, 3323–3326.

# Dienone–phenol rearrangement

Acid-promoted rearrangement of 4,4-disubstituted cyclohexadienones to 3,4-disubstituted phenols.

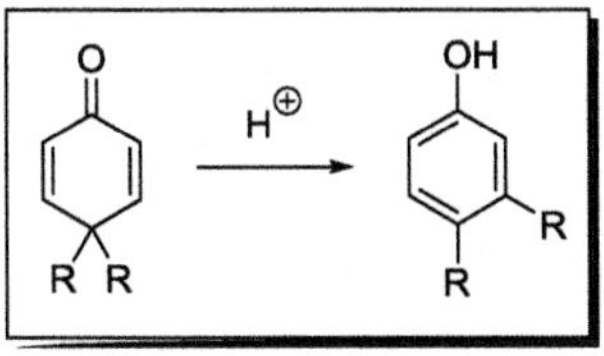

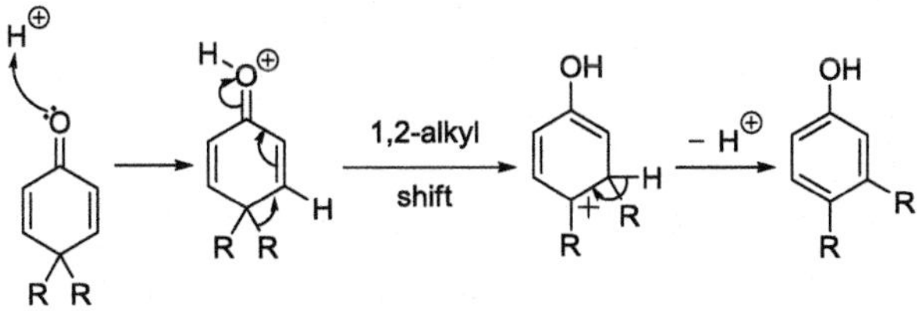

Example 1[4]

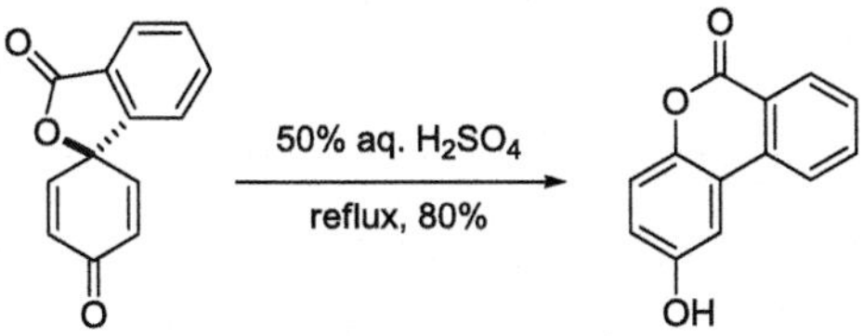

Example 2[5]

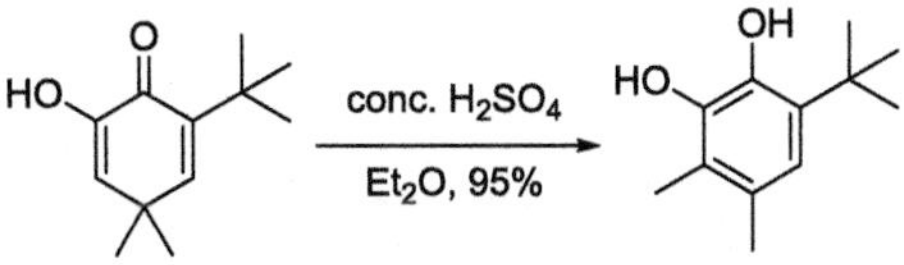

Example 3[9]

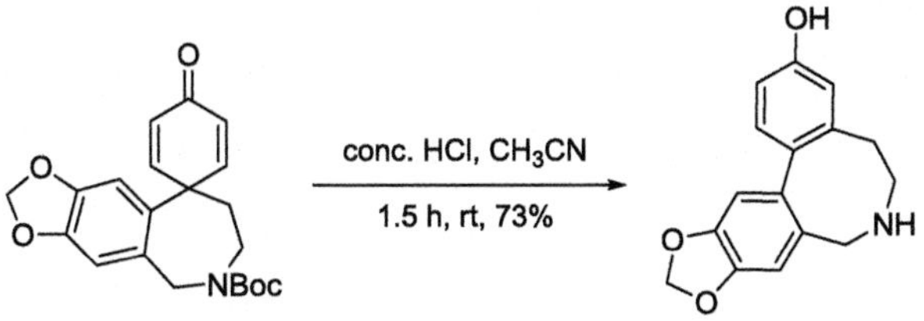

J.J. Li, *Name Reactions: A Collection of Detailed Mechanisms and Synthetic Applications*,
DOI 10.1007/978-3-319-03979-4_89, © Springer International Publishing Switzerland 2014

Example 4[10]

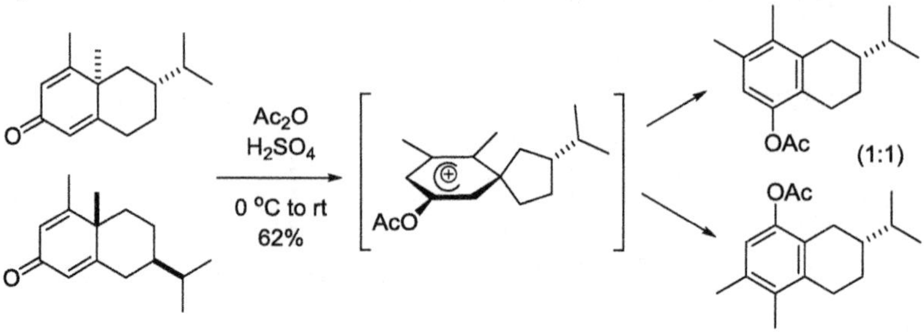

**References**

1.  Shine, H. J. In *Aromatic Rearrangements;* Elsevier: New York, **1967,** pp 55–68. (Review).
2.  Schultz, A. G.; Hardinger, S. A. *J. Org. Chem.* **1991,** *56,* 1105–1111.
3.  Schultz, A. G.; Green, N. J. *J. Am. Chem. Soc.* **1992,** *114,* 1824–1829.
4.  Hart, D. J.; Kim, A.; Krishnamurthy, R.; Merriman, G. H.; Waltos, A.-M. *Tetrahedron* **1992,** *48,* 8179–8188.
5.  Frimer, A. A.; Marks, V.; Sprecher, M.; Gilinsky-Sharon, P. *J. Org. Chem.* **1994,** *59,* 1831–1834.
6.  Oshima, T.; Nakajima, Y.-i.; Nagai, T. *Heterocycles* **1996,** *43,* 619–624.
7.  Draper, R. W.; Puar, M. S.; Vater, E. J.; Mcphail, A. T. *Steroids* **1998,** *63,* 135–140.
8.  Kodama, S.; Takita, H.; Kajimoto, T.; Nishide, K.; Node, M. *Tetrahedron* **2004,** *60,* 4901–4907.
9.  Bru, C.; Guillou, C. *Tetrahedron* **2006,** *62,* 9043–9048.
10. Sauer, A. M.; Crowe, W. E.; Henderson, G.; Laine, R. A. *Tetrahedron Lett.* **2007,** *48,* 6590–6593.

# Doebner quinoline synthesis

Three-component coupling of an aniline, pyruvic acid, and an aldehyde to provide a quinoline-4-carboxylic acid.

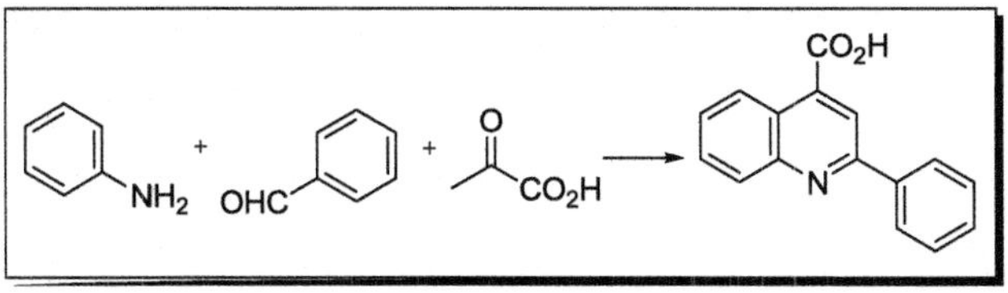

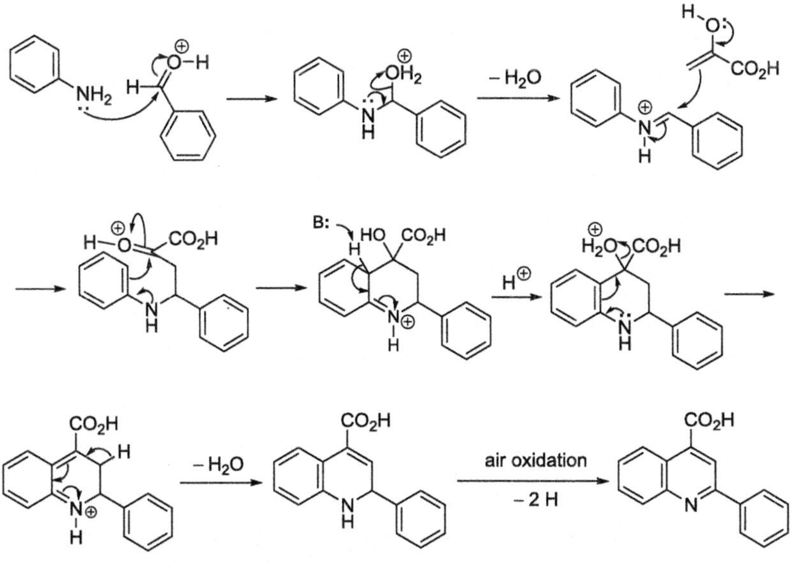

Example 1[2]

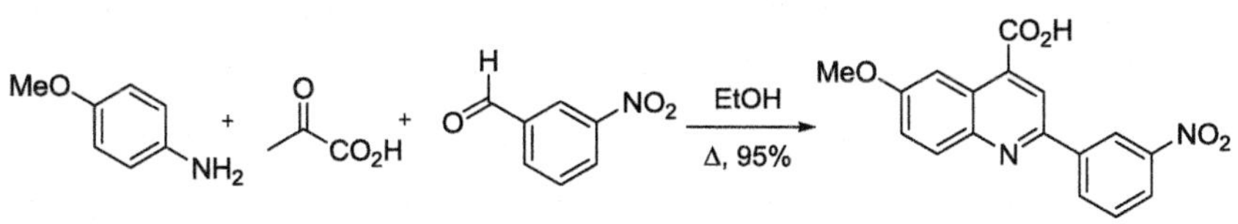

Example 2[6]

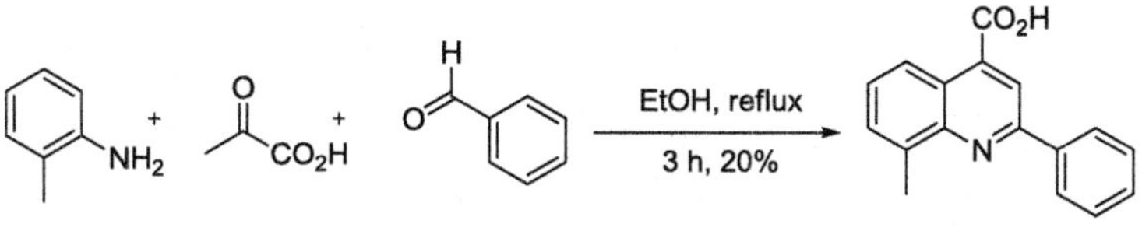

J.J. Li, *Name Reactions: A Collection of Detailed Mechanisms and Synthetic Applications*, DOI 10.1007/978-3-319-03979-4_90, © Springer International Publishing Switzerland 2014

Example 3, Combinatorial Doebner reaction[7]

Example 4, Ytterbium perfluorooctanoate-catalyzed Doebner reaction in water[9]

## References

1.  Doebner, O. G. *Ann.* **1887,** *242,* 265. Oscar Gustav Doebner (1850–1907) was born in Meininggen, Germany. After studying under Liebig, he actively took part in the Franco-Prussian War. He apprenticed with Otto and Hofmann for a few years after the war, then began his independent researches at the University at Halle.
2.  Mathur, F. C.; Robinson, R. *J. Chem. Soc.* **1934,** 1520–1523.
3.  Elderfield, R. C. *Heterocyclic Compounds*; Elderfield, R. C., Ed.; Wiley: New York, **1952,** *Vol. 4, Quinoline, Isoquinoline and Their Benzo Derivatives*, pp. 25–29. (Review).
4.  Jones, G. In *Chemistry of Heterocyclic Compounds*, Jones, G., ed.; Wiley: New York, **1977,** *Vol. 32*; Quinolines, pp. 125–131. (Review).
5.  Atwell, G. J.; Baguley, B. C.; Denny, W. A. *J. Med. Chem.* **1989,** *32,* 396–401.
6.  Herbert, R. B.; Kattah, A. E.; Knagg, E. *Tetrahedron* **1990,** *46,* 7119–7138.
7.  Gopalsamy, A.; Pallai, P. V. *Tetrahedron Lett.* **1997,** *38,* 907–910.
8.  Pflum, D. A. *Doebner Quinoline Synthesis.* In *Name Reactions in Heterocyclic Chemistry*; Li, J. J., Ed.; Wiley: Hoboken, NJ, **2005,** pp 407–410. (Review).
9.  Wang, L.-M.; Hu, L.; Chen, H.-J.; Sui, Y.-Y.; Shen, W. *J. Fluorine Chem.* **2009,** *130,* 406–409.

# Doebner–von Miller reaction

Doebner–von Miller reaction is a variant of the Skraup quinoline synthesis. Therefore, the mechanism for the Skraup reaction is also operative for the Doebner–von Miller reaction. The following mechanism is favored by Denmark's mechnistic study using [13]C-labelled α,β-unsaturated ketones.[9]

Example 1[5]

J.J. Li, *Name Reactions: A Collection of Detailed Mechanisms and Synthetic Applications*, DOI 10.1007/978-3-319-03979-4_91, © Springer International Publishing Switzerland 2014

Example 2[6]

Example 3, A novel variant[10]

Example 4, Similar to Example 1[11]

## References

1.  Doebner, O.; von Miller, W. *Ber.* **1883**, *16*, 2464.
2.  Corey, E. J.; Tramontano, A. *J. Am. Chem. Soc.* **1981**, *103*, 5599–5600.
3.  Eisch, J. J.; Dluzniewski, T. *J. Org. Chem.* **1989**, *54*, 1269–1274.
4.  Zhang, Z. P.; Tillekeratne, L. M. V.; Hudson, R. A. *Tetrahedron Lett.* **1998**, *39*, 5133–5134.
5.  Carrigan, C. N.; Esslinger, C. S.; Bartlett, R. D.; Bridges, R. J. *Bioorg. Med. Chem. Lett.* **1999**, *9*, 2607–2712.
6.  Sprecher, A.-v.; Gerspacher, M.; Beck, A.; Kimmel, S.; Wiestner, H.; Anderson, G. P.; Niederhauser, U.; Subramanian, N.; Bray, M. A. *Bioorg. Med. Chem. Lett.* **1998**, *8*, 965–970.
7.  Fürstner, A.; Thiel, O. R.; Blanda, G. *Org. Lett.* **2000**, *2*, 3731–3734.
8.  Moore, A. *Skraup Doebner–von Miller Reaction*. In *Name Reactions in Heterocyclic Chemistry*; Li, J. J., Ed.; Wiley: Hoboken, NJ, **2005**, 488–494. (Review).
9.  Denmark, S. E.; Venkatraman, S. *J. Org. Chem.* **2006**, *71*, 1668–1676. Mechanistic study using $^{13}$C-labelled α,β-unsaturated ketones.
10. Horn, J.; Marsden, S. P.; Nelson, A.; House, D.; Weingarten, G. G. *Org. Lett.* **2008**, *10*, 4117–4120.
11. Laras, Y.; Hugues, V.; Chandrasekaran, Y.; Blanchard-Desce, M.; Acher, F. C.; Pietrancosta, N. *J. Org. Chem.* **2012**, *77*, 8294–8302.

# Dötz reaction

Also known as the Dötz benzannulation, the Dötz reaction is the Cr(CO)$_3$-coordinated hydroquinone from vinylic alkoxy pentacarbonyl chromium carbene (Fischer carbene) complex and alkynes.

## Example 1[5]

## Example 3[8]

## Example 3[8]

J.J. Li, *Name Reactions: A Collection of Detailed Mechanisms and Synthetic Applications*,
DOI 10.1007/978-3-319-03979-4_92, © Springer International Publishing Switzerland 2014

Example 3[9]

Example 4[10]

**References**
1.  Dötz, K. H. *Angew. Chem. Int. Ed.* **1975**, *14*, 644–645. Karl H. Dötz (1943–) was a professor at the University of Munich in Germany.
2.  Wulff, W. D. In *Advances in Metal-Organic Chemistry*; Liebeskind, L. S., Ed.; JAI Press, Greenwich, CT; **1989**; *Vol. 1.* (Review).
3.  Wulff, W. D. In *Comprehensive Organometallic Chemistry II*; Abel, E. W., Stone, F. G. A., Wilkinson, G., Eds.; Pergamon Press: Oxford, **1995**; *Vol. 12.* (Review).
4.  Torrent, M.; Solá, M.; Frenking, G. *Chem. Rev.* **2000**, *100*, 439–494. (Review).
5.  Caldwell, J. J.; Colman, R.; Kerr, W. J.; Magennis, E. J. *Synlett* **2001**, 1428–1430.
6.  Solá, M.; Duran, M.; Torrent, M. *The Dötz reaction: A chromium Fischer carbene-mediated benzannulation reaction.* In *Computational Modeling of Homogeneous Catalysis* Maseras, F.; Lledós, eds.; Kluwer Academic: Boston; **2002**, 269–287. (Review).
7.  Pulley, S. R.; Czakó, B. *Tetrahedron Lett.* **2004**, *45*, 5511–5514.
8.  White, J. D.; Smits, H. *Org. Lett.* **2005**, *7*, 235–238.
9.  Boyd, E.; Jones, R. V. H.; Quayle, P.; Waring, A. J. *Tetrahedron Lett.* **2005**, *47*, 7983–7986.
10. Fernandes, R. A.; Mulay, S. V. *J. Org. Chem.* **2010**, *75*, 7029–7032.

# Dowd–Beckwith ring expansion

Radical-mediated ring expansion of 2-halomethyl cycloalkanones.

2,2'-azobisisobutyronitrile (AIBN)

## Example 1[4]

## Example 2[9]

J.J. Li, *Name Reactions: A Collection of Detailed Mechanisms and Synthetic Applications*, DOI 10.1007/978-3-319-03979-4_93, © Springer International Publishing Switzerland 2014

Example 3, Cascade Dowd–Beckwith Ring Expansion/Cyclization[10]

**References**

1.  Dowd, P.; Choi, S.-C. *J. Am. Chem. Soc.* **1987,** *109,* 3493–3494.  Paul Dowd (1936–1996) was a professor at the University of Pittsburgh.
2.  (a) Beckwith, A. L. J.; O'Shea, D. M.; Gerba, S.; Westwood, S. W. *J. Chem. Soc., Chem. Commun.* **1987,** 666–667. Athelstan L. J. Beckwith is a professor at University of Adelaide, Adelaide, Australia. (b) Beckwith, A. L. J.; O'Shea, D. M.; Westwood, S. W. *J. Am. Chem. Soc.* **1988,** *110,* 2565–2575. (c) Dowd, P.; Choi, S.-C. *Tetrahedron* **1989,** *45,* 77–90. (d) Dowd, P.; Choi, S.-C. *Tetrahedron Lett.* **1989,** *30,* 6129–6132. (e) Dowd, P.; Choi, S.-C. *Tetrahedron* **1991,** *47,* 4847–4860.
3.  Dowd, P.; Zhang, W. *Chem. Rev.* **1993,** *93,* 2091–2115.  (Review).
4.  Banwell, M. G.; Cameron, J. M. *Tetrahedron Lett.* **1996,** *37,* 525–526.
5.  Studer, A.; Amrein, S. *Angew. Chem. Int. Ed.* **2000,** *39,* 3080–3082.
6.  Kantorowski, E. J.; Kurth, M. J. *Tetrahedron* **2000,** *56,* 4317–4353. (Review).
7.  Sugi, M.; Togo, H. *Tetrahedron* **2002,** *58,* 3171–3177.
8.  Ardura, D.; Sordo, T. L. *Tetrahedron Lett.* **2004,** *45,* 8691–8694.
9.  Ardura, D.; Sordo, T. L. *J. Org. Chem.* **2005,** *70,* 9417–9423.
10. Lupton, David W.; Hierold, J. *Org. Lett.* **2013,** *14,* 3412–3415.

# Dudley reagent

The Dudley reagents are employed for the protection of alcohols as benzyl[1] or PMB[2] ethers, respectively, under mild conditions. Carboxylic acids are readily protected as well.[3] Activation of the appropriate Dudley reagent in the presence of an alcohol furnishes the desired arylmethyl ether. The benzyl reagent is activated upon warming to approximately 80–85 °C, whereas activation of the PMB reagent occurs at room temperature upon treatment with methyl triflate (CH$_3$OTf) or protic acid.[4] Aromatic solvents, most commonly trifluorotoluene, often provide the best results. Magnesium oxide (MgO) is typically included in the reaction mixture as an acid scavenger.[5] For benzylation of carboxylic acids, triethylamine (Et$_3$N) is used in place of MgO.[3]

Preparation:[1–3]

Dudley benzyl reagent

Dudley PMB reagent

The Dudley reagents are conveniently prepared from readily available starting materials and are indefinitely stable to storage and handling under standard laboratory conditions. Alternatively, both reagents are commercially available.

Example 1[6]

Benzylation of a monoacetylated diol is shown in Example 1.[6] The Dudley benzyl reagent was uniquely effective for protection of the free alcohol without loss and/or migration of the labile acetyl group.

J.J. Li, *Name Reactions: A Collection of Detailed Mechanisms and Synthetic Applications*,
DOI 10.1007/978-3-319-03979-4_94, © Springer International Publishing Switzerland 2014

Example 2[2]

PMB-protection of a β-hydroxysilane can be accomplished without competition from the Peterson elimination (Example 2),[2] which would occur under the basic or acidic conditions required for many other alkylation reactions.

Example 3[4]

The Dudley PMB reagent can also be activated under mildly acidic conditions using catalytic camphorsulfonic acid (CSA) in lieu of $CH_3OTf$ (Example 3).[4]

Example 4, *In situ*-formation of the Dudley benzyl reagent is achieved by treating a mixture of an alcohol and 2-benzyloxypyridine with $CH_3OTf$[7]

## References

1.  Poon, K. W. C.; Dudley, G. B. *J. Org. Chem.* **2006,** *71,* 3923–3927.
2.  Nwoye, E. O.; Dudley, G. B. *Chem. Commun.* **2007,** 1436–1437.
3.  Tummatorn, J.; Albiniak, P. A.; Dudley, G. B. *J. Org. Chem.* **2007,** *72,* 8962–8964.
4.  Stewart, C. A.; Peng, X.; Paquette, L. A. *Synthesis* **2008,** 433–437.
5.  Poon, K. W. C.; Albiniak, P. A.; Dudley, G. B. *Org. Synth.* **2007,** *84,* 295–305.
6.  Schmidt, J. P.; Beltrân-Rodil, S.; Cox, R. J.; McAllister, G. D.; Reid, M.; Taylor, R. J. K. *Org. Lett.* **2007,** *9,* 4041–4044.
7.  Lopez, S. S.; Dudley, G. B. *Beilstein J. Org. Chem.* **2008,** *4,* No. 44.
8.  Chinigo, G. M.; Breder, A.; Carreira, E. M. *Org. Lett.* **2011,** *13,* 78–81.
9.  Taber, D. F.; Nelson, C. G. *J. Org. Chem.* **2011,** *76,* 1874–1882.
10. Tomioka, T.; Yabe, Y.; Takahashi, T.; Simmons, T. K. *J. Org. Chem.* **2011,** *76,* 4669–4674.

# 🖾 Erlenmeyer–Plöchl azlactone synthesis

Formation of 5-oxazolones (or "azlactones") by intramolecular condensation of acylglycines in the presence of acetic anhydride.

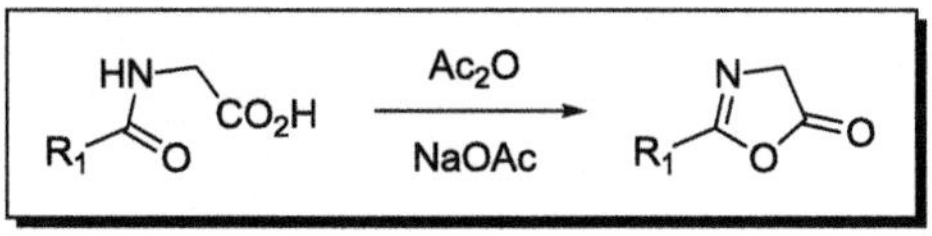

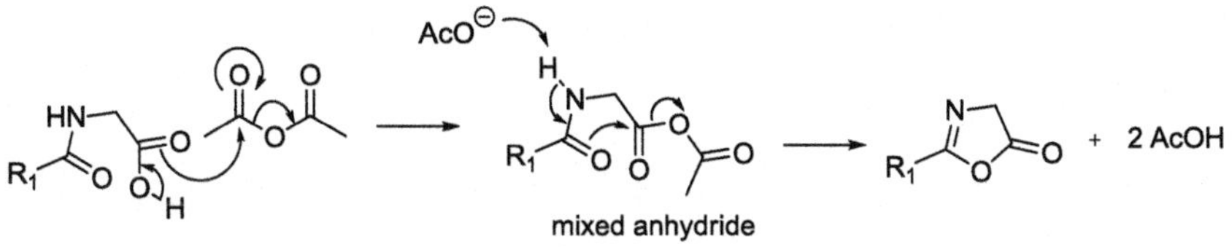

mixed anhydride

Example 1[2]

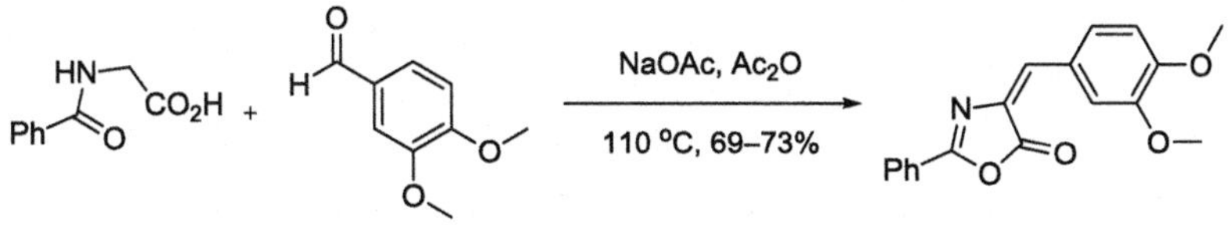

Example 2[8]

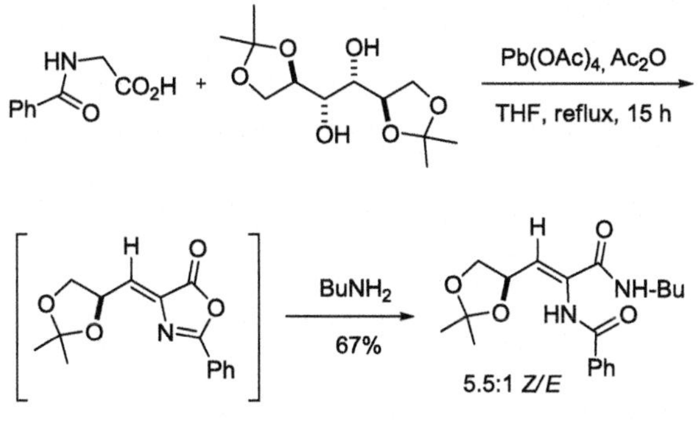

5.5:1 *Z/E*

Example 3[9]

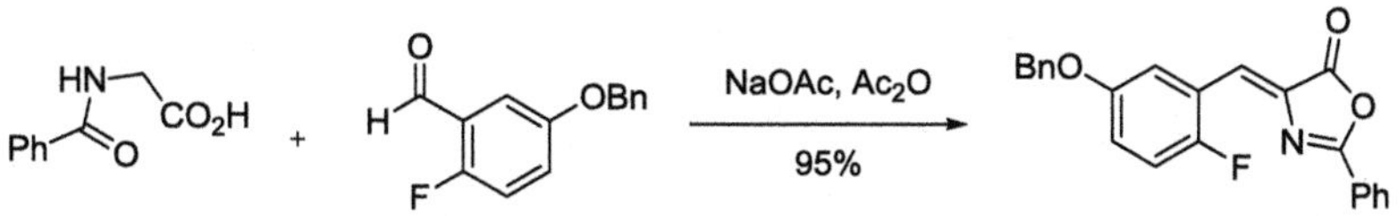

J.J. Li, *Name Reactions: A Collection of Detailed Mechanisms and Synthetic Applications*,
DOI 10.1007/978-3-319-03979-4_95, © Springer International Publishing Switzerland 2014

Example 4, The yield suffered for a more complicated substrate:[11]

**References**

1.  (a) Plöchl, J. *Ber.* **1884**, *17*, 1616–1624. (b) Erlenmeyer, E., Jr. *Ann.* **1893**, *275*, 1–3. Emil Erlenmeyer, Jr. (1864–1921) was born in Heidelberg, Germany to Emil Erlenmeyer, Sr. (1825–1909), a famous chemistry professor at the University of Heidelberg. He investigated the Erlenmeyer–Plöchl azlactone synthesis while he was a Professor of Chemistry at Strasburg. The Erlenmeyer flasks "⌂" are ubiquitous in chemistry laboratories.
2.  Buck, J. S.; Ide, W.S. *Org. Synth. Coll. II,* **1943,** 55.
3.  Carter, H. E. *Org. React.* **1946,** *3*, 198–239. (Review).
4.  Baltazzi, E. *Quart. Rev. Chem. Soc.* **1955,** *9*, 150–173. (Review).
5.  Filler, R.; Rao, Y. S. *New Development in the Chemistry of Oxazolines,* In *Adv. Heterocyclic Chem*; Katritzky, A. R.; Boulton, A. J., Eds; Academic Press, Inc: New York, **1977,** *Vol. 21*, pp 175–206. (Review).
6.  Mukerjee, A. K.; Kumar, P. *Heterocycles* **1981,** *16*, 1995–2034. (Review).
7.  Mukerjee, A. K. *Heterocycles* **1987,** *26*, 1077–1097. (Review).
8.  Combs, A. P.; Armstrong, R. W. *Tetrahedron Lett.* **1992,** *33*, 6419–6422.
9.  Konkel, J. T.; Fan, J.; Jayachandran, B.; Kirk, K. L. *J. Fluorine Chem.* **2002,** *115*, 27–32.
10. Brooks, D. A. *Erlenmeyer–Plöchl Azlactone Synthesis.* In *Name Reactions in Heterocyclic Chemistry*; Li, J. J., Ed.; Wiley: Hoboken, NJ, **2005**, pp 229–233. (Review).
11. Lee, C.-Y.; Chen, Y.-C.; Lin, H.-C.; Jhong, Y.; Chang, C.-W.; Tsai, C.-H.; Kao, C.-L.; Chien, T.-C. . *Tetrahedron* **2012,** *68*, 5898–5907.

# Eschenmoser's salt

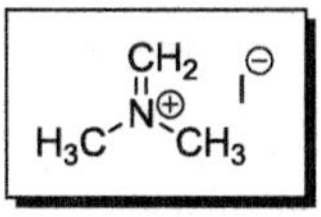

Eschenmoser's salt, dimethylmethylideneammonium iodide, is a strong dimethylaminomethylating agent, used to prepare derivatives of the type $RCH_2N(CH_3)_2$. Enolates, enolsilylethers, and even more acidic ketones undergo efficient dimethylaminomethylation—employed in the Mannich reaction.

Mechanism:

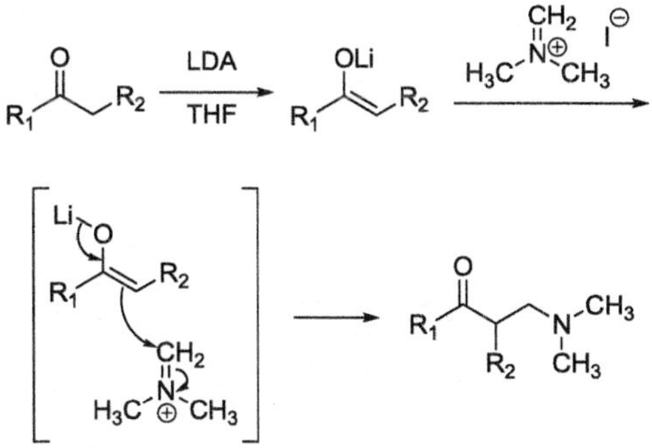

Example 1[3]

Once prepared, the resulting tertiary amines can be further methylated and then subjected to base-induced elimination to afford methylenated carbonyls.

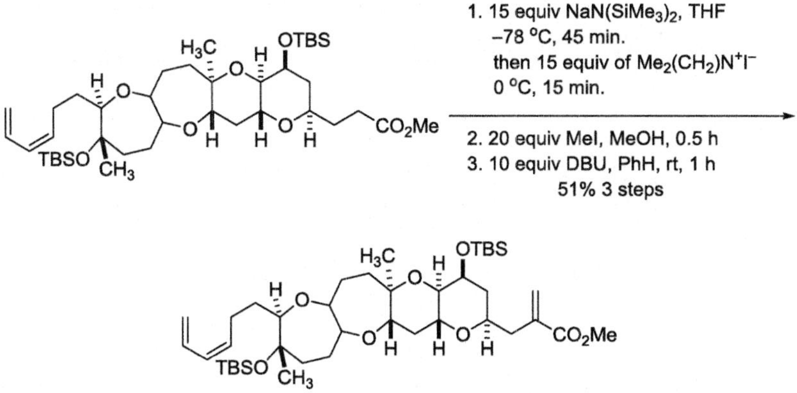

1. 15 equiv $NaN(SiMe_3)_2$, THF
   −78 °C, 45 min.
   then 15 equiv of $Me_2(CH_2)N^+I^-$
   0 °C, 15 min.

2. 20 equiv MeI, MeOH, 0.5 h
3. 10 equiv DBU, PhH, rt, 1 h
   51% 3 steps

J.J. Li, *Name Reactions: A Collection of Detailed Mechanisms and Synthetic Applications*,
DOI 10.1007/978-3-319-03979-4_96, © Springer International Publishing Switzerland 2014

Example 2[5]

Example 3[6]

## References

1.  Schreiber, J.; Maag, H.; Hashimoto, N.; Eschenmoser, A. *Angew. Chem. Int. Ed.* **1971,** *10*, 330–331. Albert Eschenmoser (Switzerland, 1925–) is best known for his work on, among many others, the monumental total synthesis of Vitamin B$_{12}$ with R. B. Woodward in 1973. He now holds joint appointments at ETH Zürich and Scripps Research Institute, La Jolla.
2.  Kleinman, E. F. *Dimethylmethyleneammonium Iodide and Chloride.* In *Encyclopedia of Reagents for Organic Synthesis* (Ed: Paquette, L. A.) 2004, WileyNew York. (Review).
3.  Nicolaou, K. C.; Reddy, K. R.; Skokotas, G.; Sato, F.; Xiao, X. Y.; Hwang, C. K. *J. Am. Chem. Soc.* **1993,** *115*, 3558–3575.
4.  Lidia Kupczyk-Subotkowska, L.; Shine, H. J. *J. Labelled Compd. Radiopharm.* **1993,** *33,* 301–304.
5.  Saczewski, J.; Gdaniec, M. *Tetrahedron Lett.* **2007,** *48*, 7624–7627.
6.  Hong, A.-W.; Cheng, T.-H.; Raghukumar, V.; Sha, C.-K. *J. Org. Chem.* **2008,** *73*, 7580–7585.
7.  Cesario, C.; Miller, M. J. *Org. Lett.* **2009,** *11*, 449–452.
8.  Crimmins, M. T.; Zuccarello, J. L.; Ellis, J. M.; McDougall P. J.; Haile, P. A.; Parrish, J. D.; Emmitte, K. A. *Org. Lett.* **2009,** *11*, 489–492.

# Eschenmoser–Tanabe fragmentation

Fragmentation of α,β-epoxyketones *via* the intermediacy of α,β-epoxy sulfonylhydrazones.

## Example 1[4]

## Example 2[7]

J.J. Li, *Name Reactions: A Collection of Detailed Mechanisms and Synthetic Applications*, DOI 10.1007/978-3-319-03979-4_97, © Springer International Publishing Switzerland 2014

Example 3[9]

1. NsNHNH₂, AcOH, THF;
   evaporation, 60 °C,
   NaBH₄, AcOH, THF, 0 °C

2. TESOTf, 2,6-lutidine
   CH₂Cl₂, rt
   60%, 2 steps

Example 4[10]

H₂NCONHNH₂•HCl, NaOAc
H₂O–EtOH, rt, 89%

Pb(OAc)₄, CH₂Cl₂
–10 °C, 60%

## References

1.  Eschenmoser, A.; Felix, D.; Ohloff, G. *Helv. Chim. Acta* **1967**, *50*, 708–713.
2.  Tanabe, M.; Crowe, D. F.; Dehn, R. L. *Tetrahedron Lett.* **1967**, 3943–3946.
3.  Felix, D.; Müller, R. K.; Horn, U.; Joos, R.; Schreiber, J.; Eschenmoser, A. *Helv. Chim. Acta* **1972**, *55*, 1276–1319.
4.  Batzold, F. H.; Robinson, C. H. *J. Org. Chem.* **1976**, *41*, 313–317.
5.  Covey, D. F.; Parikh, V. D. *J. Org. Chem.* **1982**, *47*, 5315–5318.
6.  Chinn, L. J.; Lenz, G. R.; Choudary, J. B.; Nutting, E. F.; Papaioannou, S. E.; Metcalf, L. E.; Yang, P. C.; Federici, C.; Gauthier, M. *Eur. J. Med. Chem.* **1985**, *20*, 235–240.
7.  Dai, W.; Katzenellenbogen, J. A. *J. Org. Chem.* **1993**, *58*, 1900–1908.
8.  Mück-Lichtenfeld, C. *J. Org. Chem.* **2000**, *65*, 1366–1375.
9.  Kita, Y.; Toma, T.; Kan, T.; Fukuyama, T. *Org. Lett.* **2008**, *10*, 3251–3253.
10. Nakajima, R.; Ogino, T.; Yokoshima, S.; Fukuyama, T. *J. Am. Chem. Soc.* **2010**, *132*, 1236–1237.

# Eschweiler–Clarke reductive alkylation of amines

Reductive methylation of primary or secondary amines using formaldehyde and formic acid. *Cf.* Leuckart–Wallach reaction.

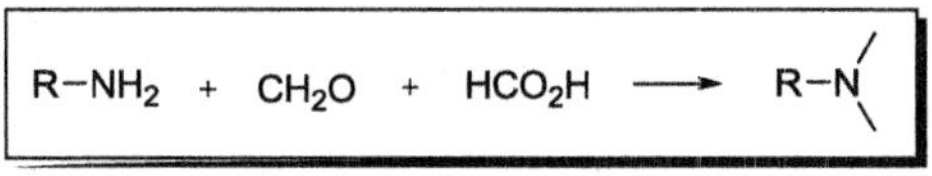
R–NH2 + CH2O + HCO2H → R–N

formic acid is the hydride source, serving as a reducing agent

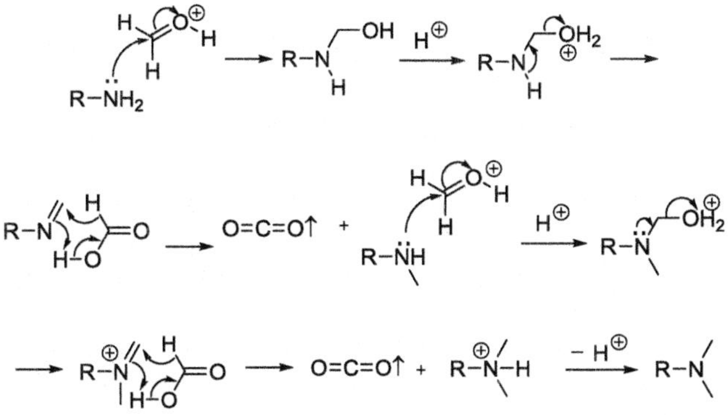

Example 1[7]

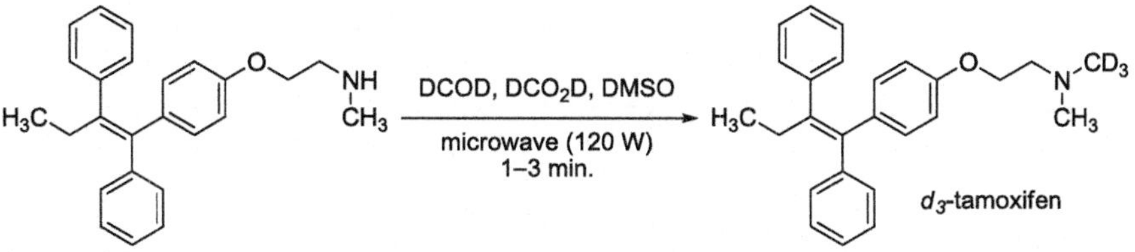
DCOD, DCO2D, DMSO
microwave (120 W)
1–3 min.
d3-tamoxifen

Example 2[9]

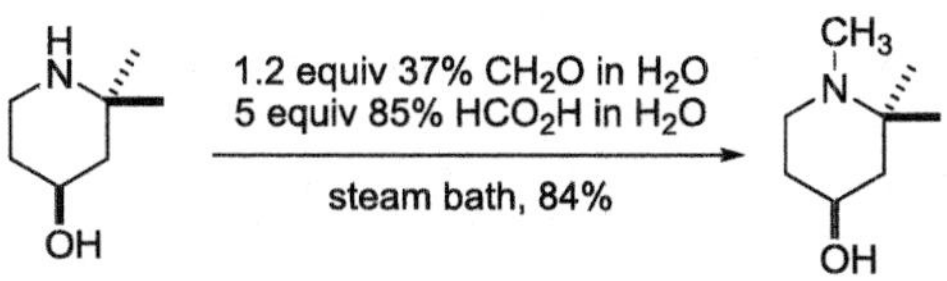
1.2 equiv 37% CH2O in H2O
5 equiv 85% HCO2H in H2O
steam bath, 84%

Example 3[10]

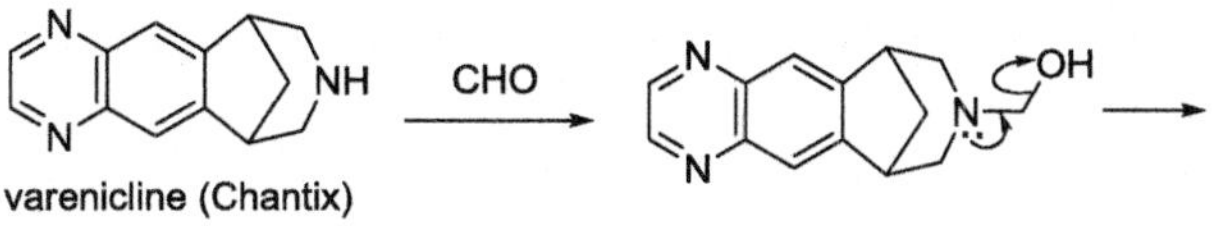
CHO
varenicline (Chantix)

J.J. Li, *Name Reactions: A Collection of Detailed Mechanisms and Synthetic Applications*,
DOI 10.1007/978-3-319-03979-4_98, © Springer International Publishing Switzerland 2014

Example 4[11]

1. NH$_2$NH$_2$·H$_2$O, EtOH, reflux, 3 h

2. HCHO, HCO$_2$H, reflux, 6 h, 73%

(S)-dapoxetine
An SSRI for PE

## References

1   (a) Eschweiler, W. *Chem. Ber.* **1905**, *38*, 880–892. Wilhelm Eschweiler (1860–1936) was born in Euskirchen, Germany. (b) Clarke, H. T.; Gillespie, H. B.; Weisshaus, S. Z. *J. Am. Chem. Soc.* **1933**, *55*, 4571–4587. Hans T. Clarke (1887–1927) was born in Harrow, England.

2   Moore, M. L. *Org. React.* **1949**, *5*, 301–330. (Review).

3   Pine, S. H.; Sanchez, B. L. *J. Org. Chem.* **1971**, *36*, 829–832.

4   Bobowski, G. *J. Org. Chem.* **1985**, *50*, 929–931.

5   Alder, R. W.; Colclough, D.; Mowlam, R. W. *Tetrahedron Lett.* **1991**, *32*, 7755–7758.

6   Bulman Page, P. C.; Heaney, H.; Rassias, G. A.; Reignier, S.; Sampler, E. P.; Talib, S. *Synlett* **2000**, 104–106.

7   Harding, J. R.; Jones, J. R.; Lu, S.-Y.; Wood, R. *Tetrahedron Lett.* **2002**, *43*, 9487–9488.

8   Brewer, A. R. E. *Eschweiler–Clarke Reductive Alkylation of Amine.* In *Name Reactions for Functional Group Transformations*; Li, J. J., Ed.; Wiley: Hoboken, NJ, **2007**, pp 86–111. (Review).

9   Weis, R.; Faist, J.; di Vora, U.; Schweiger, K.; Brandner, B.; Kungl, A. J.; Seebacher, W. *Eur. J. Med. Chem.* **2008**, *43*, 872–879.

10  Waterman, K. C.; Arikpo, W. B.; Fergione, M. B.; Graul, T. W.; Johnson, B. A.; Macdonald, B. C.; Roy, M. C.; Timpano, R. J. *J. Pharm. Sci.* **2008**, *97*, 1499–1507.

11  Sasikumar, M.; Nikalje, Milind D. *Synth. Commun.* **2012**, *42*, 3061–3067.

# Evans aldol reaction

Asymmetric aldol reaction of aldehyde and chiral acyl oxazolidinone, the Evans chiral auxiliary.

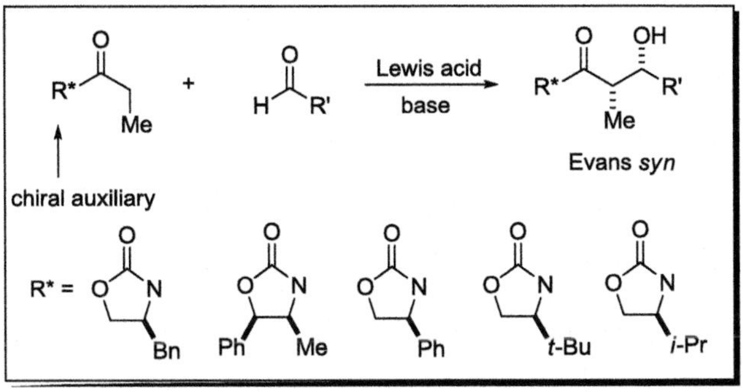

Example 1[2]

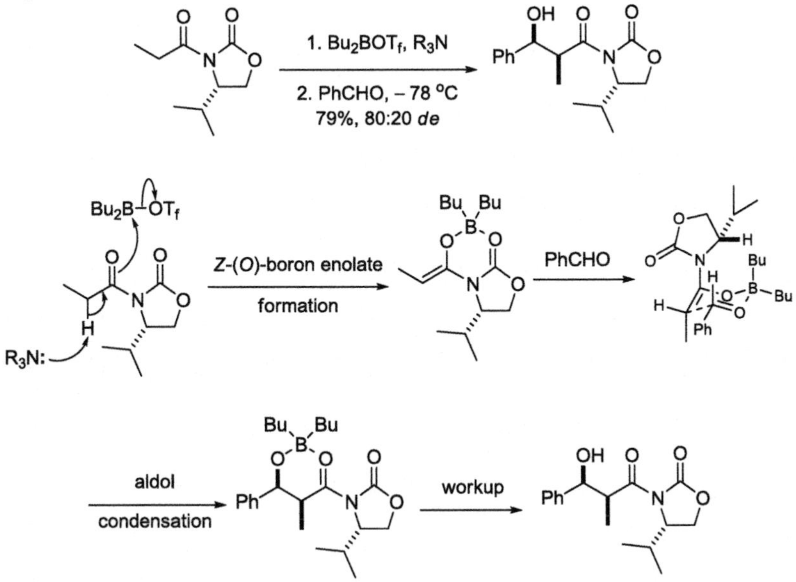

1. Bu$_2$BOT$_f$, R$_3$N

2. PhCHO, – 78 °C
79%, 80:20 de

Example 2[5]

TiCl$_4$, DIPEA
CH$_2$Cl$_2$, –78 °C, 1.5 h

then rt, overnight, 52%

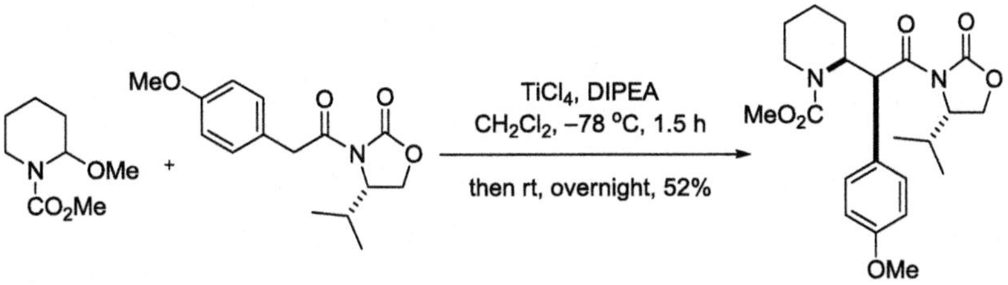

J.J. Li, *Name Reactions: A Collection of Detailed Mechanisms and Synthetic Applications*,
DOI 10.1007/978-3-319-03979-4_99, © Springer International Publishing Switzerland 2014

Example 3[9]

Example 4[10]

Example 5[12]

## References

1.  (a) Evans, D. A.; Bartroli, J.; Shih, T. L. *J. Am. Chem. Soc.* **1981**, *103*, 2127–2129. David Evans is a professor at Harvard University. (b) Evans, D. A.; McGee, L. R. *J. Am. Chem. Soc.* **1981**, *103*, 2876–2878.
2.  Danda, H.; Hansen, M. M.; Heathcock, C. H. *J. Org. Chem.* **1990**, *55*, 173–181.
3.  Ager, D. J.; Prakash, I.; Schaad, D. R. *Aldrichimica Acta* **1997**, *30*, 3–12. (Review).
4.  Braddock, D. C.; Brown, J. M. *Tetrahedron: Asymmetry* **2000**, *11*, 3591–3607.
5.  Matsumura, Y.; Kanda, Y.; Shirai, K.; Onomura, O.; Maki, T. *Tetrahedron* **2000**, *56*, 7411–7422.
6.  Williams, D. R.; Patnaik, S.; Clark, M. P. *J. Org. Chem.* **2001**, *66*, 8463–8469.
7.  Guerlavais, V.; Carroll, P. J.; Joullié, M. M. *Tetrahedron: Asymmetry* **2002**, *13*, 675–680.
8.  Li, G.; Xu, X.; Chen, D.; Timmons, C.; Carducci, M. D.; Headley, A. D. *Org. Lett.* **2003**, *5*, 329–331.
9.  Zhang, W.; Carter, R. G.; Yokochi, A. F. T. *J. Org. Chem.* **2004**, *69*, 2569–2572.
10. Ghosh, S.; Kumar, S. U.; Shashidhar, J. *J. Org. Chem.* **2008**, *73*, 1582–1585.
11. Zhang, J. *Evans Aldol Reaction.* In *Name Reactions for Homologations-Part II*; Li, J. J., Ed.; Wiley: Hoboken, NJ, **2009**, pp 532–553. (Review).
12. Siva Senkar Reddy, N.; Srinivas Reddy, A.; Yadav, J. S.; Subba Reddy, B. V. *Tetrahedron Lett.* **2012**, *53*, 6916–6918.

# Favorskii rearrangement

Transformation of enolizable α-haloketones to esters, carboxylic acids, or amides *via* alkoxide-, hydroxide-, or amine-catalyzed rearrangements, respectively.

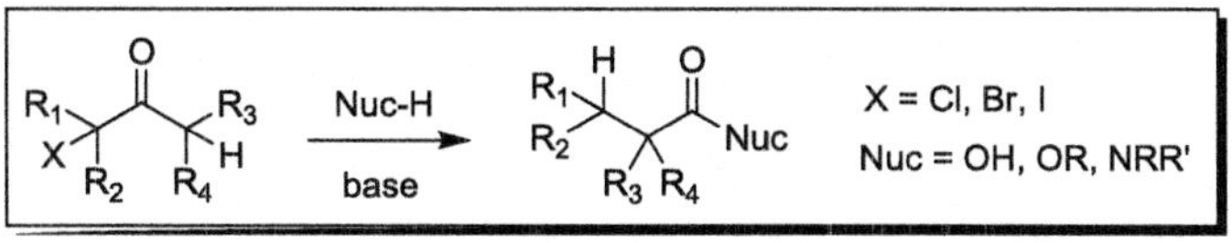

The intramolecular Favorskii Rearrangement:

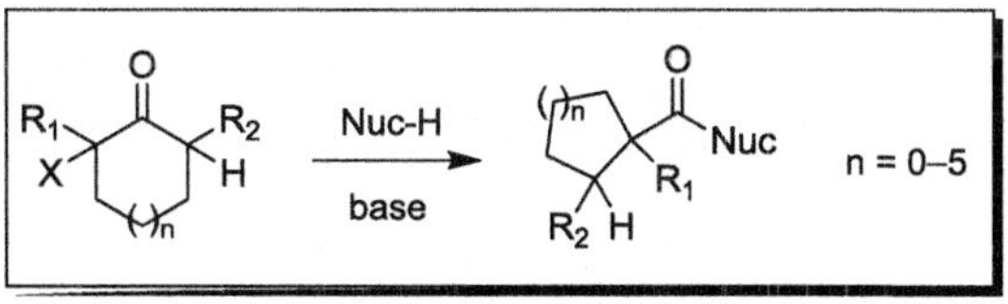

enolizable α-haloketone

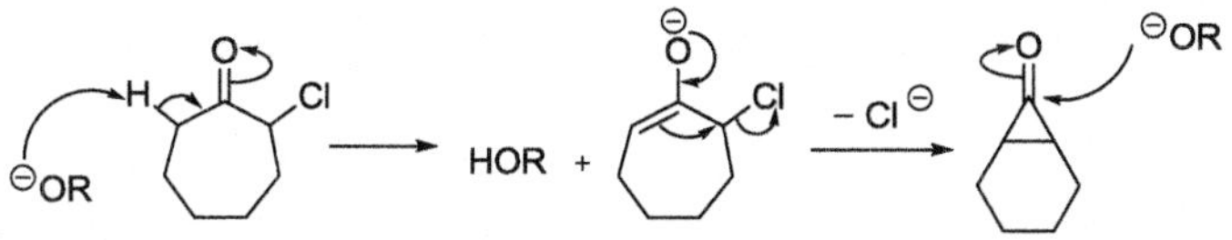

cyclopropanone intermediate

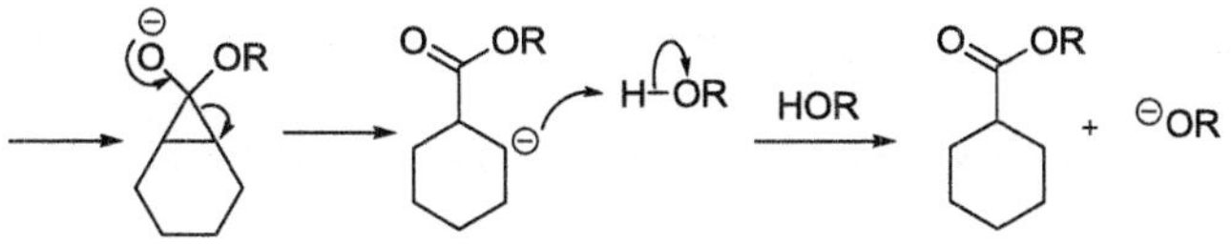

Example 1[2]

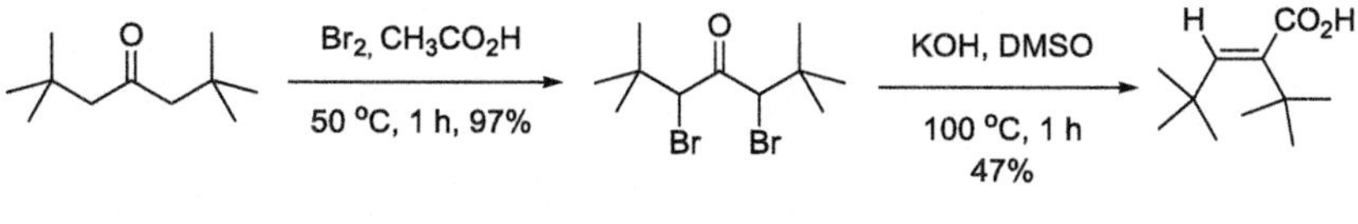

J.J. Li, *Name Reactions: A Collection of Detailed Mechanisms and Synthetic Applications*,
DOI 10.1007/978-3-319-03979-4_100, © Springer International Publishing Switzerland 2014

Example 2, Homo-Favorskii rearrangement[3]

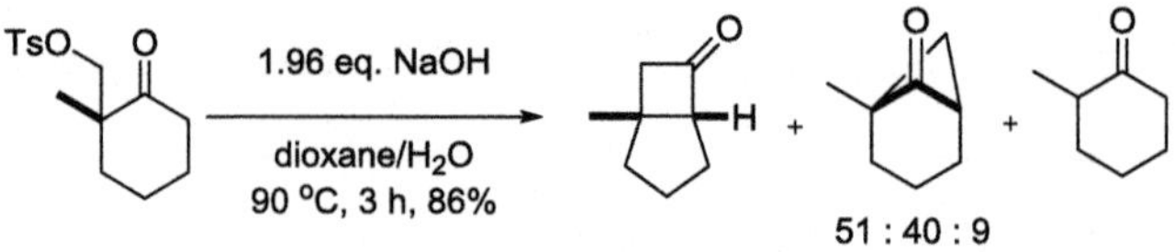

51 : 40 : 9

Example 3[6]

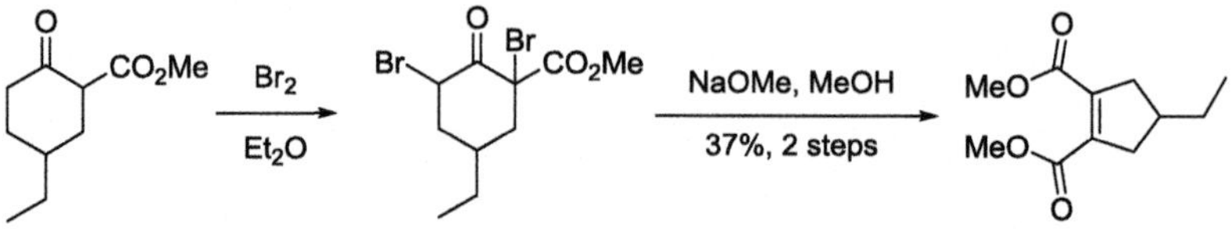

Example 4, Photo-Favorskii Rearrangement[7]

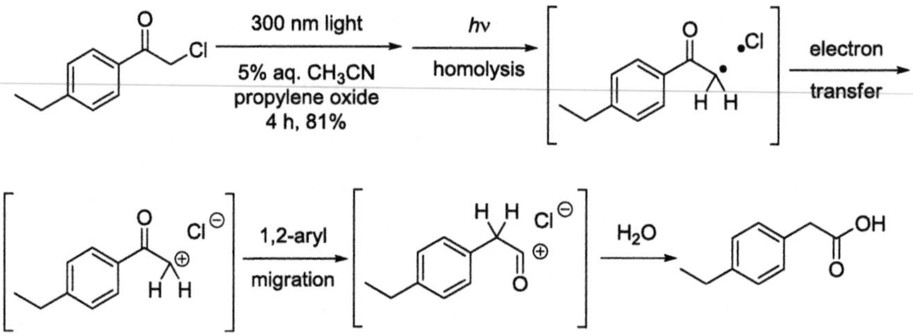

Example 5[8]

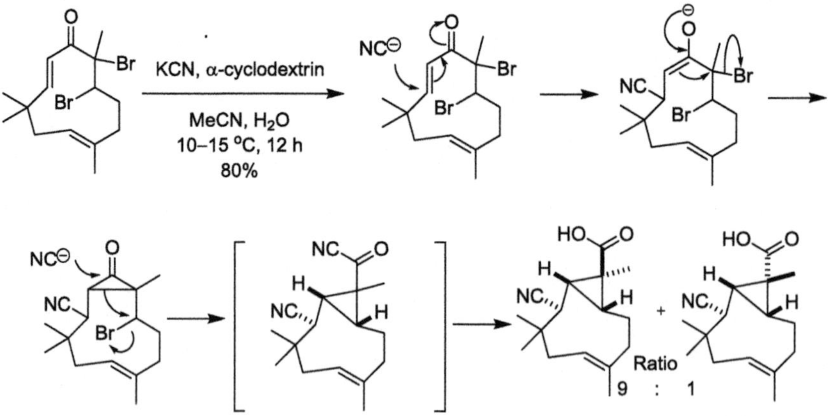

Ratio
9 : 1

Example 6[10]

Example 7[11]

## References

1.  (a) Favorskii, A. E. *J. Prakt. Chem.* **1895,** *51*, 533–563.   Aleksei E. Favorskii
    (1860–1945), born in Selo Pavlova, Russia, studied at St. Petersburg State University,
    where he became a professor since 1900.  (b) Favorskii, A. E. *J. Prakt. Chem.* **1913,**
    *88*, 658.
2.  Wagner, R. B.; Moore, J. A. *J. Am. Chem. Soc.* **1950,** *72*, 3655–3658.
3.  Wenkert, E.; Bakuzis, P.; Baumgarten, R. J.; Leicht, C. L.; Schenk, H. P. *J. Am. Chem.
    Soc.* **1971,** *93*, 3208–3216.
4.  Chenier, P. J. *J. Chem. Ed.* **1978,** *55*, 286–291. (Review).
5.  Barreta, A.; Waegell, B. In *Reactive Intermediates*; Abramovitch, R. A., ed.; Plenum
    Press: New York, **1982,** *2,* pp 527–585. (Review).
6.  White, J. D.; Dillon, M. P.; Butlin, R. J. *J. Am. Chem. Soc.* **1992,** *114*, 9673–9674.
7.  Dhavale, D. D.; Mali, V. P.; Sudrik, S. G.; Sonawane, H. R. *Tetrahedron* **1997,** *53*,
    16789–16794.
8.  Kitayama, T.; Okamoto, T. *J. Org. Chem.* **1999,** *64*, 2667–2672.
9.  Mamedov, V. A.; Tsuboi, S.; Mustakimova, L. V.; Hamamoto, H.; Gubaidullin, A. T.;
    Litvinov, I. A.; Levin, Y. A. *Chem. Heterocyclic Compd.* **2001,** *36*, 911. (Review).
10. Harmata, M.; Wacharasindhu, S. *Org. Lett.* **2005,** *7*, 2563–2565.
11. Pogrebnoi, S.; Saraber, F. C. E.; Jansen, B. J. M.; de Groot, A. *Tetrahedron* **2006,** *62*,
    1743–1748.
12. Filipski, K.J.; Pfefferkorn, J. A. *Favorskii Rearrangement*. In *Name Reactions for
    Homologations-Part II*; Li, J. J., Ed.; Wiley: Hoboken, NJ, **2009,** pp 238–252. (Re-
    view).
13. Kammath, V. B.; Šolomek, T.; Ngoy, B. P.; Heger, D.; Klán, P.; Rubina, M.; Givens,
    R. S. *J. Org. Chem.* **2013,** *78*, 1718–1729.

## *Quasi-Favorskii rearrangement*

If there are no enolizable hydrogens present, the classical Favorskii rearrangement is not possible. Instead, a semi-benzylic mechanism can lead to a rearrangement referred to as quasi-Favorskii.

Example 1, Arthur C. Cope's initial discovery[1]

non-enolizable ketone

Example 2[5]

Example 3[9]

### References

1.   Cope, A. C.; Graham, E. S. *J. Am. Chem. Soc.* **1951**, *73*, 4702–4706.
2.   Smissman, E. E.; Diebold, J. L. *J. Org. Chem.* **1965**, *30*, 4005–4007.
3.   Sasaki, T.; Eguchi, S.; Toru, T. *J. Am. Chem. Soc.* **1969**, *91*, 3390–3391.
4.   Baudry, D.; Begue, J. P.; Charpentier-Morize, M. *Tetrahedron Lett.* **1970**, 2147–2150.
5.   Stevens, C. L.; Pillai, P. M.; Taylor, K. G. *J. Org. Chem.* **1974**, *39*, 3158–3161.
6.   Harmata, M.; Wacharasindhu, S. *J. Org. Chem.* **2005**, *70*, 725–728.
7.   Filipski, K.J.; Pfefferkorn, J. A. *Favorskii Rearrangement*. In *Name Reactions for Homologations-Part II*; Li, J. J., Ed.; Wiley: Hoboken, NJ, **2009**, pp 438–452. (Review).
8.   Harmata, M.; Wacharasindhu, S. *Synthesis* **2007**, 2365–2369.
9.   Ross, A. G.; Townsend, S. D.; Danishefsky, S. J. *J. Org. Chem.* **2013**, *78*, 204–210.

# Feist–Bénary furan synthesis

α-Haloketones react with β-ketoesters in the presence of base to fashion furans.

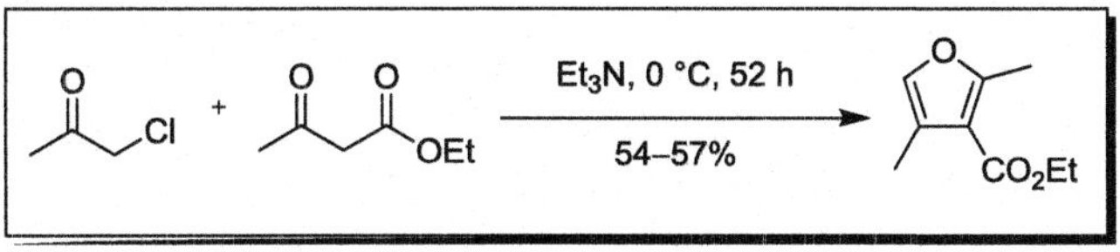

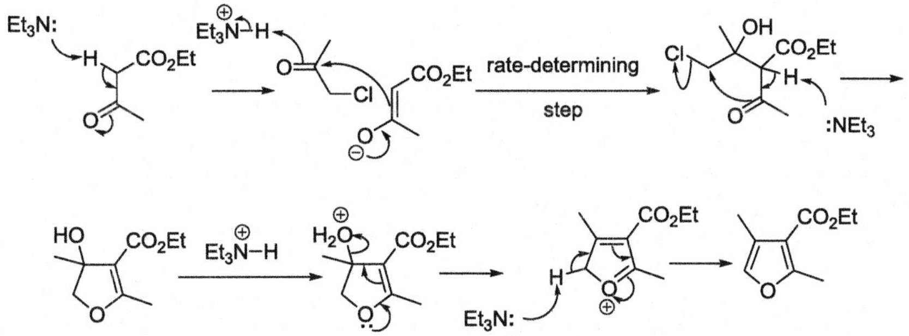

Example 1[2,3]

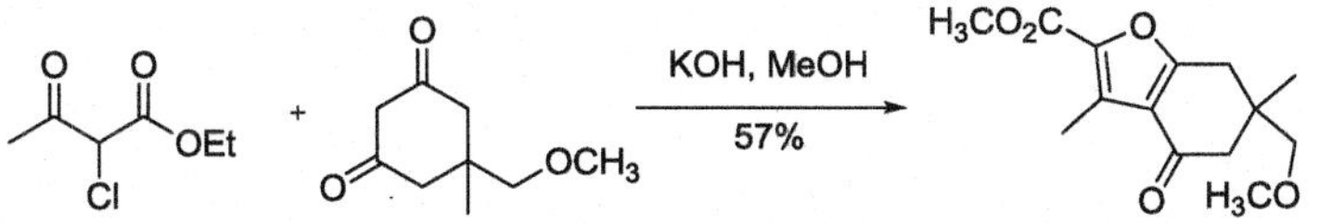

Example 2[4]

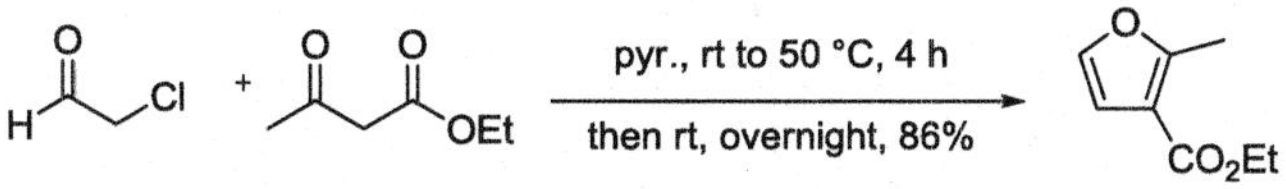

Example 3, Ionic liquid-promoted interrupted Feist–Benary reaction[10]

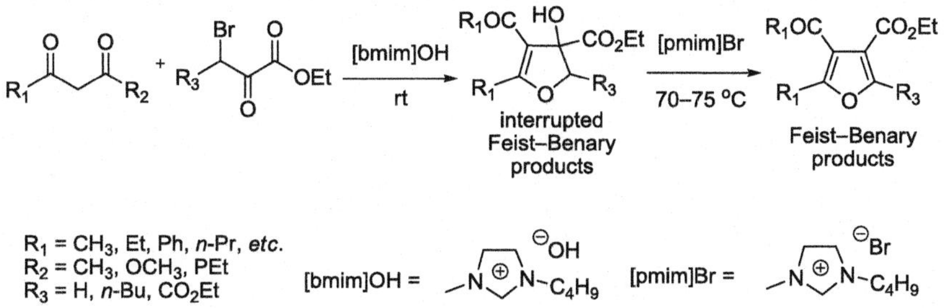

$R_1$ = $CH_3$, Et, Ph, n-Pr, etc.
$R_2$ = $CH_3$, $OCH_3$, PEt
$R_3$ = H, n-Bu, $CO_2Et$

[bmim]OH =

[pmim]Br =

J.J. Li, *Name Reactions: A Collection of Detailed Mechanisms and Synthetic Applications*,
DOI 10.1007/978-3-319-03979-4_101, © Springer International Publishing Switzerland 2014

Example 4, interrupted Feist–Benary reaction of α-tosyloxy-acetophenones[10]

Catalyst = bis(cinchona alkaloid)pyrimidines

## References

1.  (a) Feist, F. *Ber.* **1902,** *35*, 1537–1544. (b) Bénary, E. *Ber.* **1911,** *44*, 489–492.
2.  Gopalan, A.; Magnus, P. *J. Am. Chem. Soc.* **1980,** *102*, 1756–1757.
3.  Gopalan, A.; Magnus, P. *J. Org. Chem.* **1984,** *49*, 2317–2321.
4.  Padwa, A.; Gasdaska, J. R. *Tetrahedron* **1988,** *44*, 4147–4160.
5.  Dean, F. M. *Recent Advances in Furan Chemistry. Part I.* In *Advances in Heterocyclic Chemistry*, Katritzky, A. R., Ed.; Academic Press: New York, **1982**; Vol. 30, 167–238. (Review).
6.  Cambie, R. C.; Moratti, S. C.; Rutledge, P. S.; Woodgate, P. D. *Synth. Commun.* **1990,** *20*, 1923–1929.
7.  Friedrichsen, W. *Furans and Their Benzo Derivatives: Synthesis.* In *Comprehensive Heterocyclic Chemistry II*; Katritzky, A. R., Rees, C. W., Scriven, E. F. V.; Bird, C. V. Eds.; Pergamon: New York, **1996**; *Vol. 2*, 351–393. (Review).
8.  König, B. *Product Class 9: Furans.* In *Science of Synthesis: Houben–Weyl Methods of Molecular Transformations*; Maas, G., Ed.; Georg Thieme Verlag: New York, **2001**; Cat. 2, Vol. 9, 183–278. (Review).
9.  Shea, K. M. *Feist–Bénary Furan Synthesis.* In *Name Reactions in Heterocyclic Chemistry*; Li, J. J., Ed.; Wiley: Hoboken, NJ, **2005,** pp 160–167. (Review).
10. Ranu, B. C.; Adak, L.; Banerjee, S. *Tetrahedron Lett.* **2008,** *49*, 4613–4617.
11. Calter, M. A.; Korotkov, A. *Org. Lett.* **2013,** *13*, 6328–6330.

# Ferrier carbocyclization

This process (also known as the "Ferrier II Reaction") has proved to be of considerable value for the efficient, one-step conversion of 5,6-unsaturated hexopyranose derivatives into functionalized cyclohexanones useful for the preparation of such enantiomerically pure compounds as inositols and their amino, deoxy, unsaturated and selectively $O$-substituted derivatives, notably phosphate esters. In addition, the products of the carbocyclization have been incorporated into many complex compounds of interest in biological and medicinal chemistry.[1,2]

General examples:[3]

BzO BzO TsO OMe **1** HgCl₂, Me₂CO, H₂O reflux, 4.5 h, 83%

More complex products:

83%[7]    75%[7]    86% (2:1)[8]

Complex bioactive compounds made following the application of the reaction:

Paniculide A[9]    Pancratistatin[10]    Calystegine B₂[11]

J.J. Li, *Name Reactions: A Collection of Detailed Mechanisms and Synthetic Applications*,
DOI 10.1007/978-3-319-03979-4_102, © Springer International Publishing Switzerland 2014

246          Name Reactions

Modified hex-5-enopyranosides and reactions

a, Hg(OCOCF$_3$)$_2$, Me$_2$CO, H$_2$O, 0 °C; b, NaBH(OAc)$_3$, AcOH, MeCN, rt; c, *i*-Bu$_3$Al, PhMe, 40 °C; d, Ti(O*i*-Pr)Cl$_3$, CH$_2$Cl$_2$, –78 °C, 15 min. (Note: The aglycon is retained in the Al- and Ti-induced reactions).

## References
1.      Ferrier, R. J.; Middleton, S. *Chem. Rev.* **1993**, *93*, 2779–2831. (Review).
2.      Ferrier, R. J. *Top. Curr. Chem.* **2001**, *215*, 277–291 (Review).
3.      Ferrier, R. J. *J. Chem. Soc., Perkin Trans. 1* **1979**, 1455–1458. The discovery (1977) was made in the Pharmacology Department, University of Edinburgh, while R. J. Ferrier was on leave from Victoria University of Wellington, New Zealand where he was Professor of Organic Chemistry. He is now a consultant with Industrial Research Ltd., Lower Hutt, New Zealand.
4.      Blattner, R.; Ferrier, R. J.; Haines, S. R. *J. Chem. Soc., Perkin Trans. 1*, **1985**, 2413–2416.
5.      Chida, N.; Ohtsuka, M.; Ogura, K.; Ogawa, S. *Bull. Chem. Soc. Jpn.* **1991**, *64*, 2118–2121.
6.      Machado, A. S.; Olesker, A.; Lukacs, G. *Carbohydr. Res.* **1985**, *135*, 231–239.
7.      Sato, K.-i.; Sakuma, S.; Nakamura, Y.; Yoshimura, J.; Hashimoto, H. *Chem. Lett.* **1991**, 17–20.
8.      Ermolenko, M. S.; Olesker, A.; Lukacs, G. *Tetrahedron Lett.* **1994**, *35*, 711–714.
9.      Amano, S.; Takemura, N.; Ohtsuka, M.; Ogawa, S.; Chida, N. *Tetrahedron* **1999**, *55*, 3855–3870.
10.     Park, T. K.; Danishefsky, S. J. *Tetrahedron Lett.* **1995**, *36*, 195–196.
11.     Boyer, F.-D.; Lallemand, J.-Y. *Tetrahedron* **1994**, *50*, 10443–10458.
12.     Das, S. K.; Mallet, J.-M.; Sinaÿ, P. *Angew. Chem. Int. Ed.* **1997**, *36*, 493–496.
13.     Sollogoub, M.; Mallet, J.-M.; Sinaÿ, P. *Tetrahedron Lett.* **1998**, *39*, 3471–3472.
14.     Bender, S. L.; Budhu, R. J. *J. Am. Chem. Soc.* **1991**, *113*, 9883–9884.
15.     Estevez, V. A.; Prestwich, E. D. *J. Am. Chem. Soc.* **1991**, *113*, 9885–9887.
16.     Yadav, J. S.; Reddy, B. V. S.; Narasimha Chary, D.; Madavi, C.; Kunwar, A. C. *Tetrahedron Lett.* **2009**, *50*, 81–84.
17.     Chen, P.; Wang, S. *Tetrahedron* **2013**, *69*, 583–588.
18.     Chen, P.; Lin, L. *Tetrahedron* **2013**, *69*, 4524–4531.

# Ferrier glycal allylic rearrangement

In the presence of Lewis acid catalysts $O$-substituted glycal derivatives can react with $O$-, $S$-, $C$- and, less frequently, $N$-, $P$- and halide nucleophiles to give 2,3-unsaturated glycosyl products.[1,2] This allylic transformation has been termed the "Ferrier Reaction" or, to avoid complications, the "Ferrier I Reaction" or the "Ferrier Rearrangement". However, the reaction was first noted by Emil Fischer when he heated tri-$O$-acetyl-$D$-glucal in water.[3] When carbon nucleophiles are involved, the term "Carbon Ferrier Reaction" has been used,[4] although the only contribution the Ferrier group made in this area was to find that tri-$O$-acetyl-$D$-glucal dimerizes under acid catalysis to give a $C$-glycosidic product.[5] The general reaction is illustrated by the separate conversions of tri-$O$-acetyl-$D$-glucal with $O$-, $S$- and $C$-nucleophiles to the corresponding 2,3-unsaturated glycosyl derivatives. Normally, Lewis acids are used as catalysts, boron trifluoride etherate being the most common. Allyloxycarbenium ions are involved as intermediates, high yields of products are obtained, and glycosidic compounds with quasi-axial bonds (as illustrated) predominate (commonly in the α,β-ratio of about 7:1). The examples illustrated[4,6,7] are typical of a very large number of literature reports.[1]

General examples[4]

More complex products made directly from the corresponding glycols:

J.J. Li, *Name Reactions: A Collection of Detailed Mechanisms and Synthetic Applications*,
DOI 10.1007/978-3-319-03979-4_103, © Springer International Publishing Switzerland 2014

Products formed without acid catalysts:

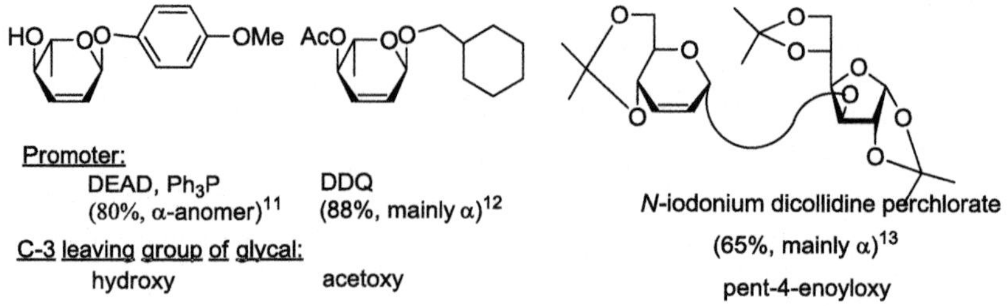

HO
O
O
OMe
AcO
O
O
Promoter:
DEAD, $Ph_3P$
(80%, α-anomer)[11]
DDQ
(88%, mainly α)[12]
N-iodonium dicollidine perchlorate
(65%, mainly α)[13]
C-3 leaving group of glycal:
hydroxy
acetoxy
pent-4-enoyloxy

Modified glycols and their reactions:

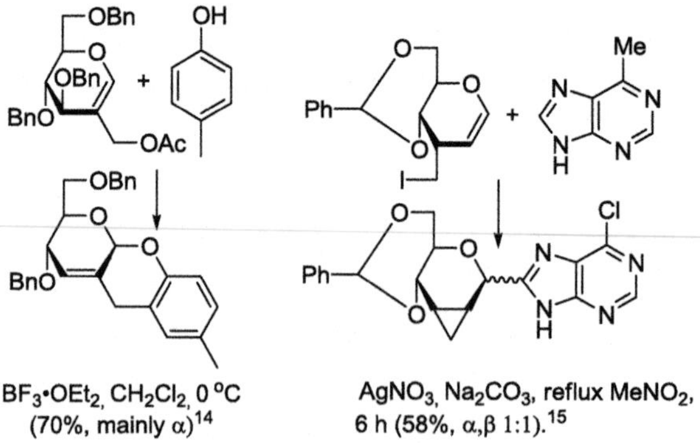

OBn
OH
O
OBn
BnO
OAc
Ph
Me
N
H
Cl
$BF_3$•$OEt_2$, $CH_2Cl_2$, 0 °C
(70%, mainly α)[14]
$AgNO_3$, $Na_2CO_3$, reflux $MeNO_2$,
6 h (58%, α,β 1:1).[15]

A variant using inexpensive Montmorillonite K-10 clay as the catalyst:

OBn
BnO
OBn OH
OH
Me
Me
cat. Mont. K10
$ClCH_2CH_2Cl$
rt, 6 h, 51%
OBn
OBn
BnO
Me
Me
OH

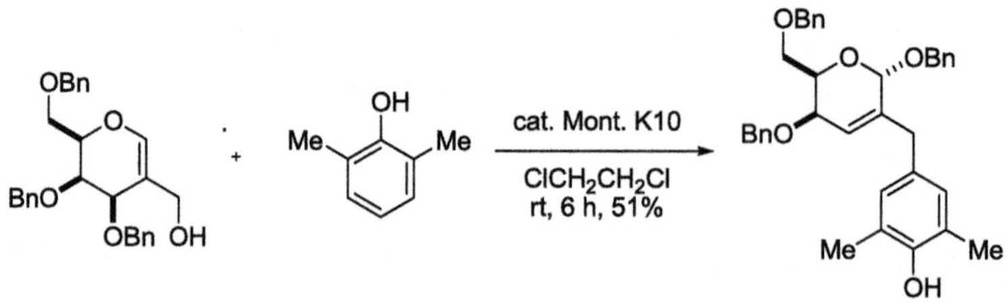

## References

1. Ferrier, R. J.; Zubkov, O. A. Transformation of glycals into 2,3-unsaturated glycosyl derivatives, In *Org. React.* **2003,** *62*, 569–736. (Review). It was almost 50 years after Fischer's seminal finding that water took part in the reaction[3] that Ann Ryan, working in George Overend's Department in Birkbeck College, University of London, found, by chance, that *p*-nitrophenol likewise participates.[16] Robin Ferrier, her immediate supervisor, who suggested her experiment, then found that simple alcohols at high temperatures also take part,[17] and with other students, notably Nagendra Prasad and George Sankey, he explored the reaction extensively. They did not apply it to make the very important *C*-glycosides.

2.  Ferrier, R. J. *Top. Curr. Chem.* **2001,** *215,* 153–175. (Review).
3.  Fischer, E. *Chem. Ber.* **1914,** *47,* 196–210.
4.  Herscovici, J.; Muleka, K.; Boumaïza, L.; Antonakis, K. *J. Chem. Soc., Perkin Trans. 1* **1990,** 1995–2009.
5.  Ferrier, R. J.; Prasad, N. *J. Chem. Soc. (C)* **1969,** 581–586.
6.  Moufid, N.; Chapleur, Y.; Mayon, P. *J. Chem. Soc., Perkin Trans. 1* **1992,** 999–1007.
7.  Whittman, M. D.; Halcomb, R. L.; Danishefsky, S. J.; Golik, J.; Vyas, D. *J. Org. Chem.* **1990,** *55,* 1979–1981.
8.  Klaffke, W.; Pudlo, P.; Springer, D.; Thiem, J. *Ann.* **1991,** 509–512.
9.  Yougai, S.; Miwa, T. *J. Chem. Soc., Chem. Commun.* **1983,** 68–69.
10.  Armstrong, P. L.; Coull, I. C.; Hewson, A. T.; Slater, M. J. *Tetrahedron Lett.* **1995,** *36,* 4311–4314.
11.  Sobti, A.; Sulikowski, G. A. *Tetrahedron Lett.* **1994,** *35,* 3661–3664.
12.  Toshima, K.; Ishizuka, T.; Matsuo, G.; Nakata, M.; Kinoshita, M. *J. Chem. Soc., Chem. Commun.* **1993,** 704–705.
13.  López, J. C.; Gómez, A. M.; Valverde, S.; Fraser-Reid, B. *J. Org. Chem.* **1995,** *60,* 3851–3858.
14.  Booma, C.; Balasubramanian, K. K. *Tetrahedron Lett.* **1993,** *34,* 6757–6760.
15.  Tam, S. Y.-K.; Fraser-Reid, B. *Can. J. Chem.* **1977,** *55,* 3996–4001.
16.  Ferrier, R. J.; Overend, W. G.; Ryan, A. E. *J. Chem. Soc. (C)* **1962,** 3667–3670.
17.  Ferrier, R. J. *J. Chem. Soc.* **1964,** 5443–5449.
18.  De, K.; Legros, J.; Crousse, B.; Bonnet-Delpon, D. *Tetrahedron* **2008,** *64,* 10497–10500.
19.  Kumaran, E.; Santhi, M., Balasubramanian, K. K.; Bhagavathy, S. *Carbohydr Res.* **2011,** *346,* 1654–1661.
20.  Okazaki, H.; Hanaya, K.; Shoji, M.; Hada, N.; Sugai, T. *Tetrahedron* **2013,** *69,* 7931–7935.

## Fiesselmann thiophene synthesis

Condensation reaction of thioglycolic acid derivatives with α,β-acetylenic esters, which upon treatment with base result in the formation of 3-hydroxy-2-thiophenecarboxylic acid derivatives.

Example 1[5]

J.J. Li, *Name Reactions: A Collection of Detailed Mechanisms and Synthetic Applications*,
DOI 10.1007/978-3-319-03979-4_104, © Springer International Publishing Switzerland 2014

Example 2[6]

Example 3[7]

Example 4[9]

## References

1. Fiesselmann, H.; Schipprak, P. *Ber.* **1954,** *87,* 835–841; Fiesselmann, H.; Schipprak, P.; Zeitler, L. *Ber.* **1954,** *87,* 841–848; Fiesselmann, H.; Pfeiffer, G. *Ber.* **1954,** *87,* 848; Fiesselmann, H.; Thoma, F. *Ber.* **1956,** *89,* 1907–1912; Fiesselmann, H.; Schipprak, P. *Ber.* **1956,** *89,* 1897–1902.
2. Gronowitz, S. In *Thiophene and Its Derivatives*, Part 1, Gronowitz, S., Ed.; Wiley: New York, **1985,** 88–125. (Review).
3. Nicolaou, K. C.; Skokotas, G.; Furuya, S.; Suemune, H.; Nicolaou, D. C. *Angew. Chem. Int. Ed.* **1990,** *29,* 1064–1068.
4. Mullican, M. D.; Sorenson, R. J.; Connor, D. T.; Thueson, D. O.; Kennedy, J. A.; Conroy, M. C. *J. Med. Chem.* **1991,** *34,* 2186–2194.
5. Donoso, R.; Jordan de Urries, P.; Lissavetzky, J. *Synthesis* **1992,** 526–528.
6. Ram, V. J.; Goel, A.; Shukla, P. K.; Kapil, A. *Bioorg. Med. Chem. Lett.* **1997,** *7,* 3101–3106.
7. Showalter, H. D. H.; Bridges, A. J.; Zhou, H.; Sercel, A. D.; McMichael, A.; Fry, D. W. *J. Med. Chem.* **1999,** *42,* 5464–5474.
8. Shkinyova, T. K.; Dalinger, I. L.; Molotov, S. I.; Shevelev, S. A. *Tetrahedron Lett.* **2000,** *41,* 4973–4975.
9. Redman, A. M.; Johnson, J. S.; Dally, R.; Swartz, S.; Wild, H.; Paulsen, H.; Caringal, Y.; Gunn, D.; Renick, J.; Osterhout, M. *Bioorg. Med. Chem. Lett.* **2001,** *11,* 9–12.
10. Migianu, E.; Kirsch, G. *Synthesis,* **2002,** 1096.
11. Mullins, R. J.; Williams, D. R. *Fiesselmann Thiophene Synthesis.* In *Name Reactions in Heterocyclic Chemistry*; Li, J. J., Ed.; Wiley: Hoboken, NJ, **2005,** pp 184–192. (Review).
12. Bezboruah, P.; Gogoi, P.; Junali Gogoi, J.; Boruah, R. C. *Synthesis* **2013,** *45,* 1341–1348.

## Fischer–Speier esterification

Esterification by refluxing a carboxylic acid and an alcohol in the presence of an acid catalyst. Often known as simply "Fischer esterification."

### References

1.   Fischer, E.; Speier, A. *Ber. Dtsch. Chem. Ges.* **1895**, *28*, 3252–3258.
2.   Hardy, J. P.; Kerrin, S. L.; Manatt, S. L. *J. Org. Chem.* **1973**, *38*, 4196–4200.
3.   Fujii, T.; Yoshifuji, S. *Chem. Pharm. Bull.* **1978**, *26*, 2253–2257.
4.   Pcolinski, M. J.; O'Mathuna, D. P.; Doskotch, R. W. *J. Nat. Prod.* **1995**, *58*, 209–216.
5.   Kai, T.; Sun, X.-L.; Tanaka, M.; Takayanagi, H.; Furuhata, K. *Chem. Pharm. Bull.* **1996**, *44*, 208–211.
6.   Birney, D. M.; Starnes, S. *J. Chem. Educ.* **1996**, *76*, 1560–1561.
7.   Cole, A. C.; Jensen, J. L.; Ntai, I.; Tran, K. L. T.; Weaver, K. J.; Forbes, D. C.; Davis, J. H., Jr. *J. Am. Chem. Soc.* **2002**, *124*, 5962–5963.
8.   Li, J. in *Name Reactions for Functional Group Transformations*, Li, J. J., Ed., Wiley: Hoboken, NJ, 2007. pp 458–461.
9.   Saavedra, H. M.; Thompson, C. M.; Hohman, J. N.; Crespi, V. H.; Weiss, P. S. *J. Am. Chem. Soc.* **2009**, *131*, 2252–2259.

J.J. Li, *Name Reactions: A Collection of Detailed Mechanisms and Synthetic Applications*,
DOI 10.1007/978-3-319-03979-4_105, © Springer International Publishing Switzerland 2014

# Fischer indole synthesis

Cyclization of arylhydrazones to indoles.

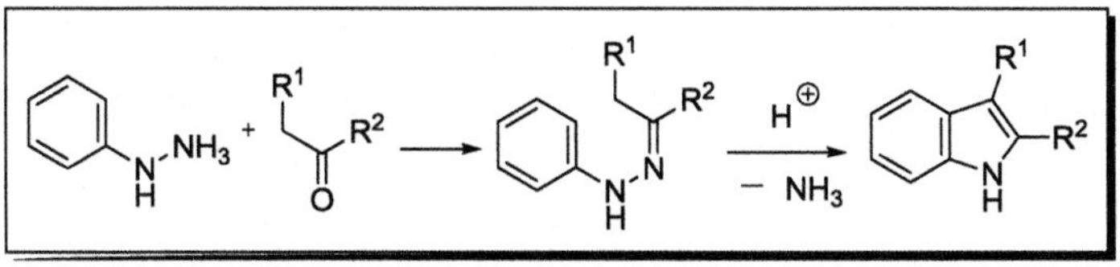
H⊕
− NH3

phenylhydrazine phenylhydrazine

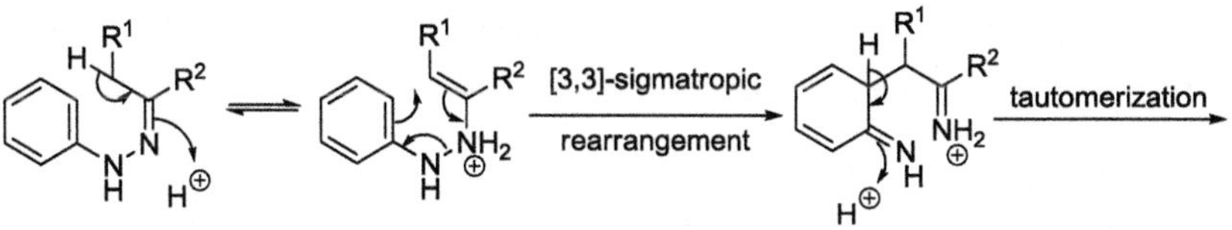
[3,3]-sigmatropic
rearrangement
tautomerization

protonation ene-hydrazine double imine

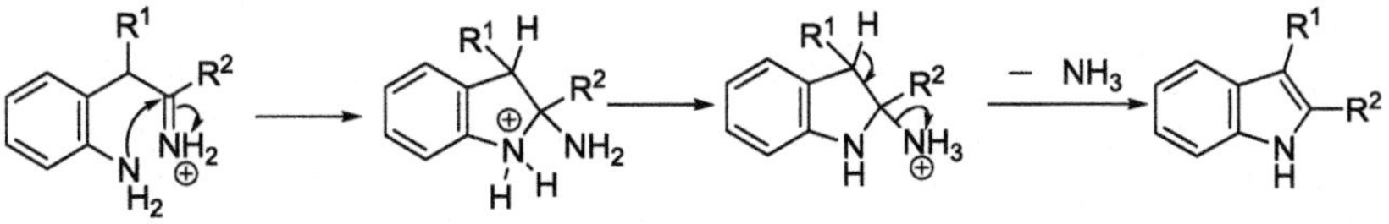
− NH3

Example 1[3]

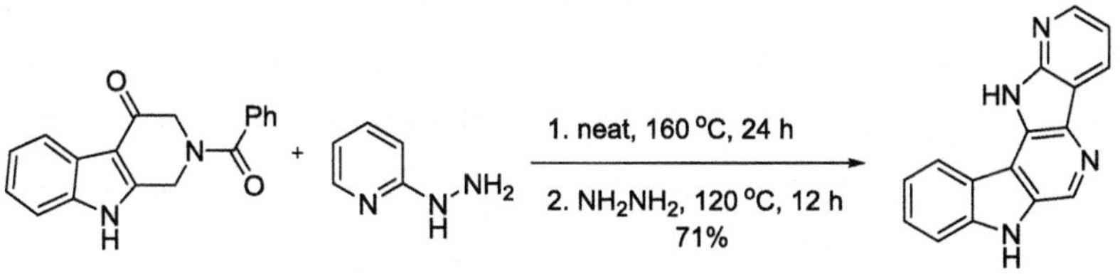
1. neat, 160 °C, 24 h
2. NH2NH2, 120 °C, 12 h
71%

Example 2[3]

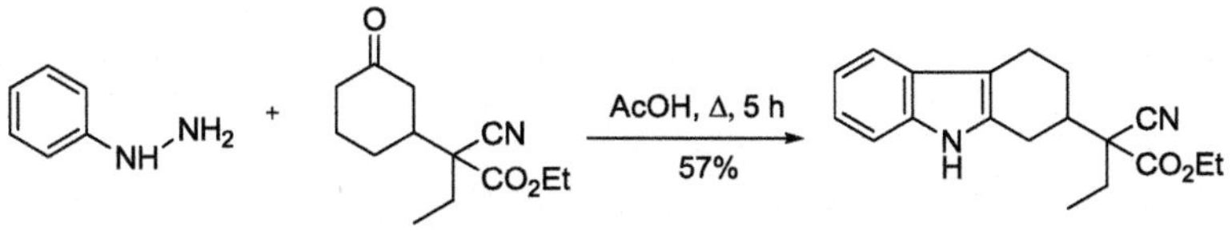
AcOH, Δ, 5 h
57%

J.J. Li, *Name Reactions: A Collection of Detailed Mechanisms and Synthetic Applications*,
DOI 10.1007/978-3-319-03979-4_106, © Springer International Publishing Switzerland 2014

Example 3[10]

Example 4[12]

## References

1.  (a) Fischer, E.; Jourdan, F. *Ber.* **1883,** *16,* 2241–2245. H. Emil Fischer (1852–1919) is arguably the greatest organic chemist ever. He was born in Euskirchen, near Bonn, Germany. When he was a boy, his father, Lorenz, said about him: "The boy is too stupid to go in to business; so in God's name, let him study." Fischer studied at Bonn and then Strassburg under Adolf von Baeyer. Fischer won the Nobel Prize in Chemistry in 1902 (three years ahead of his master, von Baeyer) for his synthetic studies in the area of sugar and purine groups. Sadly, Fischer committed suicide after WWI after his son died during the war and his fortunes completely gone. (b) Fischer, E.; Hess, O. *Ber.* **1884,** *17,* 559.
2.  Robinson, B. *The Fisher Indole Synthesis,* Wiley: New York, NY, **1982.** (Book).
3.  Martin, M. J.; Trudell, M. L.; Arauzo, H. D.; Allen, M. S.; LaLoggia, A. J.; Deng, L.; Schultz, C. A.; Tan, Y.; Bi, Y.; Narayanan, K.; Dorn, L. J.; Koehler, K. F.; Skolnick, P.; Cook, J. M. *J. Med. Chem.* **1992,** *35,* 4105–4117.
4.  Hughes, D. L. *Org. Prep. Proc. Int.* **1993,** *25,* 607–632. (Review).
5.  Bosch, J.; Roca, T.; Armengol, M.; Fernández-Forner, D. *Tetrahedron* **2001,** *57,* 1041–1048.
6.  Ergün, Y.; Patir, S.; Okay, G. *J. Heterocycl. Chem.* **2002,** *39,* 315–317.
7.  Pete, B.; Parlagh, G. *Tetrahedron Lett.* **2003,** *44,* 2537–2539.
8.  Li, J.; Cook, J. M. *Fischer Indole Synthesis.* In *Name Reactions in Heterocyclic Chemistry*; Li, J. J., Ed.; Wiley: Hoboken, NJ, **2005,** pp 116–127. (Review).
9.  Borregán, M.; Bradshaw, B.; Valls, N.; Bonjoch, J. *Tetrahedron: Asymmetry* **2008,** *19,* 2130–2134.
10. Boal, B. W.; Schammel A. W.; Garg, N. K. *Org. Lett.* **2013,** *11,* 3458–3461.
11. Donald, J. R.; Taylor, R. J. K. *Synlett* **2009,** 59–62.
12. Adams, G. L.; Carroll, P. J.; Smith, A. B. III *J. Am. Chem. Soc.* **2013,** *135,* 519–523.

# Fischer oxazole synthesis

Oxazoles from the condensation of equimolar amounts of aldehyde cyanohydrins and aromatic aldehydes in dry ether in the presence of dry hydrochloric acid.

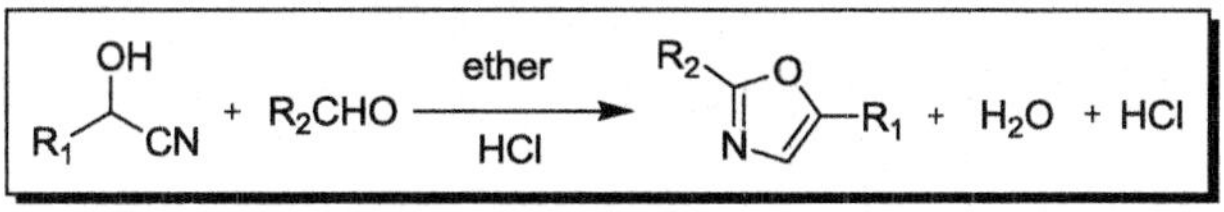

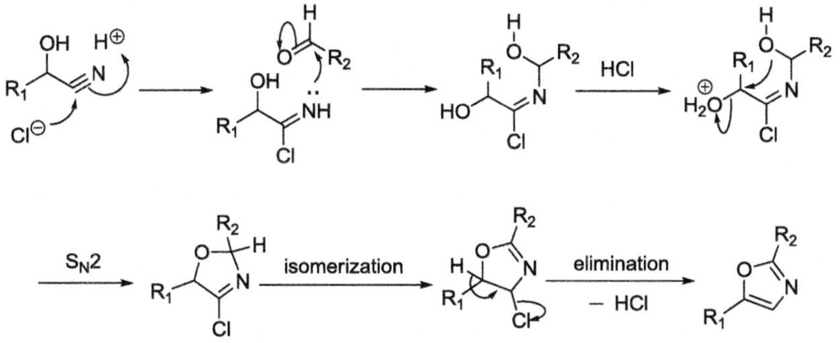

Example 1[4]

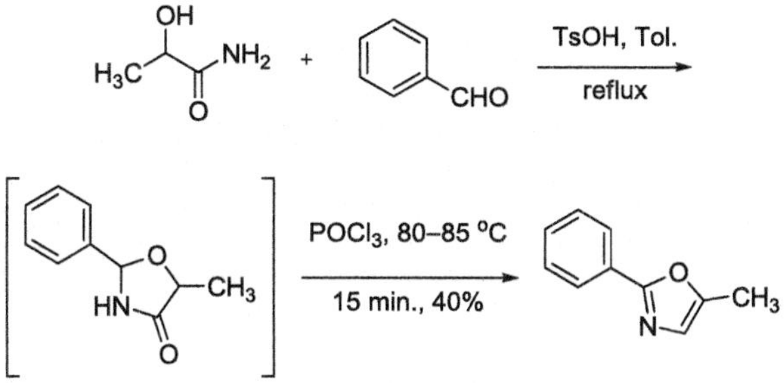

Example 2[8]

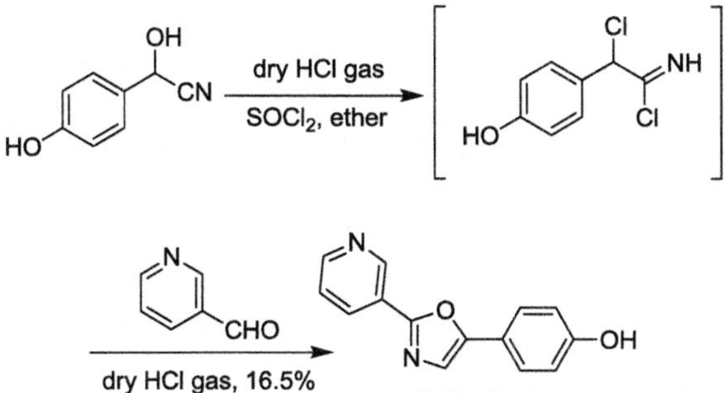

halfordinal

J.J. Li, *Name Reactions: A Collection of Detailed Mechanisms and Synthetic Applications*, DOI 10.1007/978-3-319-03979-4_107, © Springer International Publishing Switzerland 2014

## References

1.  Fischer, E. *Ber.* **1896,** *29,* 205.
2.  Ladenburg, K.; Folkers, K.; Major, R. T. *J. Am. Chem. Soc.* **1936,** *58*, 1292–1294.
3.  Wiley, R. H. *Chem. Rev.* **1945,** *37*, 401–442. (Review).
4.  Cornforth, J. W.; Cornforth, R. H. *J. Chem. Soc.* **1949,** 1028–1030.
5.  Cornforth, J. W. In *Heterocyclic Compounds 5*; Elderfield, R. C., Ed.; Wiley: New York, **1957,** *5,* 309–312. (Review).
6.  Crow, W. D.; Hodgkin, J. H. *Tetrahedron Lett.* **1963,** *2*, 85–89.
7.  Brossi, A.; Wenis, E. *J. Heterocycl. Chem.* **1965,** *2*, 310–312.
8.  Onaka, T. *Tetrahedron Lett.* **1971,** 4393–4394.
9.  Brooks, D. A. *Fisher Oxazole Synthesis.* In *Name Reactions in Heterocyclic Chemistry*; Li, J. J., Ed.; Wiley: Hoboken, NJ, **2005,** pp 234–236. (Review).

# Fleming–Kumada oxidation

Stereoselective oxidation of alkyl-silanes into the corresponding alkyl-alcohols using peracids.

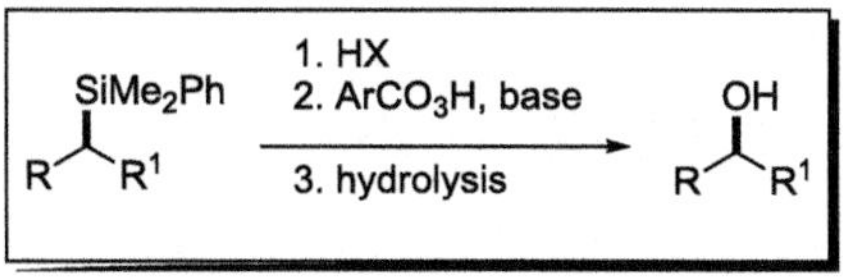

*retention* of configuration

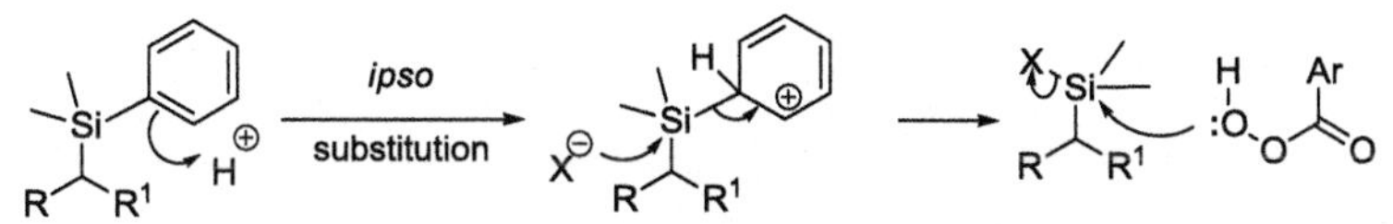

the β-carbocation is stabilized by the silicon group

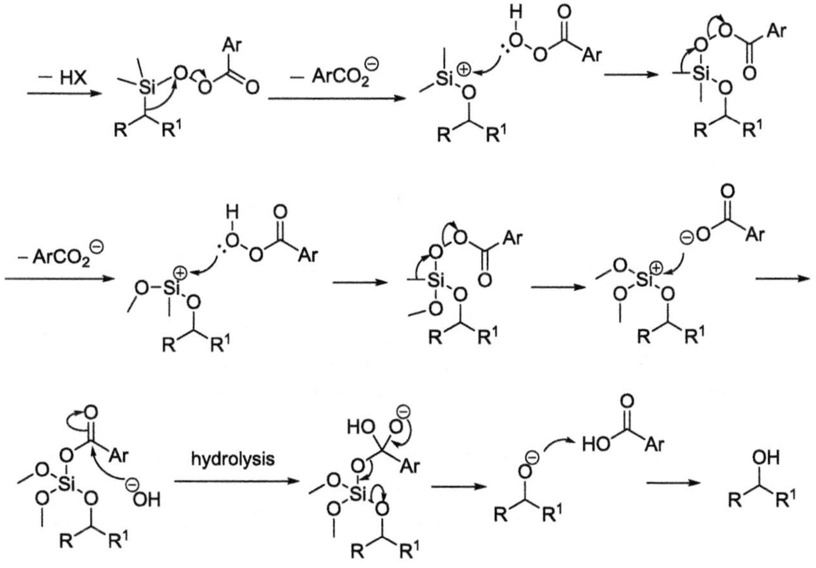

Example 1[4]

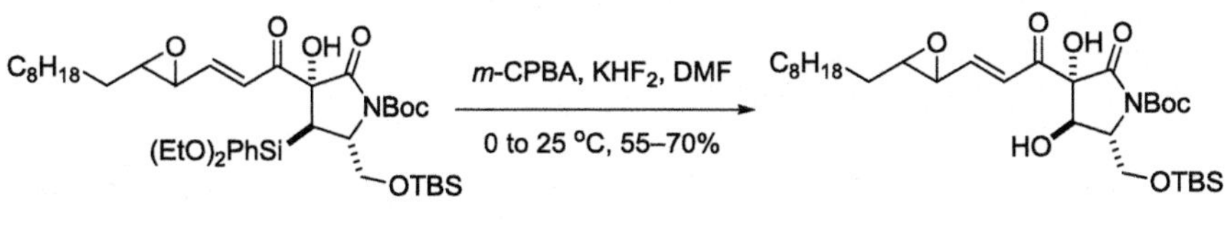

J.J. Li, *Name Reactions: A Collection of Detailed Mechanisms and Synthetic Applications*,
DOI 10.1007/978-3-319-03979-4_108, © Springer International Publishing Switzerland 2014

Example 2[5]

Example 3[8]

Example 4[9]

## References

1.    (a) Fleming, I.; Henning, R.; Plaut, H. *J. Chem. Soc., Chem. Commun.* **1984**, 29–31.
      (b) Fleming, I.; Sanderson, P. E. J. *Tetrahedron Lett.* **1987**, *28*, 4229–4232. (c) Flem-
      ing, I.; Dunoguès, J.; Smithers, R. *Org. React.* **1989**, *37*, 57–576. (Review).
2.    Hunt, J. A.; Roush, W. R. *J. Org. Chem.* **1997**, *62*, 1112–1124.
3.    Knölker, H.-J.; Jones, P. G.; Wanzl, G. *Synlett* **1997**, 613–616.
4.    Barrett, A. G. M.; Head, J.; Smith, M. L.; Stock, N. S.; White, A. J. P.; Williams, D. J.
      *J. Org. Chem.* **1999**, *64*, 6005–6018.
5.    Denmark, S.; Cottell, J. *J. Org. Chem.* **2001**, *66*, 4276–4284.
6.    Lee, T. W.; Corey, E. J. *Org. Lett.* **2001**, *3*, 3337–3339.
7.    Jung, M. E.; Piizzi, G. *J. Org. Chem.* **2003**, *68*, 2572–2582.
8.    Paquette, L. A.; Yang, J.; Long, Y. O. *J. Am. Chem. Soc.* **2003**, *125*, 1567–1574.
9.    Clive, D. L. J.; Cheng, H.; Gangopadhyay, P.; Huang, X.; Prabhudas, B. *Tetrahedron*
      **2004**, *60*, 4205–4221.
10.   Mullins, R. J.; Jolley, S. L.; Knapp, A. R. *Tamao–Kumada–Fleming Oxidation*. In
      *Name Reactions for Functional Group Transformations*; Li, J. J., Ed.; Wiley: Hobo-
      ken, NJ, **2007**, pp 237–247. (Review).

## *Tamao–Kumada oxidation*

Oxidation of alkyl fluorosilanes to the corresponding alcohols. A variant of the Fleming–Kumada oxidation.

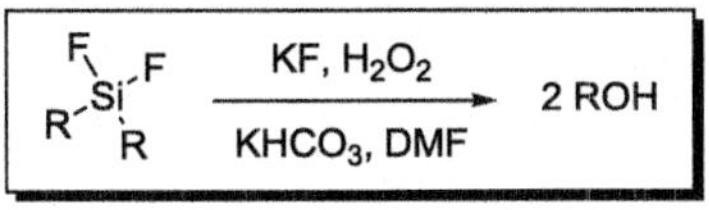

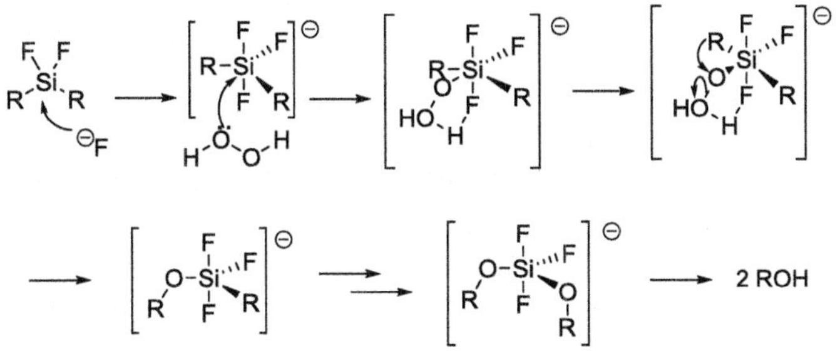

Example 1[3]

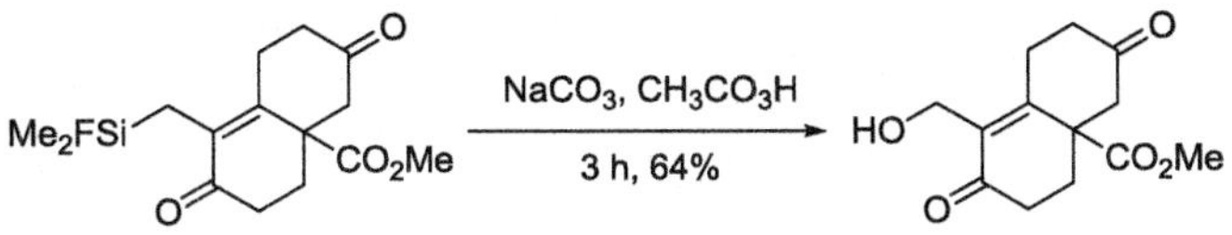

Example 2[4]

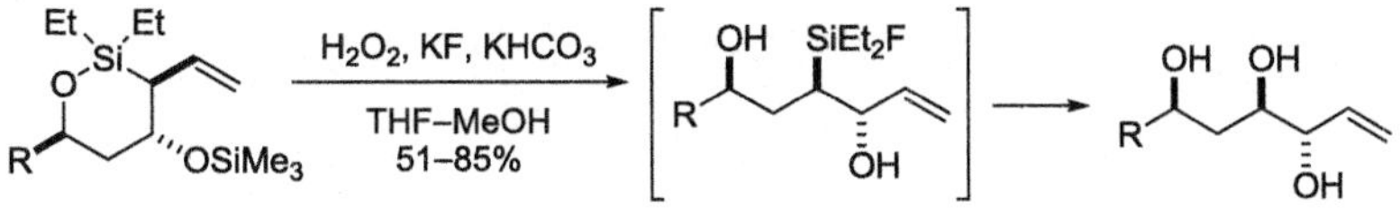

## References

1.  Tamao, K.; Ishida, N.; Kumada, M. *J. Org. Chem.* **1983**, *48*, 2120–2122.
2.  Fleming, I.; Dunoguès, J.; Smithers, R. *Org. React.* **1989**, *37*, 57–576. (Review).
3.  Kim, S.; Emeric, G.; Fuchs, P. L. *J. Org. Chem.* **1992**, *57*, 7362–7364.
4.  Mullins, R. J.; Jolley, S. L.; Knapp, A. R. *Tamao–Kumada–Fleming Oxidation*. In *Name Reactions for Functional Group Transformations*; Li, J. J., Ed.; Wiley: Hoboken, NJ, **2007**, pp 237–247. (Review).
5.  Beignet, J.; Jervis, P. J.; Cox, L. R. *J. Org. Chem.* **2008**, *73*, 5462–5475.
6.  Cardona, F.; Parmeggiani, C.; Faggi, E.; Bonaccini, C.; Gratteri, P.; Sim, L.; Gloster, T. M.; Roberts, S.; Davies, G. J.; Rose, D. R.; Goti, A. *Chem. Eur. J.* **2009**, *15*, 1627–1636.
7.  Terauchi, T.; Machida, S.; Komba, S. *Tetrahedron Lett.* **2010**, *51*, 1497–1499.

# Friedel–Crafts reaction

## *Friedel–Crafts acylation reaction:*

Introduction of an acyl group onto an aromatic substrate by treating the substrate with an acyl halide or anhydride in the presence of a Lewis acid.

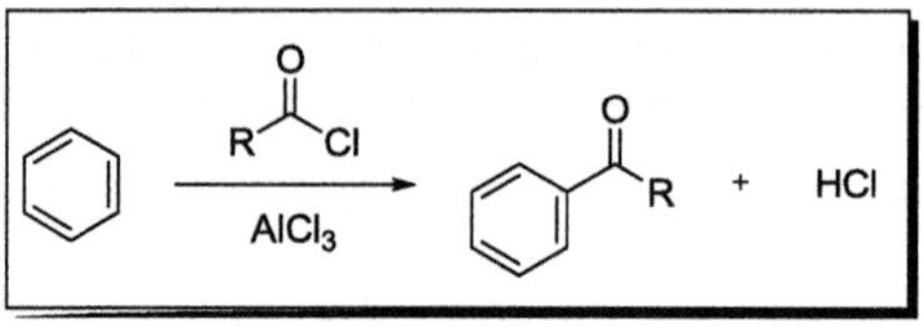
AlCl3
R
Cl
+ HCl

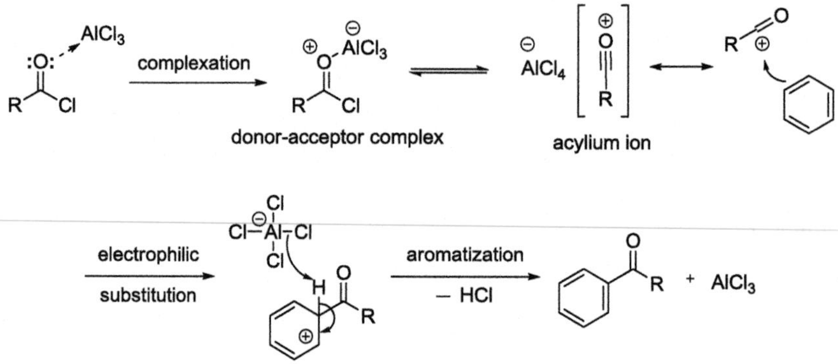
complexation
donor-acceptor complex
acylium ion
electrophilic
substitution
aromatization
– HCl
+ AlCl3

Example 1, Intermolecular Friedel–Crafts acylation[6]

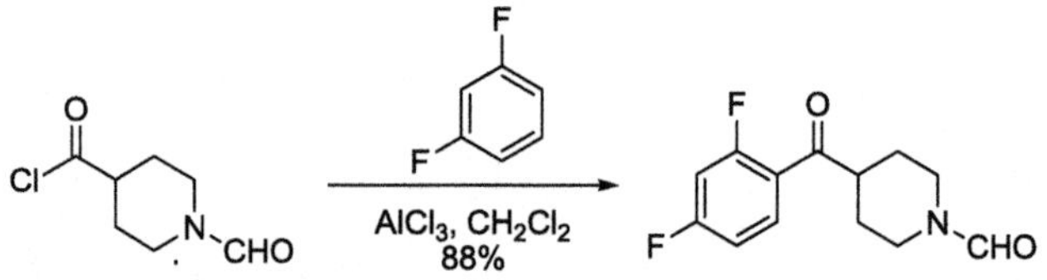
AlCl3, CH2Cl2
88%
CHO

Example 2, Intramolecular Friedel–Crafts acylation[7]

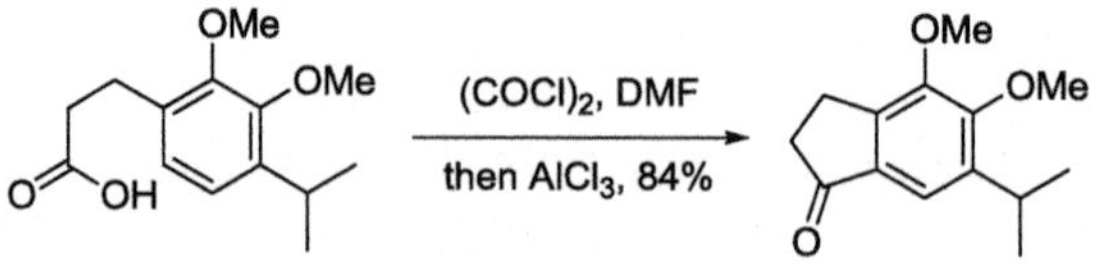
(COCl)2, DMF
then AlCl3, 84%
OMe
OMe
OH

Example 3, Intramolecular Friedel–Crafts acylation[8]

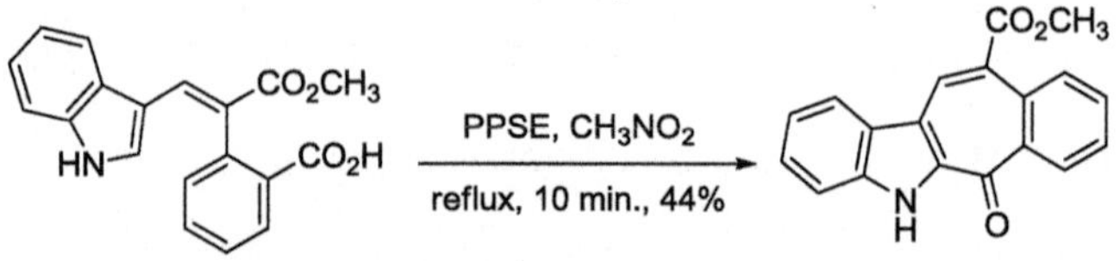
PPSE, CH3NO2
reflux, 10 min., 44%
CO2CH3
CO2H

PPSE = Trimethylsilyl polyphosphate

J.J. Li, *Name Reactions: A Collection of Detailed Mechanisms and Synthetic Applications*,
DOI 10.1007/978-3-319-03979-4_109, © Springer International Publishing Switzerland 2014

Example 4, Intramolecular Friedel–Crafts acylation[9]

Example 5, "Kinetic Capture" of Acylium Ion[11]

donor-acceptor complex                    acylium ion

## References

1. Friedel, C.; Crafts, J. M. *Compt. Rend.* **1877**, *84*, 1392–1395. Charles Friedel (1832–1899) was born in Strasbourg, France. He earned his Ph.D. In 1869 under Wurtz at Sorbonne and became a professor and later chair (1884) of organic chemistry at Sorbonne. Friedel was one of the founders of the French Chemical Society and served as its president for four terms. James Mason Crafts (1839–1917) was born in Boston, Massachusetts. He studied under Bunsen and Wurtz in his youth and became a professor at Cornell and MIT. From 1874 to 1891, Crafts collaborated with Friedel at École de Mines in Paris, where they discovered the Friedel–Crafts reaction. He returned to MIT in 1892 and later served as its president. The discovery of the Friedel–Crafts reaction was the fruit of serendipity and keen observation. In 1877, both Friedel and Crafts were working in Charles A. Wurtz's laboratory. In order to prepare amyl iodide, they treated amyl chloride with aluminum and iodide using benzene as the solvent. Instead of amyl iodide, they ended up with amylbenzene! Unlike others before them who may have simply discarded the reaction, they thoroughly investigated the Lewis acid-catalyzed alkylations and acylations and published more than 50 papers and patents on the Friedel–Crafts reaction, which has become one of the most useful organic reactions.

2. Pearson, D. E.; Buehler, C. A. *Synthesis* **1972**, 533–542. (Review).

3.  Hermecz, I.; Mészáros, Z. *Adv. Heterocyclic Chem.* **1983,** *33,* 241–330. (Review).
4.  Metivier, P. *Friedel-Crafts Acylation.* In *Friedel-Crafts Reaction* Sheldon, R. A.; Bekkum, H., eds.; Wiley-VCH: New York. **2001,** pp 161–172. (Review).
5.  Basappa; Mantelingu, K.; Sadashira, M. P.; Rangappa, K. S. *Indian J. Chem. B.* **2004,** *43B,* 1954–1957.
6.  Olah, G. A.; Reddy, V. P.; Prakash, G. K. S. *Chem. Rev.* **2006,** *106,* 1077–1104. (Review).
7.  Simmons, E.M.; Sarpong, R. *Org. Lett.* **2006,** *8,* 2883–2886.
8.  Bourderioux, A.; Routier, S.; Beneteau, V.; Merour, J.-Y. *Tetrahedron* **2007,** *63,* 9465–9475.
9.  Fillion, E.; Dumas, A. M. *J. Org. Chem.* **2008,** *73,* 2920–2923.
10. de Noronha, R. G.; Fernandes, A. C.; Romao, C. C. *Tetrahedron Lett.* **2009,** *50,* 1407–1410.
11. Huang, Z.; Jin, L.; Han, H.; Lei, A. *Org. Biomol. Chem.* **2013,** *11,* 1810–1814.

## *Friedel–Crafts alkylation reaction:*

Introduction of an alkyl group onto an aromatic substrate by treating the substrate with an alkylating agent such as alkyl halide, alkene, alkyne and alcohol in the presence of a Lewis acid.

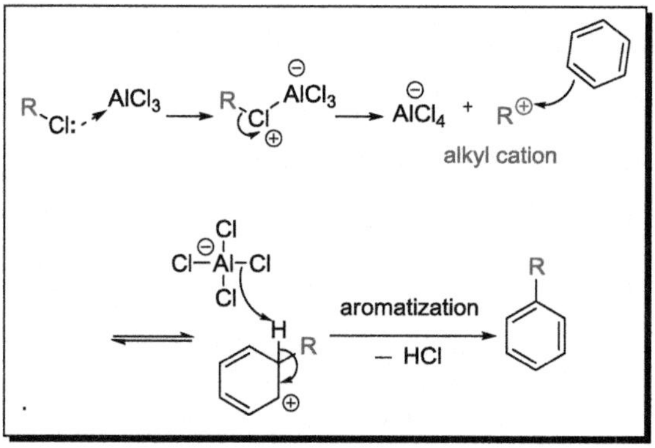

Example 1[1]

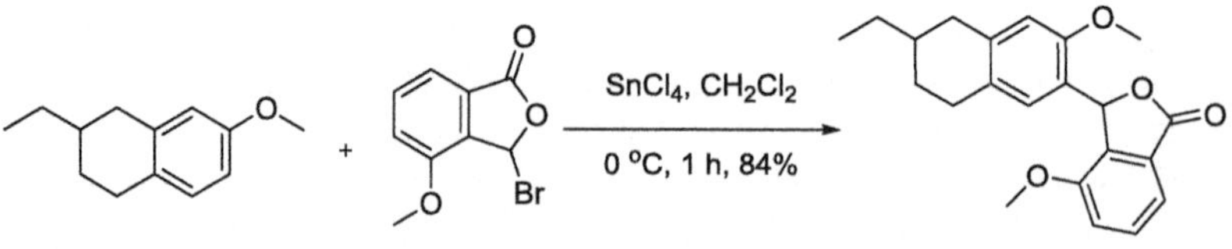

Example 2, An intramolecular Friedel–Crafts cyclization[6]

## References

1.  Patil, M. L.; Borate, H. B.; Ponde, D. E.; Bhawal, B. M.; Deshpande, V. H. *Tetrahedron Lett.* **1999,** *40*, 4437–4438.

2.  Meima, G. R.; Lee, G. S.; Garces, J. M. *Friedel-Crafts Alkylation.* In *Friedel–Crafts Reaction* Sheldon, R. A.; Bekkum, H., eds.; Wiley-VCH: New York. **2001,** pp 550–556. (Review).

3.  Bandini, M.; Melloni, A.; Umani-Ronchi, A. *Angew. Chem. Int. Ed.* **2004,** *43*, 550–556. (Review).

4.  Poulsen, T. B.; Jorgensen, K. A. *Chem. Rev.* **2008,** *108*, 2903–2915. (Review).

5.  Silvanus, A. C.; Heffernan, S. J.; Liptrot, D. J.; Kociok-Kohn, G.; Andrews, B. I.; Carbery, D. R. *Org. Lett.* **2009,** *11*, 1175–1178.

6.  Kargbo, R. B.; Sajjadi-Hashemi, Z.; Roy, S.; Jin, X.; Herr, R. J. *Tetrahedron Lett.* **2013,** *54*, 2018–2021.

# Friedländer quinoline synthesis

Also known as the Friedländer condensation, it combines an α-amino aldehyde or ketone with another aldehyde or ketone with at least one methylene α adjacent to the carbonyl to furnish a substituted quinoline. The reaction can be promoted by either acid, base, or heat.

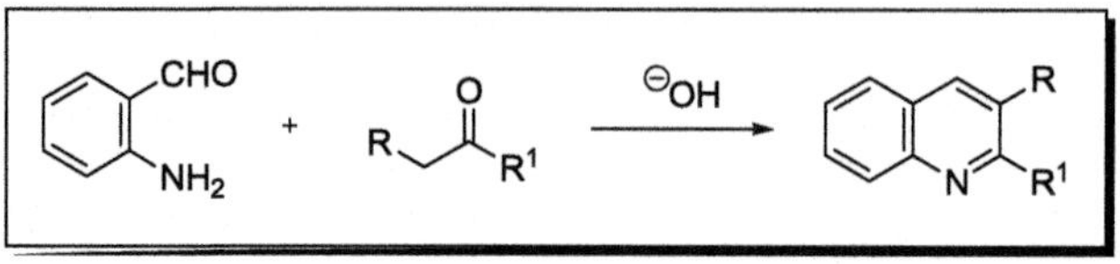

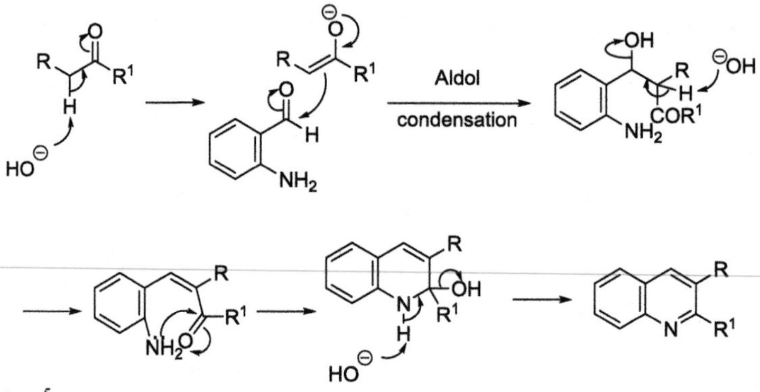

Example 1[5]

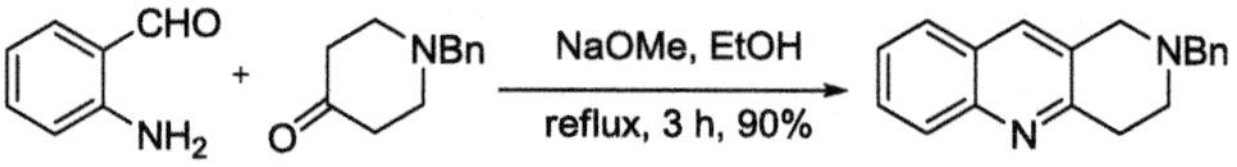

Example 2[7]

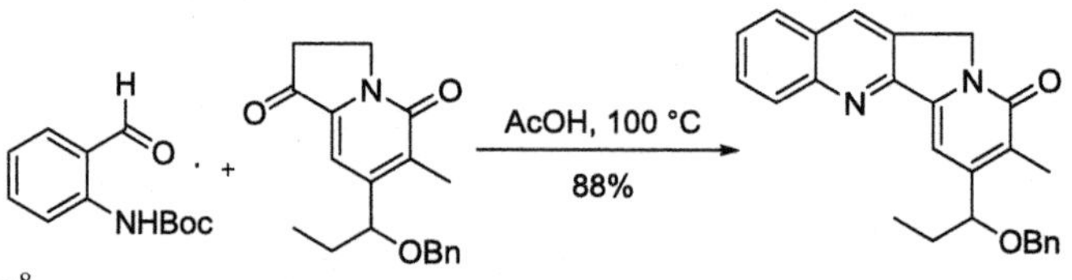

Example 3[8]

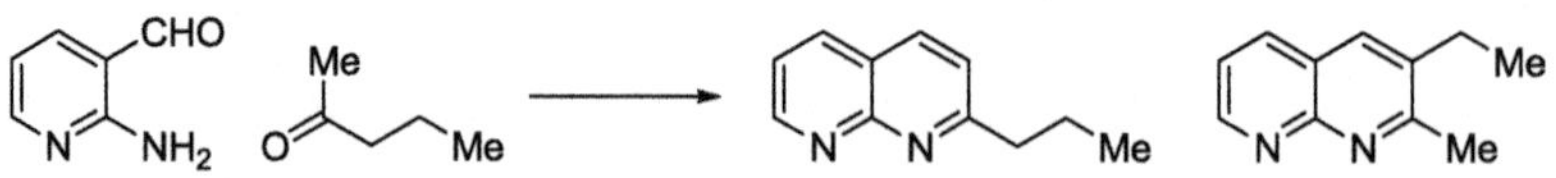

| Conditions | Conversion | Ratio |
|---|---|---|
| NaOH, rt | > 99% | 37:63 |
| pyrrolidine, 5% $H_2SO_4$, rt | 97% | 86:14 |
| **TBAO**, 5% $H_2SO_4$, rt | > 99% | 87:13 |
| **TBAO**, 5% $H_2SO_4$, slow addition, 65 °C | > 99% | 94:6 |

**TBAO** = 1,3,3-trimethyl-6-azabicyclo[3.2.1]octane

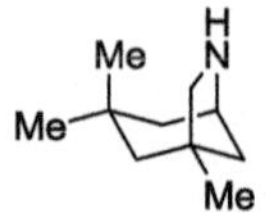

J.J. Li, *Name Reactions: A Collection of Detailed Mechanisms and Synthetic Applications*, DOI 10.1007/978-3-319-03979-4_110, © Springer International Publishing Switzerland 2014

Example 4[10]

Example 5, Using propylphosphonic anhydride (T3P) as the coupling agent[11]

## References

1.  Friedländer, P. *Ber.* **1882**, *15*, 2572–2575. Paul Friedländer (1857–1923), born in Königsberg, Prussia, apprenticed under Carl Graebe and Adolf von Baeyer. He was interested in music and was an accomplished pianist.
2.  Elderfield, R. C. In *Heterocyclic Compounds*, Elderfield, R. C., ed.; Wiley: New York, **1952**, *4*, *Quinoline, Isoquinoline and Their Benzo Derivatives*, 45–47. (Review).
3.  Jones, G. In *Heterocyclic Compounds*, Quinolines, vol. 32, **1977**; Wiley: New York, pp 181–191. (Review).
4.  Cheng, C.-C.; Yan, S.-J. *Org. React.* **1982**, *28*, 37–201. (Review).
5.  Shiozawa, A.; Ichikawa, Y.-I.; Komuro, C.; Kurashige, S.; Miyazaki, H.; Yamanaka, H.; Sakamoto, T. *Chem. Pharm. Bull.* **1984**, *32*, 2522–2529.
6.  Gladiali, S.; Chelucci, G.; Mudadu, M. S.; Gastaut, M.-A.; Thummel, R. P. *J. Org. Chem.* **2001**, *66*, 400–405.
7.  Henegar, K. E.; Baughman, T. A. *J. Heterocycl. Chem.* **2003**, *40*, 601–605.
8.  Dormer, P. G.; Eng, K. K.; Farr, R. N.; Humphrey, G. R.; McWilliams, J. C.; Reider, P. J.; Sager, J. W.; and Volante, R. P. *J. Org. Chem.* **2003**, *68*, 467–477.
9.  Pflum, D. A. *Friedländer Quinoline Synthesis*. In *Name Reactions in Heterocyclic Chemistry*; Li, J. J., Ed.; Wiley: Hoboken, NJ, **2005**, 411–415. (Review).
10. Vander Mierde, H.; Van Der Voot, P.; De Vos, D.; Verpoort, F. *Eur. J. Org. Chem.* **2008**, 1625–1631.
11. Augustine, J. K.; Bombrun, A.; Venkatachaliah, S. *Tetrahedron Lett.* **2011**, *52*, 6814–6818.

# Fries rearrangement

Lewis acid-catalyzed rearrangement of phenol esters and lactams to 2- or 4-ketophenols. Also known as the Fries–Finck rearrangement.

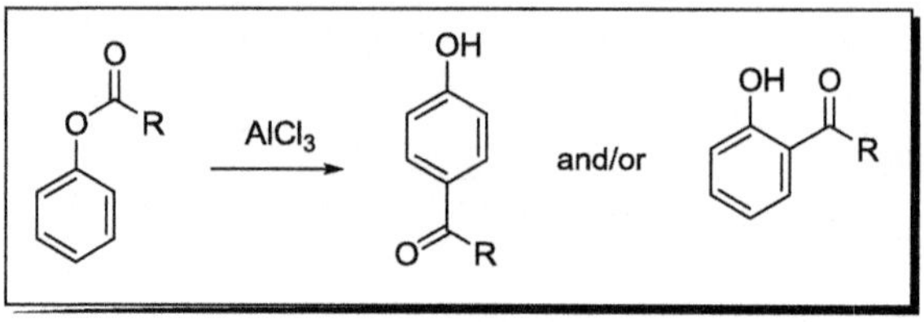

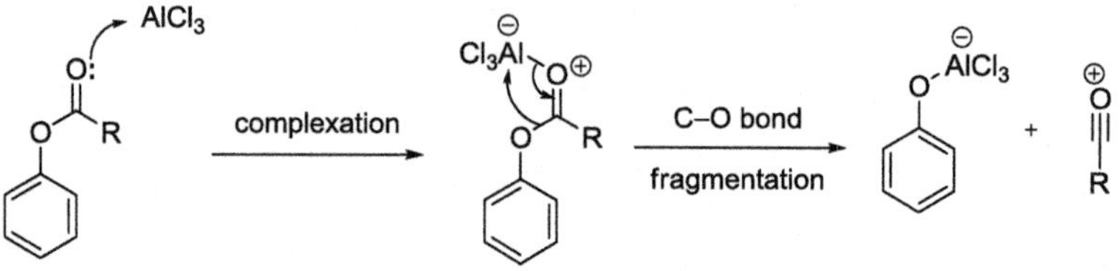

aluminum phenolate, acylium ion

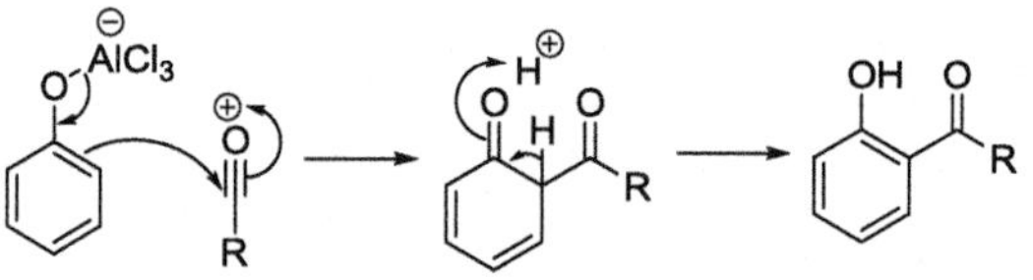

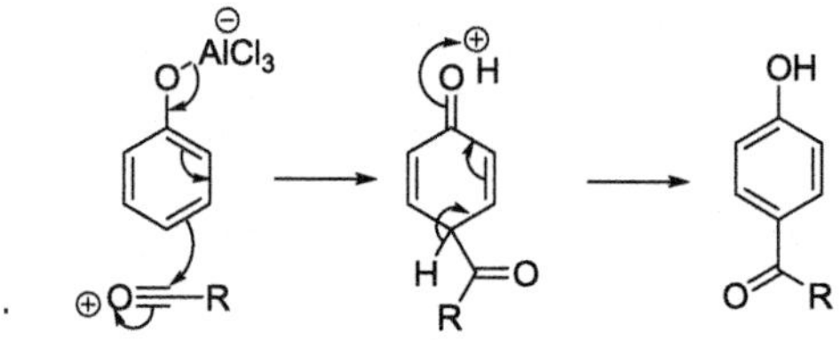

Example 1[5]

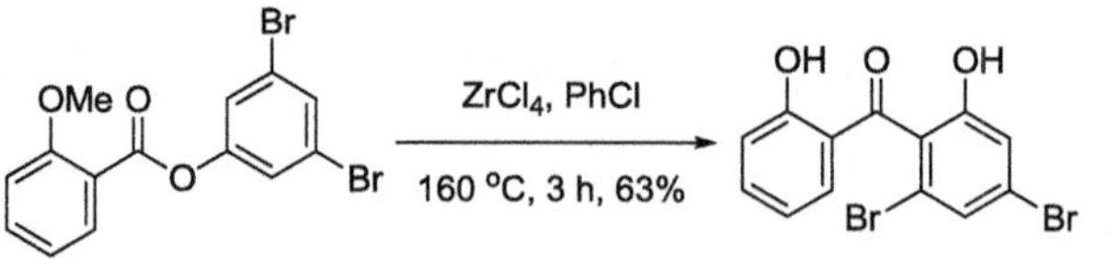

ZrCl$_4$, PhCl
160 °C, 3 h, 63%

Example 2[6]

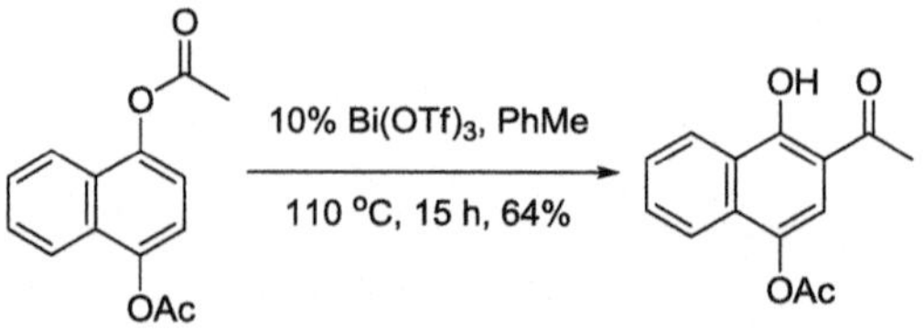

10% Bi(OTf)$_3$, PhMe
110 °C, 15 h, 64%

J.J. Li, *Name Reactions: A Collection of Detailed Mechanisms and Synthetic Applications*,
DOI 10.1007/978-3-319-03979-4_111, © Springer International Publishing Switzerland 2014

Example 3, Photo-Fries rearrangement[7]

Example 4, *ortho*-Fries rearrangement[8]

Example 5, Thia-Fries rearrangement[9]

Example 6, Remote Anionic Thia-Fries rearrangement[10]

## References

1. Fries, K.; Finck, G. *Ber.* **1908**, *41*, 4271–4284. Karl Theophil Fries (1875–1962) was born in Kiedrich near Wiesbaden on the Rhine. He earned his doctorate under Theodor Zincke. Although G. Finck co-discovered the rearrangement of phenolic esters, somehow his name has been forgotten by history. In all fairness, the Fries rearrangement should really be the Fries–Finck rearrangement.

2. Martin, R. *Org. Prep. Proced. Int.* **1992**, *24*, 369–435. (Review).

3. Boyer, J. L.; Krum, J. E.; Myers, M. C.; Fazal, A. N.; Wigal, C. T. *J. Org. Chem.* **2000**, *65*, 4712–4714.

4. Guisnet, M.; Perot, G. *The Fries rearrangement*. In *Fine Chemicals through Heterogeneous Catalysis* **2001**, 211–216. (Review).

5. Tisserand, S.; Baati, R.; Nicolas, M.; Mioskowski, C. *J. Org. Chem.* **2004**, *69*, 8982–8983.

6. Ollevier, T.; Desyroy, V.; Asim, M.; Brochu, M.-C. *Synlett* **2004**, 2794–2796.

7. Ferrini, S.; Ponticelli, F.; Taddei, M. *Org. Lett.* **2007**, *9*, 69–72.

8. Macklin, T. K.; Panteleev, J.; Snieckus, V. *Angew. Chem. Int. Ed.* **2008**, *47*, 2097–2101.

9. Dyke, A. M.; Gill, D. M.; Harvey, J. N.; Hester, A. J.; Lloyd-Jones, G. C.; Munoz, M. P.; Shepperson, I. R. *Angew. Chem. Int. Ed.* **2008**, *47*, 5067–5070.

10. Xu, X.-H.; Taniguchi, M.; Azuma, A.; Liu, G. K.; Tokunaga, E.; Shibata, N. *Org. Lett.* **2013**, *15*, 686–689.

## Fukuyama amine synthesis

Transformation of a primary amine to a secondary amine using 2,4-dinitro-benzenesulfonyl chloride and an alcohol. Also known as the Fukuyama–Mitsunobu procedure.

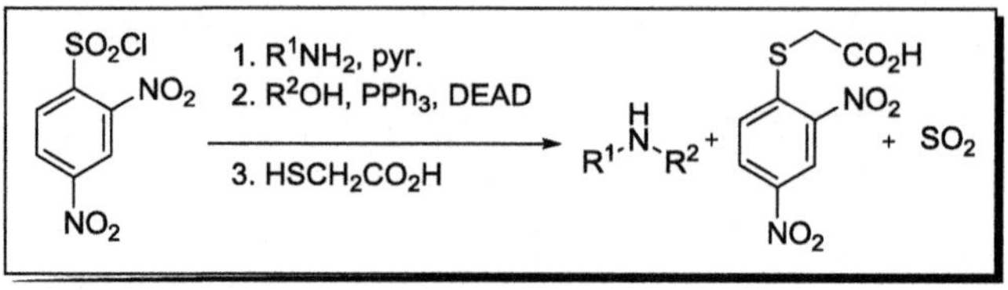

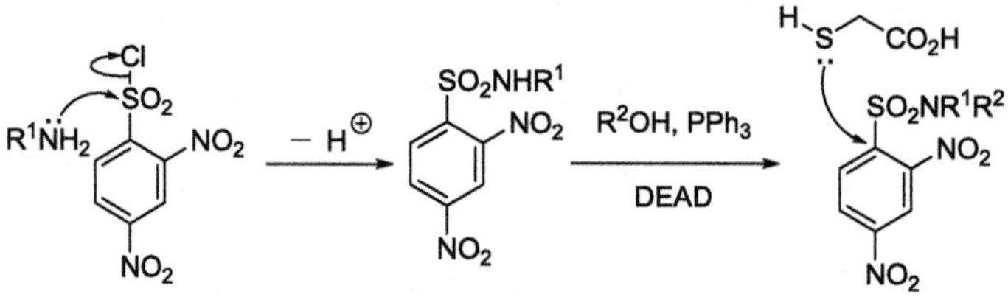

See the Mitsunobu reaction for mechanism.

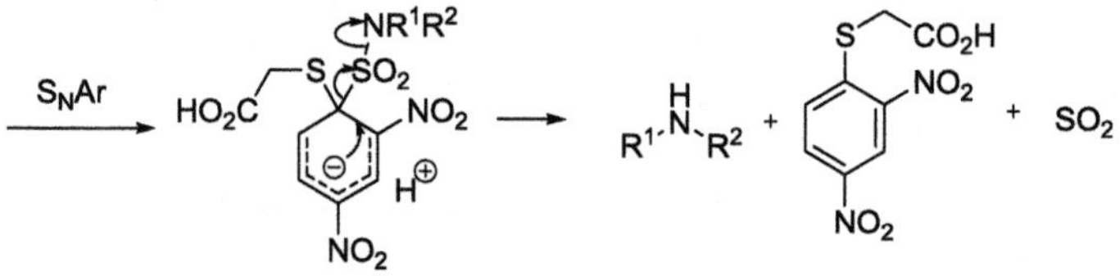

Meisenheimer complex

Example 1[6]

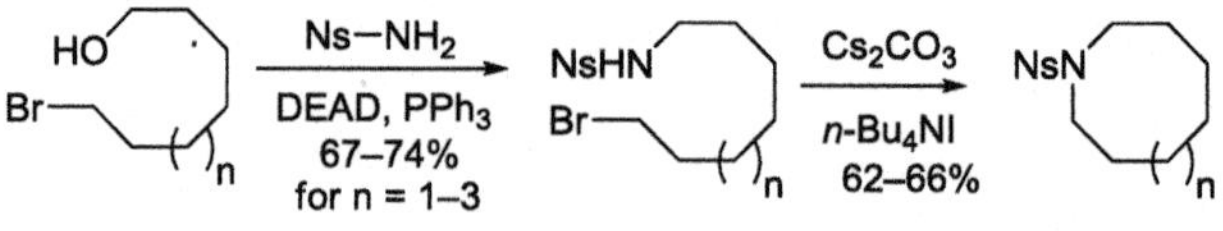

Example 2[7]

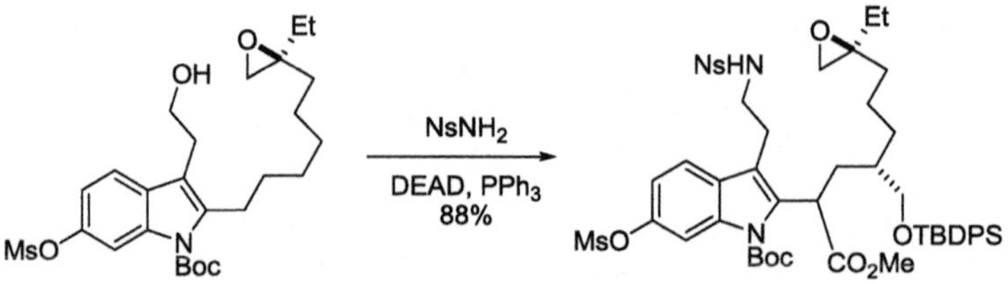

J.J. Li, *Name Reactions: A Collection of Detailed Mechanisms and Synthetic Applications*, DOI 10.1007/978-3-319-03979-4_112, © Springer International Publishing Switzerland 2014

Example 3[8]

PyPh$_2$P = diphenyl 2-pyridylphosphine; DTBAD = di-*tert*-butylazodicarbonate

## References

1.  (a) Fukuyama, T.; Jow, C.-K.; Cheung, M. *Tetrahedron Lett.* **1995**, *36*, 6373–6374. Tohru Fukuyama earned his Ph.D. at Harvard under the tutelage of Kishi. He started his independent academic career at Rice University but moved to the University of Tokyo from in 1995. (b) Fukuyama, T.; Cheung, M.; Jow, C.-K.; Hidai, Y.; Kan, T. *Tetrahedron Lett.* **1997**, *38*, 5831–5834.

2.  Piscopio, A. D.; Miller, J. F.; Koch, K. *Tetrahedron Lett.* **1998**, *39*, 2667–2670.

3.  Bolton, G. L.; Hodges, J. C. *J. Comb. Chem.* **1999**, *1*, 130–133.

4.  Lin, X.; Dorr, H.; Nuss, J. M. *Tetrahedron Lett.* **2000**, *41*, 3309–3313.

5.  Olsen, C. A.; Jørgensen, M. R.; Witt, M.; Mellor, I. R.; Usherwood, P. N. R.; Jaroszewski, J. W.; Franzyk, H. *Eur. J. Org. Chem.* **2003**, 3288–3299.

6.  Kan, T.; Fujiwara, A.; Kobayashi, H.; Fukuyama, T. *Tetrahedron* **2002**, *58*, 6267–6276.

7.  Yokoshima, S.; Ueda, T.; Kobayashi, S.; Sato, A.; Kuboyama, T.; Tokuyama, H.; Fukuyama, T. *Pure Appl. Chem.* **2003**, *75*, 29–38.

8.  Guisado, C.; Waterhouse, J. E.; Price, W. S.; Jørgensen, M. R.; Miller, A. D. *Org. Biomol. Chem.* **2005**, *3*, 1049–1057.

9.  Olsen, C. A.; Witt, M.; Hansen, S. H.; Jaroszewski, J. W.; Franzyk, H. *Tetrahedron* **2005**, *61*, 6046–6055.

10. Janey, J. M. *Fukuyama Amine Synthesis.* In *Name Reactions for Functional Group Transformations*; Li, J. J., Ed.; Wiley: Hoboken, NJ, **2007**, pp 424–437. (Review).

11. Hahn, F.; Schepers, U. *Synlett* **2009**, 2755–2760. (Review).

# Fukuyama reduction

Aldehyde synthesis through reduction of thiol esters with $Et_3SiH$ in the presence of Pd/C catalyst.

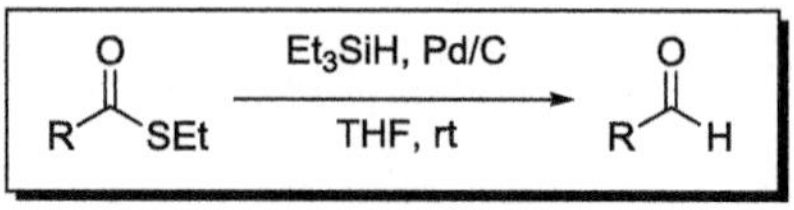

Path A:

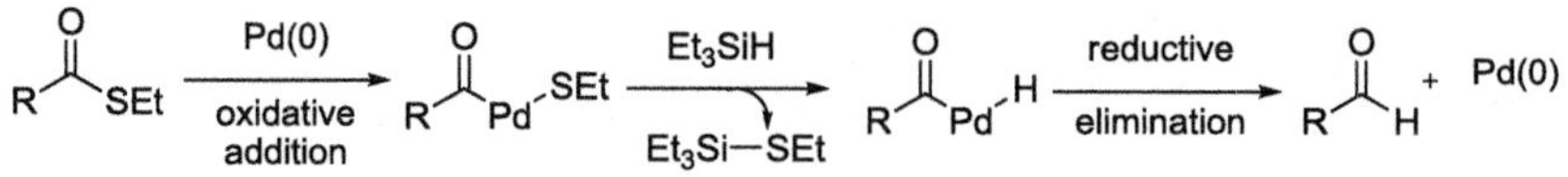

Path B:

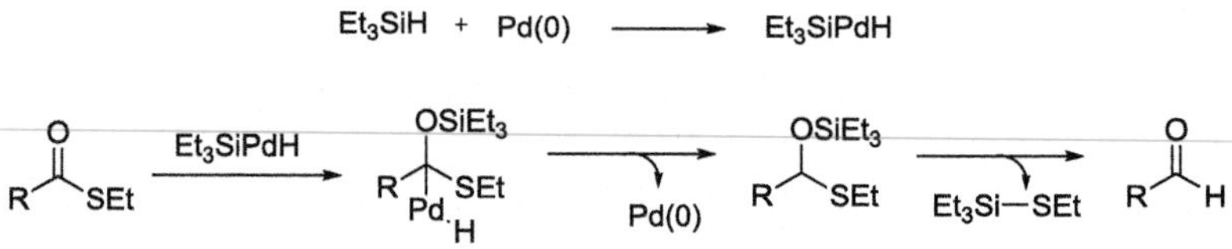

$$Et_3SiH \ + \ Pd(0) \longrightarrow Et_3SiPdH$$

Example 1[1]

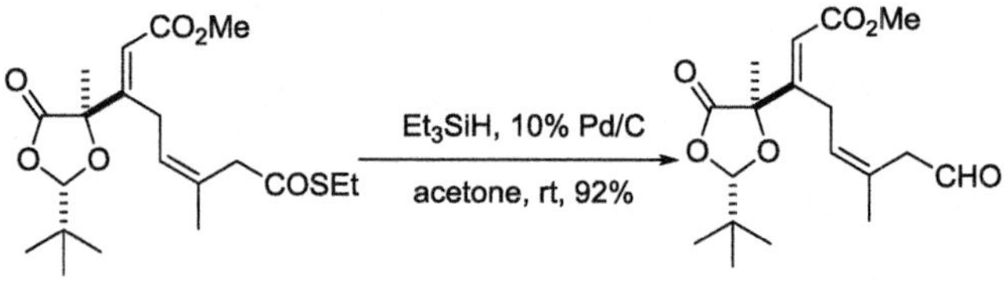

Example 2[3]

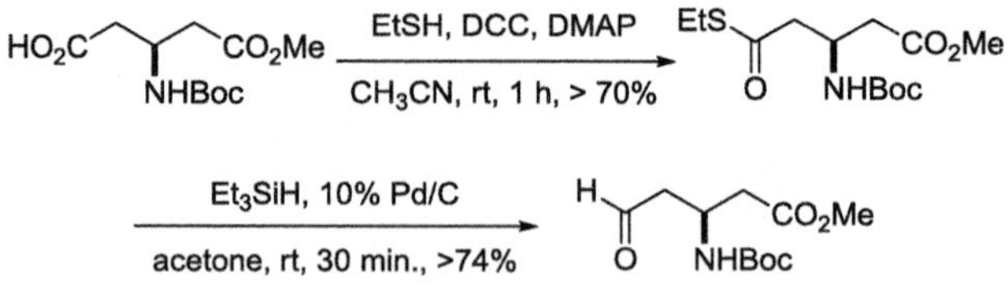

Example 3[8]

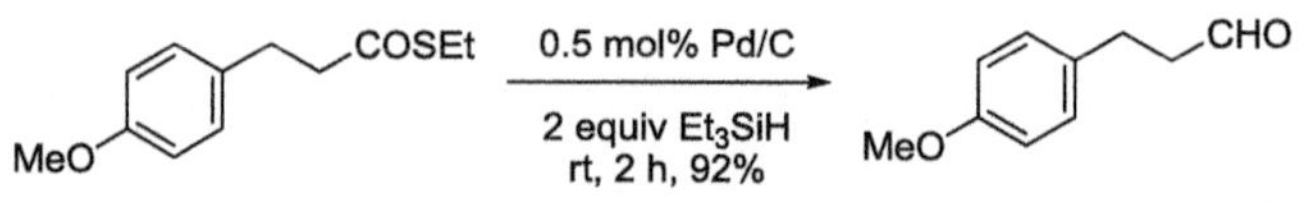

J.J. Li, *Name Reactions: A Collection of Detailed Mechanisms and Synthetic Applications*,
DOI 10.1007/978-3-319-03979-4_113, © Springer International Publishing Switzerland 2014

# References

1.  Fukuyama, T.; Lin, S.-C.; Li, L. *J. Am. Chem. Soc.* **1990,** *112*, 7050–7051.
2.  Kanda, Y.; Fukuyama, T. *J. Am. Chem. Soc.* **1993,** *115*, 8451–8452.
3.  Fujiwara, A.; Kan, T.; Fukuyama, T. *Synlett* **2000,** 1667–1673.
4.  Tokuyama, H.; Yokoshima, S.; Lin, S.-C.; Li, L.; Fukuyama, T. *Synthesis* **2002,** 1121–1123.
5.  Evans, D. A.; Rajapakse, H. A.; Stenkamp, D. *Angew. Chem. Int. Ed.* **2002,** *41*, 4569–4573.
6.  Shimada, K.; Kaburagi, Y.; Fukuyama, T. *J. Am. Chem. Soc.* **2003,** *125*, 4048–4049.
7.  Kimura, M.; Seki, M. *Tetrahedron Lett.* **2004,** *45*, 3219–3223. (Possible mechanisms were proposed in this paper).
8.  Miyazaki, T.; Han-ya, Y.; Tokuyama, H.; Fukuyama, T. *Synlett* **2004,** 477–480.
9.  Weerasinghe, L. P.; Garner, P. P.; Youngs, W. J.; Wright, B. *Abstracts of Papers, 243rd ACS National Meeting & Exposition*, San Diego, CA, March 25-29, (2012),ORGN-306.

# Gabriel synthesis

Synthesis of primary amines using potassium phthalimide and alkyl halides.

## Example 1[2]

## Example 2[6]

## Example 3[8z]

J.J. Li, *Name Reactions: A Collection of Detailed Mechanisms and Synthetic Applications*,
DOI 10.1007/978-3-319-03979-4_114, © Springer International Publishing Switzerland 2014

Example
4[9]

## References

1. Gabriel, S. *Ber.* **1887**, *20*, 2224–2226. Siegmund Gabriel (1851–1924), born in Berlin, Germany, studied under Hofmann at Berlin and Bunsen in Heidelberg. He taught at Berlin, where he discovered the Gabriel synthesis of amines. Gabriel, a good friend of Emil Fischer, often substituted for Fischer in his lectures.
2. Sheehan, J. C.; Bolhofer, V. A. *J. Am. Chem. Soc.* **1950,** *72*, 2786–2788.
3. Han, Y.; Hu, H. *Synthesis* **1990**, 122–124.
4. Ragnarsson, U.; Grehn, L. *Acc. Chem. Res.* **1991**, *24*, 285–289. (Review).
5. Toda, F.; Soda, S.; Goldberg, I. *J. Chem. Soc., Perkin Trans. 1* **1993**, 2357–2361.
6. Sen, S. E.; Roach, S. L. *Synthesis*, **1995**, 756–758.
7. Khan, M. N. *J. Org. Chem.* **1996**, *61*, 8063–8068.
8. Iida, K.; Tokiwa, S.; Ishii, T.; Kajiwara, M. *J. Labelled. Compd. Radiopharm.* **2002**, *45*, 569–570.
9. Tanyeli, C.; Özçubukçu, S. *Tetrahedron Asymmetry* **2003**, *14,* 1167–1170.
10. Ahmad, N. M. *Gabriel synthesis.* In *Name Reactions for Functional Group Transformations*; Li, J. J., Ed.; Wiley: Hoboken, NJ, **2007**, pp 438–450. (Review).
11. Al-Mousawi, S. M.; El-Apasery, M. A.; Al-Kanderi, N. H. *ARKIVOC* **2008**, *(16),* 268–278.
12. Richter, J. M. *Name Reactions in Heterocyclic Chemistry-II*, Li, J. J., Ed.; Wiley: Hoboken, NJ, 2011, pp 11–20. (Review).
13. Cytlak, T.; Marciniak, B.; Koroniak, H. In *Efficient Preparations of Fluorine Compounds*; Roesky, H. W., ed.; Wiley: Hoboken, NJ, (2013), pp 375–378. (Review).

## *Ing–Manske procedure*

A variant of Gabriel amine synthesis where hydrazine is used to release the amine from the corresponding phthalimide:

Example 1[6]

## References

1.  Ing, H. R.; Manske, R. H. F. *J. Chem. Soc.* **1926**, 2348–2351. H. R. Ing was a professor of pharmacological chemistry at Oxford. R. H. F. Manske, Ing's collaborator at Oxford, was of German origin but trained in Canada before studying at Oxford. Manske left England to return to Canada, eventually to become Director of Research in the Union Rubber Company, Guelph, Ontario, Canada.
2.  Ueda, T.; Ishizaki, K. *Chem. Pharm. Bull.* **1967**, *15*, 228–237.
3.  Khan, M. N. *J. Org. Chem.* **1995**, *60*, 4536–4541.
4.  Hearn, M. J.; Lucas, L. E. *J. Heterocycl. Chem.* **1984**, *21*, 615–622.
5.  Khan, M. N. *J. Org. Chem.* **1996**, *61*, 8063–8063.
6.  Tanyeli, C.; Özçubukçu, S. *Tetrahedron: Asymmetry* **2003**, *14*, 1167–1170.
7.  Ariffin, A.; Khan, M. N.; Lan, L. C.; May, F. Y.; Yun, C. S. *Synth. Commun.* **2004**, *34*, 4439–4445.
8.  Ali, M. M.; Woods, M.; Caravan, P.; Opina, A. C. L.; Spiller, M.; Fettinger, J. C.; Sherry, A. D. *Chem. Eur. J.* **2008**, *14*, 7250–7258.
9.  Nagarapu, L.; Apuri, S.; Gaddam, C.; Bantu, R. *Org. Prep. Proc. Int.* **2009**, *41*, 243–247.

# Gabriel–Colman rearrangement

Reaction of the enolate of a maleimidyl acetate to provide isoquinoline 1,4-diol.

G = CO, SO$_2$

Example 1[6]

4 equiv i-PrONa

i-PrOH, reflux
5 min., 85%

Example 2[9]

NaOMe, MeOH

reflux, 24 h
91%

## References

1. (a) Gabriel, S.; Colman, J. *Ber.* **1900**, *33*, 980–995. (b) Gabriel, S.; Colman, J. *Ber.* **1900**, *33*, 2630–2634. (c) Gabriel, S.; Colman, J. *Ber.* **1902**, *35*, 1358–1368.
2. Allen, C. F. H. *Chem. Rev.* **1950**, *47*, 275–305. (Review).
3. Gensler, W. J. *Heterocyclic Compounds*, Vol. 4, R. C. Elderfield, Ed., Wiley & Sons., New York, N.Y., **1952**, 378. (Review).
4. Hill, J. H. M. *J. Org. Chem.* **1965**, *30*, 620–622. (Mechanism).
5. Lombardino, J. G.; Wiseman, E. H.; McLamore, W. M. *J. Med. Chem.* **1971**, *14*, 1171–1175.
6. Schapira, C. B.; Perillo, I. A.; Lamdan, S. *J. Heterocycl. Chem.* **1980**, *17*, 1281–1288.
7. Lazer, E. S.; Miao, C. K.; Cywin, C. L.; *et al. J. Med. Chem.* **1997**, *40*, 980–989.
8. Pflum, D. A. *Gabriel–Colman Rearrangement*. In *Name Reactions in Heterocyclic Chemistry*; Li, J. J., Ed.; Wiley: Hoboken, NJ, **2005**, pp 416–422. (Review).
9. Kapatsina, E.; Lordon, M.; Baro, A.; Laschat, S. *Synthesis* **2008**, 2551–2560.

J.J. Li, *Name Reactions: A Collection of Detailed Mechanisms and Synthetic Applications*,
DOI 10.1007/978-3-319-03979-4_115, © Springer International Publishing Switzerland 2014

## Gassman indole synthesis

The Gassman indole synthesis involves a one-pot process in which a hypohalite, a β-carbonyl sulfide derivative, and a base are added sequentially to an aniline or a substituted aniline to provide 3-thioalkoxyindoles.    The mechanism of the Gassman indole synthesis involves a [2,3]-sigmatropic rearrangement (Sommelet–Hauser).   The sulfur can be easily removed by hydrogenolysis or Raney nickel.

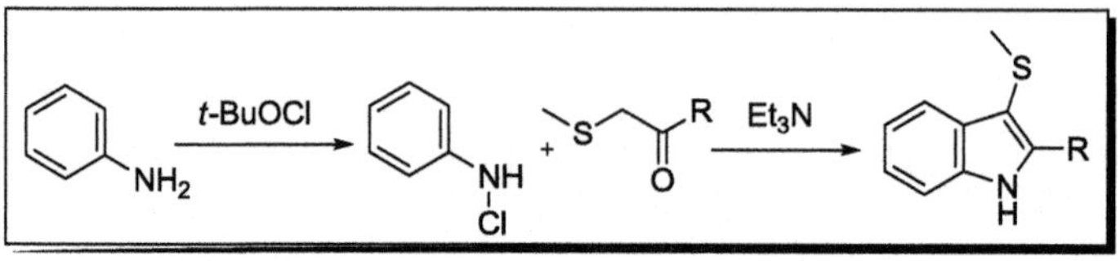

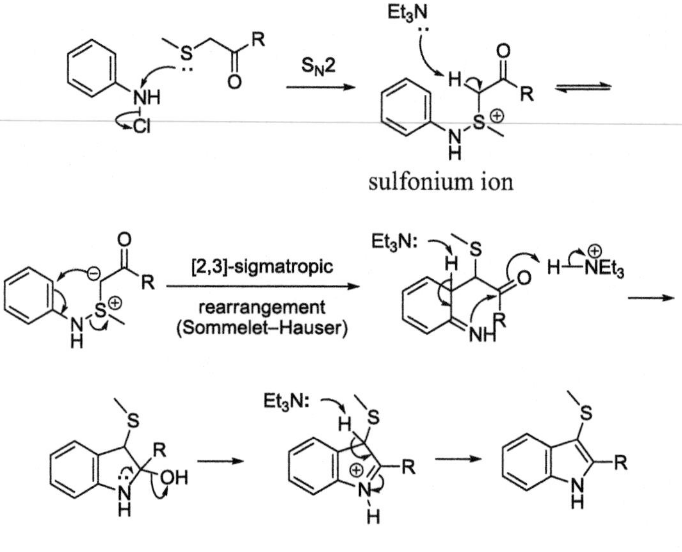

sulfonium ion

[2,3]-sigmatropic

rearrangement
(Sommelet–Hauser)

Example 1[1]

1. *t*-BuOCl

2. 

3. Et₃N, 69%

1. *t*-BuOCl

2. 

3. Et₃N

LiAlH₄, Et₂O

0 °C, 48% overall

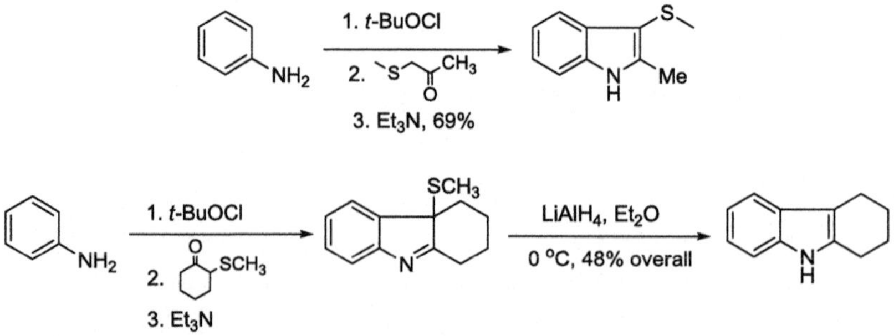

J.J. Li, *Name Reactions: A Collection of Detailed Mechanisms and Synthetic Applications*,
DOI 10.1007/978-3-319-03979-4_116, © Springer International Publishing Switzerland 2014

# Example 2[2]

## References

1. (a) Gassman, P. G.; van Bergen, T. J.; Gilbert, D. P.; Cue, B. W., Jr. *J. Am. Chem. Soc.* **1974**, *96*, 5495–5508. Paul G. Gassman (1935–1993) was a professor at the University of Minnesota (1974–1993). (b) Gassman, P. G.; van Bergen, T. J. *J. Am. Chem. Soc.* **1974**, *96*, 5508–5512. (c) Gassman, P. G.; Gruetzmacher, G.; van Bergen, T. J. *J. Am. Chem. Soc.* **1974**, *96*, 5512–5517.

2. Wierenga, W. *J. Am. Chem. Soc.* **1981**, *103*, 5621–5623.

3. Ishikawa, H.; Uno, T.; Miyamoto, H.; Ueda, H.; Tamaoka, H.; Tominaga, M.; Nakagawa, K. *Chem. Pharm. Bull.* **1990**, *38*, 2459–2462.

4. Smith, A. B., III; Sunazuka, T.; Leenay, T. L.; Kingery-Wood, J. *J. Am. Chem. Soc.* **1990**, *112*, 8197–8198.

5. Smith, A. B., III; Kingery-Wood, J.; Leenay, T. L.; Nolen, E. G.; Sunazuka, T. *J. Am. Chem. Soc.* **1992**, *114*, 1438–1449.

6. Savall, B. M.; McWhorter, W. W.; Walker, E. A. *J. Org. Chem.* **1996**, *61*, 8696–8697.

7. Li, J.; Cook, J. M. *Gassman Indole Synthesis.* In *Name Reactions in Heterocyclic Chemistry*; Li, J. J., Ed.; Wiley: Hoboken, NJ, **2005**, pp 128–131. (Review).

8. Barluenga, J.; Valdes, C. In *Modern Heterocyclic Chemistry;* Alvarez-Builla, J.; Vaquero, J. J.; Barluenga, J. eds.; Wiley-VCH: Weinheim, Germany; (2011), *1*, 377–531. (Review).

# Gattermann–Koch reaction

Formylation of arenes using carbon monoxide and hydrogen chloride in the presence of aluminum chloride under high pressure.

acylium ion

Example 1, A more practical variant[4]

orcinol

## References

1.  Gattermann, L.; Koch, J. A. *Ber.***1897**, *30*, 1622–1624.   Ludwig Gattermann (1860–1920) was born in Freiburg, Germany.   His textbook, *"Die Praxis de organischen Chemie"* (1894) was well-known all over the world. Some of his peers derided it fondly as "Gattermanns Kochbuch" ("Gattermann's cookbook", his father was a baker).
2.  Crounse, N. N. *Org. React.* **1949**, *5*, 290–300. (Review).
3.  Truce, W. E. *Org. React.* **1957**, *9*, 37–72. (Review).
4.  Solladié, G.; Rubio, A.; et al. *Tetrahedron: Asymmetry* **1990**, *1*, 187–198.
5.  (a) Tanaka, M.; Fujiwara, M.; Ando, H. *J. Org. Chem.* **1995**, *60*, 2106–2111. (b) Tanaka, M.; Fujiwara, M.; Ando, H.; Souma, Y. *Chem. Commun.* **1996**, 159–160. (c) Tanaka, M.; Fujiwara, M.; Xu, Q.; Souma, Y.; Ando, H.; Laali, K. K. *J. Am. Chem. Soc.* **1997**, *119*, 5100–5105. (d) Tanaka, M.; Fujiwara, M.; Xu, Q.; Ando, H.; Raeker, T. J. *J. Org. Chem.* **1998**, *63*, 4408–4412.
6.  Kantlehner, W.; Vettel, M.; et al. Haas, R. *J. Prakt. Chem.* **2000**, *342*, 297–310.

J.J. Li, *Name Reactions: A Collection of Detailed Mechanisms and Synthetic Applications*, DOI 10.1007/978-3-319-03979-4_117, © Springer International Publishing Switzerland 2014

# Gewald aminothiophene synthesis

Base-promoted aminothiophene formation from ketone, $\alpha$-active methylene nitrile and elemental sulfur.

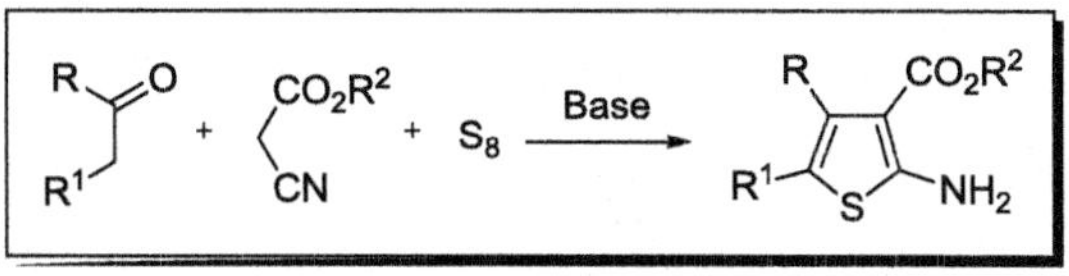

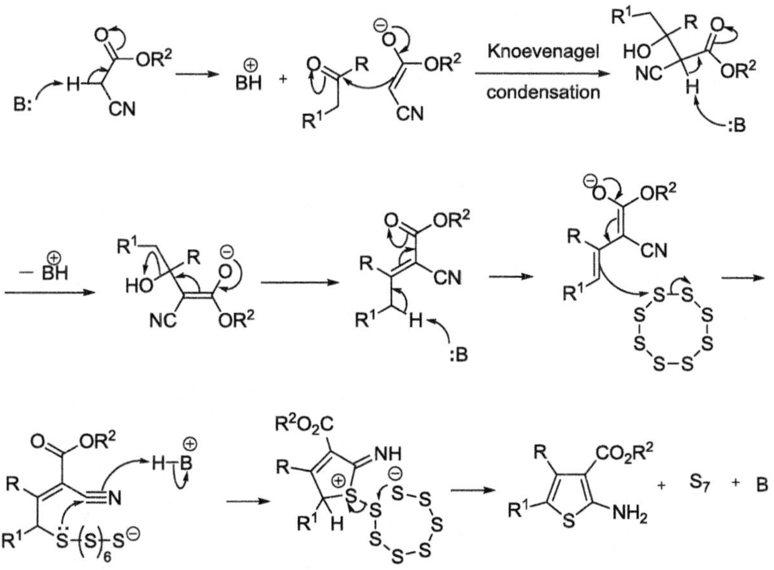

ylidene-sulfur adduct

Example 1[4]

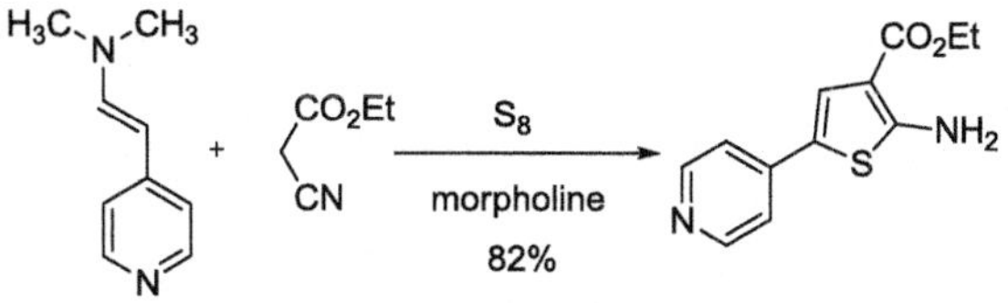

Example 2[7]

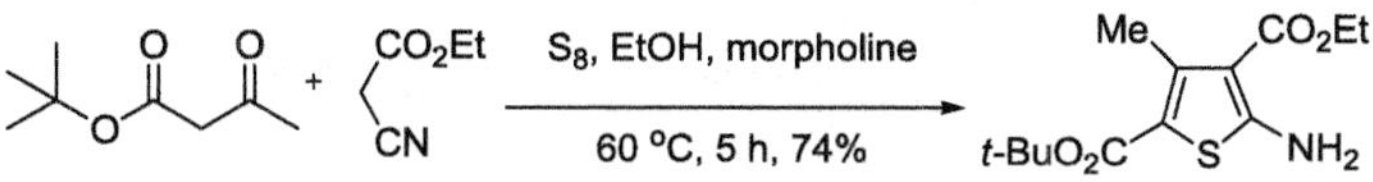

J.J. Li, *Name Reactions: A Collection of Detailed Mechanisms and Synthetic Applications*, DOI 10.1007/978-3-319-03979-4_118, © Springer International Publishing Switzerland 2014

Example 3[9]

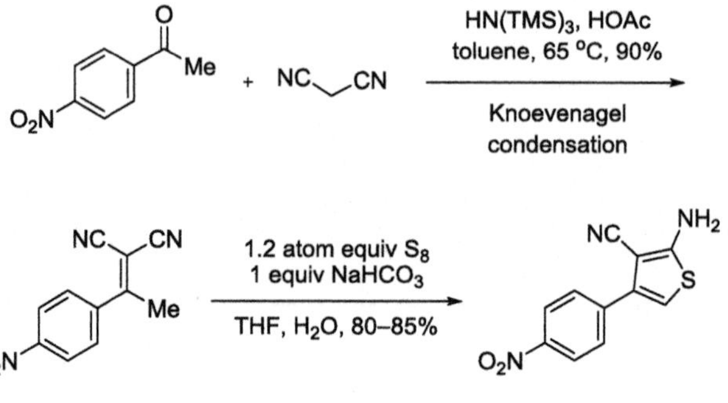

HN(TMS)$_3$, HOAc
toluene, 65 °C, 90%

Knoevenagel
condensation

1.2 atom equiv S$_8$
1 equiv NaHCO$_3$

THF, H$_2$O, 80–85%

Example 4[10]

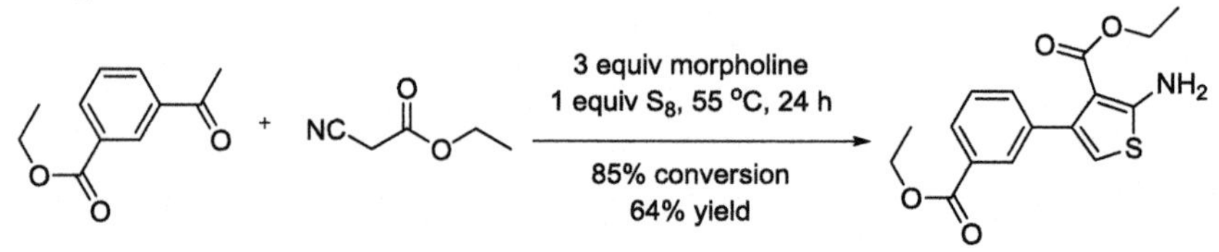

3 equiv morpholine
1 equiv S$_8$, 55 °C, 24 h

85% conversion
64% yield

Example 5[11]

3 equiv morpholine
2 equiv S$_8$, MeOH

20–45 °C, 24 h
72%

*Condensation*

MPS =
morpholine-
polysulfide ·

*Addition
of sulfur*

ylidene

ylidene-sulfur
adduct

*Dimerization*

*Cyclization*

*Re-cyclization*

dimer

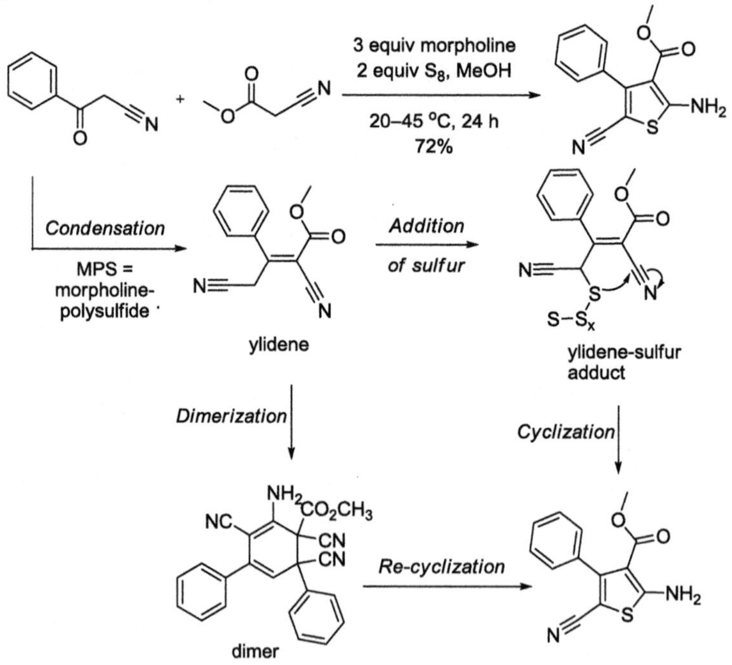

Example 6, *N*-Methylpiperazine-Functionalized Polyacrylonitrile Fiber Catalyst[12]

### References

1. (a) Gewald, K. *Z. Chem.* **1962**, *2*, 305–306. (b) Gewald, K.; Schinke, E.; Böttcher, H. *Chem. Ber.* **1966**, *99*, 94–100. (c) Gewald, K.; Neumann, G.; Böttcher, H. *Z. Chem.* **1966**, *6*, 261. (d) Gewald, K.; Schinke, E. *Chem. Ber.* **1966**, *99*, 271–275. Karl Gewald(1930–) is a professor at Technical University of Dresden.
2. Mayer, R.; Gewald, K. *Angew. Chem. Int. Ed.* **1967**, *6*, 294–306. (Review).
3. Gewald, K. *Chimia* **1980**, *34*, 101–110. (Review).
4. Bacon, E. R.; Daum, S. J. *J. Heterocycl. Chem.* **1991**, *28*, 1953-1955.
5. Sabnis, R. W. *Sulfur Reports* **1994**, *16*, 1–17. (Review).
6. Sabnis, R. W.; Rangnekar, D. W.; Sonawane, N. D. *J. Heterocycl. Chem.* **1999**, *36*, 333–345. (Review).
7. Gütschow, M.; Kuerschner, L.; Neumann, U.; Pietsch, M.; Löser, R.; Koglin, N.; Eger, K. *J. Med. Chem.* **1999**, *42*, 5437.
8. Tinsley, J. M. *Gewald Aminothiophene Synthesis.* In *Name Reactions in Heterocyclic Chemistry*; Li, J. J., Ed.; Wiley: Hoboken, NJ, **2005**, pp 193–198. (Review).
9. Barnes, D. M.; Haight, A. R.; Hameury, T.; McLaughlin, M. A.; Mei, J.; Tedrow, J. S.; Dalla Riva Toma, J. *Tetrahedron* **2006**, *62*, 11311–11319.
10. Tormyshev, V. M.; Trukhin, D. V.; Rogozhnikova, O. Yu.; Mikhalina, T. V.; Troitskaya, T. I.; Flinn, A. *Synlett* **2006**, 2559–2564.
11. Puterová, Z.; Andicsová, A.; Végh, D. *Tetrahedron* **2008**, *64*, 11262–11269.
12. Ma, L.; Yuan, L.; Xu, C.; Li, G.; Tao, M.; Zhang, W. *Synthesis* **2013**, *45*, 45–52.

## Glaser coupling

Sometimes known as the Glaser–Hay coupling, it is the oxidative homo-coupling of terminal alkynes using copper catalyst in the presence of oxygen.

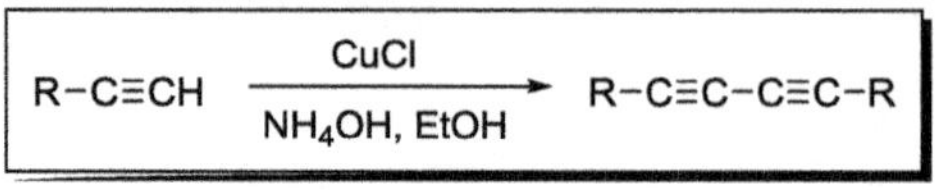

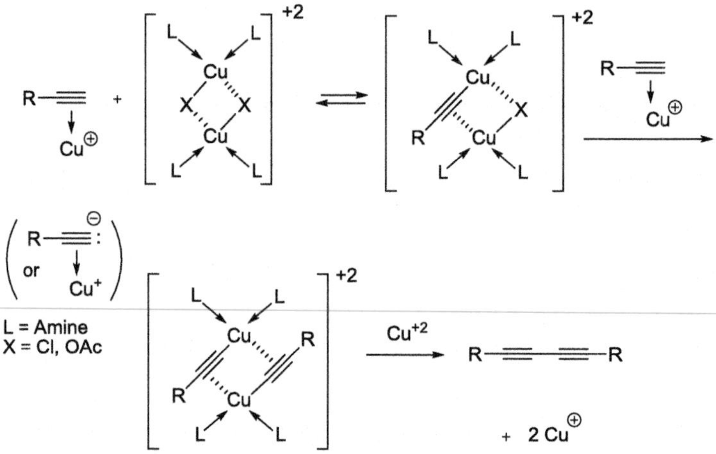

L = Amine
X = Cl, OAc

Alternatively, the radical mechanism is also operative:

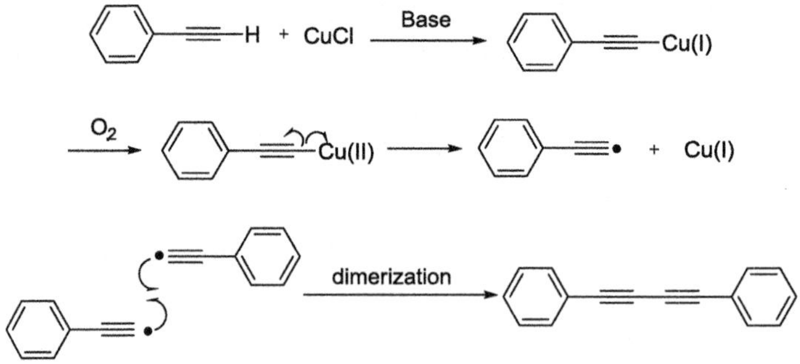

Example 1[1]

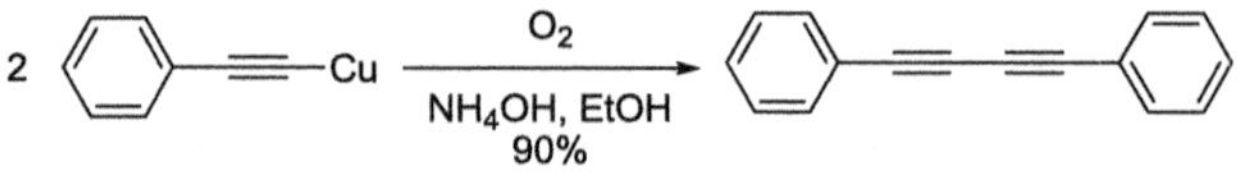

J.J. Li, *Name Reactions: A Collection of Detailed Mechanisms and Synthetic Applications*,
DOI 10.1007/978-3-319-03979-4_119, © Springer International Publishing Switzerland 2014

## Example 2, Homo-coupling[2]

## Example 3[7]

R = *n*-Hexyl

## Example 4[9]

## References

1. Glaser, C. *Ber.* **1869**, *2*, 422–424. Carl Andreas Glaser (1841–1935) studied under Justus von Liebig and Adolph Strecker. He became a professor in 1869 when the Glaser coupling was discovered. He became the Chairman of the Board of BASF after WWI.

2. Bowden, K.; Heilbron, I.; Jones, E. R. H.; Sondheimer, F. *J. Chem. Soc.* **1947**, 1583–1590.

3. Hoeger, S.; Meckenstock, A.-D.; Pellen, H. *J. Org. Chem.* **1997**, *62*, 4556–4557.

4. Siemsen, P.; Livingston, R. C.; Diederich, F. *Angew. Chem. Int. Ed.* **2000**, *39*, 2632–2657. (Review).

5. Youngblood, W. J.; Gryko, D. T.; Lammi, R. K.; Bocian, D. F.; Holten, D.; Lindsey, J. S. *J. Org. Chem.* **2002**, *67*, 2111–2117.

6. Moriarty, R. M.; Pavlovic, D. *J. Org. Chem.* **2004**, *69*, 5501–5504.

7. Andersson, A. S.; Kilsa, K.; Hassenkam, T.; Gisselbrecht, J.-P.; Boudon, C.; Gross, M.; Nielsen, M. B.; Diederich, F. *Chem. Eur. J.* **2006**, *12*, 8451–8459.

8. Gribble, G. W. *Glaser Coupling*. In *Name Reactions for Homologations-Part I*; Li, J. J., Ed.; Wiley: Hoboken, NJ, **2009**, pp 236–257. (Review).

9. Muesmann, T. W. T.; Wickleder, M. S.; Christoffers, J. *Synthesis* **2011**, 2775–2780.

## *Eglinton coupling*

Oxidative homo-coupling of terminal alkynes mediated by stoichiometric (or often excess) Cu(OAc)$_2$. A variant of the Glaser coupling reaction.

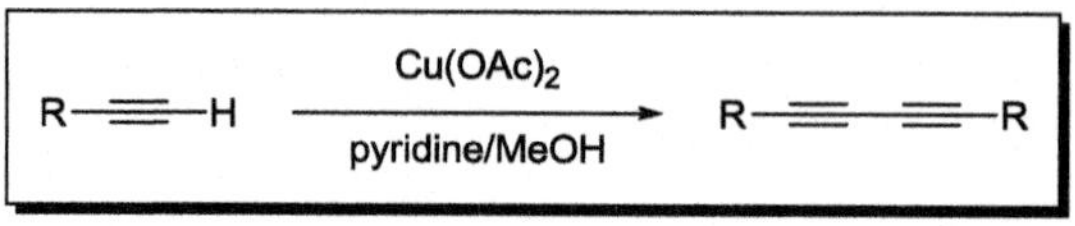

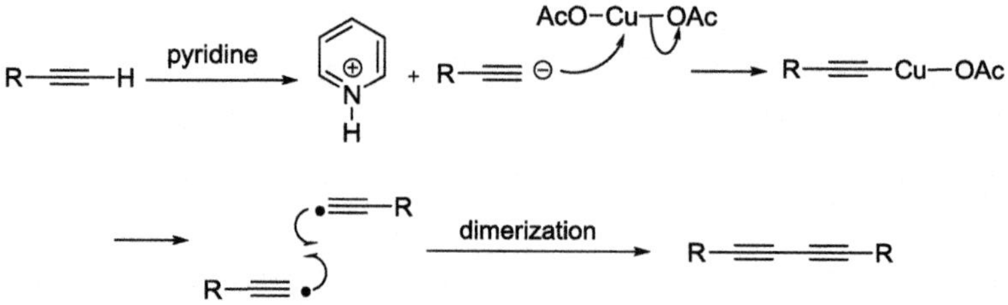

Example 1, Homo-coupling[2]

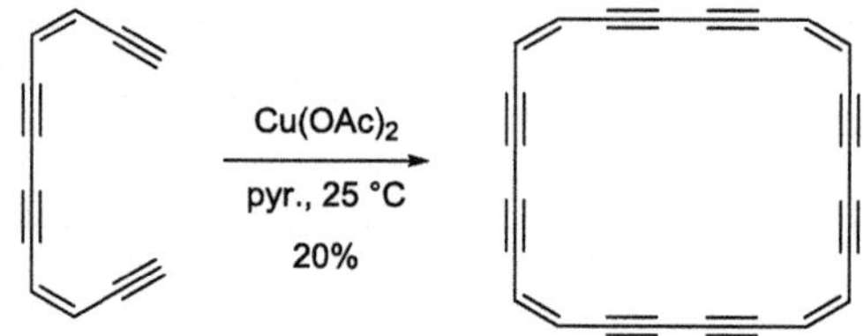

Example 2, Cross-coupling[3]

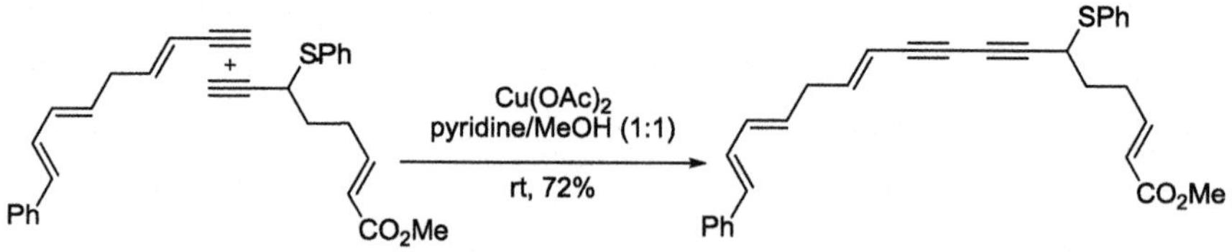

Example 3, Homo-coupling[4]

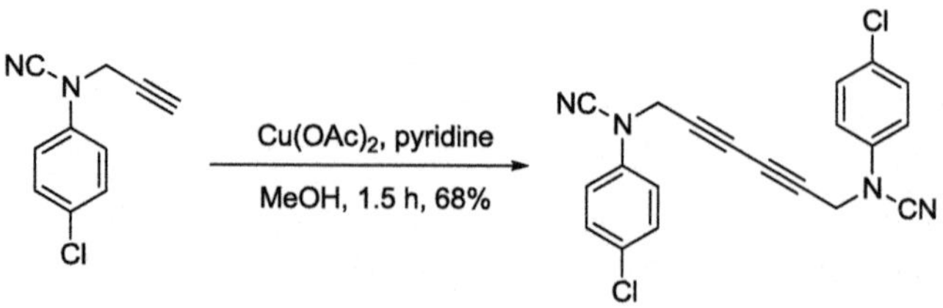

Example 4[5]

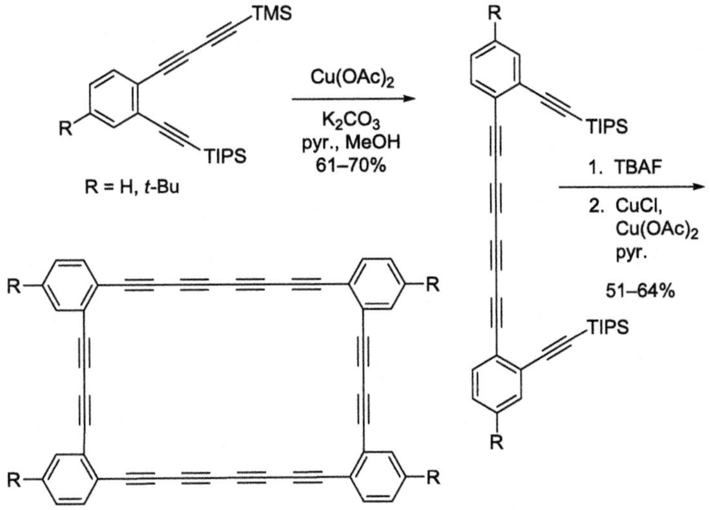

Example 5[11]

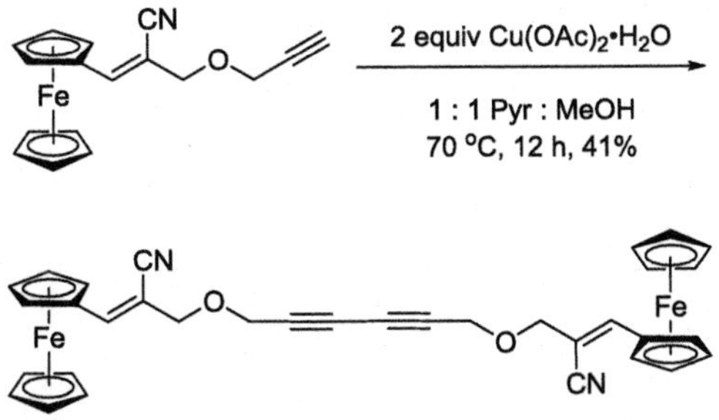

Example 6[12]

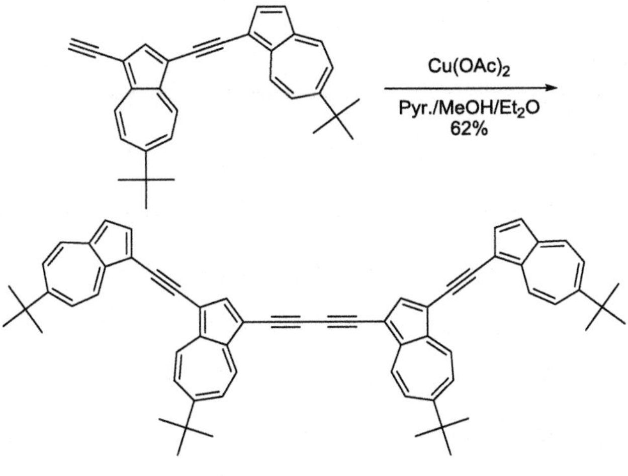

Example 7[13]

Cu(OAc)₂H₂O, CH₃CN

reflux, 54%, 12% of dimer

## References

1. (a) Eglinton, G.; Galbraith, A. R. *Chem. Ind.* **1956**, 737–738. Geoffrey Eglinton (1927–), born in Cardiff, Wales, is a Professor Emeritus at Bristol University. (b) Behr, O. M.; Eglinton, G.; Galbraith, A. R.; Raphael, R. A. *J. Chem. Soc.* **1960**, 3614–3625. (c) Eglinton, G.; McRae, W. *Adv. Org. Chem.* **1963**, *4*, 225–328. (Review).
2. McQuilkin, R. M.; Garratt, P. J.; Sondheimer, F. *J. Am. Chem. Soc.* **1970**, *92*, 6682–6683.
3. Nicolaou, K. C.; Petasis, N. A.; Zipkin, R. E.; Uenishi, J. *J. Am. Chem. Soc.* **1982**, *104*, 5558–5560.
4. Srinivasan, R.; Devan, B.; Shanmugam, P.; Rajagopalan, K. *Indian J. Chem., Sect. B* **1997**, *36B*, 123–125.
5. Haley, M. M.; Bell, M. L.; Brand, S. C.; Kimball, D. B.; Pak, J. J.; Wan, W. B. *Tetrahedron Lett.* **1997**, *38*, 7483–7486.
6. Nakanishi, H.; Sumi, N.; Aso, Y.; Otsubo, T. *J. Org. Chem.* **1998**, *63*, 8632–8633.
7. Kaigtti-Fabian, K. H. H.; Lindner, H.-J.; Nimmerfroh, N.; Hafner, K. *Angew. Chem. Int. Ed.* **2001**, *40*, 3402–3405.
8. Siemsen, P.; Livingston, R. C.; Diederich, F. *Angew. Chem. Int. Ed.* **2000**, *39*, 2632–2657. (Review).
9. Inouchi, K.; Kabashi, S.; Takimiya, K.; Aso, Y.; Otsubo, T. *Org. Lett.* **2002**, *4*, 2533–2536. ·
10. Xu, G.-L.; Zou, G.; Ni, Y.-H.; DeRosa, M. C.; Crutchley, R. J.; Ren, T. *J. Am. Chem. Soc.* **2003**, *125*, 10057–10065.
11. Shanmugam, P.; Vaithiyananthan, V.; Viswambharan, B.; Madhavan, S. *Tetrahedron Lett.* **2007**, *48*, 9190–9194.
12. Miljanic, O. S.; Dichtel, W. R.; Khan, S. I.; Mortezaei, S.; Heath, J. R.; Stoddart, J. F. *J. Am. Chem. Soc.* **2007**, *129*, 8236–8246.
13. White, N. G.; Beer, P. D. *Beilstein J. Org. Chem.* **2012**, *8*, 246–252.

# Gomberg–Bachmann reaction

Base-promoted radical coupling between an aryl diazonium salt and an arene to form a diaryl compound.

## Example 1[5]

## Example 2[6]

J.J. Li, *Name Reactions: A Collection of Detailed Mechanisms and Synthetic Applications*, DOI 10.1007/978-3-319-03979-4_120, © Springer International Publishing Switzerland 2014

Example 3, With Diazotate[7]

11 equiv

## References

1. Gomberg, M.; Bachmann, W. E. *J. Am. Chem. Soc.* **1924**, *46*, 2339–2343. Moses Gomberg (1866–1947) was born in Elizabetgrad, Russia. He discovered the triphenylmethyl stable radical at the University of Michigan in Ann Arbor, Michigan. In this article, Gomberg declared that he had reserved the field of radical chemistry for himself! Werner Bachmann (1901–1951), Gomberg's Ph.D. student, was born in Detroit, Michigan. After his postdoctoral trainings in Europe Bachmann returned to the University of Michigan, fittingly, as the Moses Gomberg Professor of Chemistry.

2. Dermer, O. C.; Edmison, M. T. *Chem. Rev.* **1957**, *57*, 77–122. (Review).

3. Rüchardt, C.; Merz, E. *Tetrahedron Lett.* **1964**, *5*, 2431–2436. (Mechanism).

4. Beadle, J. R.; Korzeniowski, S. H.; Rosenberg, D. E.; Garcia-Slanga, B. J.; Gokel, G. W. *J. Org. Chem.* **1984**, *49*, 1594–1603.

5. McKenzie, T. C.; Rolfes, S. M. *J. Heterocycl. Chem.* **1987**, *24*, 859–861.

6. Lai, Y.-H.; Jiang, J. *J. Org. Chem.* **1997**, *62*, 4412–4417.

7. Pratsch, G.; Wallaschkowski, T.; Heinrich, M. R. *Chem. Eur. J.* **2012**, *18*, 11555–11559,

# Gould–Jacobs reaction

The Gould–Jacobs reaction is a sequence of the following reactions:

a. Substitution of an aniline with either alkoxy methylenemalonic ester or acyl malonic ester providing the anilinomethylenemalonic ester;

b. Cyclization of to the 4-hydroxy-3-carboalkoxyquinoline (4-hydroxyquinolines exist predominantly in 4-oxoform);

c. Saponification to form acid;

d. Decarboxylation to give the 4-hydroxyquinoline. Extension could lead to unsubstituted parent heterocycles with fused pyridine ring of Skraup type.

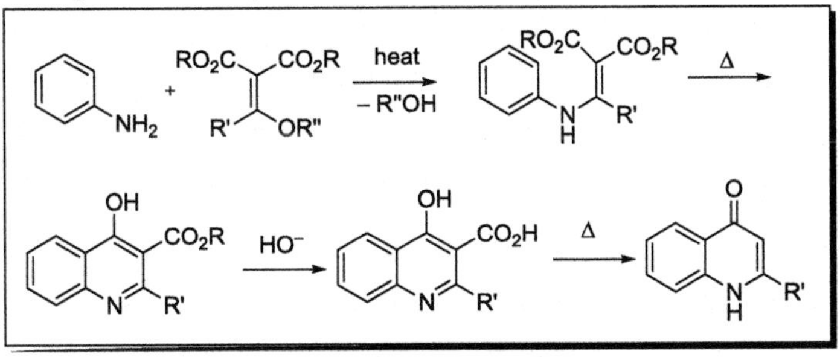

R = alkyl; R' = alkyl, aryl, or H; R" = alkyl or H

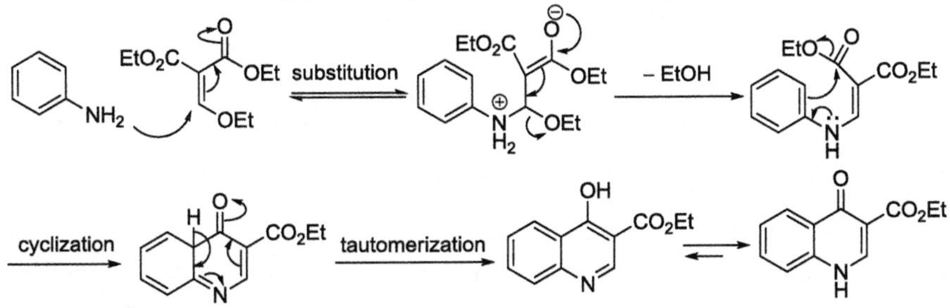

Example 1[3]

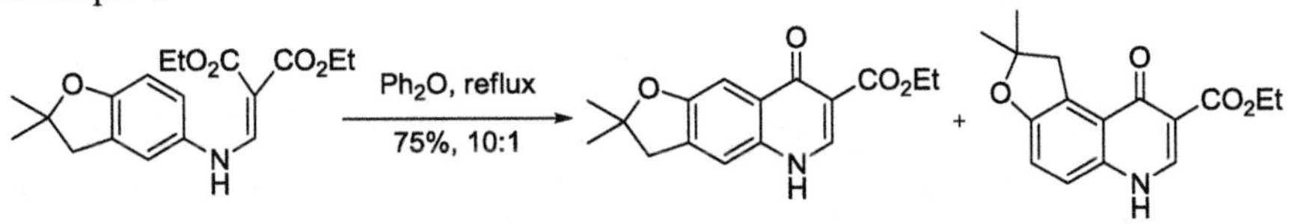

Example 2[7]

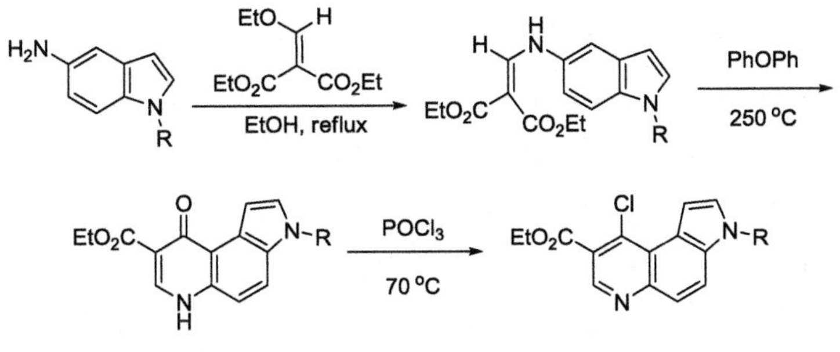

J.J. Li, *Name Reactions: A Collection of Detailed Mechanisms and Synthetic Applications*, DOI 10.1007/978-3-319-03979-4_121, © Springer International Publishing Switzerland 2014

Example 3, Microwave-assisted Gould–Jacobs reaction[8]

Example 4[9]

## References

1.  Gould, R. G.; Jacobs, W. A. *J. Am. Chem. Soc.* **1939**, *61*, 2890–2895. R. Gordon Gould was born in Chicago in 1909. He earned his Ph.D. at Harvard University in 1933. After serving as an instructor at Harvard and Iowa, Gould worked at Rockefeller Institute for Medical Research where he discovered the Gould–Jacobs reaction with his colleague Walter A. Jacobs.

2.  Reitsema, R. H. *Chem. Rev.* **1948**, *53*, 43–68. (Review).

3.  Cruickshank, P. A., Lee, F. T., Lupichuk, A. *J. Med. Chem.* **1970**, *13*, 1110–1114.

4.  Elguero J., Marzin C., Katritzky A. R., Linda P., *The Tautomerism of Heterocycles*, Academic Press, New York, **1976**, pp 87–102. (Review).

5.  Milata, V.; Claramunt, R. M.; Elguero, J.; Zálupský, P. *Targets in Heterocyclic Systems* **2000**, *4*, 167–203. (Review).

6.  Curran, T. T. *Gould–Jacobs Reaction.* In *Name Reactions in Heterocyclic Chemistry*; Li, J. J., Ed.; Wiley: Hoboken, NJ, **2005**, 423–436. (Review).

7.  Ferlin, M. G.; Chiarelotto, G.; Dall'Acqua, S.; Maciocco, E.; Mascia, M. P.; Pisu, M. G.; Biggio, G. *Bioorg. Med. Chem.* **2005**, *13*, 3531–3541.

8.  Desai, N. D. *J. Heterocycl. Chem.* **2006**, *43*, 1343–1348.

9.  Kendre, D. B.; Toche, R. B.; Jachak, M. N. *J. Heterocycl. Chem.* **2008**, *45*, 1281–1286.

10. Lengyel, L.; Nagy, T. Z.; Sipos, G.; Jones, R.; Dormán, G.; Üerge, L.; Darvas, F. *Tetrahedron Lett.* **2012**, *53*, 738–743.

# Grignard reaction

Addition of organomagnesium compounds (Grignard reagents), generated from organohalides and magnesium metal, to electrophiles.

Formation of the Grignard reagent:

Grignard reaction, ionic mechanism:

Grignard reaction, radical mechanism,

Example 1[4]

   This reaction is known as the *Hoch–Campbell aziridine synthesis*, which entails treatment of ketoximes with excess Grignard reagents and subsequent hydrolysis of the organometallic complex to produce aziridines.

J.J. Li, *Name Reactions: A Collection of Detailed Mechanisms and Synthetic Applications*,
DOI 10.1007/978-3-319-03979-4_122, © Springer International Publishing Switzerland 2014

Example 2[5]

Example 5[10]

Garner's aldehyde

Example 6[11]

52%        28%        5%

Example 7, Asymmetric Conjugate Addition[12]

97:3 *er*

$L =$ (R,S)-Rev-Josiphos

## References

1.  Grignard, V. *C. R. Acad. Sci.* **1900**, *130*, 1322–1324.  Victor Grignard (France, 1871–1935) won the Nobel Prize in Chemistry in 1912 for his discovery of the Grignard reagent.
2.  Ashby, E. C.; Laemmle, J. T.; Neumann, H. M. *Acc. Chem. Res.* **1974**, *7*, 272–280. (Review).
3.  Ashby, E. C.; Laemmle, J. T. *Chem. Rev.* **1975**, *75*, 521–546. (Review).
4.  Sasaki, T.; Eguchi, S.; Hattori, S. *Heterocycles* **1978**, *11*, 235–242.
5.  Meyers, A. I.; Flisak, J. R.; Aitken, R. A. *J. Am. Chem. Soc.* **1987**, *109*, 5446–5452.
6.  *Grignard Reagents* Richey, H. G., Jr., Ed.; Wiley: New York, **2000**. (Book).
7.  Holm, T.; Crossland, I. In *Grignard Reagents* Richey, H. G., Jr., Ed.; Wiley: New York, **2000**, Chapter 1, pp 1–26. (Review).
8.  Shinokubo, H.; Oshima, K. *Eur. J. Org. Chem.* **2004**, 2081–2091.  (Review).
9.  Graden, H.; Kann, N. *Cur. Org. Chem.* **2005**, *9*, 733–763.  (Review).
10. Babu, B. N.; Chauhan, K. R. *Tetrahedron Lett.* **2008**, *50*, 66–67.
11. Mlinaric-Majerski, K.; Kragol, G.; Ramljak, T. S. *Synlett* **2008**, 405–409.
12. Mao, B.; Fãnás-Mastral, M.; Feringa, B. L. *Org. Lett.* **2013**, *15*, 286–289.

# Grob fragmentation

The C–C bond cleavage primarily via a concerted process involving a five atom system.
General scheme:

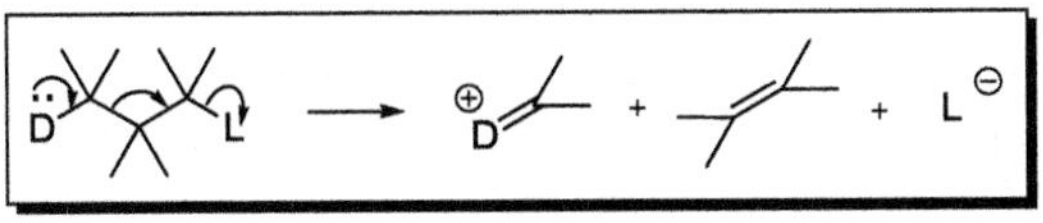

$$D = O^-, NR_2; L = OH_2^+, OTs, I, Br, Cl$$

Example 1[2]

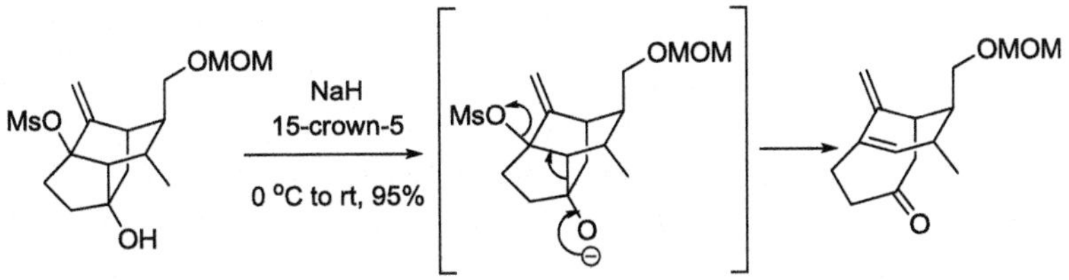

Example 2, Aza-Grob fragmentation[3]

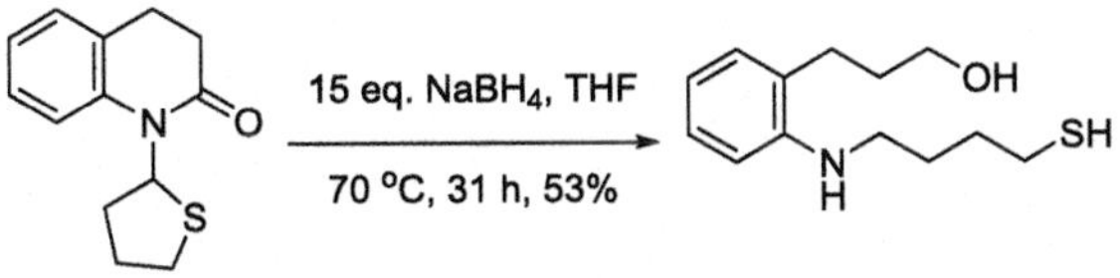

Example 3[7]

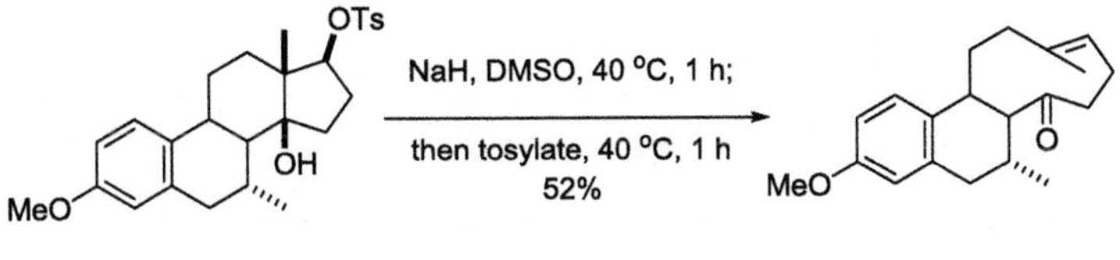

Example 4[8]

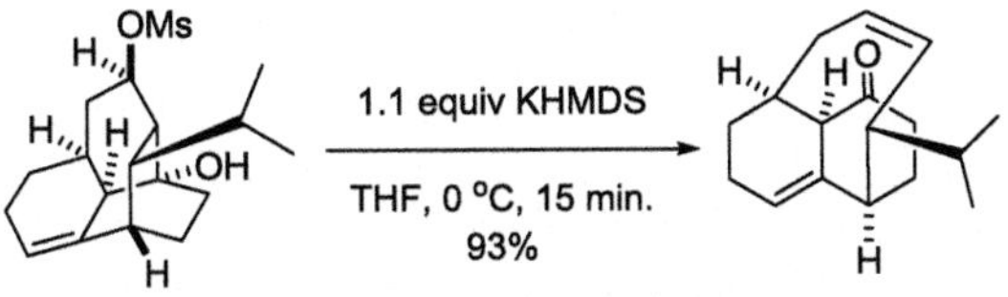

J.J. Li, *Name Reactions: A Collection of Detailed Mechanisms and Synthetic Applications*,
DOI 10.1007/978-3-319-03979-4_123, © Springer International Publishing Switzerland 2014

Example 4[8]

## References

1.  (a) Grob, C. A.; Baumann, W. *Helv. Chim. Acta* **1955**, *38*, 594–603. (b) Grob, C. A.; Schiess, P. W. *Angew. Chem. Int. Ed.* **1967**, *6*, 1–15. Cyril A. Grob (1917–2003) was born in London (UK) to Swiss parents, studied chemistry at ETH Zürich and completed his PhD in 1943 under the guidance of Leopold Ruzicka (Nobel laureate) on artificial steroidal antigens. He then moved to Basel to work with Taddeus Reichstein (another Nobel laureate) first at the pharmaceutical institute and from 1947 at the organic chemistry institute of the university, where he moved up the academic career ladder to become the director of the institute and holder of the chair there as Reichstein's successor in 1960. An investigation of the reductive elimination of bromine from 1,4-dibromides in the presence of zinc led in 1955 to the recognition of heterolytic fragmentation as a general reaction principle. The heterolytic fragmentation has now entered textbooks under his name. Experimental evidence for vinyl cations as discrete reactive intermediates was also first provided by Grob. Cyril Grob never acted impulsively, but always calmly and deliberately. He never sought attention in public, but fulfilled his social duties efficiently, reliably, and without a fuss. He died in his home in Basel (Switzerland) on December 15, 2003 at the age of 86. (Schiess, P. *Angew. Chem. Int. Ed.* **2004**, *43*, 4392.) A recent review[11] revealed that Grob was not even the first to investigate such reactions!
2.  Yoshimitsu, T.; Yanagiya, M.; Nagaoka, H. *Tetrahedron Lett.* **1999**, *40*, 5215–5218.
3.  Hu, W.-P.; Wang, J.-J.; Tsai, P.-C. *J. Org. Chem.* **2000**, *65*, 4208–4029.
4.  Molander, G. A.; Le Huerou, Y.; Brown, G. A. *J. Org. Chem.* **2001**, *66*, 4511–4516.
5.  Paquette, L. A.; Yang, J.; Long, Y. O. *J. Am. Chem. Soc.* **2002**, *124*, 6542–6543.
6.  Barluenga, J.; Alvarez-Perez, M.; Wuerth, K.; *et al. Org. Lett.* **2003**, *5*, 905–908.
7.  Khripach, V. A.; Zhabinskii, V. N.; Fando, G. P.; *et al. Steroids* **2004**, *69*, 495–499.
8.  Maimone, T. J.; Voica, A.-F.; Baran, P. S. *Angew. Chem. Int. Ed.* **2008**, *47*, 3054–3056.
9.  Yuan, D.-Y.; Tu, Y.-Q.; Fan, C.-A. *J. Org. Chem.* **2008**, *73*, 7797–7799.
10. Barbe, G.; St-Onge, M.; Charette, A. B. *Org. Lett.* **2008**, *10*, 5497–5499.
11. Mulzer, *Chem. Rev.* **2010**, *110*, 3741–4766.
12. Umland, K.-D.; Palisse, A.; Haug, T. T.; Kirsch, S. F. *Angew. Chem. Int. Ed.* **2011**, *50*, 9965–9968

# Guareschi–Thorpe condensation

2-Pyridone formation from the condensation of cyanoacetic ester with diketone in the presence of ammonia.

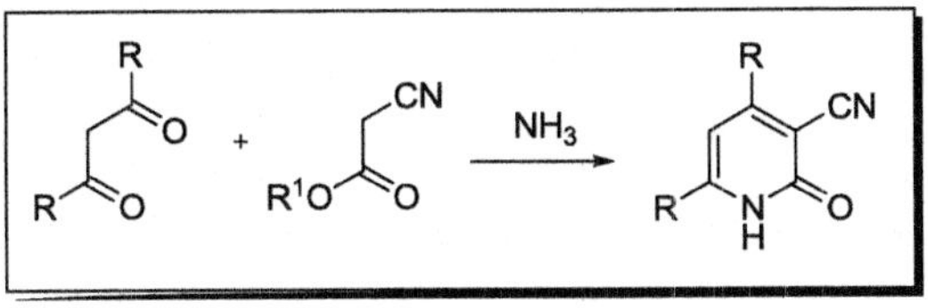

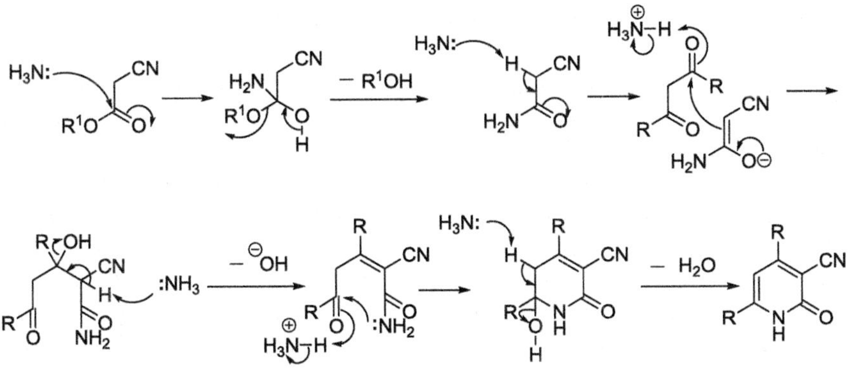

Example 1[6]

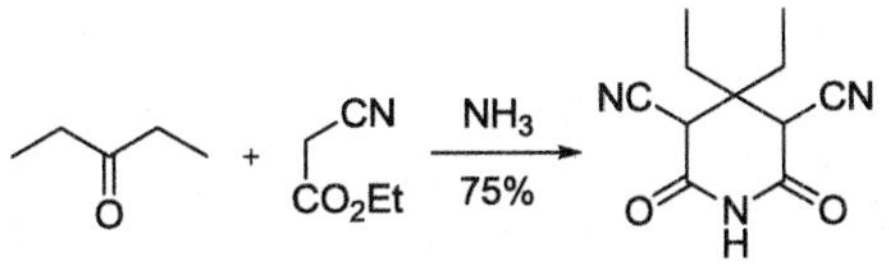

Guareschi imide

Example 2[9]

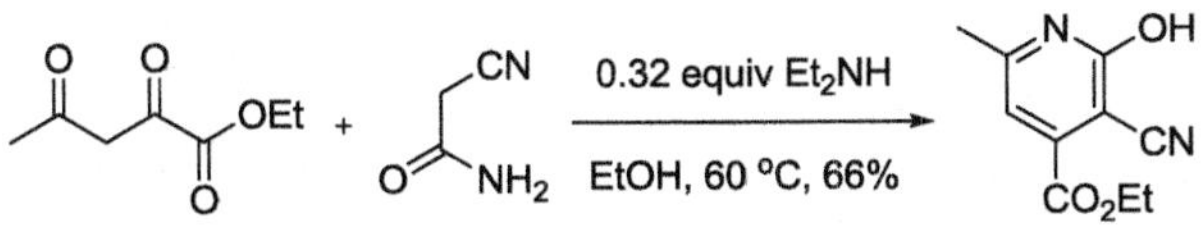

## References

1. (a) Guareschi, I. *Mem. R. Accad. Sci. Torino* **1896**, *II*, 7, 11, 25. (b) Baron, H.; Renfry, F. G. P.; Thorpe, J. F. *J. Chem. Soc.* **1904**, *85*, 1726–1961. Jocelyn F. Thorpe spent two years in Germany where he worked in the laboratory of a dyestuff manufacturer before taking a post as a lecturer at Manchester. Thorpe later became FRS (Fellow of the Royal Society) and professor of organic chemistry at Imperial College.
2. Vogel, A. I. *J. Chem. Soc.* **1934**, 1758–1765.

J.J. Li, *Name Reactions: A Collection of Detailed Mechanisms and Synthetic Applications*, DOI 10.1007/978-3-319-03979-4_124, © Springer International Publishing Switzerland 2014

3.  McElvain, S. M.; Lyle, R. E. Jr. *J. Am. Chem. Soc.* **1950,** *72*, 384–389.
4.  Brunskill, J. S. A. *J. Chem. Soc. (C)* **1968,** 960–966.
5.  Brunskill, J. S. A. *J. Chem. Soc., Perkin Trans. 1* **1972,** 2946–2950.
6.  Holder, R. W.; Daub, J. P.; Baker, W. E.; Gilbert, R. H. III; Graf, N. A. *J. Org. Chem.* **1982,** *47*, 1445–1451.
7.  Krstic, V.; Misic-Vukovic, M.; Radojkovic-Velickovic, M. *J. Chem. Res. (S)* **1991,** 82.
8.  Galatsis, P. *Guareschi–Thorpe Pyridine Synthesis.* In *Name Reactions in Heterocyclic Chemistry*; Li, J. J., Ed.; Wiley: Hoboken, NJ, **2005,** pp 307–308. (Review).
9.  Schmidt, G.; Reber, S.; Bolli, M. H.; Abele, S. *Org. Process Res. Dev.* **2012,** *16*, 595–604.

# Hajos–Wiechert reaction

Asymmetric Robinson annulation catalyzed by (*S*)-(−)-proline.

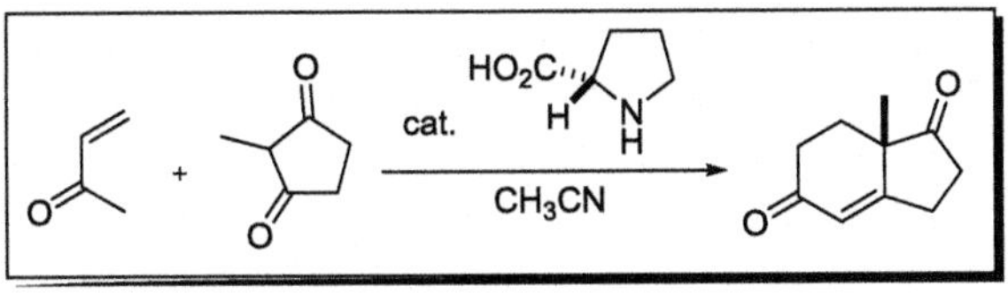
cat.
CH3CN

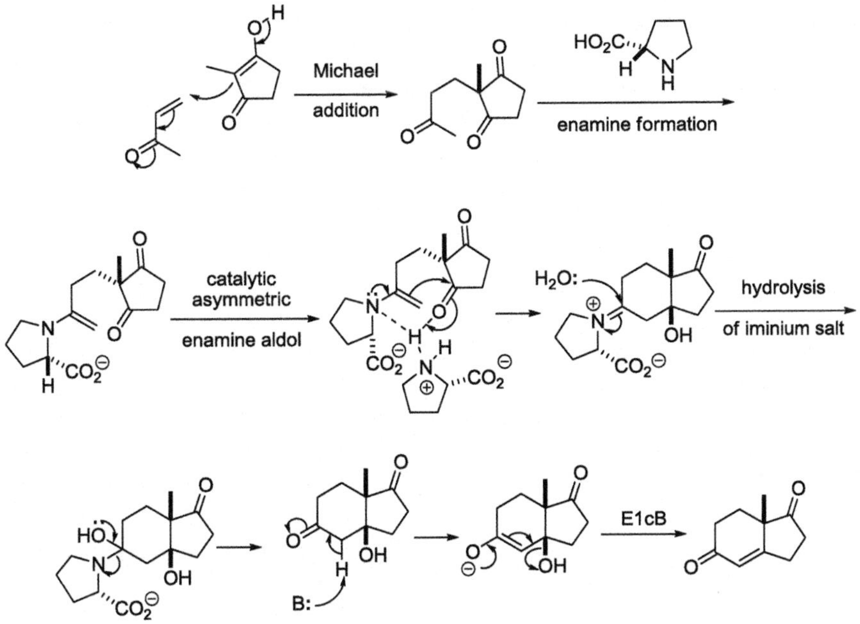
Michael
addition
enamine formation
catalytic
asymmetric
enamine aldol
hydrolysis
of iminium salt
E1cB

Example 1[1a]

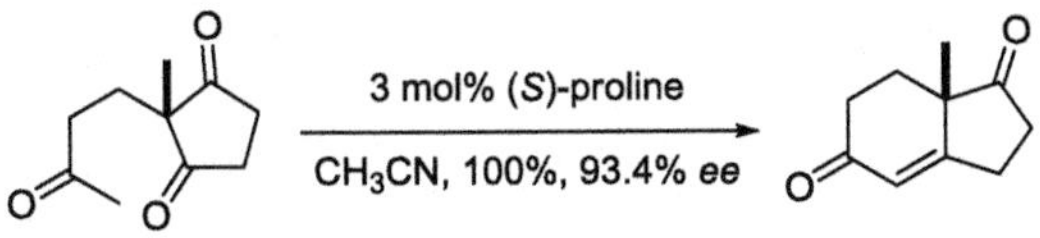
3 mol% (S)-proline
CH3CN, 100%, 93.4% ee

Example 2[3]

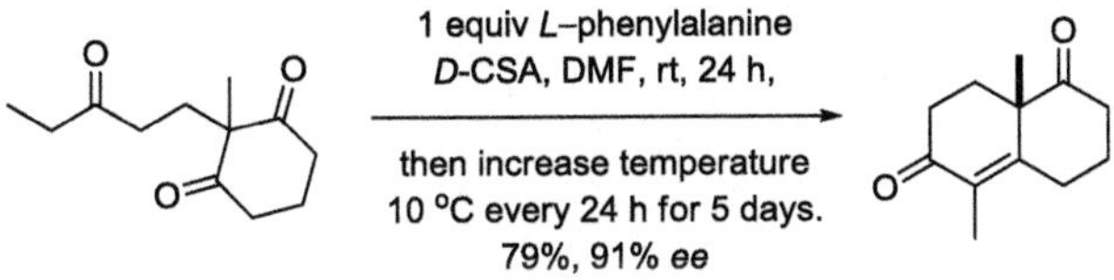
1 equiv L–phenylalanine
D-CSA, DMF, rt, 24 h,
then increase temperature
10 °C every 24 h for 5 days.
79%, 91% ee

J.J. Li, *Name Reactions: A Collection of Detailed Mechanisms and Synthetic Applications*,
DOI 10.1007/978-3-319-03979-4_125, © Springer International Publishing Switzerland 2014

Example 3[8]

L-phenylalanine, PPTS

DMSO, 50 °C, 24 h
sonication, 94%, 73% ee

Example 4[9]

1 equiv L-phenylalanine
0.5 equiv 1 N HClO$_4$

DMSO, 90 °C
86%, 48% ee

## References

1.  (a) Hajos, Z. G.; Parrish, D. R. *J. Org. Chem.* **1974**, *39*, 1615–1621. Hajos and Parrish were chemists at Hoffmann–La Roche. (b) Eder, U.; Sauer, G.; Wiechert, R. *Angew. Chem. Int. Ed.* **1971**, *10*, 496–497.

2.  Brown, K. L.; Dann, L.; Duntz, J. D.; Eschenmoser, A.; Hobi, R.; Kratky, C. *Helv. Chim. Acta* **1978**, *61*, 3108–3135.

3.  Hagiwara, H.; Uda, H. *J. Org.Chem.* **1998**, *53*, 2308–2311.

4.  Nelson, S. G. *Tetrahedron: Asymmetry* **1998**, *9*, 357–389.

5.  List, B.; Lerner, R. A.; Barbas, C. F., III. *J. Am. Chem. Soc.* **2000**, *122*, 2395–2396.

6.  List, B.; Pojarliev, P.; Castello, C. *Org. Lett.* **2001**, *3*, 573–576.

7.  Hoang, L.; Bahmanyar, S.; Houk, K. N.; List, B. *J. Am. Chem. Soc.* **2003**, *125*, 16–17.

8.  Shigehisa, H.; Mizutani, T.; Tosaki, S.-y.; Ohshima, T.; Shibasaki, M. *Tetrahedron* **2005**, *61*, 5057–5065.

9.  Nagamine, T.; Inomata, K.; Endo, Y.; Paquette, L. A. *J. Org. Chem.* **2007**, *72*, 123–131.

10. Kennedy, J. W. J.; Vietrich, S.; Weinmann, H.; Brittain, D. E. A. *J. Org. Chem.* **2009**, *73*, 5151–5154.

11. Christen, D. P. *Hajos–Wiechert Reaction*. In *Name Reactions for Homologations-Part II*; Li, J. J., Ed.; Wiley: Hoboken, NJ, **2009**, pp 554–582. (Review).

12. Zhu, H.; Clemente, F. R.; Houk, K. N.; Meyer, M. P. *J. Am. Chem. Soc.* **2009**, *131*, 1632–1633.

13. Bradshaw, B.; Bonjoch, J. *Synlett* **2012**, *23*, 337–356. (Review).

# Haller–Bauer reaction

Base-induced cleavage of non-enolizable ketones leading to carboxylic amide or acid derivative and a neutral fragment in which the carbonyl group is replaced by a hydrogen.

non-enolizable ketone

Example 1[4]

Example 2[9]

Example 3, Racemization[10]

# References

1. Haller, A.; Bauer, E. *Compt. Rend.* **1908**, *147*, 824–829.
2. Gilday, J. P.; Gallucci, J. C.; Paquette, L. A. *J. Org. Chem.* **1989**, *54*, 1399–1408.
3. Paquette, L. A.; Gilday, J. P.; Maynard, G. D. *J. Org. Chem.* **1989**, *54*, 5044–5053.
4. Paquette, L. A.; Gilday, J. P. *Org. Prep. Proc. Int.* **1990**, *22*, 167–201.
5. Mehta, G.; Praveen, M. *J. Org. Chem.* **1995**, *60*, 279–280.
6. Mehta, G.; Venkateswaran, R. V. *Tetrahedron* **2000**, *56*, 1399–1422. (Review).
7. Arjona, O.; Medel, R.; Plumet, J. *Tetrahedron Lett.* **2001**, *42*, 1287–1288.
8. Ishihara, K.; Yano, T. *Org. Lett.* **2004**, *6*, 1983–1986.
9. Patra, A.; Ghorai, S. K.; De, S. R.; Mal, D. *Synthesis* **2006**, 2556–2562.
10. Braun, I.; Rudroff, F.; Mihovilovic, M. D.; Bach, T. *Synthesis* **2007**, *24*, 3896–3906.
11. Krief, A.; Kremer, A. *Tetrahedron Lett.* **2010**, *51*, 4306–4309.

J.J. Li, *Name Reactions: A Collection of Detailed Mechanisms and Synthetic Applications*,
DOI 10.1007/978-3-319-03979-4_126, © Springer International Publishing Switzerland 2014

## Hantzsch dihydropyridine synthesis

1,4-Dihydropyridine from the condensation of aldehyde, β-ketoester and ammonia.
Hantzsch 1,4-dihydropyridines are popular reducing reagents in organo-catalysis.

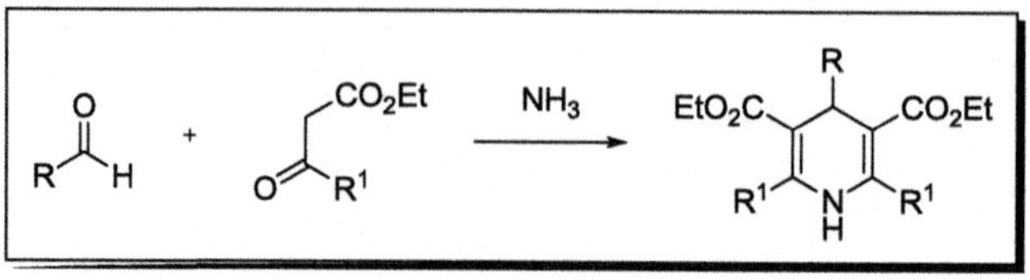

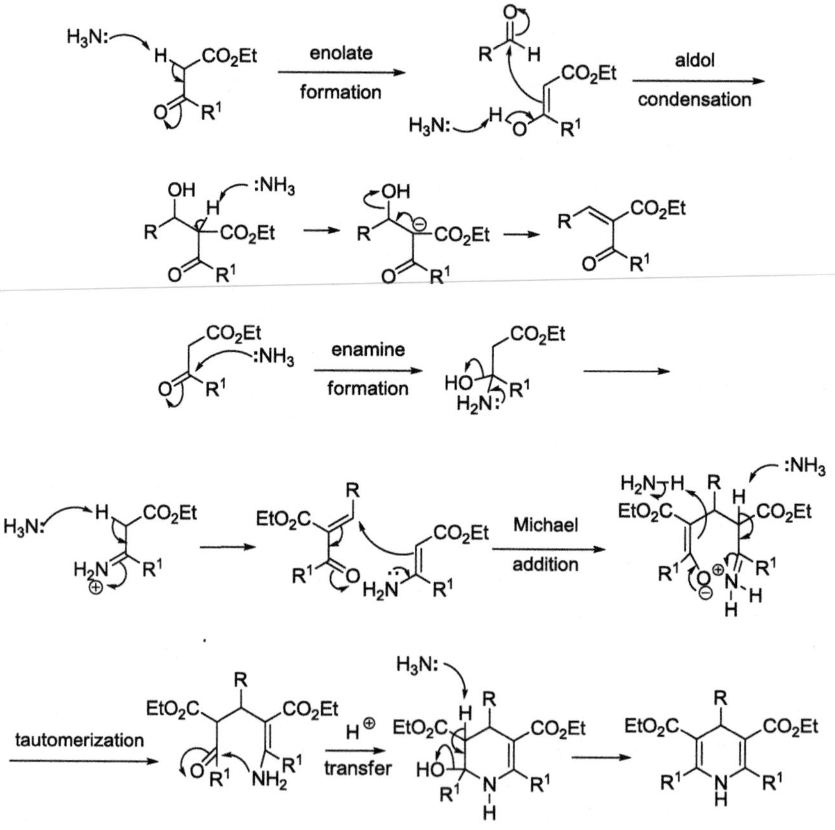

Example 1[2]

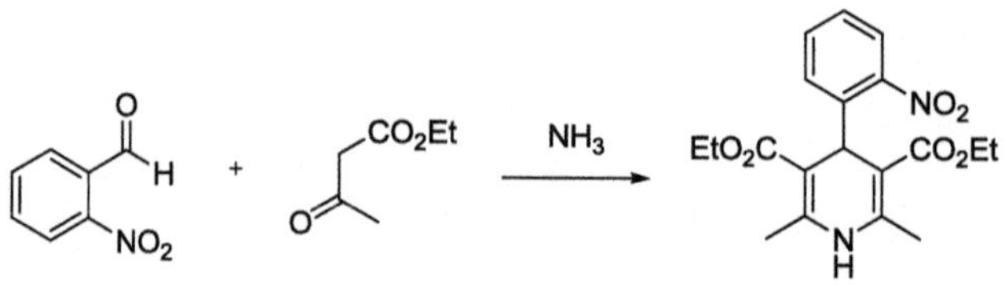

nifedipine

J.J. Li, *Name Reactions: A Collection of Detailed Mechanisms and Synthetic Applications*,
DOI 10.1007/978-3-319-03979-4_127, © Springer International Publishing Switzerland 2014

Example 2[10]

Example 3[10]

## References

1.   Hantzsch, A. *Ann.* **1882,** *215,* 1–83.
2.   Bossert, F.; Vater, W. *Naturwissenschaften* **1971,** *58,* 578–585.
3.   Balogh, M.; Hermecz, I.; Naray-Szabo, G.; Simon, K.; Meszaros, Z. *J. Chem. Soc., Perkin Trans. 1* **1986,** 753–757.
4.   Katritzky, A. R.; Ostercamp, D. L.; Yousaf, T. I. *Tetrahedron* **1987,** *43,* 5171–5187.
5.   Menconi, I.; Angeles, E.; Martinez, L.; Posada, M. E.; Toscano, R. A.; Martinez, R. *J. Heterocycl. Chem.* **1995,** *32,* 831–833.
6.   Raboin, J.-C.; Kirsch, G.; Beley, M. *J. Heterocycl. Chem.* **2000,** *37,* 1077–1080.
7.   Sambongi, Y.; Nitta, H.; Ichihashi, K.; Futai, M.; Ueda, I. *J. Org. Chem.* **2002,** *67,* 3499–3501.
8.   Wang, L.-M.; Sheng, J.; Zhang, L.; Han, J.-W.; Fan, Z.-Y.; Tian, H.; Qian, C.-T. *Tetrahedron* **2005,** *61,* 1539–1543.
9.   Galatsis, P. *Hantzsch Dihydro-Pyridine Synthesis.* In *Name Reactions in Heterocyclic Chemistry*; Li, J. J., Ed.; Wiley: Hoboken, NJ, **2005,** pp 304–307. (Review).
10.  Gupta, R.; Gupta, R.; Paul, S.; Loupy, A. *Synthesis* **2007,** 2835–2838.
11.  Snyder, N. L.; Boisvert, C. J. *Hantzsch Synthesis*, in *Name Reactions in Heterocyclic Chemistry II,* Li, J. J., Ed.; Wiley: Hoboken, NJ, **2011,** pp 591–644. (Review).
12.  Ghosh, S.; Saikh, F.; Das, J.; Pramanik, A. K. *Tetrahedron Lett.* **2013,** *54,* 58–62.

# Hantzsch pyrrole synthesis

Reaction of α-chloromethyl ketones with β-ketoesters and ammonia to assemble pyrroles.

## Example 1[4]

## Example 2[7]

## Example 3[9]

J.J. Li, *Name Reactions: A Collection of Detailed Mechanisms and Synthetic Applications*, DOI 10.1007/978-3-319-03979-4_128, © Springer International Publishing Switzerland 2014

CAN (5%), AgNO$_3$ (1 equiv)

HSVM (20 Hz), rt, 60 min., 94%

HSVM = high-speed vibration milling

## References

1. Hantzsch, A. *Ber.* **1890,** *23,* 1474–1483.

2. Katritzky, A. R.; Ostercamp, D. L.; Yousaf, T. I. *Tetrahedron* **1987,** *43,* 5171–5186.

3. Kirschke, K.; Costisella, B.; Ramm, M.; Schulz, B. *J. Prakt. Chem.* **1990,** *332,* 143–147.

4. Kameswaran, V.; Jiang, B. *Synthesis* **1997,** 530–532.

5. Trautwein, A. W.; Süßmuth, R. D.; Jung, G. *Bioorg. Med. Chem. Lett.* **1998,** *8,* 2381–2384.

6. Ferreira, V. F.; De Souza, M. C. B. V.; Cunha, A. C.; Pereira, L. O. R.; Ferreira, M. L. G. *Org. Prep. Proced. Int.* **2001,** *33,* 411–454. (Review).

7. Matiychuk, V. S.; Martyak, R. L.; Obushak, N. D.; Ostapiuk, Yu. V.; Pidlypnyi, N. I. *Chem. Heterocycl. Compounds* **2004,** *40,* 1218–1219.

8. Snyder, N. L.; Boisvert, C. J. *Hantzsch Synthesis,* in *Name Reactions in Heterocyclic Chemistry II,* Li, J. J., Ed.; Wiley: Hoboken, NJ, 2011, pp 591–644. (Review).

9. Estevez, Veronica; Villacampa, M.; Menendez, J. C. *Chem. Commun.* **2013,** *49,* 591–593.

# Heck reaction

The palladium-catalyzed alkenylation or arylation of olefins.

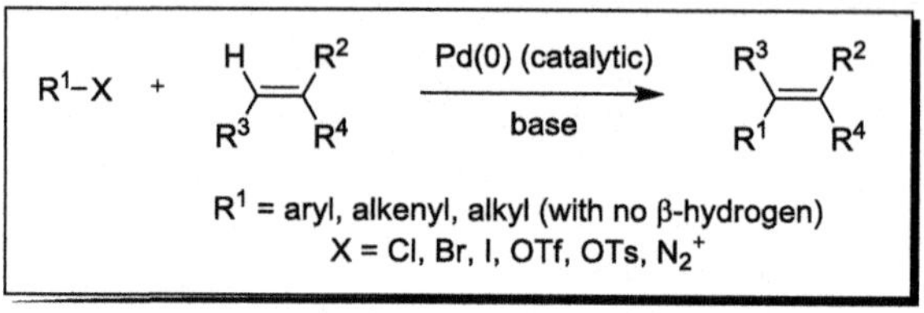
R1–X +
Pd(0) (catalytic)
base
R1 = aryl, alkenyl, alkyl (with no β-hydrogen)
X = Cl, Br, I, OTf, OTs, N2+

The catalytic cycle:

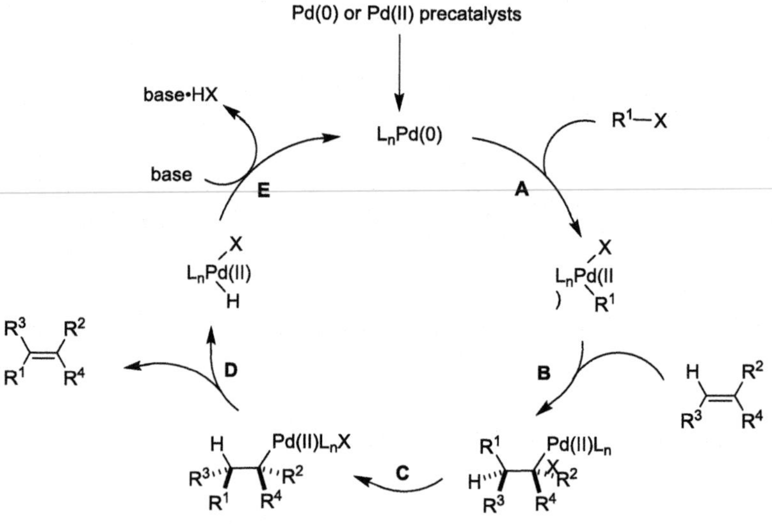
Pd(0) or Pd(II) precatalysts
LnPd(0)
base•HX
base
R1—X
LnPd(II)
Pd(II)LnX
Pd(II)Ln

A: Oxidative addition
B: Migratory insertion (*syn*)
C: C–C bond rotation
D: *syn*-β-elimination
E: Reductive elimination

Example 1, Asymmetric intermolecular Heck reaction[6]

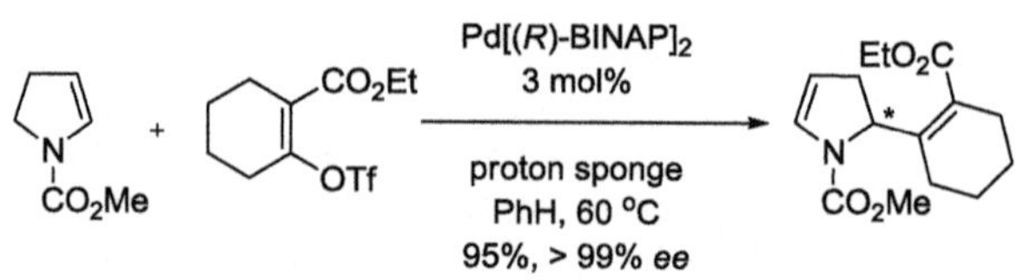
Pd[(R)-BINAP]2
3 mol%
proton sponge
PhH, 60 °C
95%, > 99% ee

J.J. Li, *Name Reactions: A Collection of Detailed Mechanisms and Synthetic Applications*,
DOI 10.1007/978-3-319-03979-4_129, © Springer International Publishing Switzerland 2014

## Example 2, Intramolecular Heck[7]

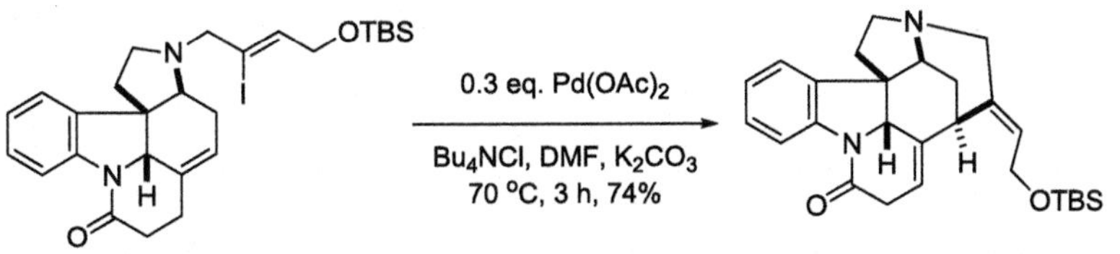

0.3 eq. Pd(OAc)$_2$

Bu$_4$NCl, DMF, K$_2$CO$_3$
70 °C, 3 h, 74%

## Example 3[8]

1.5% Pd$_2$(dba)$_3$, 6% P($t$-Bu)$_3$
1.1. eq. Cs$_2$CO$_3$, dioxane
120 °C, 24 h, 82%

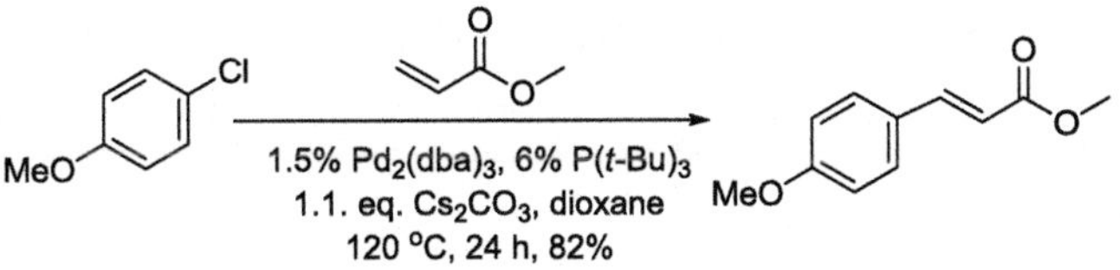

## Example 4, Intramolecular Heck[9]

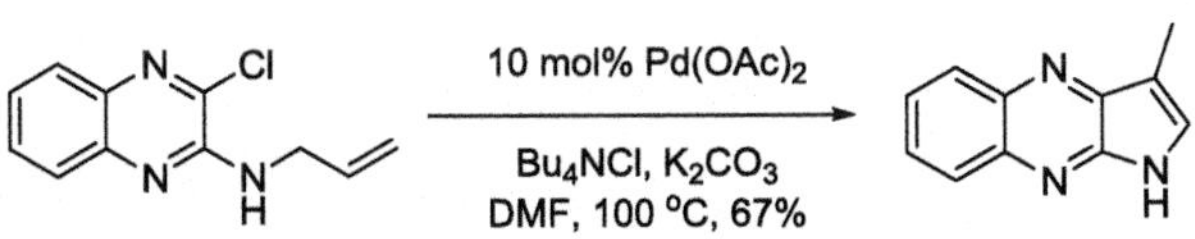

10 mol% Pd(OAc)$_2$

Bu$_4$NCl, K$_2$CO$_3$
DMF, 100 °C, 67%

## Example 5, Intramolecular Heck[13]

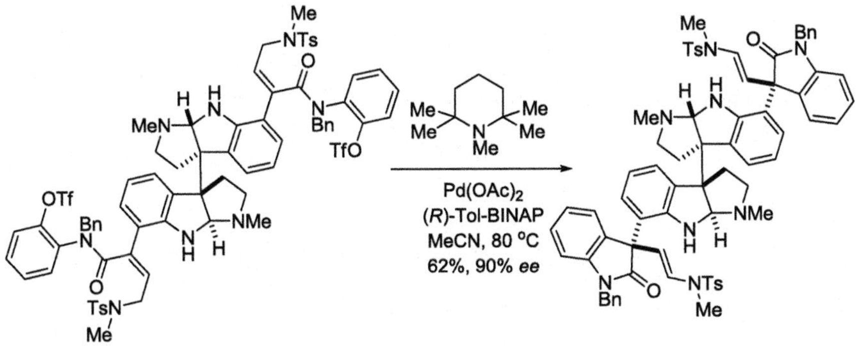

Pd(OAc)$_2$
(R)-Tol-BINAP
MeCN, 80 °C
62%, 90% ee

## Example 6, Reductive Heck reaction[17]

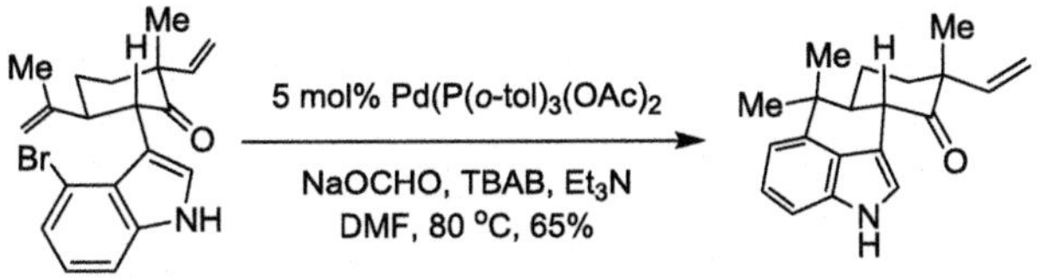

5 mol% Pd(P($o$-tol)$_3$(OAc)$_2$

NaOCHO, TBAB, Et$_3$N
DMF, 80 °C, 65%

Example 7, Intramolecular Heck[20]

## References

1. Heck, R. F.; Nolley, J. P., Jr. *J. Am. Chem. Soc.* **1968**, *90*, 5518–5526. Richard Heck discovered the Heck reaction when he was Hercules Corp. Heck won Nobel Prize in 2010 along with Akira Suzuki and Ei-ichi Negishi "for palladium-catalyzed cross couplings in organic synthesis".
2. Heck, R. F. *Acc. Chem. Res.* **1979**, *12*, 146–151. (Review).
3. Heck, R. F. *Org. React.* **1982**, *27*, 345–390. (Review).
4. Heck, R. F. *Palladium Reagents in Organic Synthesis,* Academic Press, London, **1985**. (Book).
5. Hegedus, L. S. *Transition Metals in the Synthesis of Complex Organic Molecule* **1994**, University Science Books: Mill Valley, CA, pp 103–113. (Book).
6. Ozawa, F.; Kobatake, Y.; Hayashi, T. *Tetrahedron Lett.* **1993**, *34*, 2505–2508.
7. Rawal V. H.; Iwasa, H. *J. Org. Chem.* **1994**, *59*, 2685–2686.
8. Littke, A. F.; Fu, G. C. *J. Org. Chem.* **1999**, *64*, 10–11.
9. Li, J. J. *J. Org. Chem.* **1999**, *64*, 8425–8427.
10. Beletskaya, I. P.; Cheprakov, A. V. *Chem. Rev.* **2000**, *100*, 3009–3066. (Review).
11. Amatore, C.; Jutand, A. *Acc. Chem. Res.* **2000**, *33*, 314–321. (Review).
12. Link, J. T. *Org. React.* **2002**, *60*, 157–534. (Review).
13. Lebsack, A. D.; Link, J. T.; Overman, L. E.; Stearns, B. A. *J. Am. Chem. Soc.* **2002**, *124*, 9008–9009.
14. Dounay, A. B.; Overman, L. E. *Chem. Rev.* **2003**, *103*, 2945–2963. (Review).
15. Beller, M.; Zapf, A.; Riermeier, T. H. *Transition Metals for Organic Synthesis* (2nd edn.) **2004**, *1*, 271–305. (Review).
16. Oestreich, M. *Eur. J. Org. Chem.* **2005**, 783–792. (Review).
17. Baran, P. S.; Maimone, T. J.; Richter, J. M. *Nature* **2007**, *446*, 404–406.
18. Fuchter, M. J. *Heck Reaction.* In *Name Reactions for Homologations-Part I*; Li, J. J., Ed.; Wiley: Hoboken, NJ, **2009**, pp 2–32. (Review).
19. *The Mizoroki–Heck Reaction*; Oestreich, M., Ed.; Wiley: Hoboken, NJ, **2009**.
20. Bennasar, M.-L.; Solé, D.; Zulaica, E.; Alonso, S. *Tetrahedron* **2013**, *69*, 2534–2541.

## Heteroaryl Heck reaction

Intermolecular or intramolecular Heck reaction that occurs onto a heteroaryl recipient.

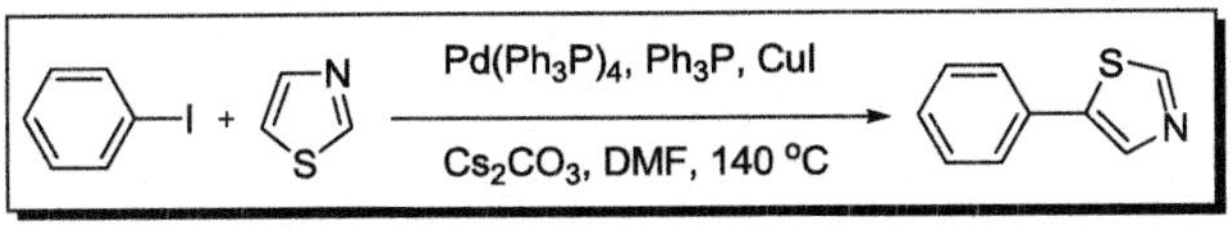

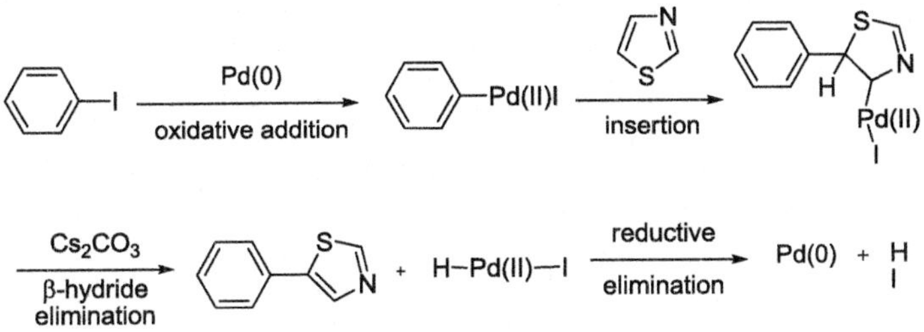

Example 1[2]

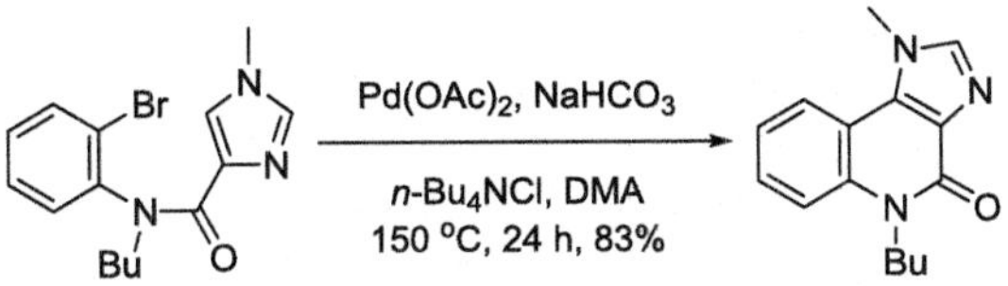

Example 2[3]

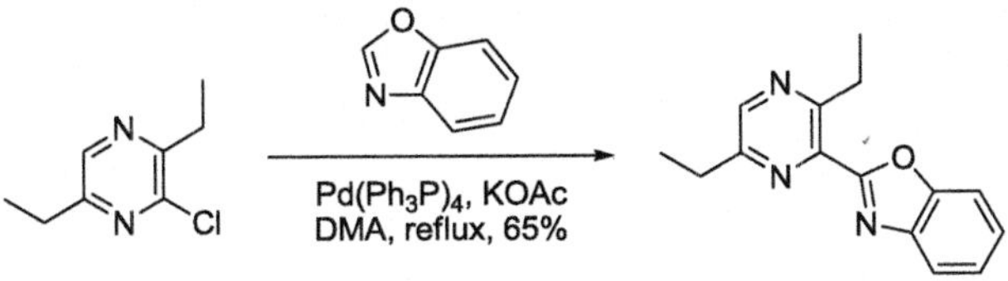

Example 3[7]

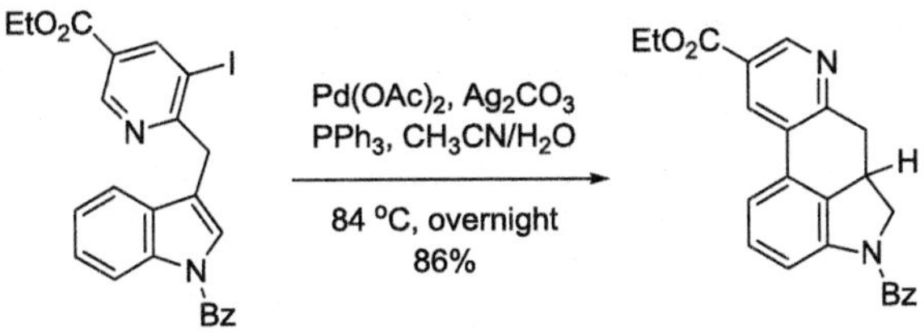

# References

1.  Ohta, A.; Akita, Y.; Ohkuwa, T.; Chiba, M.; Fukunaka, R.; Miyafuji, A.; Nakata, T.; Tani, N. Aoyagi, Y. *Heterocycles* **1990**, *31*, 1951–1958.
2.  Kuroda, T.; Suzuki, F. *Tetrahedron Lett.* **1991**, *32*, 6915–6918.
3.  Aoyagi, Y.; Inoue, A.; Koizumi, I.; Hashimoto, R.; Tokunaga, K.; Gohma, K.; Komatsu, J.; Sekine, K.; Miyafuji, A.; Kunoh, J.; Honma, R.; Akita, Y.; Ohta, A. *Heterocycles* **1992**, *33*, 257–272.
4.  Proudfoot, J. R.; Patel, U. R.; Kapadia, S. R.; Hargrave, K. D. *J. Med. Chem.* **1995**, *38*, 1406–1410.
5.  Pivsa-Art, S.; Satoh, T.; Kawamura, Y.; Miura, M.; Nomura, M. *Bull. Chem. Soc. Jpn.* **1998**, *71*, 467–473.
6.  Li, J. J.; Gribble, G. W. In *Palladium in Heterocyclic Chemistry;* 2$^{nd}$ ed.; **2007**, Elsevier: Oxford, UK. (Review).
7.  Burley, S. D.; Lam, V. V.; Lakner, F. J.; Bergdahl, B. M.; Parker, M. A. *Org. Lett.* **2013**, *15*, 2598–2600.

# Hegedus indole synthesis

Stoichiometric Pd(II)-mediated oxidative cyclization of alkenyl anilines to indoles. *Cf.* Wacker oxidation.

Example 1[1a]

Example 2[1d]

## References

1.  (a) Hegedus, L. S.; Allen, G. F.; Waterman, E. L. *J. Am. Chem. Soc.* **1976**, *98*, 2674–2676. Lou Hegedus is a professor at Colorado State University. (b) Hegedus, L. S.; Allen, G. F.; Bozell, J. J.; Waterman, E. L. *J. Am. Chem. Soc.* **1978**, *100*, 5800–5807. (c) Hegedus, L. S.; Winton, P. M.; Varaprath, S. *J. Org. Chem.* **1981**, *46*, 2215–2221. (d) Harrington, P. J.; Hegedus, L. S. *J. Org. Chem.* **1984**, *49*, 2657–2662. (e) Hegedus, L. S. *Angew. Chem. Int. Ed.* **1988**, *27*, 1113–1126. (Review).
2.  Brenner, M.; Mayer, G.; Terpin, A.; Steglich, W. *Chem. Eur. J.* **1997**, *3*, 70–74.
3.  Osanai, Y. Y.; Kondo, K.; Murakami, Y. *Chem. Pharm. Bull.* **1999**, *47*, 1587–1590.
4.  Kondo, T.; Okada, T.; Mitsudo, T. *J. Am. Chem. Soc.* **2002**, *124*, 186–187. A ruthenium variant.
5.  Johnston, J. N. *Hegedus Indole Synthesis*. In *Name Reactions in Heterocyclic Chemistry*; Li, J. J., Ed.; Wiley: Hoboken, NJ, **2005**, pp 135–139. (Review).

J.J. Li, *Name Reactions: A Collection of Detailed Mechanisms and Synthetic Applications*,
DOI 10.1007/978-3-319-03979-4_130, © Springer International Publishing Switzerland 2014

# Hell–Volhard–Zelinsky reaction

α-Halogenation of carboxylic acids using $X_2/PBr_3$.

α-bromoacid

Example 1[5]

Example 2[6]

## References

1.    (a) Hell, C. *Ber.* **1881**, *14*, 891–893. Carl M. von Hell (1849–1926) was born in
      Stuttgart, Germany. He studied under Fehling and Erlenmeyer. Hell became a profes-
      sor at Stuttgart in 1883 where he discovered the Hell–Volhard–Zelinsky reaction. (b)
      Volhard, J. *Ann.* **1887**, *242*, 141–163. Jacob Volhard (1849–1909) was born in Darm-
      stadt, Germany. He apprenticed under Liebig, Will, Bunsen, Hofmann, Kolbe, and
      von Baeyer. He improved Hell's original procedure in preparing α-bromo-acid during
      his research in thiophenes. (c) Zelinsky, N. D. *Ber.* **1887**, *20*, 2026. Nikolay D.
      Zelinsky (1861–1953) was born in Tiraspol, Russia. He studied in Germany, receiv-
      ing his degree in 1891. In 1885, Zelinsky was the first to prepare mustard gas uninten-
      tionally while exploring polymerization of sulfur dichloride. Zelinsky returned to Rus-

J.J. Li, *Name Reactions: A Collection of Detailed Mechanisms and Synthetic Applications*,
DOI 10.1007/978-3-319-03979-4_131, © Springer International Publishing Switzerland 2014

sia and became a professor at the University of Moscow, where he discovered activated charcoal gas mask. He was awarded the Order of Lenin in 1934.

2.   Watson, H. B. *Chem. Rev.* **1930**, *7,* 173–201. (Review).

3.   Sonntag, N. O. V. *Chem. Rev.* **1953**, *52,* 237–246. (Review).

4.   Harwood, H. J. *Chem. Rev.* **1962**, *62,* 99–154. (Review).

5.   Jason, E. F.; Fields, E. K. US Patent 3,148,209 (**1964**).

6.   Chow, A. W.; Jakas, D. R.; Hoover, J. R. E. *Tetrahedron Lett.* **1966**, *7,* 5427–5431.

7.   Liu, H.-J.; Luo, W. *Synth. Commun.* **1991**, *21*, 2097–2102.

8.   Zhang, L. H.; Duan, J.; Xu, Y.; Dolbier, W. R., Jr. *Tetrahedron Lett.* **1998**, *39,* 9621–9622.

9.   Sharma, A.; Chattopadhyay, S. *J. Org. Chem.* **1999**, *64*, 8059–8062.

10.  Stack, D. E.; Hill, A. L.; Diffendaffer, C. B.; Burns, N. M. *Org. Lett.* **2002**, *4*, 4487–4490.

11.  Sun, Z.; Peng, X.; Dong, X.; Shi, W. *Asian J. Chem.* **2012**, *24*, 929–930.

# Henry nitroaldol reaction

The nitroaldol condensation reaction involving aldehydes and nitronates, derived from deprotonation of nitroalkanes by bases.

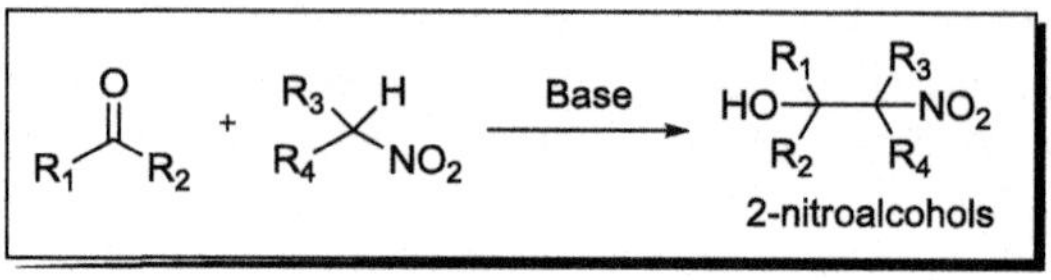

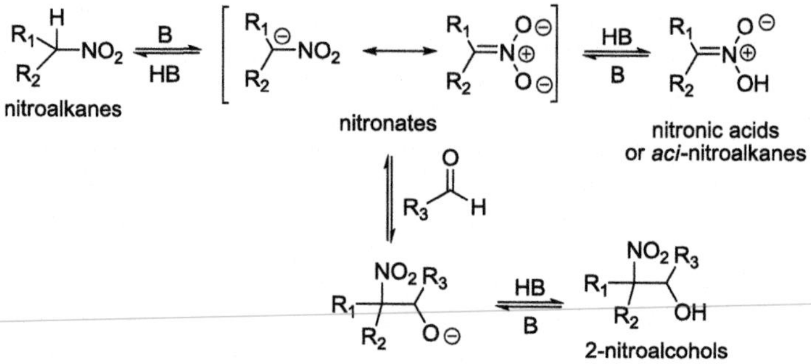

2-nitroalcohols

nitroalkanes

nitronates

nitronic acids or *aci*-nitroalkanes

2-nitroalcohols

Example 1[4]

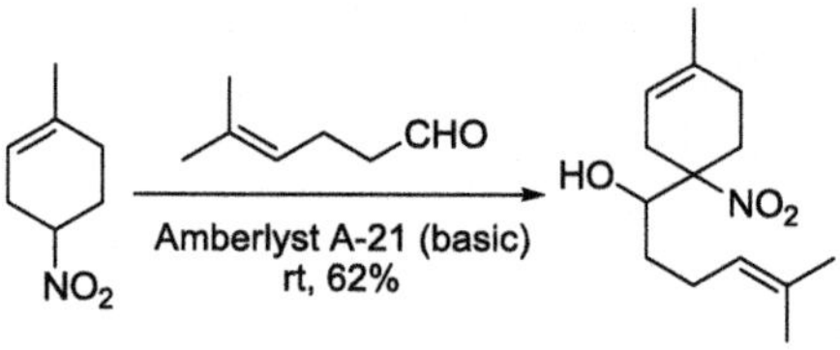

Amberlyst A-21 (basic)
rt, 62%

Example 2, Retro-Henry reaction[5]

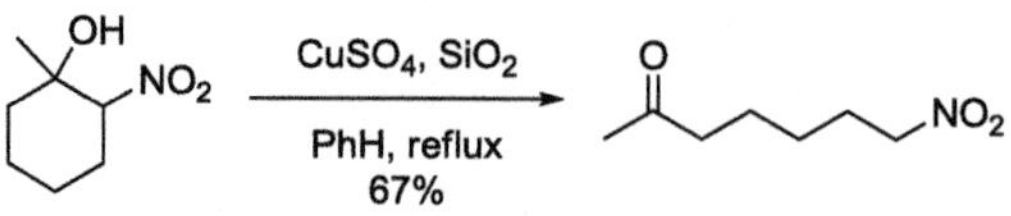

CuSO$_4$, SiO$_2$

PhH, reflux
67%

Example 3, Aza-Henry reaction[8]

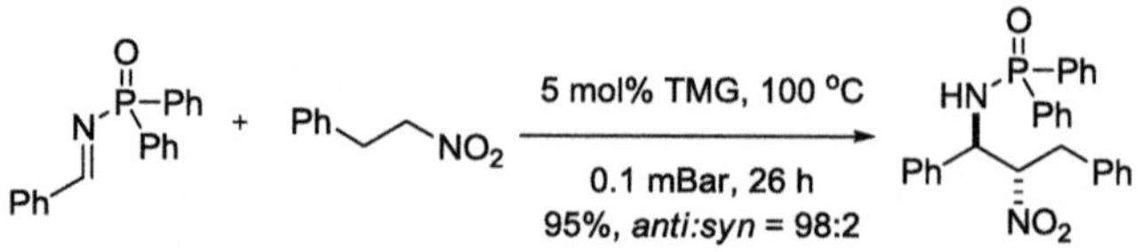

5 mol% TMG, 100 °C

0.1 mBar, 26 h
95%, *anti:syn* = 98:2

J.J. Li, *Name Reactions: A Collection of Detailed Mechanisms and Synthetic Applications*, DOI 10.1007/978-3-319-03979-4_132, © Springer International Publishing Switzerland 2014

Example 4, Intramolecular Henry reaction[10]

Example 4, A highly asymmetric Henry reaction catalyzed by chiral copper(II) complexes[12]

## References

1. Henry, L. *Compt. Rend.* **1895,** *120,* 1265–1268.
2. Barrett, A. G. M.; Robyr, C.; Spilling, C. D. *J. Org. Chem.* **1989,** *54,* 1233–1234.
3. Rosini, G. In *Comprehensive Organic Synthesis;* Trost, B. M.; Fleming, I., Eds.; Pergamon, **1991,** *2,* 321–340. (Review).
4. Chen, Y.-J.; Lin, W.-Y. *Tetrahedron Lett.* **1992,** *33,* 1749–1750.
5. Saikia, A. K.; Hazarika, M. J.; Barua, N. C.; Bezbarua, M. S.; Sharma, R. P.; Ghosh, A. C. *Synthesis* **1996,** 981–985.
6. Luzzio, F. A. *Tetrahedron* **2001,** *57,* 915–945. (Review).
7. Westermann, B. *Angew. Chem. Int. Ed.* **2003,** *42,* 151–153. (Review on aza-Henry reaction).
8. Bernardi, L.; Bonini, B. F.; Capito, E.; Dessole, G.; Comes-Franchini, M.; Fochi, M.; Ricci, A. *J. Org. Chem.* **2004,** *69,* 8168–8171.
9. Palomo, C.; Oiarbide, M.; Laso, A. *Angew. Chem. Int. Ed.* **2005,** *44,* 3881–3884.
10. Kamimura, A.; Nagata, Y.; Kadowaki, A.; Uchidaa, K.; Uno, H. *Tetrahedron* **2007,** *63,* 11856–11861.
11. Wang, A. X. *Henry Reaction.* In *Name Reactions for Homologations-Part I*; Li, J. J., Ed.; Wiley: Hoboken, NJ, **2009,** pp 404–419. (Review).
12. Ni, B.; He, J. *Tetrahedron Lett.* **2013,** *54,* 462–465.

# Hinsberg synthesis of thiophenes

Condensation of diethyl thiodiglycolate and α-diketones under basic conditions, which provides 3,4-disubstituted thiophene-2,5-dicarbonyls upon hydrolysis of the crude ester product with aqueous acid.

## Example 1[2]

## Example 2[4]

J.J. Li, *Name Reactions: A Collection of Detailed Mechanisms and Synthetic Applications*, DOI 10.1007/978-3-319-03979-4_133, © Springer International Publishing Switzerland 2014

Example 3[5]

Example 4[6]

Example 5, Polymer-support Hinsberg thiophene synthesis[9]

## References

1.  Hinsberg, O. *Ber.* **1910**, *43*, 901–906.
2.  Miyahara, Y.; Inazu, T.; Yoshino, T. *Bull. Chem. Soc. Jpn.* **1980**, *53*, 1187–1188.
3.  Gronowitz, S. In *Thiophene and Its Derivatives*, Part 1, Gronowitz, S., ed.; Wiley-Interscience: New York, **1985**, pp 34–41. (Review).
4.  Miyahara, Y.; Inazu, T.; Yoshino, T. *J. Org. Chem.* **1984**, *49*, 1177–1182.
5.  Christl, M.; Krimm, S.; Kraft, A. *Angew. Chem. Int. Ed.* **1990**, *29*, 675–677.
6.  Beye, N.; Cava, M. P. *J. Org. Chem.* **1994**, *59*, 2223–2226.
7.  Vogel, E.; Pohl, M.; Herrmann, A.; Wiss, T.; König, C.; Lex, J.; Gross, M.; Gisselbrecht, J. P. *Angew. Chem. Int. Ed.* **1996**, *35*, 1520–1525.
8.  Mullins, R. J.; Williams, D. R. *Hinsberg Synthesis of Thiophene Derivatives.* In *Name Reactions in Heterocyclic Chemistry*; Li, J. J., Ed.; Wiley: Hoboken, NJ, **2005**, pp 199–206. (Review).
9.  Traversone, A.; Brill, W. K.-D. *Tetrahedron Lett.* **2007**, *48*, 3535–3538.
10. Jimenez, R. P.; Parvez, M.; Sutherland, T. C.; Viccars, J. *Eur. J. Org. Chem.* **2009**, 5635–5646.

# Hiyama cross-coupling reaction

Palladium-catalyzed cross-coupling reaction of organosilicons with organic halides, triflates, *etc*. In the presence of an activating agent such as fluoride or hydroxide (transmetallation is reluctant to occur without the effect of an activating agent). For the catalytic cycle, see the Kumada coupling.

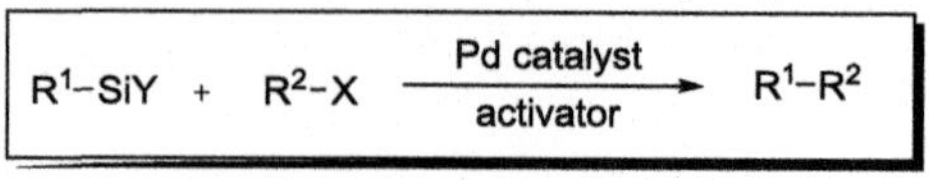

$$R^1\text{-SiY} + R^2\text{-X} \xrightarrow[\text{activator}]{\text{Pd catalyst}} R^1\text{-}R^2$$

$R^1$ = alkenyl, aryl, alkynyl, alkyl
$R^2$ = aryl, alkyl, alkenyl
Y = $(OR)_3$, $Me_3$, $Me_2OH$, $Me_{(3-n)}F_{(n+3)}$
X = Cl, Br, I, OTf
activator = TBAF, base

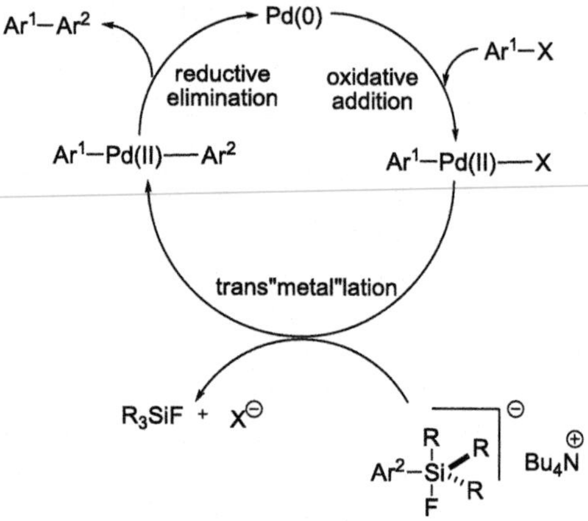

Example 1[1a]

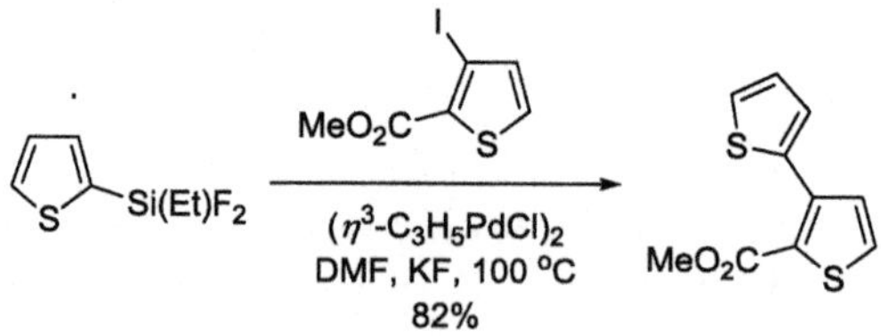

MeO_2C, ($\eta^3$-C_3H_5PdCl)_2, DMF, KF, 100 °C, 82%

Example 2[2]

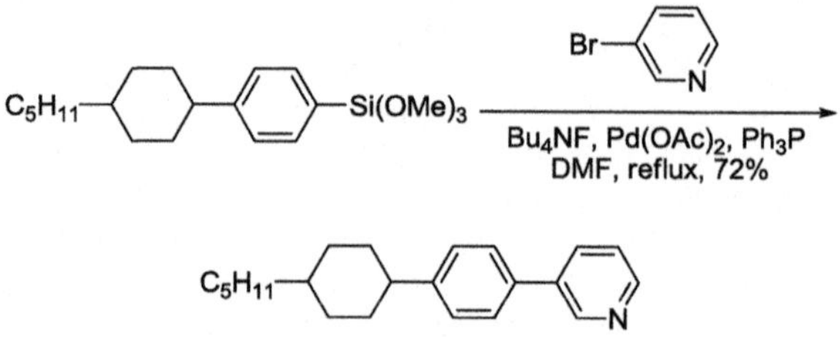

C_5H_11, Si(OMe)_3, Bu_4NF, Pd(OAc)_2, Ph_3P, DMF, reflux, 72%

J.J. Li, *Name Reactions: A Collection of Detailed Mechanisms and Synthetic Applications*,
DOI 10.1007/978-3-319-03979-4_134, © Springer International Publishing Switzerland 2014

Example 3[7]

Example 4[9]

Example 5, Reusable polystyrene-supported palladium catalyst[11]

## References

1. (a) Hatanaka, Y.; Fukushima, S.; Hiyama, T. *Heterocycles* **1990**, *30*, 303–306. (b) Hiyama, T.; Hatanaka, Y. *Pure Appl. Chem.* **1994**, *66*, 1471–1478. (c) Matsuhashi, H.; Kuroboshi, M.; Hatanaka, Y.; Hiyama, T. *Tetrahedron Lett.* **1994**, *35*, 6507–6510.

2. Shibata, K.; Miyazawa, K.; Goto, Y. *Chem. Commun.* **1997**, 1309–1310.

3. Hiyama, T. In *Metal-Catalyzed Cross-Coupling Reactions;* **1998**, Diederich, F.; Stang, P. J., Eds.; Wiley–VCH: Weinheim, Germany, pp 421–53. (Review).

4. Denmark, S. E.; Wang, Z. *J. Organomet. Chem.* **2001**, *624*, 372–375.

5. Hiyama, T. *J. Organomet. Chem.* **2002**, *653*, 58–61.

6. Pierrat, P.; Gros, P.; Fort, Y. *Org. Lett.* **2005**, *7*, 697–700.

7. Denmark, S. E.; Yang, S.-M. *J. Am. Chem. Soc.* **2004**, *126*, 12432–12440.

8. Domin, D.; Benito-Garagorri, D.; Mereiter, K.; Froehlich, J.; Kirchner, K. *Organometallics* **2005**, *24*, 3957–3965.

9. Anzo, T.; Suzuki, A.; Sawamura, K.; Motozaki, T.; Hatta, M.; Takao, K.-i.; Tadano, K.-i. *Tetrahedron Lett.* **2007**, *48*, 8442–8448.

10. Yet L. *Hiyama Cross-Coupling Reaction.* In *Name Reactions for Homologations-Part I*; Li, J. J., Ed.; Wiley: Hoboken, NJ, **2009**, pp 33–416. (Review).

11. Diebold, C.; Derible, A.; Becht, J.-M.; Drian, C. L. *Tetrahedron* **2013**, *69*, 264–267.

# Hofmann elimination

Elimination reaction of alkyl trimethyl amines proceeds with *anti*-stereochemistry, furnishing the least highly substituted olefins.

Example 1, Amine released from the resin by Hofmann elimination[10]

Example 2[11]

## References

1.  Hofmann, A. W. *Ber.* **1881**, *14*, 659–669.
2.  Eubanks, J. R. I.; Sims, L. B.; Fry, A. *J. Am. Chem. Soc.* **1991**, *113*, 8821–8829.
3.  Bach, R. D.; Braden, M. L. *J. Org. Chem.* **1991**, *56*, 7194–7195.
4.  Lai, Y. H.; Eu, H. L. *J. Chem. Soc., Perkin Trans. 1* **1993**, 233–237.
5.  Sepulveda-Arques, J.; Rosende, E. G.; Marmol, D. P.; Garcia, E. Z.; Yruretagoyena, B.; Ezquerra, J. *Monatsh. Chem.* **1993**, *124*, 323–325.
6.  Woolhouse, A. D.; Gainsford, G. J.; Crump, D. R. *J. Heterocycl. Chem.* **1993**, *30*, 873–880.
7.  Bhonsle, J. B. *Synth. Commun.* **1995**, *25*, 289–300.
8.  Berkes, D.; Netchitailo, P.; Morel, J.; Decroix, B. *Synth. Commun.* **1998**, *28*, 949–956.
9.  Morphy, J. R.; Rankovic, Z.; York, M. *Tetrahedron Lett.* **2002**, *43*, 6413–6415.
10. Liu, Z.; Medina-Franco, J. L.; Houghten, R. A.; Giulianotti, M. A. *Tetrahedron Lett.* **2010**, *51*, 5003–5004.
11. Arava, V. R.; Malreddy, S.; Thummala, S. R. *Synth. Commun.* **2012**, *42*, 3545–3552.

J.J. Li, *Name Reactions: A Collection of Detailed Mechanisms and Synthetic Applications*,
DOI 10.1007/978-3-319-03979-4_135, © Springer International Publishing Switzerland 2014

# Hofmann rearrangement

Upon treatment of primary amides with hypohalites, primary amines with one less carbon are obtained *via* the intermediacy of isocyanate. Also know as the Hofmann degradation reaction.

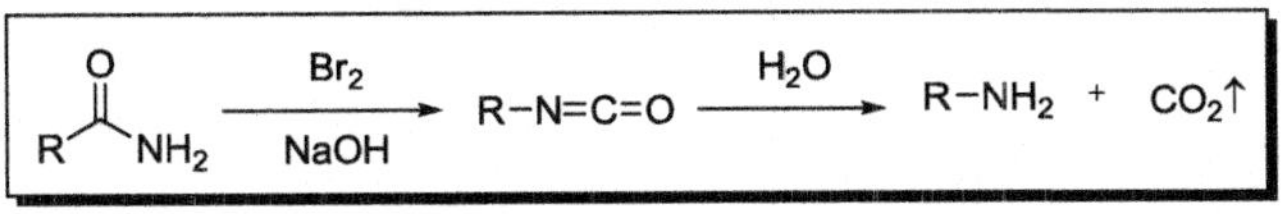

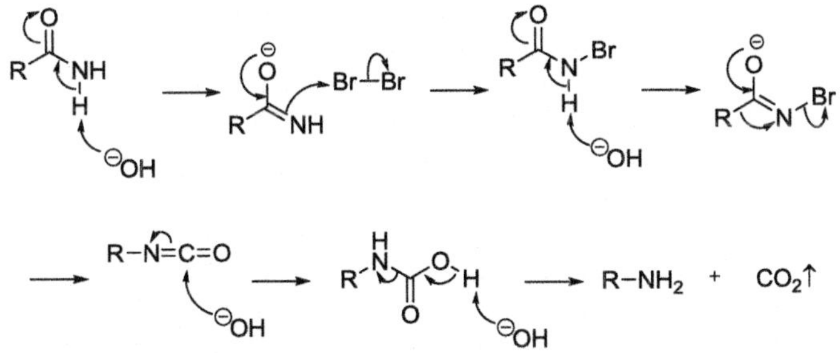

isocyanate intermediate

Example 1, An NBS variant[2]

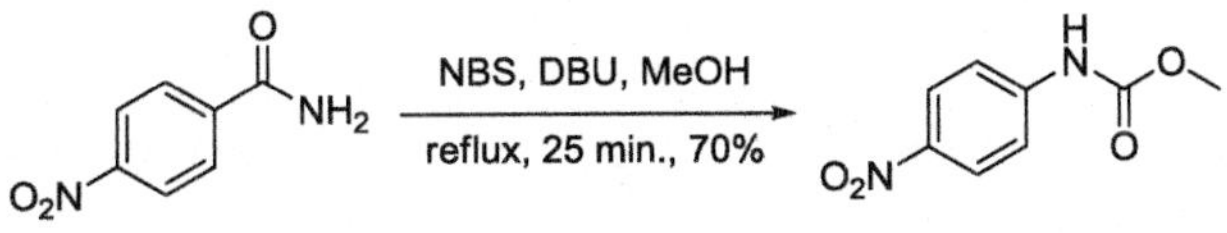

Example 2, Iodosobenzene diacetate[5]

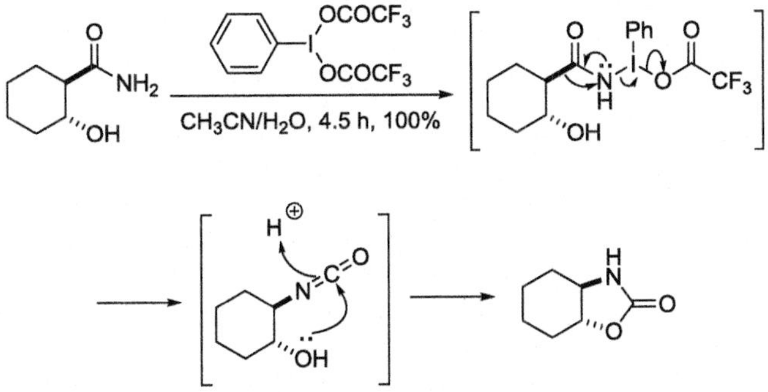

J.J. Li, *Name Reactions: A Collection of Detailed Mechanisms and Synthetic Applications*, DOI 10.1007/978-3-319-03979-4_136, © Springer International Publishing Switzerland 2014

Example 3, Bromine and alkoxide[6]

Example 4, Sodium hypochlorite[7]

Example 5, The original conditions, bromine and hydroxide[9]

Example 6, Lead tetraacetate[10]

## References

1. Hofmann, A. W. *Ber.* **1881**, *14*, 2725–2736.
2. Jew, S.-s.; Kang, M.-h. *Arch. Pharmacol Res.* **1994**, *17*, 490–491.
3. Huang, X.; Seid, M.; Keillor, J. W. *J. Org. Chem.* **1997**, *62*, 7495–7496.
4. Togo, H.; Nabana, T.; Yamaguchi, K. *J. Org. Chem.* **2000**, *65*, 8391–8394.
5. Yu, C.; Jiang, Y.; Liu, B.; Hu, L. *Tetrahedron Lett.* **2001**, *42*, 1449–1452.
6. Jiang, X.; Wang, J.; Hu, J.; Ge, Z.; Hu, Y.; Hu, H.; Covey, D. F. *Steroids* **2001**, *66*, 655–662.
7. Stick, R. V.; Stubbs, K. A. *J. Carbohydr. Chem.* **2005**, *24*, 529–547.
8. Moriarty, R. M. *J. Org. Chem.* **2005**, *70*, 2893–2903. (Review).
9. El-Mariah, F.; Hosney, M.; Deeb, A. *Phosphorus, Sulfur Silicon Relat. Elem.* **2006**, *181*, 2505–2517.
10. Jia, Y.-M.; Liang, X.-M.; Chang, L.; Wang, D.-Q. *Synthesis* **2007**, 744–748.
11. Gribble, G. W. *Hofmann rearrangement.* In *Name Reactions for Homologations-Part II*; Li, J. J., Ed.; Wiley: Hoboken, NJ, **2009**, pp 164–199. (Review).
12. Yoshimura, A.; Luedtke, M. W.; Zhdankin, V. V. *J. Org. Chem.* **2012**, *77*, 2087–2091.

# Hofmann–Löffler–Freytag reaction

Formation of pyrrolidines or piperidines by thermal or photochemical decomposition of protonated *N*-haloamines.

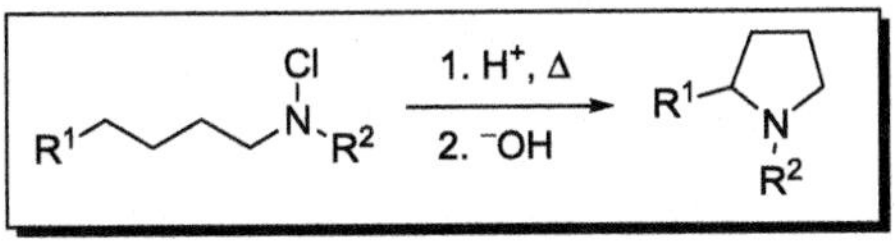

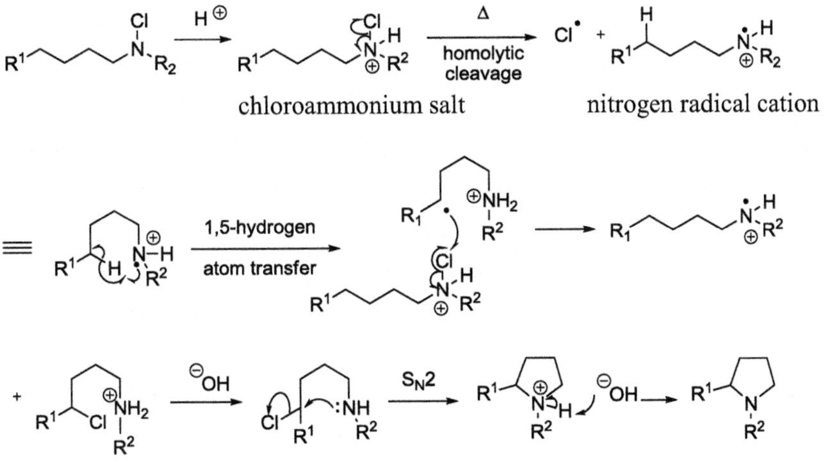

chloroammonium salt               nitrogen radical cation

1,5-hydrogen atom transfer

$S_N2$

Example 1[2]

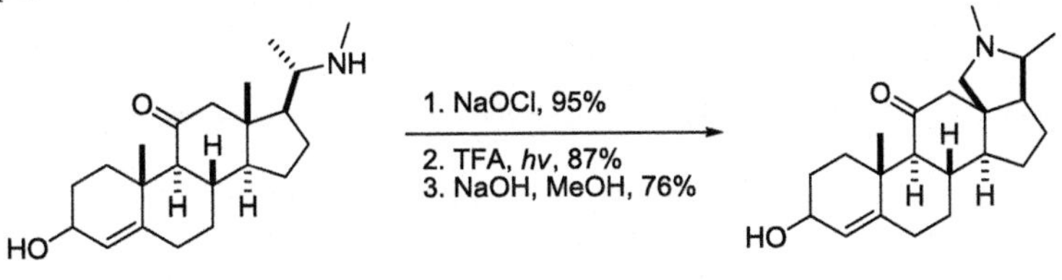

1. NaOCl, 95%
2. TFA, *hv*, 87%
3. NaOH, MeOH, 76%

Example 2[4]

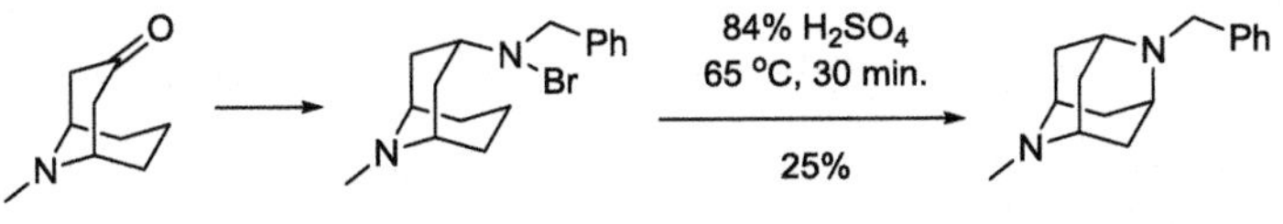

84% $H_2SO_4$
65 °C, 30 min.

25%

Example 3[5]

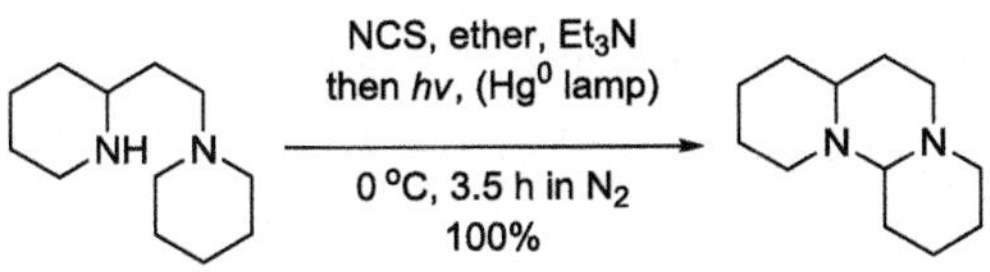

NCS, ether, $Et_3N$
then *hv*, ($Hg^0$ lamp)

0 °C, 3.5 h in $N_2$
100%

J.J. Li, *Name Reactions: A Collection of Detailed Mechanisms and Synthetic Applications*,
DOI 10.1007/978-3-319-03979-4_137, © Springer International Publishing Switzerland 2014

Example 4[7]

Example 5[12]

Example 6[13]

## References

1.  (a) Hofmann, A. W. *Ber.* **1883**, *16*, 558–560.  (b) Löffler, K.; Freytag, C. *Ber.* **1909**, *42*, 3727.
2.  Wolff, M. E.; Kerwin, J. F.; Owings, F. F.; Lewis, B. B.; Blank, B.; Magnani, A.; Karash, C.; Georgian, V. *J. Am. Chem. Soc.* **1960**, *82*, 4117–4118.
3.  Wolff, M. E. *Chem. Rev.* **1963**, *63*, 55–64. (Review).
4.  Dupeyre, R.-M.; Rassat, A. *Tetrahedron Lett.* **1973**, 2699–2701.
5.  Kimura, M.; Ban, Y. *Synthesis* **1976**, 201–202.
6.  Stella, L. *Angew. Chem. Int. Ed.* **1983**, *22*, 337–422. (Review).
7.  Betancor, C.; Concepcion, J. I.; Hernandez, R.; Salazar, J. A.; Suarez, E. *J. Org. Chem.* **1983**, *48*, 4430–4432.
8.  Majetich, G.; Wheless, K. *Tetrahedron* **1995**, *51*, 7095–7129. (Review).
9.  Togo, H.; Katohgi, M. *Synlett* **2001**, 565–581. (Review).
10. Pellissier, H.; Santelli, M. *Org. Prep. Proced. Int.* **2001**, *33*, 455–476. (Review).
11. Li, J. J. *Hofmann–Löffler–Freytag Reaction.* In *Name Reactions in Heterocyclic Chemistry*; Li, J. J., Ed.; Wiley: Hoboken, NJ, **2005**, pp 89–97. (Review).
12. Chen, K.; Richter, J. M.; Baran, P. S. *J. Am. Chem. Soc.* **2008**, *130*, 17247–17249.
13. Lechel, T.; Podolan, G.; Brusilowskij, B.; Schalley, C. A.; Reissig, H.-U. *Eur. J. Org. Chem.* **2012**, 5685–5692.

## Horner–Wadsworth–Emmons reaction

Olefin formation from aldehydes and phosphonates. Workup is more advantageous than the corresponding Wittig reaction because the phosphate by-product can be washed away with water. Typically gives the *trans*- rather than the *cis*-olefins.

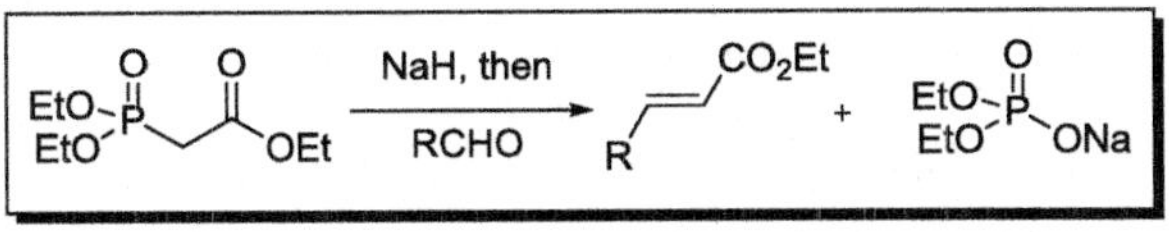
NaH, then
RCHO
CO2Et
R
EtO
ONa

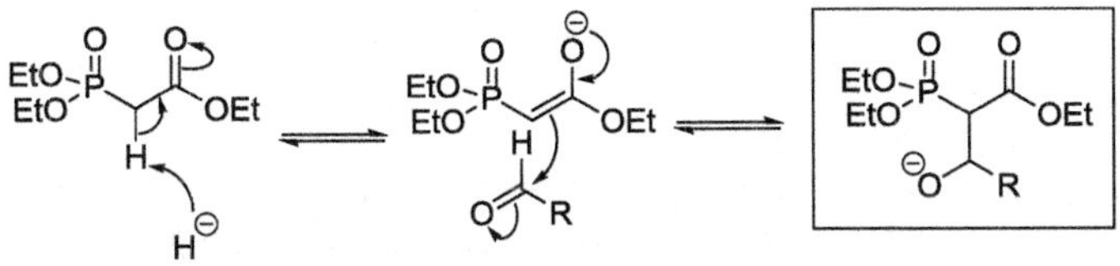

The stereochemical outcome: *erythro* (kinetic) or *threo* (thermodynamic)

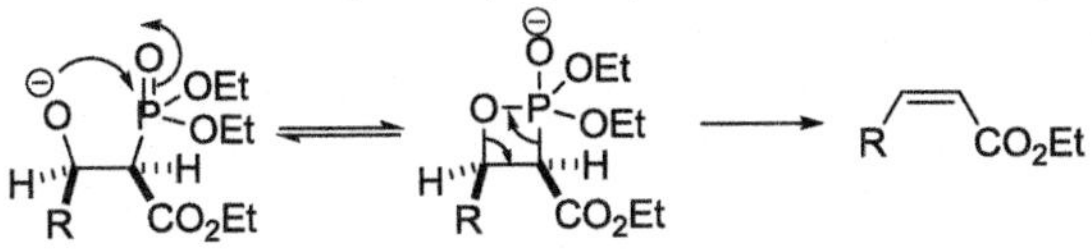

*erythro*, kinetic adduct

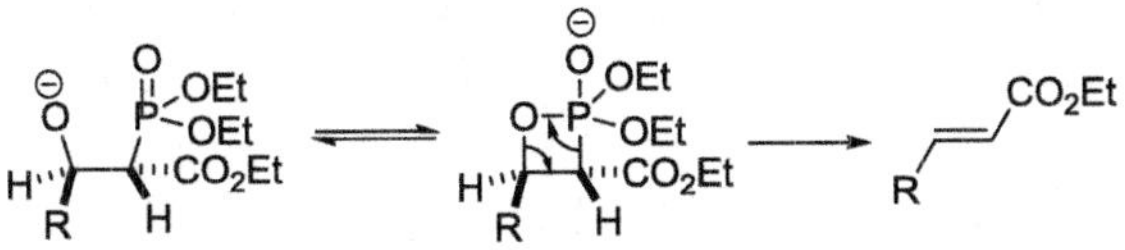

*threo*, thermodynamic adduct

Example 1[3]

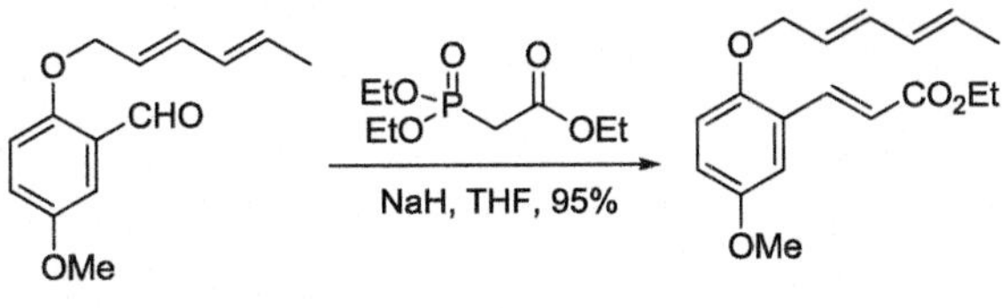
NaH, THF, 95%

Example 2[4]

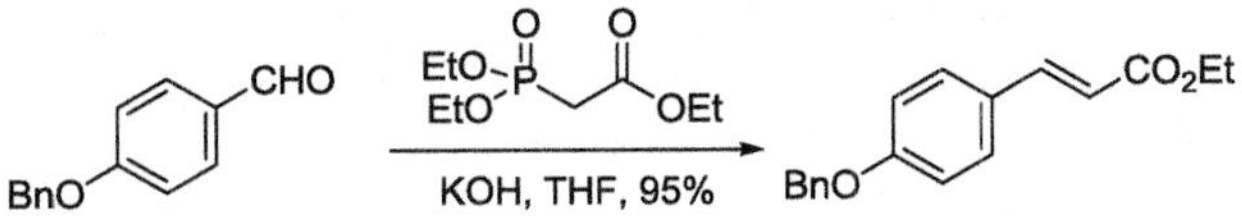
KOH, THF, 95%

Example 3[7]

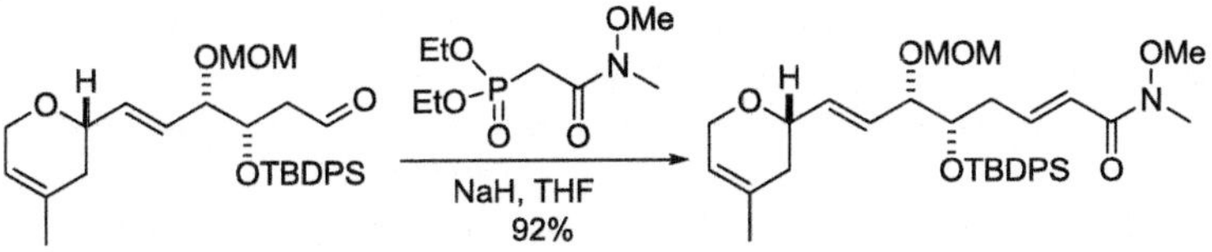
NaH, THF
92%

J.J. Li, *Name Reactions: A Collection of Detailed Mechanisms and Synthetic Applications*,
DOI 10.1007/978-3-319-03979-4_138, © Springer International Publishing Switzerland 2014

Example 4, Intramolecular Horner–Wadsworth–Emmons[9]

$K_2CO_3$
18-crown-6

Toluene
–20 °C
67%
$Z/E = 5:1$

Example 4[11]

$(EtO)_2P(O)CHMeCOMe$

NaH, THF, 0 °C, 80%

## References

1.  (a) Horner, L.; Hoffmann, H.; Wippel, H. G.; Klahre, G. *Chem. Ber.* **1959**, *92*, 2499–2505. (b) Wadsworth, W. S., Jr.; Emmons, W. D. *J. Am. Chem. Soc.* **1961**, *83*, 1733–1738. (c) Wadsworth, D. H.; Schupp, O. E.; Seus, E. J.; Ford, J. A., Jr. *J. Org. Chem.* **1965**, *30*, 680–685.
2.  Maryanoff, B. E.; Reitz, A. B. *Chem. Rev.* **1989**, *89*, 863–927. (Review).
3.  Shair, M. D.; Yoon, T. Y.; Mosny, K. K.; Chou, T. C.; Danishefsky, S. J. *J. Am. Chem. Soc.* **1996**, *118*, 9509–9525.
4.  Nicolaou, K. C.; Boddy, C. N. C.; Li, H.; Koumbis, A. E.; Hughes, R. J.; Natarajan, S.; Jain, N. F.; Ramanjulu, J. M.; Bräse, S.; Solomon, M. E. *Chem. Eur. J.* **1999**, *5*, 2602–2621.
5.  Comins, D. L.; Ollinger, C. G. *Tetrahedron Lett.* **2001**, *42*, 4115–4118.
6.  Lattanzi, A.; Orelli, L. R.; Barone, P.; Massa, A.; Iannece, P.; Scettri, A. *Tetrahedron Lett.* **2003**, *44*, 1333–1337.
7.  Ahmed, A.; Hoegenauer. E. K.; Enev, V. S.; Hanbauer, M.; Kaehlig, H.; Öhler, E.; Mulzer, J. *J. Org. Chem.* **2003**, *68*, 3026–3042.
8.  Blasdel, L. K.; Myers, A. G. *Org. Lett.* **2005**, *7*, 4281–4283.
9.  Li, D.-R.; Zhang, D.-H.; Sun, C.-Y.; Zhang, J.-W.; Yang, L.; Chen, J.; Liu, B.; Su, C.; Zhou, W.-S.; Lin, G.-Q. *Chem. Eur. J.* **2006**, *12*, 1185–1204.
10. Rong, F. *Horner–Wadsworth–Emmons reaction* In *Name Reactions for Homologations-Part I*; Li, J. J., Ed.; Wiley: Hoboken, NJ, **2009**, pp 420–466. (Review).
11. Okamoto, R.; Takeda, K.; Tokuyama, H.; Ihara, M.; Toyota, M. *J. Org. Chem.* **2013**, *78*, 93–103.

# Houben–Hoesch reaction

Acid-catalyzed acylation of phenols as well as phenolic ethers using nitriles.

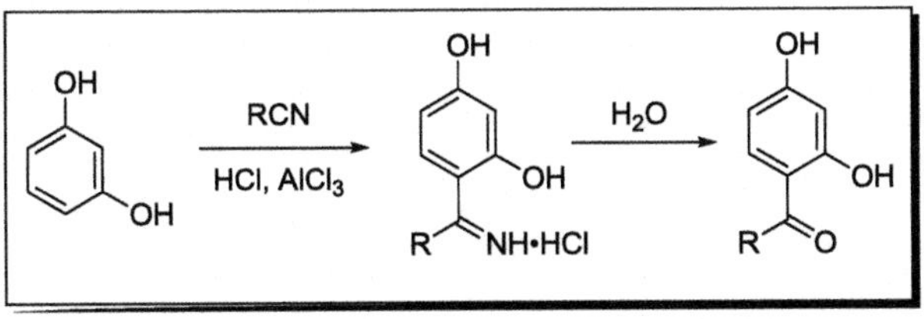

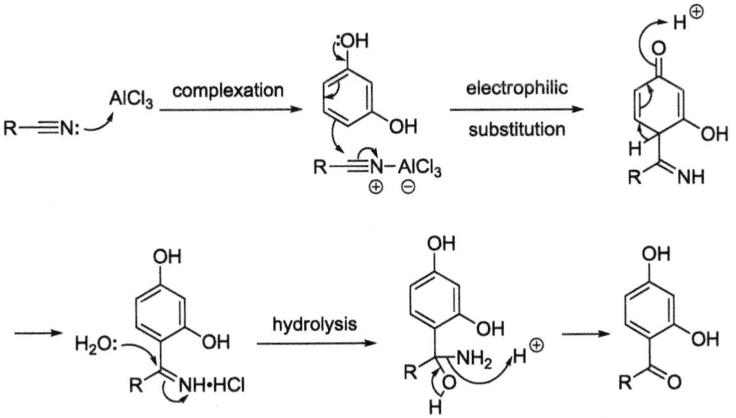

Example 1, Intramolecular Houben–Hoesch reaction[3]

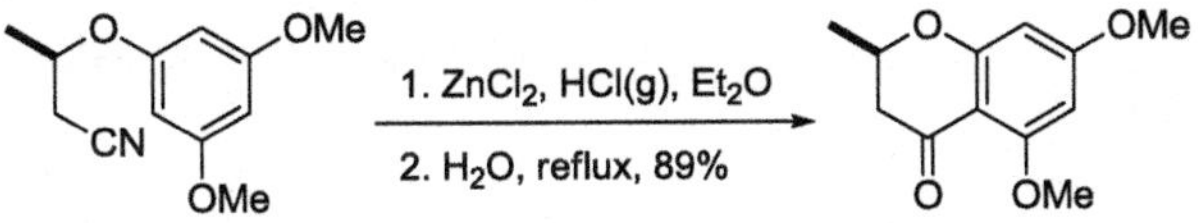

Example 2[6]

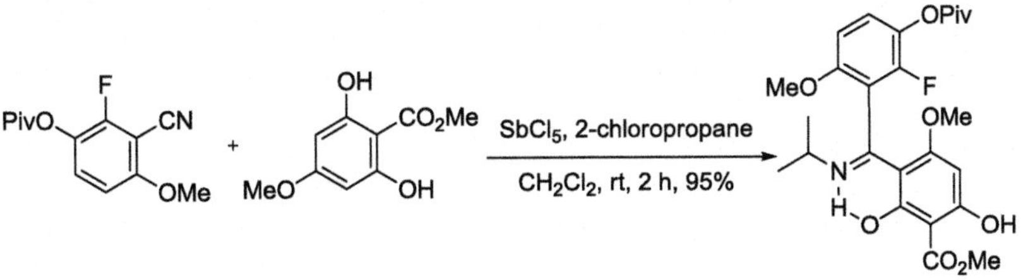

Example 3[8]

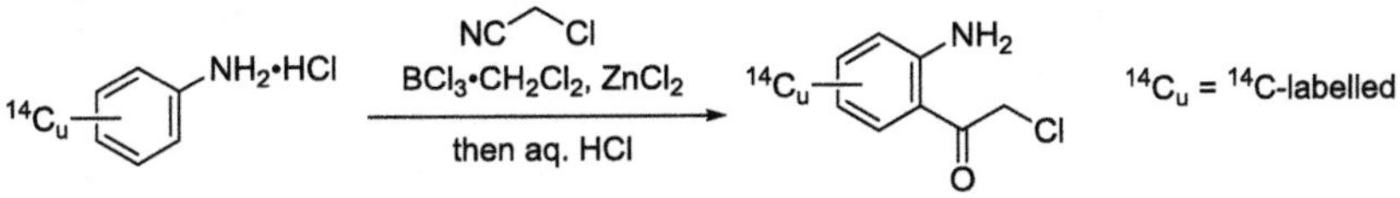

$^{14}C_u$ = $^{14}C$-labelled

J.J. Li, *Name Reactions: A Collection of Detailed Mechanisms and Synthetic Applications*,
DOI 10.1007/978-3-319-03979-4_139, © Springer International Publishing Switzerland 2014

Example 4[9]

Example 5[10]

## References

1. (a) Hoesch, K. *Ber.* **1915**, *48*, 1122–1133. Kurt Hoesch (1882–1932) was born in Krezau, Germany. He studied at Berlin under Emil Fischer. During WWI, Hoesch was Professor of Chemistry at the University of Istanbul, Turkey. After the war he gave up his scientific activities to devote himself to the management of a family business. (b) Houben, J. *Ber.* **1926**, *59*, 2878–2891.
2. Yato, M.; Ohwada, T.; Shudo, K. *J. Am. Chem. Soc.* **1991**, *113*, 691–692.
3. Rao, A. V. R.; Gaitonde, A. S.; Prakash, K. R. C.; Rao, S. P. *Tetrahedron Lett.* **1994**, *35*, 6347–6350.
4. Sato, Y.; Yato, M.; Ohwada, T.; Saito, S.; Shudo, K. *J. Am. Chem. Soc.* **1995**, *117*, 3037–3043.
5. Kawecki, R.; Mazurek, A. P.; Kozerski, L.; Maurin, J. K. *Synthesis* **1999**, 751–753.
6. Udwary, D. W.; Casillas, L. K.; Townsend, C. A. *J. Am. Chem. Soc.* **2002**, *124*, 5294–5303.
7. Sanchez-Viesca, F.; Gomez, M. R.; Berros, M. *Org. Prep. Proc. Int.* **2004**, *36*, 135–140.
8. Wager, C. A. B.; Miller, S. A. *J. Labelled Compd. Radiopharm.* **2006**, *49*, 615–622.
9. Black, D. St. C.; Kumar, N.; Wahyuningsih, T. D. *ARKIVOC* **2008**, *(6)*, 42–51.
10. Zhao, B.; Hao, X.-Y.; Zhang, J.-X.; Liu, S.; Hao, X.-J. *J. Org. Chem.* **2013**, *15*, 528–530.

# Hunsdiecker–Borodin reaction

Conversion of silver carboxylate to halide by treatment with halogen.

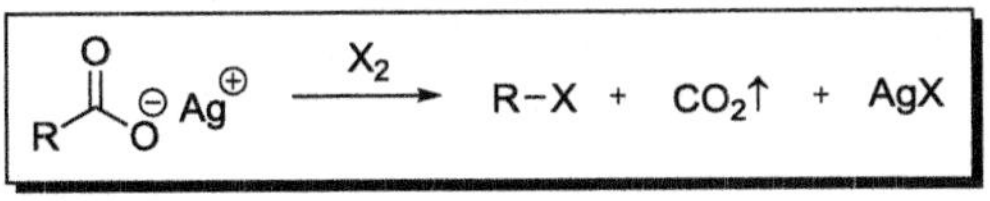

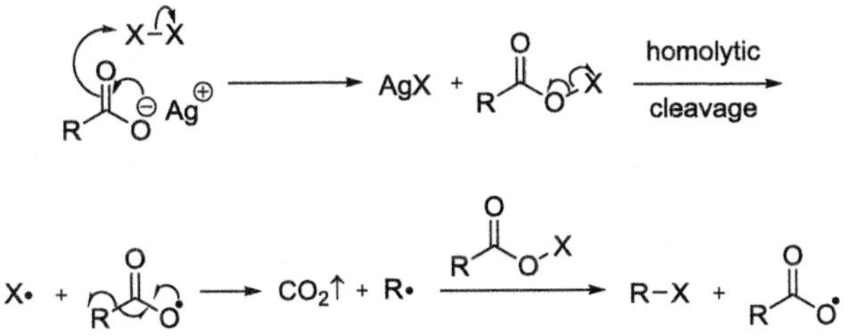

R–X + CO₂↑ + AgX (scheme with X₂)

homolytic cleavage

X• + R–C(O)–O → CO₂↑ + R• → R–X + R–C(O)–O•

Example 1[5]

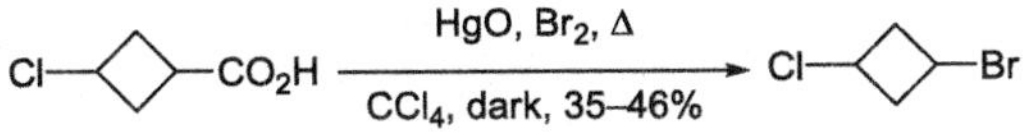

Cl—◻—CO₂H → HgO, Br₂, Δ / CCl₄, dark, 35–46% → Cl—◻—Br

Example 2[6]

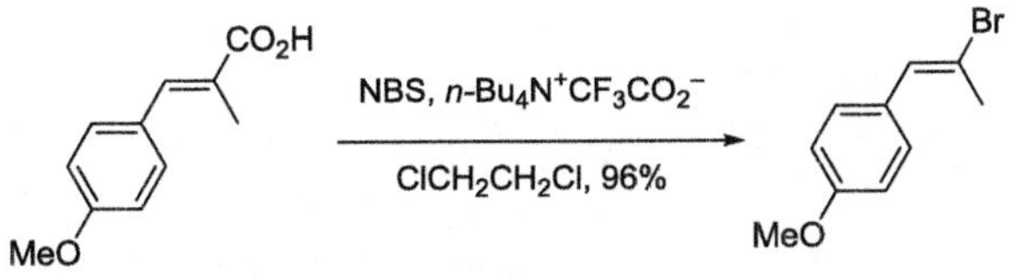

(4-MeO-phenyl) substituted 2-methylacrylic acid → NBS, n-Bu₄N⁺CF₃CO₂⁻ / ClCH₂CH₂Cl, 96% → vinyl bromide product

Example 3[8]

(4-HO-phenyl) cinnamic acid → "Select fluor" (N-chloromethyl-N'-fluoro-DABCO bis(tetrafluoroborate)), 2 BF₄⁻ / KBr, CH₃CN, 82% → (E)-vinyl bromide product

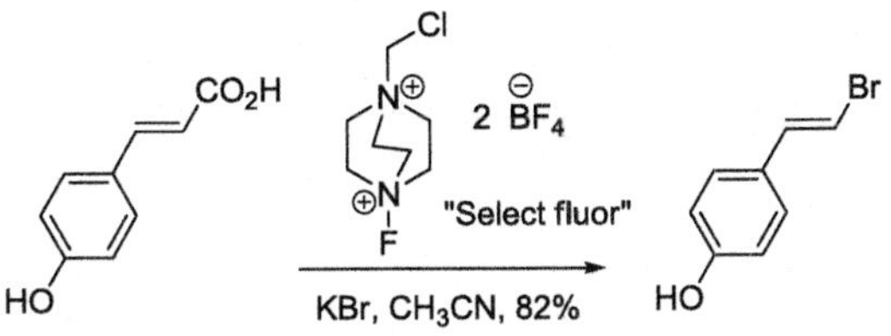

Example 4, One-pot microwave-Hunsdiecker–Borodin followed by Suzuki[10]

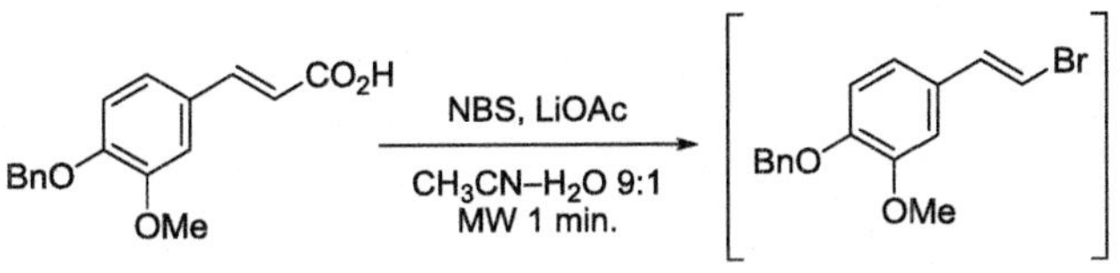

(BnO, OMe substituted) cinnamic acid → NBS, LiOAc / CH₃CN–H₂O 9:1, MW 1 min. → [vinyl bromide intermediate]

J.J. Li, *Name Reactions: A Collection of Detailed Mechanisms and Synthetic Applications*, DOI 10.1007/978-3-319-03979-4_140, © Springer International Publishing Switzerland 2014

PhB(OH)$_2$, K$_2$CO$_3$
Pd(PPh$_3$)$_4$
$\longrightarrow$
CH$_3$CN/H$_2$O 2:1
microwave, 5 min.
64% 2 steps

Example 5[11]

5 mol% Ag(Phen)$_2$OTf
150 mol% t-BuOCl
$\longrightarrow$
CH$_3$CN, rt, 3 h, 86%

## References

1. (a) Borodin, A. *Ann.* **1861**, *119*, 121–123. Aleksandr Porfirevič Borodin (1833–1887) was born in St Petersburg, the illegitimate son of a prince. He prepared methyl bromide from silver acetate in 1861, but another eighty years elapsed before Heinz and Cläre Hunsdiecker converted Borodin's synthesis into a general method, the Hunsdiecker or Hunsdiecker–Borodin reaction. Borodin was also an accomplished composer and is now best known for his musical masterpiece, opera Prince Igor. He kept a piano outside his laboratory. (b) Hunsdiecker, H.; Hunsdiecker, C. *Ber.* **1942**, *75*, 291–297. Cläre Hunsdiecker was born in 1903 and educated in Cologne. She developed the bromination of silver carboxylate alongside her husband, Heinz.
2. Sheldon, R. A.; Kochi, J. K. *Org. React.* **1972**, *19*, 326–421. (Review).
3. Barton, D. H. R.; Crich, D.; Motherwell, W. B. *Tetrahedron Lett.* **1983**, *24*, 4979–4982.
4. Crich, D. In *Comprehensive Organic Synthesis;* Trost, B. M.; Steven, V. L., Eds.; Pergamon, **1991**, *Vol. 7*, pp 723–734. (Review).
5. Lampman, G. M.; Aumiller, J. C. *Org. Synth.* **1988**, *Coll. Vol. 6*, 179.
6. Naskar, D.; Chowdhury, S.; Roy, S. *Tetrahedron Lett.* **1998**, *39*, 699–702.
7. Das, J. P.; Roy, S. *J. Org. Chem.* **2002**, *67*, 7861–7864.
8. Ye, C.; Shreeve, J. M. *J. Org. Chem.* **2004**, *69*, 8561–8563.
9. Li, J. J. *Hunsdiecker Reaction.* In *Name Reactions for Functional Group Transformations*; Li, J. J., Corey, E. J., Eds., Wiley: Hoboken, NJ, **2007**, pp 623–629. (Review).
10. Bazin, M.-A.; El Kihel, L.; Lancelot, J.-C.; Rault, S. *Tetrahedron Lett.* **2007**, *48*, 4347–4351.
11. Wang, Z.; Zhu, L.; Yin, F.; Su, Z.; Li, Z.; Li, C. *J. Am. Chem. Soc.* **2012**, *134*, 4258–4263.

# Jacobsen–Katsuki epoxidation

Mn(III)salen-catalyzed asymmetric epoxidation of (Z)-olefins.

1. Concerted oxygen transfer (*cis*-epoxide):

2. Oxygen transfer *via* radical intermediate (*trans*-epoxide):

3. Oxygen transfer *via* manganaoxetane intermediate (*cis*-epoxide):

Example 1[2]

J.J. Li, *Name Reactions: A Collection of Detailed Mechanisms and Synthetic Applications*, DOI 10.1007/978-3-319-03979-4_141, © Springer International Publishing Switzerland 2014

Example 2[5]

*cat.*, NaOCl

58% yield, 89% *ee*

Example 3[6]

*cat.* NaOCl

88%

88% *ee*

indinavir (Crixivan)

## References

1.  (a) Zhang, W.; Loebach, J. L.; Wilson, S. R.; Jacobsen, E. N. *J. Am. Chem. Soc.* **1990,** *112,* 2801–2903. (b) Irie, R.; Noda, K.; Ito, Y.; Matsumoto, N.; Katsuki, T. *Tetrahedron Lett.* **1990,** *31*, 7345–7348. (c) Irie, R.; Noda, K.; Ito, Y.; Katsuki, T. *Tetrahedron Lett.* **1991,** *32*, 1055–1058. (d) Deng, L.; Jacobsen, E. N. *J. Org. Chem.* **1992,** 57, 4320–4323. (e) Palucki, M.; McCormick, G. J.; Jacobsen, E. N. *Tetrahedron Lett.* **1995,** *36*, 5457–5460.
2.  Zhang, W.; Jacobsen, E. N. *J. Org. Chem.* **1991,** *56*, 2296–2298.
3.  Jacobsen, E. N. In *Catalytic Asymmetric Synthesis;* Ojima, I., Ed.; VCH: Weinheim, New York, **1993,** Ch. 4.2. (Review).
4.  Jacobsen, E. N. In *Comprehensive Organometallic Chemistry II*, Eds. G. W. Wilkinson, G. W.; Stone, F. G. A.; Abel, E. W.; Hegedus, L. S., Pergamon, New York, **1995,** vol 12, Chapter 11.1. (Review).
5.  Lynch, J. E.; Choi, W.-B.; Churchill, H. R. O.; Volante, R. P.; Reamer, R. A.; Ball, R. G. *J. Org. Chem.* **1997,** *62*, 9223–9228.
6.  Senananyake, C. H. *Aldrichimica Acta* **1998,** *31*, 3–15. (Review).
7.  Jacobsen, E. N.; Wu, M. H. In *Comprehensive Asymmetric Catalysis*, Jacobsen, E. N.; Pfaltz, A.; Yamamoto, H. Eds.; Springer: New York; 1999, Chapter 18.2. (Review).
8.  Katsuki, T. In *Catalytic Asymmetric Synthesis;* 2nd edn.; Ojima, I., Ed.; Wiley-VCH: New York, **2000,** 287. (Review).
9.  Katsuki, T. *Synlett* **2003,** 281–297. (Review).
10. Palucki, M. *Jacobsen–Katsuki epoxidation.* In *Name Reactions in Heterocyclic Chemistry*; Li, J. J., Ed.; Wiley: Hoboken, NJ, **2005,** pp 29–43. (Review).
11. Engelhardt, U.; Linker, T. *Chem. Commun.* **2005,** 1152–1154.
12. Fernandez de la Pradilla, R.; Castellanos, A.; Osante, I.; Colomer, I.; Sanchez, M. I. *J. Org. Chem.* **2009,** *74,* 170–181.
13. Olson, J. A.; Shea, K. M. *Acc. Chem. Res.* **2011,** *44*, 311–321. (Review).

# Japp–Klingemann hydrazone synthesis

Hydrazones from β-ketoesters and diazonium salts with the acid or base.

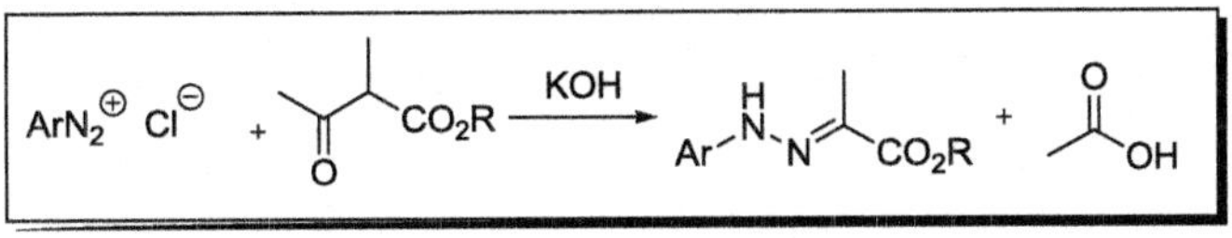

Diazonium salt    β-keto-ester       hydrazone

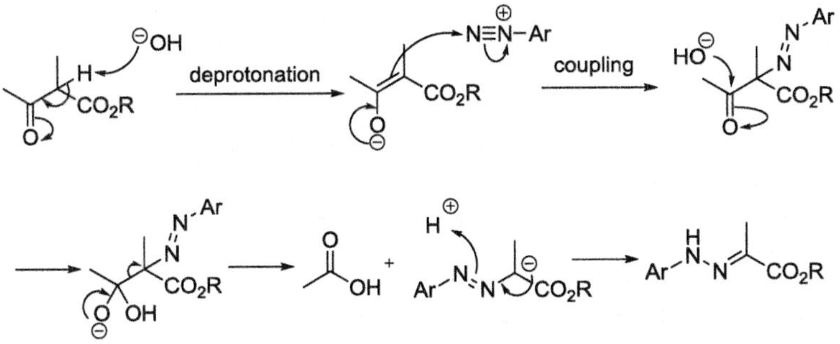

Example 1[4]

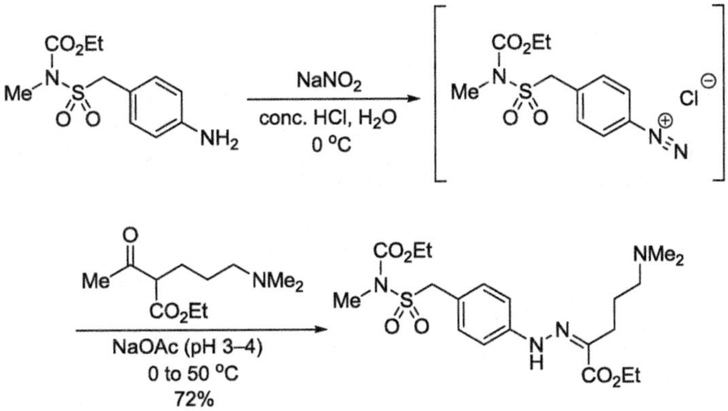

Example 2[6]

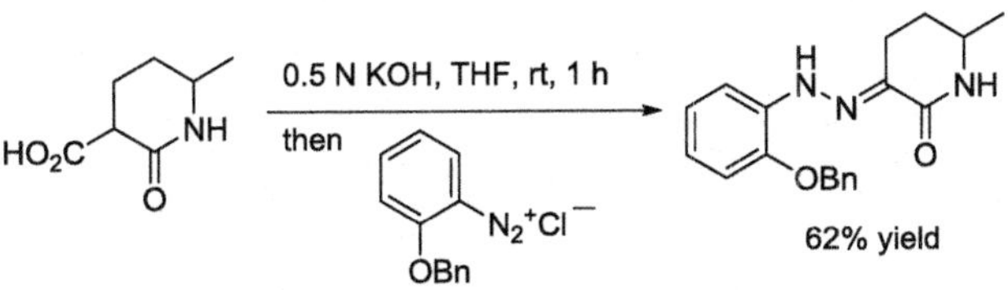

J.J. Li, *Name Reactions: A Collection of Detailed Mechanisms and Synthetic Applications*,
DOI 10.1007/978-3-319-03979-4_142, © Springer International Publishing Switzerland 2014

Example 3[10]

Example 4, A Japp–Klingemann cleavage[11]

## References

1.  (a) Japp, F. R.; Klingemann, F. *Ber.* **1887**, *20*, 2942–2944.    (b) Japp, F. R.;
    Klingemann, F. *Ber.* **1887**, *21*, 2934–2936. (c) Japp, F. R.; Klingemann, F. *Ber.* **1887**,
    *20*, 3398–3401. (d) Japp, F. R.; Klingemann, F. *Ann.* **1888**, *247*, 190–225. (e) Japp,
    F. R.; Klingemann, F. *J. Chem. Soc.* **1888**, *53*, 519–544.
2.  Phillips, R. R. *Org. React.* **1959**; *10*, 143–178. (Review).
3.  Loubinoux, B.; Sinnes, J.-L.; O'Sullivan, A. C.; Winkler, T. *J. Org. Chem.* **1995**, *60*,
    953–959.
4.  Pete, B.; Bitter, I.; Harsanyi, K.; Toke, L. *Heterocycles* **2000**, *53*, 665–673.
5.  Atlan, V.; Kaim, L. E.; Supiot, C. *Chem. Commun.* **2000**, 1385–1386.
6.  Dubash, N. P.; Mangu, N. K.; Satyam, A. *Synth. Commun.* **2004**, *34*, 1791–1799.
7.  He, W.; Zhang, B.-L.; Li, Z.-J.; Zhang, S.-Y. *Synth. Commun.* **2005**, *35*, 1359–1368.
8.  Li, J. *Japp–Klingemann hydrazone synthesis.* In *Name Reactions for Functional
    Group Transformations*; Li, J. J., Ed.; Wiley: Hoboken, NJ, **2007**, pp 630–634. (Re-
    view).
9.  Chen, Y.; Shibata, M.; Rajeswaran, M.; Srikrishnan, T.; Dugar, S.; Pandey, R. K. *Tet-
    rahedron Lett.* **2007**, *48*, 2353–2356.
10. Pete, B. *Tetrahedron Lett.* **2008**, *49*, 2835–2838.
11. Frohberg, P.; Schulze, I.; Donner, C.; Krauth, F. *Tetrahedron Lett.* **2013**, *53*,
    4507–4509.

# Jones oxidation

The **Collins/Sarett oxidation** (chromium trioxide-pyridine complex), and **Corey's PCC** (pyridinium chlorochromate) and **PDC** (pyridinium dichromate) **oxidations** follow a similar pathway as the **Jones oxidation** (chromium trioxide and sulfuric acid in acetone). All these oxidants have a chromium (VI), normally orange or yellow, which is reduced to Cr(III), often green.

| $CrO_3/H_2SO_4$ | $CrO_3$ | $CrO_3Cl^{\ominus}$ | $Cr_2O_7^{=}$ |
|---|---|---|---|
| Jones | Collins/Sarett | PCC | PDC |

## *Jones oxidation*

By the Jones oxidation, the primary alcohols are oxidized to the corresponding aldehyde or carboxylic acids, whereas the secondary alcohols are oxidized to the corresponding ketones.

$$CrO_3 + H_2O \longrightarrow H_2CrO_4$$

Cr(VI)

clear orange solution

chromate ester

Cr(III)

green

The intramolecular mechanism is also operative:

J.J. Li, *Name Reactions: A Collection of Detailed Mechanisms and Synthetic Applications*, DOI 10.1007/978-3-319-03979-4_143, © Springer International Publishing Switzerland 2014

Example 1[6]

1. Jones reagent
   acetone, 20 min.

2. HCO$_2$H, rt, 1 h
   96% 2 steps

(–)-CP-263114

Example 2[7]

CrO$_3$, H$_2$SO$_4$
acetone/H$_2$O

rt, 74%

Example 3[9]

CrO$_3$, H$_2$SO$_4$

acetone, 0 °C
1–2 h, 86%

## References

1. Bowden, K.; Heilbron, I. M., Jones, E. R. H.; Weedon, B. C. L. *J. Chem. Soc.* **1946,** 39–45. Ewart R. H. (Tim) Jones worked with Ian M. Heilbron at Imperial College. Jones later succeeded Robert Robinson to become the prestigious Chair of Organic Chemistry at Manchester. *The recipe for the Jones reagent: 25 g CrO$_3$, 25 mL conc. H$_2$SO$_4$, and 70 mL H$_2$O.*
2. Ratcliffe, R. W. *Org. Synth.* **1973,** *53,* 1852.
3. Vanmaele, L.; De Clerq, P.; Vandewalle, M. *Tetrahedron Lett.* **1982,** *23,* 995–998.
4. Luzzio, F. A. *Org. React.* **1998,** *53,* 1–222. (Review).
5. Zhao, M.; Li, J.; Song, Z.; Desmond, R. J.; Tschaen, D. M.; Grabowski, E. J. J.; Reider, P. J. *Tetrahedron Lett.* **1998,** *39,* 5323–5326. (Catalytic CrO$_3$ oxidation).
6. Waizumi, N.; Itoh, T.; Fukuyama, T. *J. Am. Chem. Soc.* **2000,** *122,* 7825–7826.
7. Hagiwara, H.; Kobayashi, K.; Miya, S.; Hoshi, T.; Suzuki, T.; Ando, M. *Org. Lett.* **2001,** *3,* 251–254.
8. Fernandes, R. A.; Kumar, P. *Tetrahedron Lett.* **2003,** *44,* 1275–1278.
9. Hunter, A. C.; Priest, S.-M. *Steroids* **2006,** *71,* 30–33.
10. Kim, D.-S.; Bolla, K.; Lee, S.; Ham, J. *Tetrahedron* **2013,** *67,* 1062–1070.
10. Marshall, A. J.; Lin, J.-M.; Grey, A.; Reid, I. R; Cornish, J.; Denny, W. A *Bioorg. Med. Chem.* **2013,** *21,* 4112–4119.

## Collins oxidation

Different from the Jones oxidation, the Collins oxidation, also known as the Collins–Sarett oxidation, converts primary alcohols to the corresponding aldehydes. $CrO_3 \cdot 2Pyr$ is known as the **Collins reagent**.

Example 1[5]

$$\text{8 eq } CrO_3 \cdot 2Pyr$$
$$CH_2Cl_2, \text{ 15 min., 86\%}$$

Example 2[7]

$$\text{7 equiv } CrO_3 \cdot Py_2$$
$$CH_2Cl_2, \text{ 30 min., 75\%}$$

Example 3[9]

$$CrO_3, Pyr.$$
$$50–60 \ ^\circ C, 48 \text{ h}$$
$$50–65\%$$

## References

1.  Poos, G. I.; Arth, G. E.; Beyler, R. E.; Sarett, L. H. *J. Am. Chem. Soc.* **1953,** *75,* 422–429.
2.  Collins, J. C; Hess, W. W.; Frank, F. J. *Tetrahedron Lett.* **1968,** 3363–3366. J. C. Collins was a chemist at Sterling-Winthrop in Rensselaer, New York.
3.  Collins, J. C; Hess, W. W. *Org. Synth.* **1972,** *Coll. Vol. V,* 310.
4.  Hill, R. K.; Fracheboud, M. G.; Sawada, S.; Carlson, R. M.; Yan, S.-J. *Tetrahedron Lett.* **1978,** 945–948.
5.  Krow, G. R.; Shaw, D. A.; Szczepanski, S.; Ramjit, H. *Synth. Commun.* **1984,** *14,* 429–433.
6.  Li, M.; Johnson, M. E. *Synth. Commun.* **1995,** *25,* 533–537.
7.  Harris, P. W. R.; Woodgate, P. D. *Tetrahedron* **2000,** *56,* 4001–4015.
8.  Nguyen-Trung, N. Q.; Botta, O.; Terenzi, S.; Strazewski, P. *J. Org. Chem.* **2003,** *68,* 2038–2041.
9.  Arumugam, N.; Srinivasan, P. C. *Synth. Commun.* **2003,** *33,* 2313–2320.

## PCC oxidation

Alcohols are oxidized by pyridinium chlorochromate (PCC) to the corresponding aldehydes or ketones. They are not further oxidized to the corresponding carboxylic acids because the reaction was done in organic solvents, not in water. If water existed, the carbonyls would form *aldehyde hydrates* or *ketone hydrates*, which are then oxidized to acids.

Example 1, One-pot PCC–Wittig reactions[2]

Example 2[3]

Example 3, Allylic oxidation[4]

Example 4, Hemiacetol oxidation[5]

## References

1. Corey, E. J.; Suggs, W. *Tetrahedron Lett.* **1975**, *16*, 2647–2650.
2. Bressette, A. R.; Glover, L. C., IV *Synlett* **2004**, 738–740.
3. Breining, S. R.; Bhatti, B. S.; Hawkins, G. D.; Miao, L. WO2005037832 (**2005**).
4. Srikanth, G. S. C.; Krishna, U. M. *Tetrahedron* **2006**, *62*, 11165–11171.
5. Kim, S.-G. *Tetrahedron Lett.* **2008**, *49*, 6148–6151.
6. Mehta, G.; Bera, M. K. *Tetrahedron* **2013**, *69*, 1815–1821.
7. Fowler, K. J.; Ellis, J. L.; Morrow, G. W. *Synth. Commun.* **2013**, *43*, 1676–1682.

## PDC oxidation

Pyridinium dichromate (PDC) may oxidize alcohols all the way to the corresponding carboxylic acids instead of aldehydes and ketones as PCC does.

Example 1[2]

Example 2, Cleavage of primary carbon–boron bond[3]

Example 3, Hemiacetal as an intermediate[5]

Example 4[2]

## References

1    Corey, E. J.; Schmidt, G. *Tetrahedron Lett.* **1979**, 399–402.
2    Terpstra, J. W.; Van Leusen, A. M. *J. Org. Chem.* **1986**, *51*, 230–208.
3    Brown, H. C.; Kulkarni, S. V.; Khanna, V. V.; Patil, V. D.; Racherla, U. S. *J. Org. Chem.* **1992**, *57*, 6173–6177.
4    Nakamura, M.; Inoue, J.; Yamada, T. *Bioorg. Med. Chem. Lett.* **2000**, *10*, 2807–2810.
5    Chênevert, R. Courchene, G.; Caron, D. *Tetrahedron: Asymmetry* **2003**, 2567–2571.
6    Jordão, A. K *Synlett* **2006**, 3364–3365. (Review).
7    Xu, G.; Hou, A.-J.; Wang, R.-R.; Liang, G.-Y.; Zheng, Y.-T.; Liu, Z.-Y.; Li, X.-L.; Zhao, Y.; Huang, S.-X.; Peng, L.-Y.; et al. *Org. Lett.* **2006**, *8*, 4453–4456.
8    Morzycki, J. W; Perez-Diaz, J. O. H; Santillan, R.; Wojtkielewicz, A. *Steroids* **2010**, *75*, 70–76.
9    Cai, Q.; You, S.-L. *Org. Lett.* **2012**, *14*, 3040–3043.

## Julia–Kocienski olefination

Modified one-pot Julia olefination to give predominantly (*E*)-olefins from heteroarylsulfones and aldehydes. A sulfone reduction step is *not* required.

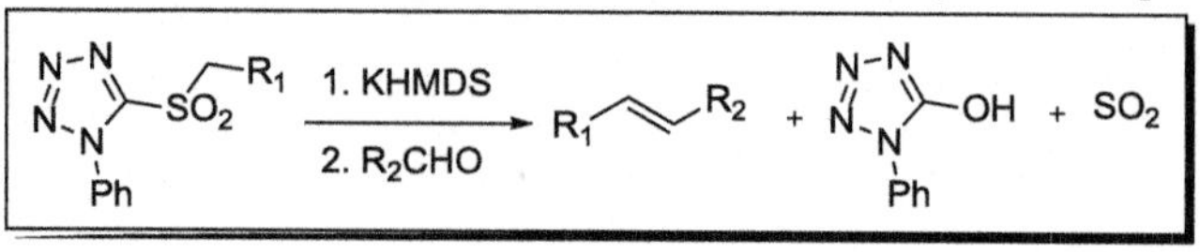

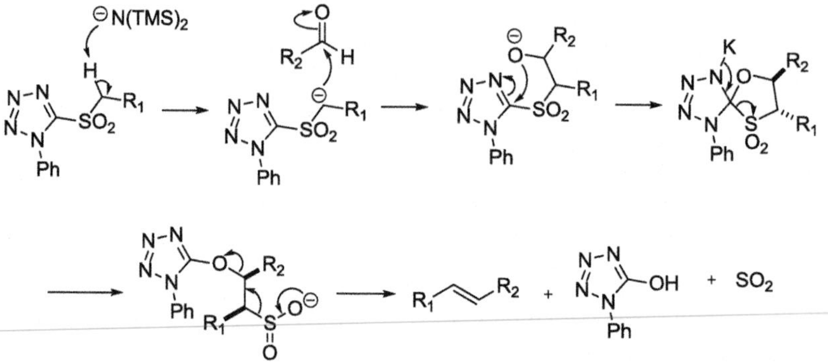

Alternatives to tetrazole:

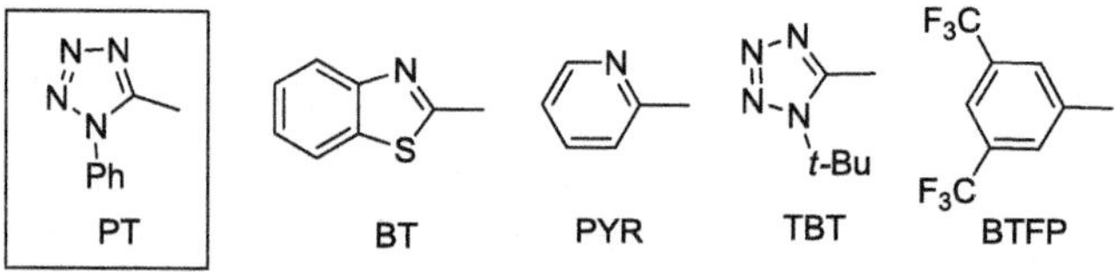

The use of larger counterion (such as $K^+$) and polar solvents (such as DME) favors an open transition state (PT = phenyltetrazolyl):

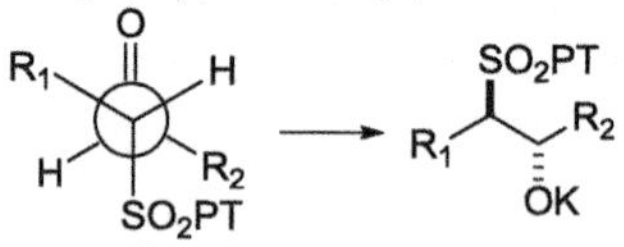

Example 1, (BT = benzothiazole)[2]

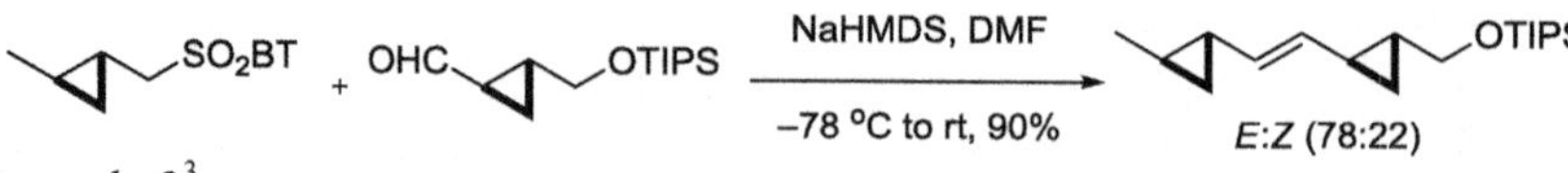

Example 2[3]

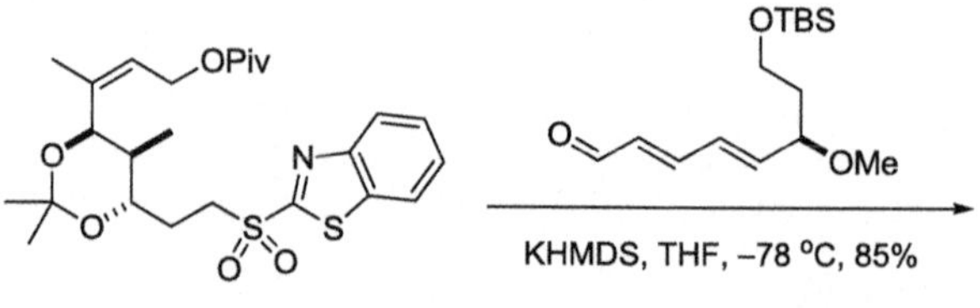

J.J. Li, *Name Reactions: A Collection of Detailed Mechanisms and Synthetic Applications*,
DOI 10.1007/978-3-319-03979-4_144, © Springer International Publishing Switzerland 2014

Example 3[7]

88 : 12

Example 4[8]

74%, yield
$E/Z = 98/2$

## References

1. (a) Baudin, J. B.; Hareau, G.; Julia, S. A.; Ruel, O. *Tetrahedron Lett.* **1991**, *32*, 1175–1178. (b) Baudin, J. B.; Hareau, G.; Julia, S. A.; Ruel, O. *Bull. Soc. Chim. Fr.* **1993**, *130*, 336–357. (c) Baudin, J. B.; Hareau, G.; Julia, S. A.; Loene, R.; Ruel, O. *Bull. Soc. Chim. Fr.* **1993**, *130*, 856–878. (d) Blakemore, P. R.; Cole, W. J.; Kocienski, P. J.; Morely, A. *Synlett* **1998**, 26–28.

2. Charette, A. B.; Lebel, H. *J. Am. Chem. Soc.* **1996**, *118*, 10327–10328.

3. Blakemore, P. R.; Kocienski, P. J.; Morley, A.; Muir, K. *J. Chem. Soc., Perkin Trans. 1* **1999**, 955–968.

4. Williams, D. R.; Brooks, D. A.; Berliner, M. A. *J. Am. Chem. Soc.* **1999**, *121*, 4924–4925.

5. Kocienski, P. J.; Bell, A.; Blakemore, P. R. *Synlett* **2000**, 365–366.

6. Liu, P.; Jacobsen, E. N. *J. Am. Chem. Soc.* **2001**, *123*, 10772–10773.

7. Charette, A. B.; Berthelette, C.; St-Martin, D. *Tetrahedron Lett.* **2001**, *42*, 5149–5153.

8. Alonso, D. A.; Najera, C.; Varea, M. *Tetrahedron Lett.* **2004**, *45*, 573–577.

9. Alonso, D. A.; Fuensanta, M.; Najera, C.; Varea, M. *J. Org. Chem.* **2005**, *70*, 6404.

10. Rong, F. *Julia–Lythgoe olefination*. In *Name Reactions for Homologations-Part I*; Li, J. J., Ed.; Wiley: Hoboken, NJ, **2009**, pp 447–473. (Review).

11. Davies, S. G.; Fletcher, A. M.; Foster, E. M.; Lee, J. A.; Roberts, P. M.; Thomson, J. E. *J. Org. Chem.* **2013**, *78*, 2500–2510.

# Julia–Lythgoe olefination

(*E*)-Olefins from sulfones and aldehydes.

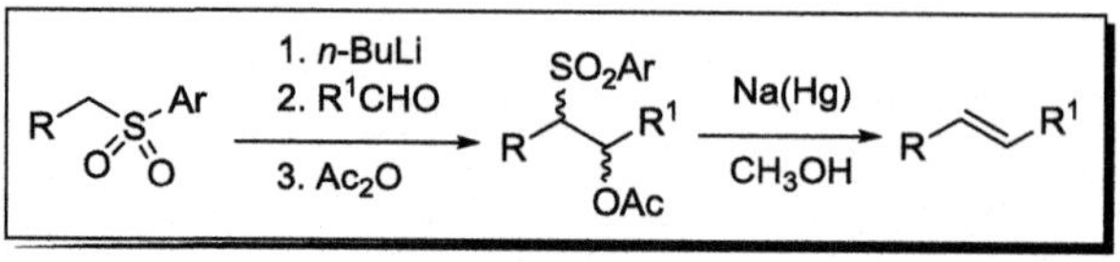

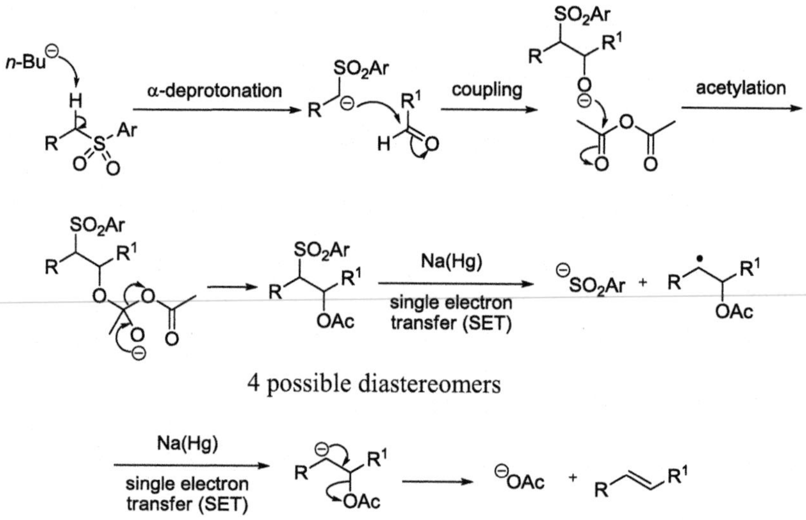

4 possible diastereomers

Example 1[2]

E:Z (99:1)

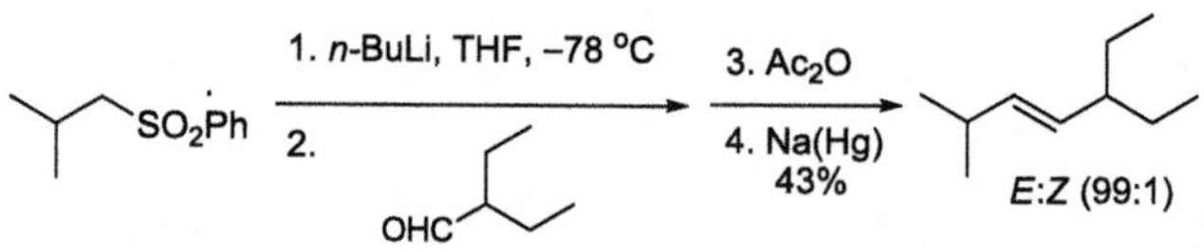

Example 2[3]

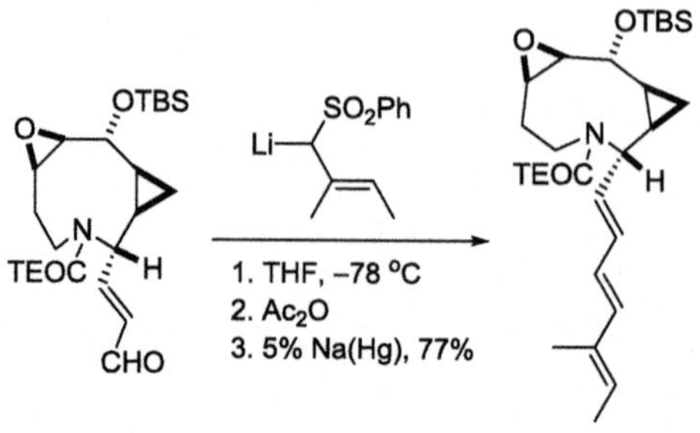

J.J. Li, *Name Reactions: A Collection of Detailed Mechanisms and Synthetic Applications*,
DOI 10.1007/978-3-319-03979-4_145, © Springer International Publishing Switzerland 2014

## Example 3[7]

## Example 4[8]

## References

1.  (a) Julia, M.; Paris, J. M. *Tetrahedron. Lett.* **1973**, 4833–4836.  (b) Lythgoe, B. *J. Chem. Soc., Perkin Trans. 1* **1978**, 834–837.
2.  Kocienski, P. J.; Lythgoe, B.; Waterhause, I. *J. Chem. Soc., Perkin Trans. 1* **1980**, 1045–1050.
3.  Kim, G.; Chu-Moyer, M. Y.; Danishefsky, S. J. *J. Am. Chem. Soc.* **1990**, *112*, 2003–2005.
4.  Keck, G. E.; Savin, K. A.; Weglarz, M. A. *J. Org. Chem.* **1995**, *60*, 3194–3204.
5.  Breit, B. *Angew. Chem. Int. Ed.* **1998**, 37, 453–456.
6.  Marino, J. P.; McClure, M. S.; Holub, D. P.; Comasseto, J. V.; Tucci, F. C. *J. Am. Chem. Soc.* **2002**, *124*, 1664–1668.
7.  Bernard, A. M.; Frongia, A.; Piras, P. P.; Secci, F. *Synlett* **2004**, *6*, 1064–1068.
8.  Pospíšil, J.; Pospíšil, T, Markó, I. E. *Org. Lett.* **2005**, *7*, 2373–2376.
9.  Gollner, A.; Mulzer, J. *Org. Lett.* **2008**, *10*, 4701–4704.
10. Rong, F. *Julia–Lythgoe olefination*. In *Name Reactions for Homologations-Part I*; Li, J. J., Ed.; Wiley: Hoboken, NJ, **2009**, pp 447–473. (Review).
11. Dams, I.; Chodynski, M.; Krupa, M.; Pietraszek, A.; Zezula, M.; Cmoch, P.; Kosinska, M.; Kutner, A. *Tetrahedron* **2013**, *69*, 1634–1648.

# Kahne glycosidation

Diastereoselective glycosidation of a sulfoxide at the anomeric center as the glycosyl acceptor. The sulfoxide activation is achieved using Tf$_2$O.

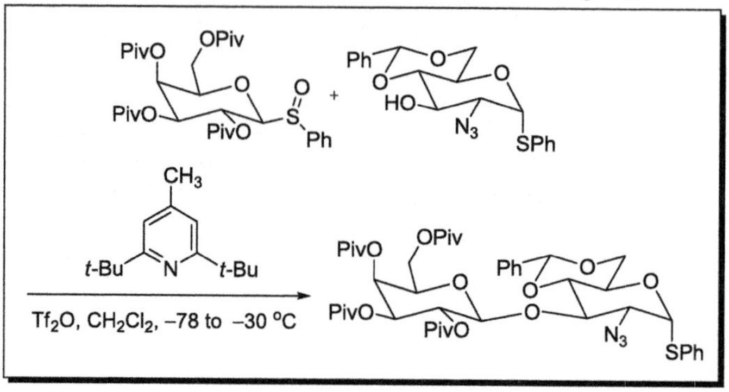

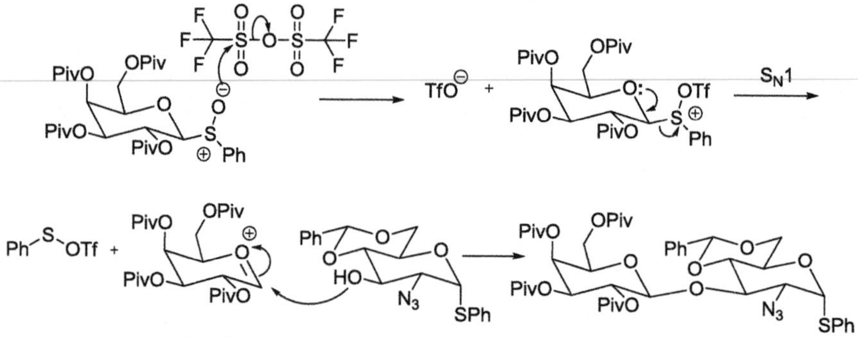

Example 1[1d]

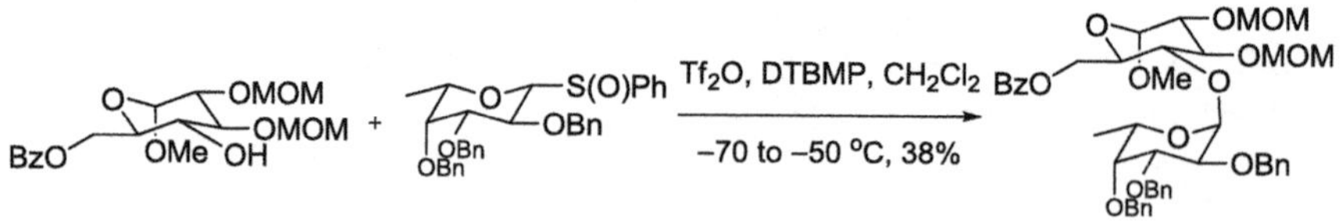

Example 2[4]

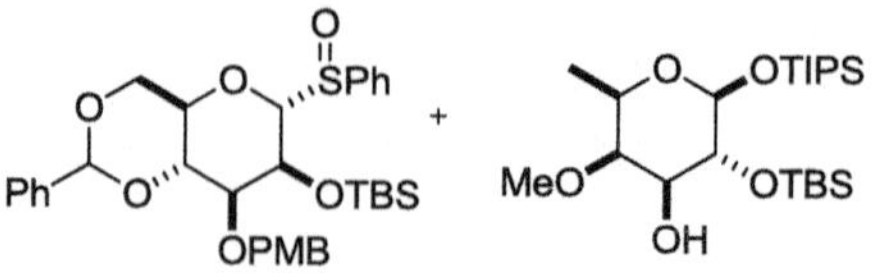

J.J. Li, *Name Reactions: A Collection of Detailed Mechanisms and Synthetic Applications*,
DOI 10.1007/978-3-319-03979-4_146, © Springer International Publishing Switzerland 2014

Example 3, Reverse Kahne-type glycosylation[6]

NuH = ROH, ArOH, ArNH$_2$, CH$_2$=CHCH$_2$TMS

## References

1.  (a) Kahne, D.; Walker, S.; Cheng, Y.; Van Engen, D. *J. Am. Chem. Soc.* **1989**, *111*, 6881–6882. (b) Yan, L.; Taylor, C. M.; Goodnow, R., Jr.; Kahne, D. *J. Am. Chem. Soc.* **1994**, *116*, 6953–6954. (c) Yan, L.; Kahne, D. *J. Am. Chem. Soc.* **1996**, *118*, 9239–9248. (d) Gildersleeve, J.; Pascal, R. A.; Kahne, D. *J. Am. Chem. Soc.* **1998**, *120*, 5961–5969. Daniel Kahne now teaches at Harvard University.
2.  Boeckman, R. K., Jr.; Liu, Y. *J. Org. Chem.* **1996**, *61*, 7984–7985.
3.  Crich, D.; Sun, S. *J. Am. Chem. Soc.* **1998**, *120*, 435–436.
4.  Crich, D.; Li, H. *J. Org. Chem.* **2000**, *65*, 801–805.
5.  Nicolaou, K. C.; Rodríguez, R. M.; Mitchell, H. J.; Suzuki, H.; Fylaktakidou, K. C.; Baudoin, O.; van Delft, F. L. *Chem. Eur. J.* **2000**, *6*, 3095–3115.
6.  Berkowitz, D. B.; Choi, S.; Bhuniya, D.; Shoemaker, R. K. *Org. Lett.* **2000**, *2*, 1149–1152.
7.  Crich, D.; Li, H.; Yao, Q.; Wink, D. J.; Sommer, R. D.; Rheingold, A. L. *J. Am. Chem. Soc.* **2001**, *123*, 5826–5828.
8.  Crich, D.; Lim, L. B. L. *Org. React.* **2004**, *64*, 115–251. (Review).
9.  Yu, B.; Yang, Z.; Cao, H. *Cur. Org. Chem.* **2005**, *9*, 179–194.

## Knoevenagel condensation

Condensation between carbonyl compounds and activated methylene compounds catalyzed by amines.

Example 1[3]

J.J. Li, *Name Reactions: A Collection of Detailed Mechanisms and Synthetic Applications*,
DOI 10.1007/978-3-319-03979-4_147, © Springer International Publishing Switzerland 2014

Example 2, EDDA = Ethylenediamine diacetate[5]

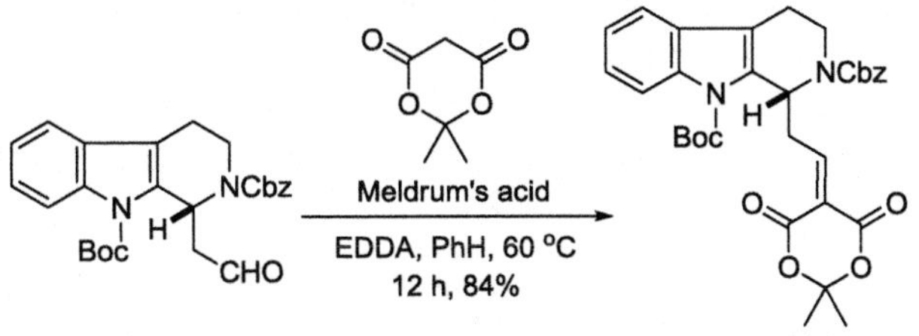

Meldrum's acid

EDDA, PhH, 60 °C
12 h, 84%

Example 3, Using ionic liquid ethylammonium nitrate (EAN) as solvent[8]

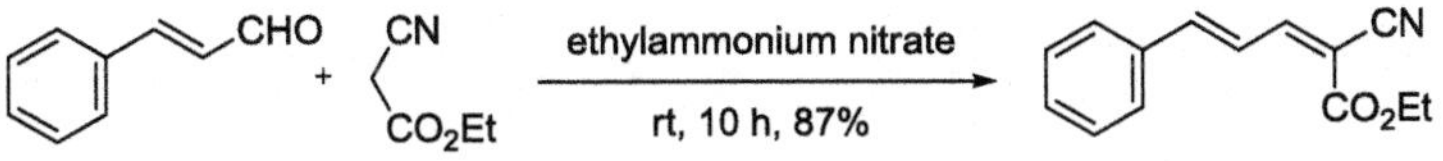

ethylammonium nitrate

rt, 10 h, 87%

Example 4[9]

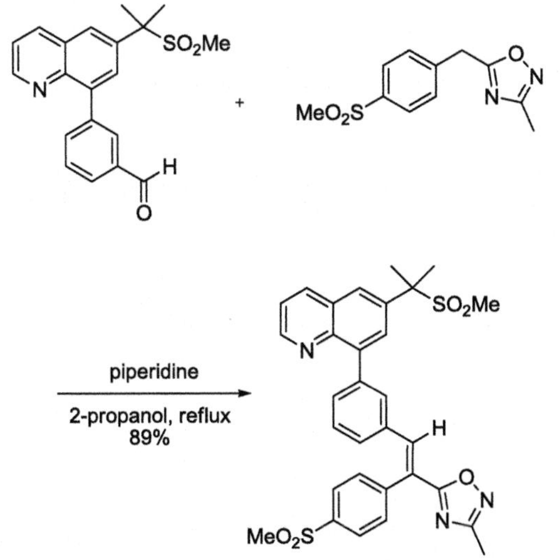

piperidine

2-propanol, reflux
89%

Example 5[11]

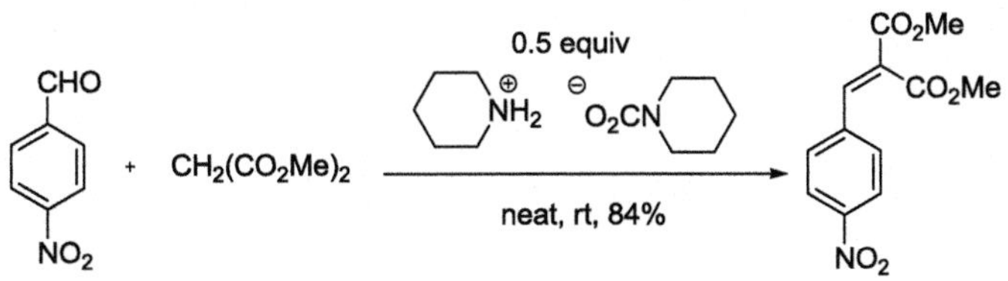

0.5 equiv

neat, rt, 84%

## References

1. Knoevenagel, E. *Ber.* **1898**, *31*, 2596–2619. Emil Knoevenagel (1865–1921) was born in Hannover, Germany. He studied at Göttingen under Victor Meyer and Gattermann, receiving a Ph.D. in 1889. He became a full professor at Heidelberg in 1900. When WWI broke out in 1914, Knoevenagel was one of the first to enlist and rose to the rank of staff officer. After the war, he returned to his academic work until his sudden death during an appendectomy.
2. Jones, G. *Org. React.* **1967,** *15*, 204–599. (Review).
3. Cantello, B. C. C.; Cawthornre, M. A.; Cottam, G. P.; Duff, P. T.; Haigh, D.; Hindley, R. M.; Lister, C. A.; Smith, S. A.; Thurlby, P. L. *J. Med. Chem.* **1994,** *37*, 3977–3985.
4. Paquette, L. A.; Kern, B. E.; Mendez-Andino, J. *Tetrahedron Lett.* **1999,** *40*, 4129–4132.
5. Tietze, L. F.; Zhou, Y. *Angew. Chem. Int. Ed.* **1999,** *38*, 2045–2047.
6. Pearson, A. J.; Mesaros, E. F. *Org. Lett.* **2002,** *4*, 2001–2004.
7. Kourouli, T.; Kefalas, P.; Ragoussis, N.; Ragoussis, V. *J. Org. Chem.* **2002,** *67,* 4615–4618.
8. Hu, Y.; Chen, J.; Le, Z.-G.; Zheng, Q.-G. *Synth. Commun.* **2005,** *35*, 739–744.
9. Conlon, D. A.; Drahus-Paone, A.; Ho, G.-J.; Pipik, B.; Helmy, R.; McNamara, J. M.; Shi, Y.-J.; Williams, J. M.; MacDonald, D. *Org. Process Res. Dev.* **2006,** *10*, 36–45.
10. Rong, F. *Knoevenagel Condensation.* In *Name Reactions for Homologations-Part I*; Li, J. J., Ed.; Wiley: Hoboken, NJ, **2009,** pp 474–501. (Review).
11. Mase, N.; Horibe, T. *Org. Lett.* **2013,** *15*, 1854–1857.

# Knorr pyrazole synthesis

Also known as Knorr reaction. Reaction of hydrazine or substituted hydrazine with 1,3-dicarbonyl compounds to provide the pyrazole or pyrazolone ring system. *Cf.* Paal–Knorr pyrrole synthesis.

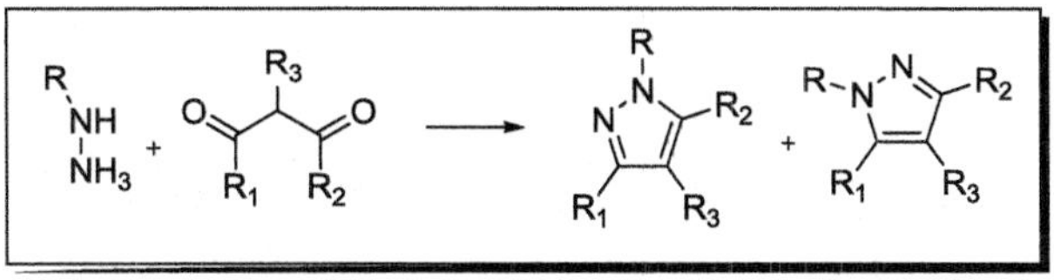

R = H, Alkyl, Aryl, Het-aryl, Acyl, *etc.*

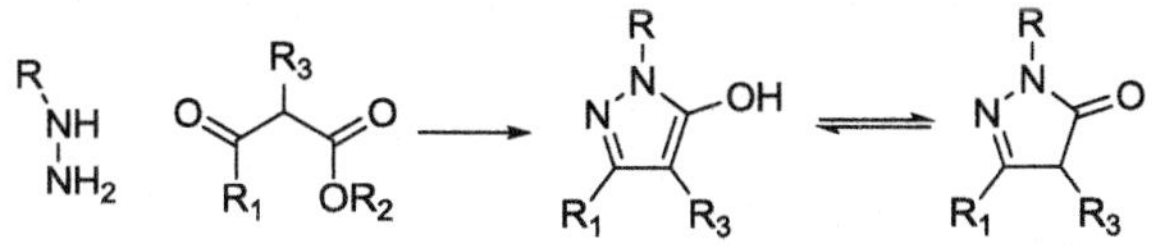

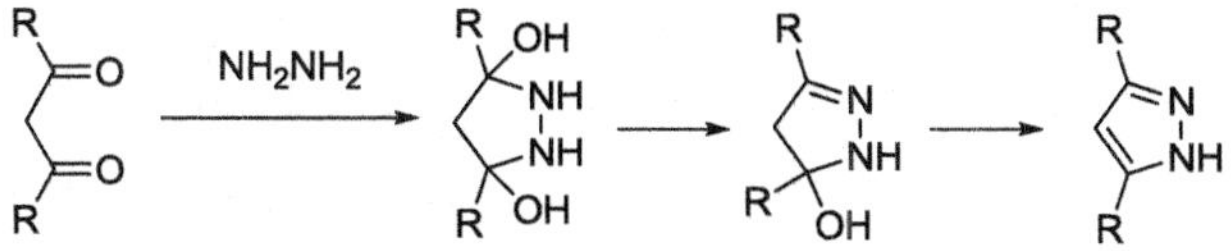

Alternatively,

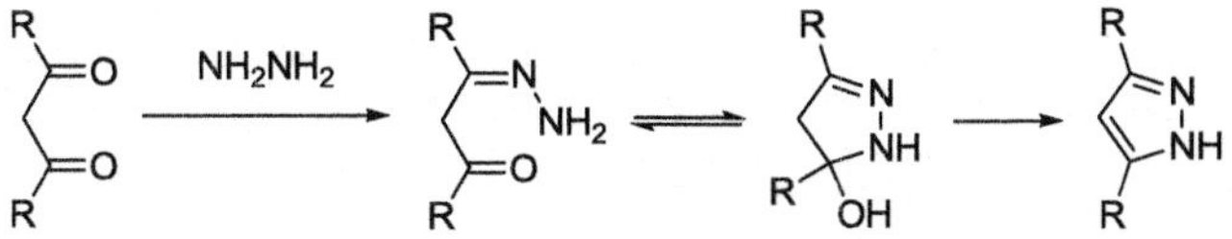

Example 1[2]

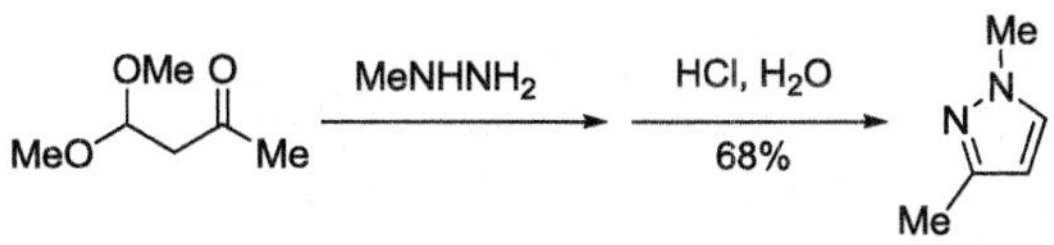

Example 2[8]

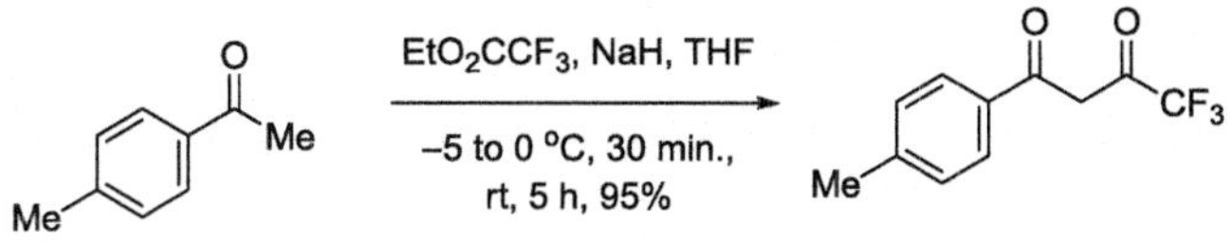

J.J. Li, *Name Reactions: A Collection of Detailed Mechanisms and Synthetic Applications*, DOI 10.1007/978-3-319-03979-4_148, © Springer International Publishing Switzerland 2014

Example 3[9]

66%                                      +                                    27%

## References

1   (a) Knorr, L. *Ber* **1883,** *16,* 2597.  Ludwig Knorr (1859–1921) was born near Munich,
    Germany.  After studying under Volhard, Emil Fischer, and Bunsen, he was appointed
    professor of chemistry at Jena.  Knorr made tremendous contributions in the synthesis
    of heterocycles in addition to discovering the important pyrazolone drug, pyrine.  (b)
    Knorr, L. *Ber* **1884,** *17,* 546, 2032. (c) Knorr, L. *Ber.* **1885,** *18,* 311. (d) Knorr, L. *Ann.*
    **1887,** *238,* 137.

2   Burness, D. M. *J. Org. Chem.* **1956,** *21,* 97–101.

3   Jacobs, T. L. in *Heterocyclic Compounds,* Elderfield, R. C., Ed.; Wiley: New York,
    **1957,** *5,* 45.  (Review).

4   *Houben–Weyl,* **1967,** *10/2,* 539, 587, 589, 590.  (Review).

5   Elguero, J., In *Comprehensive Heterocyclic Chemistry II,* Katrizky, A. R.; Rees, C.
    W.: Scriven, E. F. V., Eds; Elsevier: Oxford, **1996,** *3,* 1.  (Review).

6   Stanovnik, E.; Svete, J. In *Science of Synthesis,* **2002,** *12,* 15; Neier, R., Ed.; Thieme.
    (Review).

7   Sakya, S. M. *Knorr Pyrazole Synthesis.* In *Name Reactions in Heterocyclic Chemistry;*
    Li, J. J., Corey, E. J., Eds, Wiley: Hoboken, NJ, **2005,** pp 292–300. (Review).

8   Ahlstroem, M. M.; Ridderstroem, M.; Zamora, I.; Luthman, K. *J. Med. Chem.* **2007,**
    *50,* 4444–4452.

9   Jiang, J. A.; Huang, W. B.; Zhai, J. J.; Liu, H. W.; Cai, Q.; Xu, L. X.; Wang, W.; Ji, Y.
    F. *Tetrahedron* **2013,** *69,* 627–635.

# Koch–Haaf carbonylation

Strong acid-catalyzed tertiary carboxylic acid formation from alcohols or olefins and CO.

protonation

alkyl migration

The tertiary carbocation is thermodynamically favored

acylium ion

## References

1. Koch, H.; Haaf, W. *Ann.* **1958**, *618*, 251–266.
2. Hiraoka, K.; Kebarle, P. *J. Am. Chem. Soc.* **1977**, *99*, 366–370.
3. Takeuchi, K.; Akiyama, F.; Miyazaki, T.; Kitagawa, I.; Okamoto, K. *Tetrahedron* **1987**, *43*, 701–709.
4. Stepanov, A. G.; Luzgin, M. V.; Romannikov, V. N.; Zamaraev, K. I. *J. Am. Chem. Soc.* **1995**, *117*, 3615–3616.
5. Olah, G. A.; Prakash, G. K. S.; Mathew, T.; Marinez, E. R. *Angew. Chem. Int. Ed.* **2000**, *39*, 2547–2548.
6. Emert, J. I.; Dankworth, D. C.; Gutierrez, A. *Macromol.* **2001**, *34*, 2766–2775.
7. Li, T.; Tsumori, N.; Souma, Y.; Xu, Q. *Chem. Commun.* **2003**, 2070–2071.
8. Davis, M. C.; Liu, S. *Synth. Commun.* **2006**, *36*, 3509–3514.
9. Barton, V.; Ward, S. A.; Chadwick, J.; Hill, A. *J. Med. Chem.* **2010**, *53*, 4555–4559.

J.J. Li, *Name Reactions: A Collection of Detailed Mechanisms and Synthetic Applications*, DOI 10.1007/978-3-319-03979-4_149, © Springer International Publishing Switzerland 2014

# Koenig–Knorr glycosidation

Formation of the β-glycoside from α-halocarbohydrate under the influence of silver salt.

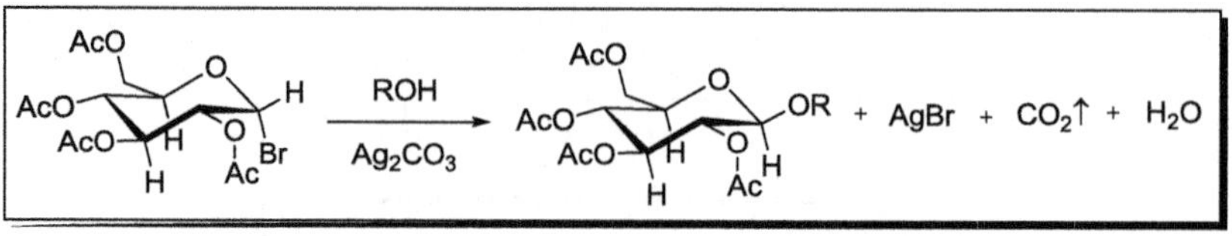
ROH
$Ag_2CO_3$
+ AgBr + $CO_2$↑ + $H_2O$

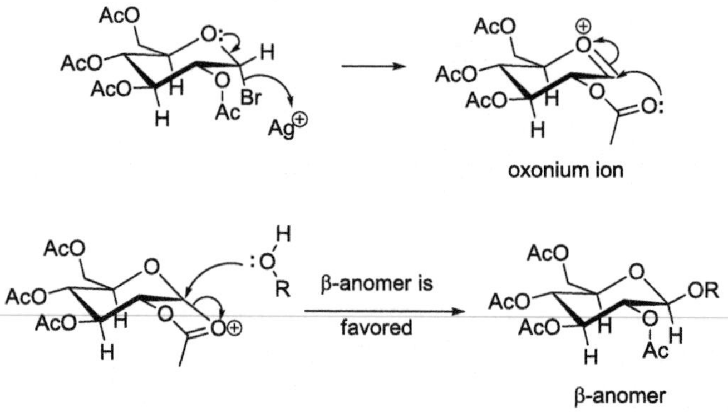
oxonium ion
β-anomer is
favored
β-anomer

Example 1[7]

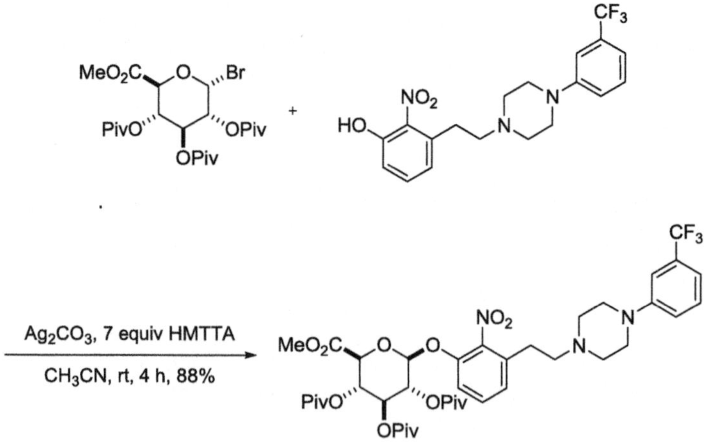
$Ag_2CO_3$, 7 equiv HMTTA
$CH_3CN$, rt, 4 h, 88%

Example 2[8]

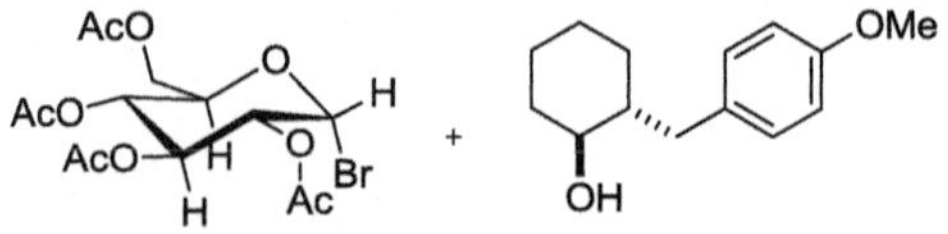

J.J. Li, *Name Reactions: A Collection of Detailed Mechanisms and Synthetic Applications*,
DOI 10.1007/978-3-319-03979-4_150, © Springer International Publishing Switzerland 2014

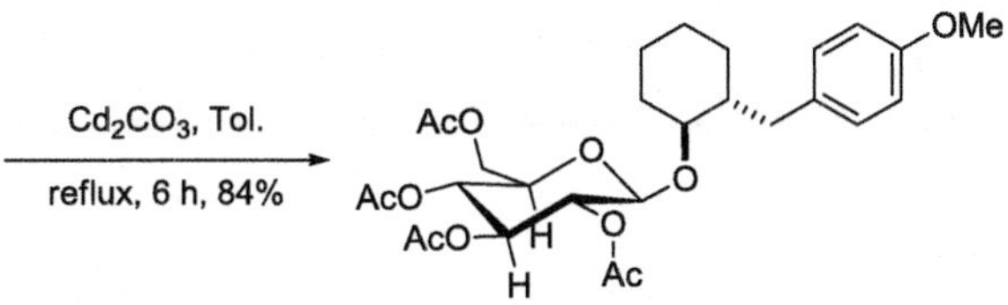

Cd$_2$CO$_3$, Tol.
reflux, 6 h, 84%

Example 3[9]

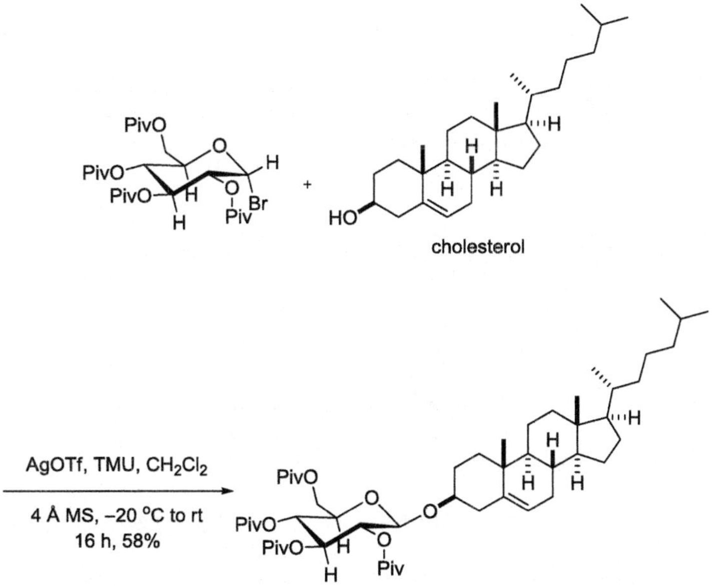

+

cholesterol

AgOTf, TMU, CH$_2$Cl$_2$

4 Å MS, −20 °C to rt
16 h, 58%

Example 4[11]

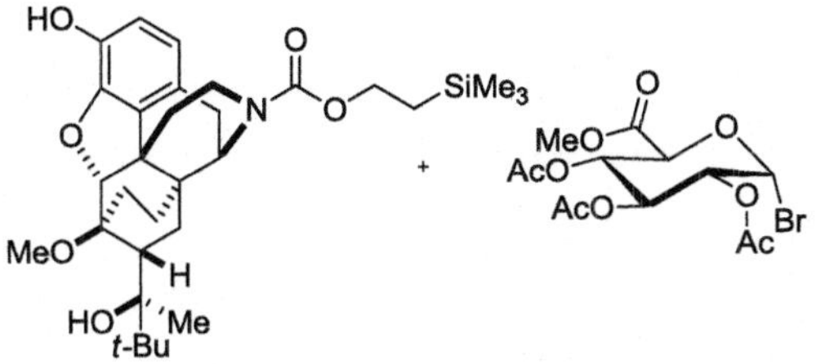

+

## References

1.  Koenig, W.; Knorr, E. *Ber.* **1901,** *34,* 957–981.
2.  Igarashi, K. *Adv. Carbohydr. Chem. Biochem.* **1977,** *34,* 243–83. (Review).
3.  Schmidt, R. R. *Angew. Chem.* **1986,** *98,* 213–236.
4.  Smith, A. B., III; Rivero, R. A.; Hale, K. J.; Vaccaro, H. A. *J. Am. Chem. Soc.* **1991,** *113,* 2092–2112.
5.  Fürstner, A.; Radkowski, K.; Grabowski, J.; Wirtz, C.; Mynott, R. *J. Org. Chem.* **2000,** *65,* 8758–8762.
6.  Yashunsky, D. V.; Tsvetkov, Y. E.; Ferguson, M. A. J.; Nikolaev, A. V. *J. Chem. Soc., Perkin Trans. 1* **2002,** 242–256.
7.  Stazi, F.; Palmisano, G.; Turconi, M.; Clini, S.; Santagostino, M. *J. Org. Chem.* **2004,** *69,* 1097–1103.
8.  Wimmer, Z.; Pechova, L.; Saman, D. *Molecules* **2004,** *9,* 902–912.
9.  Presser, A.; Kunert, O.; Pötschger, I. *Monat. Chem.* **2006,** *137,* 365–374.
10. Schoettner, E.; Simon, K.; Friedel, M.; Jones, P. G.; Lindel, T. *Tetrahedron Lett.* **2008,** *49,* 5580–5582.
11. Fan, J.; Brown, S. M.; Tu, Z.; Kharasch, E. D. *Bioconjugate Chem.* **2011,** *22,* 752–758.

# Kostanecki reaction

Also known as **Kostanecki–Robinson reaction**. Transformation **1→2** represents an **Allan–Robinson reaction**, whereas **1→3** is a **Kostanecki (acylation) reaction**:

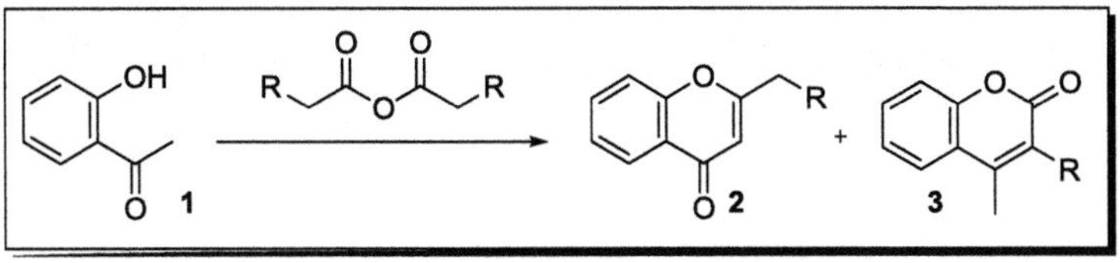

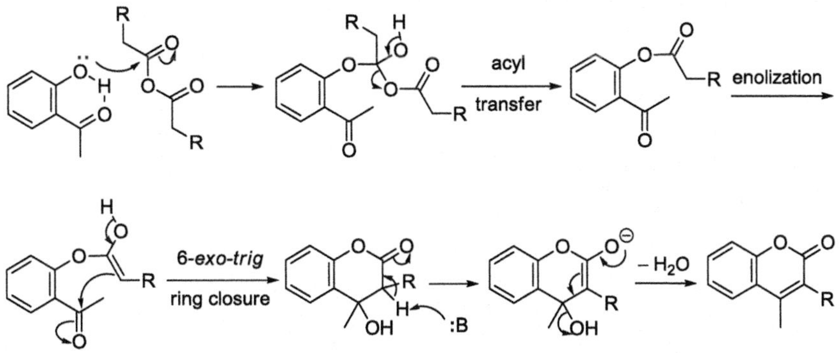

Example 1[2]

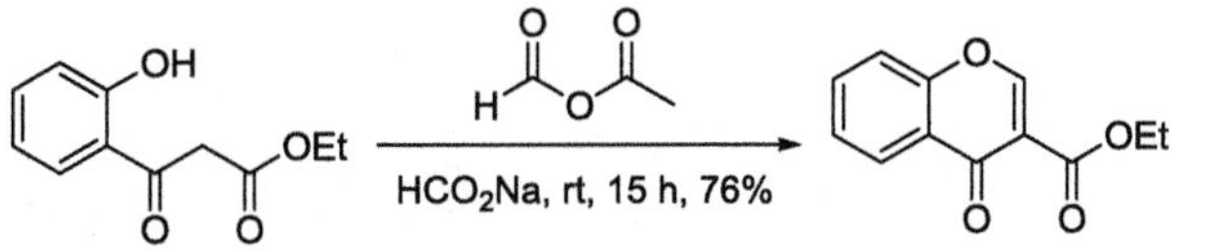

Example 2[3]

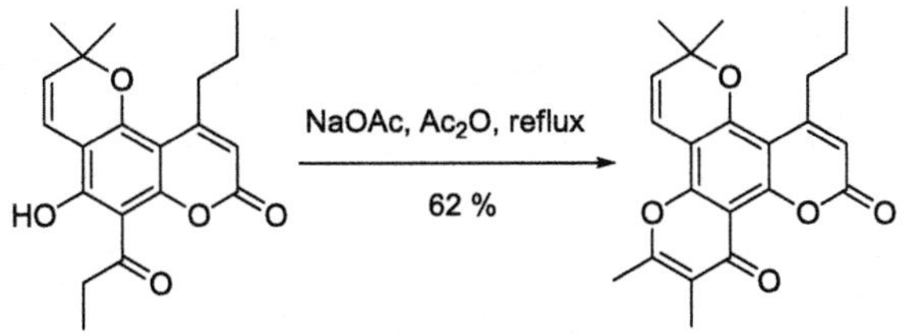

# References

1.  von Kostanecki, S.; Rozycki, A. *Ber.* **1901**, *34*, 102–109.
2.  Pardanani, N. H.; Trivedi, K. N. *J. Indian Chem. Soc.* **1972**, *49*, 599–604.
3.  Flavin, M. T.; Rizzo, J. D.; Khilevich, A.; *et al. J. Med. Chem.* **1996**, *39*, 1303–1313.
4.  Mamedov, V. A.; et al. *Chemistry of Heterocyclic Compounds* **2003**, *39*, 96–100.
5.  Limberakis, C. *Kostanecki–Robinson Reaction.* In *Name Reactions in Heterocyclic Chemistry*; Li, J. J., Ed.; Wiley: Hoboken, NJ, **2005**, pp 521–535. (Review).
6.  Hwang, I.-T.; Lee, S.-A.; Hwang, J.-S.; Lee, K.-I. *Mol.* **2011**, *16*, 6313–6321.

J.J. Li, *Name Reactions: A Collection of Detailed Mechanisms and Synthetic Applications*,
DOI 10.1007/978-3-319-03979-4_151, © Springer International Publishing Switzerland 2014

# Kröhnke pyridine synthesis

Pyridines from α-pyridinium methyl ketone salts and α,β-unsaturated ketones.

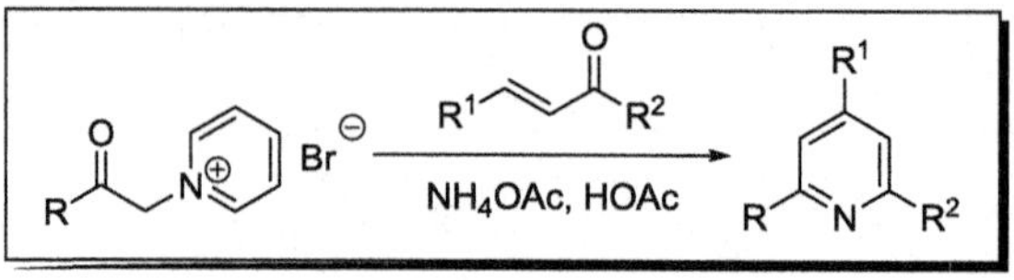

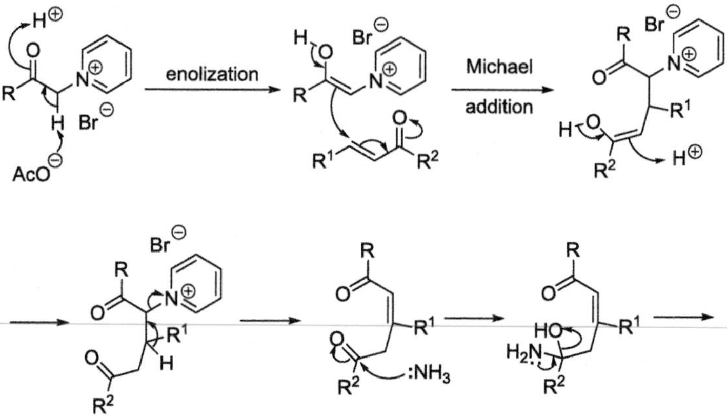

The ketone is more reactive than the enone

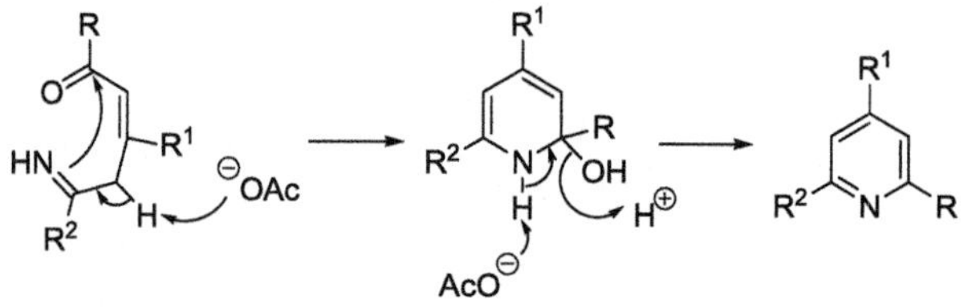

Example 1[1b]

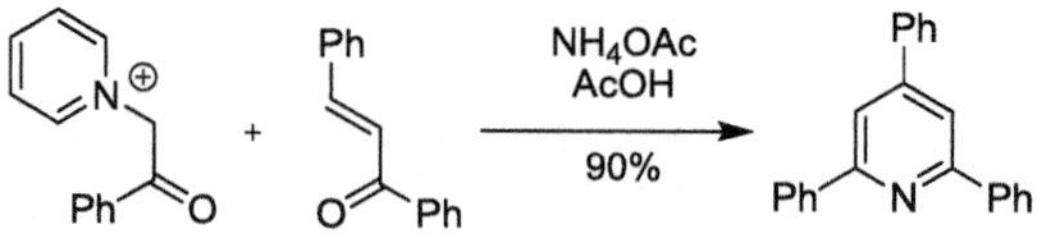

Example 2[4]

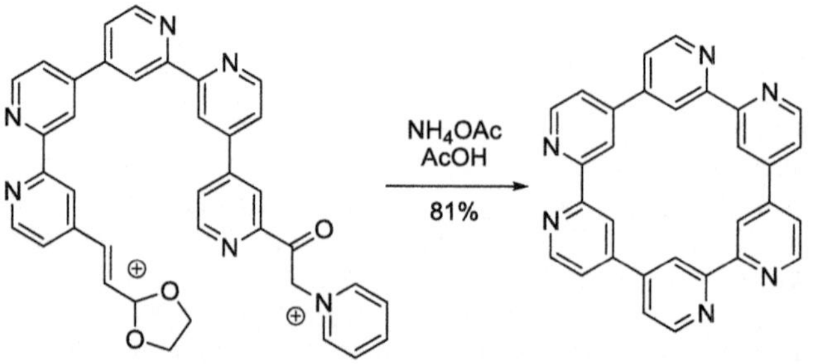

J.J. Li, *Name Reactions: A Collection of Detailed Mechanisms and Synthetic Applications*,
DOI 10.1007/978-3-319-03979-4_152, © Springer International Publishing Switzerland 2014

Example 3[6]

X = H, 65%
X = F, 83%
X = Br, 82%
X = OMe, 40%

Example 3[6]

## References

1. (a) Zecher, W.; Kröhnke, F. *Ber.* **1961,** *94,* 690–697. (b) Kröhnke, F.; Zecher, W. *Angew. Chem.* **1962,** *74,* 811–817. (c) Kröhnke, F. *Synthesis* **1976,** 1–24. (Review).
2. Potts, K. T.; Cipullo, M. J.; Ralli, P.; Theodoridis, G. *J. Am. Chem. Soc.* **1981,** *103,* 3584–3585, 3585–3586.
3. Newkome, G. R.; Hager, D. C.; Kiefer, G. E. *J. Org. Chem.* **1986,** *51,* 850–853.
4. Kelly, T. R.; Lee, Y.-J.; Mears, R. J. *J. Org. Chem.* **1997,** *62,* 2774–2781.
5. Bark, T.; Von Zelewsky, A. *Chimia* **2000,** *54,* 589–592.
6. Malkov, A. V.; Bella, M.; Stara, I. G.; Kocovsky, P. *Tetrahedron Lett.* **2001,** *42,* 3045–3048.
7. Cave, G. W. V.; Raston, C. L. *J. Chem. Soc., Perkin Trans. 1* **2001,** 3258–3264.
8. Malkov, A. V.; Bell, M.; Vassieu, M.; Bugatti, V.; Kocovsky, P. *J. Mol. Cat. A: Chem.* **2003,** *196,* 179–186.
9. Galatsis, P. *Kröhnke Pyridine Synthesis.* In *Name Reactions in Heterocyclic Chemistry*; Li, J. J., Ed.; Wiley: Hoboken, NJ, **2005,** 311–313. (Review).
10. Yan, C.-G.; Wang, Q.-F.; Cai, X.-M.; Sun, J. *Central Eur. J. Chem.* **2008,** *6,* 188–198.
11. Xu, T.; Luo, X.-L.; Yang, Y.-R. *Tetrahedron Lett.* **2013,** *54,* 2858–2860.

# Krapcho reaction

Nucleophilic decarboxylation of β-ketoesters, malonate esters, α-cyanoesters, or α-sulfonylesters.

Example 1[5]

Example 2[10]

## References

1.  Krapcho, A. P.; Glynn, G. A.; Grenon, B. J. *Tetrahedron Lett.* **1967**, 215–217. A. Paul Krapcho is a professor at the University of Vermont.
2.  Duval, O.; Gomes, L. M. *Tetrahedron* **1989**, *45*, 4471–4476.
3.  Flynn, D. L.; Becker, D. P.; Nosal, R.; Zabrowski, D. L. *Tetrahedron Lett.* **1992**, *33*, 7283–7286.
4.  Martin, C. J.; Rawson, D. J.; Williams, J. M. J. *Tetrahedron: Asymmetry* **1998**, *9*, 3723–3730.
5.  Gonzalez-Gomez, J. C.; Uriarte, E. *Synlett* **2002**, 2095–2097.
6.  Bridges, N. J.; Hines, C. C.; Smiglak, M.; Rogers, R. D. *Chem. Eur. J.* **2007**, *13*, 207–5212.
7.  Poon, P. S.; Banerjee, A. K.; Laya, M. S. *J. Chem. Res.* **2011**, *35*, 67–73. (Review).
8.  Farran, D.; Bertrand, P. *Synth. Commun.* **2012**, *42*, 989–1001.
9.  Adepu, R.; Rambabu, D.; Prasad, B.; Meda, C. L. T.; Kandale, A.; Rama Krishna, G.; Malla Reddy, C.; Chennuru, L. N. *Org. Biomol. Chem.* **2012**, *10*, 5554–5569.
10. Mason, J. D.; Murphree, S. S. *Synlett* **2013**, *24*, 1391–1394.

J.J. Li, *Name Reactions: A Collection of Detailed Mechanisms and Synthetic Applications*,
DOI 10.1007/978-3-319-03979-4_153, © Springer International Publishing Switzerland 2014

# Kumada cross-coupling reaction

The Kumada cross-coupling reaction (also known as Kumada-Tamao-Corriu coupling, also occasionally known as the Kharasch cross-coupling reaction) was originally reported as the nickel-catalyzed cross-coupling of Grignard reagents with aryl- or alkenyl halides. It has subsequently been developed to encompass the coupling of organolithium or organomagnesium compounds with aryl-, alkenyl or alkyl halides, catalyzed by nickel or palladium. The Kumada cross-coupling reaction, as well as the Negishi, Stille, Hiyama, and Suzuki cross-coupling reactions, belong to the same category of Pd-catalyzed cross-coupling reactions of organic halides, triflates and other electrophiles with organometallic reagents. These reactions follow a general mechanistic catalytic cycle as shown below. There are slight variations for the Hiyama and Suzuki reactions, for which an additional activation step is required for the transmetallation to occur.

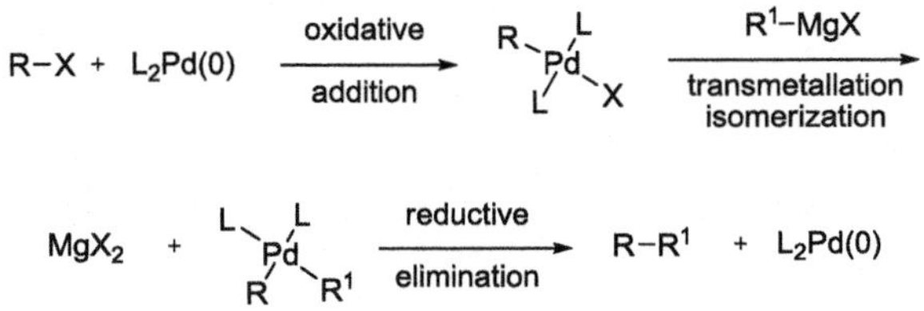

The catalytic cycle:

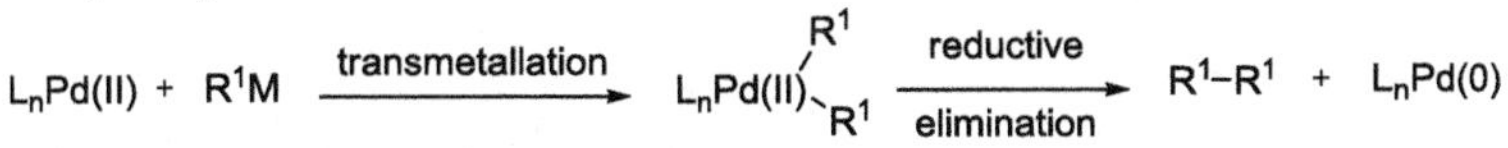

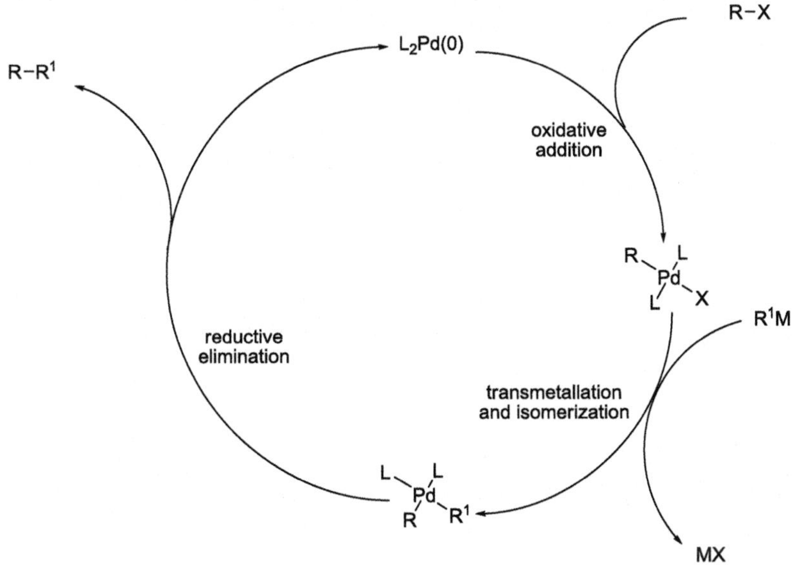

J.J. Li, *Name Reactions: A Collection of Detailed Mechanisms and Synthetic Applications*, DOI 10.1007/978-3-319-03979-4_154, © Springer International Publishing Switzerland 2014

Example 1[2]

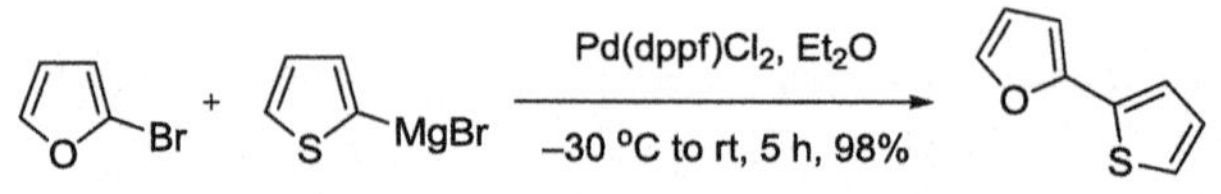

Example 2[3]

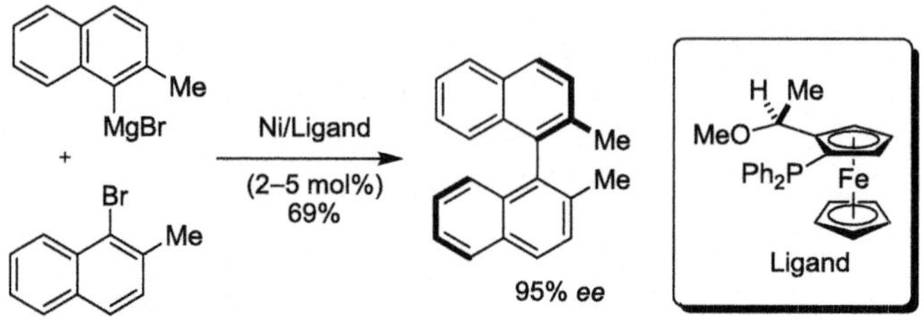

95% ee

Example 3[5]

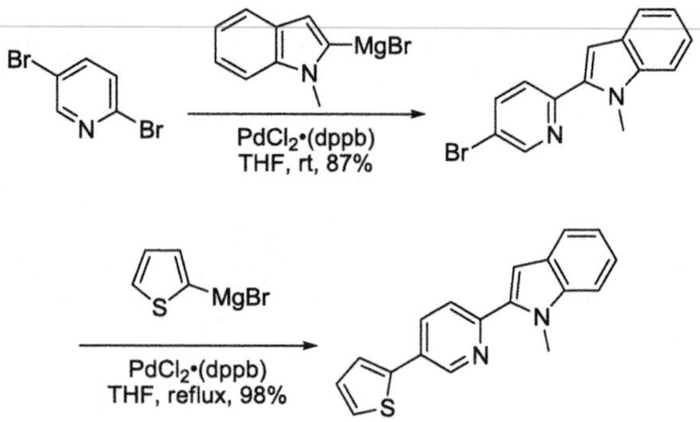

Example 4[8]

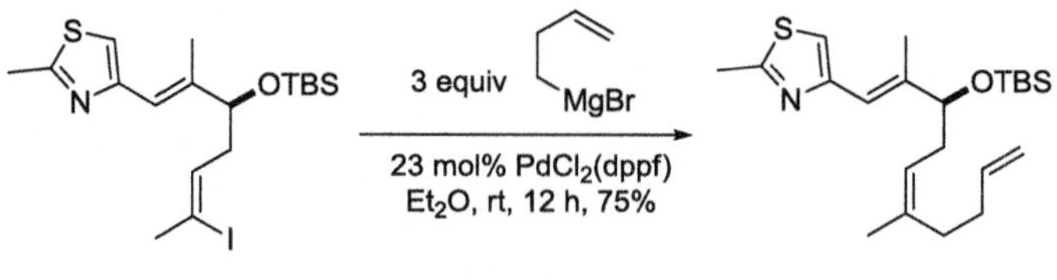

Example 5[9]

Example 6, Nickel-catalyzed Kumada reaction of tosylalkanes[11]

## References

1.   Tamao, K.; Sumitani, K.; Kiso, Y.; Zembayashi, M.; Fujioka, A.; Kodma, S.-i.; Nakajima, I.; Minato, A.; Kumada, M. *Bull. Chem. Soc. Jpn.* **1976**, *49*, 1958–1969.
2.   Carpita, A.; Rossi, R.; Veracini, C. A. *Tetrahedron* **1985**, *41*, 1919–1929.
3.   Hayashi, T.; Hayashizaki, K.; Kiyoi, T.; Ito, Y. *J. Am. Chem. Soc.* **1988**, *110*, 8153–8156.
4.   Kalinin, V. N. *Synthesis* **1992**, 413–432. (Review).
5.   Meth-Cohn, O.; Jiang, H. *J. Chem. Soc., Perkin Trans. 1* **1998**, 3737–3746.
6.   Stanforth, S. P. *Tetrahedron* **1998**, *54*, 263–303. (Review).
7.   Huang, J.; Nolan, S. P. *J. Am. Chem. Soc.* **1999**, *121*, 9889–9890.
8.   Rivkin, A.; Njardarson, J. T.; Biswas, K.; Chou, T.-C.; Danishefsky, S. J. *J. Org. Chem.* **2002**, *67*, 7737–7740.
9.   William, A. D.; Kobayashi, Y. *J. Org. Chem.* **2002**, *67*, 8771–8782.
10.  Fuchter, M. J. *Kumada Cross-Coupling Reaction.* In *Name Reactions for Homologations-Part I*; Li, J. J., Ed.; Wiley: Hoboken, NJ, **2009**, pp 47–69. (Review).
11.  Wu, J.-C.; Gong, L.-B.; Xia, Y.; Song, R.-J.; Xie, Y.-X.; Li, J.-H. *Angew. Chem. Int. Ed.* **2012**, *51*, 9909–9913.
12.  Handa, S.; Arachchige, Y. L. N. M.; Slaughter, L. M. *J. Org. Chem.* **2013**, *78*, 5694–5699.

# Lawesson's reagent

2,4-Bis(4-methoxyphenyl)-1,3-dithiadiphosphetane-2,4-disulfide transforms the carbonyl groups of aldehydes, ketones, amides, lactams, esters and lactones into the corresponding thiocarbonyl compounds. *Cf.* Knorr thiophene synthesis.

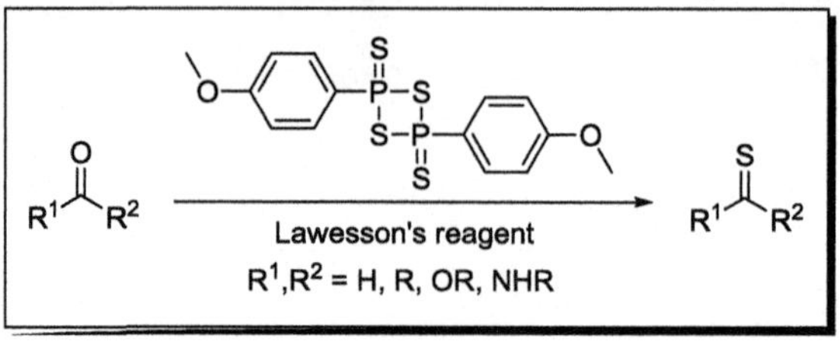

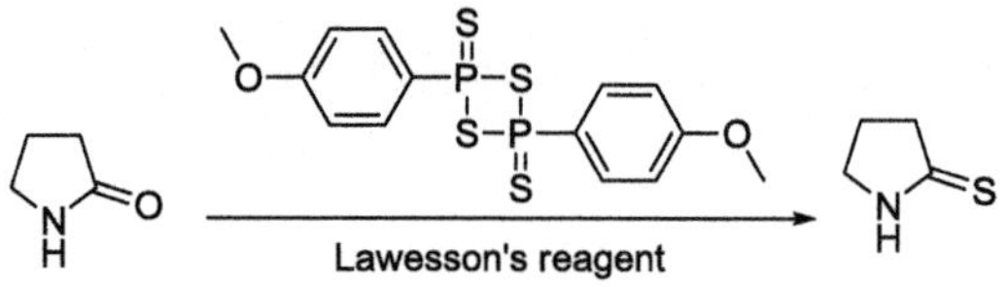

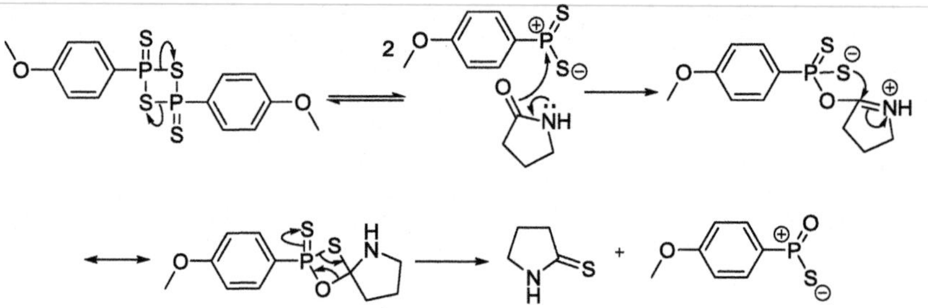

Example 1[4]

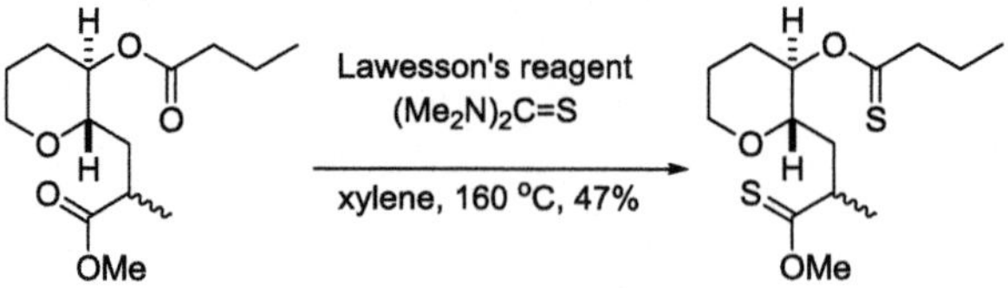

Example 2[5]

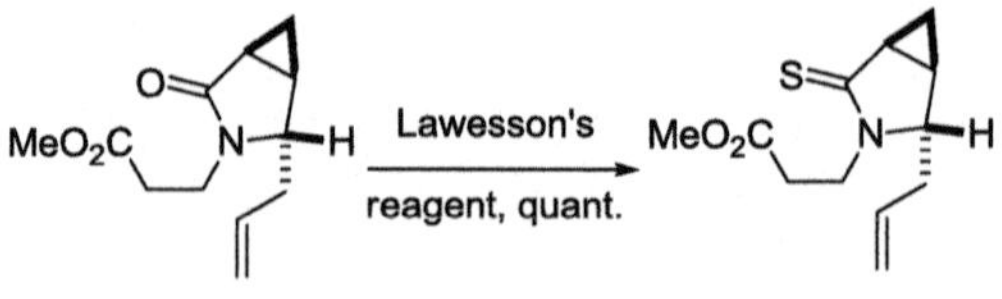

J.J. Li, *Name Reactions: A Collection of Detailed Mechanisms and Synthetic Applications*,
DOI 10.1007/978-3-319-03979-4_155, © Springer International Publishing Switzerland 2014

## Example 3, Thiophene from dione[8]

R = H
R = Me
R = OMe
R = Cl
R = Br

## Example 4[10]

## Example 5[11]

## References

1.  Scheibye, S.; Shabana, R.; Lawesson, S. O.; Rømming, C. *Tetrahedron* **1982**, *38*, 993–1001.
2.  Navech, J.; Majoral, J. P.; Kraemer, R. *Tetrahedron Lett.* **1983**, *24*, 5885–5886.
3.  Cava, M. P.; Levinson, M. I. *Tetrahedron* **1985**, *41*, 5061–5087. (Review).
4.  Nicolaou, K. C.; Hwang, C.-K.; Duggan, M. E.; Nugiel, D. A.; Abe, Y.; Bal Reddy, K.; DeFrees, S. A.; Reddy, D. R.; Awartani, R. A.; Conley, S. R.; Rutjes, F. P. J. T.; Theodorakis, E. A. *J. Am. Chem. Soc.* **1995**, *117*, 10227–10238.
5.  Kim, G.; Chu-Moyer, M. Y.; Danishefsky, S. J. *J. Am. Chem. Soc.* **1990**, *112*, 2003–2005.
6.  Luheshi, A.-B. N.; Smalley, R. K.; Kennewell, P. D.; Westwood, R. *Tetrahedron Lett.* **1990**, *31*, 123–127.
7.  Ishii, A.; Yamashita, R.; Saito, M.; Nakayama, J. *J. Org. Chem.* **2003**, *68*, 1555–1558.
8.  Diana, P.; Carbone, A.; Barraja, P.; Montalbano, A.; Martorana, A.; Dattolo, G.; Gia, O.; Dalla Via, L.; Cirrincione, G. *Bioorg. Med. Chem. Lett.* **2007**, *17*, 2342–2346.
9.  Ozturk, T.; Ertas, E.; Mert, O. *Chem. Rev.* **2007**, *107*, 5210–5278. (Review).
10. Taniguchi, T.; Ishibashi, H. *Tetrahedron* **2008**, *64*, 8773–8779.
11. de Moreira, D. R. M. *Synlett* **2008**, 463–464. (Review).
12. Kaschel, J.; Schmidt, C. D.; Mumby, M.; Kratzert, D.; Stalke, D.; Werz, D. B. *Chem. Commun.* **2013**, *49*, 4403-4405.

## Leuckart–Wallach reaction

Amine synthesis from reductive amination of a ketone and an amine in the presence of excess formic acid, which serves as the reducing reagent by delivering a hydride. When the ketone is replaced by formaldehyde, it becomes the Eschweiler–Clarke reductive alkylation of amines.

gem-aminoalcohol; iminium ion intermediate

Reduction

### Example 1[4]

### Example 2[6]

### Example 3[7]

### Example 4[8]

J.J. Li, *Name Reactions: A Collection of Detailed Mechanisms and Synthetic Applications*,
DOI 10.1007/978-3-319-03979-4_156, © Springer International Publishing Switzerland 2014

An unexpected intramolecular transamidation *via* a Wagner–Meerwein shift after the Leuckart–Wallach reaction

## References

1.  Leuckart, R. *Ber.* **1885,** *18*, 2341–2344. Carl L. R. A. Leuckart (1854–1889) was born in Giessen, Germany. After studying under Bunsen, Kolbe, and von Baeyer, he became an assistant professor at Göttingen. Unfortunately, chemistry lost a brilliant contributor by his sudden death at age 35 as a result of a fall in his parent's house.
2.  Wallach, O. *Ann.* **1892,** *272*, 99. Otto Wallach (1847–1931), born in Königsberg, Prussia, studied under Wöhler and Hofmann. He was the director of the Chemical Institute at Göttingen from 1889 to 1915. His book "Terpene und Kampfer" served as the foundation for future work in terpene chemistry. Wallach was awarded the Nobel Prize in Chemistry in 1910 for his work on alicyclic compounds.
3.  Moore, M. L. *Org. React.* **1949,** *5*, 301–330. (Review).
4.  DeBenneville, P. L.; Macartney, J. H. *J. Am. Chem. Soc.* **1950,** *72*, 3073–3075.
5.  Lukasiewicz, A. *Tetrahedron* **1963,** *19*, 1789–1799. (Mechanism).
6.  Bach, R. D. *J. Org. Chem.* **1968,** *33*, 1647–1649.
7.  Musumarra, G.; Sergi, C. *Heterocycles* **1994,** *37*, 1033–1039.
8.  Martínez, A. G.; Vilar, E. T.; Fraile, A. G.; Ruiz, P. M.; San Antonio, R. M.; Alcazar, M. P. M. *Tetrahedron: Asymmetry* **1999,** *10*, 1499–1505.
9.  Kitamura, M.; Lee, D.; Hayashi, S.; Tanaka, S.; Yoshimura, M. *J. Org. Chem.* **2002,** *67*, 8685–8687.
10. Brewer, A. R. E. *Leuckart–Wallach reaction.* In *Name Reactions for Functional Group Transformations*; Li, J. J., Ed.; Wiley: Hoboken, NJ, **2007,** pp 451–455. (Review).
11. Muzalevskiy, V. M.; Nenajdenko, V. G.; Shastin, A. V.; Balenkova, E. S.; Haufe, G. *J. Fluorine Chem.* **2008,** *129*, 1052–1055.

# Li A$^3$ reaction

The Li A$^3$-reaction is the direct dehydrative condensation reaction of aldehyde-alkyne-amine catalyzed by various transition-metals to generate propargyl amines, often in water.[1–4] Many catalyztic systems such as [Ru]/[Cu],[5] [Au],[6] [Ag],[7] and iron[8,9] were effective for such reactions. The catalytic cycle of the reaction involves the in situ generation an alkynylmetal intermediate as well as an imine (or iminium) intermediate, which react together to give the propargylamine products. Carbohydrates[10–11] can be used directly to generate the propagyl amine products. Multi A$^3$-reactions are also successful,[12] which allows for the site-specific functionalization of amino acids and peptides under physiological conditions.[13] Highly efficient asymmetric A$^3$-reactions involving both primary amines[14,15] and secondary amines[16] have also been succeeded. The reaction is also amenable for flow chemistry.[17]

Catalytic cycle of the Li A$^3$ reaction

Example 1[5]

Example 2[6]

J.J. Li, *Name Reactions: A Collection of Detailed Mechanisms and Synthetic Applications*, DOI 10.1007/978-3-319-03979-4_157, © Springer International Publishing Switzerland 2014

Example 3[11]

83% aldehyde conversion
50 µL liquor

[Au(CN)Cl$_2$]

40 °C, 2 h
H$_2$O

Example 4[13]

≡—R'
aq. CH$_2$O
cat. CuCl

35 °C

R= quanidine, disulfide, thioether, phenol, alcohol
R= aryl, alkyl, TMS

Example 5[14]

$$R^1\text{-CHO} + Ar\text{-NH}_2 + R^2\text{-}\!\!\equiv$$

cat. Cu(OTf)/ligand **1**

toluene or water

### References

1. Yoo, W.-J.; Zhao, L.; Li, C.-J. *Aldrichimica Acta*, **2011**, *44*, 43–51. (Review).
2. Wei, C.; Li, Z.; Li, C.-J. *Synlett* **2004**, 1472.
3. Zani, L.; Bolm, C. *Chem. Commun.* **2006**, 4263.
4. Peshkov, V. A.; Pereshivko, O. P.; Van der Eycken, E. V. *Chem. Soc. Rev.* **2012**, *41*, 3702 (Review).
5. Li, C. J.; Wei, C. M. *Chem. Commun.* **2002**, 268–269.
6. Wei, C.; Li, C.-J. *J. Am. Chem. Soc.* **2003**, *125*, 9584.
7. Wei, C. M.; Li, Z. G.; Li, C. J. *Org. Lett.* **2003**, *5*, 4473–4475.
8. Chen, W.-W.; Nguyen, R. V.; Li, C.-J. *Tetrahedron Lett.* **2009**, *50*, 2895.
9. Li, P.; Zhang, Y.; Wang, L. *Chem. Eur. J.* **2009**, *15*, 2045.
10. Roy, B.; Raj, R.; Mukhopadhya, B. *Tetrahedron Lett.* **2009**, *50*, 5838–5841.
11. Kung, K. K. Y.; Li, G. L.; Zou, L.; Chong, H. C.; Leung, Y. C.; Wong, K. H.; Lo, V. K. Y.; Che, C.-M.; Wong, M.-K. *Org. Biomol. Chem.*, **2012**, *10*, 925–930.
12. Bonfield, E. R.; Li, C.-J. *Org. Biomol. Chem.* **2007**, *5*, 435.

13. Uhlig, N.; Li, C.-J. *Org. Lett.* **2012,** *14*, 3000–3003.
14. Wei, C.; Li, C.-J. *J. Am. Chem. Soc.* **2002,** *124*, 5638.
15. Wei, C.; Mague, J. T.; Li, C.-J. *Proc. Natl. Acad. Sci. U.S.A.* **2004,** *101*, 5749.
16. Gommermann, N.; Koradin, C.; Polborn, K.; Knochel, P. *Angew. Chem. Int. Ed.* **2003,** *42*, 5763.
17. Shore, G.; Yoo, W.-J.; Li, C.-J.; Organ, M. G. *Chem. Eur. J.* **2010,** 16, 126–133.

# Lossen rearrangement

The Lossen rearrangement involves the generation of an isocyanate via thermal or base-mediated rearrangement of an activated hydroxamate which can be generated from the corresponding hydroxamic acid. Activation of the hydroxamic acid can be achieved through O-acylation, O-arylation, chlorination, or O-sulfonylation. Such hydroxamic acids can also be activated using polyphosphoric acid, carbodiimide, Mitsunobu conditions, or silyation. The product of the Lossen rearrangement, an isocyanate can be subsequently converted to a urea or an amine resulting in the net loss of one carbon atom relative to the starting hydroxamic acid.

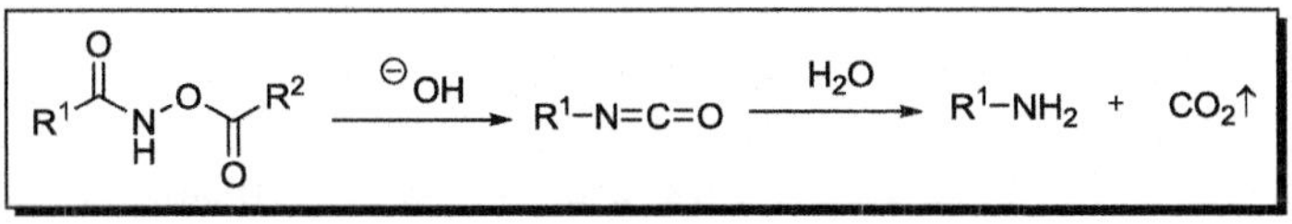

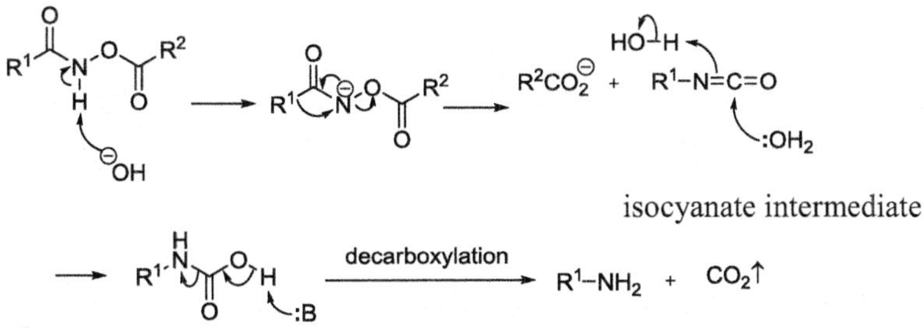

isocyanate intermediate

Example 1[6]

BnOH, CH₃CN, 85 °C, 78%

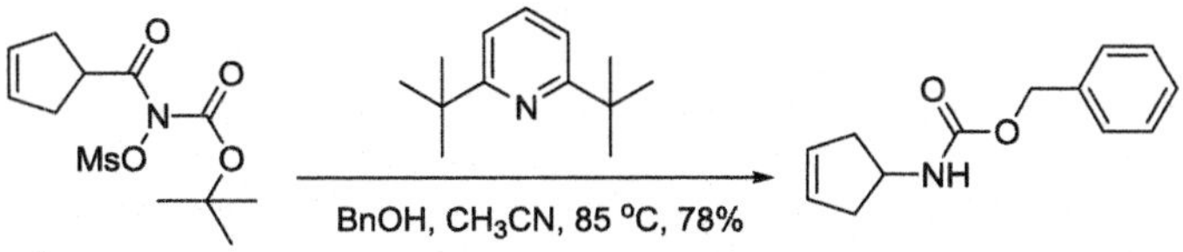

Example 2[7]

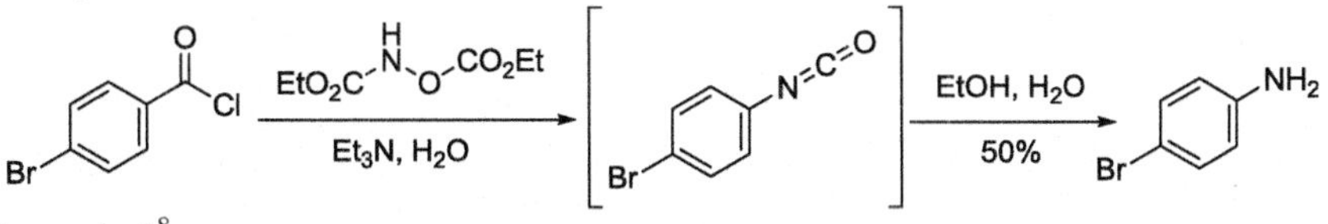

Example 3[8]

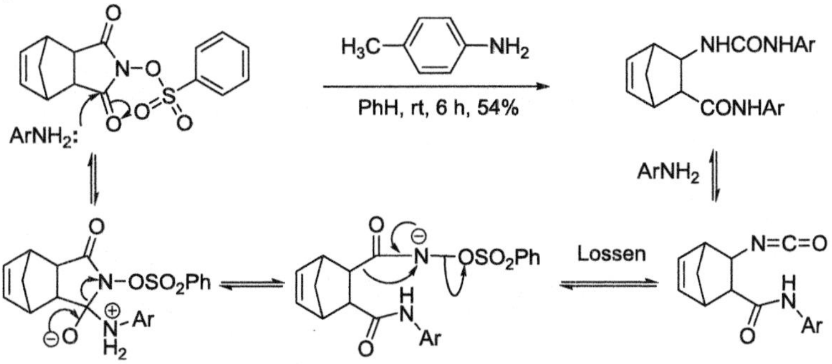

J.J. Li, *Name Reactions: A Collection of Detailed Mechanisms and Synthetic Applications*, DOI 10.1007/978-3-319-03979-4_158, © Springer International Publishing Switzerland 2014

Example 4[9]

Example 5[11]

## References

1.  Lossen, W. *Ann.* **1872,** *161,* 347.  Wilhelm C. Lossen (1838–1906) was born in Kreuznach, Germany.  After his Ph.D. studies at Göttingen in 1862, he embarked on his independent academic career, and his interests centered on hydroxyamines.
2.  Bauer, L.; Exner, O. *Angew. Chem. Int. Ed.* **1974,** *13,* 376.
3.  Lipczynska-Kochany, E. *Wiad. Chem.* **1982,** *36,* 735–756.
4.  Casteel, D. A.; Gephart, R. S.; Morgan, T. *Heterocycles* **1993,** *36,* 485–495.
5.  Zalipsky, S. *Chem. Commun.* **1998,** 69–70.
6.  Stafford, J. A.; Gonzales, S. S.; Barrett, D. G.; Suh, E. M.; Feldman, P. L. *J. Org. Chem.* **1998,** 63, 10040–10044.
7.  Anilkumar, R.; Chandrasekhar, S.; Sridhar, M. *Tetrahedron Lett.* **2000,** *41,* 5291–5293.
8.  Abbady, M. S.; Kandeel, M. M.; Youssef, M. S. K. *Phosphorous, Sulfur and Silicon* **2000,** *163,* 55–64.
9.  Ohmoto, K.; Yamamoto, T.; Horiuchi, T.; Kojima, T.; Hachiya, K.; Hashimoto, S.; Kawamura, M.; Nakai, H.; Toda, M. *Synlett* **2001,** 299–301.
10. Choi, C.; Pfefferkorn, J. A. *Lossen rearrangement.* In *Name Reactions for Homologations-Part II*; Li, J. J., Ed.; Wiley: Hoboken, NJ, **2009,** pp 200–209. (Review).
11. Yoganathan, S.; Miller, S. J. *Org. Lett.* **2013,** *15,* 602–605.

# McFadyen–Stevens reduction

Treatment of acylbenzenesulfonylhydrazines with base delivers the corresponding aldehydes.

Example 1[5]

Example 2[7]

## References

1. McFadyen, J. S.; Stevens, T. S. *J. Chem. Soc.* **1936**, 584–587. Thomas S. Stevens (1900–2000) was born in Renfrew, Scotland. After earning his Ph.D. under W. H. Perkin at Oxford University, he became a reader at the University of Sheffield. J. S. McFadyen (1908–) was born in Toronto, Canada. After studying under Stevens at the University of Glasgow, he worked for ICI for 15 years before returning to Canada where he worked for the Canadian Industries, Ltd., Montreal.

2. Graboyes, H.; Anderson, E. L.; Levinson, S. H.; Resnick, T. M. *J. Heterocycl. Chem.* **1975**, *12*, 1225–1231.

3. Eichler, E.; Rooney, C. S.; Williams, H. W. R. *J. Heterocycl. Chem.* **1976**, *13*, 841–844.

4. Nair, M.; Shechter, H. *J. Chem. Soc., Chem. Commun.* **1978**, 793–796.

5. Dudman, C. C.; Grice, P.; Reese, C. B. *Tetrahedron Lett.* **1980**, *21*, 4645–4648.

6. Manna, R. K.; Jaisankar, P.; Giri, V. S. *Synth. Commun.* **1998**, *28*, 9–16.

7. Jaisankar, P.; Pal, B.; Giri, V. S. *Synth. Commun.* **2002**, *32*, 2569–2573.

8. Ma, B.; Banerjee, B.; Litvinov, D. N.; He, L.; Castle, S. L. *J. Am. Chem. Soc.* **2010**, *132*, 1159–1171.

9. Iwai, Y.; Ozaki, T.; Takita, R.; Uchiyama, M.; Shimokawa, J.; Fukuyama, T. *Chem. Sci.* **2013**, *4*, 1111–1119.

J.J. Li, *Name Reactions: A Collection of Detailed Mechanisms and Synthetic Applications*,
DOI 10.1007/978-3-319-03979-4_159, © Springer International Publishing Switzerland 2014

# McMurry coupling

Olefination of carbonyls with low-valent titanium such as Ti(0) derived from TiCl$_3$/LiAlH$_4$. A single-electron process.

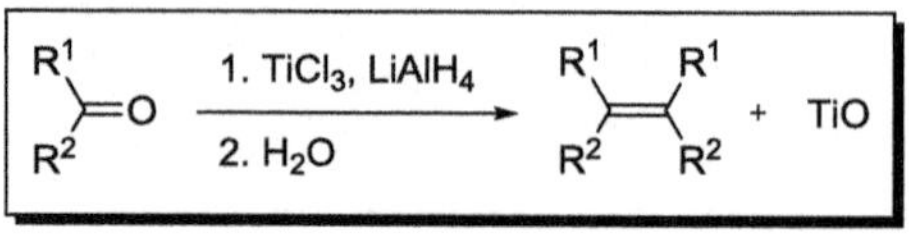

$$Ti(III)Cl_3 + LiAlH_4 \longrightarrow Ti(0)$$

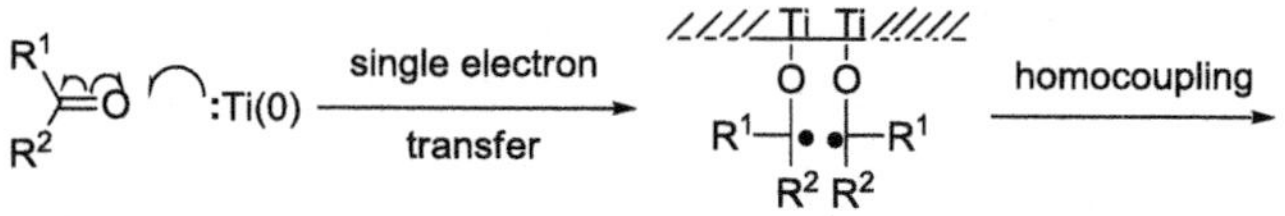

radical anion intermediate

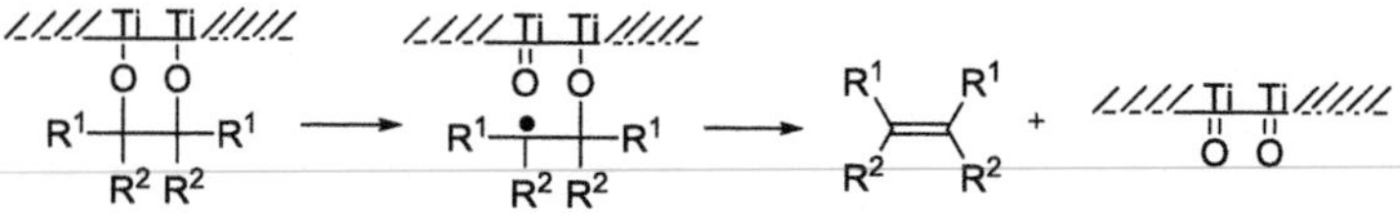

oxide-coated titanium surface

Example 1, Cross-McMurry coupling[7]

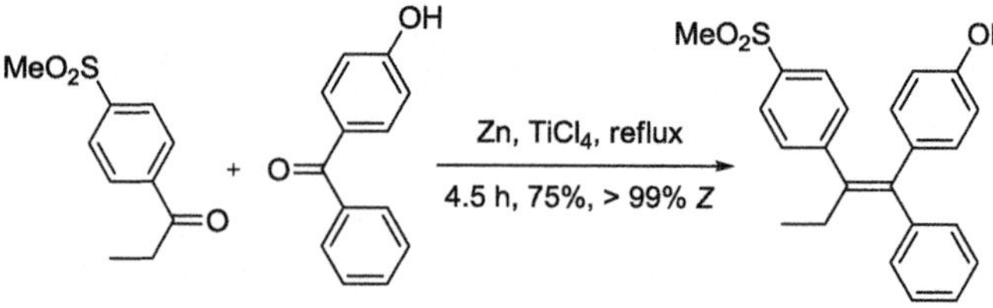

Example 2, Homo-McMurry coupling[8]

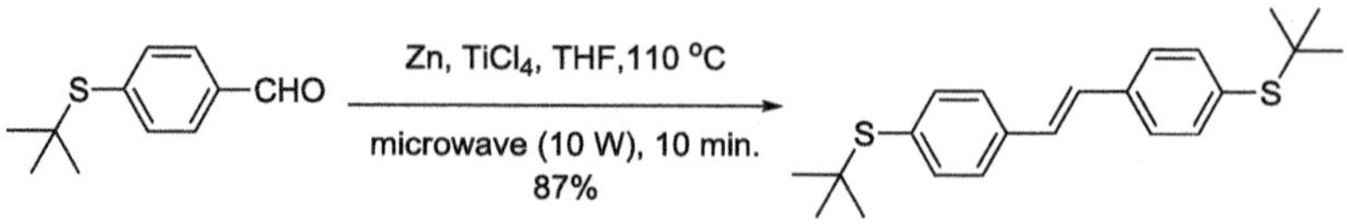

J.J. Li, *Name Reactions: A Collection of Detailed Mechanisms and Synthetic Applications*, DOI 10.1007/978-3-319-03979-4_160, © Springer International Publishing Switzerland 2014

Example 3, Cross-McMurry coupling[9]

Example 4, Cross-McMurry coupling[10]

Example 5[12]

## References

1. (a) McMurry, J. E.; Fleming, M. P. *J. Am. Chem. Soc.* **1974**, *96*, 4708–4712.   (b) McMurry, J. E. *Chem. Rev.* **1989**, *89*, 1513–1524. (Review).
2. Hirao, T. *Synlett* **1999**, 175–181.
3. Sabelle, S.; Hydrio, J.; Leclerc, E.; Mioskowski, C.; Renard, P.-Y. *Tetrahedron Lett.* **2002**, *43*, 3645–3648.
4. Williams, D. R.; Heidebrecht, R. W., Jr. *J. Am. Chem. Soc.* **2003**, *125*, 1843–1850.
5. Honda, T.; Namiki, H.; Nagase, H.; Mizutani, H. *Tetrahedron Lett.* **2003**, *44*, 3035–3038.
6. Ephritikhine, M.; Villiers, C. In *Modern Carbonyl Olefination* Takeda, T., Ed.; Wiley-VCH: Weinheim, Germany, **2004**, 223–285. (Review).
7. Uddin, M. J.; Rao, P. N. P.; Knaus, E. E. *Synlett* **2004**, 1513–1516.
8. Stuhr-Hansen, N. *Tetrahedron Lett.* **2005**, *46*, 5491–5494.
9. Zeng, D. X.; Chen, Y. *Synlett* **2006**, 490–492.
10. Duan, X.-F.; Zeng, J.; Zhang, Z.-B.; Zi, G.-F. *J. Org. Chem.* **2007**, *72*, 10283–10286.
11. Debroy, P.; Lindeman, S. V.; Rathore, R. *J. Org. Chem.* **2009**, *74*, 2080–2087.
12. Kumar, A. S.; Nagarajan, R. *Synthesis* **2013**, *45*, 1235–1246.

# MacMillan catalyst

Highly enantioselective and general asymmetric organocatalytic Diels–Alder reaction using α-amino acid-derived imidazolidinones (of type **1**) as catalysts. The first generation of MacMillan catalyst (**1**) has been employed in a variety of organocatalytic enantioselective reactions. Typical examples are: Diels–Alder reaction;[1] nitrone cycloaddition,[2] pyrrole Friedel–Crafts reaction,[3] indole addition,[4] vinylogous Michael addition;[5] α-chlorination;[6] hydride addition;[7] cyclopropanation;[8] α-fluorination.[9] The second generation MacMillan catalyst (**2**) was used to catalyze 1,4-addition of indoles to α,β-unsaturated aldehydes.

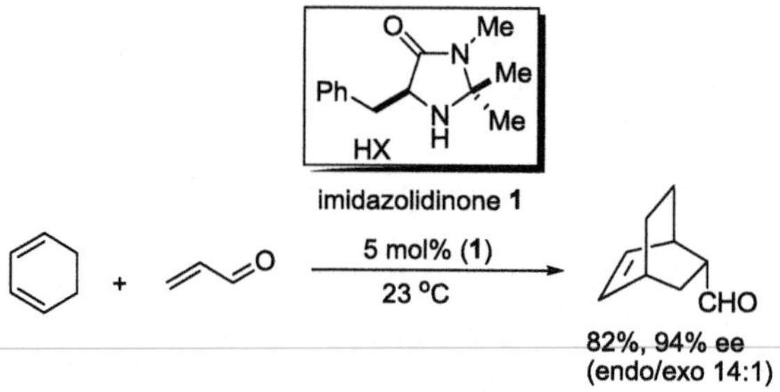
Me
Me
Me
Ph
HX
imidazolidinone 1
5 mol% (1)
23 °C
CHO
82%, 94% ee
(endo/exo 14:1)

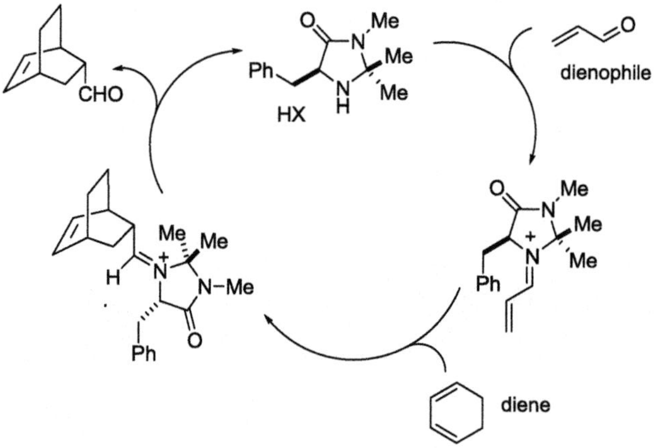
CHO
Ph
HX
Me
dienophile
N–Me
diene

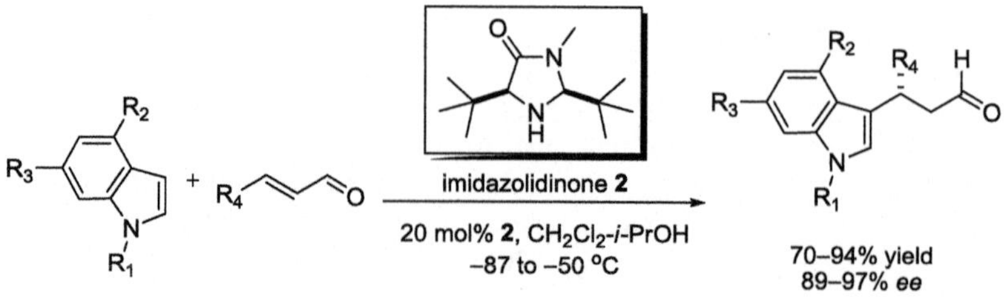
R2
R3
R1
R4
imidazolidinone 2
20 mol% 2, CH2Cl2-i-PrOH
–87 to –50 °C
70–94% yield
89–97% ee

J.J. Li, *Name Reactions: A Collection of Detailed Mechanisms and Synthetic Applications*,
DOI 10.1007/978-3-319-03979-4_161, © Springer International Publishing Switzerland 2014

Example 1[11]

Example 2[10]

78% yield
90% ee

(−)-flustramine B

## References

1. Ahrendt, K.; Borths, C.; MacMillan, D. W. C. *J. Am. Chem. Soc.* **2000,** *122*, 4243.
2. Jen, W.; Wiener, J.; MacMillan, D. W. C. *J. Am. Chem. Soc.* **2000,** *122*, 9874.
3. Paras, N.; MacMillan, D. W. C. *J. Am. Chem. Soc.* **2001,** *123*, 4370.
4. Austin, J. F.; MacMillan, D. W. C. *J. Am. Chem. Soc.* **2002,** *124*, 1172.
5. Brown, S. P.; Goodwin, N. C.; MacMillan, D. W. C. *J. Am. Chem. Soc.* **2003,** *125*, 1192.
6. Brochu, M. P.; Brown, S. P.; MacMillan, D. W. C. *J. Am. Chem. Soc.* **2004,** *126*, 4108.
7. Ouellet, S. G.; Tuttle, J. B.; MacMillan, D. W. C. *J. Am. Chem. Soc.* **2005,** *127*, 32.
8. Kunz, R. K; MacMillan, D. W. C. *J. Am. Chem. Soc.* **2005,** *127*, 3240.
9. Beeson, T. D.; MacMillan, D. W. C. *J. Am. Chem. Soc.* **2005,** *127*, 8826.
10. Austin, J. F.; Kim, S.-G.; Sinz, C. J.; Xiao, W.-J.; MacMillan, D. W. C. *Proc. Nat. Acad. Sci. USA,* **2004,** *101*, 5482.
11. Kim, S.-G.; Kim, J.; Jung, H. *Tetrahedron Lett.* **2005,** *46*, 2437.
12. Riente, P.; Yadav, J.; Pericas, M. A. *Org. Lett.* **2012,** *14,* 3668–3671.
13. Zhang, Y.; Wang, S.-Y.; Xu, X.-P.; Jiang, R.; Ji, S.-J. *Org. Biomol. Chem.* **2013,** *11*, 1933–1937.

# Mannich reaction

Three-component aminomethylation from amine, aldehyde and a compound with an acidic methylene moiety.

When R = Me, the $^+Me_2N=CH_2$ salt is known as *Eschenmoser's salt*

The Mannich reaction can also operate under basic conditions:

Mannich Base

Example 1, Asymmetric Mannich reaction[2]

Example 2, Asymmetric Mannich-type reaction[9]

Example 3, Asymmetric Mannich-type reaction[10]

J.J. Li, *Name Reactions: A Collection of Detailed Mechanisms and Synthetic Applications*,
DOI 10.1007/978-3-319-03979-4_162, © Springer International Publishing Switzerland 2014

Example 4[11]

Example 5, Vinylogous Mannich Reaction (VMR)[13]

## References

1. Mannich, C.; Krösche, W. *Arch. Pharm.* **1912,** *250*, 647–667. Carl U. F. Mannich (1877–1947) was born in Breslau, Germany. After receiving a Ph.D. at Basel in 1903, he served on the faculties of Göttingen, Frankfurt and Berlin. Mannich synthesized many esters of *p*-aminobenzoic acid as local anesthetics.
2. List, B. *J. Am. Chem. Soc.* **2000,** *122*, 9336–9337.
3. Schlienger, N.; Bryce, M. R.; Hansen, T. K. *Tetrahedron* **2000,** *56*, 10023–10030.
4. Bur, S. K.; Martin, S. F. *Tetrahedron* **2001,** *57*, 3221–3242. (Review).
5. Martin, S. F. *Acc. Chem. Res.* **2002,** *35*, 895–904. (Review).
6. Padwa, A.; Bur, S. K.; Danca, D. M.; Ginn, J. D.; Lynch, S. M. *Synlett* **2002,** 851–862. (Review).
7. Notz, W.; Tanaka, F.; Barbas, C. F., III. *Acc. Chem. Res.* **2004,** *37*, 580–591. (Review).
8. Córdova, A. *Acc. Chem. Res.* **2004,** *37*, 102–112. (Review).
9. Harada, S.; Handa, S.; Matsunaga, S.; Shibasaki, M. *Angew. Chem. Int. Ed.* **2005,** *44*, 4365–4368.
10. Lou, S.; Dai, P.; Schaus, S. E. *J. Org. Chem.* **2007,** *72*, 9998–10008.
11. Hahn, B. T.; Fröhlich, R.; Harms, K.; Glorius, F. *Angew. Chem. Int. Ed.* **2008,** *47*, 9985–9988.
12. Galatsis, P. *Mannich reaction.* In *Name Reactions for Homologations-Part II*; Li, J. J., Ed.; Wiley: Hoboken, NJ, **2009,** pp 653–670. (Review).
13. Liu, X.-K.; Ye, J.-L.; Ruan, Y.-P.; Li, Y.-X.; Huang, P.-Q. *J. Org. Chem.* **2013,** *78*, 35–41.

## Markovnikov's rule

For addition of HX to olefins, Markovnikov's Rule predicts the regiochemistry of HX addition to unsymmetrically substituted alkenes: The halide component of HX bonds preferentially at the more highly substituted carbon, whereas the hydrogen prefers the carbon which already contains more hydrogen atoms.

The intermediate is the secondary cation, and the formal charge is on one carbon.

Exception to Markovnikov's Rule:

Example 1[3]

Example 2, Markovnikov-selective hydrothiolation of styrenes[4]

J.J. Li, *Name Reactions: A Collection of Detailed Mechanisms and Synthetic Applications*,
DOI 10.1007/978-3-319-03979-4_163, © Springer International Publishing Switzerland 2014

## References

1      Markownikoff, W. *Ann. Pharm.* **1870**, *153*, 228–259. Vladimir Vasilyevich Markovnikov (1838–1904) fomulated the rule for addition to alkenes at Moscow University. He was one of the most eminant Russian organic chemists in the 19th century. He was a prickly individual with strong opinions, and he had no fear in expressing them. His tactless outspokenness led to his ouster from professorship at Kazan' and Moscow. (Lewis, D. E. *Early Russian Organic Chemists and Their Legacy*, Springer: Heldelberg, Germany, 2012, p 71.).

2      Oparina, L. A.; Artem'ev, A. V.; Vysotskaya, O. V.; Kolyvanov, N. A.; Bagryanskaya, Y. I.; Doronina, E. P.; Gusarova, N. K. *Tetrahedron* **2013**, *69*, 6185–6195.

3      Ziyaei Halimehjani, A.; Pasha Zanussi, H. *Synthesis* **2013**, *45*, 1483–1488

4      Savolainen, M. A.; Wu, J. *Org. Lett.* **2013**, *15*, 3802–3804.

## *Anti*-Markovnikov

Some reactions do not follow Markovnikov's Rule, and *anti*-Markovnikov products are isolated. The outcome of the regioselectivity may be explained by the relative stability of the radical intermediates.

Radical mechanism:

Initiation:

Propagation:

Termination:

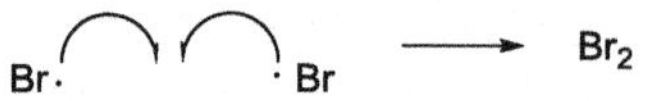
Br·
· Br
Br2

Example 1, Anti-Markovnikov oxidation of allylic esters[1]

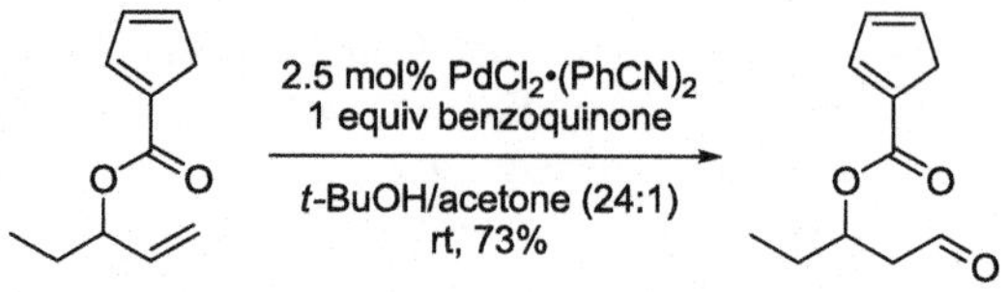
2.5 mol% PdCl2•(PhCN)2
1 equiv benzoquinone
t-BuOH/acetone (24:1)
rt, 73%

Example 2, Anti-Markovnikov hydroamination[3]

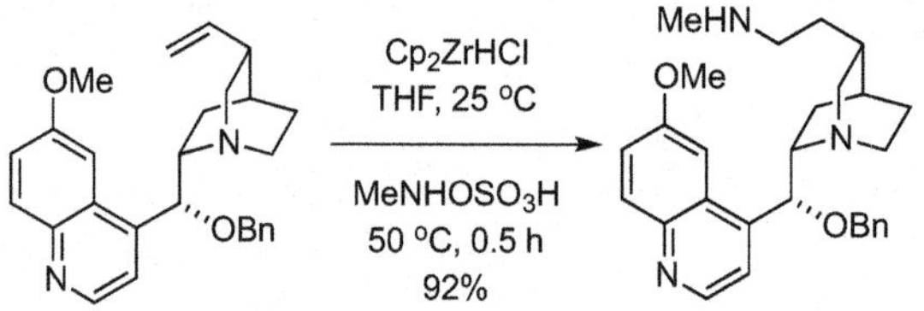
OMe
N
OBn
Cp2ZrHCl
THF, 25 °C
MeNHOSO3H
50 °C, 0.5 h
92%
MeHN

## References


1. Nishizawa, M.; Asai, Y.; Imagawa, H. *Org. Lett.* **2006,** *8,* 5793–5796.
2. Dong, J. J.; Fañanás-Mastral, M.; Alsters, P. L.; Browne, W. R.; Feringa, B. L. *Angew. Chem. Int. Ed.* **2008,** *47*, 5561–5565.
3. Strom, A. E.; Hartwig, J. F. *J. Org. Chem.* **2013,** *78*, 8909–8914.

# Martin's sulfurane dehydrating reagent

Dehydrates secondary and tertiary alcohols to give olefins, but forms ethers with primary alcohols. *Cf.* Burgess dehydrating reagent.

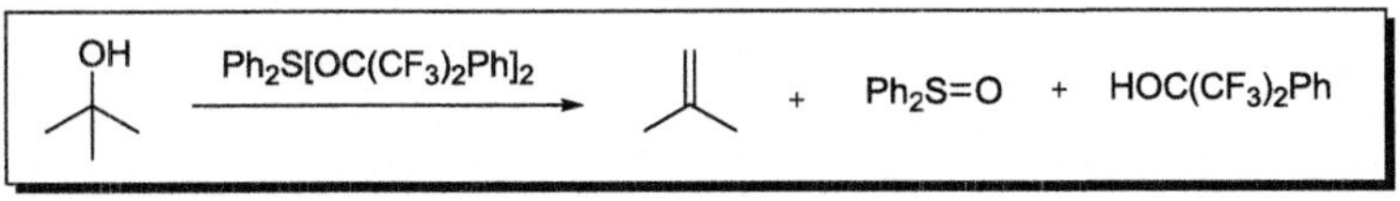

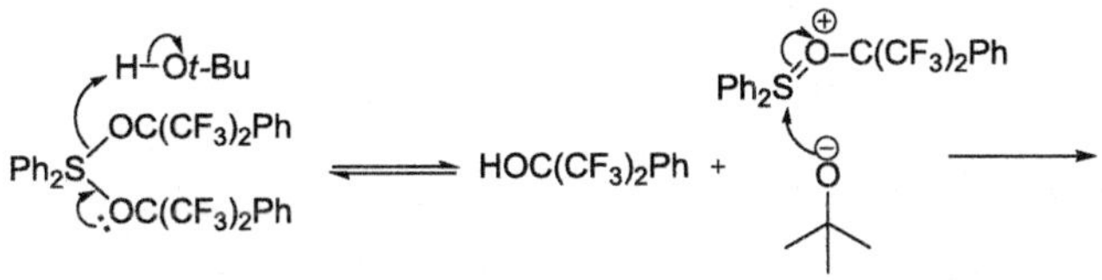

The alcohol is acidic

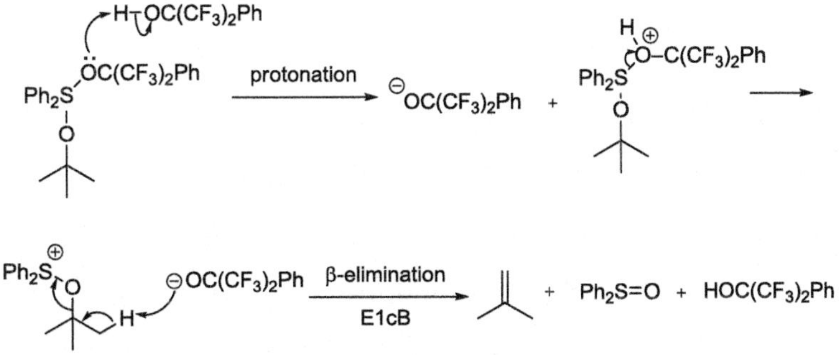

Example 1[5]

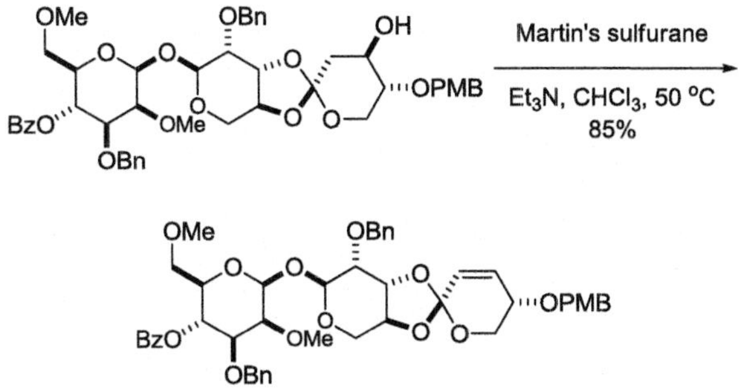

J.J. Li, *Name Reactions: A Collection of Detailed Mechanisms and Synthetic Applications,*
DOI 10.1007/978-3-319-03979-4_164, © Springer International Publishing Switzerland 2014

Example 2[6]

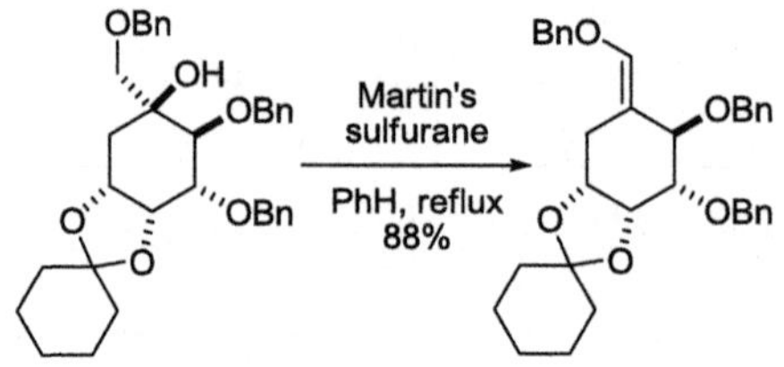

Example 3[7]

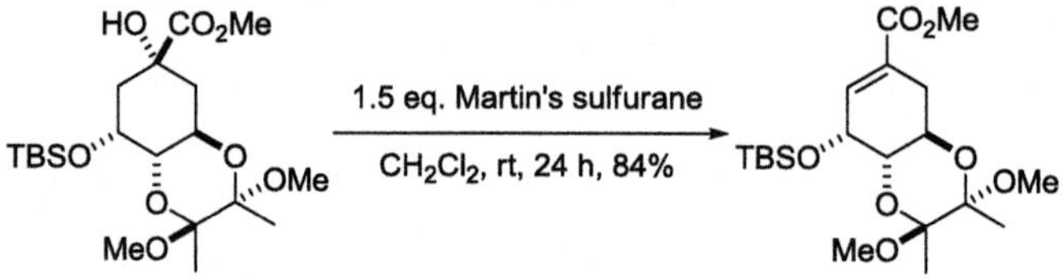

Example 4[9]

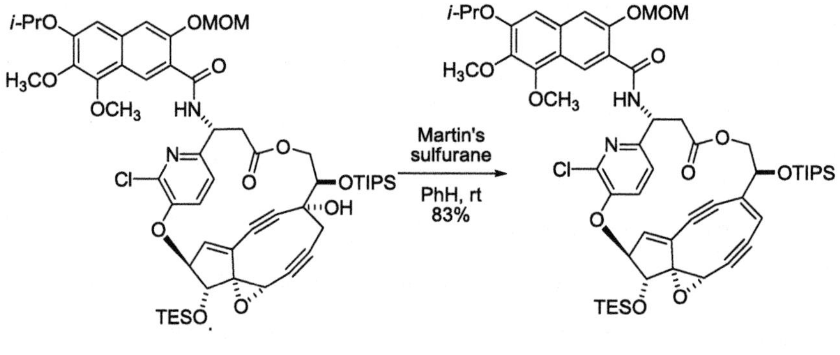

Example 5[12]

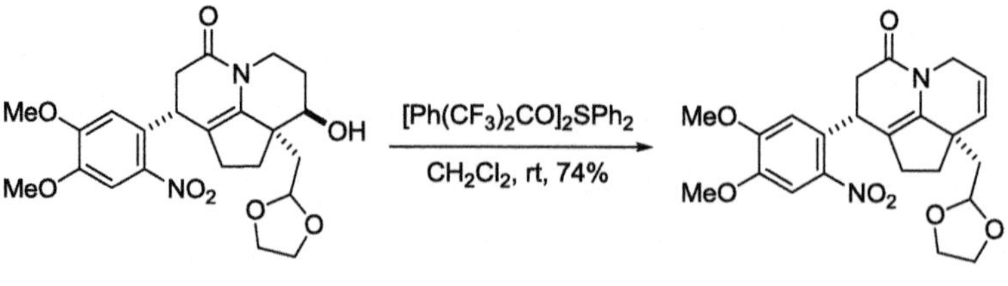

## References

1.    (a) Martin, J. C.; Arhart, R. J. *J. Am. Chem. Soc.* **1971**, *93*, 2339–2341; (b) Martin, J.
      C.; Arhart, R. J. *J. Am. Chem. Soc.* **1971**, *93*, 2341–2342; (c) Martin, J. C.; Arhart, R.

J. *J. Am. Chem. Soc.* **1971,** *93*, 4327–4329. (d) Martin, J. C.; Arhart, R. J.; Franz, J. A.; Perozzi, E. F.; Kaplan, L. J. *Org. Synth.* **1977,** *57*, 22–26.

2. Gallagher, T. F.; Adams, J. L. *J. Org. Chem.* **1992,** *57*, 3347–3353.

3. Tse, B.; Kishi, Y. *J. Org. Chem.* **1994,** *59*, 7807–7814.

4. Winkler, J. D.; Stelmach, J. E.; Axten, J. *Tetrahedron Lett.* **1996,** *37*, 4317–4320.

5. Nicolaou, K. C.; Rodríguez, R. M.; Fylaktakidou, K. C.; Suzuki, H.; Mitchell, H. J. *Angew. Chem. Int. Ed.* **1999,** *38*, 3340–3345.

6. Kok, S. H. L.; Lee, C. C.; Shing, T. K. M. *J. Org. Chem.* **2001,** *66,* 7184–7190.

7. Box, J. M.; Harwood, L. M.; Humphreys, J. L.; Morris, G. A.; Redon, P. M.; Whitehead, R. C. *Synlett* **2002,** 358–360.

8. Myers, A. G.; Glatthar, R.; Hammond, M.; Harrington, P. M.; Kuo, E. Y.; Liang, J.; Schaus, S. E.; Wu, Y.; Xiang, J.-N. *J. Am. Chem. Soc.* **2002,** *124*, 5380–5401.

9. Myers, A. G.; Hogan, P. C.; Hurd, A. R.; Goldberg, S. D. *Angew. Chem. Int. Ed.* **2002,** *41,* 1062–1067.

10. Shea, K. M. *Martin's sulfurane dehydrating reagent.* In *Name Reactions for Functional Group Transformations*; Li, J. J., Ed.; Wiley: Hoboken, NJ, **2007,** pp 248–264. (Review).

11. Sparling, B. A.; Moslin, R. M.; Jamison, T. F. *Org. Lett.* **2008,** *10,* 1291–1294.

12. Miura, Y.; Hayashi, N.; Yokoshima, S.; Fukuyama, T. *J. Am. Chem. Soc.* **2012,** *134*, 11995–11997.

## Masamune–Roush conditions for the Horner–Emmons reaction

Applicable to base-sensitive aldehydes and phosphonates for the Horner–Wadsworth–Emmons reaction to prepare olefins. α-Keto or α-alkoxycarbonyl phosphonate required.

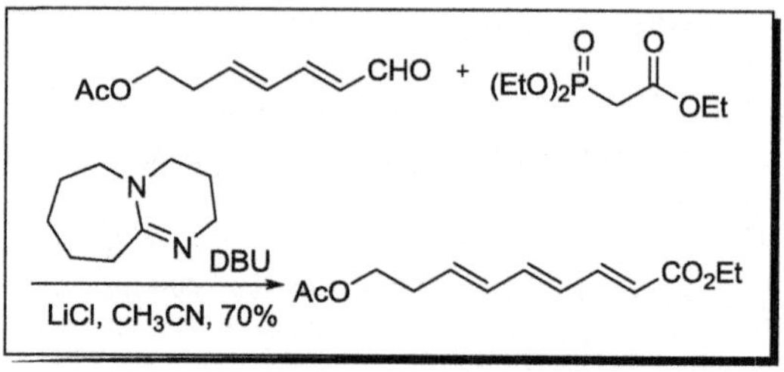

DBU = 1,8-diazabicyclo[5.4.0]undec-7-ene

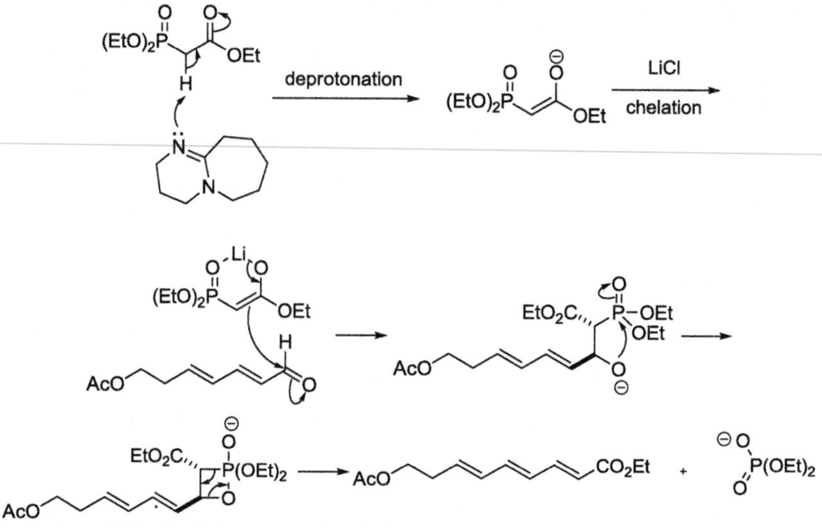

Formation of the P=O is thermodynamically favored, which is the driving force of this reaction.

Example 1[5]

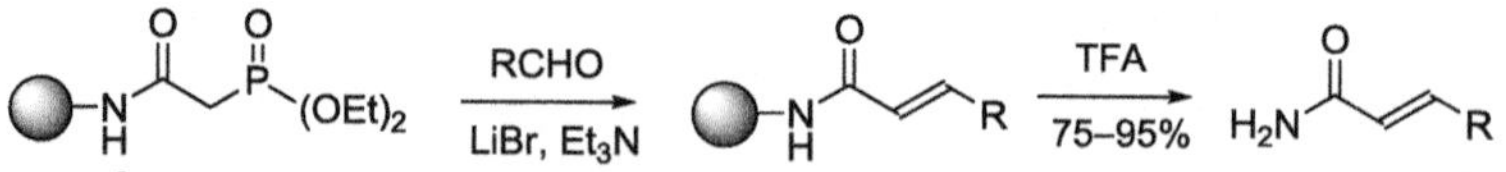

Example 2[6]

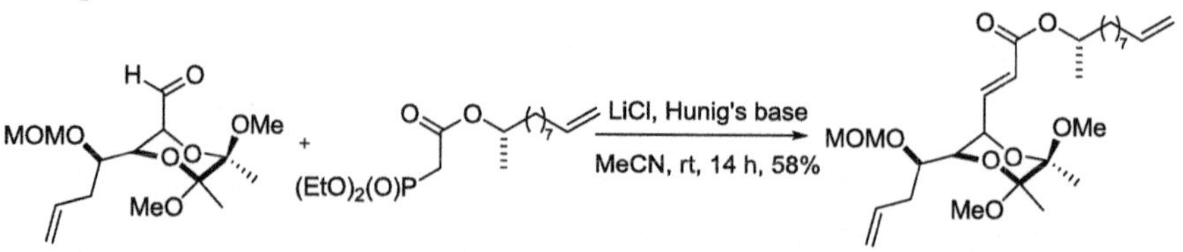

J.J. Li, *Name Reactions: A Collection of Detailed Mechanisms and Synthetic Applications*, DOI 10.1007/978-3-319-03979-4_165, © Springer International Publishing Switzerland 2014

Example 3[7]

Example 4[8]

Example 5[10]

## References

1. Blanchette, M. A.; Choy, W.; Davis, J. T.; Essenfeld, A. P.; Masamune, S.; Roush, W. R.; Sakai, T. *Tetrahedron Lett.* **1984**, *25*, 2183–2186.
2. Rathke, M. W.; Nowak, M. *J. Org. Chem.* **1985**, *50*, 2624–2636.
3. Tius, M. A.; Fauq, A. H. *J. Am. Chem. Soc.* **1986**, *108*, 1035–1039, and 6389–6391.
4. Marshall, J. A.; DuBay, W. J. *J. Org. Chem.* **1994**, *59*, 1703–1708.
5. Johnson, C. R.; Zhang, B. *Tetrahedron Lett.* **1995**, *36*, 9253–9256.
6. Rychnovsky, S. D.; Khire, U. R.; Yang, G. *J. Am. Chem. Soc.* **1997**, *119*, 2058–2059.
7. Dixon, D. J.; Foster, A. C.; Ley, S. V. *Org. Lett.* **2000**, *2*, 123–125.
8. Simoni, D.; Rossi, M.; Rondannin, R.; Mazzali, A.; Baruchello, R.; Malagutti, C.; Roberti, M.; Invidiata, F. P. *Org. Lett.* **2000**, *2*, 3765–3768.
9. Crackett, P.; Demont, E.; Eatherton, A.; Frampton, C. S.; Gilbert, J.; Kahn, I.; Redshaw, S.; Watson, W. *Synlett* **2004**, 679–683.
10. Ordonez, M.; Hernandez-Fernandez, E.; Montiel-Perez, M.; Bautista, R.; Bustos, P.; Rojas-Cabrera, H.; Fernandez-Zertuche, M.; Garcia-Barradas, O. *Tetrahedron: Asymmetry* **2007**, *18*, 2427–2436.
11. Zanato, C.; Pignataro, L.; Hao, Z.; Gennari, C. *Synthesis* **2008**, 2158–2162.
12. Paterson, I.; Fink, S. J.; Blakey, S. B. *Org. Lett.* **2013**, *15*, 3188–3121.

# Meerwein's salt

Meerwein's salts, also known as the Meerwein reagents, refer to trimethyloxonium tetrafluoroborate $(Me_3O^+BF_4^-)$ and triethyloxonium tetrafluoroborate $(Et_3O^+BF_4^-)$. Named after the inventor Hans Meerwein,[1] these trialkyloxonium salts are powerful alkylating agents.

Preparation:[2]

Example 1, The Meerwein reagent is an excellent $O$-alkylating agent:[5]

Transforming an amide into its corresponding ethyl or methyl esters

Example 2, *Metal*-methylation[4]

Example 3, $N$-Alkylation, the product is an ionic liquid[8]

J.J. Li, *Name Reactions: A Collection of Detailed Mechanisms and Synthetic Applications*,
DOI 10.1007/978-3-319-03979-4_166, © Springer International Publishing Switzerland 2014

Example 4, *N*-Methylation[9]

1. Me$_3$O·BF$_4$, slow addition CH$_2$Cl$_2$, rt
2. BuMeNH, CH$_3$CN, Δ, quant.

**References**

1. (a) Meerwein, H.; Hinz, G.; Hofmann, P.; Kroning, E.; Pfeil, E. *J. Prakt. Chem.* **1937**, *147*, 257–285. (b) Meerwein, H.; Bettenberg, E.; Pfeil, E.; Willfang, G. *J. Prakt. Chem.* **1939**, *154*, 83–156.
2. (a) Meerwein, H. *Org. Synth.*; *Coll. Vol. V* **1973**, 1080. Triethyloxonium tetrafluoroborate. (b) Curphey, T. J. *Org. Synth.*; *Coll. Vol. VI*, **1988**, 1019. Trimethyloxonium tetrafluoroborate.
3. Chen, F. M. F.; Benoiton, N. L. *Can. J. Chem.* **1977**, *55*, 1433–1534.
4. Dötz, K. H.; Möhlemeier, J.; Schubert, U.; Orama, O. *J. Organomet. Chem.* **1983**, *247*, 187–201.
5. Downie, I. M.; Heaney, H.; Kemp, G.; King, D.; Wosley, M. *Tetrahedron* **1992**, *48*, 4005–4016.
6. Kiessling, A. J.; McClure, C. K. *Synth. Commun.* **1997**, *27*, 923–937.
7. Pichlmair, S. *Synlett* **2004**, 195–196. (Review).
8. Egashira, M.; Yamamoto, Y.; Fukutake, T.; Yoshimoto, N.; Morita, M. *J. Fluorine Chem.* **2006**, *127*, 1261–1264.
9. Delest, B.; Nshimyumukiza, P.; Fasbender, O.; Tinant, B.; Marchand-Brynaert, J.; Darro, F.; Robiette, R. *J. Org. Chem.* **2008**, *73*, 6816–6823.
10. Perst, H.; Seapy, D. G. *Triethyloxonium Tetrafluoroborate* In *Encyclopedia of Reagents for Organic Synthesis* Wiley: New York, **2008**,
11. Hari, D. P., König, B. *Angew. Chem. Int. Ed.* **2013**, *52*, 4734–4743. (Review).

# Meerwein–Ponndorf–Verley reduction

Reduction of ketones to the corresponding alcohols using Al(O$i$-Pr)$_3$ in iso-propanol. Reverse of the Oppernauer oxidation.

cyclic transition state

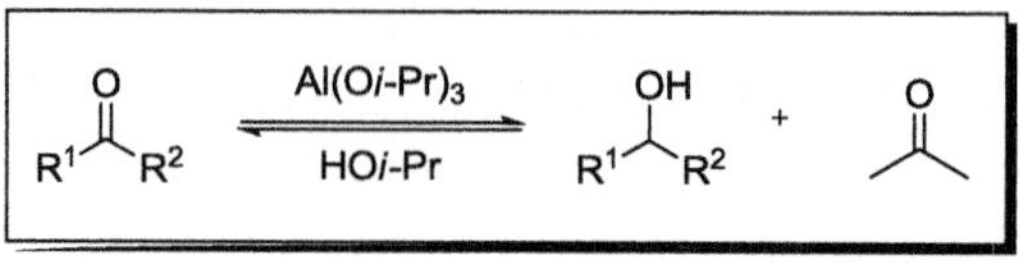

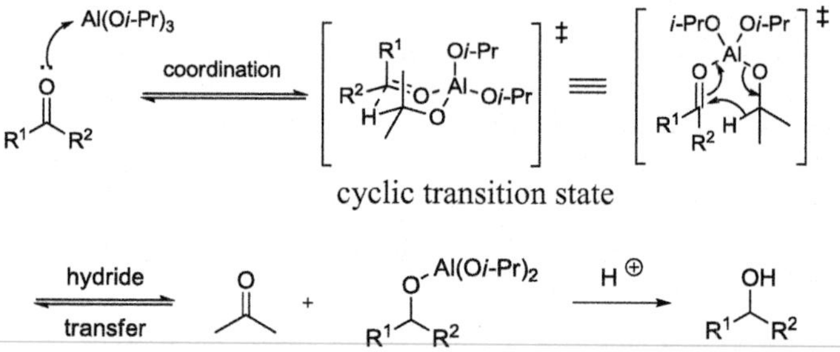

Example 1[2]

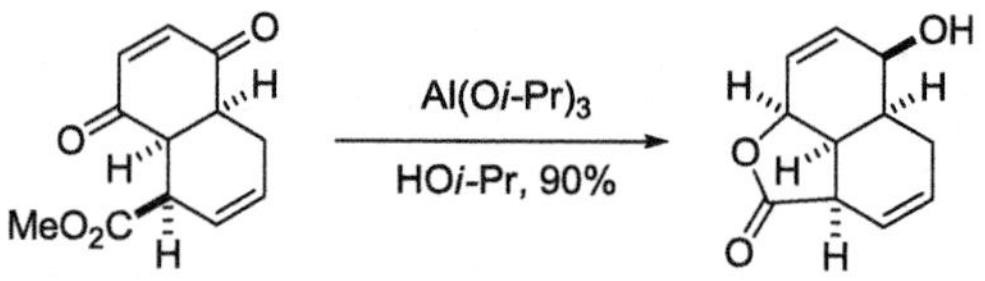

Example 2[4]

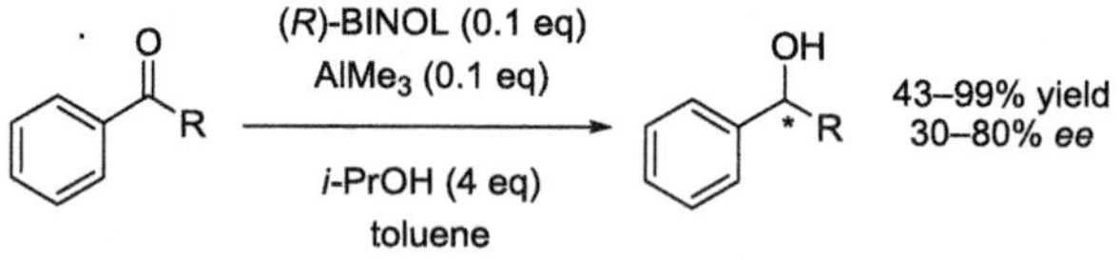

43–99% yield
30–80% *ee*

Example 3[7]

84%

15%

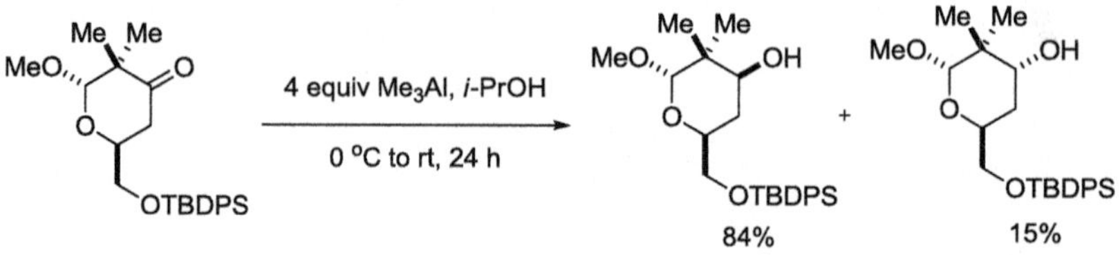

J.J. Li, *Name Reactions: A Collection of Detailed Mechanisms and Synthetic Applications*,
DOI 10.1007/978-3-319-03979-4_167, © Springer International Publishing Switzerland 2014

Example 4[9]

Example 5[10]

## References

1. Meerwein, H.; Schmidt, R. *Ann.* **1925**, *444*, 221–238. Hans L. Meerwein, born in Hamburg Germany in 1879, received his Ph.D. at Bonn in 1903. In his long and productive academic career, Meerwein made many notable contributions in organic chemistry.

2. Woodward, R. B.; Bader, F. E.; Bickel, H.; Frey, A. J.; Kierstead, R. W. *Tetrahedron* **1958**, *2*, 1–57.

3. de Graauw, C. F.; Peters, J. A.; van Bekkum, H.; Huskens, J. *Synthesis* **1994**, 1007–1017. (Review).

4. Campbell, E. J.; Zhou, H.; Nguyen, S. T. *Angew. Chem. Int. Ed.* **2002**, *41*, 1020–1022.

5. Sominsky, L.; Rozental, E.; Gottlieb, H.; Gedanken, A.; Hoz, S. *J. Org. Chem.* **2004**, *69*, 1492–1496.

6. Cha, J. S. *Org. Proc. Res. Dev.* **2006**, *10*, 1032–1053.

7. Manaviazar, S.; Frigerio, M.; Bhatia, G. S.; Hummersone, M. G.; Aliev, A. E.; Hale, K. J. *Org. Lett.* **2006**, *8*, 4477–4480.

8. Clay, J. M. *Meerwein–Ponndorf–Verley reduction.* In *Name Reactions for Functional Group Transformations*; Li, J. J., Ed.; Wiley: Hoboken, NJ, **2007**, pp 123–128. (Review).

9. Dilger, A. K.; Gopalsamuthiram, V.; Burke, S. D. *J. Am. Chem. Soc.* **2007**, *129*, 16273–16277.

10. Flack, K.; Kitagawa, K.; Pollet, P.; Eckert, C. A.; Richman, K.; Stringer, J.; Dubay, W.; Liotta, C. L. *Org. Process Res. Dev.* **2012**, *16*, 1301–1306.

11. Lenze, M.; Bauer, E. B. *Chem. Commun.* **2013**, *49*, 5889–5891.

## Meisenheimer complex

Also known as **the Meisenheimer–Jackson salt**, the stable intermediate for certain $S_NAr$ reactions.

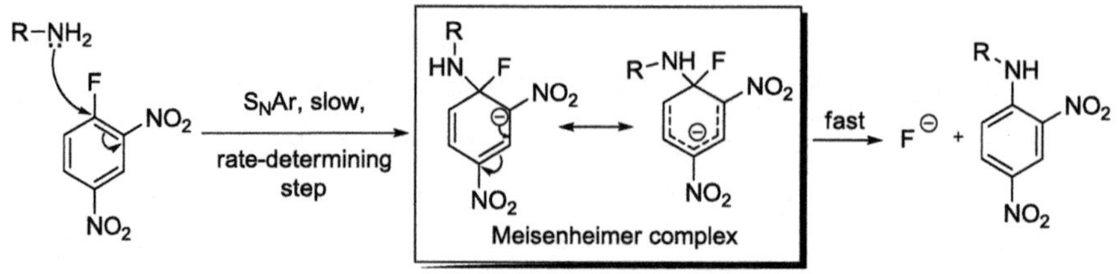

Sanger's reagent, *ipso* attack          *ipso* substitution

Example 1[7]

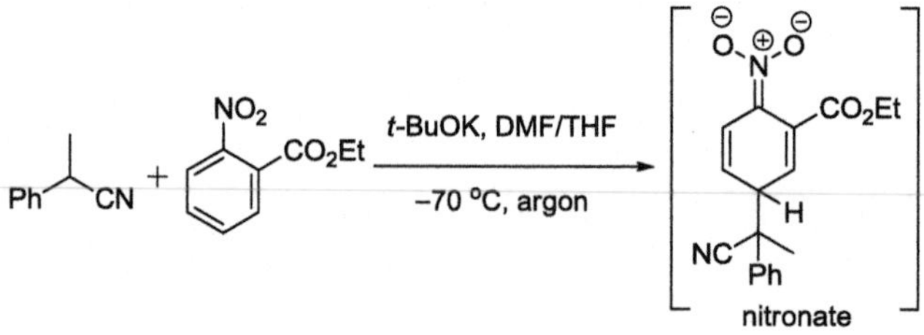

Example 2[9]

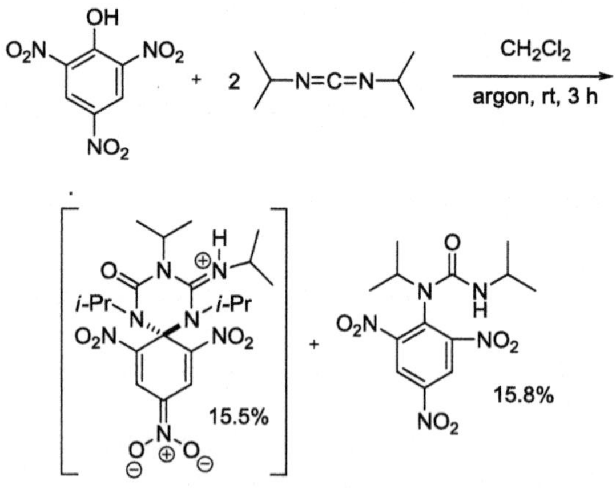

The reaction using Sanger's reagent is faster than using the corresponding chloro-, bromo-, and iododinitrobenzene—the fluoro-Meisenheimer complex is the most stabilized because F is the most electron-withdrawing. The reaction rate does not depend upon the capacity of the leaving group.

J.J. Li, *Name Reactions: A Collection of Detailed Mechanisms and Synthetic Applications*, DOI 10.1007/978-3-319-03979-4_168, © Springer International Publishing Switzerland 2014

Example 3[10]

## References

1.  Meisenheimer, J. *Ann.* **1902,** *323*, 205–214.
2.  Strauss, M. J. *Acc. Chem. Res.* **1974,** *7*, 181–188. (Review).
3.  Bernasconi, C. F. *Acc. Chem. Res.* **1978,** *11*, 147–152. (Review).
4.  Terrier, F. *Chem. Rev.* **1982,** *82*, 77–152. (Review).
5.  Manderville, R. A.; Buncel, E. *J. Org. Chem.* **1997,** *62*, 7614–7620.
6.  Hoshino, K.; Ozawa, N.; Kokado, H.; Seki, H.; Tokunaga, T.; Ishikawa, T. *J. Org. Chem.* **1999,** *64*, 4572–4573.
7.  Adam, W.; Makosza, M.; Zhao, C.-G.; Surowiec, M. *J. Org. Chem.* **2000,** *65*, 1099–1101.
8.  Gallardo, I.; Guirado, G.; Marquet, J. *J. Org. Chem.* **2002,** *67*, 2548–2555.
9.  Al-Kaysi, R. O.; Guirado, G.; Valente, E. J. *Eur. J. Org. Chem.* **2004,** 3408–3411.
10. Um, I.-H.; Min, S.-W.; Dust, J. M. *J. Org. Chem.* **2007,** *72*, 8797–8803.
11. Han, T. Y.-J.; Pagoria, P. F.; Gash, A. E.; Maiti, A.; Orme, C. A.; Mitchell, A. R.; Fried, L. E. *New J. Chem.* **2009,** *33*, 50–56.
12. Campodónico, P. R.; Tapia, R. A.; Contreras, R.; Ormazábal-Toledo, R. *Org. Biomol. Chem.* **2013,** *11*, 2302–2309.

# [1,2]-Meisenheimer rearrangement

[1,2]-Sigmatropic rearrangement of tertiary amine *N*-oxides to substituted hydroxylamines.

Example 1[7]

Example 2[9]

## References

1. Meisenheimer, J. *Ber.* **1919**, *52*, 1667–1677.
2. Castagnoli, N., Jr.; Craig, J. C.; Melikian, A. P.; Roy, S. K. *Tetrahedron* **1970**, *26*, 4319–4327.
3. Johnstone, R. A. W. *Mech. Mol. Migr.* **1969**, *2*, 249–266. (Review).
4. Kurihara, T.; Sakamoto, Y.; Tsukamoto, K.; Ohishi, H.; Harusawa, S.; Yoneda, R. *J. Chem. Soc., Perkin Trans. 1*, **1993**, 81–87.
5. Yoneda, R.; Sakamoto, Y.; Oketo, Y.; Minami, K.; Harusawa, S.; Kurihara, T. *Tetrahedron Lett.* **1994**, *35*, 3749–3752.
6. Kurihara, T.; Sakamoto, Y.; Takai, M.; Ohishi, H.; Harusawa, S.; Yoneda, R. *Chem. Pharm. Bull.* **1995**, *43*, 1089–1095.
7. Yoneda, R.; Sakamoto, Y.; Oketo, Y.; Harusawa, S.; Kurihara, T. *Tetrahedron* **1996**, *52*, 14563–14576.
8. Yoneda, R.; Araki, L.; Harusawa, S.; Kurihara, T. *Chem. Pharm. Bull.* **1998**, *46*, 853–856.
9. Menguy, L.; Drouillat, B.; Marrot, J.; Couty, F. *Tetrahedron Lett.* **2012**, *53*, 4697–4699.

J.J. Li, *Name Reactions: A Collection of Detailed Mechanisms and Synthetic Applications*, DOI 10.1007/978-3-319-03979-4_169, © Springer International Publishing Switzerland 2014

# [2,3]-Meisenheimer rearrangement

[2,3]-Sigmatropic rearrangement of allylic tertiary amine-*N*-oxides to give *O*-allyl hydroxylamines:

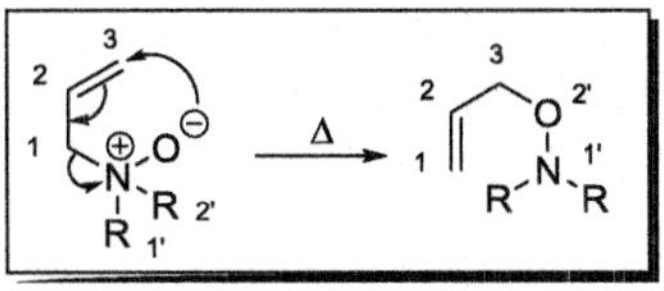

Example 1[7]

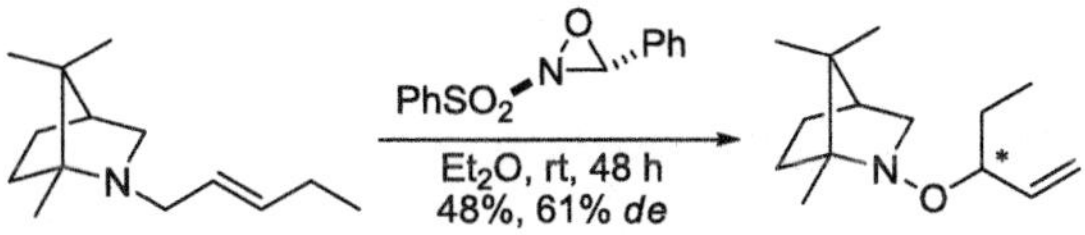

Example 2[8]

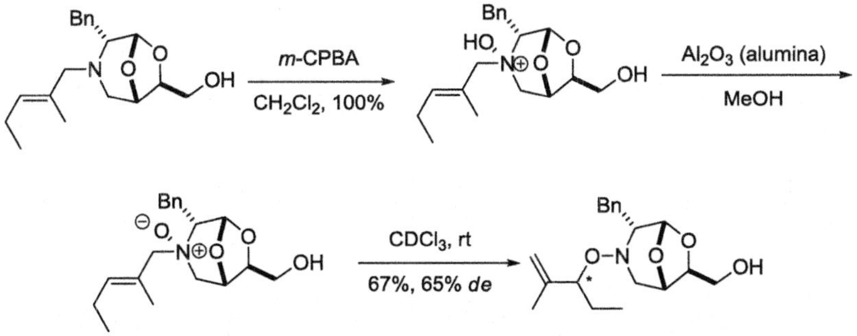

Example 3[8]

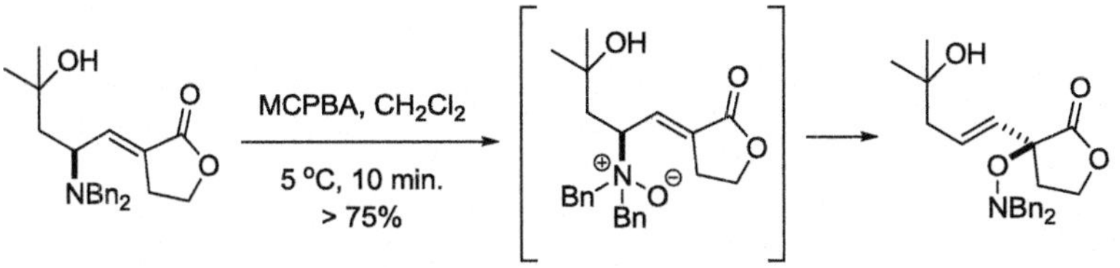

## References

1. Meisenheimer, J. *Ber.* **1919**, *52*, 1667–1677.
2. Yamamoto, Y.; Oda, J.; Inouye, Y. *J. Org. Chem.* **1976**, *41*, 303–306.
3. Johnstone, R. A. W. *Mech. Mol. Migr.* **1969**, *2*, 249–266. (Review).
4. Kurihara, T.; Sakamoto, Y.; Matsumoto, H.; Kawabata, N.; Harusawa, S.; Yoneda, R. *Chem. Pharm. Bull.* **1994**, *42*, 475–480.

J.J. Li, *Name Reactions: A Collection of Detailed Mechanisms and Synthetic Applications*,
DOI 10.1007/978-3-319-03979-4_170, © Springer International Publishing Switzerland 2014

5.  Blanchet, J.; Bonin, M.; Micouin, L.; Husson, H.-P. *Tetrahedron Lett.* **2000,** *41*, 8279–8283.
6.  Enders, D.; Kempen, H. *Synlett* **1994,** 969–971.
7.  Buston, J. E. H.; Coldham, I.; Mulholland, K. R. *Synlett* **1997,** 322–324.
8.  Guarna, A.; Occhiato, E. G.; Pizzetti, M.; Scarpi, D.; Sisi, S.; van Sterkenburg, M. *Tetrahedron: Asymmetry* **2000,** *11*, 4227–4238.
9.  Mucsi, Z.; Szabó, A.; Hermecz, I.; Kucsman, Á.; Csizmadia, I. G. *J. Am. Chem. Soc.* **2005,** *127*, 7615–7621.
10. Bourgeois, J.; Dion, I.; Cebrowski, P. H.; Loiseau, F.; Bedard, A.-C.; Beauchemin, A. M. *J. Am. Chem. Soc.* **2009,** *131*, 874–875.
11. Yang, H.; Sun, M.; Zhao, S.; Zhu, M.; Xie, Y.; Niu, C.; Li, C. *J. Org. Chem.* **2013,** *78*, 339–346.

# Meyers oxazoline method

Chiral oxazolines employed as activating groups and/or chiral auxiliaries in nucleophilic addition and substitution reactions that lead to the asymmetric construction of carbon–carbon bonds.

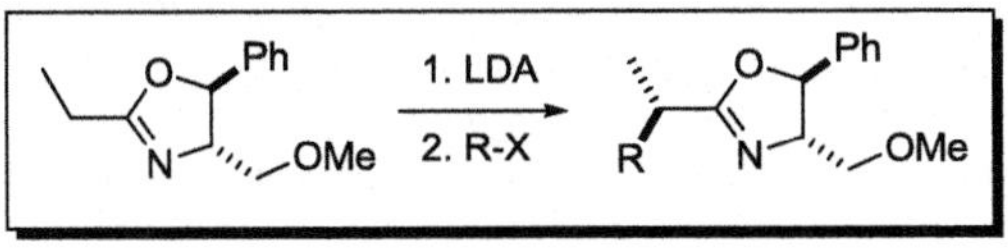

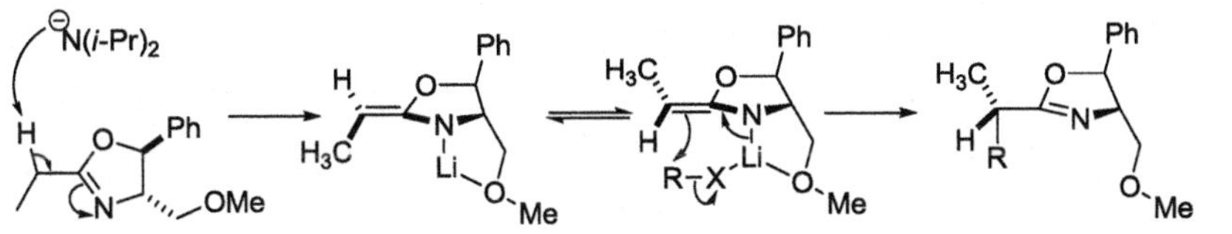

Example 1[2]

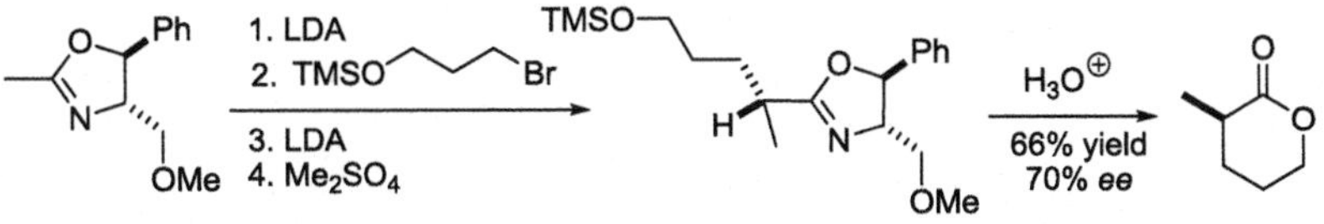

Example 2[5]

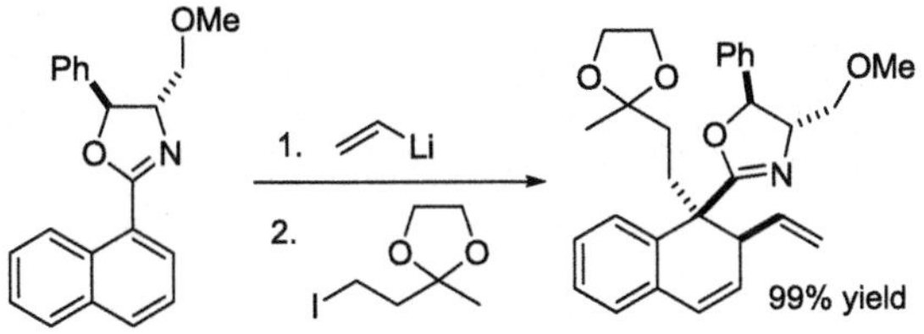

Example 3[9]

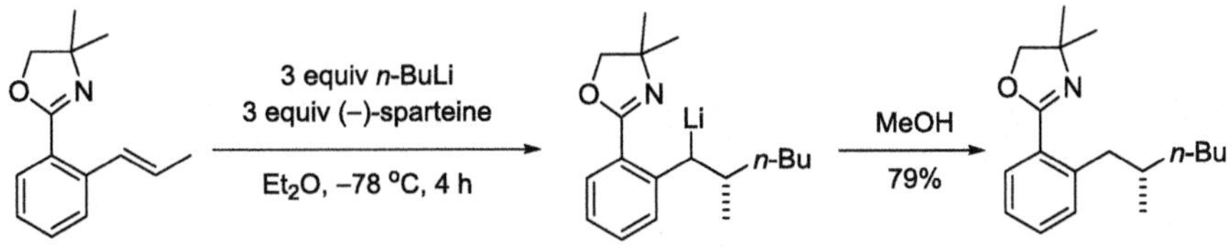

er (S:R) = 86:14

J.J. Li, *Name Reactions: A Collection of Detailed Mechanisms and Synthetic Applications*, DOI 10.1007/978-3-319-03979-4_171, © Springer International Publishing Switzerland 2014

## References

1.  (a) Meyers, A. I.; Knaus, G.; Kamata, K. *J. Am. Chem. Soc.* **1974**, *96*, 268–270. While Albert I. Meyers was an assistant professor at Wayne State University, neighboring pharmaceutical firm Parke–Davis (Drs. George Moersch and Harry Crooks) donated several kilograms of (1*S*,2*S*)-(+)-2-amino-1-phenyl-1,3-propanediol (Meyers referred to it as the Parke–Davis diol), from which his chemistry with chiral oxazolines began. He taught at Colorado State University since 1972. Meyers passed away in 2007. (b) Meyers, A. I.; Knaus, G. *J. Am. Chem. Soc.* **1974,** *96*, 6508–6510. (c) Meyers, A. I.; Knaus, G. *Tetrahedron Lett.* **1974,** *15,* 1333–1336. (d) Meyers, A. I.; Whitten, C. E. *J. Am. Chem. Soc.* **1975,** *97*, 6266–6267. (e) Meyers, A. I.; Mihelich, E. D. *J. Org. Chem.* **1975,** *40,* 1186–1187. (f) Meyers, A. I.; Mihelich, E. D. *Angew. Chem. Int. Ed.* **1976,** *15*, 270–271. (Review). (g) Meyers, A. I. *Acc. Chem. Res.* **1978,** *11*, 375–381. (Review).

2.  Meyers, A. I.; Yamamoto, Y.; Mihelich, E. D.; Bell, R. A. *J. Org. Chem.* **1980,** *45*, 2792–2796.

3.  Meyers, A. I., Lutomski, K. A. In *Asymmetric Synthesis*, Morrison, J. D. Ed.; Vol III, Part B, Chapter 3, Academic Press, **1983**. (Review).

4.  Reuman, M.; Meyers, A. I. *Tetrahedron* **1985,** *41*, 837–860. (Review).

5.  Robichaud, A. J.; Meyers, A. I. *J. Org. Chem.* **1991,** *56*, 2607–2609.

6.  Gant, T. G.; Meyers, A. I. *Tetrahedron* **1994,** 50, 2297–2360. (Review).

7.  Meyers, A. I. *J. Heterocycl. Chem.* **1998,** 35, 991–1002. (Review).

8.  Wolfe, J. P. *Meyers Oxazoline Method.* In *Name Reactions in Heterocycl. Chemistry*; Li, J. J., Ed.; Wiley: Hoboken, NJ, **2005,** pp 237–248. (Review).

9.  Hogan, A.-M. L.; Tricotet, T.; Meek, A.; Khokhar, S. S.; O'Shea, D. F. *J. Org. Chem.* **2008,** *73*, 6041–6044.

# Meyer–Schuster rearrangement

The isomerization of secondary and tertiary α-acetylenic alcohols to α,β-unsaturated carbonyl compounds *via* 1,3-shift. When the acetylenic group is terminal, the products are aldehydes, whereas the internal acetylenes give ketones. *Cf.* Rupe rearrangement.

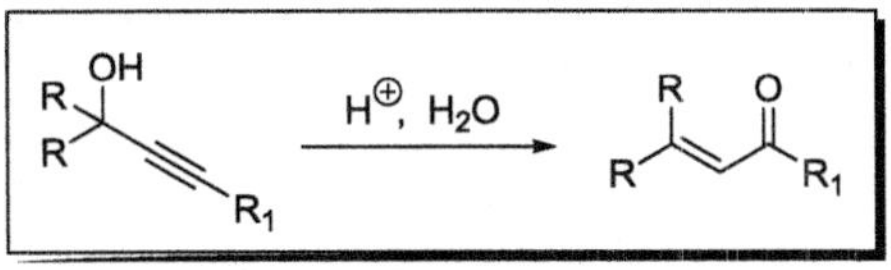

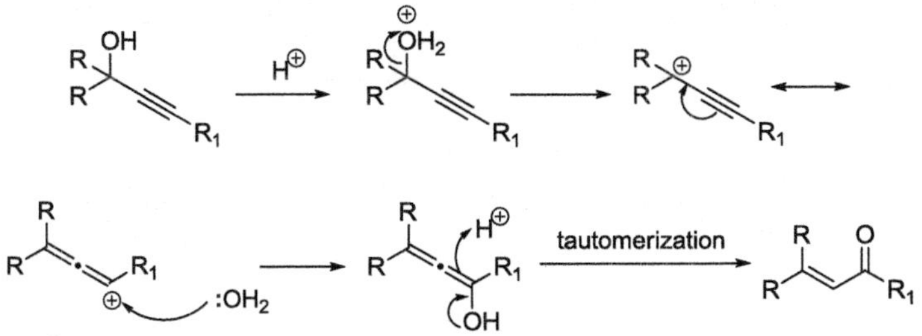

Example 1[6]

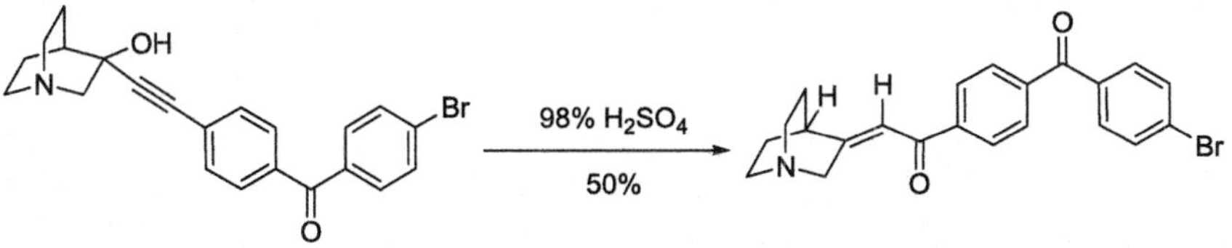

Example 2[7]

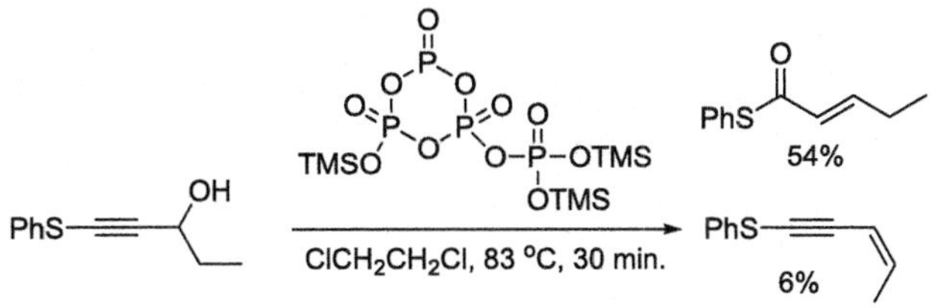

Example 3[8]

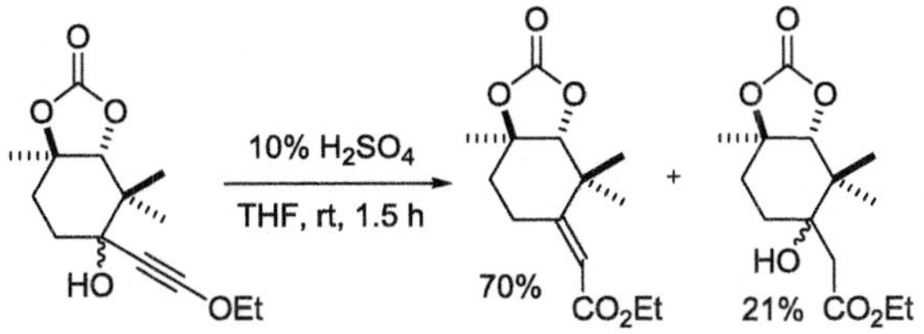

J.J. Li, *Name Reactions: A Collection of Detailed Mechanisms and Synthetic Applications*, DOI 10.1007/978-3-319-03979-4_172, © Springer International Publishing Switzerland 2014

Example 4[9]

Example 5[11]

## References

1.  Meyer, K. H.; Schuster, K. *Ber.* **1922**, *55*, 819–823.
2.  Swaminathan, S.; Narayanan, K. V. *Chem. Rev.* **1971**, *71*, 429–438. (Review).
3.  Edens, M.; Boerner, D.; Chase, C. R.; Nass, D.; Schiavelli, M. D. *J. Org. Chem.* **1977**, *42*, 3403–3408.
4.  Andres, J.; Cardenas, R.; Silla, E.; Tapia, O. *J. Am. Chem. Soc.* **1988**, *110*, 666–674.
5.  Tapia, O.; Lluch, J. M.; Cardenas, R.; Andres, J. *J. Am. Chem. Soc.* **1989**, *111*, 829–835.
6.  Brown, G. R.; Hollinshead, D. M.; Stokes, E. S.; Clarke, D. S.; Eakin, M. A.; Foubister, A. J.; Glossop, S. C.; Griffiths, D.; Johnson, M. C.; McTaggart, F.; Mirrlees, D. J.; Smith, G. J.; Wood, R. *J. Med. Chem.* **1999**, *42*, 1306–1311.
7.  Yoshimatsu, M.; Naito, M.; Kawahigashi, M.; Shimizu, H.; Kataoka, T. *J. Org. Chem.* **1995**, *60*, 4798–4802.
8.  Crich, D.; Natarajan, S.; Crich, J. Z. *Tetrahedron* **1997**, *53*, 7139–7158.
9.  Williams, C. M.; Heim, R.; Bernhardt, P. V. *Tetrahedron* **2005**, *61*, 3771–3779.
10. Mullins, R. J.; Collins, N. R. *Meyer–Schuster Rearrangement*. In *Name Reactions for Homologations-Part II*; Li, J. J., Ed.; Wiley: Hoboken, NJ, **2009**, pp 305–318. (Review).
11. Collins, B. S. L.; Suero, M. G.; Gaunt, M. J. *Angew. Chem.* **2013**, *125*, 5911–5914.

# Michael addition

Also known as conjugate addition, Michael addition is the 1,4-addition of a nucleophile to an α,β-unsaturated system.

Example 1, Asymmetric Michael addition[2]

Example 2, Thia-Michael addition[3]

Example 3, Phospha-Michael addition[7]

Example 4, Asymmetric aza-Michael addition[9]

J.J. Li, *Name Reactions: A Collection of Detailed Mechanisms and Synthetic Applications*, DOI 10.1007/978-3-319-03979-4_173, © Springer International Publishing Switzerland 2014

Example 5, Intramolecular Michael addition[10]

Example 6, Intramolecular Michael addition[11]

## References

1. Michael, A. *J. Prakt. Chem.* **1887,** *35,* 349. Arthur Michael (1853–1942) was born in Buffalo, New York. He studied under Robert Bunsen, August Hofmann, Adolphe Wurtz, and Dimitri Mendeleev, but never bothered to take a degree. Back to the United States, Michael became a Professor of Chemistry at Tufts University, where he married one of his students, Helen Abbott, one of the few female organic chemists in this period. Since he failed miserably as an administrator, Michael and his wife set up their own private laboratory at Newton Center, Massachusetts, where the 1,4-addition was discovered.
2. Hunt, D. A. *Org. Prep. Proced. Int.* **1989,** *21,* 705–749.
3. D'Angelo, J.; Desmaële, D.; Dumas, F.; Guingant, A. *Tetrahedron: Asymmetry* **1992,** *3,* 459–505.
4. Lipshutz, B. H.; Sengupta, S. *Org. React.* **1992,** *41,* 135–631. (Review).
5. Hoz, S. *Acc. Chem. Res.* **1993,** *26,* 69–73. (Review).
6. Ihara, M.; Fukumoto, K. *Angew. Chem. Int. Ed.* **1993,** *32,* 1010–1022. (Review).
7. Simoni, D.; Invidiata, F. P.; Manferdini, M.; Lampronti, I.; Rondanin, R.; Roberti, M.; Pollini, G. P. *Tetrahedron Lett.* **1998,** *39,* 7615–7618.
8. Enders, D.; Saint-Dizier, A.; Lannou, M.-I.; Lenzen, A. *Eur. J. Org. Chem.* **2006,** 29–49. (Review on the phospha-Michael addition).
9. Chen, L.-J.; Hou, D.-R. *Tetrahedron: Asymmetry* **2008,** *19,* 715–720.
10. Sakaguchi, H.; Tokuyama, H.; Fukuyama, T. *Org. Lett.* **2008,** *10,* 1711–1714.
11. Kwan, E. E.; Scheerer, J. R.; Evans, D. A. *J. Org. Chem.* **2013,** *78,* 175–203.

# Michaelis–Arbuzov phosphonate synthesis

Aliphatic phosphonate synthesis from the reaction of alkyl halides with phosphites.

General scheme:

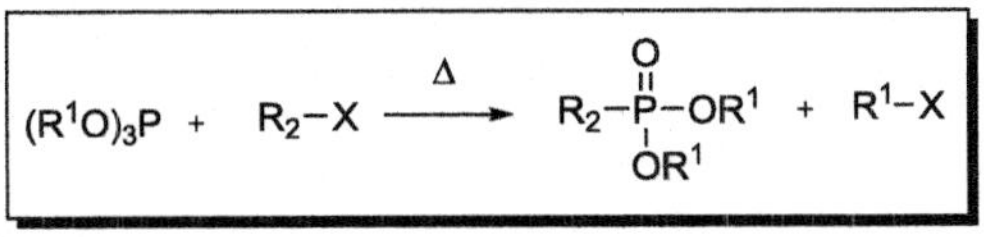

$$(R^1O)_3P \ + \ R_2-X \ \xrightarrow{\Delta} \ R_2-\overset{\overset{\displaystyle O}{\|}}{\underset{\underset{\displaystyle OR^1}{|}}{P}}-OR^1 \ + \ R^1-X$$

$R^1$ = alkyl, *etc.*;  $R_2$ = alkyl, acyl, *etc.*;  X = Cl, Br, I

For instance:

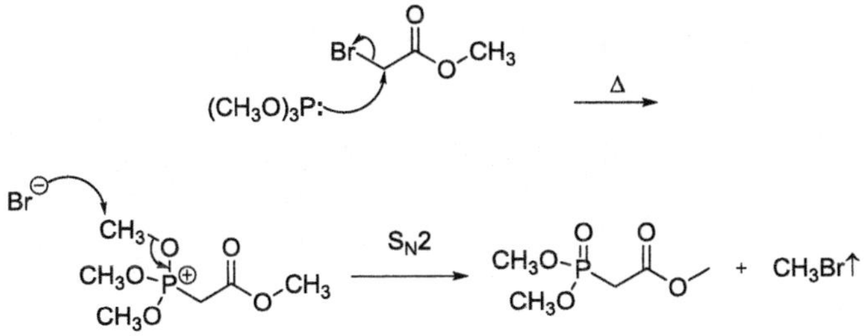

Example 1[2]

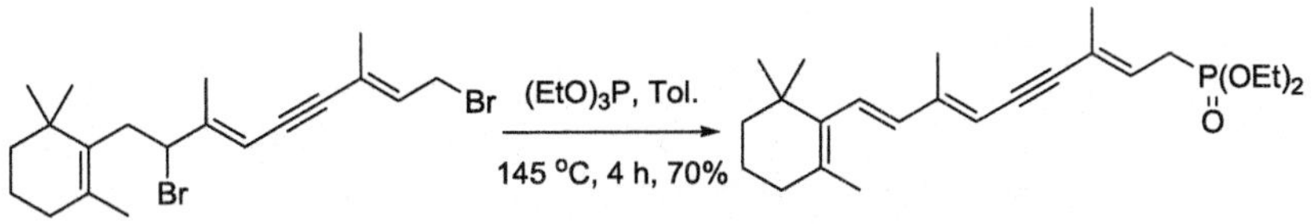

Example 2[6]

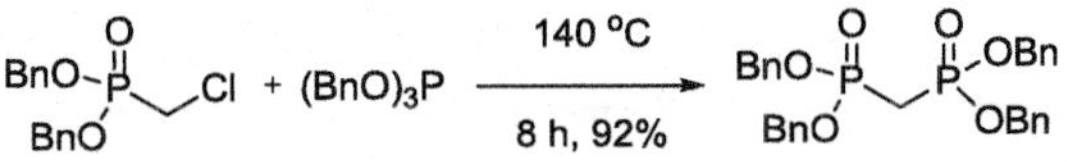

Example 3, Transition-metal catalyzed coupling, not via $S_N2$[7]

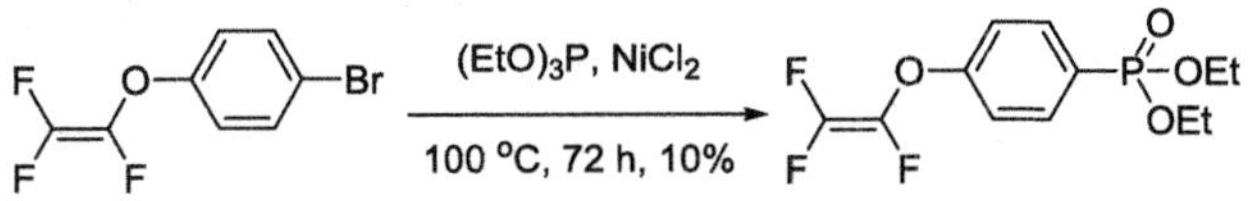

J.J. Li, *Name Reactions: A Collection of Detailed Mechanisms and Synthetic Applications*,
DOI 10.1007/978-3-319-03979-4_174, © Springer International Publishing Switzerland 2014

Example 4[9]

Example 5[10]

Example 6, An Approach to prepare aromatic phosphonates[11]

## References

1.  (a) Michaelis, A.; Kaehne, R. *Ber.* **1898**, *31,* 1048–1055. (b) Arbuzov, A. E. *J. Russ. Phys. Chem. Soc.* **1906**, *38*, 687.
2.  Surmatis, J. D.; Thommen, R. *J. Org. Chem.* **1969**, *34*, 559–560.
3.  Gillespie, P.; Ramirez, F.; Ugi, I.; Marquarding, D. *Angew. Chem. Int. Ed.* **1973**, *12*, 91–119. (Review).
4.  Waschbüsch, R.; Carran, J.; Marinetti, A.; Savignac, P. *Synthesis* **1997**, 727–743.
5.  Bhattacharya, A. K.; Stolz, F.; Schmidt, R. R. *Tetrahedron Lett.* **2001**, *42*, 5393–5395.
6.  Erker, T.; Handler, N. *Synthesis* **2004**, 668–670.
7.  Souzy, R.; Ameduri, B.; Boutevin, B.; Virieux, D. *J. Fluorine Chem.* **2004**, *125*, 1317–1324.
8.  Kadyrov, A. A.; Sîlaev, D. V.; Makarov, K. N.; Gervits, L. L.; Röschenthaler, G.-V. *J. Fluorine Chem.* **2004**, *125*, 1407–1410.
9.  Ordonez, M.; Hernandez-Fernandez, E.; Montiel-Perez, M.; Bautista, R.; Bustos, P.; Rojas-Cabrera, H.; Fernandez-Zertuche, M.; Garcia-Barradas, O. *Tetrahedron: Asymmetry* **2007**, *18*, 2427–2436.
10. Piekutowska, M.; Pakulski, Z. *Carbohydrate Res.* **2008**, *343*, 785–792.
11. Dhokale, R. A.; Mhaske, S. B. *Org. Lett.* **2013**, *15*, 2218–2221.

# Midland reduction

Asymmetric reduction of ketones using Alpine-borane®.
Alpine-borane® = B-isopinocampheyl-9-borabicyclo[3.3.1]nonane.

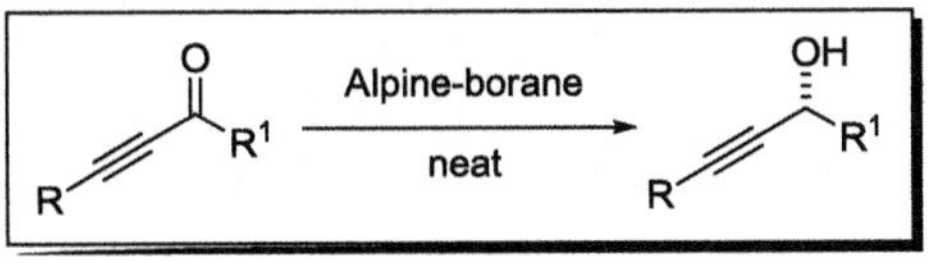

Preparation:

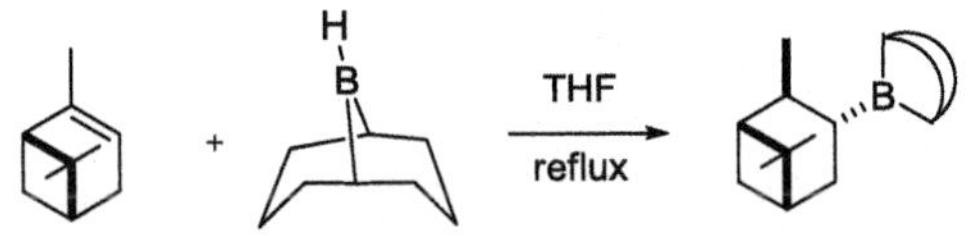

(1R)-(+)-α-pinene      9-BBN                    (R)-Alpine-borane
9-BBN = 9-borabicyclo[3.3.1]nonane

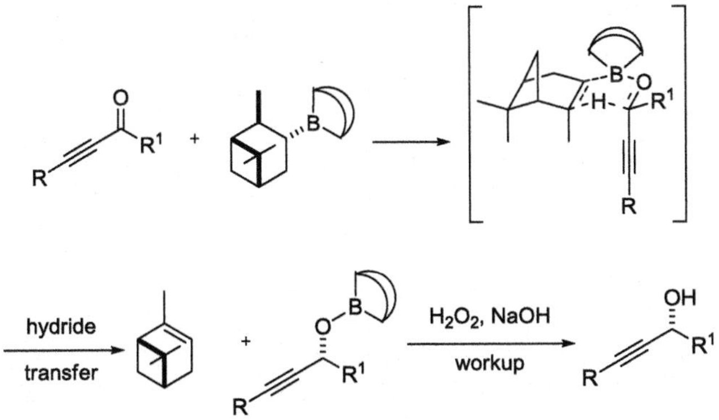

Example 1[6]

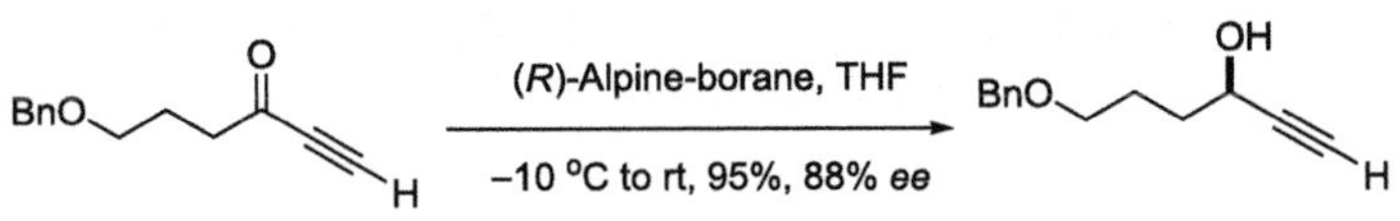

Example 2[7]

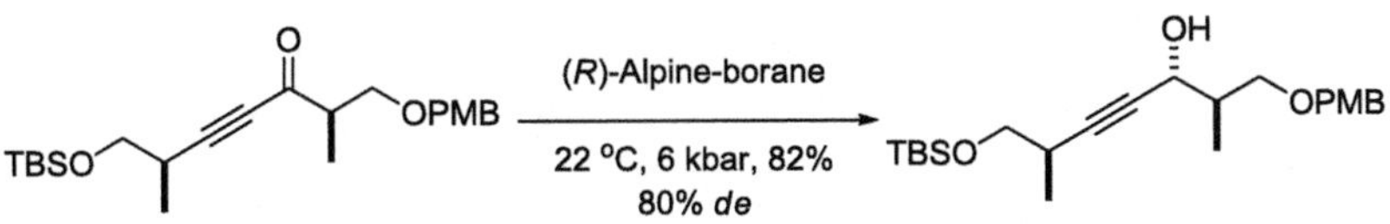

J.J. Li, *Name Reactions: A Collection of Detailed Mechanisms and Synthetic Applications*,
DOI 10.1007/978-3-319-03979-4_175, © Springer International Publishing Switzerland 2014

Example 3[8]

Example 4[10]

## References

1. Midland, M. M.; Greer, S.; Tramontano, A.; Zderic, S. A. *J. Am. Chem. Soc.* **1979,** *101*, 2352–2355. M. Mark Midland was a professor at the University of California, Riverside.
2. Midland, M. M.; McDowell, D. C.; Hatch, R. L.; Tramontano, A. *J. Am. Chem. Soc.* **1980,** *102*, 867–869.
3. Brown, H. C.; Pai, G. G. *J. Org. Chem.* **1982,** *47*, 1606–1608.
4. Brown, H. C.; Pai, G. G.; Jadhav, P. K. *J. Am. Chem. Soc.* **1984,** *106*, 1531–1533.
5. Singh, V. K. *Synthesis* **1992,** 605–617. (Review).
6. Williams, D. R.; Fromhold, M. G.; Earley, J. D. *Org. Lett.* **2001,** *3*, 2721–2724.
7. Mulzer, J.; Berger, M. *J. Org. Chem.* **2004,** *69*, 891–898.
8. Kiewel, K.; Luo, Z.; Sulikowski, G. A. *Org. Lett.* **2005,** *7*, 5163–5165.
9. Clay, J. M. *Midland reduction.* In *Name Reactions for Functional Group Transformations*; Li, J. J., Ed.; Wiley: Hoboken, NJ, **2007,** pp 40–45. (Review).
10. Ramesh, D.; Shekhar, V.; Chantibabu, D.; Rajaram, S.; Ramulu, U.; Venkateswarlu, Y. *Tetrahedron Lett.* **2012,** *53*, 1258–1260.

# Minisci reaction

Radical-based carbon–carbon bond formation with electron-deficient heteroaromatics. The reaction entails an intermolecular addition of a nucleophilic *radical* to protonated heteroaromatic nucleus.

$$R-CO_2H + \text{(pyridine)} \xrightarrow[\text{H}_2\text{SO}_4]{\substack{2\ AgNO_3 \\ (NH_4)_2S_2O_8}} \text{(2-R-pyridine)}$$

$$R-CO_2H \xrightarrow[\substack{\text{silver-promoted} \\ \text{oxidative decarboxylation}}]{2\ AgNO_3,\ (NH_4)_2S_2O_8,\ H_2SO_4} CO_2 + R\bullet$$

Example 1[4]

$$S_2O_8^= + CH_3OH \longrightarrow \bullet CH_2OH + H^+ + SO_4^= + SO_4^{\bullet-}$$

$$\xrightarrow[\substack{\text{reflux, 1 h, 40\%}}]{\substack{(NH_4)_2S_2O_8 \\ MeOH,\ H_2O}}$$

Example 2[5]

$$\xrightarrow[\substack{\text{acetone, rt, 1.5 h} \\ 75\%}]{1.6\ \text{equiv}\ m\text{-CPBA}} \xrightarrow[\substack{CH_2Cl_2,\ rt \\ 90\ min.}]{(CH_3)_3O\bullet BF_4}$$

Meerwein's methylating reagent

$$\xrightarrow[\substack{\text{reflux, 1 h} \\ 40\%,\ 2\ steps}]{\substack{(NH_4)_2S_2O_8 \\ MeOH,\ H_2O}}$$

$$\xrightarrow[\substack{\text{acetone, rt, 1.5 h} \\ 78\%}]{1.6\ \text{equiv}\ m\text{-CPBA}}$$

$$\xrightarrow[\substack{CH_2Cl_2,\ rt \\ 90\ min.,\ 85\%}]{(CH_3)_3O\bullet BF_4} \xrightarrow[\substack{\text{reflux, 1 h, 41\%}}]{\substack{(NH_4)_2S_2O_8 \\ MeOH,\ H_2O}}$$

J.J. Li, *Name Reactions: A Collection of Detailed Mechanisms and Synthetic Applications*, DOI 10.1007/978-3-319-03979-4_176, © Springer International Publishing Switzerland 2014

Example 3, Intramolecular Minisci reaction[6]

Example 4[7]

Example 5[10]

Example 6[12]

## References

1. Minisci, F, Bernardi. R, Bertini, F, Galli, R, Perchinummo, M. *Tetrahedron* **1971**, *27*, 3575–3579.
2. Minisci, F. *Synthesis* **1973**, 1–24. (Review).
3. Minisci, F. *Acc. Chem. Res.* **1983**, *16*, 27–32. (Review).
4. Katz, R. B.; Mistry, J.; Mitchell, M. B. *Synth. Commun.* **1989**, *19*, 317–325.
5. Biyouki, M. A. A.; Smith, R. A. J.; Bedford, J. J.; Leader, J. P. *Synth. Commun.* **1998**, *28*, 3817–3825.
6. Doll, M. K. H. *J. Org. Chem.* **1999**, *64*, 1372–1374.
7. Cowden, C. J. *Org. Lett.* **2003**, *5*, 4497–4499.
8. Kast, O.; Bracher, F. *Synth. Commun.* **2003**, *33*, 3843–3850.
9. Benaglia, M.; Puglisi, A.; Holczknecht, O.; Quici, S.; Pozzi, G. *Tetrahedron* **2005**, *61*, 12058–12064.
10. Palde, P. B.; McNaughton, B. R.; Ross, N. T.; Gareiss, P. C.; Mace, C. R.; Spitale, R. C.; Miller, B. L. *Synthesis* **2007**, 2287–2290.
11. Brebion, F.; Nàjera, F.; Delouvrié, B.; Lacôte, E.; Fensterbank, L.; Malacria, M. *J. Heterocycl. Chem.* **2008**, *45*, 527–532.
12. Presset, M.; Fleury-Brégeot, N.; Oehlrich, D.; Rombouts, F.; Molander, G. A. *J. Org. Chem.* **2013**, *78*, 4615–4619.

# Mislow–Evans rearrangement

[2,3]-Sigmatropic rearrangement followed by reduction of allylic sulfoxide to allylic alcohol.

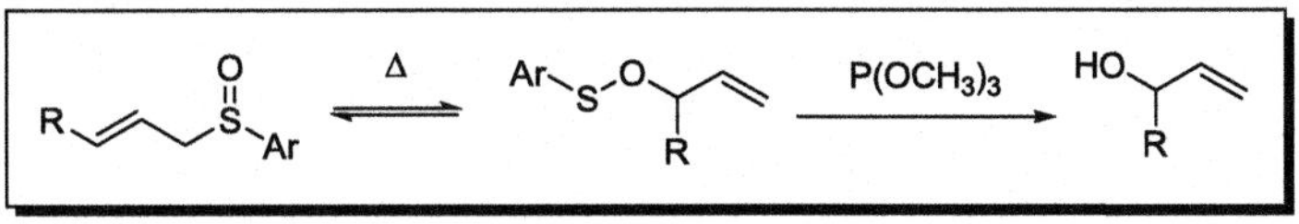

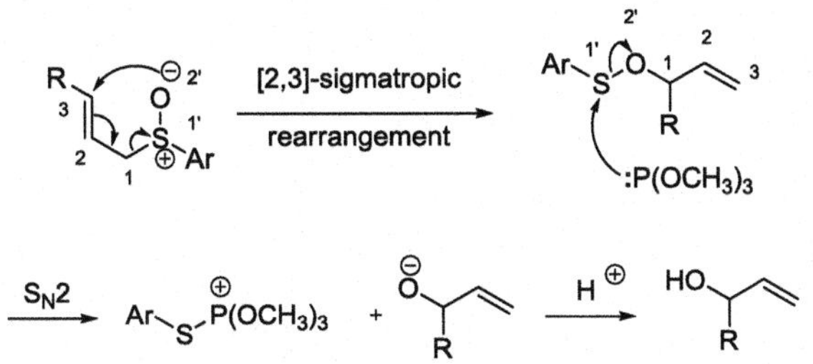

Example 1[2]

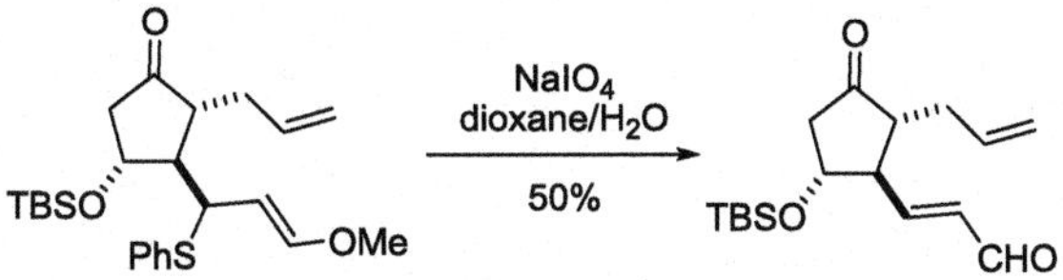

Example 2[7]

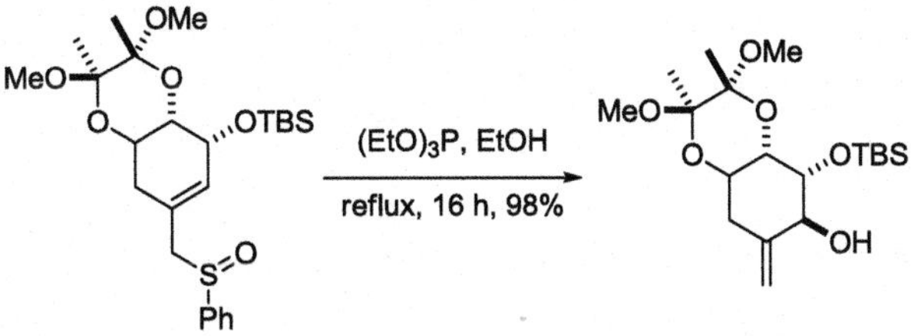

Example 3, Seleno-Mislow–Evans[8]

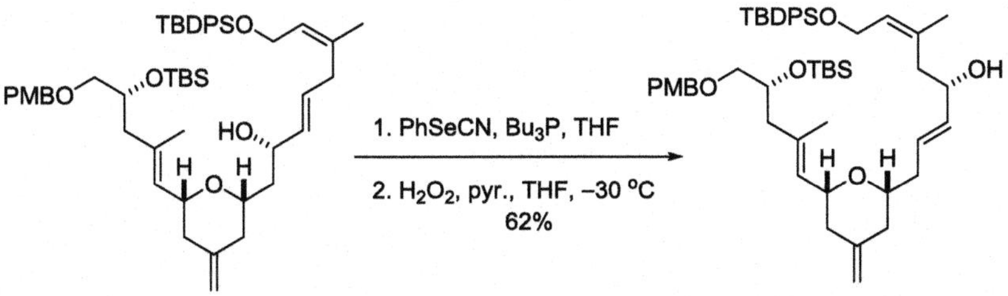

J.J. Li, *Name Reactions: A Collection of Detailed Mechanisms and Synthetic Applications*, DOI 10.1007/978-3-319-03979-4_177, © Springer International Publishing Switzerland 2014

Example 4[12]

**References**

1.  (a) Tang, R.; Mislow, K. *J. Am. Chem. Soc.* **1970**, *92*, 2100–2104.  (b) Evans, D. A.; Andrews, G. C.; Sims, C. L. *J. Am. Chem. Soc.* **1971**, *93*, 4956–4957.  (c) Evans, D. A.; Andrews, G. C. *J. Am. Chem. Soc.* **1972**, *94*, 3672–3674.  (d) Evans, D. A.; Andrews, G. C. *Acc. Chem. Res.* **1974**, *7*, 147–155. (Review).
2.  Sato, T.; Shima, H.; Otera, J. *J. Org. Chem.* **1995**, *60*, 3936–3937.
3.  Jones-Hertzog, D. K.; Jorgensen, W. L. *J. Am. Chem. Soc.* **1995**, *117*, 9077–9078.
4.  Jones-Hertzog, D. K.; Jorgensen, W. L. *J. Org. Chem.* **1995**, *60*, 6682–6683.
5.  Mapp, A. K.; Heathcock, C. H. *J. Org. Chem.* **1999**, *64*, 23–27.
6.  Zhou, Z. S.; Flohr, A.; Hilvert, D. *J. Org. Chem.* **1999**, *64*, 8334–8341.
7.  Shinada, T.; Fuji, T.; Ohtani, Y.; Yoshida, Y.; Ohfune, Y. *Synlett* **2002**, 1341–1343.
8.  Aubele, D. L.; Wan, S.; Floreancig, P. E. *Angew. Chem. Int. Ed.* **2005**, *44*, 3485–3499.
9.  Albert, B. J.; Sivaramakrishnan, A.; Naka, T.; Koide, K. *J. Am. Chem. Soc.* **2006**, *128*, 2792–2793.
10. Pelc, M. J.; Zakarian, A. *Tetrahedron Lett.* **2006**, *47*, 7519–7523.
11. Brebion, F.; Najera, F.; Delouvrie, B.; Lacote, E.; Fensterbank, L.; Malacria, M. *Synthesis* **2007**, 2273–2278.
12. Palko, J. W.; Buist, P. H.; Manthorpe, J. M. *Tetrahedron: Assmmetry* **2013**, *24*, 165–168.

# Mitsunobu reaction

$S_N2$ inversion of an alcohol by a nucleophile using disubstituted azodicarboxylates (originally, diethyl diazodicarboxylate, or DEAD) and trisubstituted phosphines (originally, triphenylphosphine).

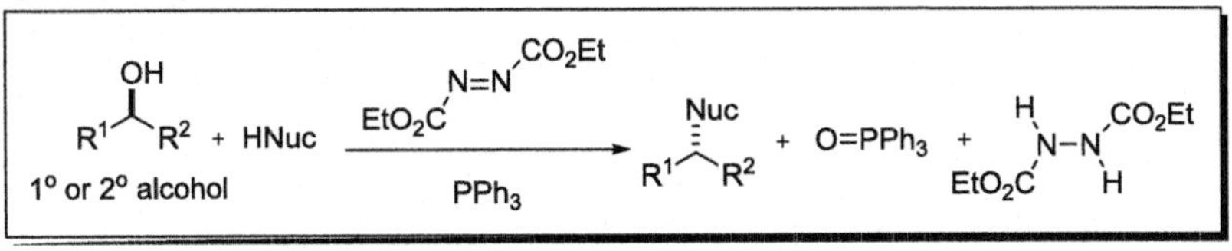

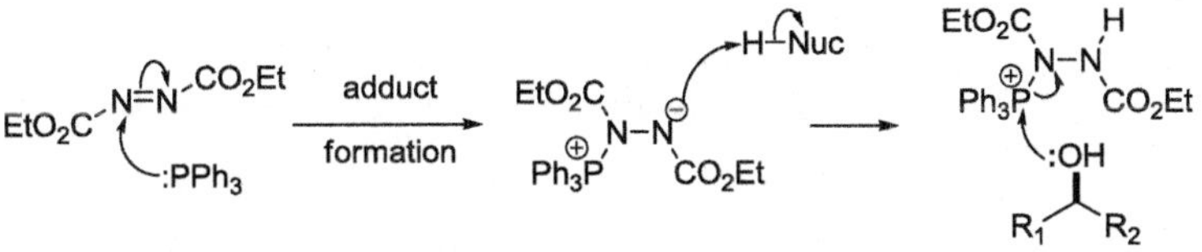

Diethyl azodicarboxylate (DEAD)

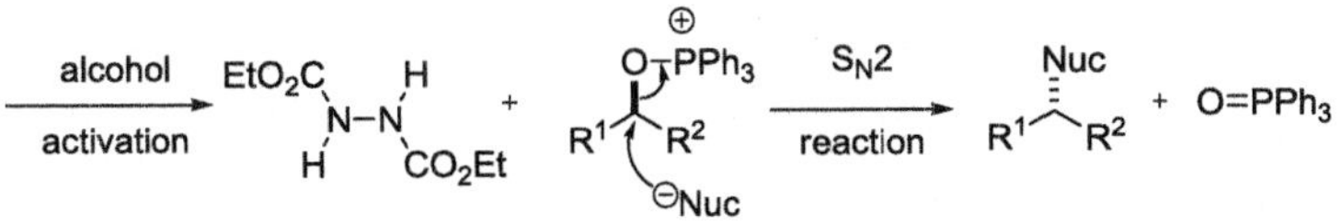

Example 1[2]

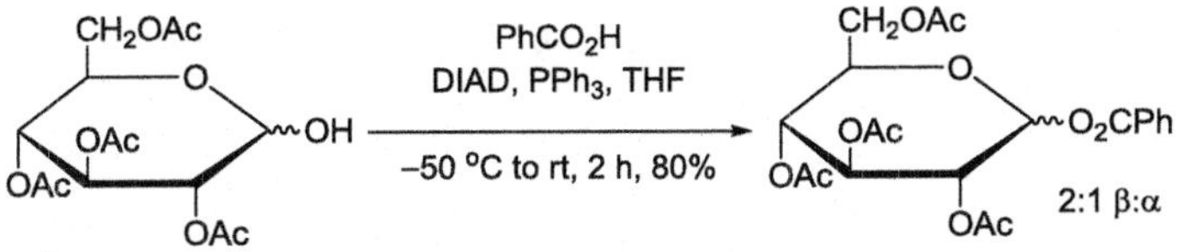

Example 2[3]

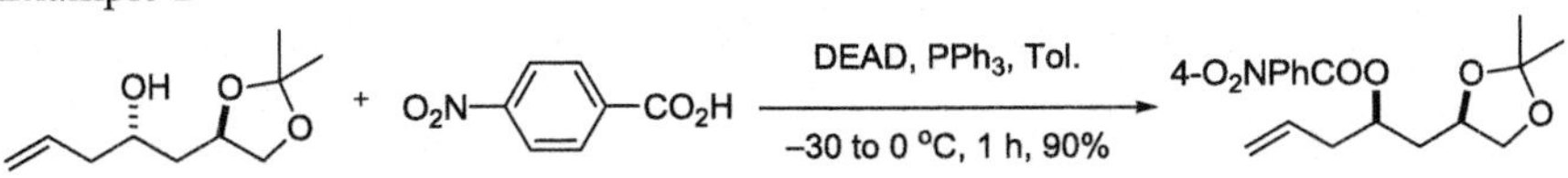

Example 3, Ether formation[6]

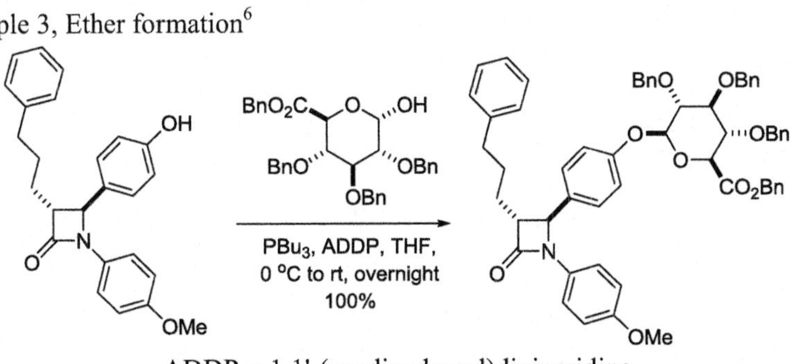

ADDP = 1,1'-(azodicarbonyl)dipiperidine

J.J. Li, *Name Reactions: A Collection of Detailed Mechanisms and Synthetic Applications*,
DOI 10.1007/978-3-319-03979-4_178, © Springer International Publishing Switzerland 2014

Example 4[7]

Example 5[8]

Example 6, Intramolecular Mitsunobu reaction[9]

## References

1.  (a) Mitsunobu, O.; Yamada, M. *Bull. Chem. Soc. Jpn.* **1967**, *40*, 2380–2382.    (b) Mitsunobu, O. *Synthesis* **1981**, 1–28. (Review).
2.  Smith, A. B., III; Hale, K. J.; Rivero, R. A. *Tetrahedron Lett.* **1986**, *27*, 5813–5816.
3.  Kocieński, P. J.; Yeates, C.; Street, D. A.; Campbell, S. F. *J. Chem. Soc., Perkin Trans. 1*, **1987**, 2183–2187.
4.  Hughes, D. L. *Org. React.* **1992**, *42*, 335–656. (Review).
5.  Hughes, D. L. *Org. Prep. Proc. Int.* **1996**, *28*, 127–164. (Review).
6.  Vaccaro, W. D.; Sher, R.; Davis, H. R., Jr. *Bioorg. Med. Chem. Lett.* **1998**, *8*, 35–40.
7.  Cevallos, A.; Rios, R.; Moyano, A.; Pericàs, M. A.; Riera, A. *Tetrahedron: Asymmetry* **2000**, *11*, 4407–4416.
8.  Mukaiyama, T.; Shintou, T.; Fukumoto, K. *J. Am. Chem. Soc.* **2003**, *125*, 10538–10539.
9.  Sumi, S.; Matsumoto, K.; Tokuyama, H.; Fukuyama, T. *Tetrahedron* **2003**, *59*, 8571–8587.
10. Christen, D. P. *Mitsunobu reaction*. In *Name Reactions for Homologations-Part II*; Li, J. J., Ed.; Wiley: Hoboken, NJ, **2009**, pp 671–748. (Review).
11. Ganesan, M.; Salunke, R. V.; Singh, N.; Ramesh, N. G. *Org. Biomol. Chem.* **2013**, *11*, 559–611.

# Miyaura borylation

Palladium-catalyzed reaction of aryl halides with diboron reagents to produce arylboronates. Also known as Hosomi–Miyaura borylation.

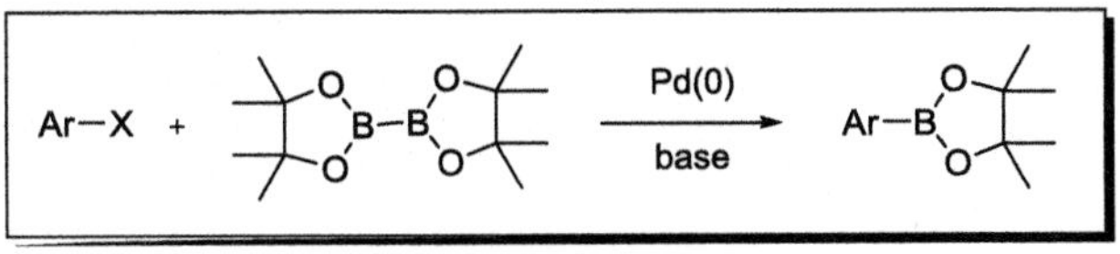

$$X = I, Br, Cl, OTf.$$

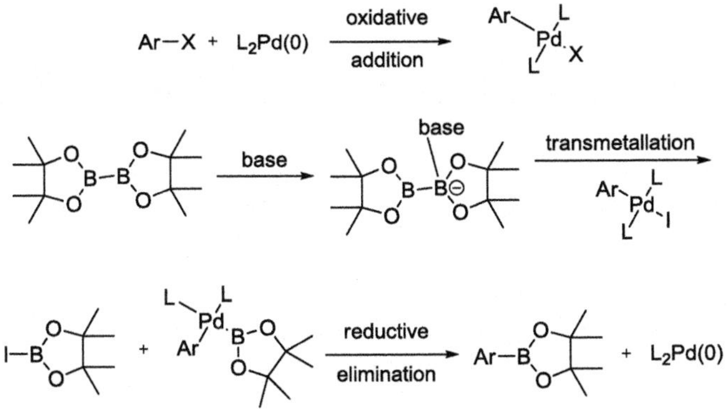

Example 1[7]

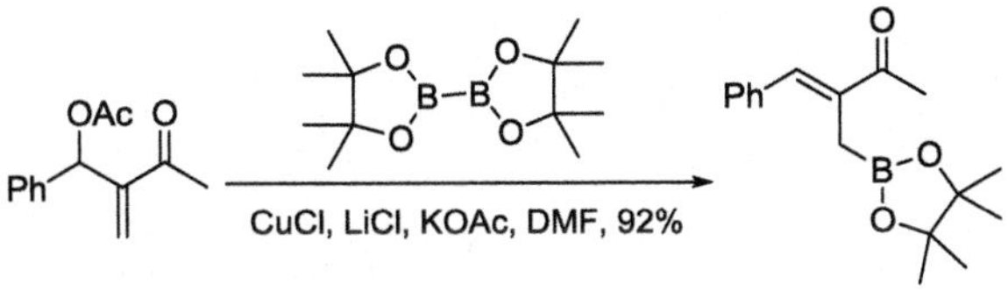

CuCl, LiCl, KOAc, DMF, 92%

Example 2[8]

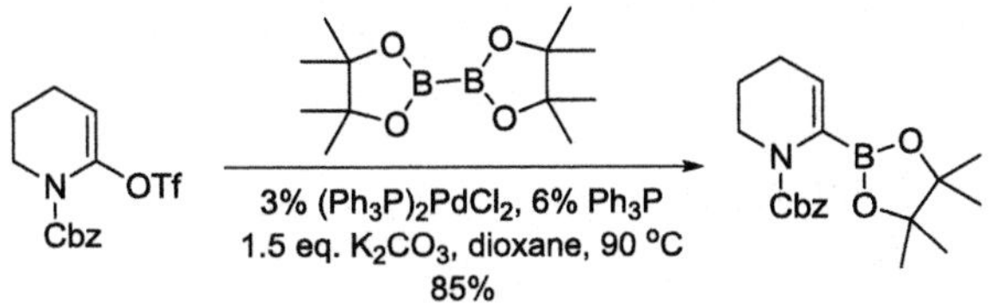

3% (Ph$_3$P)$_2$PdCl$_2$, 6% Ph$_3$P
1.5 eq. K$_2$CO$_3$, dioxane, 90 °C
85%

J.J. Li, *Name Reactions: A Collection of Detailed Mechanisms and Synthetic Applications*,
DOI 10.1007/978-3-319-03979-4_179, © Springer International Publishing Switzerland 2014

Example 3[9]

Example 4, One-pot synthesis of biindolyl[10]

## References

1.  Ishiyama, T.; Murata, M.; Miyaura, N. *J. Org. Chem.* **1995,** *60,* 7508–7510.
2.  Miyaura, N.; Suzuki, A. *Chem. Rev.* **1995,** *95,* 2457–2483. (Review).
3.  Suzuki, A. *J. Organomet. Chem.* **1995,** *576,* 147–168. (Review).
4.  Carbonnelle, A.-C.; Zhu, J. *Org. Lett.* **2000,** *2,* 3477–3480.
5.  Giroux, A. *Tetrahedron Lett.* **2003,** *44,* 233–235.
6.  Kabalka, G. W.; Yao, M.-L. *Tetrahedron Lett.* **2003,** *44,* 7885–7887.
7.  Ramachandran, P. V.; Pratihar, D.; Biswas, D.; Srivastava, A.; Reddy, M. V. R. *Org. Lett.* **2004,** *6,* 481–484.
8.  Occhiato, E. G.; Lo Galbo, F.; Guarna, A. *J. Org. Chem.* **2005,** *70,* 7324–7330.
9.  Skaff, O.; Jolliffe, K. A.; Hutton, C. A. *J. Org. Chem.* **2005,** *70,* 7353–7363.
10. Duong, H. A.; Chua, S.; Huleatt, P. B.; Chai, C. L. L. *J. Org. Chem.* **2008,** *73,* 9177–9180.
11. Jo, T. S.; Kim, S. H.; Shin, J.; Bae, C. *J. Am. Chem. Soc.* **2009,** *131,* 1656–1657.
12. Marciasini, L. D.; Richy, N.; Vaultier, M.; Pucheault, M. *Adv. Synth. Cat.* **2013,** *355,* 1083–1088.

# Moffatt oxidation

Oxidation of alcohols using DCC and DMSO, aka "Pfitzner–Moffatt oxidation".

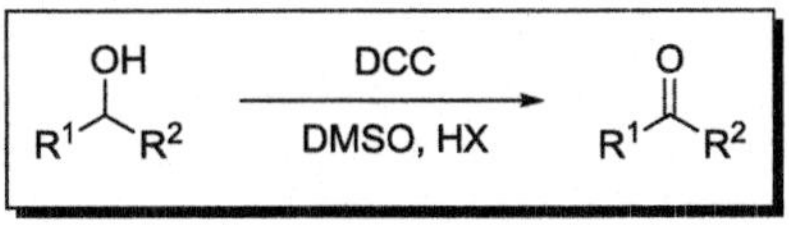

DCC =1,3-dicyclohexylcarbodiimide

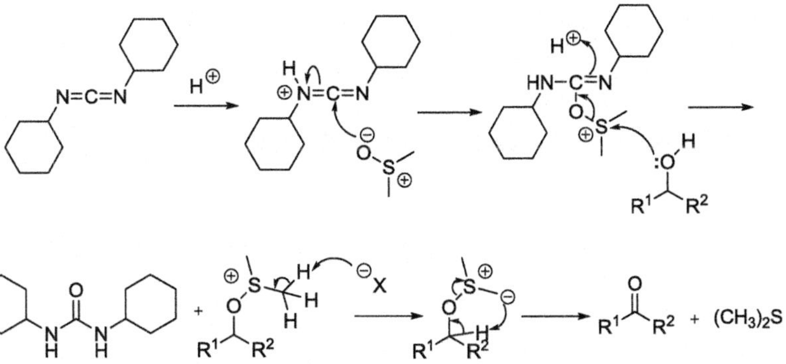

1,3-dicyclohexylurea

Example 1[2]

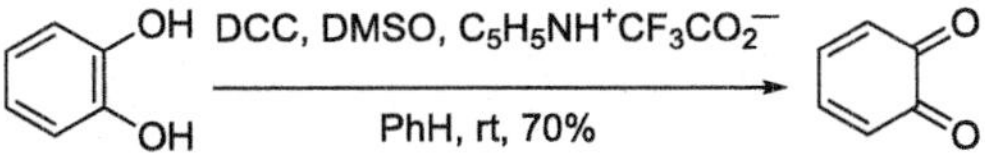

Example 2[8]

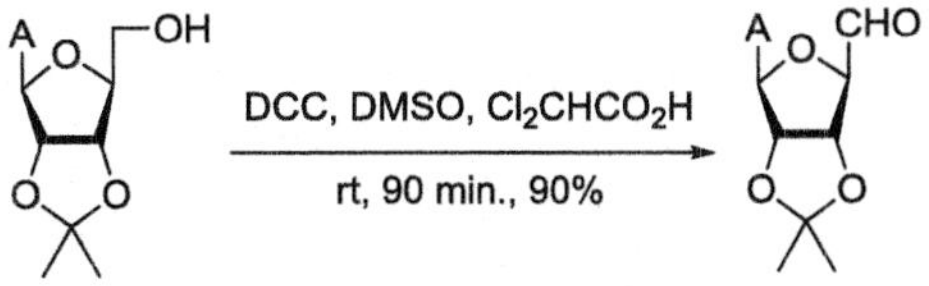

A = adenosine

Example 3[10]

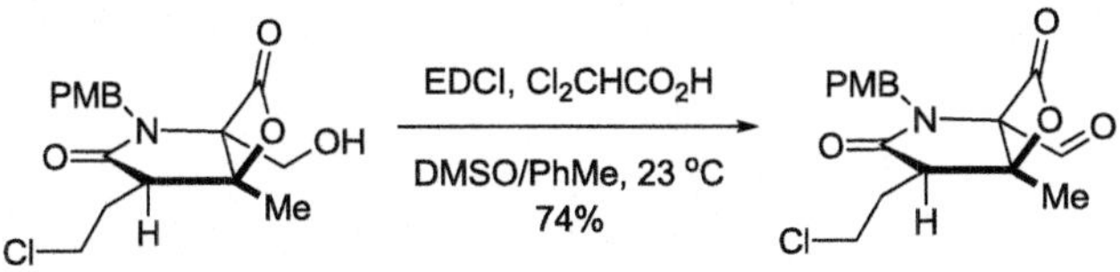

EDCI = 1-Ethyl-3-(3-dimethylaminopropyl)carbodiimide

J.J. Li, *Name Reactions: A Collection of Detailed Mechanisms and Synthetic Applications*, DOI 10.1007/978-3-319-03979-4_180, © Springer International Publishing Switzerland 2014

## References

1. Pfitzner, K. E.; Moffatt, J. G. *J. Am. Chem. Soc.* **1963,** *85*, 3027–3028.
2. Schobert, R. *Synthesis* **1987,** 741–742.
3. Liu, H. J.; Nyangulu, J. M. *Tetrahedron Lett.* **1988,** *29*, 3167–3170.
4. Tidwell, T. T. *Org. React.* **1990,** *39*, 297–572. (Review).
5. Gordon, J. F.; Hanson, J. R.; Jarvis, A. G.; Ratcliffe, A. H. *J. Chem. Soc., Perkin Trans. 1*, **1992,** 3019–3022.
6. Krysan, D. J.; Haight, A. R. *Org. Prep. Proced. Int.* **1993,** *25*, 437–443.
7. Adak, A. K. *Synlett* **2004,** 1651–1652.
8. Wang, M.; Zhang, J.; Andrei, D.; Kuczera, K.; Borchardt, R. T.; Wnuk, S. F. *J. Med. Chem.* **2005,** *48*, 3649–3653.
9. van der Linden, J. J. M.; Hilberink, P. W.; Kronenburg, C. M. P.; Kemperman, G. J. *Org. Proc. Res. Dev.* **2008,** *12*, 911–920.
10. Nguyen, H.; Ma, G.; Gladysheva, T.; Fremgen, T.; Romo, D. *J. Org. Chem.* **2011,** *76*, 2–12.

# Morgan–Walls reaction

Phenanthridine cyclization by dehydrative ring closure of acyl-*o*-aminobiphenyls with phosphorus oxychloride in boiling nitrobenzene.

Example 1[6]

## *Pictet–Hubert reaction*

The Pictet-Hubert reaction is a variant of the Morgan-Walls reaction where the phenanthridine cyclization was accomplished by dehydrative ring closure of acyl-*o*-aminobiphenyls on heating with zinc chloride at 250–300 °C.

J.J. Li, *Name Reactions: A Collection of Detailed Mechanisms and Synthetic Applications*, DOI 10.1007/978-3-319-03979-4_181, © Springer International Publishing Switzerland 2014

Example 2[4]

## References

1.  (a) Pictet, A.; Hubert, A. *Ber.* **1896,** *29*, 1182–1189.  (b) Morgan, C. T.; Walls, L. P. *J. Chem. Soc.* **1931,** 2447–2456.  (c) Morgan, C. T.; Walls, L. P. *J. Chem. Soc.* **1932,** 2225–2231.
2.  Gilman, H.; Eisch, J. *J. Am. Chem. Soc.* **1957,** *79*, 4423–4426.
3.  Hollingsworth, B. L.; Petrow, V. *J. Chem. Soc.* **1961,** 3664–3667.
4.  Fodor, G.; Nagubandi, S. *Tetrahedron* **1980,** *36*, 1279–1300.
5.  Atwell, G. J.; Baguley, B. C.; Denny, W. A. *J. Med. Chem.* **1988,** *31*, 774–779.
6.  Peytou, V.; Condom, R.; Patino, N.; Guedj, R.; Aubertin, A.-M.; Gelus, N.; Bailly, C.; Terreux, R.; Cabrol-Bass, D. *J. Med. Chem.* **1999,** *42*, 4042–4043.
7.  Holsworth, D. D. *Pictet–Hubert Reaction.* In *Name Reactions in Heterocyclic Chemistry*; Li, J. J., Ed.; Wiley: Hoboken, NJ, **2005,** 465–468. (Review).

# Mori–Ban indole synthesis

Intramolecular Heck reaction of *o*-halo-aniline with pendant olefin to prepare indole.

Reduction of $Pd(OAc)_2$ to $Pd(0)$ using $Ph_3P$:

Mori–Ban indole synthesis:

Regeneration of $Pd(0)$:

$$H-PdBrL_n + NaHCO_3 \longrightarrow Pd(0) + NaBr + H_2O + CO_2\uparrow$$

J.J. Li, *Name Reactions: A Collection of Detailed Mechanisms and Synthetic Applications*, DOI 10.1007/978-3-319-03979-4_182, © Springer International Publishing Switzerland 2014

Example 1[1a]

Example 2[4]

Example 3[7]

## References

1.  Mori–Ban indole synthesis, (a) Mori, M.; Chiba, K.; Ban, Y. *Tetrahedron Lett.* **1977,** *18,* 1037–1040; (b) Ban, Y.; Wakamatsu, T.; Mori, M. *Heterocycles* **1977,** *6,* 1711–1715.
2.  Reduction of Pd(OAc)$_2$ to Pd(0), (a) Amatore, C.; Carre, E.; Jutand, A.; M'Barki, M. A.; Meyer, G. *Organometallics* **1995,** *14,* 5605–5614; (b) Amatore, C.; Carre, E.; M'Barki, M. A. *Organometallics* **1995,** *14,* 1818–1826; (c) Amatore, C.; Jutand, A.; M'Barki, M. A. *Organometallics* **1992,** *11,* 3009–3013; (d) Amatore, C.; Azzabi, M.; Jutand, A. *J. Am. Chem. Soc.* **1991,** *113,* 8375–8384.
3.  Macor, J. E.; Ogilvie, R. J.; Wythes, M. J. *Tetrahedron Lett.* **1996,** *37,* 4289–4293.
4.  Li, J. J. *J. Org. Chem.* **1999,** *64,* 8425–8427.
5.  Gelpke, A. E. S.; Veerman, J. J. N.; Goedheijt, M. S.; Kamer, P. C. J.; van Leuwen, P. W. N. M.; Hiemstra, H. *Tetrahedron* **1999,** *55,* 6657–6670.
6.  Sparks, S. M.; Shea, K. J. *Tetrahedron Lett.* **2000,** *41,* 6721–6724.
7.  Bosch, J.; Roca, T.; Armengol, M.; Fernandez-Forner, D. *Tetrahedron* **2001,** *57,* 1041–1048.
8.  Ma, J.; Yin, W.; Zhou, H.; Liao, X.; Cook, J. M. *J. Org. Chem.* **2009,** *74,* 264–273.
9.  Platon, M.; Amardeil, R.; Djakovitch, L.; Hierso, J.-C. *Chem. Soc. Rev.* **2012,** *41,* 3929–3968. (Review).

# Mukaiyama aldol reaction

Lewis acid-catalyzed aldol condensation of aldehyde and silyl enol ether.

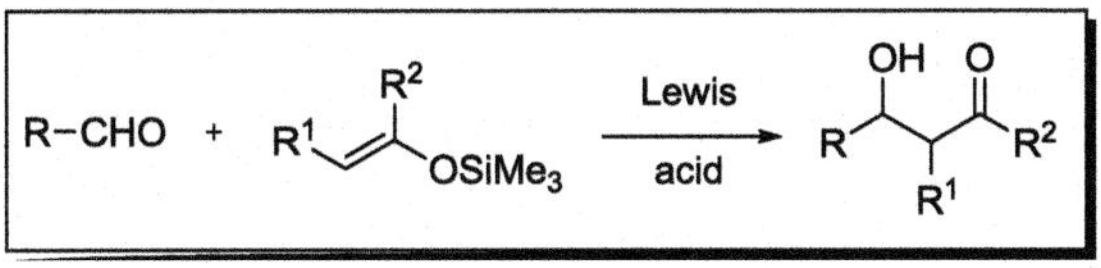

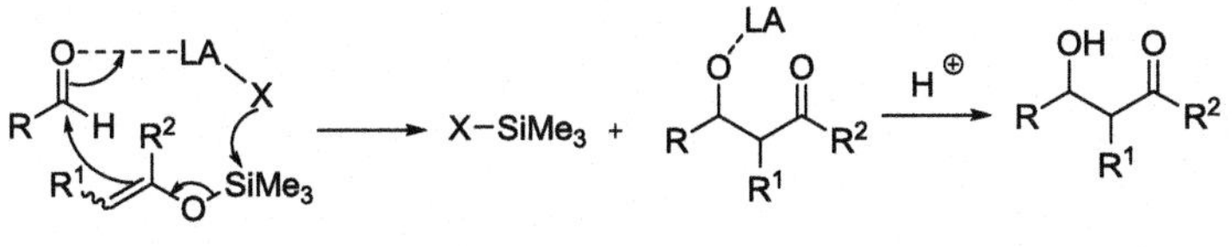

Example 1, Intramolecular Mukaiyama aldol reaction[3]

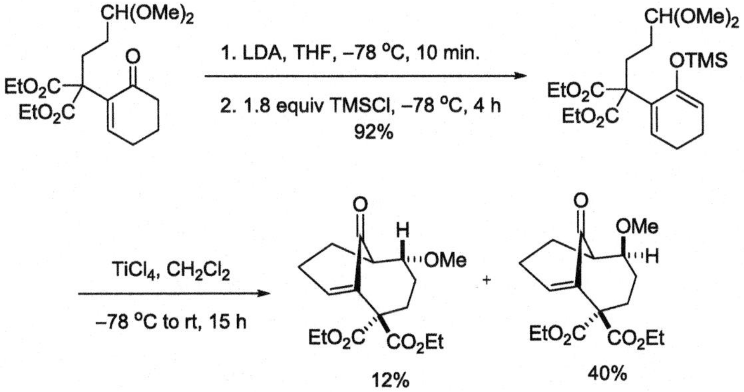

1. LDA, THF, –78 °C, 10 min.

2. 1.8 equiv TMSCl, –78 °C, 4 h
92%

TiCl$_4$, CH$_2$Cl$_2$

–78 °C to rt, 15 h

12%          40%

Example 2, Mukaiyama aldol reaction[7]

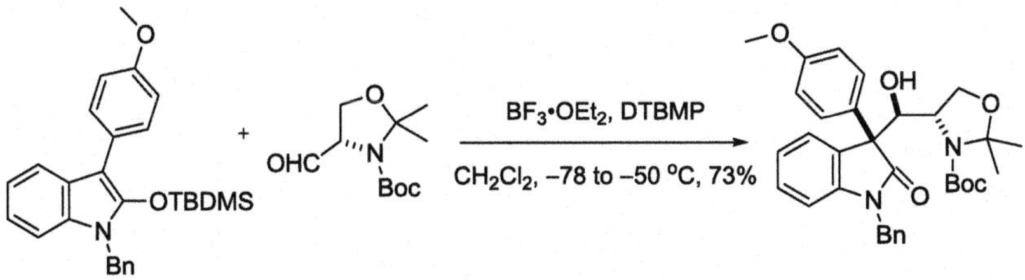

BF$_3$·OEt$_2$, DTBMP

CH$_2$Cl$_2$, –78 to –50 °C, 73%

Example 3, Vinylogous Mukaiyama aldol reaction[8]

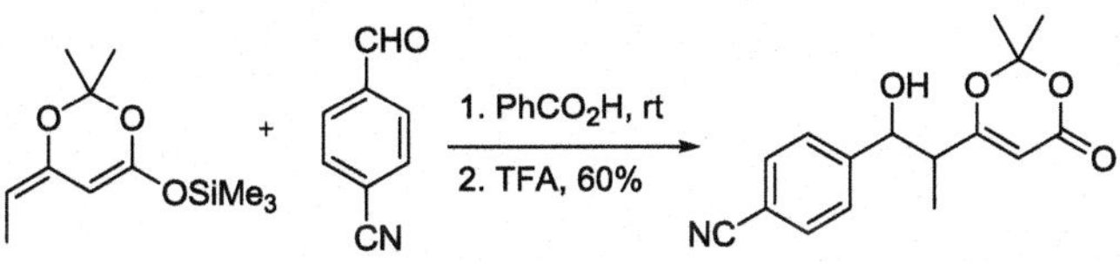

1. PhCO$_2$H, rt

2. TFA, 60%

J.J. Li, *Name Reactions: A Collection of Detailed Mechanisms and Synthetic Applications*,
DOI 10.1007/978-3-319-03979-4_183, © Springer International Publishing Switzerland 2014

Example 4, Asymmetric Mukaiyama aldol reaction[10]

Example 5[12]

10 mol% Bi(OTf)$_3$
CH$_2$Cl$_2$, –40 °C
76% out of 56%
conversion

## References

1.  (a) Mukaiyama, T.; Narasaka, K.; Banno, K. *Chem. Lett.* **1973**, 1011–1014.    (b) Mukaiyama, T.; Narasaka, K.; Banno, K. *J. Am. Chem. Soc.* **1974**, *96*, 7503–7509.
2.  Ishihara, K.; Kondo, S.; Yamamoto, H. *J. Org. Chem.* **2000**, *65*, 9125–9128.
3.  Armstrong, A.; Critchley, T. J.; Gourdel-Martin, M.-E.; Kelsey, R. D.; Mortlock, A. A. *J. Chem. Soc., Perkin Trans. 1* **2002**, 1344–1350.
4.  Clézio, I. L.; Escudier, J.-M.; Vigroux, A. *Org. Lett.* **2003**, *5*, 161–164.
5.  Ishihara, K.; Yamamoto, H. *Boron and Silicon Lewis Acids for Mukaiyama Aldol Reactions*. In *Modern Aldol Reactions* Mahrwald, R., Ed.; **2004**, 25–68. (Review).
6.  Mukaiyama, T. *Angew. Chem. Int. Ed.* **2004**, *43*, 5590–5614. (Review).
7.  Adhikari, S.; Caille, S.; Hanbauer, M.; Ngo, V. X.; Overman, L. E. *Org. Lett.* **2005**, *7*, 2795–2797.
8.  Acocella, M. R.; Massa, A.; Palombi, L.; Villano, R.; Scettri, A. *Tetrahedron Lett.* **2005**, *46*, 6141–6144.
9.  Jiang, X.; Liu, B.; Lebreton, S.; De Brabander, J. K. *J. Am. Chem. Soc.* **2007**, *129*, 6386–6387.
10. Webb, M. R.; Addie, M. S.; Crawforth, C. M.; Dale, J. W.; Franci, X.; Pizzonero, M.; Donald, C.; Taylor, R. J. K. *Tetrahedron* **2008**, *64*, 4778–4791.
11. Frings, M.; Atodiresei, I.; Runsink, J.; Raabe, G.; Bolm, C. *Chem. Eur. J.* **2009**, *15*, 1566–1569.
12. Gao, S.; Wang, Q.; Chen, C. *J. Am. Chem. Soc.* **2009**, *131*, 1410–1412.
13. Matsuo, J.-i.; Murakami, M. *Angew. Chem. Int. Ed.* **2013**, *52*, 9109–9118. (Review).

# Mukaiyama Michael addition

Lewis acid-catalyzed Michael addition of silyl enol ether to an α,β-unsaturated system.

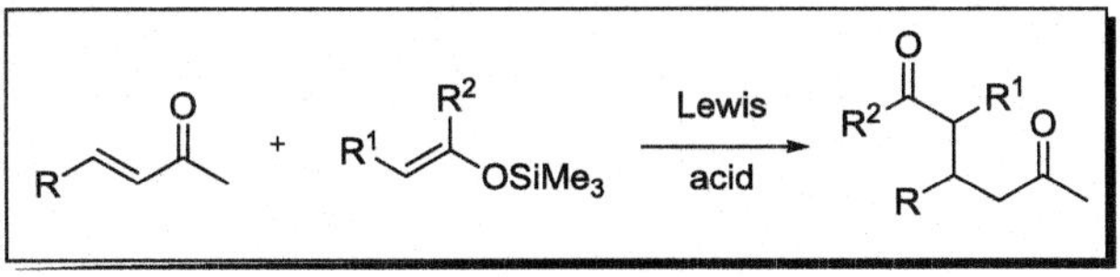

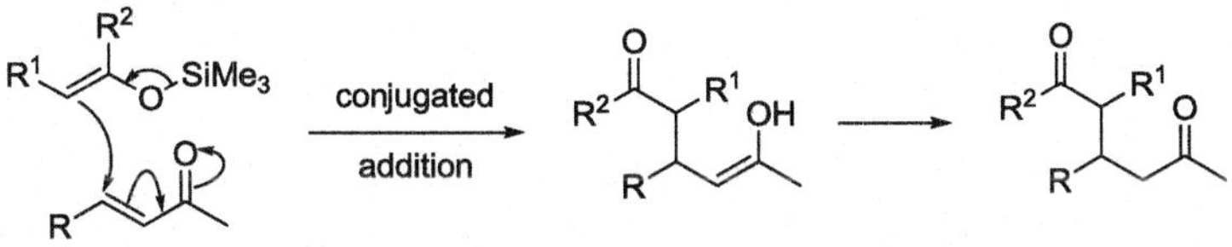

Example 1[2]

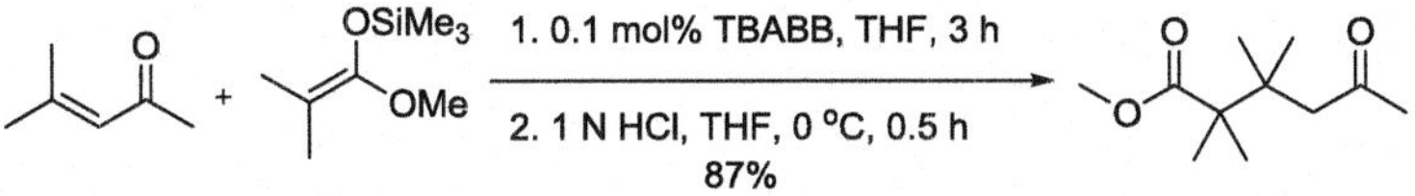

TBABB = tetra-*n*-butylammonium bibenzoate

Example 2[5]

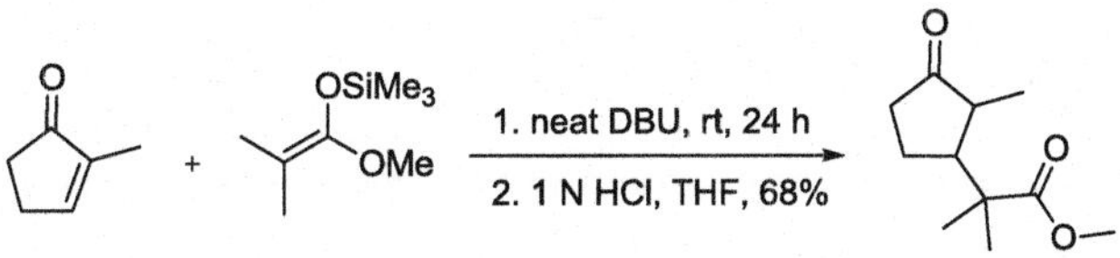

Example 3[8]

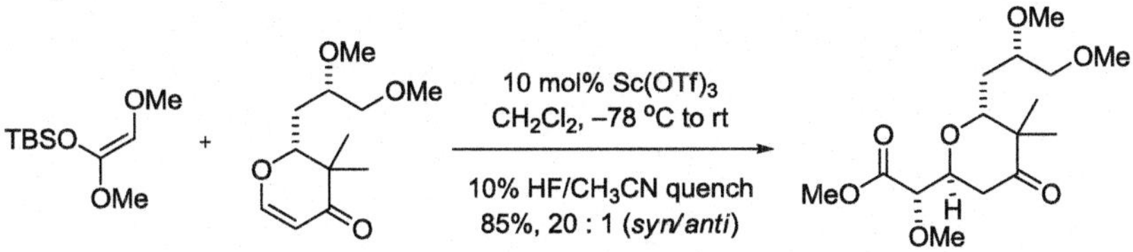

Example 4[9]

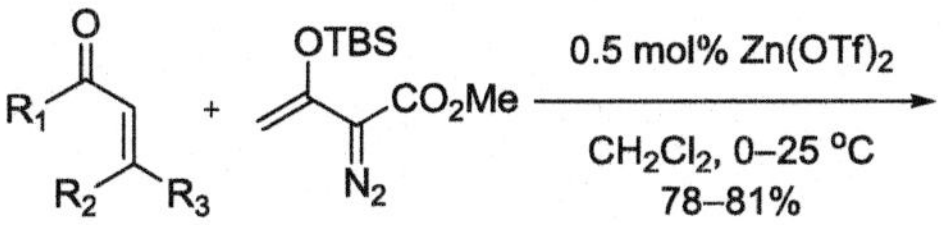

J.J. Li, *Name Reactions: A Collection of Detailed Mechanisms and Synthetic Applications*,
DOI 10.1007/978-3-319-03979-4_184, © Springer International Publishing Switzerland 2014

Example 5, Enantioselective Mukaiyama–Michael reaction[11]

## References

1.   (a) Mukaiyama, T.; Narasaka, K.; Banno, K. *Chem. Lett.* **1973**, 1011–1014. (b) Mukaiyama, T.; Narasaka, K.; Banno, K. *J. Am. Chem. Soc.* **1974**, *96*, 7503–7509. (c) Mukaiyama, T. *Angew. Chem. Int. Ed.* **2004**, *43*, 5590–5614. (Review).
2.   Gnaneshwar, R.; Wadgaonkar, P. P.; Sivaram, S. *Tetrahedron Lett.* **2003**, *44*, 6047–6049.
3.   Wang, X.; Adachi, S.; Iwai, H.; Takatsuki, H.; Fujita, K.; Kubo, M.; Oku, A.; Harada, T. *J. Org. Chem.* **2003**, *68*, 10046–10057.
4.   Jaber, N.; Assie, M.; Fiaud, J.-C.; Collin, J. *Tetrahedron* **2004**, *60*, 3075–3083.
5.   Shen, Z.-L.; Ji, S.-J.; Loh, T.-P. *Tetrahedron Lett.* **2005**, *46*, 507–508.
6.   Wang, W.; Li, H.; Wang, J. *Org. Lett.* **2005**, *7*, 1637–1639.
7.   Ishihara, K.; Fushimi, M. *Org. Lett.* **2006**, *8*, 1921–1924.
8.   Jewett, J. C.; Rawal, V. H. *Angew. Chem. Int. Ed.* **2007**, *46*, 6502–6504.
9.   Liu, Y.; Zhang, Y.; Jee, N.; Doyle, M. P. *Org. Lett.* **2008**, *10*, 1605–1608.
10.   Takahashi, A.; Yanai, H.; Taguchi, T. *Chem. Commun.* **2008**, 2385–2387.
11.   Rout, S.; Ray, S. K.; Singh, V. K. *Org. Biomol. Chem.* **2013**, *11*, 4537–4545.

# Mukaiyama reagent

Pyridinium halide reagent for esterification or amide formation.
General scheme:

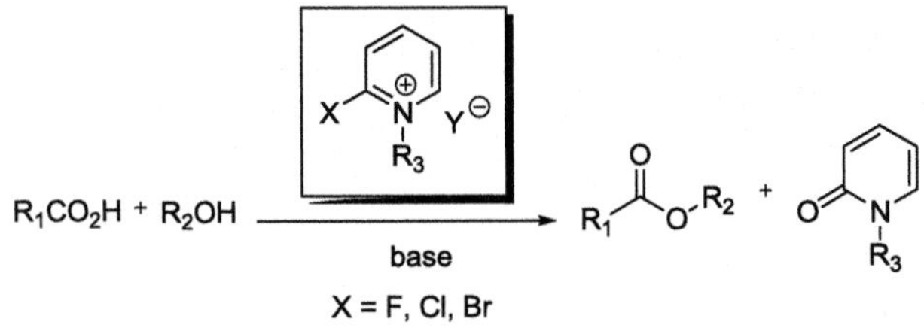
$R_1CO_2H$ + $R_2OH$
base
X = F, Cl, Br

Example 1[1c]

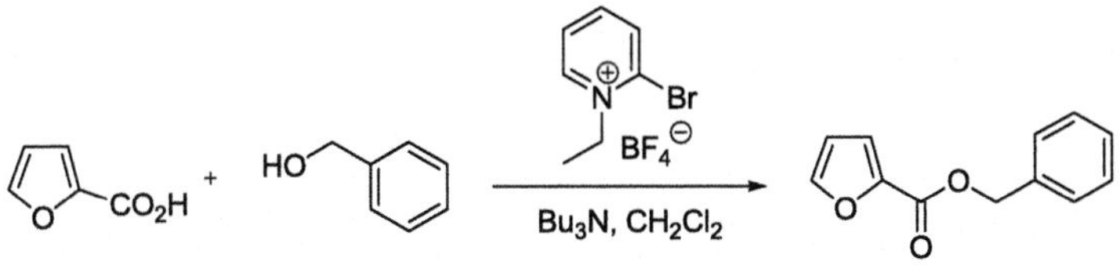
$Bu_3N$, $CH_2Cl_2$

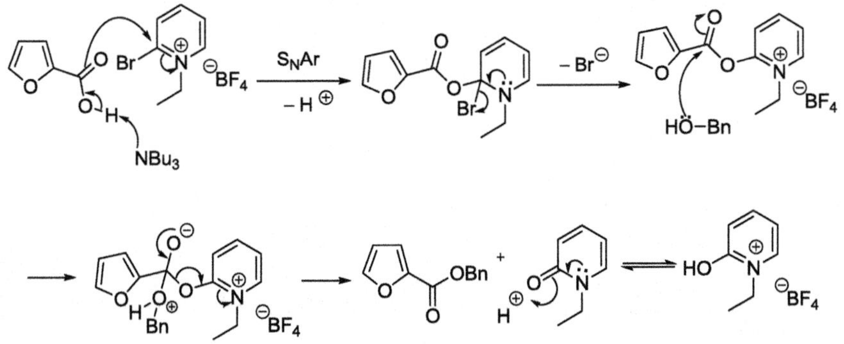
$S_NAr$
$- H^{\oplus}$
$- Br^{\ominus}$
$NBu_3$

Amide formation using the Mukaiyama reagent follows a similar mechanistic pathway.[1d]
Example 2, Polymer-supported Mukaiyama reagent[5]

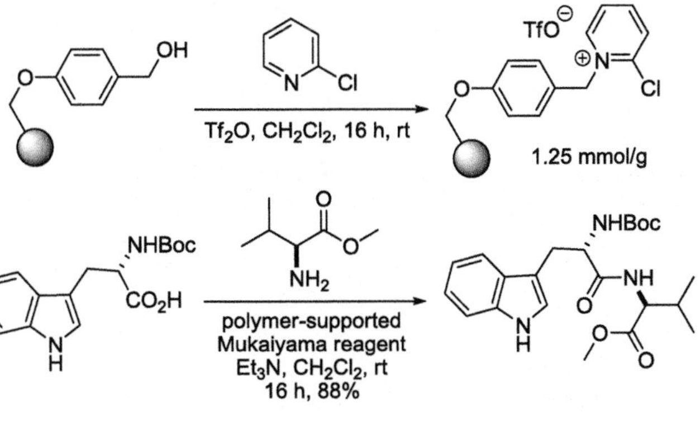
$Tf_2O$, $CH_2Cl_2$, 16 h, rt
1.25 mmol/g
polymer-supported
Mukaiyama reagent
$Et_3N$, $CH_2Cl_2$, rt
16 h, 88%

J.J. Li, *Name Reactions: A Collection of Detailed Mechanisms and Synthetic Applications*,
DOI 10.1007/978-3-319-03979-4_185, © Springer International Publishing Switzerland 2014

Example 3[9]

Example 4, Fluorous Mukaiyama reagent[10]

$$RCO_2H + R^1NH_2 \text{ or } R^2OH \xrightarrow[\text{2. H}_2\text{O, rt, 5 min., 87–100\%}]{\begin{array}{c}\text{1. Fluorous Mukaiyama reagent}\\ \text{1 equiv DMAP, 3 equiv Et}_3\text{N}\\ \text{dry DMF, rt, 1h}\end{array}} RCONHR^1 \text{ or } RCO_2R^2$$

Fluorous Mukaiyama reagent

## References

1.  (a) Mukaiyama, T.; Usui, M.; Shimada, E.; Saigo, K. *Chem. Lett.* **1975**, 1045–1048. (b) Hojo, K.; Kobayashi, S.; Soai, K.; Ikeda, S.; Mukaiyama, T. *Chem. Lett.* **1977**, 635–636. (c) Mukaiyama, T. *Angew. Chem. Int. Ed.* **1979**, *18*, 707–708. (d) For amide formation, see: Huang, H.; Iwasawa, N.; Mukaiyama, T. *Chem. Lett.* **1984**, 1465–1466.
2.  Nicolaou, K. C.; Bunnage, M. E.; Koide, K. *J. Am. Chem. Soc.* **1994**, *116*, 8402–8403.
3.  Yong, Y. F.; Kowalski, J. A.; Lipton, M. A. *J. Org. Chem.* **1997**, *62*, 1540–1542.
4.  Folmer, J. J.; Acero, C.; Thai, D. L.; Rapoport, H. *J. Org. Chem.* **1998**, *63*, 8170–8182.
5.  Crosignani, S.; Gonzalez, J.; Swinnen, D. *Org. Lett.* **2004**, *6*, 4579–4582.
6.  Mashraqui, S. H.; Vashi, D.; Mistry, H. D. *Synth. Commun.* **2004**, *34*, 3129–3134.
7.  Donati, D.; Morelli, C.; Taddei, M. *Tetrahedron Lett.* **2005**, *46*, 2817–2819.
8.  Vandromme, L.; Monchaud, D.; Teulade-Fichou, M.-P. *Synlett* **2006**, 3423–3426.
9.  Ren, Q.; Dai, L.; Zhang, H.; Tan, W.; Xu, Z.; Ye, T. *Synlett* **2008**, 2379–2383.
10. Matsugi, M.; Suganuma, M.; Yoshida, S.; Hasebe, S.; Kunda, Y.; Hagihara, K.; Oka, S. *Tetrahedron Lett.* **2008**, *49*, 6573–6574.
11. Novosjolova, I. *Synlett* **2013**, *24*, 135–136. (Review).

# Myers–Saito cyclization

Construction of substituted arenes through the thermal or photochemical cycloaromatization of *allenyl enynes* in the presence of a H· donor such as 1,4-cyclohexadiene. *Cf.* Bergman cyclization and Schmittel cyclization.

allenyl enyne                  diradical

Example 1[3]

Example 2, Aza-Myers–Saito reaction[8]

# References

1. (a) Myers, A. G.; Proteau, P. J.; Handel, T. M. *J. Am. Chem. Soc.* **1988**, *110*, 7212–7214. (b) Myers, A. G.; Dragovich, P. S.; Kuo, E. Y. *J. Am. Chem. Soc.* **1992**, *114*, 9369–9386.
2. Schmittel, M.; Strittmatter, M.; Kiau, S. *Tetrahedron Lett.* **1995**, *36*, 4975–4978.
3. Schmittel, M.; Steffen, J.-P.; Auer, D.; Maywald, M. *Tetrahedron Lett.* **1997**, *38*, 6177–6180.
4. Bruckner, R.; Suffert, J. *Synlett* **1999**, 657–679. (Review).
5. Stahl, F.; Moran, D.; Schleyer, P. von R.; Prall, M.; Schreiner, P. R. *J. Org. Chem.* **2002**, *67*, 1453–1461.
6. Musch, P. W; Remenyi, C.; Helten, H.; Engels, B. *J. Am. Chem. Soc.* **2002**, *124*, 1823–1828.
7. Bui, B. H.; Schreiner, P. R. *Org. Lett.* **2003**, *5*, 4871–4874.
8. Feng, L.; Kumar, D.; Birney, D. M.; Kerwin, S. M. *Org. Lett.* **2004**, *6*, 2059–2062.
9. Schmittel, M.; Mahajan, A. A.; Bucher, G. *J. Am. Chem. Soc.* **2005**, *127*, 5324–5325.
10. Karpov, G.; Kuzmin, A.; Popik, V. V. *J. Am. Chem. Soc.* **2008**, *130*, 11771–11777.
11. Schmittel, M.; Strittmatter, M.; Vollmann, K. *Tetrahedron Lett.* **2013**, *37*, 999–1002.

J.J. Li, *Name Reactions: A Collection of Detailed Mechanisms and Synthetic Applications*, DOI 10.1007/978-3-319-03979-4_186, © Springer International Publishing Switzerland 2014

# Nazarov cyclization

Acid-catalyzed electrocyclic formation of cyclopentenone from di-vinyl ketone.

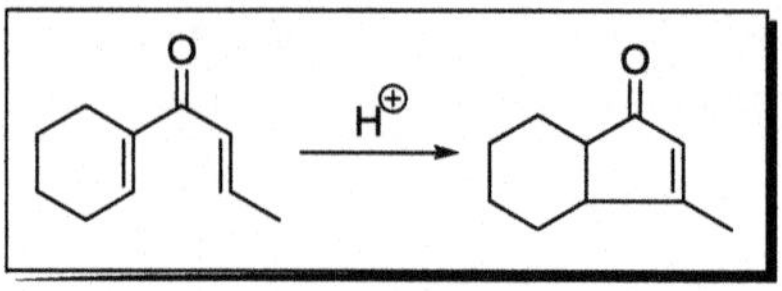

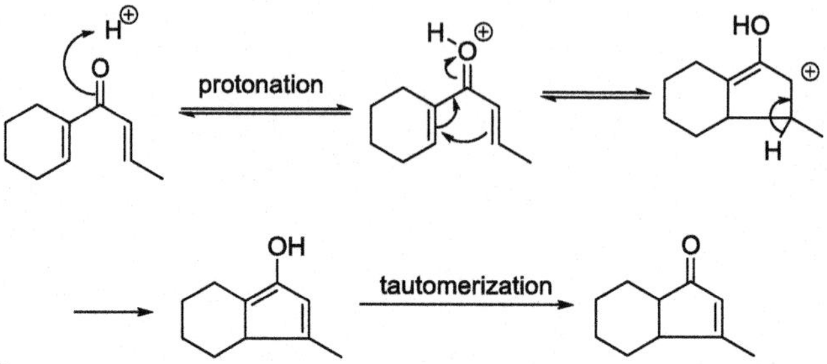

Example 1[2]

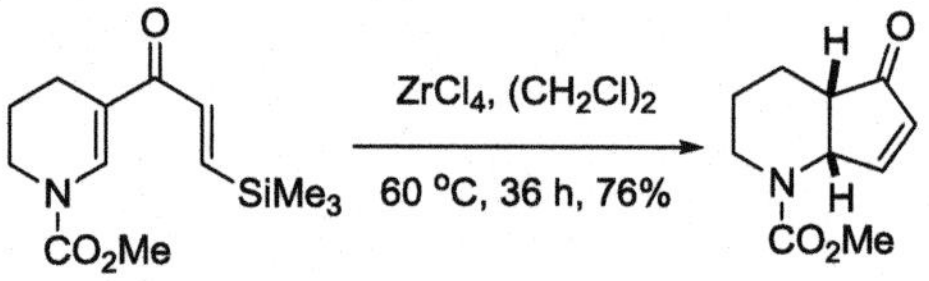

Example 2[6]

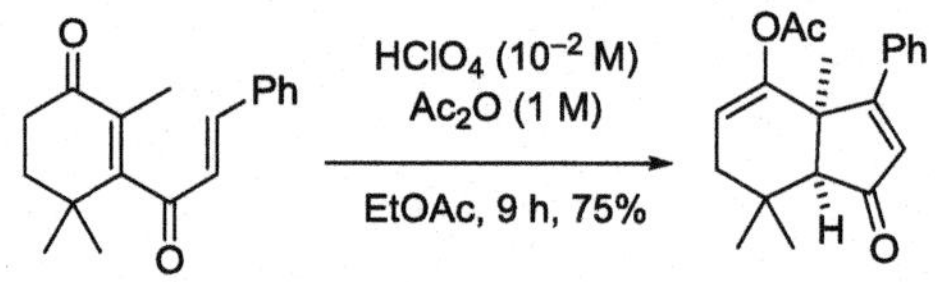

Example 3[9]

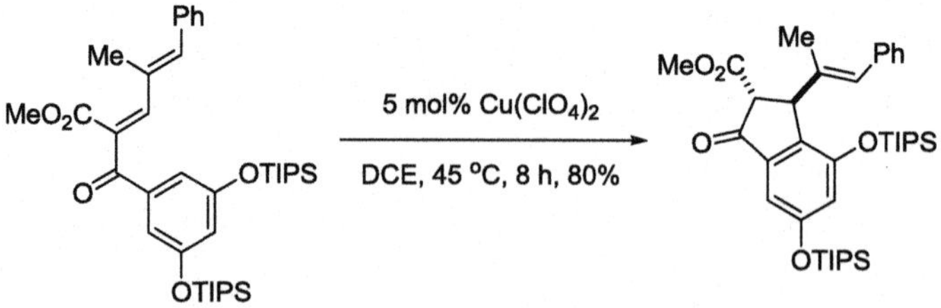

Example 4[10]

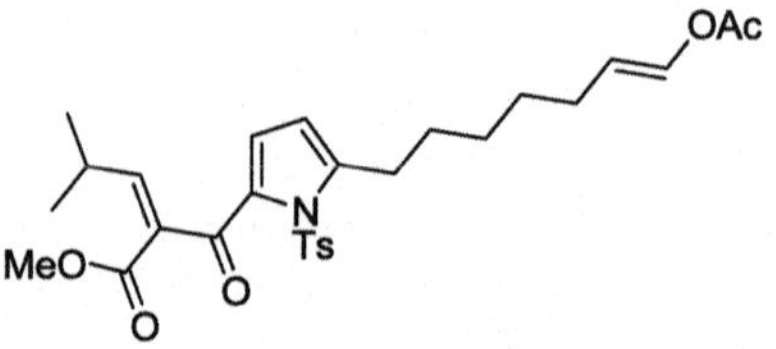

J.J. Li, *Name Reactions: A Collection of Detailed Mechanisms and Synthetic Applications*,
DOI 10.1007/978-3-319-03979-4_187, © Springer International Publishing Switzerland 2014

10 mol% Sc(OTf)$_3$
LiClO$_4$, 80 °C

ClCH$_2$CH$_2$Cl (0.3 M)
65%

Example 5, an example with a different mechanism[11]

1. $h\nu$ (350 nm), CH$_3$CN, rt

2. $i$-Pr$_2$NH, MeOH, 50 °C
60%, 2 steps

Example 6[12]

2 mol% In(OTf)$_3$
4 mol% (PhO)$_2$P(O)OH

CH$_2$Cl$_2$, 40 °C, 48 h
81%

## References

1. Nazarov, I. N.; Torgov, I. B.; Terekhova, L. N. *Bull. Acad. Sci. (USSR)* **1942**, 200. I. N. Nazarov (1900–1957), a Soviet Union Scientist, discovered this reaction in 1942. It was said that almost as many young synthetic chemists have been lost in the pursuit of an asymmetric Nazarov cyclization as of the Bayliss–Hillman reaction.
2. Denmark, S. E.; Habermas, K. L.; Hite, G. A. *Helv. Chim. Acta* **1988**, *71*, 168–194; 195–208.
3. Habermas, K. L.; Denmark, S. E.; Jones, T. K. *Org. React.* **1994**, *45*, 1–158. (Review).
4. Kim, S.-H.; Cha, J. K. *Synthesis* **2000**, 2113–2116.
5. Giese, S.; West, F. G. *Tetrahedron* **2000**, *56*, 10221–10228.
6. Mateos, A. F.; de la Nava, E. M. M.; González, R. R. *Tetrahedron* **2001**, *57*, 1049–1057.
7. Harmata, M.; Lee, D. R. *J. Am. Chem. Soc.* **2002**, *124*, 14328–14329.
8. Leclerc, E.; Tius, M. A. *Org. Lett.* **2003**, *5*, 1171–1174.
9. Marcus, A. P.; Lee, A. S.; Davis, R. L.; Tantillo, D. J.; Sarpong, R. *Angew. Chem. Int. Ed.* **2008**, *47*, 6379–6383.
10. Bitar, A. Y.; Frontier, A. J. *Org. Lett.* **2009**, *11*, 49–52.
11. Gao, S.; Wang, Q.; Chen, C. *J. Am. Chem. Soc.* **2009**, *131*, 1410–1412.
12. Xi, Z.-G.; Zhu, L.; Luo, S.; Cheng, J.-P. *J. Org. Chem.* **2013**, *78*, 606–613.

# Neber rearrangement

α-Aminoketone from tosyl ketoxime and base. The net conversion of a ketone into an α-aminoketone *via* the oxime.

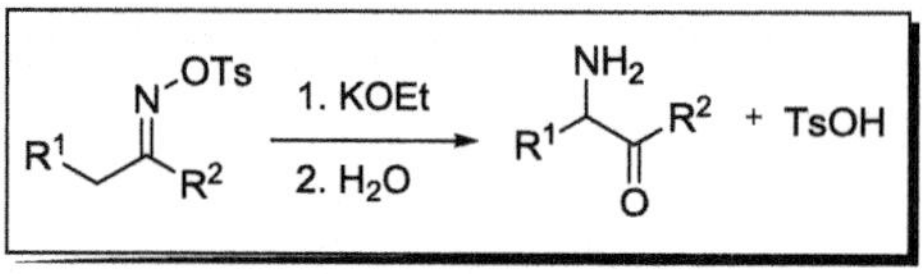

ketoxime                    α-aminoketone

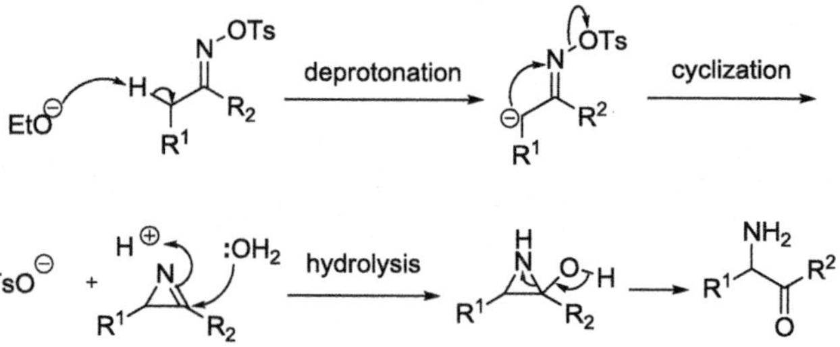

azirine intermediate

Example 1[3]

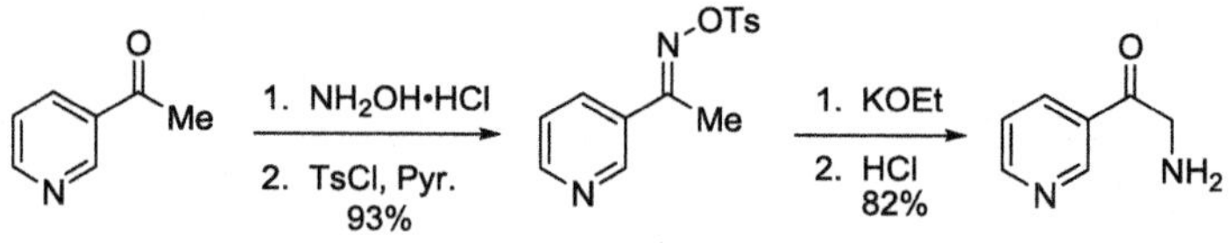

Example 2, A variant using iminochloride[5]

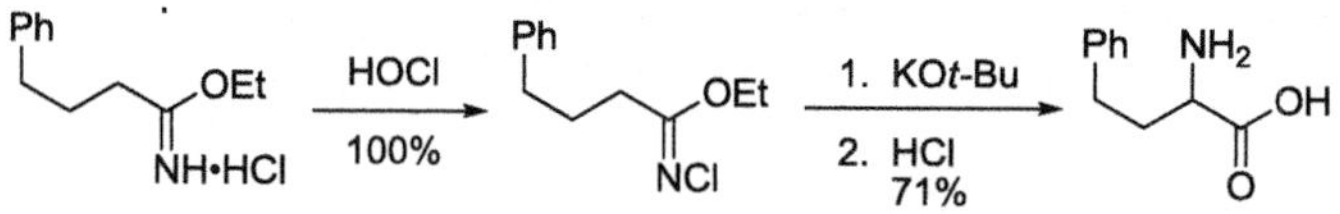

Example 3[8]

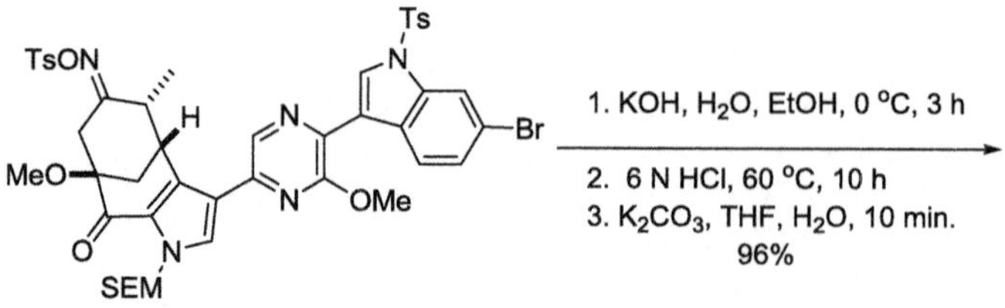

1. KOH, H$_2$O, EtOH, 0 °C, 3 h

2. 6 N HCl, 60 °C, 10 h
3. K$_2$CO$_3$, THF, H$_2$O, 10 min.
   96%

J.J. Li, *Name Reactions: A Collection of Detailed Mechanisms and Synthetic Applications*, DOI 10.1007/978-3-319-03979-4_188, © Springer International Publishing Switzerland 2014

Example 4[9]

Example 5[11]

## References

1. Neber, P. W.; v. Friedolsheim, A. *Ann.* **1926,** *449*, 109–134.
2. O'Brien, C. *Chem. Rev.* **1964,** *64*, 81–89. (Review).
3. LaMattina, J. L.; Suleske, R. T. *Synthesis* **1980,** 329–330.
4. Verstappen, M. M. H.; Ariaans, G. J. A.; Zwanenburg, B. *J. Am. Chem. Soc.* **1996,** *118*, 8491–8492.
5. Oldfield, M. F.; Botting, N. P. *J. Labeled Compd. Radiopharm.* **1998,** *16*, 29–36.
6. Palacios, F.; Ochoa de Retana, A. M.; Gil, J. I. *Tetrahedron Lett.* **2002,** *41*, 5363–5366.
7. Ooi, T.; Takahashi, M.; Doda, K.; Maruoka, K. *J. Am. Chem. Soc.* **2002,** *124*, 7640–7641.
8. Garg, N. K.; Caspi, D. D.; Stoltz, B. M. *J. Am. Chem. Soc.* **2005,** *127*, 5970–5978.
9. Taber, D. F.; Tian, W. *J. Am. Chem. Soc.* **2006,** *128*, 1058–1059.
10. Richter, J. M. *Neber Rearrangement.* In *Name Reactions for Homologations-Part I*; Li, J. J., Ed.; Wiley: Hoboken, NJ, **2009,** pp 464–473. (Review).
11. Cardoso, A. L.; Gimeno, L.; Lemos, A.; Palacios, F.; Teresa, M. V. D.; e Melo, P. *J. Org. Chem.* **2013,** *78*, 6983–6991.

# Nef reaction

Conversion of a primary or secondary nitroalkane into the corresponding carbonyl compound.

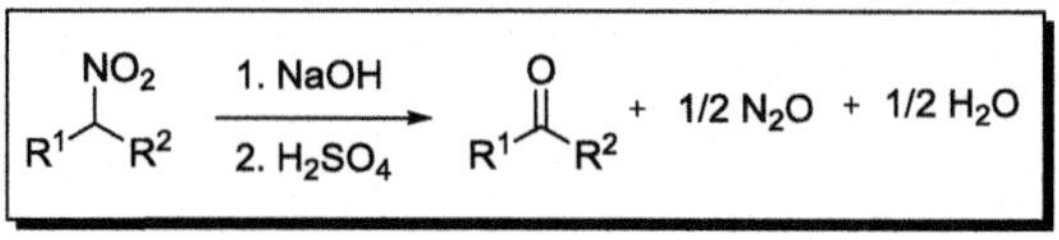

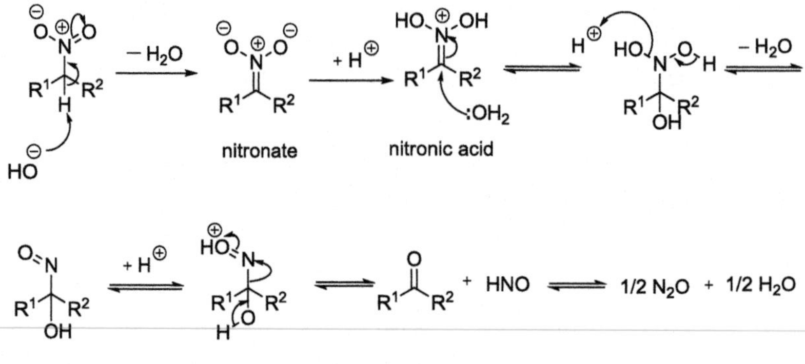

Example 1[4]

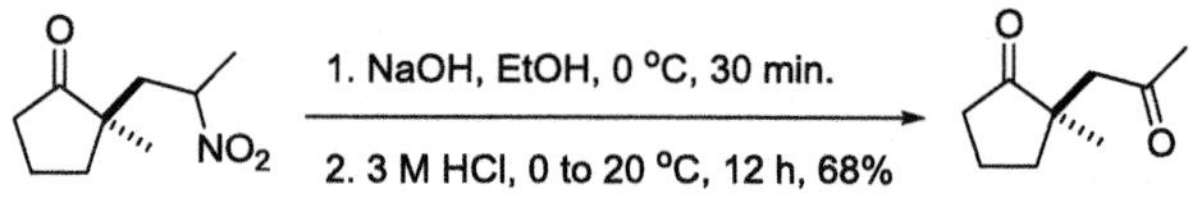

1. NaOH, EtOH, 0 °C, 30 min.

2. 3 M HCl, 0 to 20 °C, 12 h, 68%

Example 2[7]

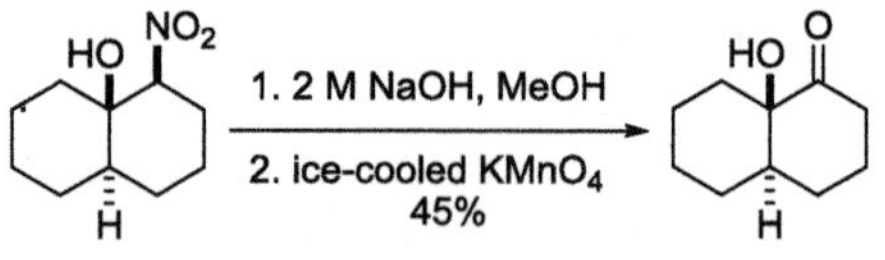

1. 2 M NaOH, MeOH

2. ice-cooled KMnO₄
45%

Example 3[9]

2.2 equiv PMe₃, THF, rt, 30 min.

H₂O, rt, 5 min.

94%, 2 steps

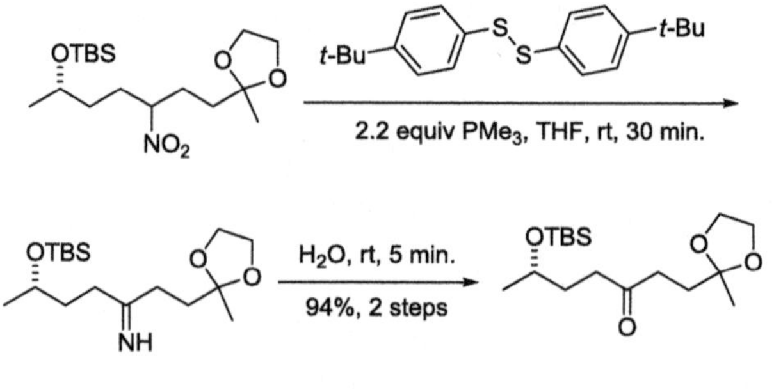

J.J. Li, *Name Reactions: A Collection of Detailed Mechanisms and Synthetic Applications*, DOI 10.1007/978-3-319-03979-4_189, © Springer International Publishing Switzerland 2014

Example 4[10]

Example 5[11]

## References

1.  Nef, J. U. *Ann.* **1894**, *280*, 263–342. John Ulrich Nef (1862–1915) was born in Switzerland and immigrated to the US at the age of four with his parents. He went to Munich, Germany to study with Adolf von Baeyer, earning a Ph.D. In 1886. Back to the States, he served as a professor at Purdue University, Clark University, and the University of Chicago. The Nef reaction was discovered at Clark University in Worcester, Massachusetts. Nef was temperamental and impulsive, suffering from a couple of mental breakdowns. He was also highly individualistic, and had never published with a coworker save for three early articles.
2.  Pinnick, H. W. *Org. React.* **1990**, *38*, 655–792. (Review).
3.  Adam, W.; Makosza, M.; Saha-Moeller, C. R.; Zhao, C.-G. *Synlett* **1998**, 1335–1336.
4.  Thominiaux, C.; Rousse, S.; Desmaele, D.; d'Angelo, J.; Riche, C. *Tetrahedron: Asymmetry* **1999**, *10*, 2015–2021.
5.  Capecchi, T.; de Koning, C. B.; Michael, J. P. *J. Chem. Soc., Perkin Trans. 1* **2000**, 2681–2688.
6.  Ballini, R.; Bosica, G.; Fiorini, D.; Petrini, M. *Tetrahedron Lett.* **2002**, *43*, 5233–5235.
7.  Chung, W. K.; Chiu, P. *Synlett* **2005**, 55–58.
8.  Wolfe, J. P. *Nef reaction*. In *Name Reactions for Functional Group Transformations*; Li, J. J., Ed.; Wiley: Hoboken, NJ, **2007**, pp 645–652. (Review).
9.  Burés, J.; Vilarrasa, J. *Tetrahedron Lett.* **2008**, *49*, 441–444.
10. Felluga, F.; Pitacco, G.; Valentin, E.; Venneri, C. D. *Tetrahedron: Asymmetry* **2008**, *19*, 945–955.
11. Chinmay Bhat, C.; Tilve, S. G. *Tetrahedron* **2013**, *69*, 6129–6143.

# Negishi cross-coupling reaction

The Negishi cross-coupling reaction is the nickel- or palladium-catalyzed coupling of organozinc compounds with various halides or triflates (aryl, alkenyl, alkynyl, and acyl).

$R^1$ = aryl, alkenyl, alkynyl, acyl
$R^2$ = aryl, heteroaryl, alkenyl, allyl, Bn, homoallyl, homopropargyl
X = Cl, Br, I, OTf
Y = Cl, Br, I
$L_n$ = PPh$_3$, dba, dppe

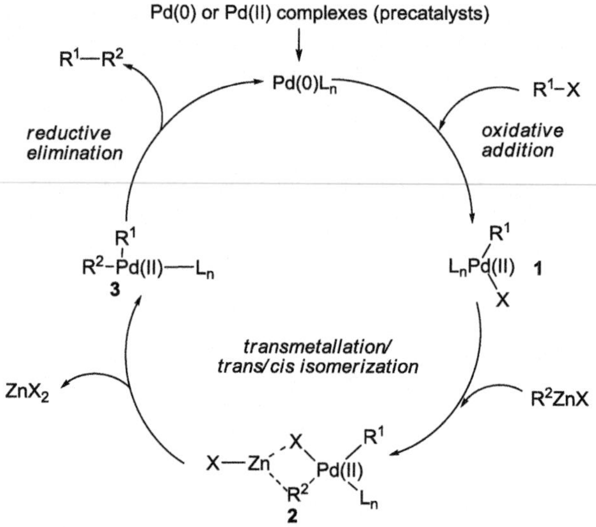

Example 1[3]

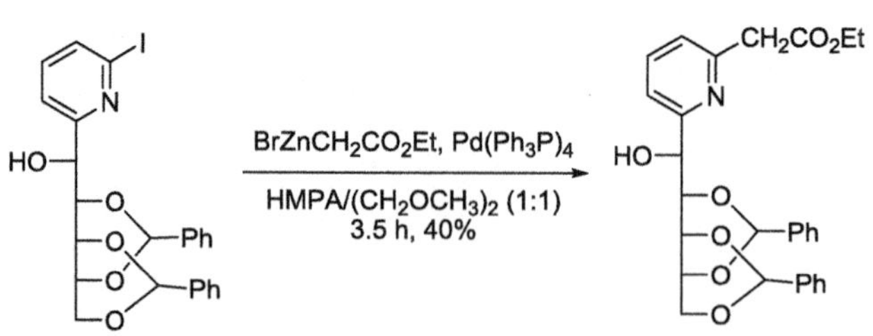

BrZnCH$_2$CO$_2$Et, Pd(Ph$_3$P)$_4$
HMPA/(CH$_2$OCH$_3$)$_2$ (1:1)
3.5 h, 40%

Example 2[4]

activated
Zn/Cu couple

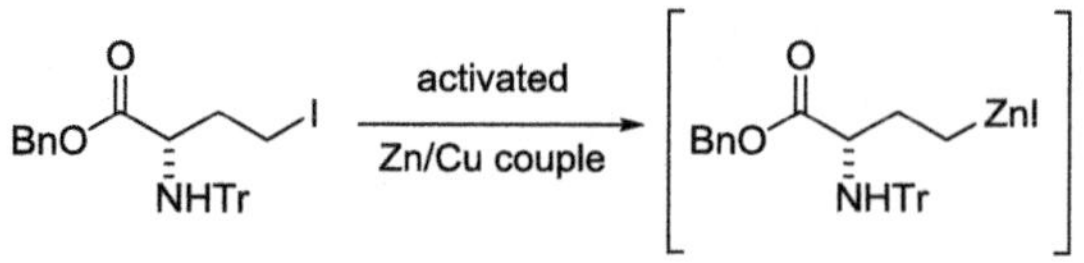

J.J. Li, *Name Reactions: A Collection of Detailed Mechanisms and Synthetic Applications*,
DOI 10.1007/978-3-319-03979-4_190, © Springer International Publishing Switzerland 2014

Example 3[8]

Example 4[9]

Example 5[11]

## References

1.  (a) Negishi, E.-I.; Baba, S. *J. Chem. Soc., Chem. Commun.* **1976**, 596–597.   (b)
    Negishi, E.-I.; King, A. O.; Okukado, N. *J. Org. Chem.* **1977**, *42*, 1821–1823.   (c)
    Negishi, E.-I. *Acc. Chem. Res.* **1982**, *15*, 340–348. (Review). Negishi is a professor at
    Purdue University. He won Nobel Prize in 2010 along with Richard F. Heck and Akira
    Suzuki "for palladium-catalyzed cross couplings in organic synthesis".
2.  Erdik, E. *Tetrahedron* **1992**, *48*, 9577–9648. (Review).
3.  De Vos, E.; Esmans, E. L.; Alderweireldt, F. C.; Balzarini, J.; De Clercq, E. *J. Hetero-
    cycl. Chem.* **1993**, *30*, 1245–1252.
4.  Evans, D. A.; Bach, T. *Angew. Chem. Int. Ed.* **1993**, *32*, 1326–1327.
5.  Negishi, E.-I.; Liu, F. In *Metal-Catalyzed Cross-Coupling Reactions;* Diederich, F.;
    Stang, P. J., Eds.; Wiley–VCH: Weinheim, Germany, **1998**, pp 1–47. (Review).
6.  Arvanitis, A. G.; Arnold, C. R.; Fitzgerald, L. W.; Frietze, W. E.; Olson, R. E.; Gilli-
    gan, P. J.; Robertson, D. W. *Bioorg. Med. Chem. Lett.* **2003**, *13*, 289–291.
7.  Ma, S.; Ren, H.; Wei, Q. *J. Am. Chem. Soc.* **2003**, *125*, 4817–4830.
8.  Corley, E. G.; Conrad, K.; Murry, J. A.; Savarin, C.; Holko, J.; Boice, G. *J. Org.
    Chem.* **2004**, *69*, 5120–5123.
9.  Inoue, M.; Yokota, W.; Katoh, T. *Synthesis* **2007**, 622–637.
10. Yet, L. *Negishi cross-coupling reaction*. In *Name Reactions for Homologations-Part I*;
    Li, J. J., Ed.; Wiley: Hoboken, NJ, **2009**, pp 70–99. (Review).
11. Dolliver, D. D.; Bhattarai, B. T.; et al. *J. Org. Chem.* **2013**, *78*, 3676–3687.

# Nenitzescu indole synthesis

5-Hydroxylindole from condensation of *p*-benzoquinone and β-aminocrotonate.

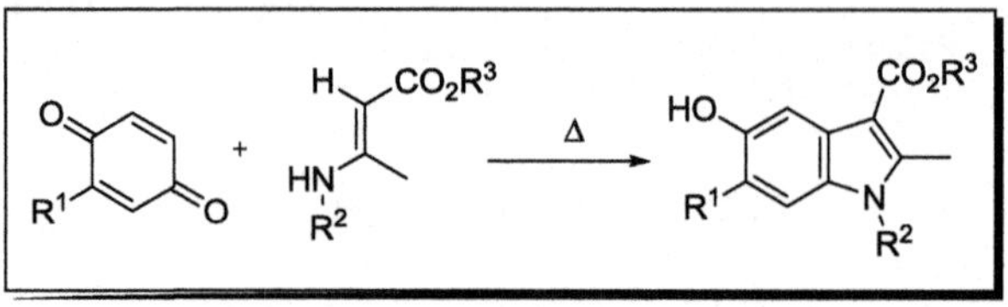

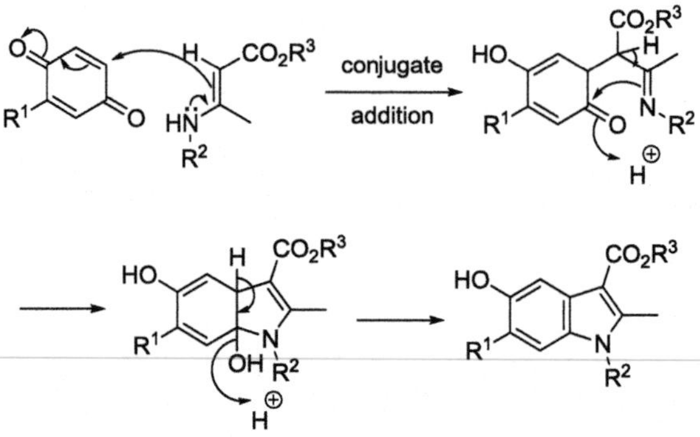

Example 1[5]

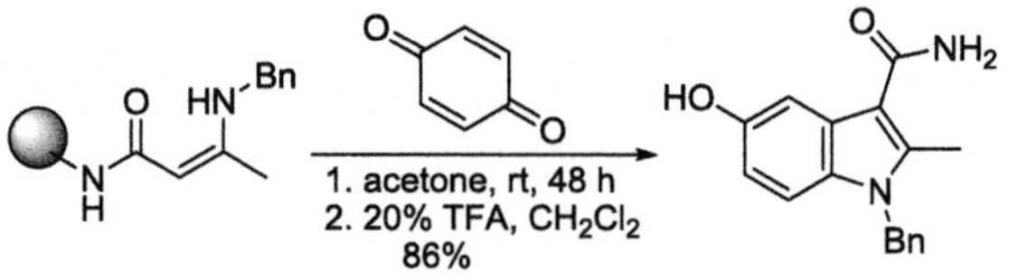

1. acetone, rt, 48 h
2. 20% TFA, CH$_2$Cl$_2$
86%

Example 2[6]

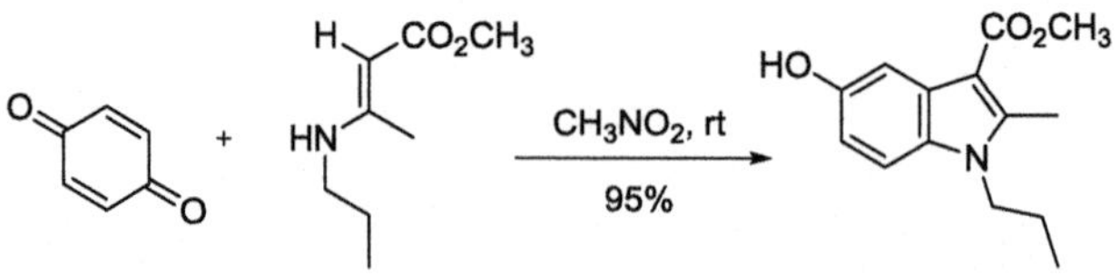

CH$_3$NO$_2$, rt

95%

Example 3[7]

R$_4$-NH$_2$
rt

HOAc, rt

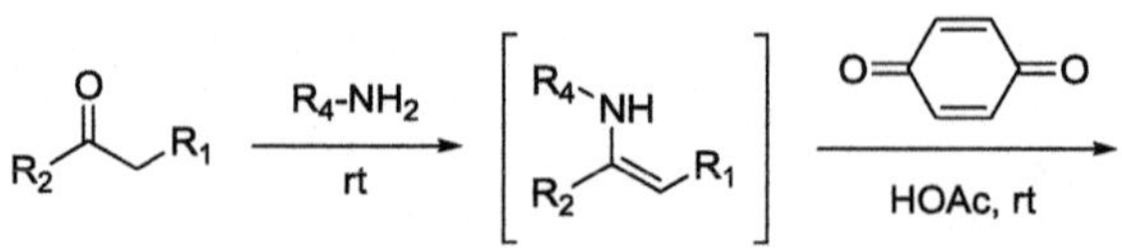

J.J. Li, *Name Reactions: A Collection of Detailed Mechanisms and Synthetic Applications*,
DOI 10.1007/978-3-319-03979-4_191, © Springer International Publishing Switzerland 2014

Example 4[10]

Example 4[12]

## References

1. Nenitzescu, C. D. *Bull. Soc. Chim. Romania* **1929**, *11*, 37–43.
2. Allen, G. R., Jr. *Org. React.* **1973**, *20*, 337–454. (Review).
3. Kinugawa, M.; Arai, H.; Nishikawa, H.; Sakaguchi, A.; Ogasa, T.; Tomioka, S.; Kasai, M. *J. Chem. Soc., Perkin Trans. 1* **1995**, 2677–2681.
4. Mukhanova, T. I.; Panisheva, E. K.; Lyubchanskaya, V. M.; Alekseeva, L. M.; Sheinker, Y. N.; Granik, V. G. *Tetrahedron* **1997**, *53*, 177–184.
5. Ketcha, D. M.; Wilson, L. J.; Portlock, D. E. *Tetrahedron Lett.* **2000**, *41*, 6253–6257.
6. Brase, S.; Gil, C.; Knepper, K. *Bioorg. Med. Chem.* **2002**, *10*, 2415–2418.
7. Böhme, T. M.; Augelli-Szafran, C. E.; Hallak, H.; Pugsley, T.; Serpa, K.; Schwarz, R. D. *J. Med. Chem.* **2002**, *45*, 3094–3102.
8. Schenck, L. W.; Sippel, A.; Kuna, K.; Frank, W.; Albert, A.; Kucklaender, U. *Tetrahedron* **2005**, *61*, 9129–9139.
9. Li, J.; Cook, J. M. *Nenitzescu indole synthesis*. In *Name Reactions in Heterocyclic Chemistry*; Li, J. J., Ed.; Wiley: Hoboken, NJ, **2005**, pp 145–153. (Review).
10. Velezheva, V. S.; Sokolov, A. I.; Kornienko, A. G.; Lyssenko, K. A.; Nelyubina, Y. V.; Godovikov, I. A.; Peregudov, A. S.; Mironov, A. F. *Tetrahedron Lett.* **2008**, *49*, 7106–7109.
11. Inman, M.; Moody, C. J. *Chem. Sci.* **2013**, *4*, 29–41. (Review).
12. Suryavanshi, P. A.; Sridharan, V.; Menendez, J. C. *Tetrahedron* **2013**, *69*, 5401–5406.

# Newman–Kwart rearrangement

Transformation of phenol to the corresponding thiophenol, a variant of the Smiles reaction.

The Newman–Kwart rearrangement is a member of a series of related rearrangements, such as the **Schönberg rearrangement** and the **Chapman rearrangement** (page 105), in which aryl groups migrate intramolecularly between nonadjacent atoms. The Schönberg rearrangement is the most similar and involves the 1,3-migration of an aryl group from oxygen to sulfur in a diarylthiocarbonate. The Chapman rearrangement involves an analogous migration but to nitrogen.

Schönberg rearrangement                          Chapman rearrangement

Example 1[5]

J.J. Li, *Name Reactions: A Collection of Detailed Mechanisms and Synthetic Applications*,
DOI 10.1007/978-3-319-03979-4_192, © Springer International Publishing Switzerland 2014

Example 2[6]

Example 3[7]

## References

1. (a) Kwart, H.; Evans, E. R. *J. Org. Chem.* **1966**, *31*, 410–413. (b) Newman, M. S.; Karnes, H. A. *J. Org. Chem.* **1966**, *31*, 3980–3984. (c) Newman, M. S.; Hetzel, F. W. *J. Org. Chem.* **1969**, *34*, 3604–3606.

2. Cossu, S.; De Lucchi, O.; Fabbri, D.; Valle, G.; Painter, G. F.; Smith, R. A. J. *Tetrahedron* **1997**, *53*, 6073–6084.

3. Lin, S.; Moon, B.; Porter, K. T.; Rossman, C. A.; Zennie, T.; Wemple, J. *Org. Prep. Proc. Int.* **2000**, *32*, 547–555.

4. Ponaras, A. A.; Zain, Ö. In *Encyclopedia of Reagents for Organic Synthesis,* Paquette, L. A., Ed.; Wiley: New York, **1995**, 2174–2176. (Review).

5. Kane, V. V.; Gerdes, A.; Grahn, W.; Ernst, L.; Dix, I.; Jones, P. G.; Hopf, H. *Tetrahedron Lett.* **2001**, *42*, 373–376.

6. Albrow, V.; Biswas, K.; Crane, A.; Chaplin, N.; Easun, T.; Gladiali, S.; Lygo, B.; Woodward, S. *Tetrahedron: Asymmetry* **2003**, *14*, 2813–2819.

7. Bowden, S. A.; Burke, J. N.; Gray, F.; McKown, S.; Moseley, J. D.; Moss, W. O.; Murray, P. M.; Welham, M. J.; Young, M. J. *Org. Proc. Res. Dev.* **2004**, *8*, 33–44.

8. Nicholson, G.; Silversides, J. D.; Archibald, S. J. *Tetrahedron Lett.* **2006**, *47*, 6541–6544.

9. Gilday, J. P.; Lenden, P.; Moseley, J. D.; Cox, B. G. *J. Org. Chem.* **2008**, *73*, 3130–3134.

10. Lloyd-Jones, G. C.; Moseley, J. D.; Renny, J. S. *Synthesis* **2008**, 661–689.

11. Tilstam, U.; Defrance, T.; Giard, T.; Johnson, M. D. *Org. Proc. Res. Dev.* **2009**, *13*, 321–323.

12. Das, J.; Le Cavelier, F.; Rouden, J.; Blanchet, J. *Synthesis* **2012**, *44*, 1349–1352.

# Nicholas reaction

Hexacarbonyldicobalt-stabilized propargyl cation is captured by a nucleophile. Subsequent oxidative demetallation then gives the propargylated product.

propargyl cation intermediate (stabilized by the hexacarbonyldicobalt complex).

Example 1, A chromium variant of the Nicholas reaction[3]

J.J. Li, *Name Reactions: A Collection of Detailed Mechanisms and Synthetic Applications*, DOI 10.1007/978-3-319-03979-4_193, © Springer International Publishing Switzerland 2014

Example 2, A Nicholas-Pauson–Khand sequence[4]

Example 3, Intramolecular Nicholas reaction using chromium[7]

Example 4[9]

**References**
1.  Nicholas, K. M.; Pettit, R. *J. Organomet. Chem.* **1972**, *44*, C21–C24.
2.  Nicholas, K. M. *Acc. Chem. Res.* **1987**, *20*, 207–214. (Review).
3.  Corey, E. J.; Helal, C. J. *Tetrahedron Lett.* **1996**, *37*, 4837–4840.
4.  Jamison, T. F.; Shambayati, S.; Crowe, W. E.; Schreiber, S. L. *J. Am. Chem. Soc.* **1997**, *119*, 4353–4363.
5.  Teobald, B. J. *Tetrahedron* **2002**, *58*, 4133–4170. (Review).
6.  Takase, M.; Morikawa, T.; Abe, H.; Inouye, M. *Org. Lett.* **2003**, *5*, 625–628.
7.  Ding, Y.; Green, J. R. *Synlett* **2005**, 271–274.
8.  Pinacho Crisóstomo, F. R.; Carrillo, R.; Martin, T.; Martin, V. S. *Tetrahedron Lett.* **2005**, *46*, 2829–2832.
9.  Hamajima, A.; Isobe, M. *Org. Lett.* **2006**, *8*, 1205–1208.
10. Shea, K. M. *Nicholas Reaction.* In *Name Reactions for Homologations-Part I*; Li, J. J., Ed.; Wiley: Hoboken, NJ, **2009**, pp 284–298. (Review).
11. Mukai, C.; Kojima, T.; Kawamura, T.; Inagaki, F. *Tetrahedron* **2013**, *69*, 7659–7669.

# Nicolaou IBX dehydrogenation

α,β-Unsaturation of aldehydes and ketones mediated by stoichiometric amounts of
o-iodoxybenzoic acid (IBX), alternative to the Saegusa oxidation.

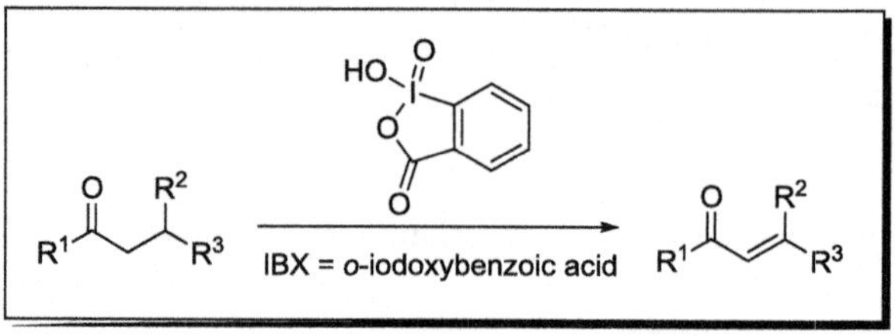

A SET mechanism has also been proposed. Additionally, silyl enol ethers are
also viable substrates.

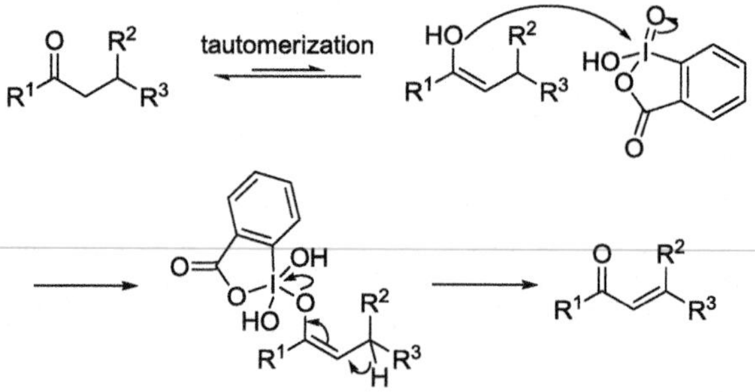

Example 1[1a]

Example 2[3]

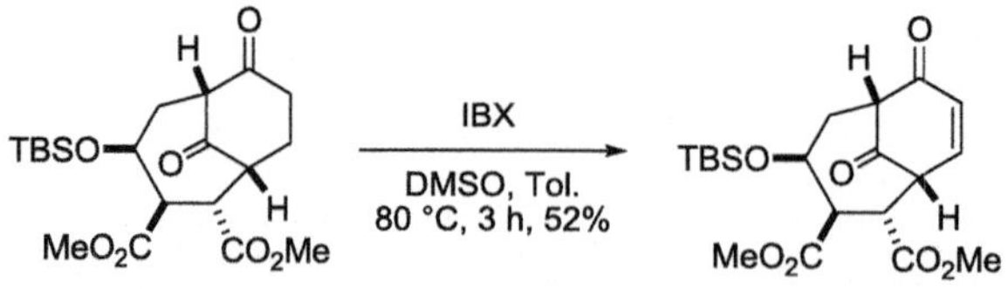

Example 3[7]

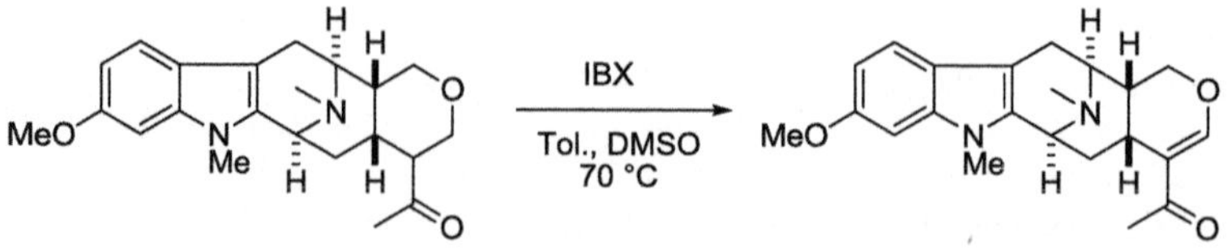

J.J. Li, *Name Reactions: A Collection of Detailed Mechanisms and Synthetic Applications*,
DOI 10.1007/978-3-319-03979-4_194, © Springer International Publishing Switzerland 2014

Example 4, *o*-Methyl-IBX (Me-IBX)[9]

$$R_1 \overset{S}{\diagup} R_2 \quad \xrightarrow[\substack{CH_3CN, \ reflux \\ 40–90\%}]{} \quad R_1 \overset{O}{\underset{\parallel}{\diagup}} \overset{S}{\underset{}{}} R_2$$

Example 5, Stabilized IBX (SIBX)[10]

## References

1.  (a) Nicolaou, K. C.; Zhong, Y.-L.; Baran, P. S. *J. Am. Chem. Soc.* **2000,** *122,* 7596–7597.  (b) Nicolaou, K. C.; Montagnon, T.; Baran, P. S. *Angew. Chem. Int. Ed.* **2002,** *41,* 993–996.  (c) Nicolaou, K. C.; Gray, D. L.; Montagnon, T.; Harrison, S. T. *Angew. Chem. Int. Ed.* **2002,** *41,* 996–1000.
2.  Nagata, H.; Miyazawa, N.; Ogasawara, K. *Org. Lett.* **2001,** *3,* 1737–1740.
3.  Ohmori, N. *J. Chem. Soc., Perkin Trans. 1* **2002,** 755–767.
4.  Hayashi, Y.; Yamaguchi, J.; Shoji, M. *Tetrahedron* **2002,** *58,* 9839–9846.
5.  Shimokawa, J.; Shirai, K.; Tanatani, A.; Hashimoto, Y.; Nagasawa, K. *Angew. Chem. Int. Ed.* **2004,** *43,* 1559–1562.
6.  Smith, N. D.; Hayashida. J.; Rawal, V. H. *Org. Lett.* **2005,** *7,* 4309–4312.
7.  Liu, X.; Deschamp, J. R.; Cook, J. M. *Org. Lett.* **2002,** *4,* 3339–3342.
8.  Herzon, S. B.; Myers, A. G. *J. Am. Chem. Soc.* **2005,** *127,* 5342–5344.
9.  Moorthy, J. N.; Singhal, N.; Senapati, K. *Tetrahedron Lett.* **2008,** *49,* 80–84.
10. Pouységu, L.; Marguerit, M.; Gagnepain, J.; Lyvinec, G.; Eatherton, A. J.; Quideau, S. *Org. Lett.* **2008,** *10,* 5211–5214.
11. Raghavan, S.; Babu, Vaddela S. *Tetrahedron* **2011,** *69,* 2044–2050.

# Noyori asymmetric hydrogenation

Asymmetric reduction of carbonyls and alkenes *via* hydrogenation, catalyzed by a ruthenium(II) BINAP complex.

$(R)$-BINAP-Ru =

$$[RuCl_2(binap)(solv)_2] \xrightarrow[- HCl]{H_2} [RuHCl(binap)(solv)_2]$$

The catalytic cycle:

J.J. Li, *Name Reactions: A Collection of Detailed Mechanisms and Synthetic Applications*,
DOI 10.1007/978-3-319-03979-4_195, © Springer International Publishing Switzerland 2014

Example 1[1b]

Ru[(S)-BINAP](CF₃CO₂)₂

30 atm H₂, rt, 92% ee

Example 2[1c]

Ru[(R)-BINAP]Cl₂

100 atm H₂, rt, 92% ee

Example 3[9]

5 bar H₂
3.2 mol% Ru(II)-(+)-(R)-BINAP

MeOH, 70 °C, 24 h, 90%

Example 4[10]

100 atm H₂
Ru[(S)-BINAP]Cl₂

EtOH, rt, 75%
98% ee

Example 5[11]

IPA/35%HCl/LiCl
H₂ (85–90 psi), 65 °C
93%

96% ee; 94% de

## References

1.  (a) Noyori, R.; Ohta, M.; Hsiao, Y.; Kitamura, M.; Ohta, T.; Takaya, H. *J. Am. Chem. Soc.* **1986,** *108*, 7117–7119.  Ryoji Noyori (Japan, 1938–) and William S. Knowles (USA, 1917–) shared half of the Nobel Prize in Chemistry in 2001 for their work on chirally catalyzed hydrogenation reactions. K. Barry Sharpless (USA, 1941–) shared the other half for his work on chirally catalyzed oxidation reactions. (b) Takaya, H.; Ohta, T.; Sayo, N.; Kumobayashi, H.; Akutagawa, S.; Inoue, S.; Kasahara, I.; Noyori, R. *J. Am. Chem. Soc.* **1987,** *109*, 1596–1598. (c) Kitamura, M.; Ohkuma, T.; Inoue, S.; Sayo, N.; Kumobayashi, H.; Akutagawa, S.; Ohta, T.; Takaya, H.; Noyori, R. *J. Am. Chem. Soc.* **1988,** *110*, 629–631. (d) Noyori, R.; Ohkuma, T.; Kitamura, H.; Ta-

kaya, H.; Sayo, H.; Kumobayashi, S.; Akutagawa, S. *J. Am. Chem. Soc.* **1987,** *109*, 5856–5858.   (e) Noyori, R.; Ohkuma, T. *Angew. Chem. Int. Ed.* **2001,** *40*, 40–73. (Review). (f) Noyori, R. *Angew. Chem. Int. Ed.* **2002,** *41*, 2008–2022. (Review, Nobel Prize Address).

2.   Noyori, R. In *Asymmetric Catalysis in Organic Synthesis;* Ojima, I., ed.; Wiley: New York, **1994,** Chapter 2. (Review).

3.   Chung, J. Y. L.; Zhao, D.; Hughes, D. L.; McNamara, J. M.; Grabowski, E. J. J.; Reider, P. J. *Tetrahedron Lett.* **1995,** *36*, 7379–7382.

4.   Bayston, D. J.; Travers, C. B.; Polywka, M. E. C. *Tetrahedron: Asymmetry* **1998,** *9*, 2015–2018.

5.   Berkessel, A.; Schubert, T. J. S.; Mueller, T. N. *J. Am. Chem. Soc.* **2002,** *124*, 8693–8698.

6.   Fujii, K.; Maki, K.; Kanai, M.; Shibasaki, M. *Org. Lett.* **2003,** *5*, 733–736.

7.   Ishibashi, Y.; Bessho, Y.; Yoshimura, M.; Tsukamoto, M.; Kitamura, M. *Angew. Chem. Int. Ed.* **2005,** *44*, 7287–7290.

8.   Lall, M. S. *Noyori Asymmetric Hydrogenation*, In *Name Reactions for Functional Group Transformations*; Li, J. J., Ed.; Wiley: Hoboken, NJ, **2007,** pp 46–66. (Review).

9.   Bouillon, M. E.; Meyer, H. H. *Tetrahedron* **2007,** *63*, 2712–2723.

10.   Case-Green, S. C.; Davies, S. G.; Roberts, P. M.; Russell, A. J.; Thomson, J. E. *Tetrahedron: Asymmetry* **2008,** *19*, 2620–2631.

11.   Magnus, N. A.; Astleford, B. A.; Laird, D. L. T.; Maloney, T. D.; McFarland, A. D.; Rizzo, J. R.; Ruble, J. C.; Stephenson, G. A.; Wepsiec, J. P. *J. Org. Chem.* **2013,** *78*, 5768–5774.

# Nozaki–Hiyama–Kishi reaction

Cr–Ni bimetallic catalyst-promoted redox addition of vinyl- or propargyl-halides to aldehydes.

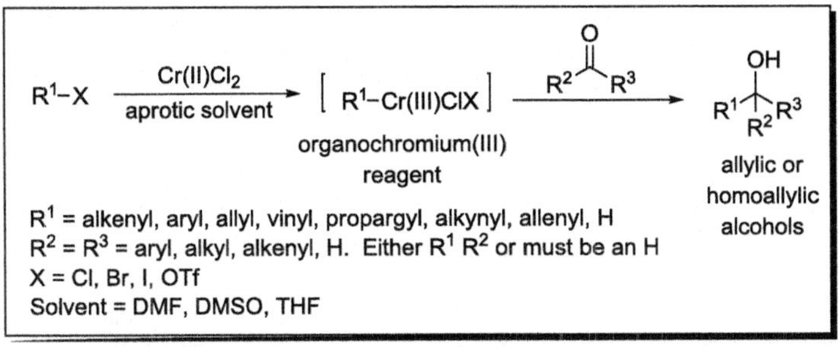

The catalytic cycle:[2]

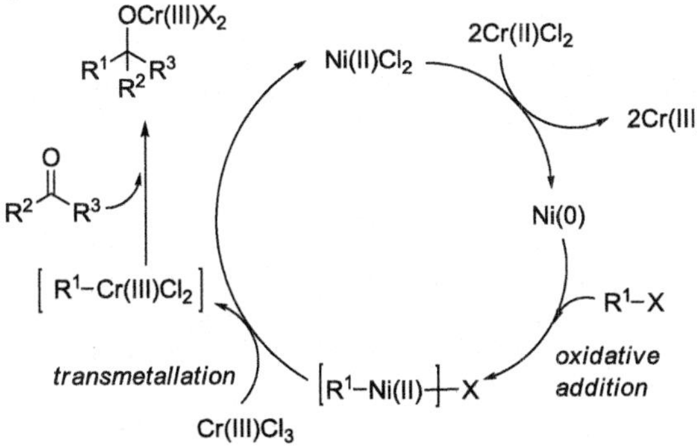

Example 1[3]

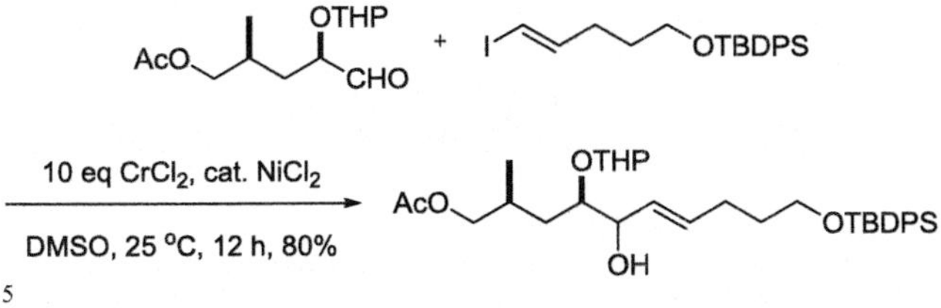

Example 2[5]

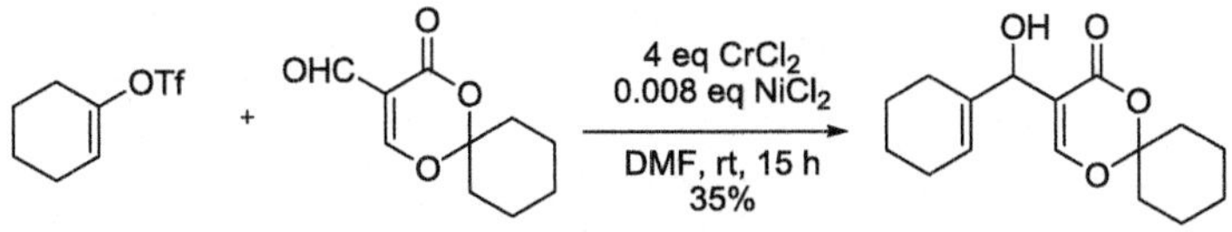

Example 3, Intramolecular Nozaki–Hiyama–Kishi reaction[8]

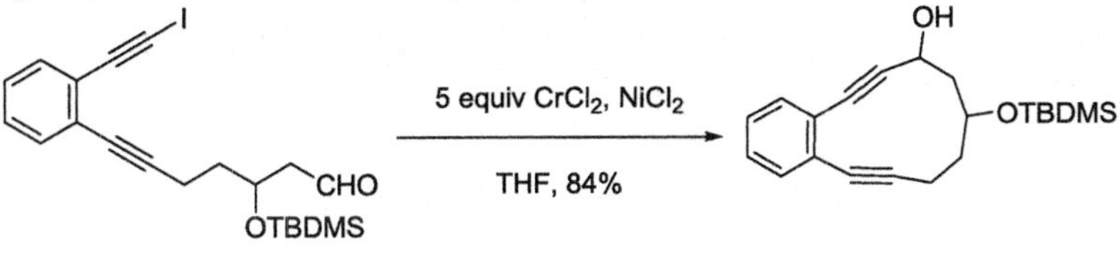

J.J. Li, *Name Reactions: A Collection of Detailed Mechanisms and Synthetic Applications*,
DOI 10.1007/978-3-319-03979-4_196, © Springer International Publishing Switzerland 2014

Example 4, Intramolecular Nozaki–Hiyama–Kishi reaction[9]

1. HCl, THF, 0.003 M dark

2. CrCl₂, NiCl₂, DMSO
0.0025 M, 50 °C
37%, 2 steps

Example 5, Asymmetric Nozaki–Hiyama–Kishi reaction[11]

cat.

NiCl₂, CrCl₂
Et₃N, THF, 25 °C
> 60%

## References

1. (a) Okude, C. T.; Hirano, S.; Hiyama, T.; Nozaki, H. *J. Am. Chem. Soc.* **1977**, *99*, 3179–3181. Hitosi Nozaki and T. Hiyama are professors at the Japanese Academy. (b) Takai, K.; Kimura, K.; Kuroda, T.; Hiyama, T.; Nozaki, H. *Tetrahedron Lett.* **1983**, *24*, 5281–5284. Kazuhiko Takai was Prof. Nozaki's student during the discovery of the reaction and is a professor at Okayama University. (c) Jin, H.; Uenishi, J.; Christ, W. J.; Kishi, Y. *J. Am. Chem. Soc.* **1986**, *108*, 5644–5646. Yoshito Kishi at Harvard independently discovered the catalytic effect of nickel during his total synthesis of polytoxin. (d) Takai, K.; Tagahira, M.; Kuroda, T.; Oshima, K.; Utimoto, K.; Nozaki, H. *J. Am. Chem. Soc.* **1986**, *108*, 6048–6050. (e) Kress, M. H.; Ruel, R.; Miller, L. W. H.; Kishi, Y. *Tetrahedron Lett.* **1993**, *34*, 5999–6002.
2. Fürstner, A.; Shi, N. *J. Am. Chem. Soc.* **1996**, *118*, 12349–12357. (The catalytic cycle).
3. Chakraborty, T. K.; Suresh, V. R. *Chem. Lett.* **1997**, 565–566.
4. Fürstner, A. *Chem. Rev.* **1999**, *99*, 991–1046. (Review).
5. Blaauw, R. H.; Benningshof, J. C. J.; van Ginkel, A. E.; van Maarseveen, J. H.; Hiemstra, H. *J. Chem. Soc., Perkin Trans. 1* **2001**, 2250–2256.
6. Berkessel, A.; Menche, D.; Sklorz, C. A.; Schroder, M.; Paterson, I. *Angew. Chem. Int. Ed.* **2003**, *42*, 1032–1035.
7. Takai, K. *Org. React.* **2004**, *64*, 253–612. (Review).
8. Karpov, G. V.; Popik, V. V. *J. Am. Chem. Soc.* **2007**, *129*, 3792–3793.
9. Valente, C.; Organ, M. G. *Chem. Eur. J.* **2008**, *14*, 8239–8245.
10. Yet, L. *Nozaki–Hiyama–Kishi reaction.* In *Name Reactions for Homologations-Part I*; Li, J. J., Ed.; Wiley: Hoboken, NJ, **2009**, pp 299–318. (Review).
11. Austad, B. C.; Benayoud, F.; Calkins, T. L.; et al. *Synlett* **2013**, *17*, 327–332.

# Nysted reagent

The Nysted reagent, cyclo-dibromodi-μ-methylene(μ-tetrahydrofuran)trizinc, is used for the olefination of ketones and aldehydes.

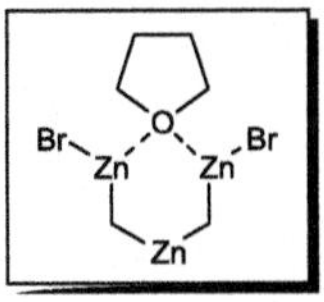

Example 1, The Wittig reagent opened the lactone:[6]

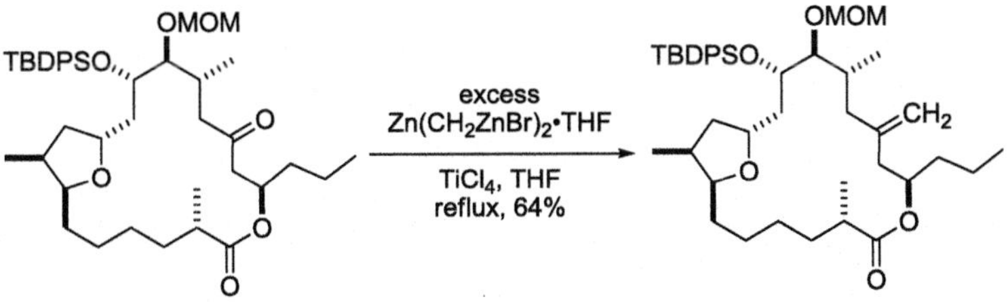

excess
Zn(CH₂ZnBr)₂·THF

TiCl₄, THF
reflux, 64%

Example 2[8]

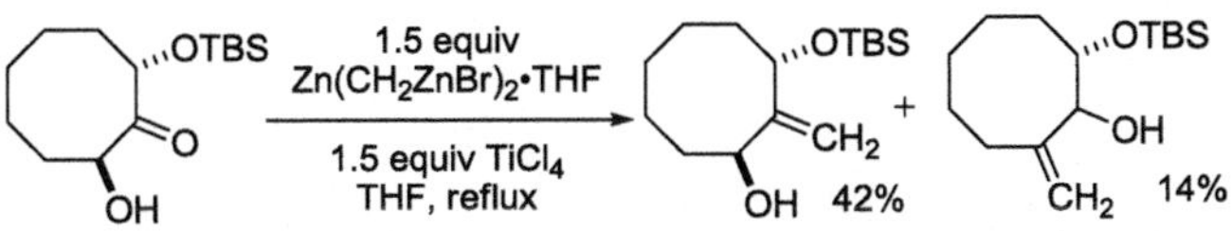

1.5 equiv
Zn(CH₂ZnBr)₂·THF

1.5 equiv TiCl₄
THF, reflux

Example 3[9]

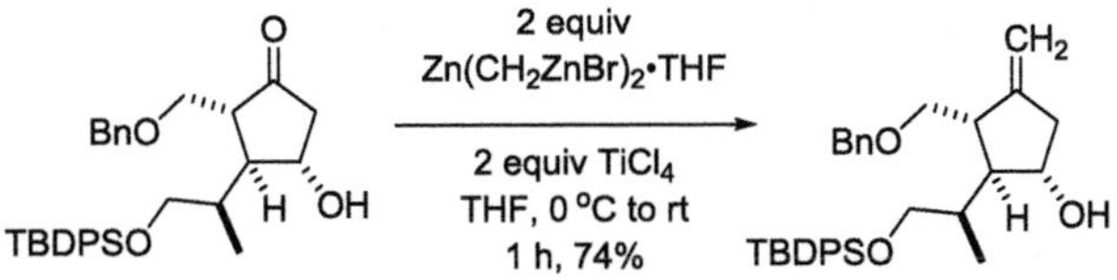

2 equiv
Zn(CH₂ZnBr)₂·THF

2 equiv TiCl₄
THF, 0 °C to rt
1 h, 74%

Example 4[11]

3 equiv Nysted reagent
2.5 equiv Ti(Oi-Pr)₂Cl₂

THF, 0–15 °C, 10–20 min.
70%

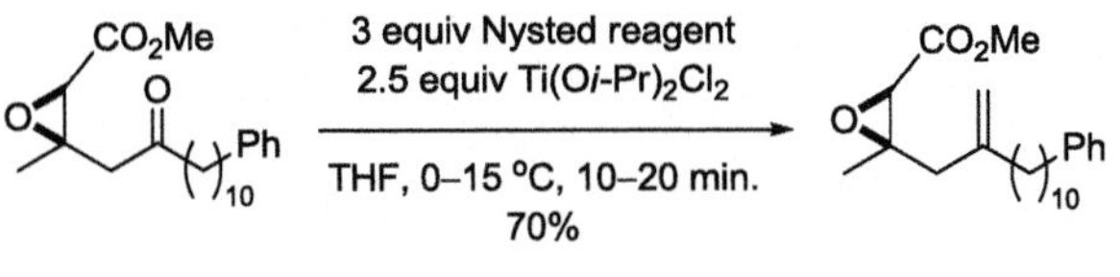

# References

1.    Nysted, L. N. US Patent 3,865,848 (**1975**).
2.    Tochtermann, W.; Bruhn, S.; Meints, M.; Wolff, C.; Peters, E.-M.; Peters, K.; von Schnering, H. G. *Tetrahedron* **1995**, *51*, 1623–1630.

J.J. Li, *Name Reactions: A Collection of Detailed Mechanisms and Synthetic Applications*,
DOI 10.1007/978-3-319-03979-4_197, © Springer International Publishing Switzerland 2014

3.      Matsubara, S.; Sugihara, M.; Utimoto, K. *Synlett* **1998,** 313–315.
4.      Tanaka, M.; Imai, M.; Fujio, M.; Sakamoto, E.; Takahashi, M.; Eto-Kato, Y.;
        Wu, X. M.; Funakoshi, K.; Sakai, K.; Suemune, H. *J. Org. Chem.* **2000,** *65,*
        5806–5816.
5.      Tarraga, A.; Molina, P.; Lopez, J. L.; Velasco, M. D. *Tetrahedron Lett.* **2001,** *42,*
        8989–8992.
6.      Aïssa, C.; Riveiros, R.; Ragot, J.; Fürstner, A. *J. Am. Chem. Soc.* **2003,** *125,*
        15512–15520.
7.      Clark, J. S.; Marlin, F.; Nay, B.; Wilson, C. *Org. Lett.* **2003,** *5,* 89–92.
8.      Paquette, L. A.; Hartung, R. E.; Hofferberth, J. E.; Vilotijevic, I.; Yang, J. *J. Org.
        Chem.* **2004,** *69,* 2454–2460.
9.      Hanessian, S.; Mainetti, E.; Lecomte, F. *Org. Lett.* **2006,** *8,* 4047–4049.
10.     Haahr, A.; Rankovic, Z.; Hartley, R. C. *Tetrahedron Lett.* **2011,** *52,* 3020–3022.
11.     Barnych, B.; Fenet, B.; Vatele, J.-M. *Tetrahedron* **2013,** *69,* 334–340.

# Oppenauer oxidation

Alkoxide-catalyzed oxidation of secondary alcohols. Reverse of the Meerwein–Ponndorf–Verley reduction.

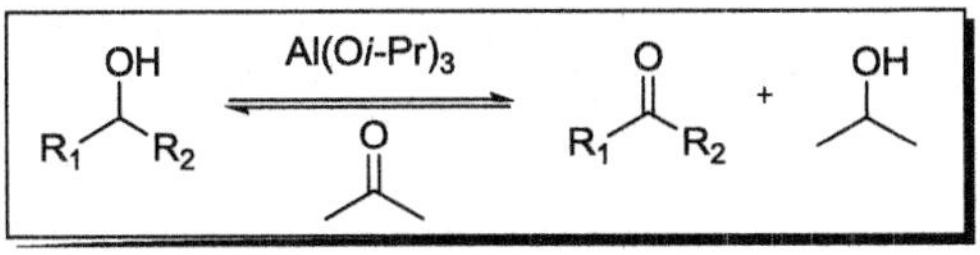

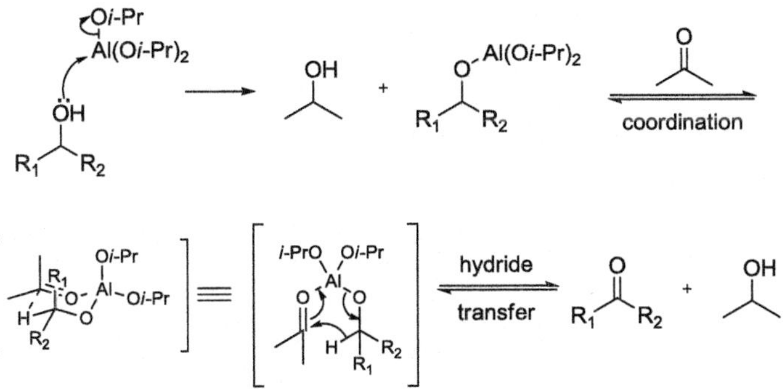

cyclic transition state

Example 1, Mg-Oppenauer oxidation[3]

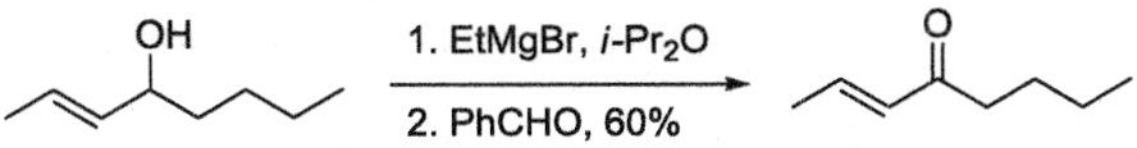

Example 2[6]

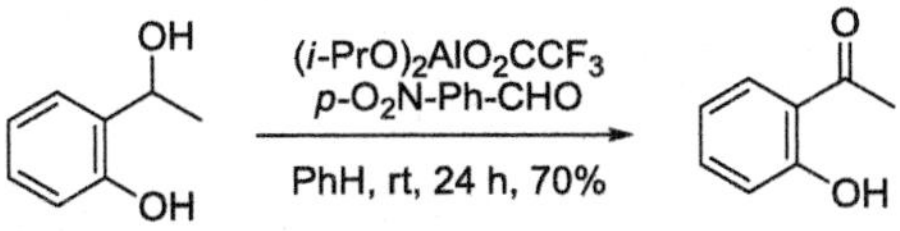

Example 3, Mg-Oppenauer oxidation[8]

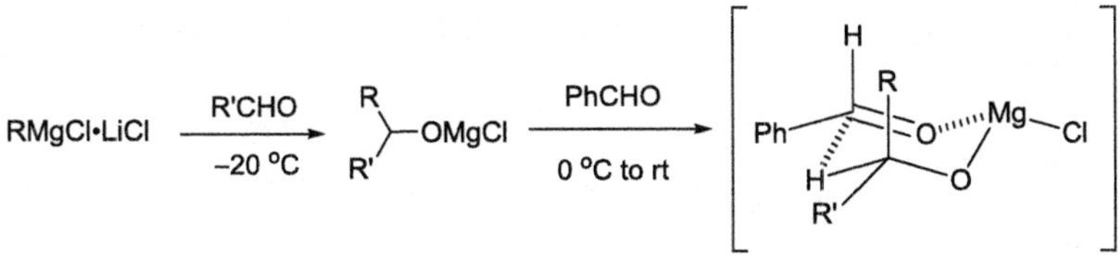

J.J. Li, *Name Reactions: A Collection of Detailed Mechanisms and Synthetic Applications*,
DOI 10.1007/978-3-319-03979-4_198, © Springer International Publishing Switzerland 2014

Example 4[10]

Example 5, Tandem nucleophilic addition–Oppenauer oxidation[12]

## References

1.  Oppenauer, R. V. *Rec. Trav. Chim.* **1937,** *56*, 137–144.   Rupert V. Oppenauer (1910–), born in Burgstall, Italy, studied at ETH in Zurich under Ruzicka and Reichstei, both Nobel laureates. After a string of academic appointments around Europe and a stint at Hoffman–La Roche, Oppenauer worked for the Ministry of Public Health in Buenos Aires, Argentina.
2.  Djerassi, C. *Org. React.* **1951,** *6*, 207–235. (Review).
3.  Byrne, B.; Karras, M. *Tetrahedron Lett.* **1987,** *28*, 769–772.
4.  Ooi, T.; Otsuka, H.; Miura, T.; Ichikawa, H.; Maruoka, K. *Org. Lett.* **2002,** *4*, 2669–2672.
5.  Suzuki, T.; Morita, K.; Tsuchida, M.; Hiroi, K. *J. Org. Chem.* **2003,** *68*, 1601–1602.
6.  Auge, J.; Lubin-Germain, N.; Seghrouchni, L. *Tetrahedron Lett.* **2003,** *44*, 819–822.
7.  Hon, Y.-S.; Chang, C.-P.; Wong, Y.-C. Byrne, B.; Karras, M. *Tetrahedron Lett.* **2004,** *45*, 3313–3315.
8.  Kloetzing, R. J.; Krasovskiy, A.; Knochel, P. *Chem. Eur. J.* **2007,** *13*, 215–227.
9.  Fuchter, M. J. *Oppenauer Oxidation.* In *Name Reactions for Functional Group Transformations*; Li, J. J., Ed.; Wiley: Hoboken, NJ, **2007,** pp 265–373. (Review).
10. Mello, R.; Martinez-Ferrer, J.; Asensio, G.; Gonzalez-Nunez, M. E. *J. Org. Chem.* **2008,** *72*, 9376–9378.
11. Borzatta, V.; Capparella, E.; Chiappino, R.; Impala, D.; Poluzzi, E.; Vaccari, A. *Cat. Today* **2009,** *140*, 112–116.
12. Fu, Y.; Yang, Y.; Hügel, H. M.; Du, Z.; Wang, K.; Huang, D.; Hu, Y. *Org. Biomol. Chem.* **2013,** *11*, 4429–4432.

## Overman rearrangement

Stereoselective transformation of allylic alcohol to allylic trichloroacetamide *via* trichloroacetimidate intermediate.

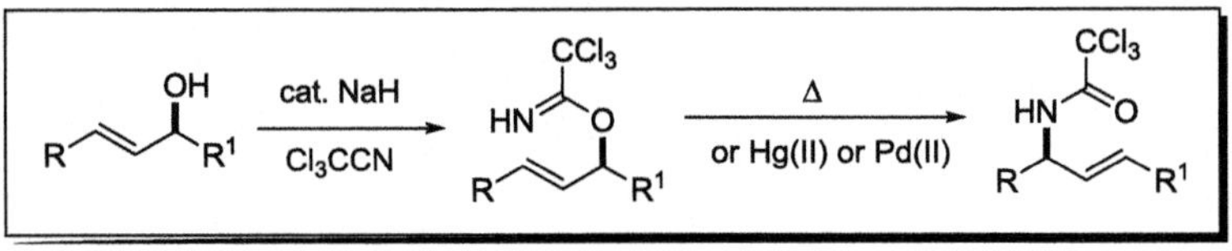

trichloroacetimidate

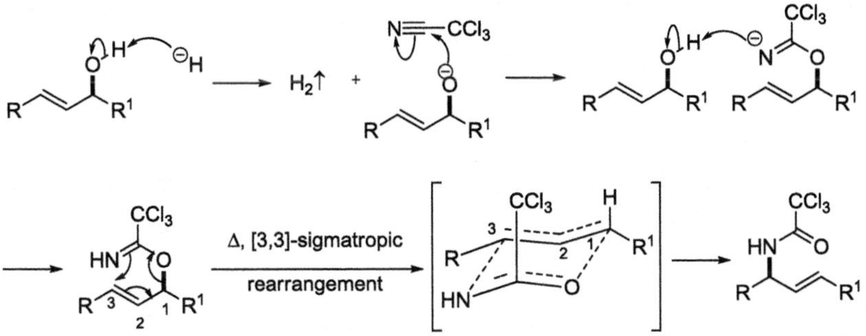

Example 1[5]

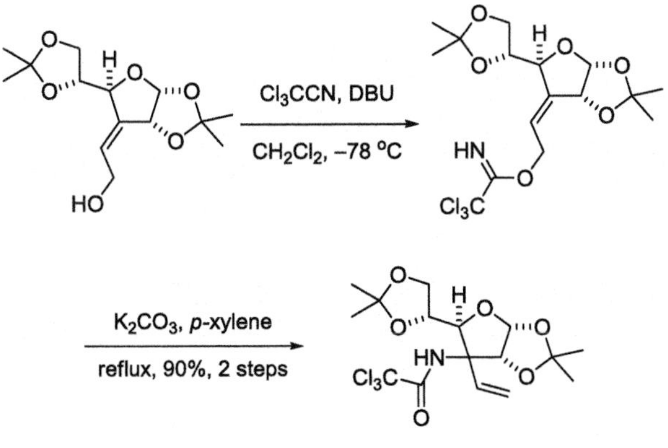

Example 2[6]

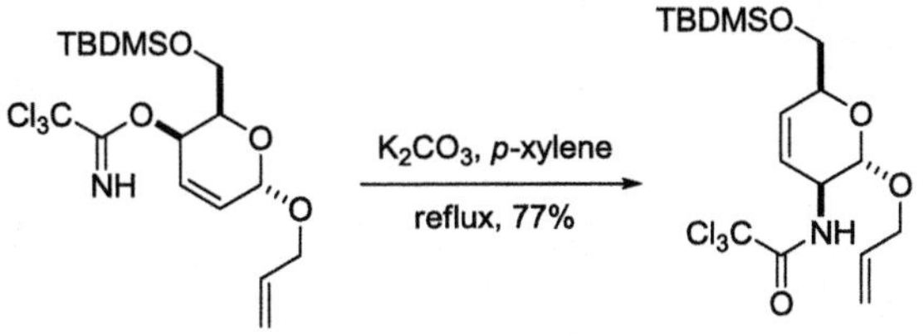

Example 3[7]

J.J. Li, *Name Reactions: A Collection of Detailed Mechanisms and Synthetic Applications*, DOI 10.1007/978-3-319-03979-4_199, © Springer International Publishing Switzerland 2014

Example 4[9]

Example 5, Cascade-type Overman rearrangement[11]

## References

1. (a) Overman, L. E. *J. Am. Chem. Soc.* **1974**, *96*, 597–599. (b) Overman, L. E. *J. Am. Chem. Soc.* **1976**, *98*, 2901–2910. (c) Overman, L. E. *Acc. Chem. Res.* **1980**, *13*, 218–224. (Review).
2. Demay, S.; Kotschy, A.; Knochel, P. *Synthesis* **2001**, 863–866.
3. Oishi, T.; Ando, K.; Inomiya, K.; Sato, H.; Iida, M.; Chida, N. *Org. Lett.* **2002**, *4*, 151–154.
4. Reilly, M.; Anthony, D. R.; Gallagher, C. *Tetrahedron Lett.* **2003**, *44*, 2927–2930.
5. Tsujimoto, T.; Nishikawa, T.; Urabe, D.; Isobe, M. *Synlett* **2005**, 433–436.
6. Montero, A.; Mann, E.; Herradon, B. *Tetrahedron Lett.* **2005**, *46*, 401–405.
7. Hakansson, A. E.; Palmelund, A.; Holm, H.; Madsen, R. *Chem. Eur. J.* **2006**, *12*, 3243–3253.
8. Bøjstrup, M.; Fanejord, M.; Lundt, I. *Org. Biomol. Chem.* **2007**, *5*, 3164–3171.
9. Lamy, C.; Hifmann, J.; Parrot-Lopez, H.; Goekjian, P. *Tetrahedron Lett.* **2007**, *48*, 6177–6180.
10. Wu, Y.-J. *Overman Rearrangement*. In *Name Reactions for Homologations-Part II*; Li, J. J., Ed.; Wiley: Hoboken, NJ, 2009, pp 210–225. (Review).
11. Nakayama, Y.; Sekiya, R.; Oishi, H.; Hama, N.; Yamazaki, M.; Sato, T.; Chida, N. *Chem. Eur. J.* **2013**, *19*, 12052–12058.

# Paal thiophene synthesis

Thiophene synthesis from addition of a sulfur atom to 1,4-diketones and subsequent dehydration.

The reaction now is frequently carried out using the Lawesson's reagent. For the mechanism of carbonyl to thiocarbonyl transformation, see Lawesson's reagent.

Example 1[2]

Example 2[3]

## References

1.  (a) Paal, C. *Ber.* **1885,** *18*, 2251–2254. (b) Paal, C. *Ber.* **1885,** *18*, 367–371.
2.  Thomsen, I.; Pedersen, U.; Rasmussen, P. B.; Yde, B.; Andersen, T. P.; Lawesson, S.-O. *Chem. Lett.* **1983,** 809–810.
3.  Parakka, J. P.; Sadannandan, E. V.; Cava, M. P. *J. Org. Chem.* **1994,** *59*, 4308–4310.
4.  Kikuchi, K.; Hibi, S.; Yoshimura, H.; Tokuhara, N.; Tai, K.; Hida, T.; Yamauchi, T.; Nagai, M. *J. Med. Chem.* **2000,** *43*, 409–423.
5.  Sonpatki, V. M.; Herbert, M. R.; Sandvoss, L. M.; Seed, A. J. *J. Org. Chem.* **2001,** *66*, 7283–7286.
6.  Kiryanov, A. A.; Sampson, P.; Seed, A. J. *J. Org. Chem.* **2001,** *66*, 7925–7929.
7.  Mullins, R. J.; Williams, D. R. *Paal Thiophene Synthesis.* In *Name Reactions in Heterocyclic Chemistry*; Li, J. J., Ed.; Wiley: Hoboken, NJ, **2005,** 207–217. (Review).
8.  Kaniskan, N.; Elmali, D.; Civcir, P. U. *ARKIVOC* **2008,** 17–29.

J.J. Li, *Name Reactions: A Collection of Detailed Mechanisms and Synthetic Applications*, DOI 10.1007/978-3-319-03979-4_200, © Springer International Publishing Switzerland 2014

## Paal–Knorr furan synthesis

Acid-catalyzed cyclization of 1,4-diketones to form furans.

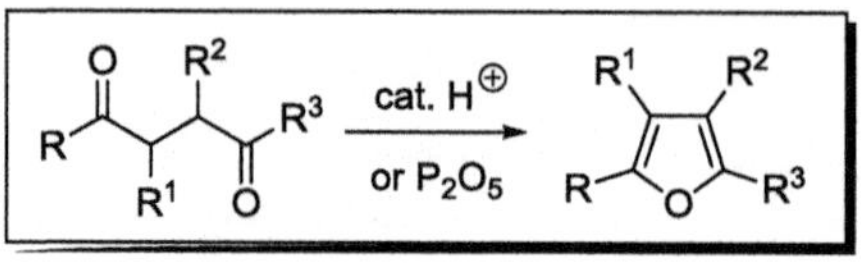

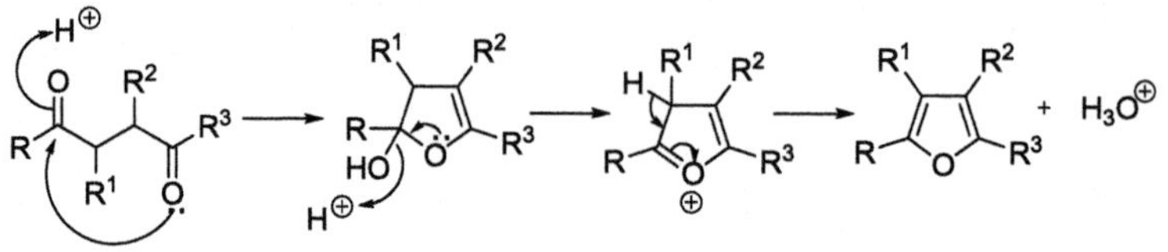

Example 1[3]

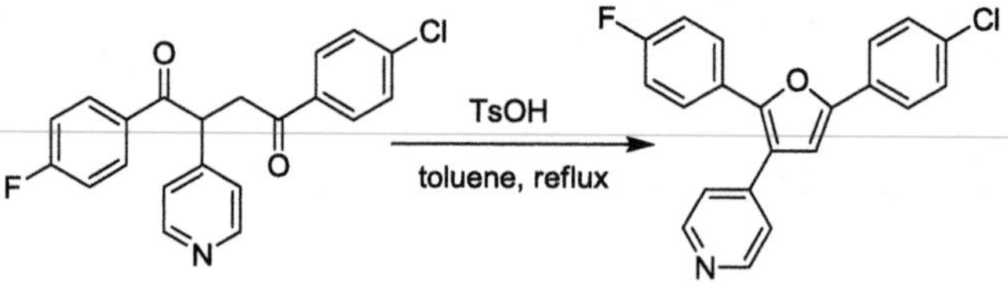

Example 2[6]

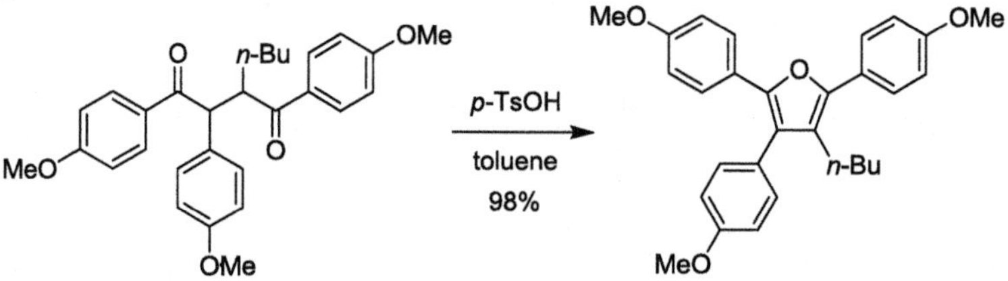

Example 3[9]

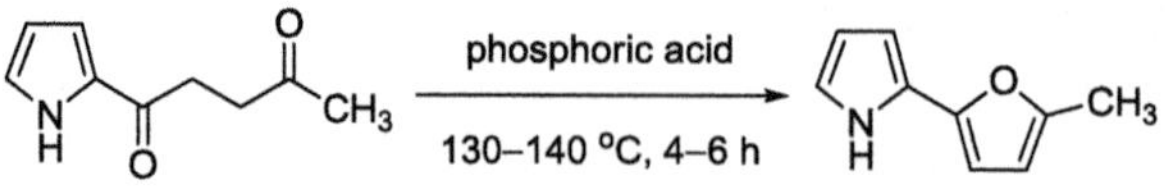

Example 4[10]

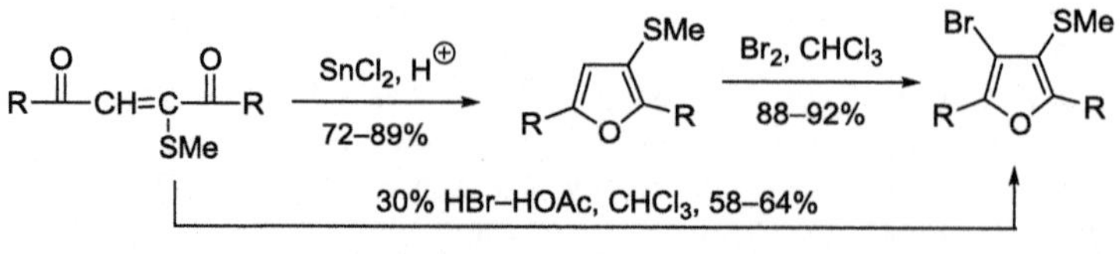

J.J. Li, *Name Reactions: A Collection of Detailed Mechanisms and Synthetic Applications*,
DOI 10.1007/978-3-319-03979-4_201, © Springer International Publishing Switzerland 2014

Example 5, Concurrent debromination along with furan formation[10]

## References

1. (a) Paal, C. *Ber.* **1884,** *17,* 2756–2767. (b) Knorr, L. *Ber.* **1885,** *17,* 2863–2870. (c) Paal, C. *Ber.* **1885,** *18,* 367–371.

2. Friedrichsen, W. Furans and Their Benzo Derivatives: Synthesis. In *Comprehensive Heterocyclic Chemistry II*; Katritzky, A. R., Rees, C. W., Scriven, E. F. V., Eds.; Pergamon: New York, **1996**; *Vol. 2,* 351–393. (Review).

3. de Laszlo, S. E.; Visco, D.; Agarwal, L.; *et al. Bioorg. Med. Chem. Lett.* **1998,** *8,* 2689–2694.

4. Gupta, R. R.; Kumar, M.; Gupta, V. *Heterocyclic Chemistry*, Springer: New York, **1999**; Vol. 2, 83–84. (Review).

5. Joule, J. A.; Mills, K. *Heterocyclic Chemistry*, 4th ed.; Blackwell Science: Cambridge, **2000**; 308–309. (Review).

6. Mortensen, D. S.; Rodriguez, A. L.; Carlson, K. E.; Sun, J.; Katzenellenbogen, B. S.; Katzenellenbogen, J. A. *J. Med. Chem.* **2001,** *44,* 3838–3848.

7. König, B. *Product Class 9: Furans.* In *Science of Synthesis: Houben–Weyl Methods of Molecular Transformations*; Maas, G., Ed.; Georg Thieme Verlag: New York, **2001**; *Cat. 2, Vol.* 9, 183–278. (Review).

8. Shea, K. M. *Paal–Knorr Furan Synthesis.* In *Name Reactions in Heterocyclic Chemistry*; Li, J. J., Ed.; Wiley: Hoboken, NJ, **2005**, pp 168–181. (Review).

9. Kaniskan, N.; Elmali, D.; Civcir, P. U. *ARKIVOC* **2008**, 17–29.

10. Yin, G.; Wang, Z.; Chen, A.; Gao, M.; Wu, A.; Pan, Y. *J. Org. Chem.* **2008,** *73,* 3377–3383.

11. Wang, G.; Guan, Z.; Tang, R.; He, Y. *Synth. Commun.* **2010,** *40,* 370–377.

## Paal–Knorr pyrrole synthesis

Reaction between 1,4-diketones and primary amines (or ammonia) to give pyrroles. A variation of the Knorr pyrazole synthesis.

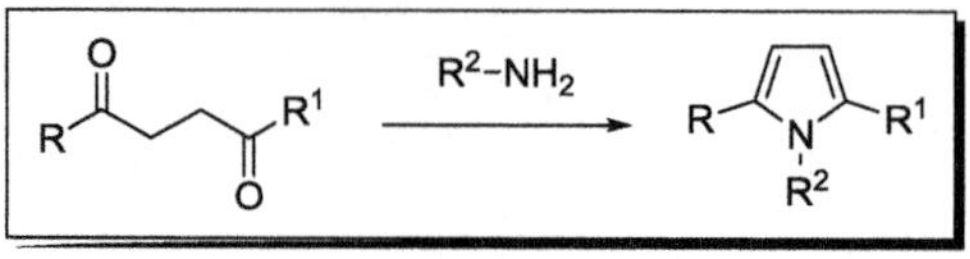

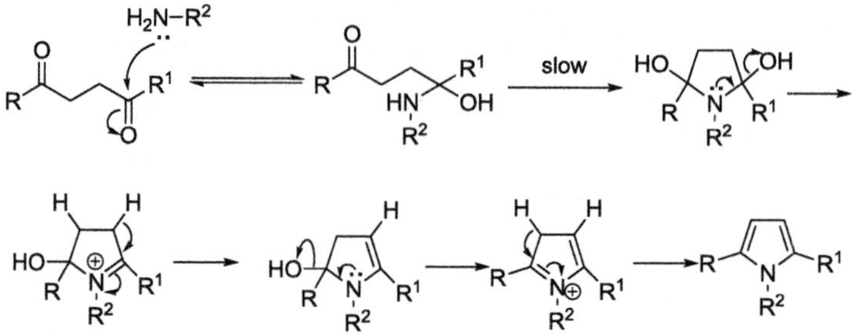

Example 1[4]

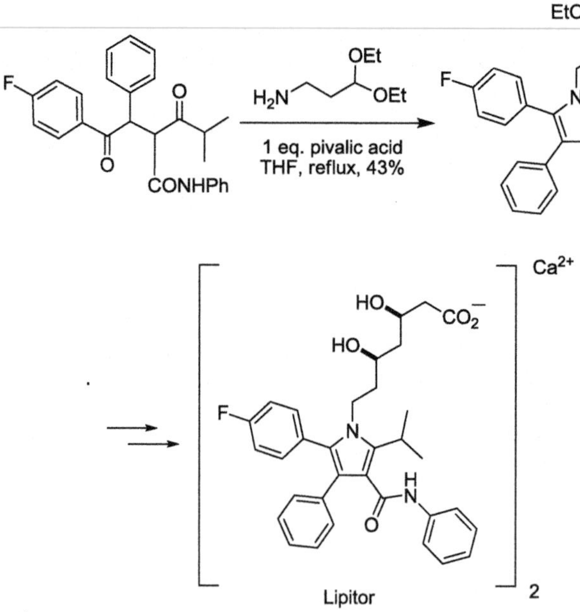

Example 2[5]

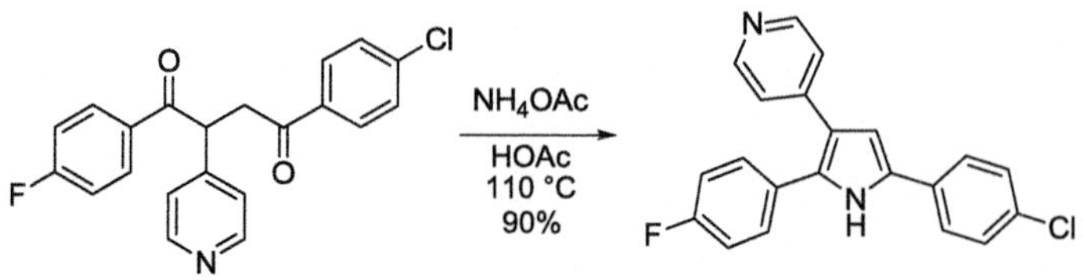

J.J. Li, *Name Reactions: A Collection of Detailed Mechanisms and Synthetic Applications*,
DOI 10.1007/978-3-319-03979-4_202, © Springer International Publishing Switzerland 2014

Example 3[9]

Example 4[10]

Example 5, Furan ring opening–pyrrole ring closure[10]

## References

1.  (a) Paal, C. *Ber.* **1885**, *18*, 367–371. (b) Paal, C. *Ber.* **1885**, *18*, 2251–2254. (c) Knorr, L. *Ber.* **1885**, *18*, 299–311.
2.  Corwin, A. H. *Heterocyclic Compounds Vol. 1*, Wiley, NY, **1950**; Chapter 6. (Review).
3.  Jones, R. A.; Bean, G. P. *The Chemistry of Pyrroles*, Academic Press, London, **1977**, pp 51–57, 74–79. (Review).
4.  (a) Brower, P. L.; Butler, D. E.; Deering, C. F.; Le, T. V.; Millar, A.; Nanninga, T. N.; Roth, B. D. *Tetrahedron Lett.* **1992**, *33*, 2279-2282. (b) Baumann, K. L.; Butler, D. E.; Deering, C. F.; Mennen, K. E.; Millar, A.; Nanninga, T. N.; Palmer, C. W.; Roth, B. D. *Tetrahedron Lett.* **1992**, *33*, 2279, 2283–2284.
5.  de Laszlo, S. E.; Visco, D.; Agarwal, L.; *et al. Bioorg. Med. Chem. Lett.* **1998**, *8*, 2689–2694.
6.  Braun, R. U.; Zeitler, K.; Müller, T. J. J. *Org. Lett.* **2001**, *3*, 3297–3300.
7.  Quiclet-Sire, B.; Quintero, L.; Sanchez-Jimenez, G.; Zard, Z. *Synlett* **2003**, 75–78.
8.  Gribble, G. W. *Knorr and Paal–Knorr Pyrrole Syntheses*. In *Name Reactions in Heterocyclic Chemistry*; Li, J. J., Corey, E. J., Eds, Wiley: Hoboken, NJ, **2005**, 77–88. (Review).
9.  Salamone, S. G.; Dudley, G. B. *Org. Lett.* **2005**, *7*, 4443–4445.
10. Fu, L.; Gribble, G. W. *Tetrahedron Lett.* **2008**, *49*, 7352–7354.
11. Trushkov, I. V.; Nevolina, T. A.; Shcherbinin, V. A.; Sorotskaya, L. N.; Butin, A. V. *Tetrahedron Lett.* **2013**, *54*, 3974–3976.

## Parham cyclization

The Parham cyclization is the generation by halogen–lithium exchange of aryllithiums and heteroaryllithiums, and their subsequent intramolecular cyclization onto an electrophilic site.

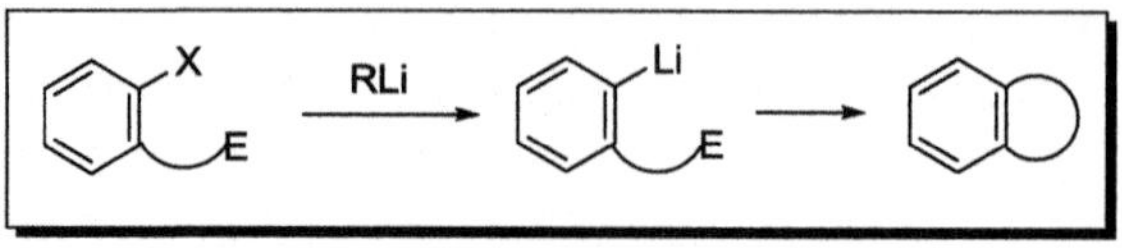

Example 1

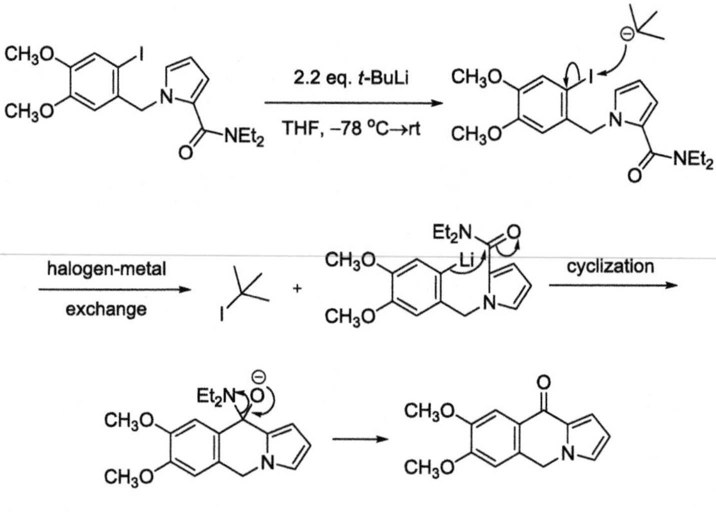

The fate of the second equivalent of *t*-BuLi:

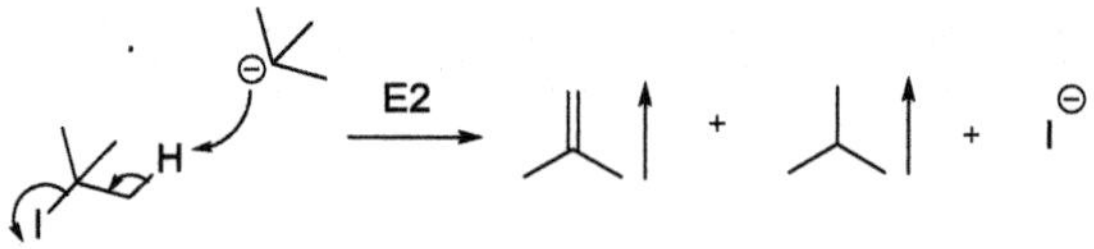

Example 2[2]

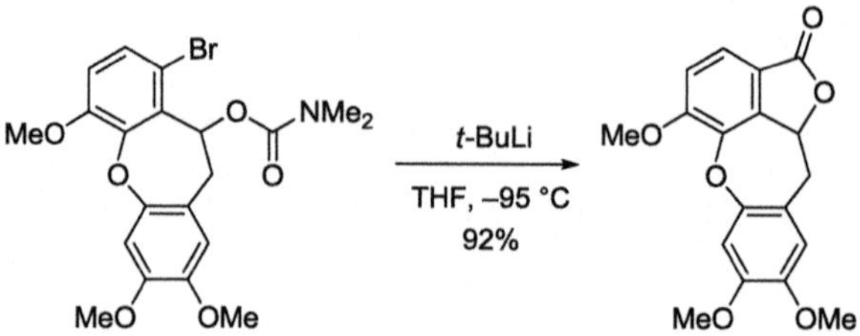

J.J. Li, *Name Reactions: A Collection of Detailed Mechanisms and Synthetic Applications*,
DOI 10.1007/978-3-319-03979-4_203, © Springer International Publishing Switzerland 2014

## Example 3[4]

$$n\text{-BuLi}$$
$$\text{Et}_2\text{O}, -78\ ^\circ\text{C}$$
$$64\%$$

## Example 4[5]

$$2\ \text{eq. } n\text{-BuLi}$$
$$\text{THF}, -78\ ^\circ\text{C}$$
$$2.5\ \text{h}, 83\%$$

## Example 5[9]

$$t\text{-BuLi, THF}, -100\ ^\circ\text{C}$$
$$\text{Ar, 30 min.}, 55\%$$

## References

1.  (a) Parham, W. E.; Jones, L. D.; Sayed, Y. *J. Org. Chem.* **1975**, *40*, 2394–2399. William E. Parham was a professor at Duke University. (b) Parham, W. E.; Jones, L. D.; Sayed, Y. *J. Org. Chem.* **1976**, *41*, 1184–1186. (c) Parham, W. E.; Bradsher, C. K. *Acc. Chem. Res.* **1982**, *15*, 300–305. (Review).
2.  Paleo, M. R.; Lamas, C.; Castedo, L.; Domínguez, D. *J. Org. Chem.* **1992**, *57*, 2029–2033.
3.  Gray, M.; Tinkl, M.; Snieckus, V. In *Comprehensive Organometallic Chemistry II*; Abel, E. W., Stone, F. G. A., Wilkinson, G., Eds.; Pergamon: Exeter, **1995**; Vol. 11; p 66. (Review).
4.  Gauthier, D. R., Jr.; Bender, S. L. *Tetrahedron Lett.* **1996**, *37*, 13–16.
5.  Collado, M. I.; Manteca, I.; Sotomayor, N.; Villa, M.-J.; Lete, E. *J. Org. Chem.* **1997**, *62*, 2080–2092.
6.  Mealy, M. M.; Bailey, W. F. *J. Organomet. Chem.* **2002**, *646*, 59–67. (Review).
7.  Sotomayor, N.; Lete, E. *Current Org. Chem.* **2003**, *7*, 275–300. (Review).
8.  González-Temprano, I.; Osante, I.; Lete, E.; Sotomayor, N. *J. Org. Chem.* **2004**, *69*, 3875–3885.
9.  Moreau, A.; Couture, A.; Deniau, E.; Grandclaudon, P.; Lebrun, S. *Org. Biomol. Chem.* **2005**, *3*, 2305–2309.
10. Gribble, G. W. *Parham cyclization.* In *Name Reactions for Homologations-Part II;* Li, J. J., Ed.; Wiley: Hoboken, NJ, **2009**, pp 749–764. (Review).
11. Aranzamendi, E.; Sotomayor, N.; Lete, E. *J. Org. Chem.* **2012**, *77*, 2986–2991.
12. Huard, K.; Bagley, S. W.; Menhaji-Klotz, E.; et al. *J. Org. Chem.* **2012**, *77*, 10050–10057.

# Passerini reaction

Three-component condensation (3CC) of carboxylic acids, *C*-isocyanides, and carbonyl compounds to afford α-acyloxycarboxamides. Also known as three-component reaction (3CR). *Cf.* Ugi reaction.

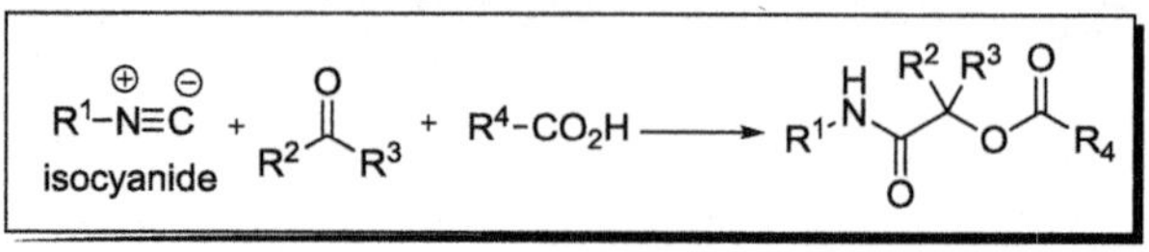

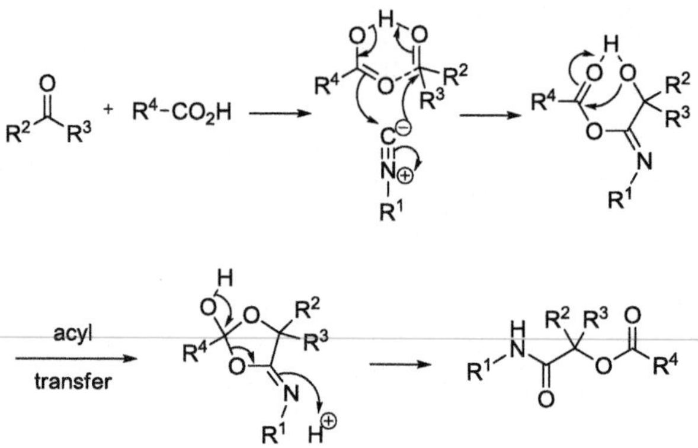

Example 1[3]

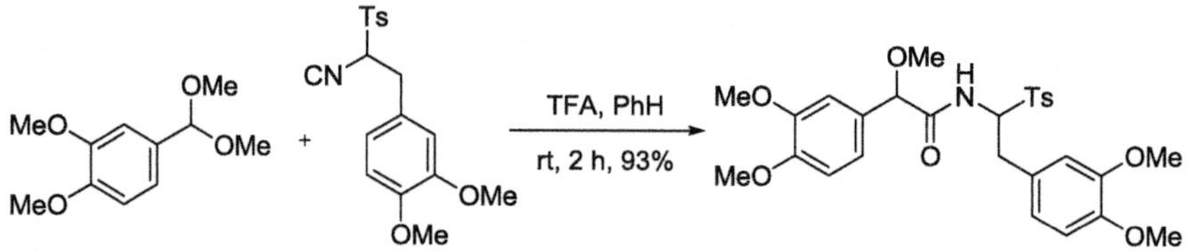

Example 2[5]

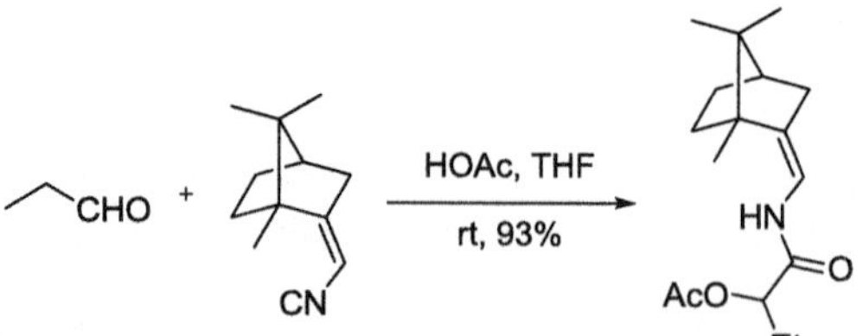

Example 3[6]

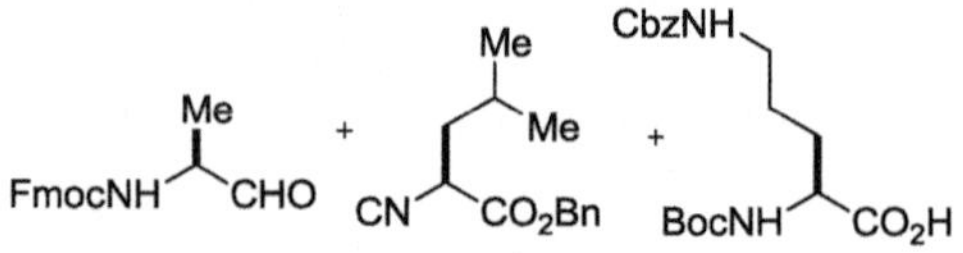

J.J. Li, *Name Reactions: A Collection of Detailed Mechanisms and Synthetic Applications*,
DOI 10.1007/978-3-319-03979-4_204, © Springer International Publishing Switzerland 2014

CH₂Cl₂, 0 °C → rt

3–5 days, 80%

Example 4[7]

CH₂Cl₂, 0 °C → rt

2 days, 59%

## References

1. Passerini, M. *Gazz. Chim. Ital.* **1921**, *51*, 126–129. (b) Passerini, M. *Gazz. Chim. Ital.* **1921**, *51*, 181–188. Mario Passerini (b, 1891) was born in Scandicci, Italy. He obtained his Ph.D. in chemistry and pharmacy at the University of Florence, where he was a professor for most of his career.
2. Ferosie, I. *Aldrichimica Acta* **1971**, *4*, 21. (Review).
3. Barrett, A. G. M.; Barton, D. H. R.; Falck, J. R.; Papaioannou, D.; Widdowson, D. A. *J. Chem. Soc., Perkin Trans. 1* **1979**, 652–661.
4. Ugi, I.; Lohberger, S.; Karl, R. In *Comprehensive Organic Synthesis*; Trost, B. M.; Fleming, I., Eds.; Pergamon: Oxford, **1991**, *Vol. 2*, p.1083. (Review).
5. Bock, H.; Ugi, I. *J. Prakt. Chem.* **1997**, *339*, 385–389.
6. Banfi, L.; Guanti, G.; Riva, R. *Chem. Commun.* **2000**, 985–986.
7. Owens, T. D.; Semple, J. E. *Org. Lett.* **2001**, *3*, 3301–3304.
8. Xia, Q.; Ganem, B. *Org. Lett.* **2002**, *4*, 1631–1634.
9. Banfi, L.; Riva, R. *Org. React.* **2005**, *65*, 1–140. (Review).
10. Klein, J. C.; Williams, D. R. *Passerini Reaction*. In *Name Reactions for Homologations-Part II*; Li, J. J., Ed.; Wiley: Hoboken, NJ, **2009**, pp 765–785. (Review).
11. Sato, K.; Ozu, T.; Takenaga, N. *Tetrahedron Lett.* **2013**, *54*, 661–664.

## Paternó–Büchi reaction

Photoinduced electrocyclization of a carbonyl with an alkene to form polysubstituted oxetane ring systems

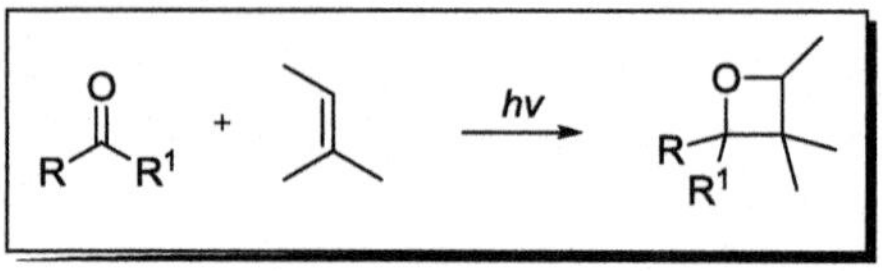

Oxetanes

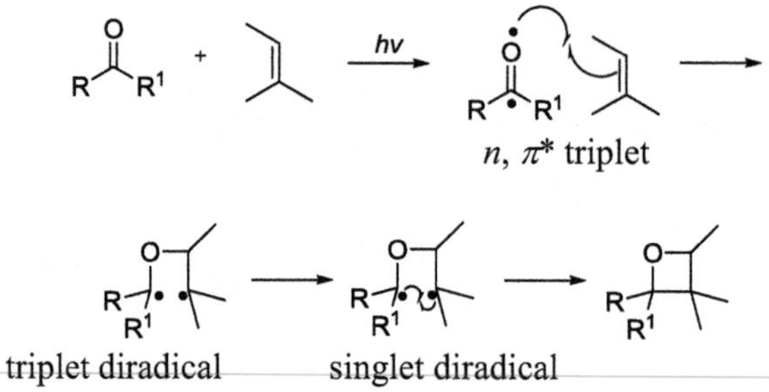

$n, \pi^*$ triplet

triplet diradical          singlet diradical

Example 1[2]

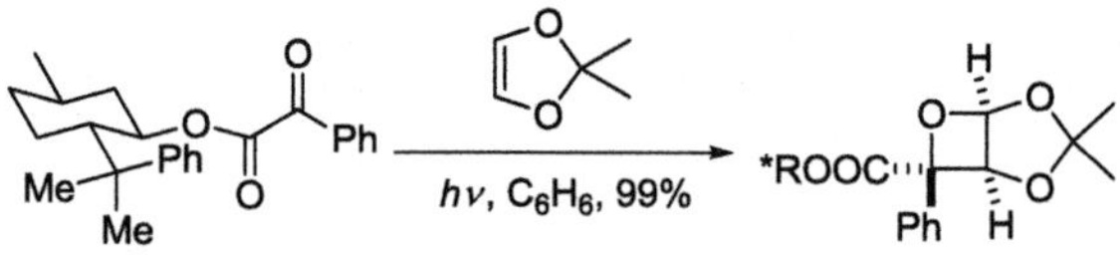

Example 2[4]

$(E/Z = 6/1)$

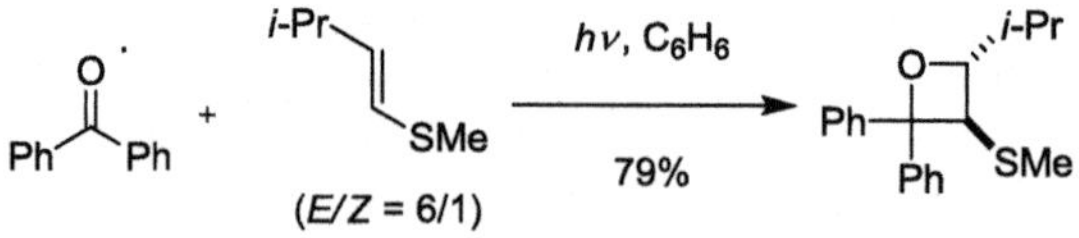

Example 3[6]

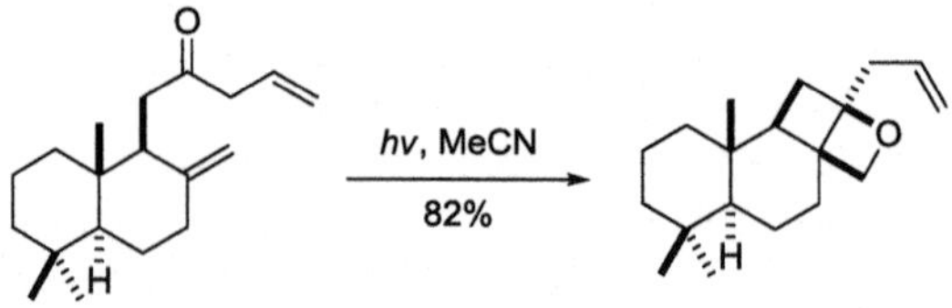

J.J. Li, *Name Reactions: A Collection of Detailed Mechanisms and Synthetic Applications*, DOI 10.1007/978-3-319-03979-4_205, © Springer International Publishing Switzerland 2014

Example 4[8]

Example 5[9]

## References

1.  (a) Paternó, E.; Chieffi, G. *Gazz. Chim. Ital.* **1909**, *39*, 341–361. Emaubuele Paternó (1847–1935) was born in Palermo, Sicily, Italy. It was 104 years ago when he first described photoinduced oxetane formation. (b) Büchi, G.; Inman, C. G.; Lipinsky, E. S. *J. Am. Chem. Soc.* **1954**, *76*, 4327–4331. George H. Büchi (1921–1998) was born in Baden, Switzerland. He was a professor at MIT when he elucidated the structure of oxetanes, the products from the light-catalyzed addition of carbonyl compounds to ole-fins, which had been observed by E. Paternó in 1909. Büchi died of heart failure while hiking with his wife in his native Switzerland.
2.  Koch, H.; Runsink, J.; Scharf, H.-D. *Tetrahedron Lett.* **1983**, *24*, 3217–3220.
3.  Carless, H. A. J. In *Synthetic Organic Photochemistry*; Horspool, W. M., Ed.; Plenum Press: New York, **1984**, 425. (Review).
4.  Morris, T. H.; Smith, E. H.; Walsh, R. *J. Chem. Soc., Chem. Commun.* **1987**, 964–965.
5.  Porco, J. A., Jr.; Schreiber, S. L. In *Comprehensive Organic Synthesis;* Trost, B. M.; Fleming, I., Eds.; Pergamon: Oxford, **1991**, *Vol. 5*, 151–192. (Review).
6.  de la Torre, M. C.; Garcia, I.; Sierra, M. A. *J. Org. Chem.* **2003**, *68*, 6611–6618.
7.  Griesbeck, A. G.; Mauder, H.; Stadtmüller, S. *Acc. Chem. Res.* **1994**, *27*, 70–75. (Review).
8.  D'Auria, M.; Emanuele, L.; Racioppi, R. *Tetrahedron Lett.* **2004**, *45*, 3877–3880.
9.  Liu, C. M. *Paternó–Büchi Reaction*. In *Name Reactions in Heterocyclic Chemistry*; Li, J. J., Ed.; Wiley: Hoboken, NJ, **2005**, pp 44–49. (Review).
10. Cho, D. W.; Lee, H.-Y.; Oh, S. W.; Choi, J. H.; Park, H. J.; Mariano, P. S.; Yoon, U. C. *J. Org. Chem.* **2008**, *73*, 4539–4547.
11. D'Annibale, A.; D'Auria, M.; Prati, F.; Romagnoli, C.; Stoia, S.; Racioppi, R.; Viggiani, L. *Tetrahedron* **2013**, *69*, 3782–3795.

## Pauson–Khand reaction

Formal [2 + 2 +1] cycloaddition of an alkene, alkyne, and carbon monoxide mediated by octacarbonyl dicobalt to form cyclopentenones.

hexacarbonyldicobalt complex

*exo* complex                          sterically-favored isomer

### Example 1[3]

### Example 2, A catalytic version[6]

J.J. Li, *Name Reactions: A Collection of Detailed Mechanisms and Synthetic Applications*,
DOI 10.1007/978-3-319-03979-4_206, © Springer International Publishing Switzerland 2014

Example 3, Intramolecular Pauson–Khand reaction[9]

Co$_2$(CO)$_8$, DMSO

THF, 65 °C, 94%

Example 4, Intramolecular Pauson–Khand reaction[10]

Co$_2$(CO)$_8$
5 equiv Me$_3$NO•2H$_2$O

THF/H$_2$O (3:1), 0 °C to rt
7 h, 71%

Example 5[12]

1. Co(CO)$_8$, rt, 2 h
2. CO, NMO, rt, 20 h

72–96% yield
single diastereomers

## References

1. (a) Pauson, P. L.; Khand, I. U.; Knox, G. R.; Watts, W. E. *J. Chem. Soc., Chem. Commun.* **1971**, 36. Ihsan U. Khand and Peter L. Pauson were at the University of Strathclyde, Glasgow in Scotland. (b) Khand, I. U.; Knox, G. R.; Pauson, P. L.; Watts, W. E.; Foreman, M. I. *J. Chem. Soc., Perkin Trans. 1* **1973**, 975–977. (c) Bladon, P.; Khand, I. U.; Pauson, P. L. *J. Chem. Res. (S)*, **1977**, 9. (d) Pauson, P. L. *Tetrahedron* **1985**, *41*, 5855–5860. (Review).
2. Schore, N. E. *Chem. Rev.* **1988**, *88*, 1081–1119. (Review).
3. Billington, D. C.; Kerr, W. J.; Pauson, P. L.; Farnocchi, C. F. *J. Organomet. Chem.* **1988**, *356*, 213–219.
4. Schore, N. E. In *Comprehensive Organic Synthesis*; Paquette, L. A.; Fleming, I.; Trost, B. M., Eds.; Pergamon: Oxford, **1991**, *Vol. 5*, p.1037. (Review).
5. Schore, N. E. *Org. React.* **1991**, *40*, 1–90. (Review).
6. Jeong, N.; Hwang, S. H.; Lee, Y.; Chung, J. *J. Am. Chem. Soc.* **1994**, *116*, 3159–3160.
7. Brummond, K. M.; Kent, J. L. *Tetrahedron* **2000**, *56*, 3263–3283. (Review).
8. Tsujimoto, T.; Nishikawa, T.; Urabe, D.; Isobe, M. *Synlett* **2005**, 433–436.
9. Miller, K. A.; Martin, S. F. *Org. Lett.* **2007**, *9*, 1113–1116.
10. Kaneda, K.; Honda, T. *Tetrahedron* **2008**, *64*, 11589–11593.
11. Torres, R. R. *The Pauson-Khand Reaction: Scope, Variations and Applications*, Wiley: Hoboken, NJ, 2012. (Review).
12. McCormack, M. P.; Waters, S. P. *J. Org. Chem.* **2013**, *78*, 1127–1137.

## Payne rearrangement

The isomerization of 2,3-epoxy alcohol under the influence of a base to 1,2-epoxy-3-ol is referred to as the Payne rearrangement. Also known as epoxide migration.

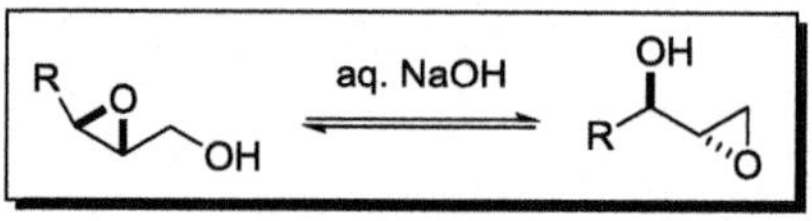

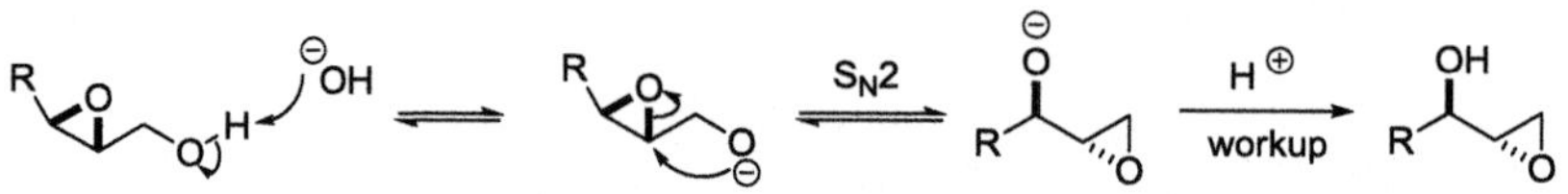

Example 1[2]

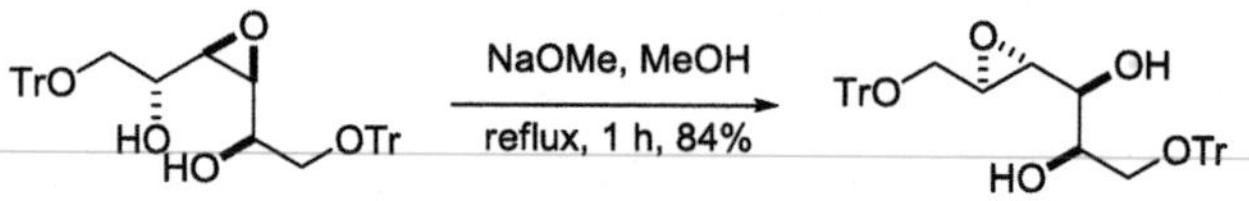

Example 2[3]

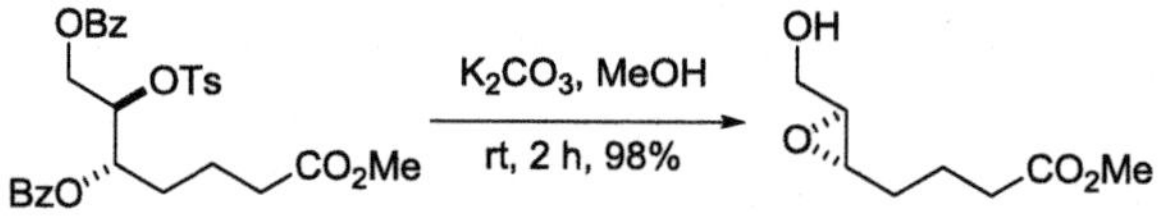

Example 3, Aza-Payne rearrangement[8]

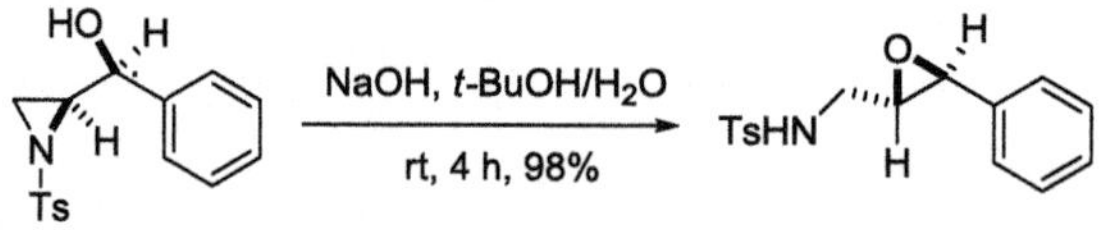

Example 4, Aza-Payne rearrangement[9]

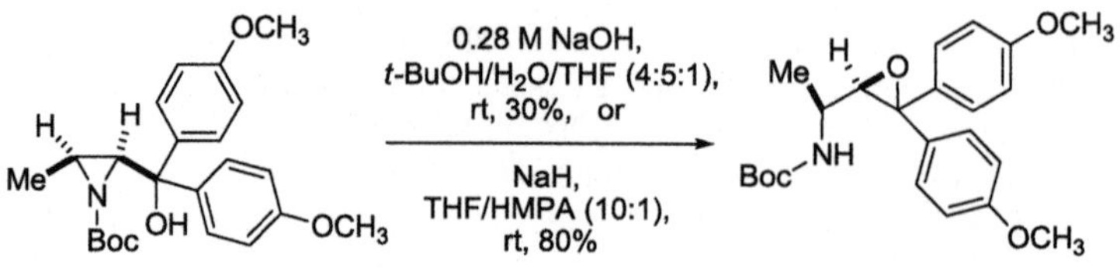

J.J. Li, *Name Reactions: A Collection of Detailed Mechanisms and Synthetic Applications*,
DOI 10.1007/978-3-319-03979-4_207, © Springer International Publishing Switzerland 2014

Example 5, Lipase-mediated dynamic kinetic resolution via a *vinylogous* Payne rearrangement[11]

## References

1. Payne, G. B. *J. Org. Chem.* **1962**, *27*, 3819–3822. George B. Payne was a chemist at Shell Development Co. in Emeryville, CA.
2. Buchanan, J. G.; Edgar, A. R. *Carbohydr. Res.* **1970**, *10*, 295–302.
3. Corey, E. J.; Clark, D. A.; Goto, G.; Marfat, A.; Mioskowski, C.; Samuelsson, B.; Hammerstrom, S. *J. Am. Chem. Soc.* **1980**, *102*, 1436–1439, and 3663–3665.
4. Ibuka, T. *Chem. Soc. Rev.* **1998**, *27*, 145–154. (Review).
5. Hanson, R. M. *Org. React.* **2002**, *60*, 1–156. (Review).
6. Yamazaki, T.; Ichige, T.; Kitazume, T. *Org. Lett.* **2004**, *6*, 4073–4076.
7. Bilke, J. L.; Dzuganova, M.; Froehlich, R.; Wuerthwein, E.-U. *Org. Lett.* **2005**, *7*, 3267–3270.
8. Feng, X.; Qiu, G.; Liang, S.; Su, J.; Teng, H.; Wu, L.; Hu, X. *Russ. J. Org. Chem.* **2006**, *42*, 514–500.
9. Feng, X.; Qiu, G.; Liang, S.; Teng, H.; Wu, L.; Hu, X. *Tetrahedron: Asymmetry* **2006**, *17*, 1394–1401.
10. Kumar, R. R.; Perumal, S. *Payne Rearrangement*. In *Name Reactions for Homologations-Part II*; Li, J. J., Ed.; Wiley: Hoboken, NJ, **2009**, pp 474–488. (Review).
11. Hoye, T. R.; Jeffrey, C. S.; Nelson, D. P. *Org. Lett.* **2010**, *12*, 52–55.
12. Kulshrestha, A.; Salehi Marzijarani, N.; Dilip Ashtekar, K.; Staples, R.; Borhan, B. *Org. Lett.* **2012**, *14*, 3592–3595.

# Pechmann coumarin synthesis

Lewis and Brønsted acid-mediated condensation of phenol with β-ketoester to produce coumarin. Some call it the von Pechmann cyclization.

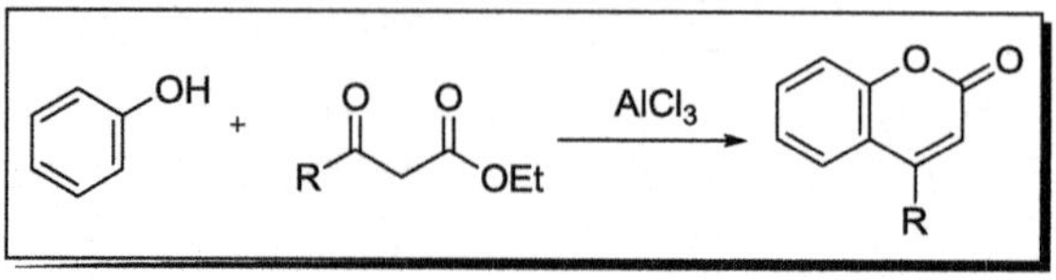

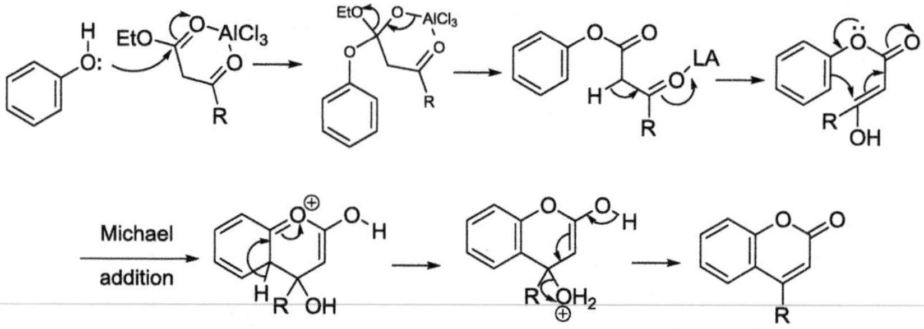

Example 1[6]

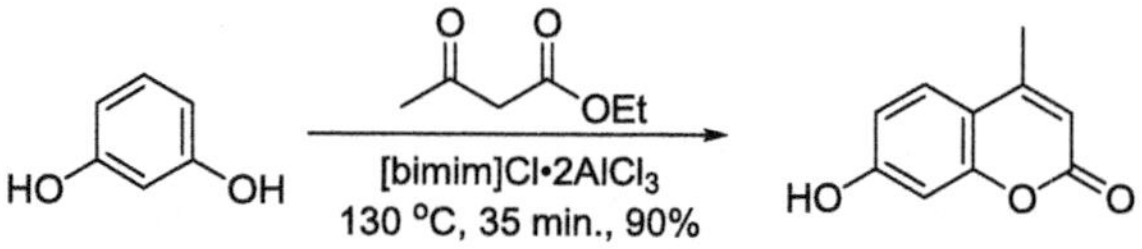

Example 2[8]

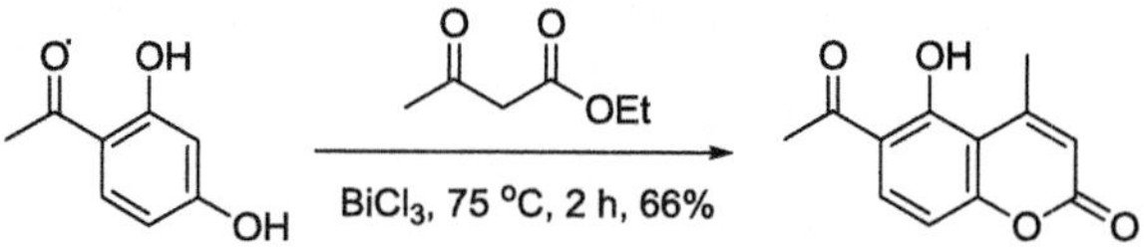

Example 3[11]

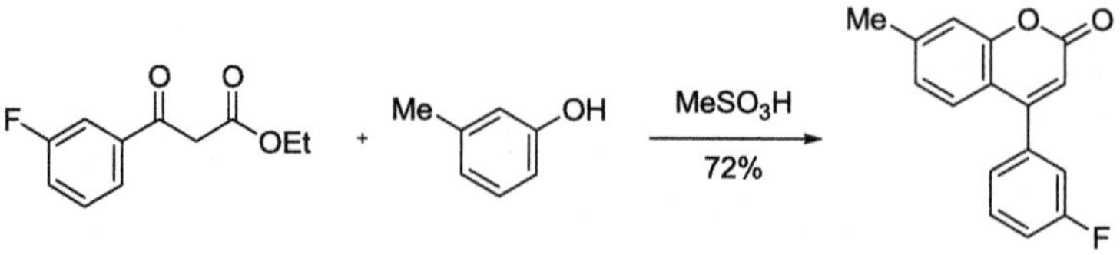

J.J. Li, *Name Reactions: A Collection of Detailed Mechanisms and Synthetic Applications*,
DOI 10.1007/978-3-319-03979-4_208, © Springer International Publishing Switzerland 2014

# References

1. von Pechmann, H.; Duisberg, C. *Ber.* **1883,** *16,* 2119.  Hans von Pechmann (1850–1902) was born in Nürnberg, Germany.  After his doctorate, he worked with Frankland and von Baeyer.  Pechmann taught at Munich and Tübingen.  He committed suicide at 52 by taking cyanide.
2. Corrie, J. E. T. *J. Chem. Soc., Perkin Trans. 1* **1990,** 2151–2997.
3. Hua, D. H.; Saha, S.; Roche, D.; Maeng, J. C.; Iguchi, S.; Baldwin, C. *J. Org. Chem.* **1992,** *57,* 399–403.
4. Li, T.-S.; Zhang, Z.-H.; Yang, F.; Fu, C.-G. *J. Chem. Res., (S)* **1998,** 38–39.
5. Potdar, M. K.; Mohile, S. S.; Salunkhe, M. M. *Tetrahedron Lett.* **2001,** *42,* 9285–9287.
6. Khandekar, A. C.; Khandilkar, B. M. *Synlett.* **2002,** 152–154.
7. Smitha, G.; Sanjeeva Reddy, C. *Synth. Commun.* **2004,** *34,* 3997–4003.
8. De, S. K.; Gibbs, R. A. *Synthesis* **2005,** 1231–1233.
9. Manhas, M. S.; Ganguly, S. N.; Mukherjee, S.; Jain, A. K.; Bose, A. K. *Tetrahedron Lett.* **2006,** *47,* 2423–2425.
10. Rodriguez-Dominguez, J. C.; Kirsch, G. *Synthesis* **2006,** 1895–1897.
11. Ouellet, S. G.; Gauvreau, D.; Cameron, M.; Dolman, S.; Campeau, L.-C.; Hughes, G.; O'Shea, P. D.; Davies, I. W. *Org. Process Res. Dev.* **2012,** *16,* 214–219.

# Perkin reaction

Cinnamic acid synthesis from aryl aldehyde and acetic anhydride.

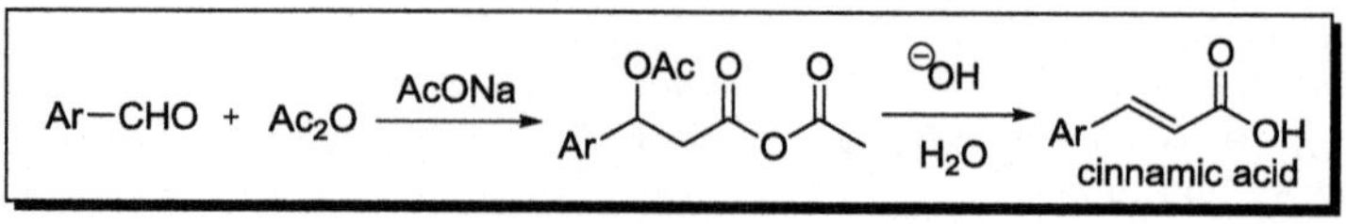

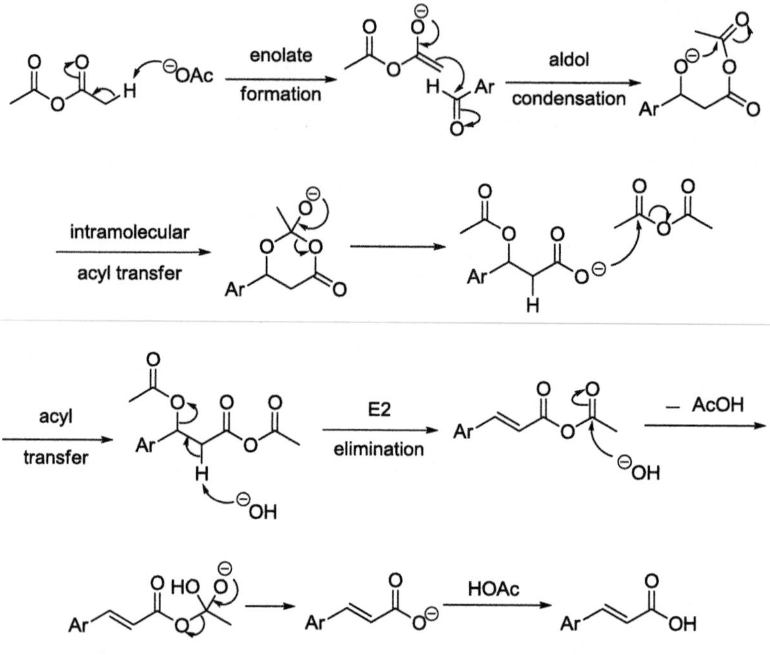

Example 1[7]

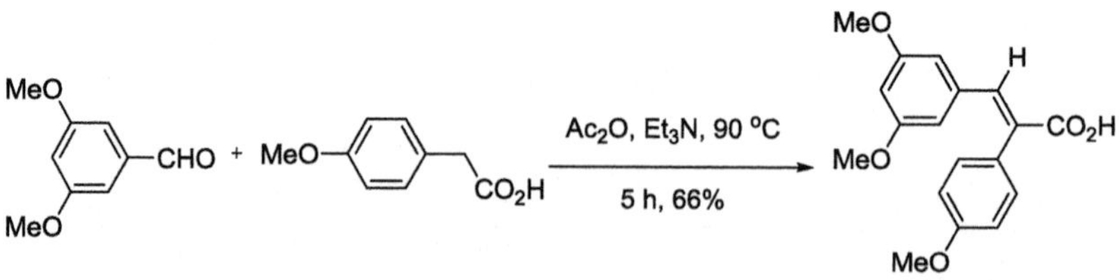

Example 2[9]

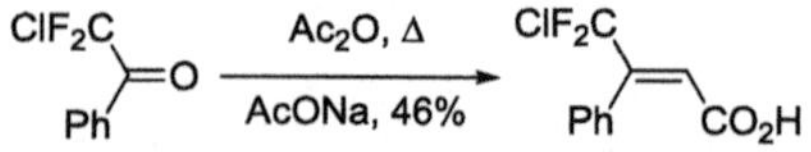

J.J. Li, *Name Reactions: A Collection of Detailed Mechanisms and Synthetic Applications*,
DOI 10.1007/978-3-319-03979-4_209, © Springer International Publishing Switzerland 2014

Example 3[12]

DES = Biodegradable deep eutectic solvent generated from choline chloride and urea.

## References

1.  Perkin, W. H. *J. Chem. Soc.* **1868,** *21,* 53.  William Henry Perkin (1838–1907), born in London, England, studied under A. W. von Hofmann at the Royal College of Chemistry.  In an attempt to synthesize quinine in his home laboratory in 1856, Perkin synthesized mauve, the purple dye.  He then started a factory to manufacture mauve and later other dyes including alizarin.  Perkin was the first person to show that organic chemistry was not just mere intellectual curiosity but could be profitable, which catapulted the discipline into a higher level.  In addition, Perkin also happened to be an exceptionally talented pianist.

2.  Gaset, A.; Gorrichon, J. P. *Synth. Commun.* **1982,** *12,* 71–79.

3.  Kinastowski, S.; Nowacki, A. *Tetrahedron Lett.* **1982,** *23,* 3723–3724.

4.  Koepp, E.; Vögtle, F. *Synthesis* **1987,** 177–179.

5.  Brady, W. T.; Gu, Y.-Q. *J. Heterocycl. Chem.* **1988,** *25,* 969–971.

6.  Pálinkó, I.; Kukovecz, A.; Török, B.; Körtvélyesi, T. *Monatsh. Chem.* **2001,** *131,* 1097–1104.

7.  Gaukroger, K.; Hadfield, J. A.; Hepworth, L. A.; Lawrence, N. J.; McGown, A. T. *J. Org. Chem.* **2001,** *66,* 8135–8138.

8.  Solladié, G.; Pasturel-Jacopé, Y.; Maignan, J. *Tetrahedron* **2003,** *59,* 3315–3321.

9.  Sevenard, D. V. *Tetrahedron Lett.* **2003,** *44,* 7119–7126.

10. Chandrasekhar, S.; Karri, P. *Tetrahedron Lett.* **2006,** *47,* 2249–2251.

11. Lacova, M.; Stankovicova, H.; Bohac, A.; Kotzianova, B. *Tetrahedron* **2008,** *64,* 9646–9653.

12. Pawar, P. M.; Jarag, K. J.; Shankarling, G. S. *Green Chem.* **2011,** *13,* 2130–2134.

# Perkow vinyl phosphate synthesis

Enol phosphate synthesis from α-halocarbonyls and trialkylphosphites.

General scheme:

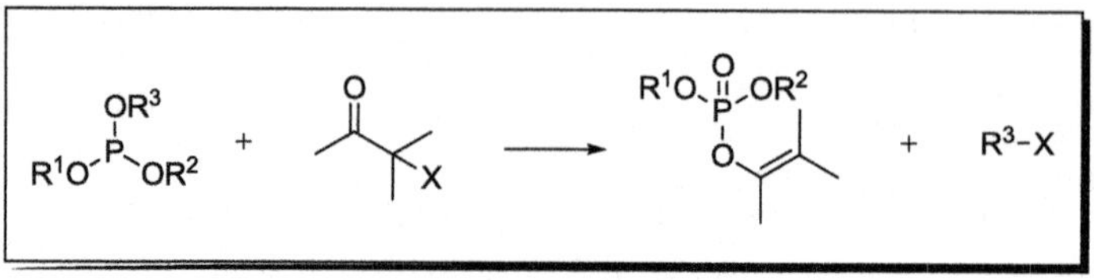

X = Cl, Br, I, secondary or tertiary halides are required to retard the
competing Michaelis–Arbuzov reaction.

Example 1.

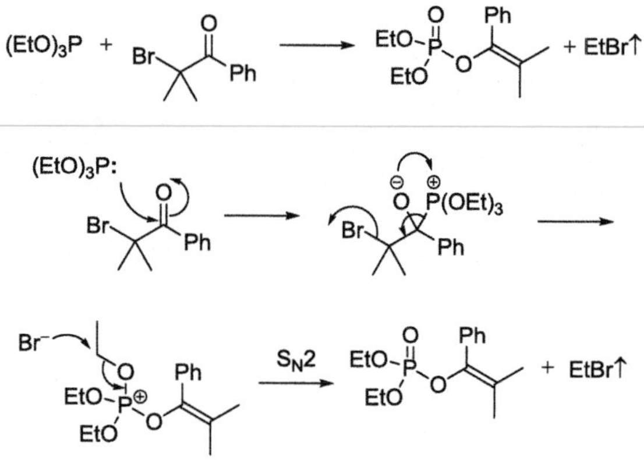

Example 2[7]

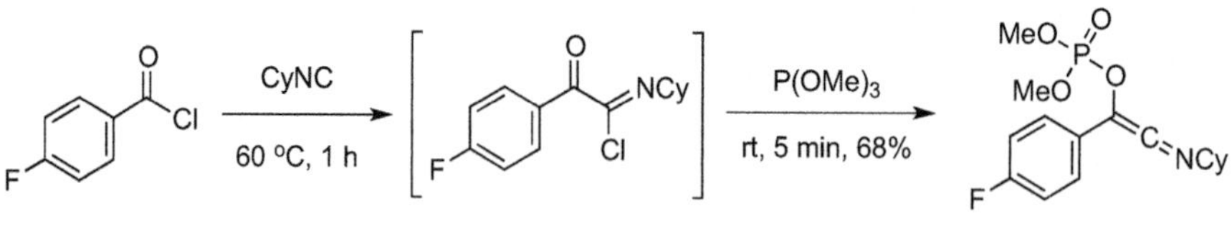

Example 3[8]

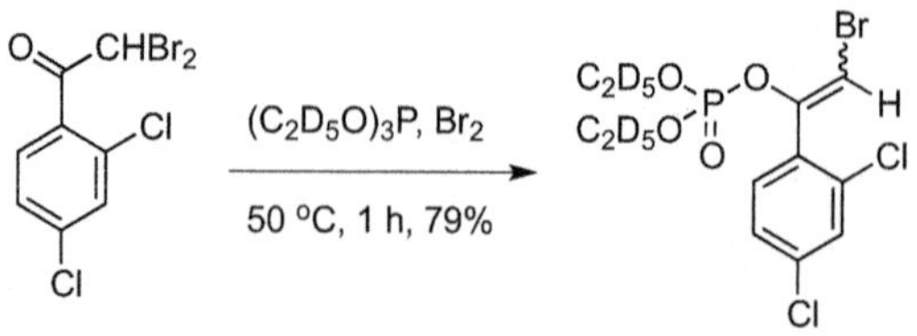

J.J. Li, *Name Reactions: A Collection of Detailed Mechanisms and Synthetic Applications*,
DOI 10.1007/978-3-319-03979-4_210, © Springer International Publishing Switzerland 2014

## References

1.  Perkow, W.; Ullrich, K.; Meyer, F. *Nasturwiss.* **1952,** *39,* 353.
2.  Perkow, W. *Ber. Dtsch. Chem. Ges.* **1954,** *87,* 755.
3.  Borowitz, G. B.; Borowitz, I. J. *Handb. Organophosphorus Chem.* **1992,** 115. (Review).
4.  Hudson, H. R.; Matthews, R. W.; McPartlin, M.; Pryce, M. A.; Shode, O. O. *J. Chem. Soc., Perkin Trans. 2* **1993,** 1433.
5.  Janecki, T.; Bodalski, R. *Heteroat. Chem.* **2000,** *11,* 115.
6.  Balasubramanian, M. *Perkow Reaction, in Name Reactions for Functional Group Transformations,* Li, J. J. Ed., Wiley: Hoboken, NJ, **2007,** pp 369–385. (Review).
7.  Coffinier, D.; El Kaim, L.; Grimaud, L. *Org. Lett.* **2009,** *11,* 1825–1827.
8.  Huras, B.; Konopski, L.; Zakrzewski, J. *J. Labeled Compd. Radiopharm.* **2011,** *54,* 399–400.
9.  Yavari, I.; Hosseinpour, R.; Pashazadeh, R.; Ghanbari, E.; Skoulika, S. *Tetrahedron* **2013,** *69,* 2462–2467.

## Petasis reaction

Benzylic or allylic amine from the three-component reaction of an aryl- or a vinyl-boronic acid, a carbonyl and an amine. Also known as boronic acid-Mannich or Petasis boronic acid-Mannich reaction. *Cf.* Mannich reaction.

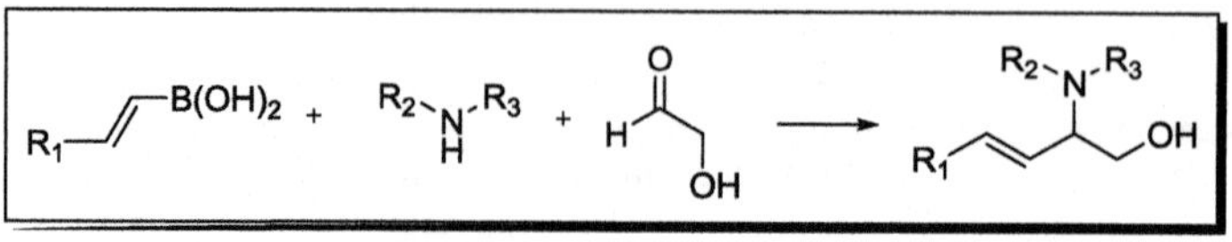

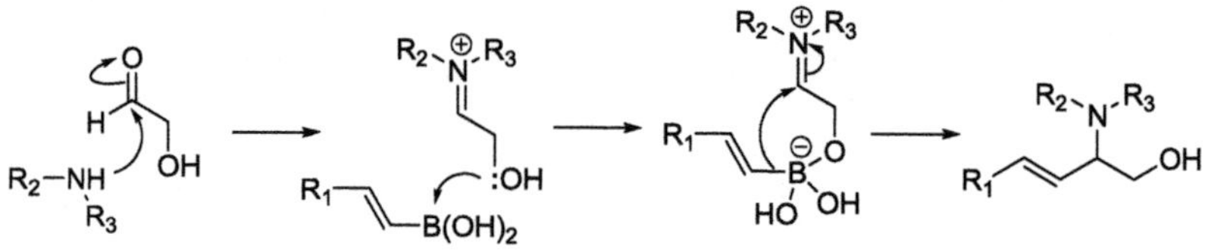

Example 1[2]

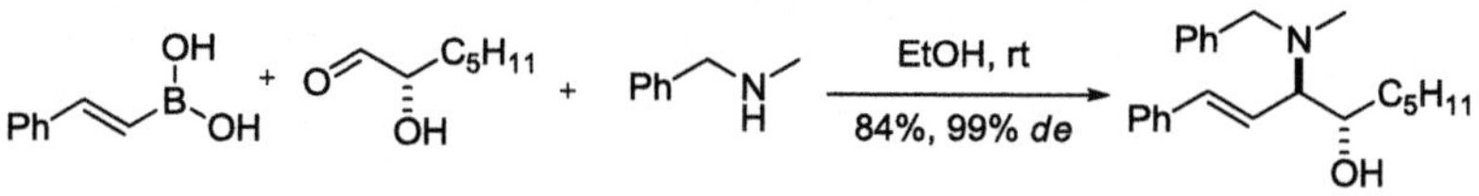

Example 2[4]

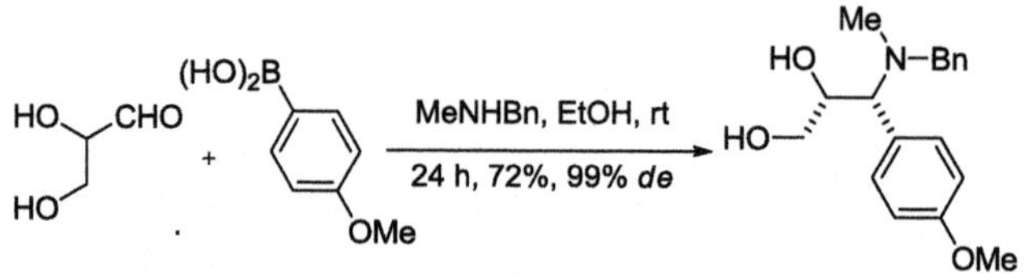

Example 3[9]

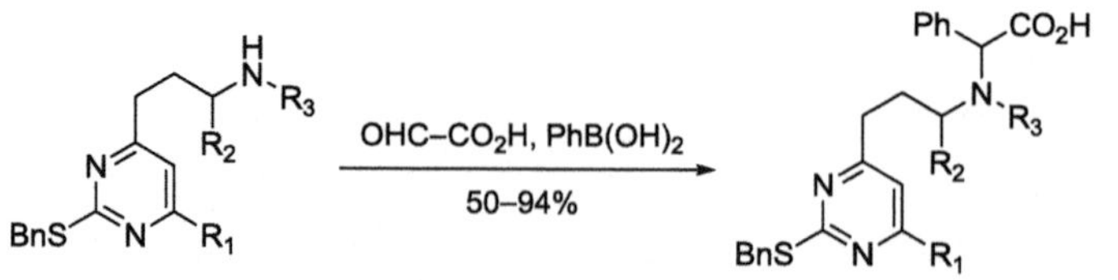

J.J. Li, *Name Reactions: A Collection of Detailed Mechanisms and Synthetic Applications*, DOI 10.1007/978-3-319-03979-4_211, © Springer International Publishing Switzerland 2014

Example 4, Asymmetric Petasis reaction[10]

$R_1$ = aryl, alkyl  $R_2$ = Bn, allyl
$R_3$ = alkyl

15 mol% (S)-VAPOL

3 A MS, –15 °C, Tol.  70–92% yield
89:11 to 98:2 er

Example 5, Asymmetric Petasis reaction[11]

20% mol% **cat.**

MTBE, 5 °C, 96 h
70%, 95% ee

**cat. =**

# References

1. (a) Petasis, N. A.; Akritopoulou, I. *Tetrahedron Lett.* **1993**, *34*, 583–586. (b) Petasis, N. A.; Zavialov, I. A. *J. Am. Chem. Soc.* **1997**, *119*, 445–446. (c) Petasis, N. A.; Goodman, A.; Zavialov, I. A. *Tetrahedron* **1997**, *53*, 16463–16470. (d) Petasis, N. A.; Zavialov, I. A. *J. Am. Chem. Soc.* **1998**, *120*, 11798–11799. Nicos A. Petasis is a professor at the University of Southern California in Los Angeles.
2. Koolmeister, T.; Södergren, M.; Scobie, M. *Tetrahedron Lett.* **2002**, *43*, 5969–5970.
3. Orru, R. V. A.; deGreef, M. *Synthesis* **2003**, 1471–1499. (Review).
4. Sugiyama, S.; Arai, S.; Ishii, K. *Tetrahedron: Asymmetry* **2004**, *15*, 3149–3153.
5. Chang, Y. M.; Lee, S. H.; Nam, M. H.; Cho, M. Y.; Park, Y. S.; Yoon, C. M. *Tetrahedron Lett.* **2005**, *46*, 3053–3056.
6. Follmann, M.; Graul, F.; Schaefer, T.; Kopec, S.; Hamley, P. *Synlett* **2005**, 1009–1011.
7. Danieli, E.; Trabocchi, A.; Menchi, G.; Guarna, A. *Eur. J. Org. Chem.* **2007**, 1659–1668.
8. Konev, A. S.; Stas, S.; Novikov, M. S.; Khlebnikov, A. F.; Abbaspour Tehrani, K. *Tetrahedron* **2007**, *64*, 117–123.
9. Font, D.; Heras, M.; Villalgordo, J. M. *Tetrahedron* **2007**, *64*, 5226–5235.
10. Lou, S.; Schaus, S. E. *J. Am. Chem. Soc.* **2008**, *130*, 6922–6923.
11. Abbaspour Tehrani, K.; Stas, S.; Lucas, B.; De Kimpe, N. *Tetrahedron* **2009**, *65*, 1957–1966.
12. Han, W.-Y.; Zuo, J.; Zhang, X.-M.; Yuan, W.-C. *Tetrahedron* **2013**, *69*, 537–541.

# Petasis reagent

The Petasis reagent (Cp$_2$TiMe$_2$, dimethyltitanocene) undergoes similar olefination reactions with ketones and aldehydes as does the Tebbe reagent. The originally proposed mechanism[5] was very different from that of Tebbe olefination. However, later experimental data seem to suggest that both Petasis and Tebbe olefination share the same mechanism, i.e., the carbene mechanism involving a four-membered titanium oxide ring intermediate.[9] Petasis reagent is easier to make than the Tebbe reagent.

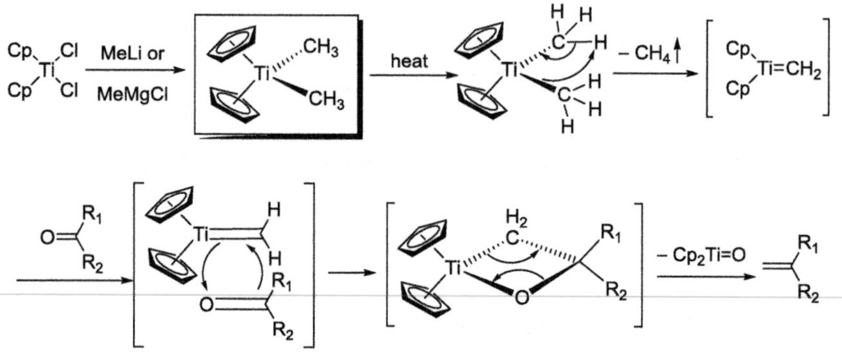

Example 1[2]

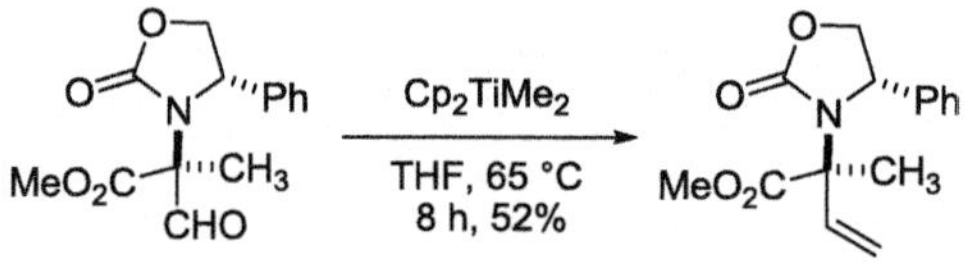

Example 2[3]

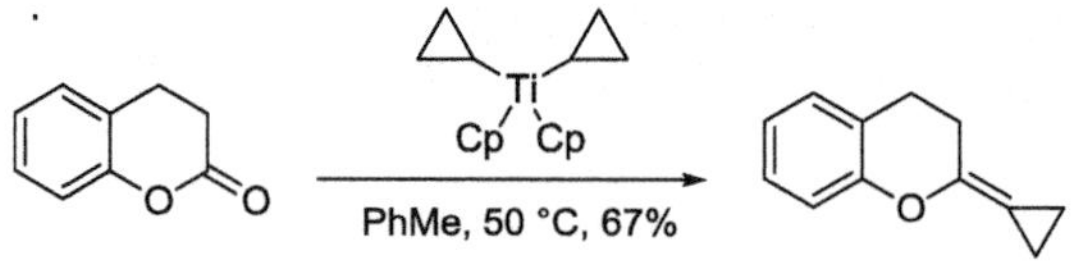

Example 3[5]

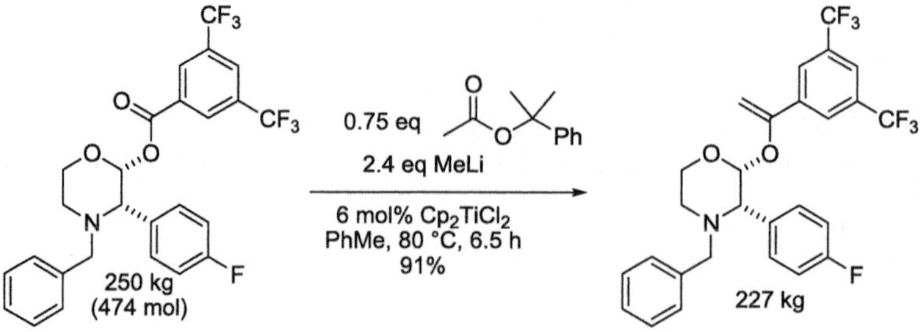

J.J. Li, *Name Reactions: A Collection of Detailed Mechanisms and Synthetic Applications*,
DOI 10.1007/978-3-319-03979-4_212, © Springer International Publishing Switzerland 2014

Example 4[8]

$$R_3 \quad R_1 \quad OMe \qquad \xrightarrow[\substack{\text{Tol., THF, microwave, 65 °C} \\ \text{3–10 min., ~ 50–60\%}}]{\text{1.8 equiv Cp}_2\text{TiMe}_2} \qquad R_3 \quad R_1 \quad OMe$$

Example 5[11]

$$\xrightarrow[\substack{\text{2. Br}_2 \\ \text{3. H}_3\text{O} \oplus}]{\text{1. Petasis reagent}}$$

85–95%

## References

1. Petasis, N. A.; Bzowej, E. I. *J. Am. Chem. Soc.* **1990**, *112*, 6392–6394. Nicos A. Petasis is a professor at the University of Southern California in Los Angeles.
2. Colson, P. J.; Hegedus, L. S. *J. Org. Chem.* **1993**, *58*, 5918–5924.
3. Petasis, N. A.; Bzowej, E. I. *Tetrahedron Lett.* **1993**, *34*, 943–946.
4. Payack, J. F.; Hughes, D. L.; Cai, D.; Cottrell, I. F.; Verhoeven, T. R. *Org. Synth.* **2002**, *79*, 19.
5. Payack, J. F.; Huffman, M. A.; Cai, D. W.; Hughes, D. L.; Collins, P. C.; Johnson, B. K.; Cottrell, I. F.; Tuma, L. D. *Org. Pro. Res. Dev.* **2004**, *8*, 256–259.
6. Cook, M. J.; Fleming, E. I. *Tetrahedron Lett.* **2005**, *46*, 297–300.
7. Morency, L.; Barriault, L. *J. Org. Chem.* **2005**, *70*, 8841–8853.
8. Adriaenssens, L. V.; Hartley, R. C. *J. Org. Chem.* **2007**, *72*, 10287–10290.
9. Naskar, D.; Neogi, S.; Roy, A.; Mandal, A. B. *Tetrahedron Lett.* **2008**, *49*, 6762–6764.
10. Zhang, J. *Tebbe Reagent.* In *Name Reactions for Homologations-Part I*, Li, J. J. Ed., Wiley: Hoboken, NJ, **2009**, pp 319–333. (Review).
11. Kobeissi, M.; Cherry, K.; Jomaa, W. *Synth. Commun.* **2013**, *43*, 2955–2965.

## Peterson olefination

Alkenes from α-silyl carbanions and carbonyl compounds. Also known as the sila-Wittig reaction.

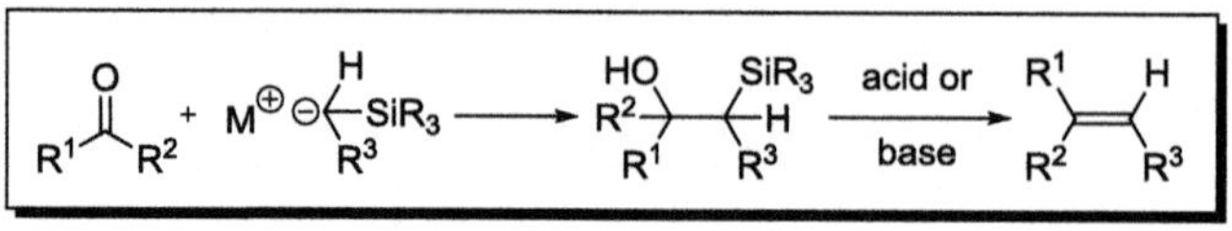

Basic conditions:

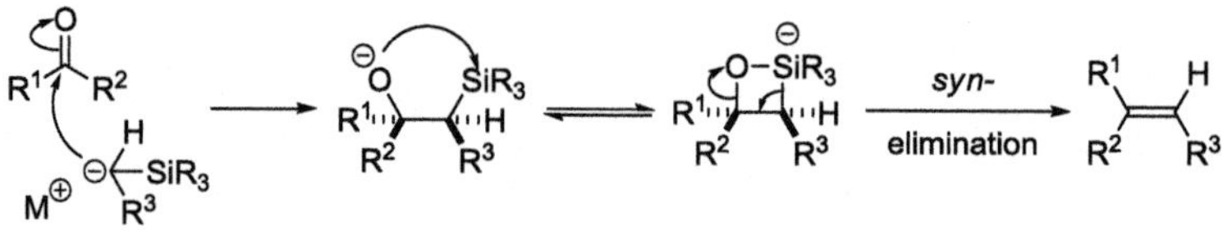

β-silylalkoxide intermediate

Acidic conditions:

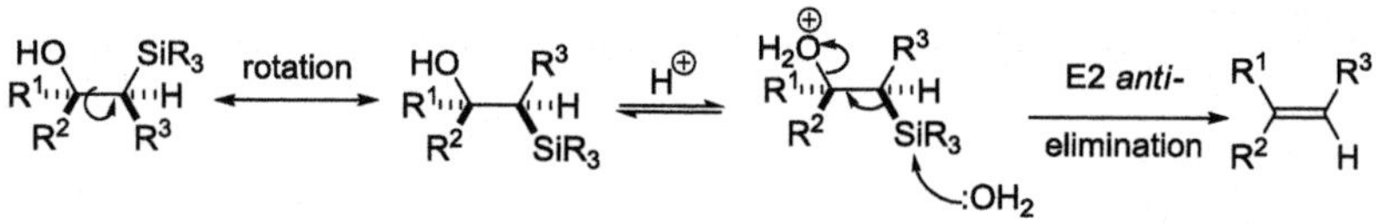

β-hydroxysilane

Example 1[6]

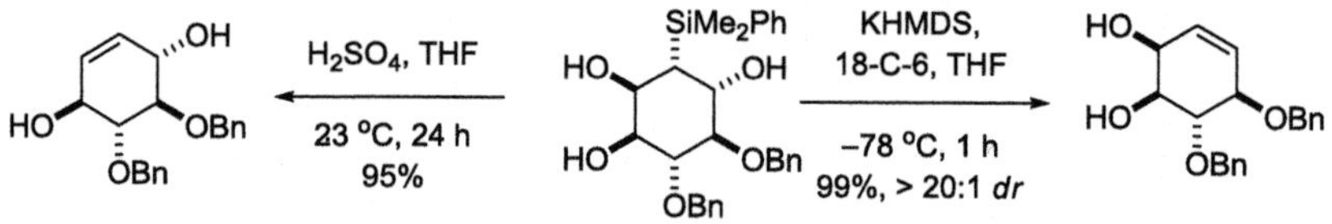

Example 2[7]

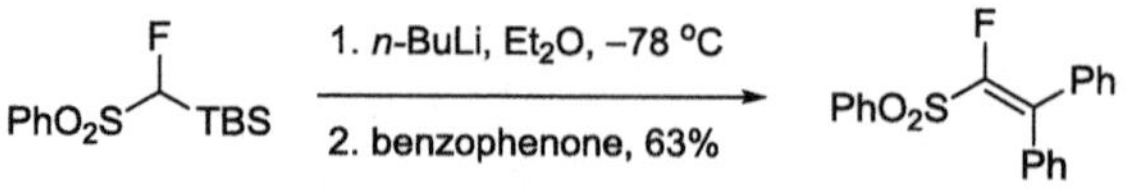

Example 3[8]

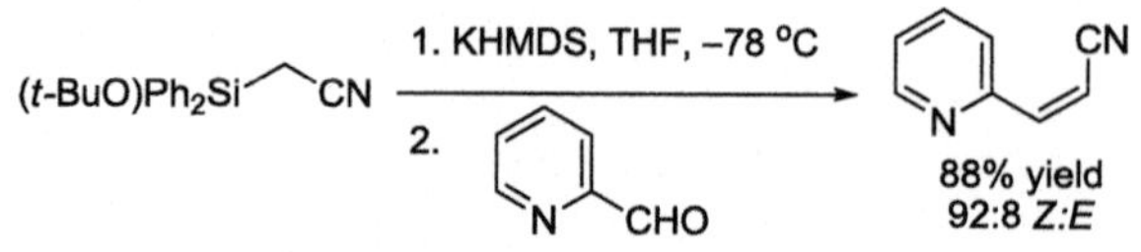

J.J. Li, *Name Reactions: A Collection of Detailed Mechanisms and Synthetic Applications*, DOI 10.1007/978-3-319-03979-4_213, © Springer International Publishing Switzerland 2014

Example 4[10]

1. LiCH$_2$TMS, THF 0 °C, 15 min.
2. KHMDS, 0 °C to rt, 1.5 h

3. HCl, MeOH/Et$_2$O, 5 min.
74%

Example 5[12]

t-BuOK, THF

45 °C, 16 h

## References

1. Peterson, D. J. *J. Org. Chem.* **1968**, *33*, 780–784.
2. Ager, D. J. *Org. React.* **1990**, *38*, 1–223. (Review).
3. Barrett, A. G. M.; Hill, J. M.; Wallace, E. M.; Flygare, J. A. *Synlett* **1991,** 764–770. (Review).
4. van Staden, L. F.; Gravestock, D.; Ager, D. J. *Chem. Soc. Rev.* **2002**, *31*, 195–200. (Review).
5. Ager, D. J. *Science of Synthesis* **2002**, *4*, 789–809. (Review).
6. Heo, J.-N.; Holson, E. B.; Roush, W. R. *Org. Lett.* **2003**, *5*, 1697–1700.
7. Asakura, N.; Usuki, Y.; Iio, H. *J. Fluorine Chem.* **2003**, *124*, 81–84.
8. Kojima, S.; Fukuzaki, T.; Yamakawa, A.; Murai, Y. *Org. Lett.* **2004**, *6,* 3917–3920.
9. Kano, N.; Kawashima, T. *The Peterson and Related Reactions* in *Modern Carbonyl Olefination;* Takeda, T., Ed.; Wiley-VCH: Weinheim, Germany, **2004**, 18–103. (Review).
10. Huang, J.; Wu, C.; Wulff, W. D. *J. Am. Chem. Soc.* **2007**, *129*, 13366.
11. Ahmad, N. M. *Peterson Olefination.* In *Name Reactions for Homologations-Part I*; Li, J. J., Ed., Wiley: Hoboken, NJ, **2009,** pp 521–538. (Review).
12. Beveridge, R. E.; Batey, R. A. *Org. Lett.* **2013**, *15,* 3086–3089.

## Pictet–Gams isoquinoline synthesis

The isoquinoline framework is derived from the corresponding acyl derivatives of
β-hydroxy-β-phenylethylamines.  Upon exposure to a dehydrating agent such as
phosphorus pentoxide, or phosphorus oxychloride, under reflux and in an inert
solvent such as decalin, isoquinoline frameworks are formed.

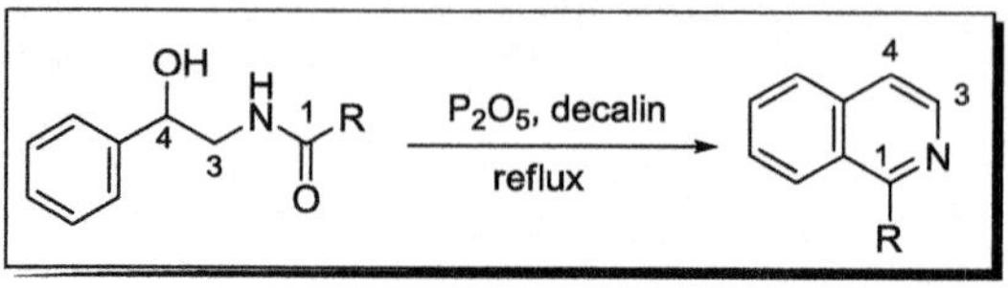

$P_2O_5$ actually exists as $P_4O_{10}$, an adamantane-like structure:

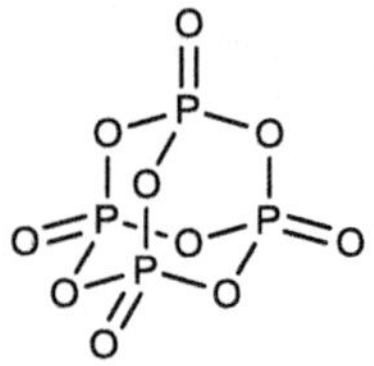

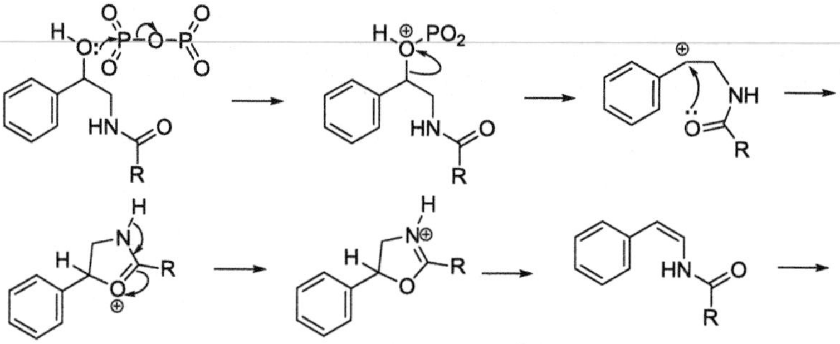

Oxazoline intermediate[2]

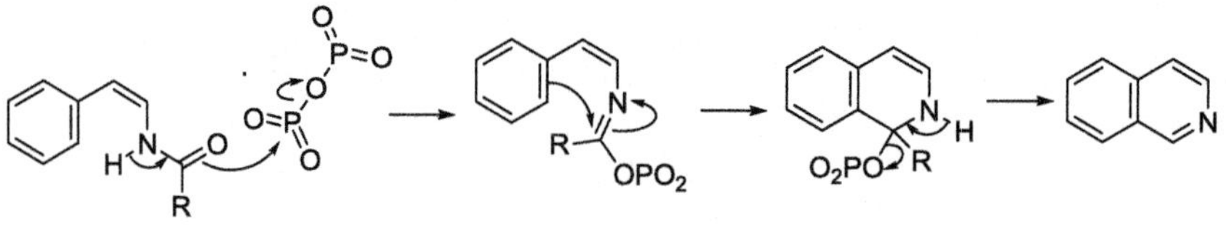

Example 1[4]

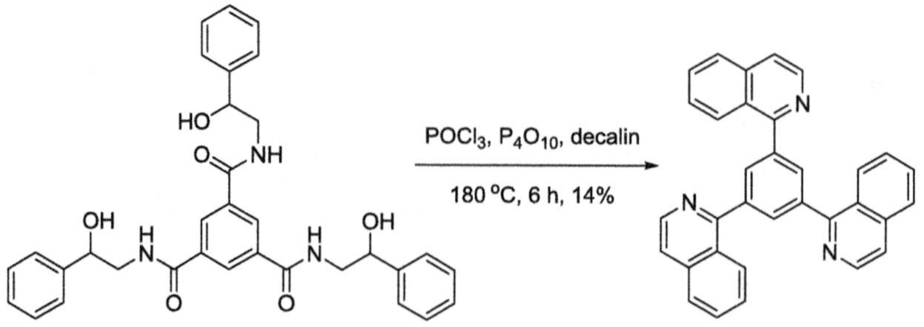

J.J. Li, *Name Reactions: A Collection of Detailed Mechanisms and Synthetic Applications*,
DOI 10.1007/978-3-319-03979-4_214, © Springer International Publishing Switzerland 2014

Example 2[7]

Example 3[9]
An Alternative to Pictet—Gams Reaction Triggered by Hendrickson Reagent:
Isoquinolines and β-Carbolines from amides:

Hendrickson reagent

## Reference

1.  (a) Pictet, A.; Kay, F. W. *Ber.* **1909**, *42*, 1973–1979. (b) Pictet, A.; Gams, A. *Ber.* **1909**, *42*, 2943–2952. Amé Pictet (1857–1937), born in Geneva, Switzerland, carried out a tremendous amount of work on alkaloids.
2.  Fritton, A. O.; Frost, J. R.; Zakaria, M. M.; Andrew, G. *J. Chem. Soc., Chem. Commun.*, **1973**, 889.
3.  (a) Ardabilchi, N.; Fitton, A. O.; Frost, J. R.; Oppong-Boachie, F. *Tetrahedron Lett.* **1977**, *18*, 4107–4110. (b) Ardabilchi, N.; Fitton, A. O.; Frost, J. R.; Oppong-Boachie, F. K.; Hadi, A. H. A.; Sharif, A. M. *J. Chem. Soc., Perkin Trans. 1* **1979**, 539–543.
4.  Dyker, G.; Gabler, M.; Nouroozian, M.; Schulz, P. *Tetrahedron Lett.* **1994**, *35*, 9697–9700.
5.  Poszávácz, L.; Simig, G. *J. Heterocycl. Chem.* **2000**, *37*, 343–348.
6.  Poszávácz, L.; Simig, G. *Tetrahedron* **2001**, *57*, 8573–8580.
7.  Manning, H. C.; Goebel, T.; Marx, J. N.; Bornhop, D. J. *Org. Lett.* **2002**, *4*, 1075–1081.
8.  Holsworth, D. D. *Pictet–Gams Isoquinoline Synthesis.* In *Name Reactions in Heterocyclic Chemistry*; Li, J. J., Ed.; Wiley: Hoboken, NJ, **2005**, 457–465. (Review).
9.  Wu, M.; Wang, S. *Synthesis* **2010**, 587–592.
10. Caille, F.; Buron, F.; Toth, E.; Suzenet, F. *E. J. Org. Chem.* **2011**, 2120–2127, S2120/1-S2120/25.
11. Blair, A.; Stevenson, L.; Sutherland, A. *Tetrahedron Lett.* **2012**, *53*, 4084–4086.

## Pictet–Spengler tetrahydroisoquinoline synthesis

Tetrahydroisoquinolines from condensation of β-arylethylamines and carbonyl compounds followed by cyclization.

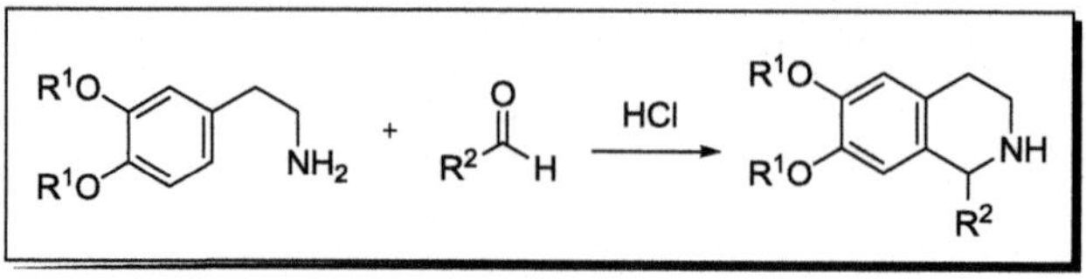

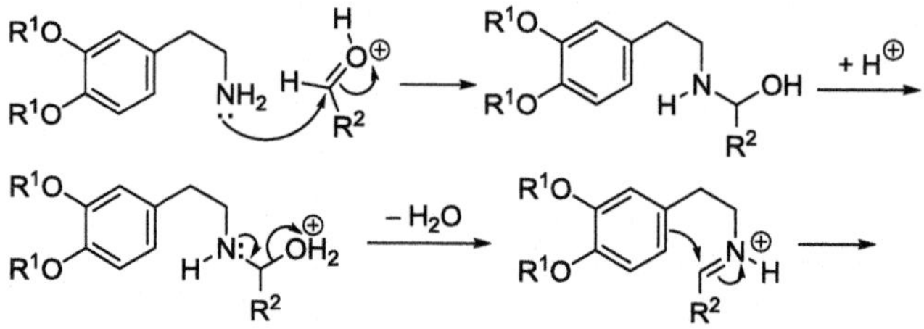

Iminium ion intermediate

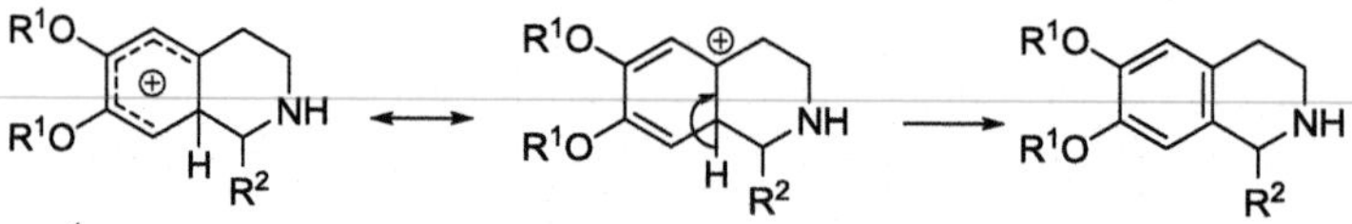

Example 1[4]

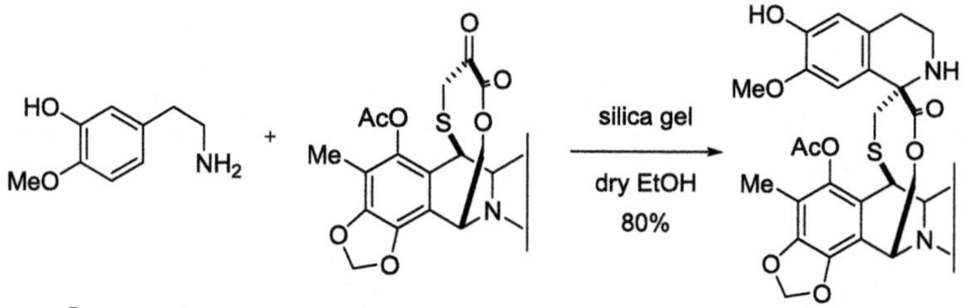

Example 2[7]

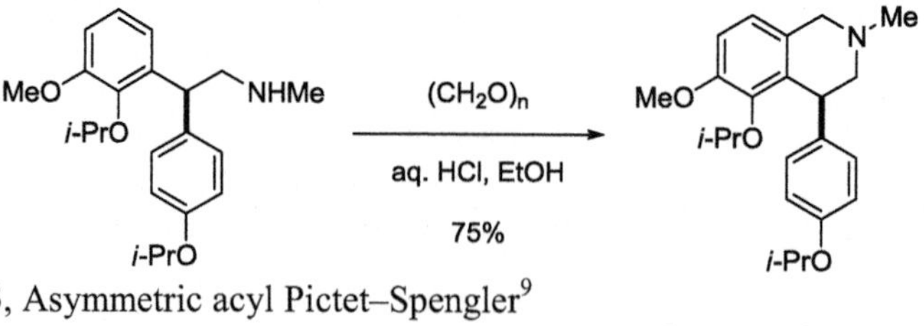

Example 3, Asymmetric acyl Pictet–Spengler[9]

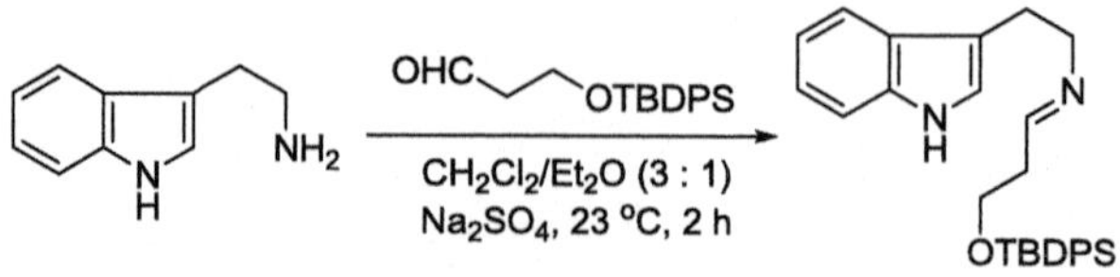

J.J. Li, *Name Reactions: A Collection of Detailed Mechanisms and Synthetic Applications*,
DOI 10.1007/978-3-319-03979-4_215, © Springer International Publishing Switzerland 2014

AcCl, 2,6-lutidine, Et$_2$O
−78 to −60 °C, 23 h

81% 2 steps
94% *ee*

Example 4, Oxa-Pictet–Spengler[10]

CHO

BF$_3$·OEt$_2$, CH$_2$Cl$_2$
0 °C to rt, 88%, 88% *de*

Example 5,

CO$_2$Me
NH$_2$

OHC

TFA, 4 Å MS

CH$_2$Cl$_2$, rt, 4 h
98%

CO$_2$Me

### References

1. Pictet, A.; Spengler, T. *Ber.* **1911**, *44*, 2030–2036.
2. Cox, E. D.; Cook, J. M. *Chem. Rev.* **1995**, *95*, 1797–1842. (Review).
3. Corey, E. J.; Gin, D. Y.; Kania, R. S. *J. Am. Chem. Soc.* **1996**, *118*, 9202–9203.
4. Zhou, B.; Guo, J.; Danishefsky, S. J. *Org. Lett.* **2002**, *4*, 43–46.
5. Yu, J.; Wearing, X. Z.; Cook, J. M. *Tetrahedron Lett.* **2003**, *44*, 543–547.
6. Tsuji, R.; Nakagawa, M.; Nishida, A. *Tetrahedron: Asymmetry* **2003**, *14*, 177–180.
7. Couture, A.; Deniau, E.; Grandclaudon, P.; Lebrun, S. *Tetrahedron: Asymmetry* **2003**, *14*, 1309–1320.
8. Tinsley, J. M. *Pictet–Spengler Isoquinoline Synthesis.* In *Name Reactions in Heterocyclic Chemistry*; Li, J. J., Ed.; Wiley: Hoboken, NJ, **2005**, 469–479. (Review).
9. Mergott, D. J.; Zuend, S. J.; Jacobsen, E. N. *Org. Lett.* **2008**, *10*, 745–748.
10. Eid, C. N.; Shim, J.; Bikker, J.; Lin, M. *J. Org. Chem.* **2009**, *74*, 423–426.
11. Pradhan, P.; Nandi, D.; Pradhan, S. D.; Jaisankar, P.; Giri, V. S. *Synlett* **2013**, *24*, 85–89.

## Pinacol rearrangement

Acid-catalyzed rearrangement of vicinal diols (pinacols) to carbonyl compounds.

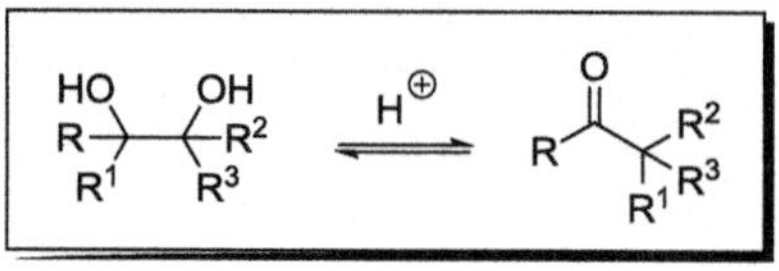

The most electron-rich alkyl group (more substituted carbon) migrates first.  The general migration order:

tertiary alkyl > cyclohexyl > secondary alkyl > benzyl > phenyl >
primary alkyl > methyl >> H.

For substituted aryls:
*p*-MeO-Ar > *p*-Me-Ar > *p*-Cl-Ar > *p*-Br-Ar > *p*-O$_2$N-Ar

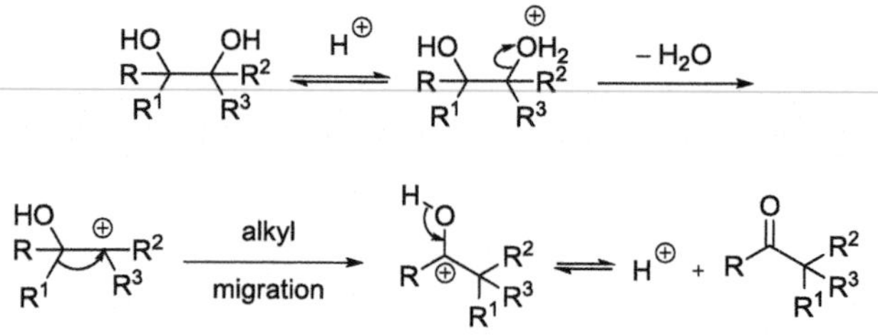

Example 1[4]

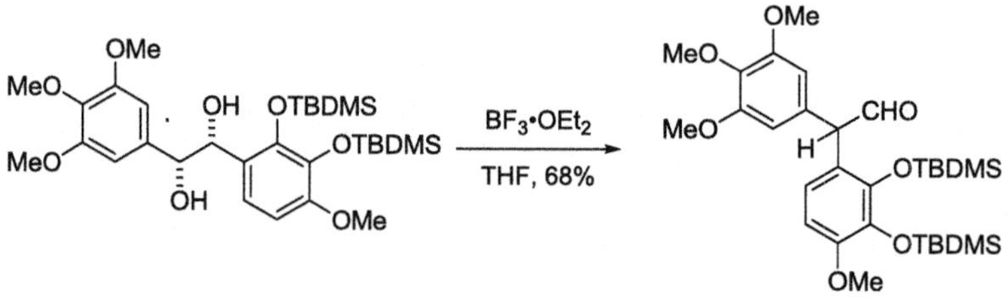

Example 2[5]

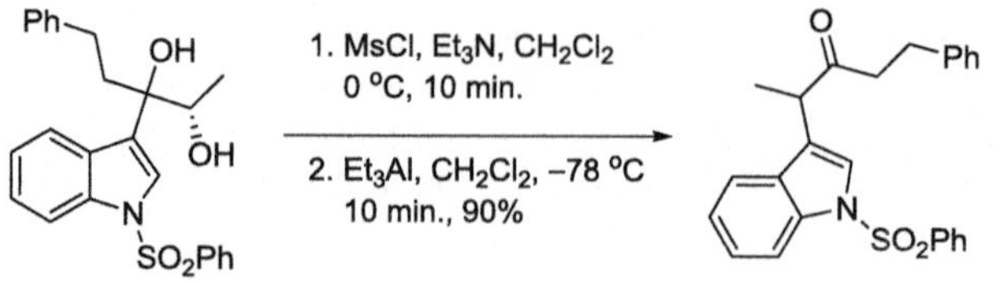

J.J. Li, *Name Reactions: A Collection of Detailed Mechanisms and Synthetic Applications*,
DOI 10.1007/978-3-319-03979-4_216, © Springer International Publishing Switzerland 2014

Example 3[7]

Example 4[9]

R = vinyl, 92%
R = allyl, 95%
R = furyl, 90%
R = prenyl, 94%
} 98% ee

Example 5, A Trivalent organophosphorus reagent induced pinacol rearrangement[11]

## References

1. Fittig, R. *Ann.* **1860,** *114*, 54–63.
2. Magnus, P.; Diorazio, L.; Donohoe, T. J.; Giles, M.; Pye, P.; Tarrant, J.; Thom, S. *Tetrahedron* **1996**, *52*, 14147–14176.
3. Razavi, H.; Polt, R. *J. Org. Chem.* **2000,** *65*, 5693–5706.
4. Pettit, G. R.; Lippert III, J. W.; Herald, D. L. *J. Org. Chem.* **2000,** *65*, 7438–7444.
5. Shinohara, T.; Suzuki, K. *Tetrahedron Lett.* **2002,** *43*, 6937–6940.
6. Overman, L. E.; Pennington, L. D. *J. Org. Chem.* **2003,** *68*, 7143–7157. (Review).
7. Mladenova, G.; Singh, G.; Acton, A.; Chen, L.; Rinco, O.; Johnston, L. J.; Lee-Ruff, E. *J. Org. Chem.* **2004,** *69*, 2017–2023.
8. Birsa, M. L.; Jones, P. G.; Hopf, H. *Eur. J. Org. Chem.* **2005**, 3263–3270.
9. Suzuki, K.; Takikawa, H.; Hachisu, Y.; Bode, J. W. *Angew. Chem. Int. Ed.* **2007**, *46*, 3252–3254.
10. Goes, B. *Pinacol Rearrangement*. In *Name Reactions for Homologations-Part I*; Li, J. J., Ed., Wiley: Hoboken, NJ, **2009**, pp 319–333. (Review).
11. Marin, L.; Zhang, Y.; Robeyns, K.; Champagne, B.; Adriaensens, P.; Lutsen, L.; Vanderzande, D.; Bevk, D.; Maes, W. *Tetrahedron Lett.* **2013**, *54*, 526–529.

# Pinner reaction

Transformation of a nitrile into an imino ether, which can be converted to either an ester or an amidine.

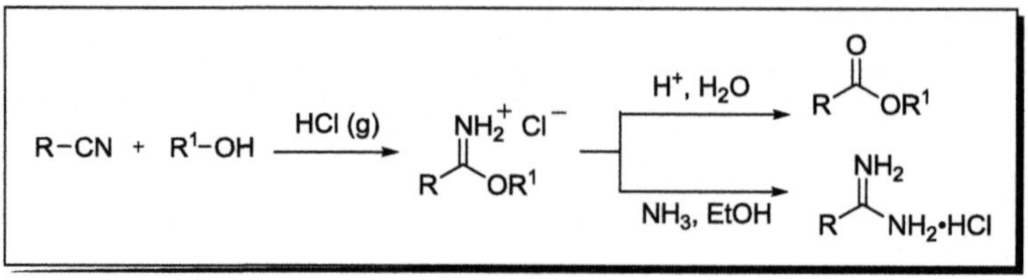

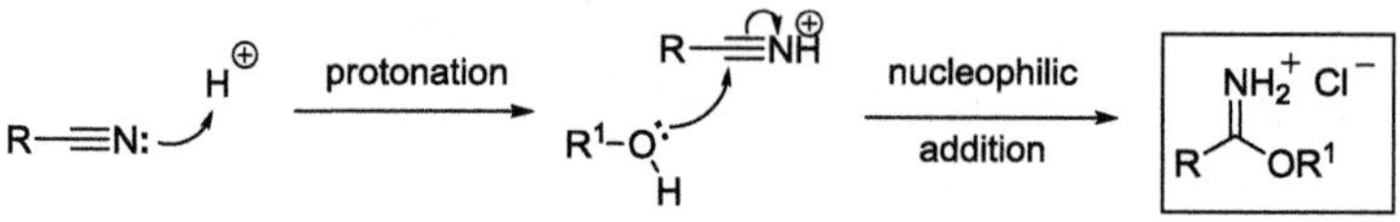

common intermediate

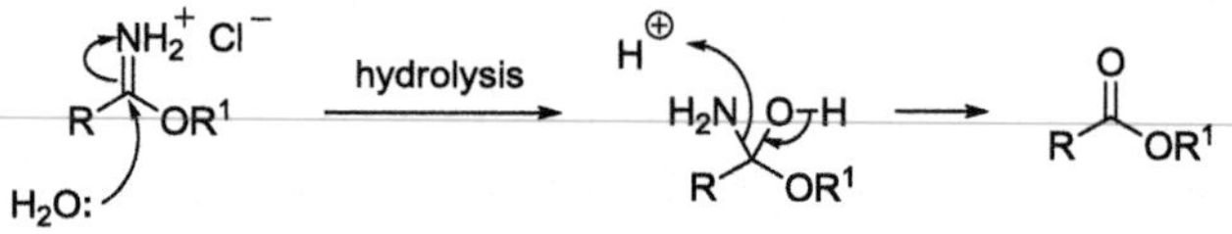

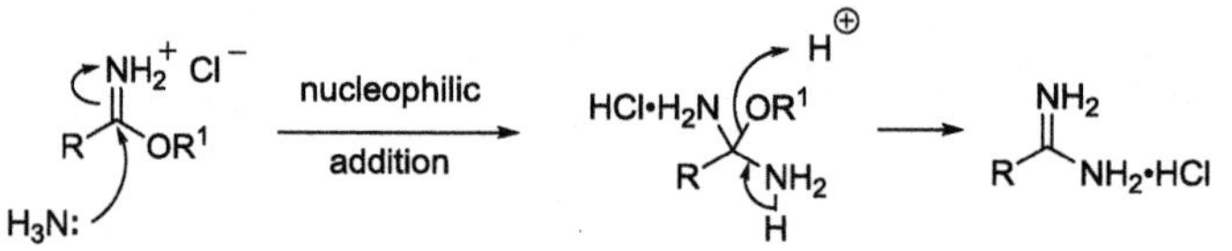

imidate hydrochloride

Example 1[2]

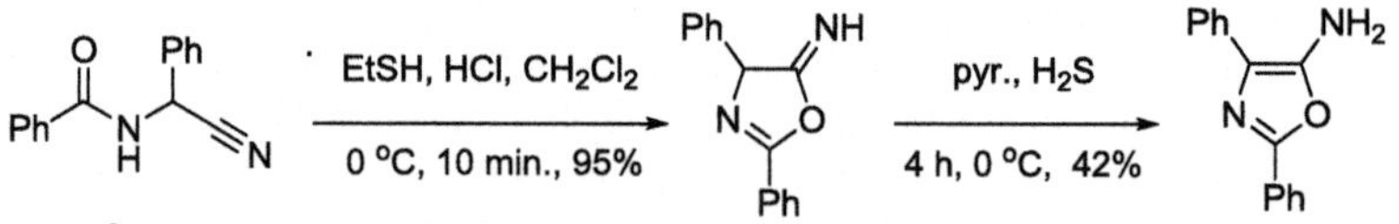

Example 2[2]

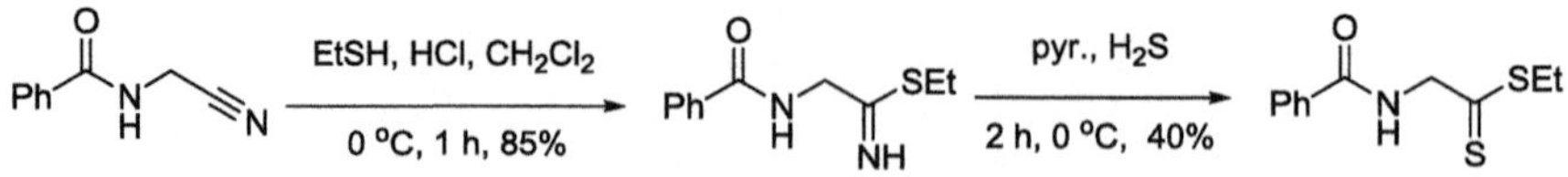

Example 3[6]

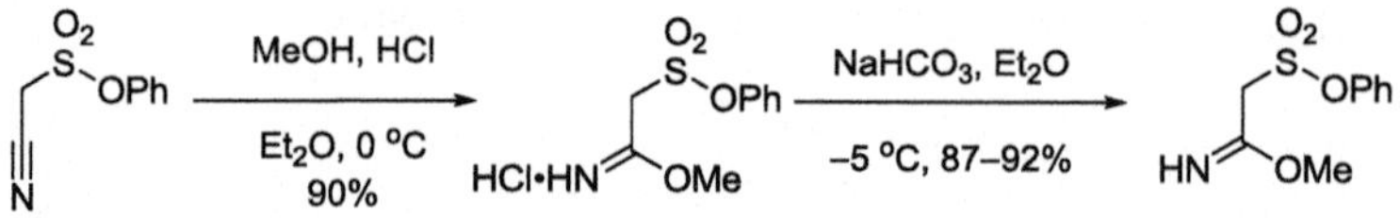

J.J. Li, *Name Reactions: A Collection of Detailed Mechanisms and Synthetic Applications*,
DOI 10.1007/978-3-319-03979-4_217, © Springer International Publishing Switzerland 2014

Example 4[10]

Example 5[11]

monaspilosin

## References

1.  (a) Pinner, A.; Klein, F. *Ber.* **1877**, *10*, 1889–1897.  (b) Pinner, A.; Klein, F. *Ber.* **1878**, *11*, 1825.
2.  Poupaert, J.; Bruylants, A.; Crooy, P. *Synthesis* **1972**, 622–624.
3.  Lee, Y. B.; Goo, Y. M.; Lee, Y. Y.; Lee, J. K. *Tetrahedron Lett.* **1990**, *31*, 1169–1170.
4.  Cheng, C. C. *Org. Prep. Proced. Int.* **1990**, *22*, 643–645.
5.  Siskos, A. P.; Hill, A. M. *Tetrahedron Lett.* **2003**, *44*, 789–794.
6.  Fischer, M.; Troschuetz, R. *Synthesis* **2003**, 1603–1609.
7.  Fringuelli, F.; Piermatti, O.; Pizzo, F. *Synthesis* **2003**, 2331–2334.
8.  Cushion, M. T.; Walzer, P. D.; Collins, M. S.; Rebholz, S.; Vanden Eynde, J. J.; Mayence, A.; Huang, T. L. *Antimicrob. Agents Chemoth.* **2004**, *48*, 4209–4216.
9.  Li, J.; Zhang, L.; Shi, D.; Li, Q.; Wang, D.; Wang, C.; Zhang, Q.; Zhang, L.; Fan, Y. *Synlett* **2008**, 233–236.
10. Racané, L.; Tralic-Kulenovic, V.; Mihalic, Z.; Pavlovic, G.; Karminski-Zamola, G. *Tetrahedron* **2008**, *64*, 11594–11602.
11. Pfaff, D.; Nemecek, G.; Podlech, J. *Beilstein J. Org. Chem.* **2013**, *9*, 1572–1577.

## Polonovski reaction

Treatment of a tertiary *N*-oxide with an activating agent such as acetic anhydride, resulting in rearrangement where an *N,N*-disubstituted acetamide and an aldehyde are generated.

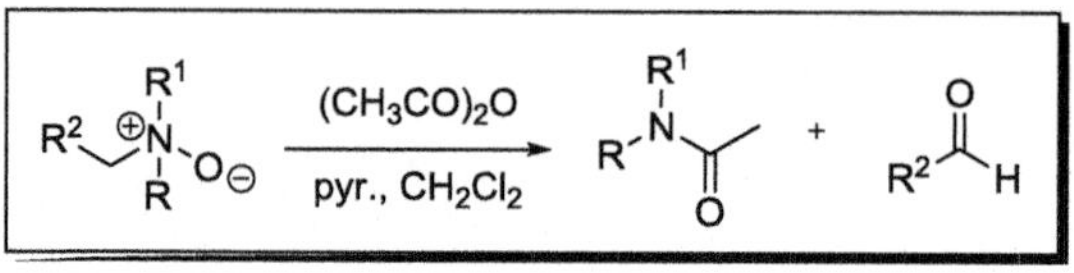

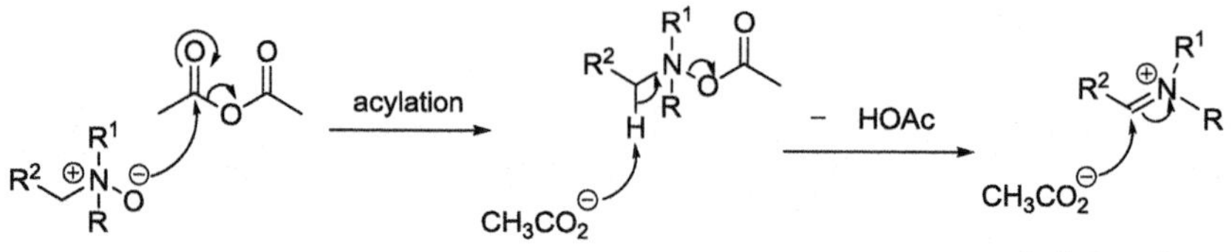

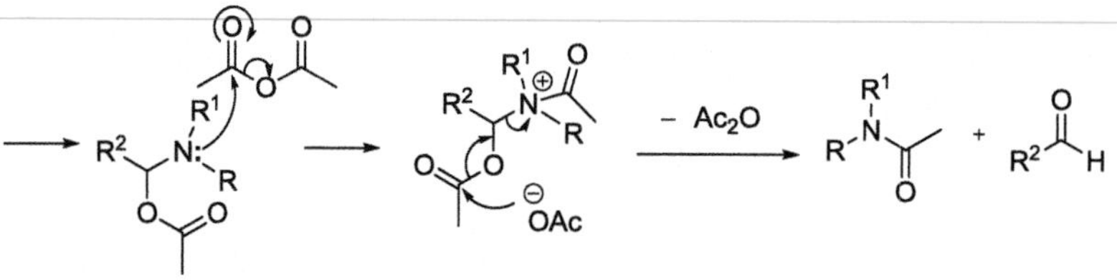

The intramolecular pathway is also operative:

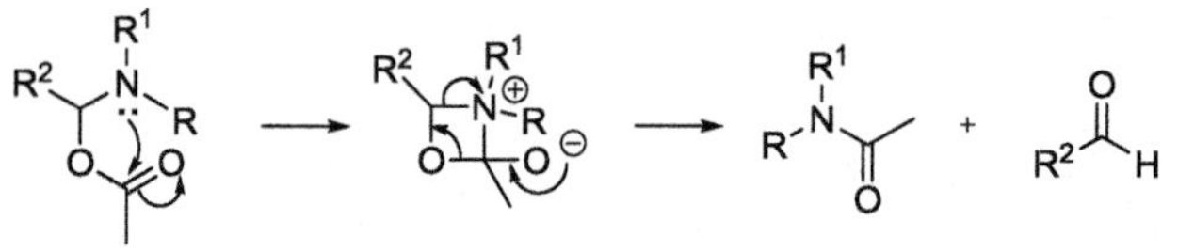

Example 1[1]

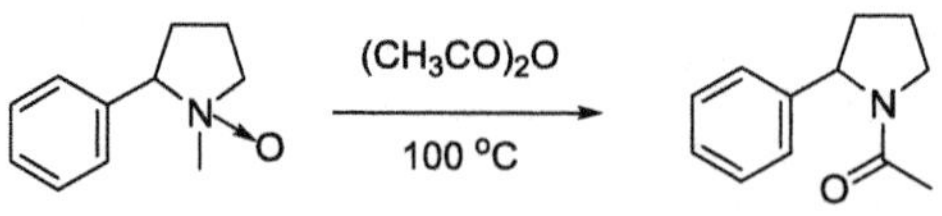

Example 2[2]

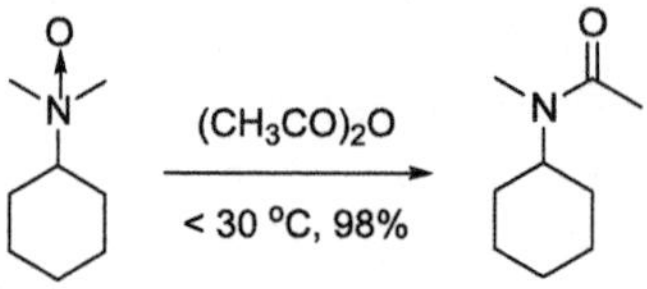

J.J. Li, *Name Reactions: A Collection of Detailed Mechanisms and Synthetic Applications*, DOI 10.1007/978-3-319-03979-4_218, © Springer International Publishing Switzerland 2014

Example 3, Iron salt-mediated Polonovski reaction[9]

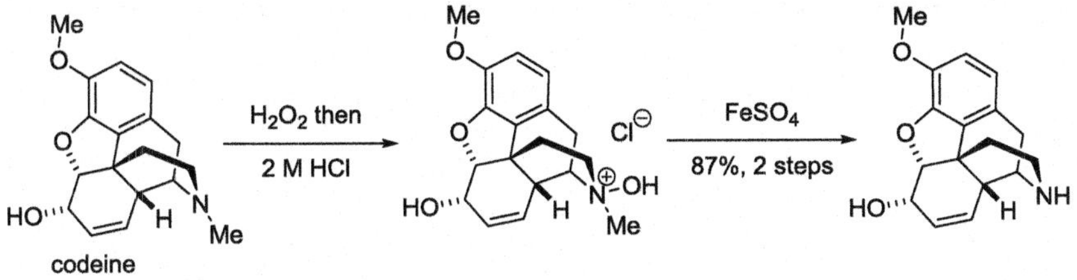

codeine

Example 4[11]

Tröger's base

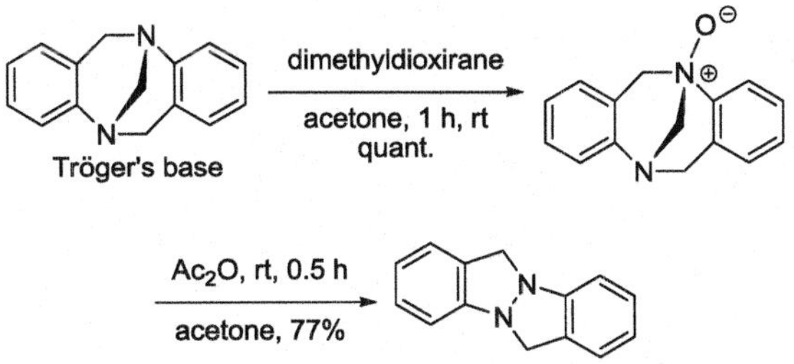

## References

1. Polonovski, M.; Polonovski, M. *Bull. Soc. Chim. Fr.* **1927,** *41*, 1190–1208.
2. Michelot, R. *Bull. Soc. Chim. Fr.* **1969,** 4377–4385.
3. Lounasmaa, M.; Karvinen, E.; Koskinen, A.; Jokela, R. *Tetrahedron* **1987,** *43*, 2135–2146.
4. Tamminen, T.; Jokela, R.; Tirkkonen, B.; Lounasmaa, M. *Tetrahedron* **1989,** *45*, 2683–2692.
5. Grierson, D. *Org. React.* **1990,** *39*, 85–295. (Review).
6. Morita, H.; Kobayashi, J. *J. Org. Chem.* **2002,** *67*, 5378–5381.
7. McCamley, K.; Ripper, J. A.; Singer, R. D.; Scammells, P. J. *J. Org. Chem.* **2003,** *68*, 9847–9850.
8. Nakahara, S.; Kubo, A. *Heterocycles* **2004,** *63*, 1849–1854.
9. Thavaneswaran, S.; Scammells, P. J. *Bioorg. Med. Chem. Lett.* **2006,** *16*, 2868–2871.
10. Volz, H.; Gartner, H. *Eur. J. Org. Chem.* **2007,** 2791–2801.
11. Pacquelet, S.; Blache, Y.; Kimny, T.; Dubois, M.-A. L.; Desbois, N. *Synth. Commun.* **2013,** *43*, 1092–1100.

## Polonovski–Potier reaction

A modification of the Polonovski reaction where trifluoroacetic anhydride is used in place of acetic anhydride. Because the reaction conditions for the Polonovski–Potier reaction are mild, it has largely replaced the Polonovski reaction.

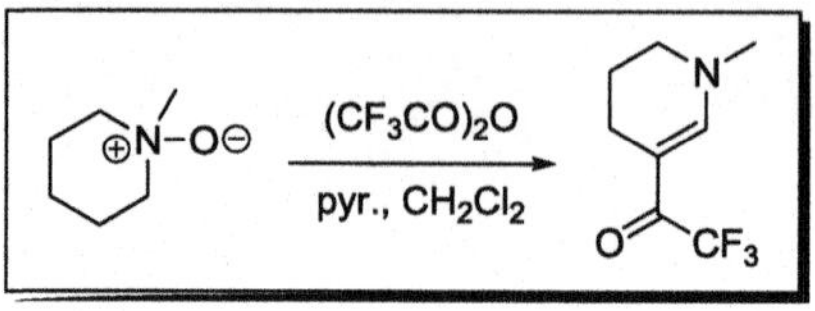

(CF3CO)2O
pyr., CH2Cl2


tertiary *N*-oxide

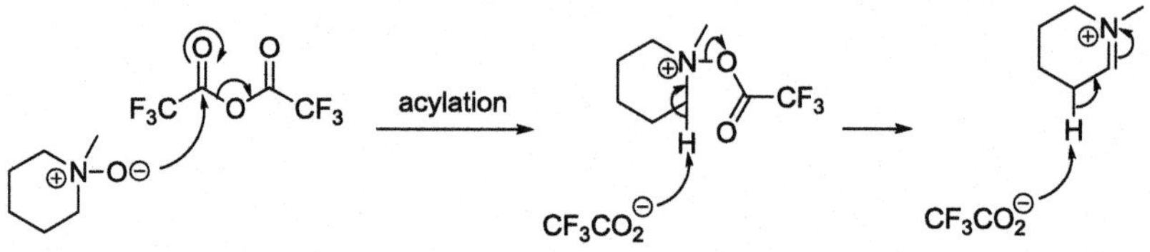

acylation


iminium ion

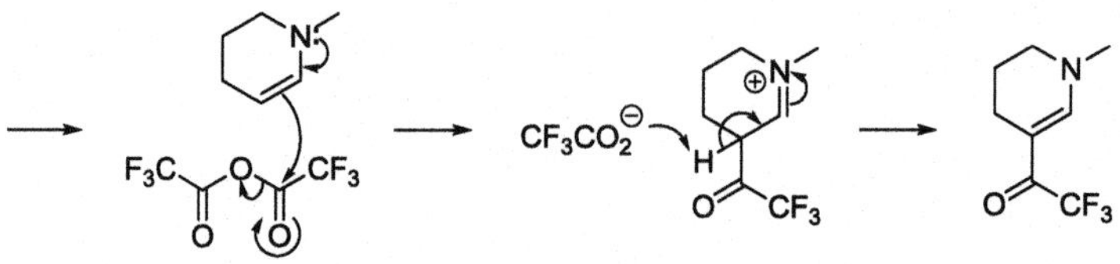

enamine

Example 1[2]

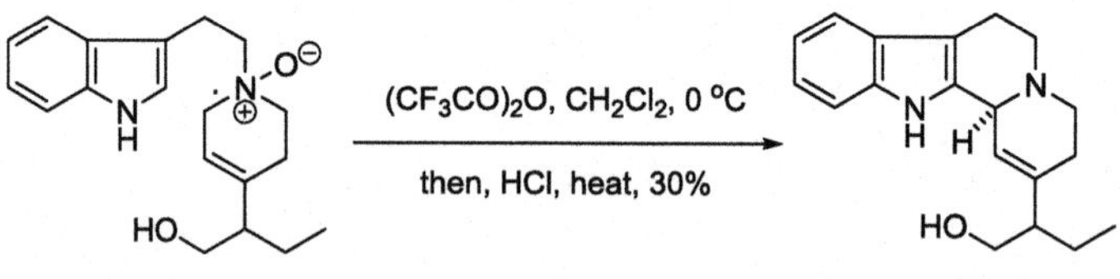

(CF3CO)2O, CH2Cl2, 0 °C
then, HCl, heat, 30%


Example 2[5]

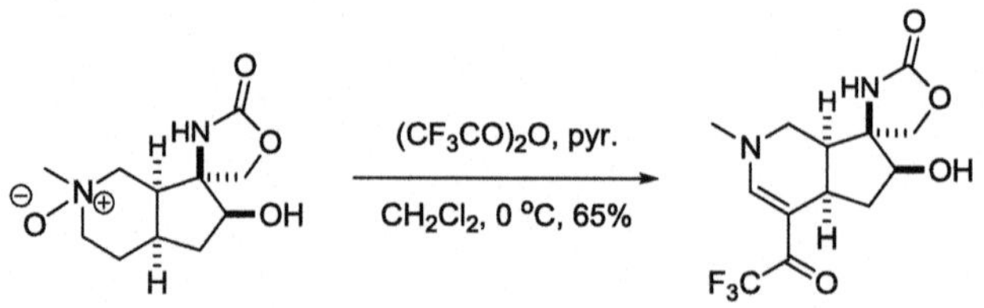

(CF3CO)2O, pyr.
CH2Cl2, 0 °C, 65%


J.J. Li, *Name Reactions: A Collection of Detailed Mechanisms and Synthetic Applications*,
DOI 10.1007/978-3-319-03979-4_219, © Springer International Publishing Switzerland 2014

Example 3[8]

Example 4, m-CPBA also concurrently oxidized the aldehyde[10]

## References

1.  Ahond, A.; Cavé, A.; Kan-Fan, C.; Husson, H.-P.; de Rostolan, J.; Potier, P. *J. Am. Chem. Soc.* **1968,** *90,* 5622–5623.
2.  Husson, H.-P.; Chevolot, L.; Langlois, Y.; Thal, C.; Potier, P. *J. Chem. Soc., Chem. Commun.* **1972,** 930–931.
3.  Grierson, D. *Org. React.* **1990,** *39,* 85–295. (Review).
4.  Sundberg, R. J.; Gadamasetti, K. G.; Hunt, P. J. *Tetrahedron* **1992,** *48,* 277–296.
5.  Kende, A. S.; Liu, K.; Brands, J. K. M. *J. Am. Chem. Soc.* **1995,** *117,* 10597–10598.
6.  Renko, D.; Mary, A.; Guillou, C.; Potier, P.; Thal, C. *Tetrahedron Lett.* **1998,** *39,* 4251–4254.
7.  Suau, R.; Nájera, F.; Rico, R. *Tetrahedron* **2000,** *56,* 9713–9720.
8.  Thomas, O. P.; Zaparucha, A.; Husson, H.-P. *Tetrahedron Lett.* **2001,** *42,* 3291–3293.
9.  Lim, K.-H.; Low, Y.-Y.; Kam, T.-S. *Tetrahedron Lett.* **2006,** *47,* 5037–5039.
10. Gazak, R.; Kren, V.; Sedmera, P.; Passarella, D.; Novotna, M.; Danieli, B. *Tetrahedron* **2007,** *63,* 10466–10478.
11. Nishikawa, Y.; Kitajima, M.; Kogure, N.; Takayama, H. *Tetrahedron* **2009,** *65,* 1608–1617.
12. Perry, M. A.; Morin, M. D.; Slafer, B. W.; Rychnovsky, S. D. **2012,** *77,* 3390–3400.

# Pomeranz–Fritsch reaction

Isoquinoline and saturated variants synthesis *via* acid-mediated cyclization of the appropriate aminoacetal intermediate.

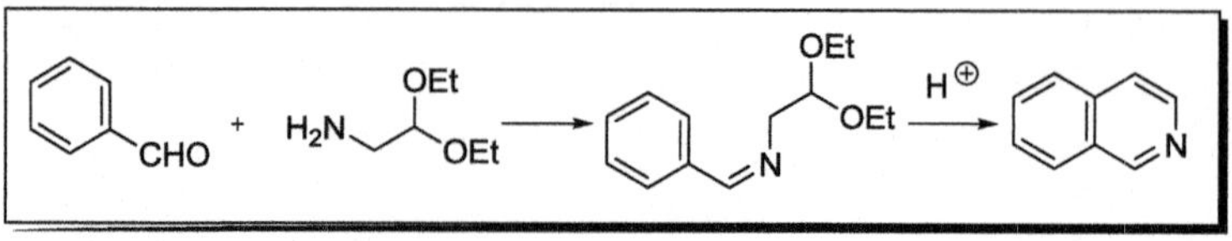

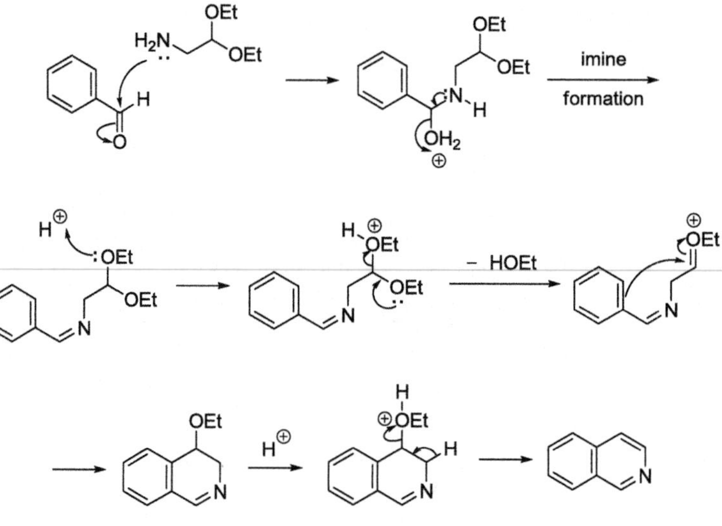

Example 1[3]

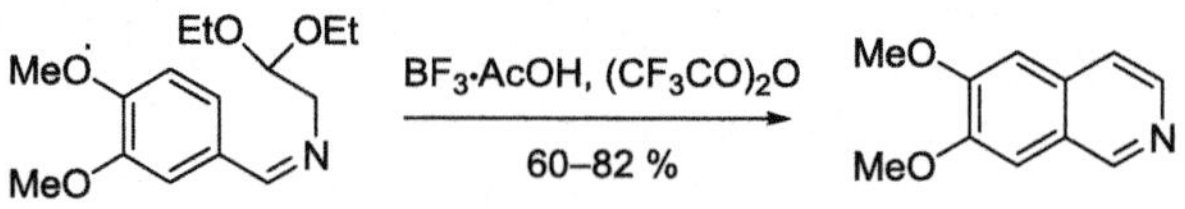

$$BF_3 \cdot AcOH, (CF_3CO)_2O$$
$$60–82\%$$

Example 2[4]

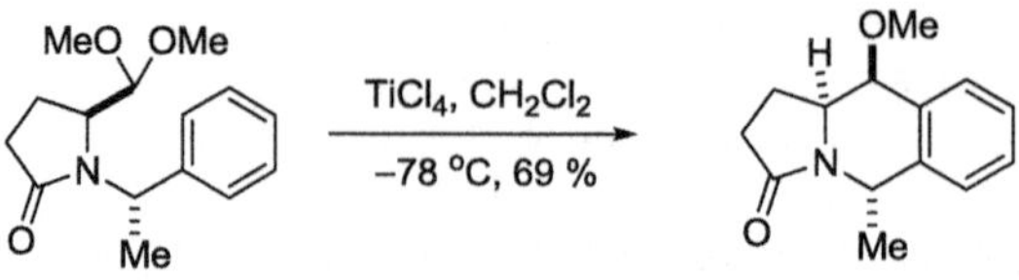

$$TiCl_4, CH_2Cl_2$$
$$-78\ ^\circ C, 69\%$$

J.J. Li, *Name Reactions: A Collection of Detailed Mechanisms and Synthetic Applications*, DOI 10.1007/978-3-319-03979-4_220, © Springer International Publishing Switzerland 2014

Example 3[9]

Example 4, **Bobbitt modification**[10]

**References**

1. (a) Pomeranz, C. *Monatsh.* **1893,** *14*, 116–119. Cesar Pomeranz (1860–1926) received his Ph.D. degree at Vienna, where he was employed as an associate professor of chemistry. (b) Fritsch, P. *Ber.* **1893,** *26*, 419–422. Paul Fritsch (1859–1913) was born in Oels, Silesia. He studied at Munich where he received his doctorate in 1884. Fritsch eventually became a professor at Marburg after several junior positions.
2. Gensler, W. J. *Org. React.* **1951,** *6*, 191–206. (Review).
3. Bevis, M. J.; Forbes, E. J.; Naik, N. N.; Uff, B. C. *Tetrahedron* **1971,** *27*, 1253–1259.
4. Ishii, H.; Ishida, T. *Chem. Pharm. Bull.* **1984,** *32*, 3248–3251.
5. Bobbitt, J. M.; Bourque, A. J. *Heterocycles* **1987,** *25*, 601–616. (Review).
6. Gluszyńska, A.; Rozwadowska, M. D. *Tetrahedron: Asymmetry* **2000,** *11*, 2359–2368.
7. Capilla, A. S.; Romero, M.; Pujol, M. D.; Caignard, D. H.; Renard, P. *Tetrahedron* **2001,** *57*, 8297–8303.
8. Hudson, A. *Pomeranz–Fritsch Reaction.* In *Name Reactions in Heterocyclic Chemistry*; Li, J. J., Ed.; Wiley: Hoboken, NJ, **2005,** 480–486. (Review).
9. Bracca, A. B. J.; Kaufman, T. S. *Eur. J. Org. Chem.* **2007,** 5284–5293.
10. Grajewska, A.; Rozwadowska, M. D. *Tetrahedron: Asymmetry* **2007,** *18*, 2910–2914.
11. Chrzanowska, M.; Grajewska, A.; Rozwadowska, M. D. *Heterocycles* **2012,** *86*, 1119–1127.

## *Schlittler–Müller modification*

Simple permutation where the amine and the aldehyde switch places for the two reactants in comparison to the Pomeranz–Fritsch reaction.

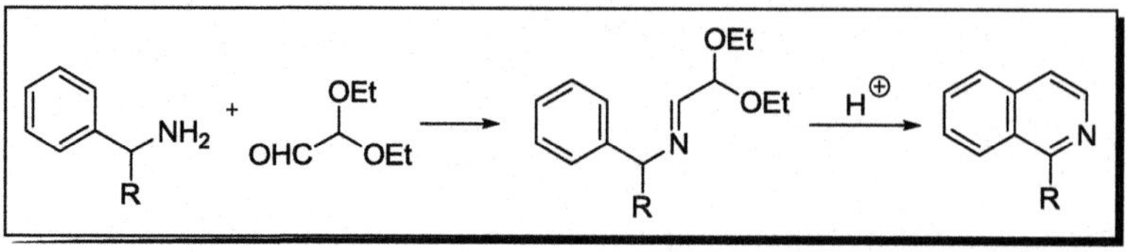

Example 1[3]

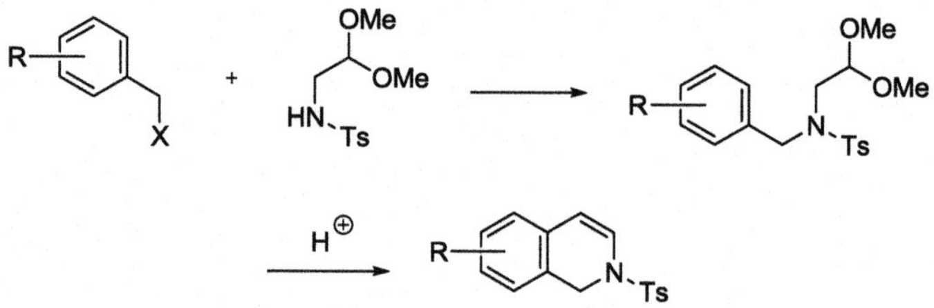

Example 2[4]

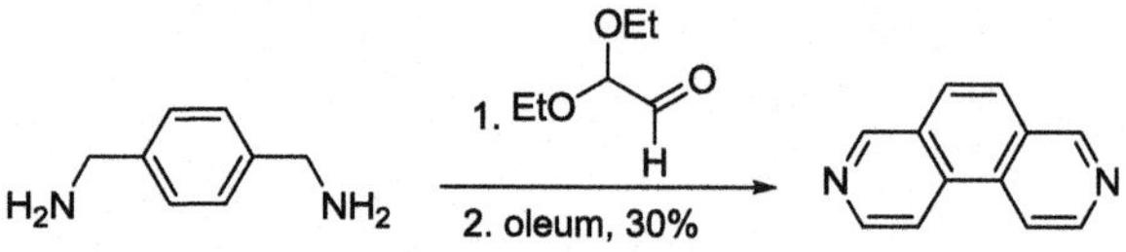

## References

1.  Schlittler, E.; Müller, J. *Helv. Chim. Acta* **1948**, *31*, 914–924, 1119–1132.
2.  Guthrie, D. A.; Frank, A. W.; Purves, C. B. *Can. J. Chem.* **1955**, *33*, 729–742.
3.  Boger, D. L.; Brotherton, C. E.; Kelley, M. D. *Tetrahedron* **1981**, *37*, 3977–3980.
4.  Gill, E. W.; Bracher, A. W. *J. Heterocycl. Chem.* **1983**, *20*, 1107–1109.
5.  Hudson, A. *Pomeranz–Fritsch Reaction*. In *Name Reactions in Heterocyclic Chemistry*; Li, J. J., Ed.; Wiley: Hoboken, NJ, **2005**, 480–486. (Review).

# Pavorov reaction

The Povarov reaction is the inverse electron-demand aza-Diels–Alder reaction, a [4 + 2] cycloaddition between an *N*-arylimine (as the diene) and an electron-rich olefin (as the dienophile), which gives tetrahydroquinolines or substituted quinolines as the product.

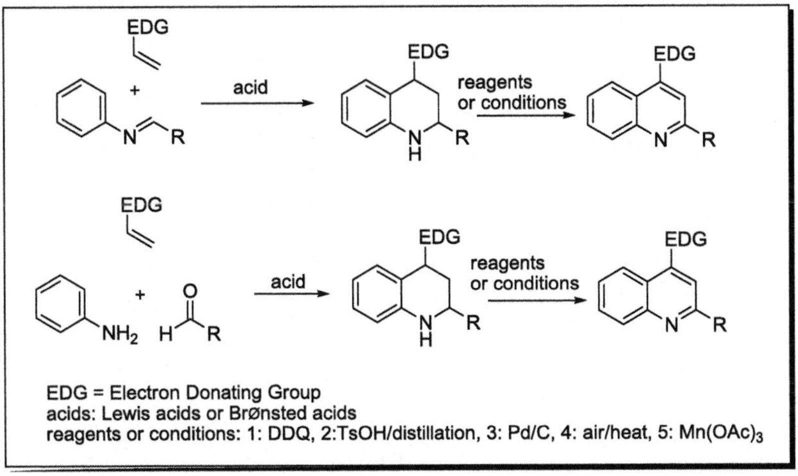
EDG
acid
reagents
or conditions
N
R
H
NH2
O
EDG = Electron Donating Group
acids: Lewis acids or Brønsted acids
reagents or conditions: 1: DDQ, 2:TsOH/distillation, 3: Pd/C, 4: air/heat, 5: Mn(OAc)3

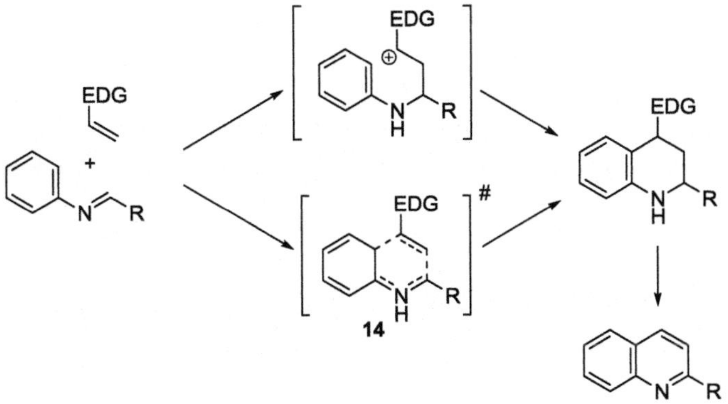
EDG
N
R
H
#
14

Example 1[2]

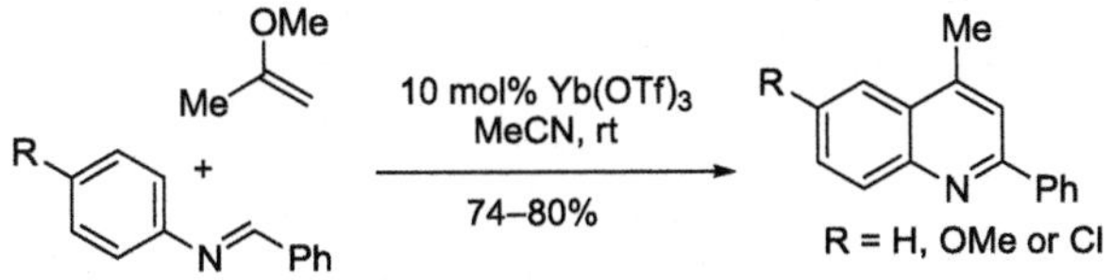
OMe
Me
R
Ph
10 mol% Yb(OTf)3
MeCN, rt
74–80%
R = H, OMe or Cl

J.J. Li, *Name Reactions: A Collection of Detailed Mechanisms and Synthetic Applications*,
DOI 10.1007/978-3-319-03979-4_221, © Springer International Publishing Switzerland 2014

Example 2, Katritzky variation[3,5]

Torcetrapib (CP-529,414)

Example 3[7]

35–91% yield
up to 98:2 *dr*

## References

1. Povarov, L. S.; Mikhailov, B. M. *Izv. Akad. Nauk SSR, Ser. Khim.* **1963,** 953–956.
2. Makioka, Y.; Shindo, T.; Taniguchi, Y.; Takaki, K.; Fujiwara, Y. *Synthesis* **1995,** *7,* 801–804.
3. Katrizky, A. R.; Belyakov, S. A. *Aldrichimica Acta* **1998,** *31,* 35–45. (Review).
4. Buonora, P.; Ólsen, J.-C.; Oh, T. *Tetrahedron* **2001,** *57,* 6099–6138. (Review).
5. Damon, D. B.; Dugger, R. W.; Magnus-Aryitey, G.; Ruggeri, R. B.; Wester, R. T.; Tu, M.; Abramov, Y. *Org. Process Res. Dev.* **2006,** *10,* 464–471.
6. Kouznetsov, V. V. *Tetrahedron* **2009,** *65,* 2721–2750. (Review).
7. Smith, C. D.; Gavrilyuk, J. I.; Lough, A. J.; Batey, R. A. *J. Org. Chem.* **2010,** *75,* 702–715.
8. Xu, H.; Zuend, S. J.; Woll, M. G.; Tao, Y.; Jacobsen, E. N. *Science* **2010,** *327,* 986–990.
9. Zhang, J. *Povarov Reaction.* In *Name Reactions in Heterocyclic Chemistry II*; Li, J. J., Ed.; Wiley: Hoboken, NJ, **2011,** 385–399. (Review).

# Prévost *trans*-dihydroxylation

*Cf.* Woodward *cis*-dihydroxylation.

cyclic iodonium ion intermediate    neighboring group assistance

Example 1[5]

AgOCOPh, I$_2$

PhH, rt, 2 h,
reflux, 10 h, 46%

Example 2[9]

O$_2$C-C$_6$H$_4$-*p*-OMe    AgOCOPh, I$_2$

CCl$_4$, 74%

O$_2$C-C$_6$H$_4$-*p*-OMe    1. KOH, H$_2$O

2. Ac$_2$O, pyr.

## References

1. Prévost, C. *Compt. Rend.* **1933**, *196,* 1129–1131.
2. Campbell, M. M.; Sainsbury, M.; Yavarzadeh, R. *Tetrahedron* **1984,** *40*, 5063–5070.
3. Ciganek, E.; Calabrese, J. C. *J. Org. Chem.* **1995**, *60*, 4439–4443.
4. Brimble, M. A.; Nairn, M. R. *J. Org. Chem.* **1996**, *61*, 4801–4805.
5. Zajc, B. *J. Org. Chem.* **1999**, *64*, 1902–1907.
6. Hamm, S.; Hennig, L.; Findeisen, M.; Muller, D. *Tetrahedron* **2000**, *56*, 1345–1348.
7. Ray, J. K.; Gupta, S.; Kar, G. K.; Roy, B. C.; Lin, J.-M. *J. Org. Chem.* **2000**, *65*, 8134–8138.
8. Sabat, M.; Johnson, C. R. *Tetrahedron Lett.* **2001**, *42*, 1209–1212.
9. Hodgson, R.; Nelson, A. *Org. Biomol. Chem.* **2004**, *2*, 373–386.
10. Emmanuvel, L.; Shaikh, T. M. A.; Sudalai, A. *Org. Lett.* **2005**, *7*, 5071–5074.

J.J. Li, *Name Reactions: A Collection of Detailed Mechanisms and Synthetic Applications*,
DOI 10.1007/978-3-319-03979-4_222, © Springer International Publishing Switzerland 2014

# Prins reaction

The Prins reaction is the acid-catalyzed addition of aldehydes to alkenes and gives different products depending on the reaction conditions.

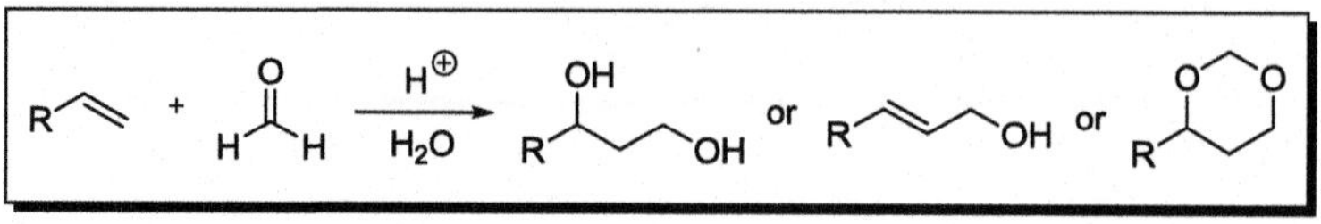
H⊕
H₂O
or
or

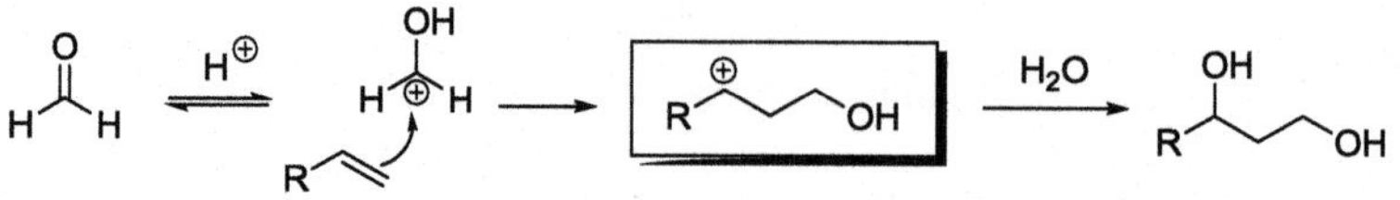
H⊕
H₂O

the common intermediate

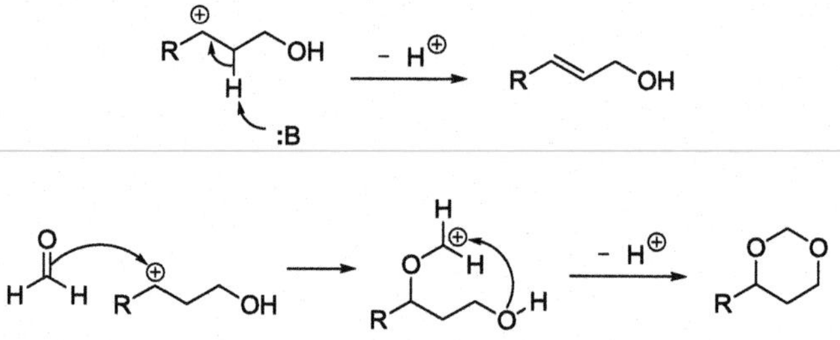
- H⊕
:B
- H⊕

Example 1[5]

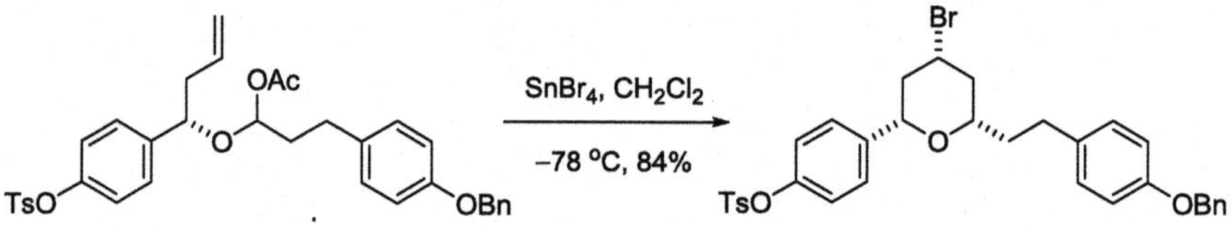
SnBr4, CH2Cl2
−78 °C, 84%

Example 2[7]

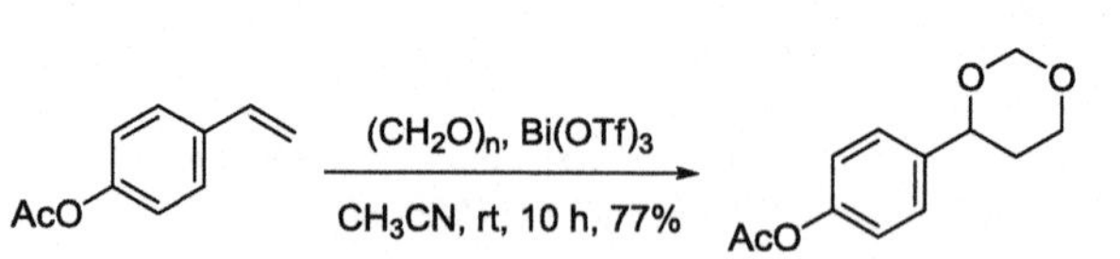
(CH2O)n, Bi(OTf)3
CH3CN, rt, 10 h, 77%

Example 3[9]

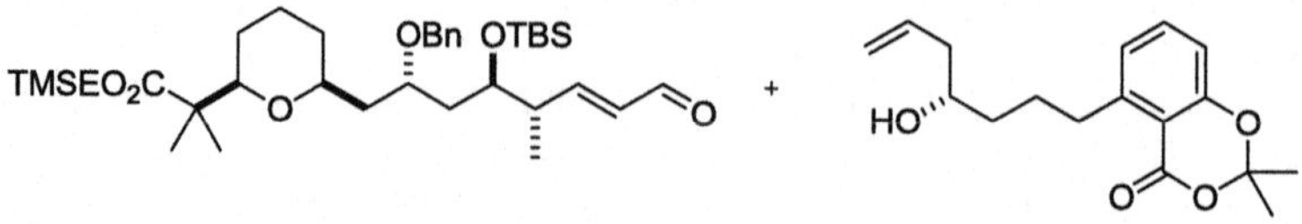

J.J. Li, *Name Reactions: A Collection of Detailed Mechanisms and Synthetic Applications*,
DOI 10.1007/978-3-319-03979-4_223, © Springer International Publishing Switzerland 2014

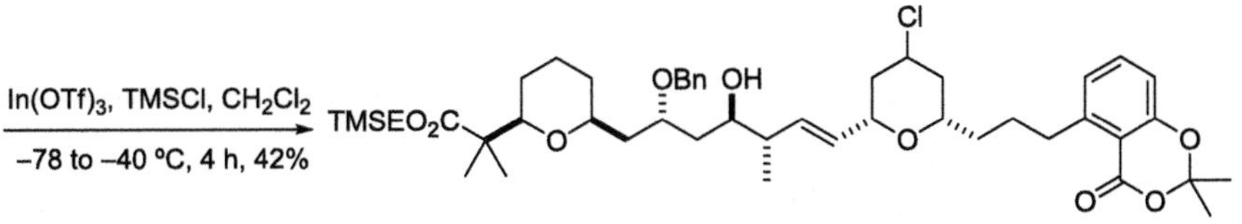

In(OTf)$_3$, TMSCl, CH$_2$Cl$_2$
$\xrightarrow{\phantom{xxxxxx}}$
−78 to −40 °C, 4 h, 42%

**Example 4**[10]

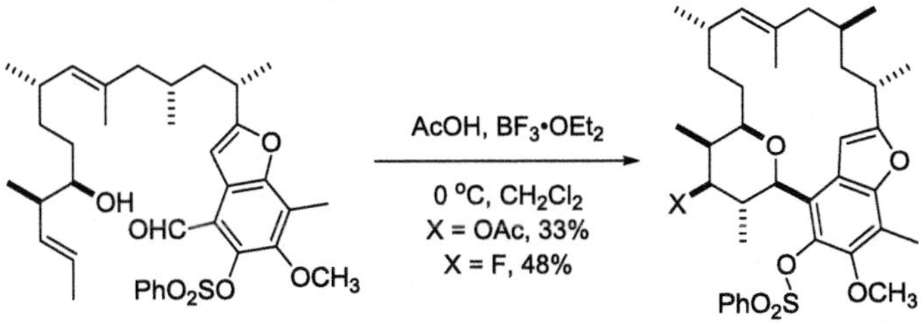

AcOH, BF$_3$·OEt$_2$

0 °C, CH$_2$Cl$_2$
X = OAc, 33%
X = F, 48%

**Example 5, A cascade of the Prins/Ritter amidation reaction**[11]

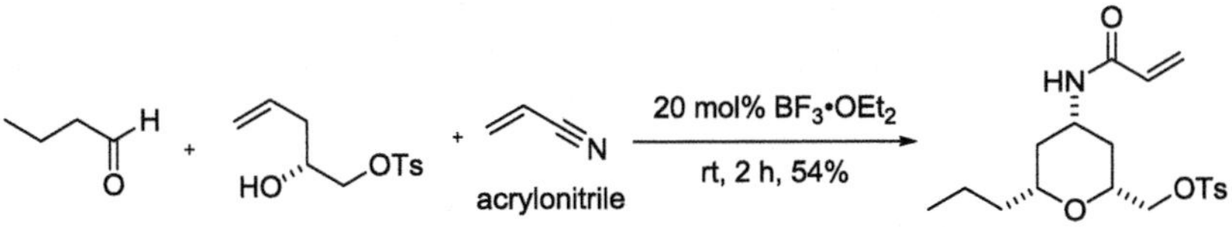

acrylonitrile

20 mol% BF$_3$·OEt$_2$

rt, 2 h, 54%

**Example 6**[12]

20 mol% Sc(OTf)$_3$

ClCH$_2$CH$_2$Cl,
4 Å MS, 80 °C

45%                              15%

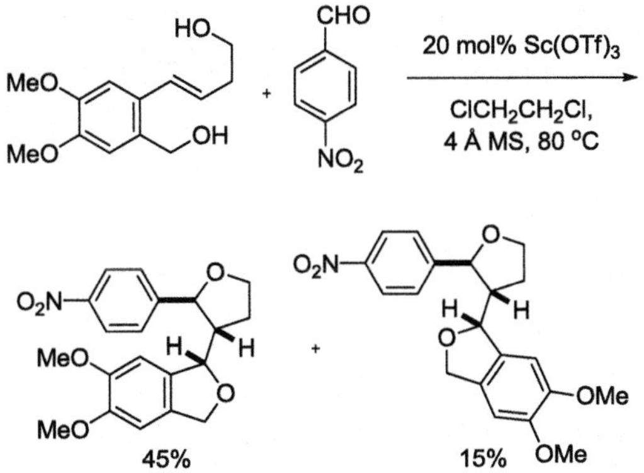

## References

1.  Prins, H. *J. Chem. Weekblad* **1919**, *16*, 1072–1023. Born in Zaandam, The Nether-
    lands, Hendrik J. Prins (1889–1958) was not even an organic chemist *per se*. After
    obtaining a doctorate in chemical engineering, Prins worked for an essential oil com-
    pany and then a company dealing with the rendering of condemned meats and carcass-
    es. But he had a small laboratory near his house where he carried out his experiments

in his spare time, which obviously was not a big distraction—for he rose to be the president-director of the firm he worked for.

2.   Adam, D. R.; Bhatnagar, S. P. *Synthesis* **1977,** 661–672. (Review).

3.   Hanaki, N.; Link, J. T.; MacMillan, D. W. C.; Overman, L. E.; Trankle, W. G.; Wurster, J. A. *Org. Lett.* **2000,** *2,* 223–226.

4.   Davis, C. E.; Coates, R. M. *Angew. Chem. Int. Ed.* **2002,** *41,* 491–493.

5.   Marumoto, S.; Jaber, J. J.; Vitale, J. P.; Rychnovsky, S. D. *Org. Lett.* **2002,** *4,* 3919–3922.

6.   Braddock, D. C.; Badine, D. M.; Gottschalk, T.; Matsuno, A.; Rodriguez-Lens, M. *Synlett* **2003,** 345–348.

7.   Sreedhar, B.; Swapna, V.; Sridhar, Ch.; Saileela, D.; Sunitha, A. *Synth. Commun.* **2005,** *35,* 1177–1182.

8.   Aubele, D. L.; Wan, S.; Floreancig, P. E. *Angew. Chem. Int. Ed.* **2005,** *44,* 3485–3488.

9.   Chan, K.-P.; Ling, Y. H.; Loh, T.-P. *Chem. Commun.* **2007,** 939–941.

10.  Bahnck, K. B.; Rychnovsky, S. D. *J. Am. Chem. Soc.* **2008,** *130,* 13177–13181.

11.  Yadav, J. S.; Reddy, Y. J.; Reddy, P. A. N.; Reddy, B. V. S. *Org. Lett.* **2013,** *15,* 546–549.

12.  Subba Reddy, B. V.; Jalal, S.; Borkar, P.; Yadav, J. S.; Gurava Reddy, P.; Sarma, A.V.S. *Tetrahedron Lett.* **2013,** *54,* 1519–1523.

# Pschorr cyclization

The intramolecular version of the Gomberg–Bachmann reaction.

Example 1[7]

J.J. Li, *Name Reactions: A Collection of Detailed Mechanisms and Synthetic Applications*,
DOI 10.1007/978-3-319-03979-4_224, © Springer International Publishing Switzerland 2014

Example 2[8]

Example 3[10]

34%                    62%

## References

1.  Pschorr, R. *Ber.* **1896,** *29*, 496–501.   Robert Pschorr (1868–1930), born in Munich, Germany, studied under von Baeyer, Bamberger, Knorr, and Fischer.  He became an assistant professor in 1899 at Berlin where he discovered the phenanthrene synthesis. During WWI, Pschorr served as a major in the German Army.
2.  Kupchan, S. M.; Kameswaran, V.; Findlay, J. W. A. *J. Org. Chem.* **1973,** *38*, 405–406.
3.  Wassmundt, F. W.; Kiesman, W. F. *J. Org. Chem.* **1995,** *60*, 196–201.
4.  Qian, X.; Cui, J.; Zhang, R. *Chem. Commun.* **2001,** 2656–2657.
5.  Hassan, J.; Sévignon, M.; Gozzi, C.; Schulz, E.; Lemaire, M. *Chem. Rev.* **2002,** *102*, 1359–1469. (Review).
6.  Karady, S.; Cummins, J. M.; Dannenberg, J. J.; del Rio, E.; Dormer, P. G.; Marcune, B. F.; Reamer, R. A.; Sordo, T. L. *Org. Lett.* **2003,** *5*, 1175–1178.
7.  Xu, Y.; Qian, X.; Yao, W.; Mao, P.; Cui, J. *Bioorg. Med. Chem.* **2003,** *11*, 5427–5433.
8.  Tapolcsányi, P.; Maes, B. U. W.; Monsieurs, K.; Lemière, G. L. F.; Riedl, Z.; Hajós, G.; Van der Driessche, B.; Dommisse, R. A.; Mátyus, P. *Tetrahedron* **2003,** *59*, 5919–5926.
9.  Mátyus, P.; Maes, B. U. W.; Riedl, Z.; Hajós, G.; Lemière, G. L. F.; Tapolcśanyi, P.; Monsieurs, K.; Éliás, O.; Dommisse, R. A.; Krajsovszky, G. *Synlett* **2004,** 1123–1139. (Review).
10. Moorthy, J. N.; Samanta, S. *J. Org. Chem.* **2007,** *72*, 9786–9789.
11. Lockner, J. W.; Dixon, D. D.; Risgaard, R.; Baran, P. S. *Org. Lett.* **2011,** *13*, 5628–5631.

# Pummerer rearrangement

The transformation of sulfoxides into α-acyloxythioethers using acetic anhydride.

## Example 1[2]

## Example 2[7]

## Example 3[8]

J.J. Li, *Name Reactions: A Collection of Detailed Mechanisms and Synthetic Applications*,
DOI 10.1007/978-3-319-03979-4_225, © Springer International Publishing Switzerland 2014

Example 4[9]

Example 5, Stereoselective Pummerer rearrangement[10,12]

## References

1. Pummerer, R. *Ber.* **1910**, *43*, 1401–1412. Rudolf Pummerer, born in Austria in 1882, studied under von Baeyer, Willstätter, and Wieland. He worked for BASF for a few years and in 1921 he was appointed head of the organic division of the Munich Laboratory, fulfilling his long-desired ambition.
2. Katsuki, T.; Lee, A. W. M.; Ma, P.; Martin, V. S.; Masamune, S.; Sharpless, K. B.; Tuddenham, D.; Walker, F. J. *J. Org. Chem.* **1982**, *47*, 1373–1378.
3. De Lucchi, O.; Miotti, U.; Modena, G. *Org. React.* **1991**, *40*, 157–406. (Review).
4. Padwa, A.; Gunn, D. E., Jr.; Osterhout, M. H. *Synthesis* **1997**, 1353–1378. (Review).
5. Padwa, A.; Waterson, A. G. *Curr. Org. Chem.* **2000**, *4*, 175–203. (Review).
6. Padwa, A.; Bur, S. K.; Danca, D. M.; Ginn, J. D.; Lynch, S. M. *Synlett* **2002**, 851–862. (Review).
7. Gámez Montaño, R.; Zhu, J. *Chem. Commun.* **2002**, 2448–2449.
8. Padwa, A.; Danca, M. D.; Hardcastle, K.; McClure, M. *J. Org. Chem.* **2003**, *68*, 929–941.
9. Suzuki, T.; Honda, Y.; Izawa, K.; Williams, R. M. *J. Org. Chem.* **2005**, *70*, 7317–7323.
10. Nagao, Y.; Miyamoto, S.; Miyamoto, M.; Takeshige, H.; Hayashi, K.; Sano, S.; Shiro, M.; Yamaguchi, K.; Sei, Y. *J. Am. Chem. Soc.* **2006**, *128*, 9722–9729.
11. Ahmad, N. M. *Pummerer Rearrangement.* In *Name Reactions for Homologations-Part II*; Li, J. J., Ed.; Wiley: Hoboken, NJ, **2009**, pp 334–352. (Review).
12. Patil, M.; Loerbroks, C.; Thiel, W. *Org. Lett.* **2013**, *15*, 1682–1685.

# Ramberg–Bäcklund reaction

Olefin synthesis *via* α-halosulfone extrusion.

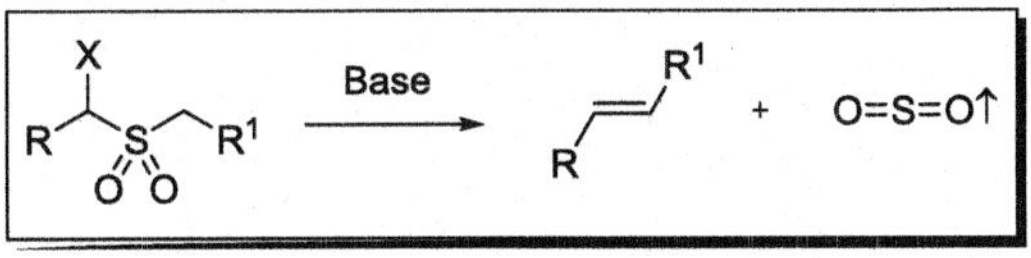
Base
+ O=S=O↑

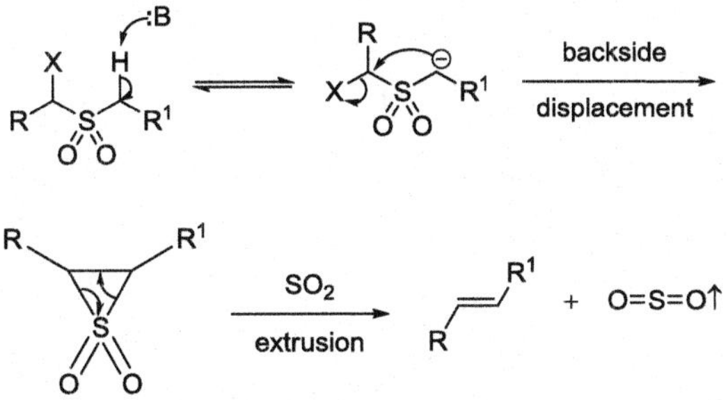
backside
displacement
SO2
extrusion
+ O=S=O↑

episulfone intermediate

Example 1[4]

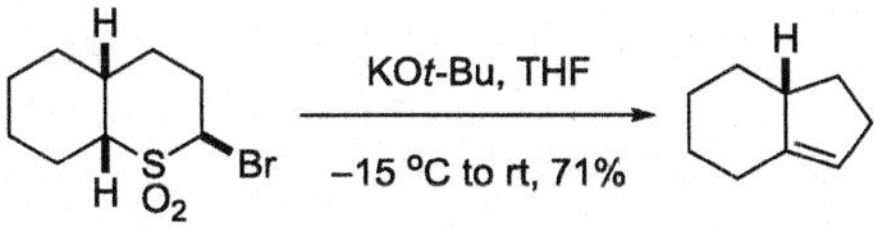
KO*t*-Bu, THF
−15 °C to rt, 71%

Example 2[5]

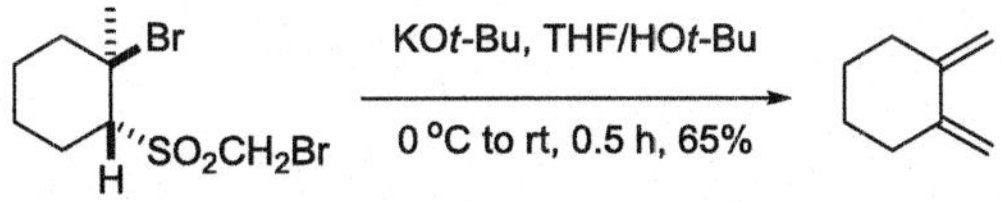
KO*t*-Bu, THF/HO*t*-Bu
0 °C to rt, 0.5 h, 65%

Example 3[6]

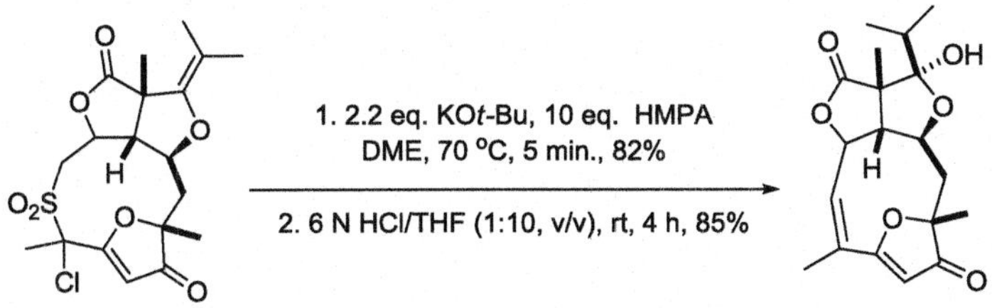
1. 2.2 eq. KO*t*-Bu, 10 eq. HMPA
DME, 70 °C, 5 min., 82%
2. 6 N HCl/THF (1:10, v/v), rt, 4 h, 85%

J.J. Li, *Name Reactions: A Collection of Detailed Mechanisms and Synthetic Applications*,
DOI 10.1007/978-3-319-03979-4_226, © Springer International Publishing Switzerland 2014

Example 4, *in situ* chlorination[7]

Example 5[8]

## References

1.  Ramberg, L.; Bäcklund, B. *Arkiv. Kemi, Mineral Geol.* **1940**, *13A*, 1–50.
2.  Paquette, L. A. *Acc. Chem. Res.* **1968**, *1*, 209–216. (Review).
3.  Paquette, L. A. *Org. React.* **1977**, *25*, 1–71. (Review).
4.  Becker, K. B.; Labhart, M. P. *Helv. Chim. Acta* **1983**, *66*, 1090–1100.
5.  Block, E.; Aslam, M.; Eswarakrishnan, V.; Gebreyes, K.; Hutchinson, J.; Iyer, R.; Laffitte, J. A.; Wall, A. *J. Am. Chem. Soc.* **1986**, *108*, 4568–4580.
6.  Boeckman, R. K., Jr.; Yoon, S. K.; Heckendorn, D. K. *J. Am. Chem. Soc.* **1991**, *113*, 9682–9684.
7.  Trost, B. M.; Shi, Z. *J. Am. Chem. Soc.* **1994**, *116*, 7459–7460.
8.  Cao, X.-P.; Chan, T.-L.; Chow, H.-F. *Tetrahedron Lett.* **1996**, *37*, 1049–1052.
9.  Taylor, R. J. K. *Chem. Commun.* **1999**, 217–227. (Review).
10. Taylor, R. J. K.; Casy, G. *Org. React.* **2003**, *62*, 357–475. (Review).
11. Li, J. J. *Ramberg–Bäcklund olefin synthesis.* In *Name Reactions for Functional Group Transformations*; Li, J. J., Ed.; Wiley: Hoboken, NJ, **2007**, pp 386–404. (Review).
12. Pal, T. K.; Pathak, T. *Carbohydrate Res.* **2008**, *343*, 2826–2829.
13. Baird, L. J.; Timmer, M. S. M.; Teesdale-Spittle, P. H.; Harvey, J. E. *J. Org. Chem.* **2009**, *74*, 2271–2277.

# Reformatsky reaction

Nucleophilic addition of organozinc reagents generated from α-haloesters to carbonyls.

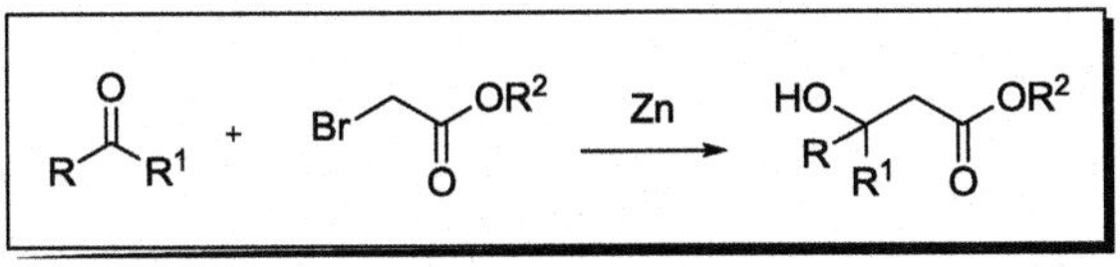

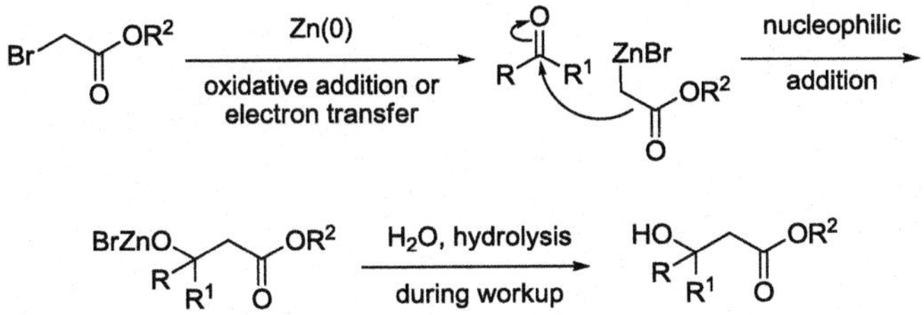

Example 1[4]

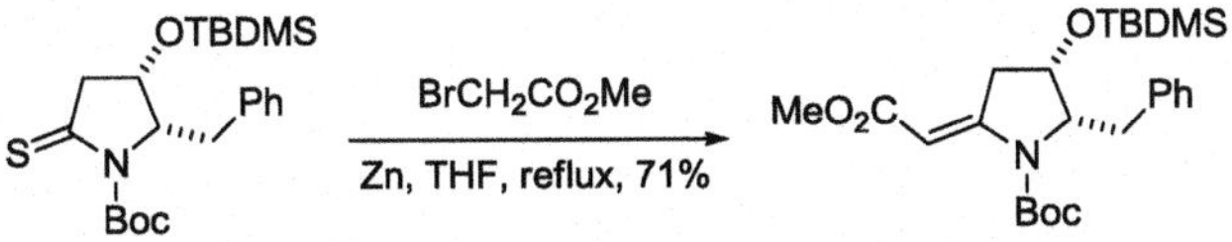

Example 2[6]

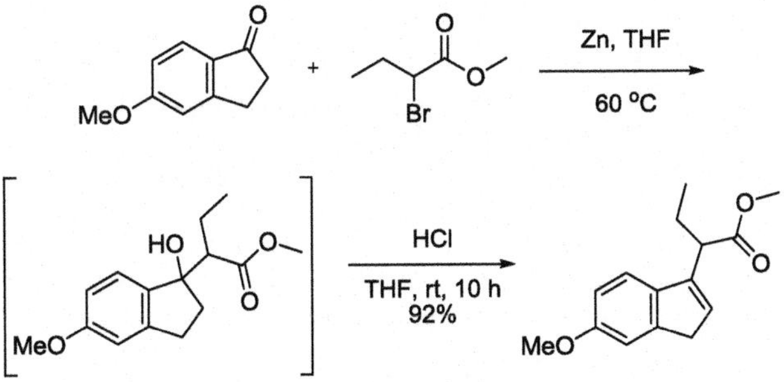

Example 3, Boron-mediated Reformatsky reaction[8]

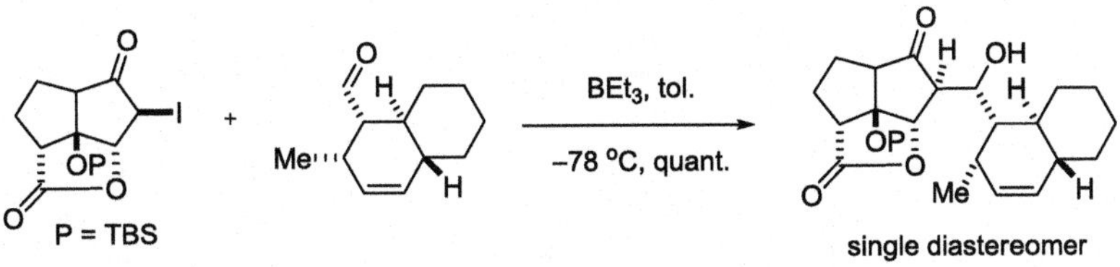

single diastereomer

J.J. Li, *Name Reactions: A Collection of Detailed Mechanisms and Synthetic Applications*, DOI 10.1007/978-3-319-03979-4_227, © Springer International Publishing Switzerland 2014

Example 4, SmI$_2$-mediated Reformatsky reaction[9]

Example 2[6]

## References

1.  Reformatsky, S. *Ber.* **1887**, *20,* 1210–1211.  Sergei Reformatsky (1860–1934) was born in Russia.  He studied at the University of Kazan in Russia, the cradle of Russian chemistry professors, where he found competent guidance of a distinguished chemist, Alexander M. Zaĭtsev.  Reformatsky then studied at Göttingen, Heidelberg, and Leipzig in Germany.  After returning to Russia, Reformatsky became the Chair of Organic Chemistry at the University of Kiev.
2.  Rathke, M. W. *Org. React.* **1975,** *22,* 423–460. (Review).
3.  Fürstner, A. *Synthesis* **1989,** 571–590. (Review).
4.  Lee, H. K.; Kim, J.; Pak, C. S. *Tetrahedron Lett.* **1999,** *40,* 2173–2174.
5.  Fürstner, A. In *Organozinc Reagents* Knochel, P., Jones, P., Eds.; Oxford University Press: New York, **1999,** pp 287–305. (Review).
6.  Zhang, M.; Zhu, L.; Ma, X. *Tetrahedron: Asymmetry* **2003,** *14,* 3447–3453.
7.  Ocampo, R.; Dolbier, W. R., Jr. *Tetrahedron* **2004,** *60,* 9325–9374. (Review).
8.  Lambert, T. H.; Danishefsky, S. J. *J. Am. Chem. Soc.* **2006,** *128,* 426–427.
9.  Moslin, R. M.; Jamison, T. F. *J. Am. Chem. Soc.* **2006,** *128,* 15106–15107.
10. Cozzi, P. G. *Angew. Chem. Int. Ed.* **2007,** *46,* 2568–2571. (Review).
11. Ke, Y.-Y.; Li, Y.-J.; Jia, J.-H.; Sheng, W.-J.; Han, L.; Gao, J.-R. *Tetrahedron Lett.* **2009,** *50,* 1389–1391.
12. Grellepois, F. *J. Org. Chem.* **2013,** *78,* 1127–1137.
13. Schulze, T. M.; Grunenberg, J.; Schulz, S. *Tetrahedron Lett.* **2013,** *54,* 921–924.

# Regitz diazo synthesis

Synthesis of 2-diazo-1,3-diketones or 2-diazo-3-oxoesters using sulfonyl azides.

tosyl amide is the by-product

When only one carbonyl is present, ethylformate can be used as an activating auxiliary:[6–9]

Alternatively, the triazole intermediate may be assembled *via* a 1,3-dipolar cycloaddition of the enol and mesyl azide:

J.J. Li, *Name Reactions: A Collection of Detailed Mechanisms and Synthetic Applications*, DOI 10.1007/978-3-319-03979-4_228, © Springer International Publishing Switzerland 2014

Example 1[5]

Example 2[10]

## References

1.   (a) Regitz, M. *Angew. Chem. Int. Ed.* **1967**, *6*, 733–741. (b) Regitz, M.; Anschütz, W.; Bartz, W.; Liedhegener, A. *Tetrahedron Lett.* **1968**, *9*, 3171–3174. (c) Regitz, M. *Synthesis* **1972**, 351–373. (Review).
2.   Pudleiner, H.; Laatsch, H. *Ann.* **1990**, 423–426.
3.   Evans, D. A.; Britton, T. C.; Ellman, J. A.; Dorow, R. L. *J. Am. Chem. Soc.* **1990**, *112*, 4011–4030.
4.   Charette, A. B.; Wurz, R. P.; Ollevier, T. *J. Org. Chem.* **2000**, *65*, 9252–9254.
5.   Hodgson, D. M.; Labande, A. H.; Pierard, F. Y. T. M.; Expósito Castro, M. A. *J. Org. Chem.* **2003**, *68*, 6153–6159.
6.   Sarpong, R.; Su, J. T.; Stoltz, B. M. *J. Am. Chem. Soc.* **2003**, *125*, 13624–13628.
7.   Mejía-Oneto, J. M.; Padwa, A. *Org. Lett.* **2004**, *6*, 3241–3244.
8.   Muroni, D.; Saba, A.; Culeddu, N. *Tetrahedron: Asymmetry* **2004**, *15*, 2609–2614.
9.   Davies, J. R.; Kane, P. D.; Moody, C. J. *Tetrahedron* **2004**, *60*, 3967–3977.
10.  Oguri, H.; Schreiber, S. L. *Org. Lett.* **2005**, *7*, 47–50.
11.  Balasubramanian, M. *Regitz diazo synthesis* In *Name Reactions for Functional Group Transformations*; Li, J. J., Ed.; Wiley: Hoboken, NJ, **2007**, pp 658–688. (Review).

# Reimer–Tiemann reaction

Synthesis of *o*-formylphenol from phenols and chloroform in alkaline medium.

a.     Carbene generation:

b.     Addition of dichlorocarbene and hydrolysis:

Example 1, Photo-Reimer–Tiemann reaction without base[7]

Example 2[8]

## References

1.  Reimer, K.; Tiemann, F. *Ber.* **1876**, *9*, 824–828.
2.  Wynberg, H.; Meijer, E. W. *Org. React.* **1982**, *28*, 1–36. (Review).
3.  Bird, C. W.; Brown, A. L.; Chan, C. C. *Tetrahedron* **1985**, *41*, 4685–4690.
4.  Neumann, R.; Sasson, Y. *Synthesis* **1986**, 569–570.
5.  Cochran, J. C.; Melville, M. G. *Synth. Commun.* **1990**, *20*, 609–616.
6.  Langlois, B. R. *Tetrahedron Lett.* **1991**, *32*, 3691–3694.
7.  Jiménez, M. C.; Miranda, M. A.; Tormos, R. *Tetrahedron* **1995**, *51*, 5825–5828.
8.  Jung, M. E.; Lazarova, T. I. *J. Org. Chem.* **1997**, *62*, 1553–1555.
9.  Bhunia, S. C.; Patra, G. C.; Pal, S. C. *Synth. Commun.* **2011**, *41*, 3678–3682.

J.J. Li, *Name Reactions: A Collection of Detailed Mechanisms and Synthetic Applications*,
DOI 10.1007/978-3-319-03979-4_229, © Springer International Publishing Switzerland 2014

# Reissert reaction

Treatment of quinoline or isoquinoline with acid chloride and KCN gives quinaldic acid, aldehyde, and $NH_3$.

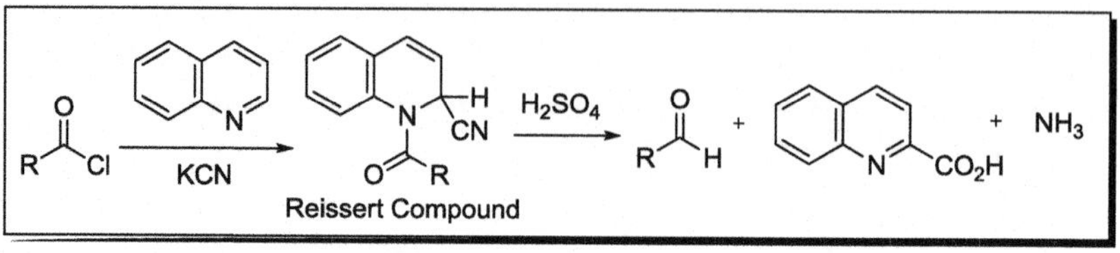

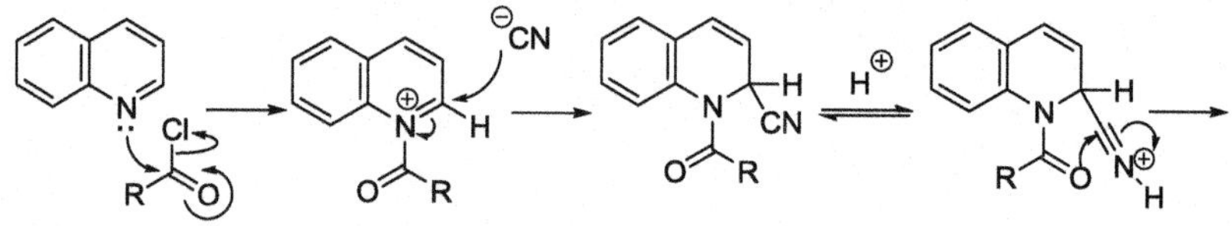

Reissert Compound

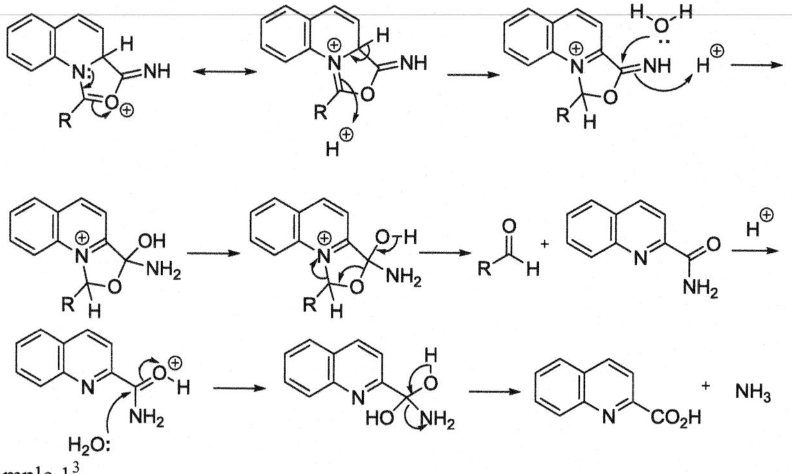

Reissert compound

Example 1[3]

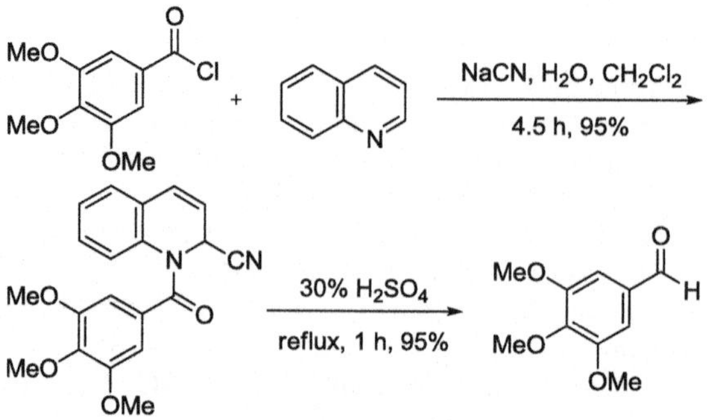

J.J. Li, *Name Reactions: A Collection of Detailed Mechanisms and Synthetic Applications*,
DOI 10.1007/978-3-319-03979-4_230, © Springer International Publishing Switzerland 2014

Example 2, Reissert compound from isoquinoline[7]

Example 3, Reissert compound from isoquinoline[10]

Reissert compounds

Example 4, Asymmetric organocatalytic allylic alkylation of Reissert compounds[12]

## References

1. (a) Reissert, A. *Ber.* **1905,** *38*, 1603–1614. (b) Reissert, A. *Ber.* **1905,** *38*, 3415–3435. Carl Arnold Reissert was born in 1860 in Powayen, Germany. He received his Ph.D. In 1884 at Berlin, where he became an assistant professor. He collaborated with Tiemann. Reissert later joined the faculty at Marburg in 1902.

2. Popp, F. D. *Adv. Heterocycl. Chem.* **1979,** *24*, 187–214. (Review).

3. Schwartz, A. *J. Org. Chem.* **1982,** *47*, 2213–2215.

4. Lorsbach, B. A.; Bagdanoff, J. T.; Miller, R. B.; Kurth, M. J. *J. Org. Chem.* **1998,** *63*, 2244–2250.

5. Perrin, S.; Monnier, K.; Laude, B.; Kubicki, M.; Blacque, O. *Eur. J. Org. Chem.* **1999,** 297–303.

6. Takamura, M.; Funabashi, K.; Kanai, M.; Shibasaki, M. *J. Am. Chem. Soc.* **2001,** *123*, 6801–6808.

7. Shibasaki, M.; Kanai, M.; Funabashi, K. *Chem. Commun.* **2002,** 1989–1999.

8. Sieck, O.; Schaller, S.; Grimme, S.; Liebscher, J. *Synlett* **2003,** 337–340.

9. Kanai, M.; Kato, N.; Ichikawa, E.; Shibasaki, M. *Synlett* **2005,** 1491–1508. (Review).

10. Gibson, H. W.; Berg, M. A. G.; Clifton Dickson, J.; Lecavalier, P. R.; Wang, H.; Merola, J. S. *J. Org. Chem.* **2007,** *72*, 5759–5770.

11. Fuchs, C.; Bender, C.; Ziemer, B.; Liebscher, J. *J. Heterocycl. Chem.* **2008,** *45*, 1651–1658.

12. Qin, T.Y.; Liao, W.-W.; Zhang, Y.-J.; Zhang, S. X.-A. *Org. Biomol. Chem.* **2013,** *11*, 984–990.

## Reissert indole synthesis

The Reissert indole synthesis involves base-catalyzed condensation of an
o-nitrotoluene derivative with an ethyl oxalate, which is followed by reductive
cyclization to an indole-2-carboxylic acid derivative.

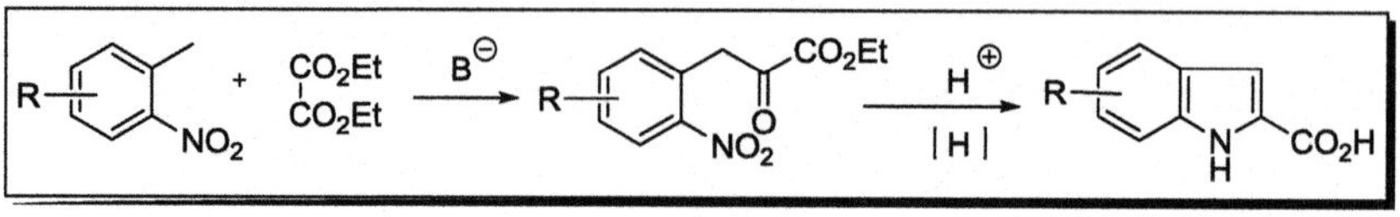

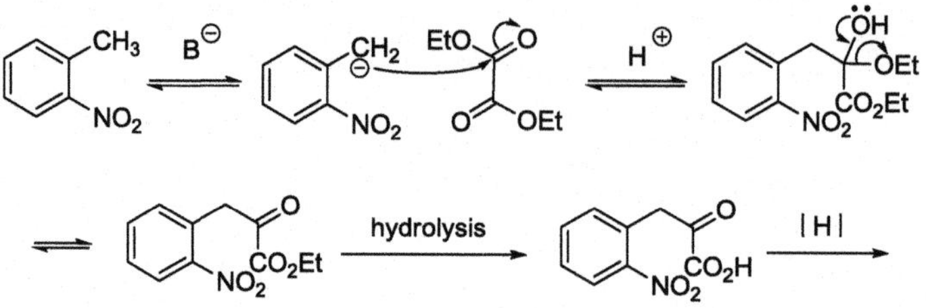

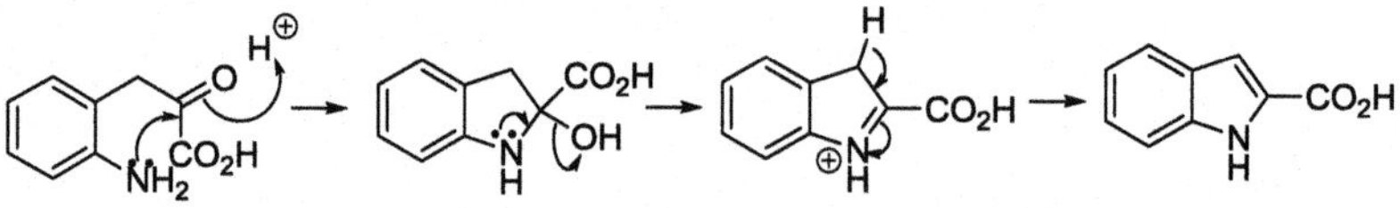

Example 1[2]

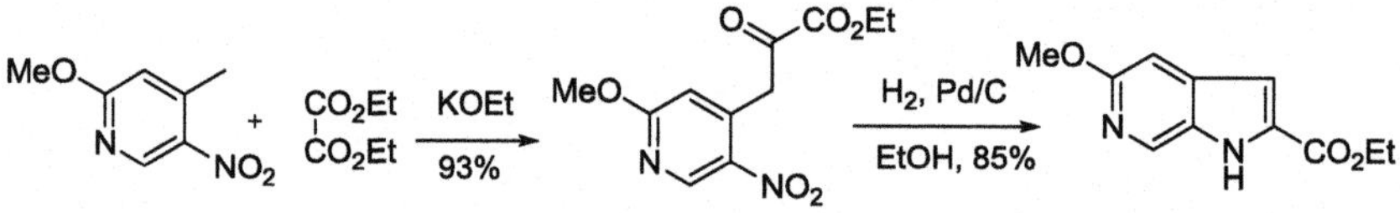

Example 2[3]

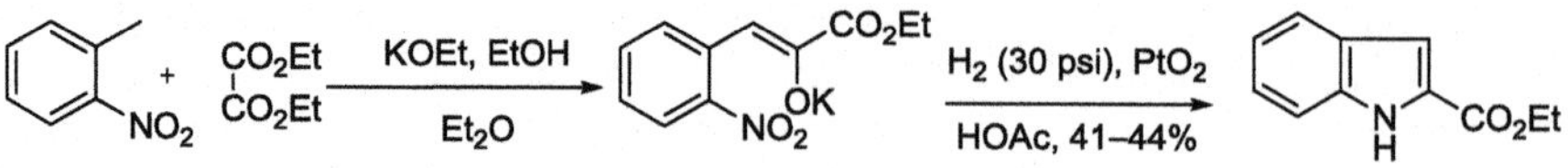

Example 3, Furan ring as the masked carbonyl[10]

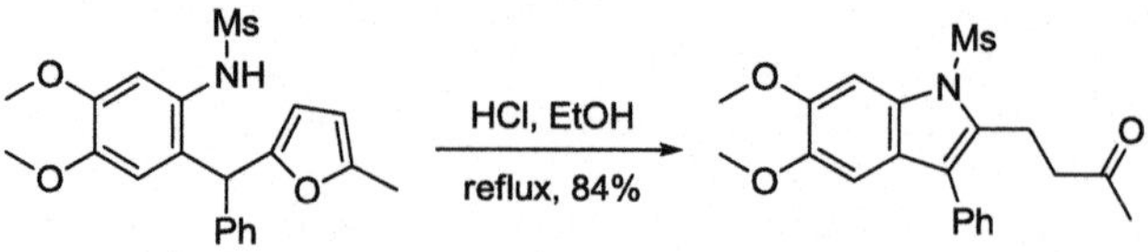

J.J. Li, *Name Reactions: A Collection of Detailed Mechanisms and Synthetic Applications*,
DOI 10.1007/978-3-319-03979-4_231, © Springer International Publishing Switzerland 2014

# References

1. Reissert, A. *Ber.* **1897,** *30*, 1030–1053.
2. Frydman, B.; Despuy, M. E.; Rapoport, H. *J. Am. Chem. Soc.* **1965,** *87*, 3530–3531.
3. Noland, W. E.; Baude, F. J. *Org. Synth.* **1973**; *Coll. Vol.* 567–571.
4. Leadbetter, G.; Fost, D. L.; Ekwuribe, N. N.; Remers, W. A. *J. Org. Chem.* **1974,** *39*, 3580–3583.
5. Cannon, J. G.; Lee, T.; Ilhan, M.; Koons, J.; Long, J. P. *J. Med. Chem.* **1984,** *27*, 386–389.
6. Suzuki, H.; Gyoutoku, H.; Yokoo, H.; Shinba, M.; Sato, Y.; Yamada, H.; Murakami, Y. *Synlett* **2000,** 1196–1198.
7. Butin, A. V.; Stroganova, T. A.; Lodina, I. V.; Krapivin, G. D. *Tetrahedron Lett.* **2001,** *42*, 2031–2036.
8. Katayama, S.; Ae, N.; Nagata, R. *J. Org. Chem.* **2001,** *66*, 3474–3483.
9. Li, J.; Cook, J. M. *Reissert Indole Synthesis.* In *Name Reactions in Heterocyclic Chemistry*; Li, J. J., Ed.; Wiley: Hoboken, NJ, **2005,** pp 154–158. (Review).
10. Butin, A. V.; Smirnov, S. K.; Stroganova, T. A.; Bender, W.; Krapivin, G. D. *Tetrahedron* **2006,** *63*, 474–491.
11. Colombo, E.; Ratel, P.; Mounier, L.; Guillier, F. *J. Flow Chem.* **2011,** *1*, 68–73.

# Ring-closing metathesis (RCM)

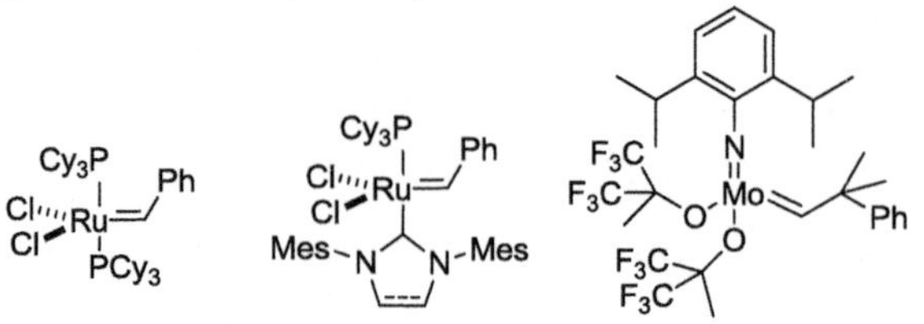

Grubbs' catalysts                          Schrock's catalyst
Mes = mesityl

All three catalysts are illustrated as "$L_nM=CHR$" in the mechanism below.

Generation of the real catalyst from the precatalysts:

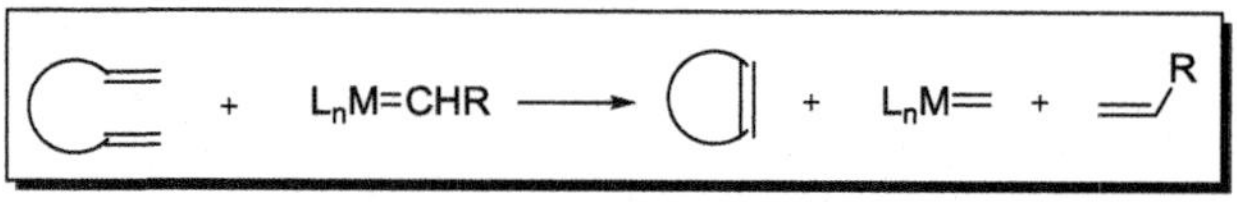

the active catalyst

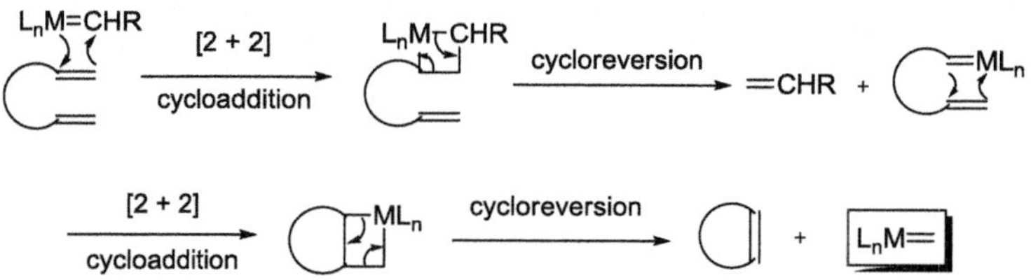

Catalytic cycle:

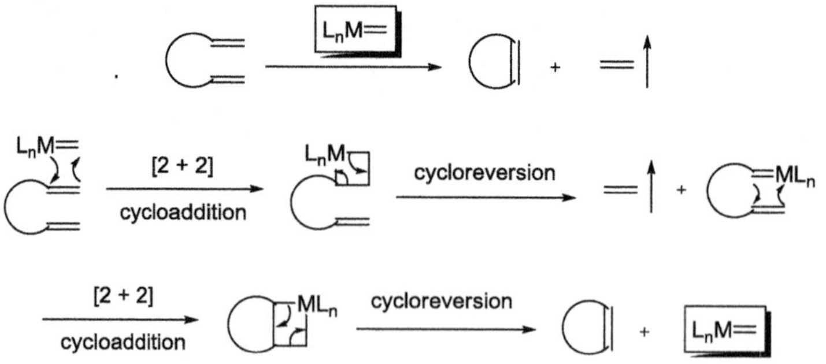

Example 1[3]

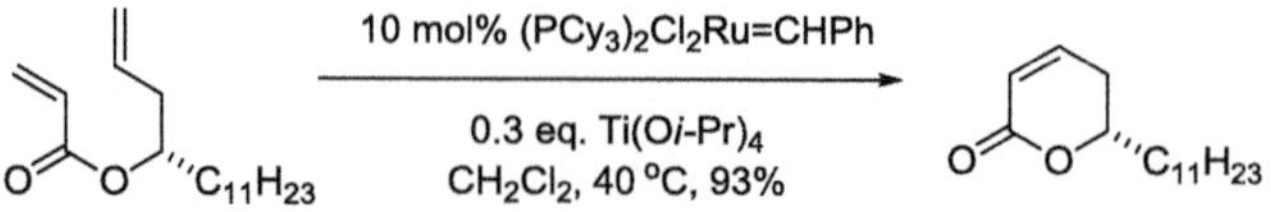

10 mol% $(PCy_3)_2Cl_2Ru=CHPh$

0.3 eq. $Ti(Oi\text{-}Pr)_4$
$CH_2Cl_2$, 40 °C, 93%

J.J. Li, *Name Reactions: A Collection of Detailed Mechanisms and Synthetic Applications*,
DOI 10.1007/978-3-319-03979-4_232, © Springer International Publishing Switzerland 2014

Example 2[4]

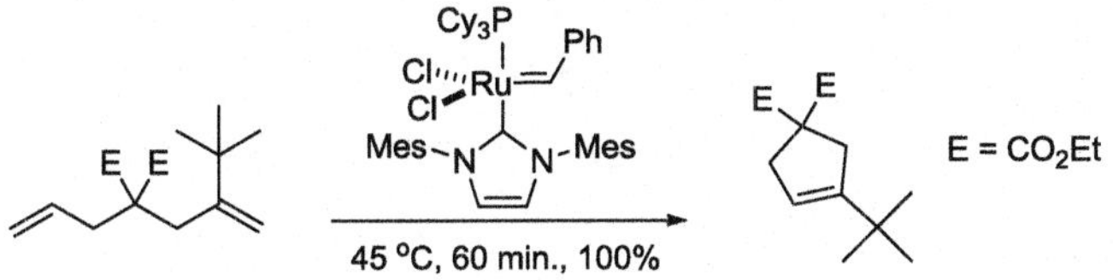

45 °C, 60 min., 100%

E = CO₂Et

Example 3[7]

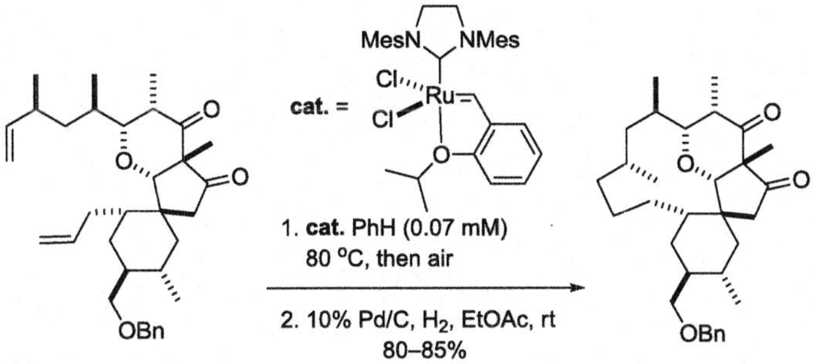

cat. =

1. **cat.** PhH (0.07 mM)
   80 °C, then air

2. 10% Pd/C, H₂, EtOAc, rt
   80–85%

Example 4[9]

5.4 mol%

CH₂Cl₂, rt, 73%

5 mol%

CH₂Cl₂, rt, 93%, > 10:1 *E:Z*

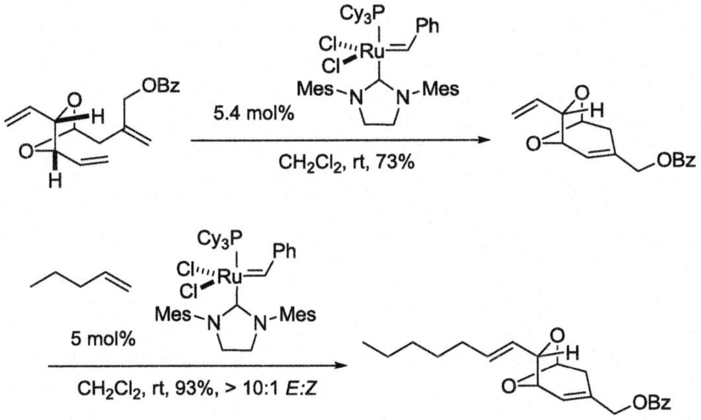

Example 5[10]

15 mol% Grubbs II

toluene, 110 °C, 78%

single
isomer

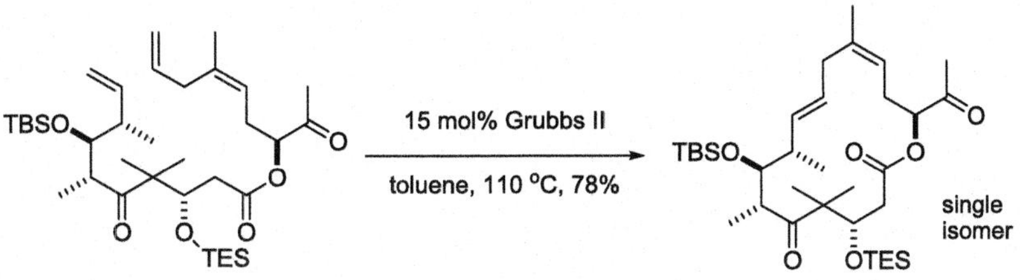

Example 6[12]

Example 7[13]

## References

1. Schrock, R. R.; Murdzek, J. S.; Bazan, G. C.; Robbins, J.; DiMare, M.; O'Regan, M. *J. Am. Chem. Soc.* **1990,** *112,* 3875–3886. Richard Schrock is a professor at MIT. He shared the 2005 Nobel Prize in Chemistry with Robert Grubbs of Caltech and Yves Chauvin of Institut Français du Pétrole in France for their contributions to metathesis.
2. Grubbs, R. H.; Miller, S. J.; Fu, G. C. *Acc. Chem. Res.* **1995,** *28,* 446–452. (Review).
3. Scholl, M.; Tunka, T. M.; Morgan, J. P.; Grubbs, R. H. *Tetrahedron Lett.* **1999,** *40,* 2247–2250.
4. Fellows, I. M.; Kaelin, D. E., Jr.; Martin, S. F. *J. Am. Chem. Soc.* **2000,** *122,* 10781–10787.
5. Timmer, M. S. M.; Ovaa, H.; Filippov, D. V.; van der Marel, G. A.; van Boom, J. H. *Tetrahedron Lett.* **2000,** *41,* 8635–8638.
6. Thiel, O. R. *Alkene and alkyne metathesis in organic synthesis.* In *Transition Metals for Organic Synthesis (2nd Edn.),* **2004,** *1,* pp 321–333. (Review).
7. Smith, A. B., III; Basu, K.; Bosanac, T. *J. Am. Chem. Soc.* **2007,** *129,* 14872–14874.
8. Hoveyda, A.H.; Zhugralin, A. R. *Nature* **2007,** *450,* 243–251. (Review).
9. Marvin, C. C.; Clemens, A. J. L.; Burke, S. D. *Org. Lett.* **2007,** *9,* 5353–5356.
10. Keck, G. E.; Giles, R. L.; Cee, V. J.; Wager, C. A.; Yu, T.; Kraft, M. B. *J. Org. Chem.* **2008,** *73,* 9675–9691.
11. Donohoe, T. J.; Fishlock, L. P.; Procopiou, P. A. *Chem. Eur. J.* **2008,** *14,* 5716–5726. (Review).
12. Sattely, E. S.; Meek, S. J.; Malcolmson, S. J.; Schrock, R. R.; Hoveyda, A. H. *J. Am. Chem. Soc.* **2009,** *131,* 943–953.
13. Moss, T. A. *Tetrahedron Lett.* **2013,** *54,* 993–997.

# Ritter reaction

Amides from nitriles and alcohols in strong acids.
General scheme:

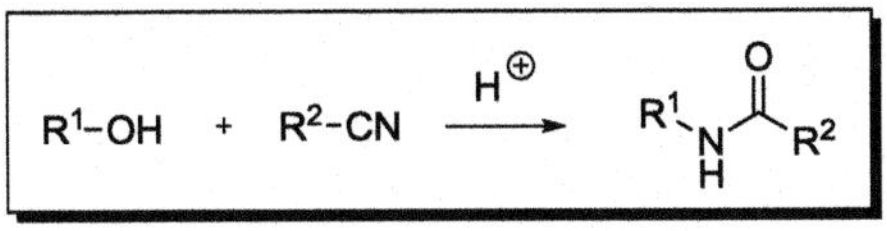

$$R^1\text{-OH} \;+\; R^2\text{-CN} \xrightarrow{\;H^{\oplus}\;} R^1\underset{H}{\overset{O}{\underset{N}{\bigwedge}}}R^2$$

Example 1

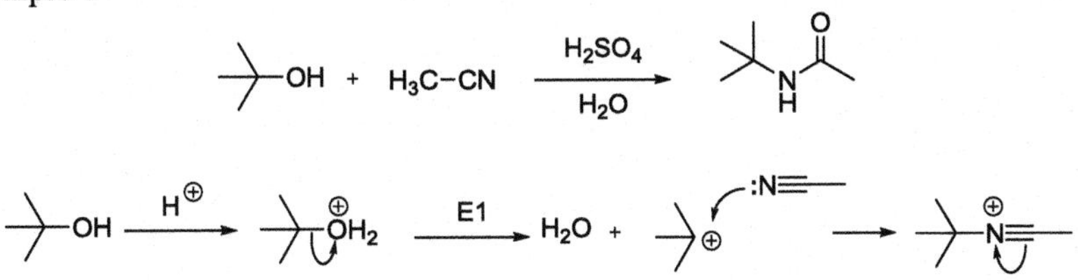

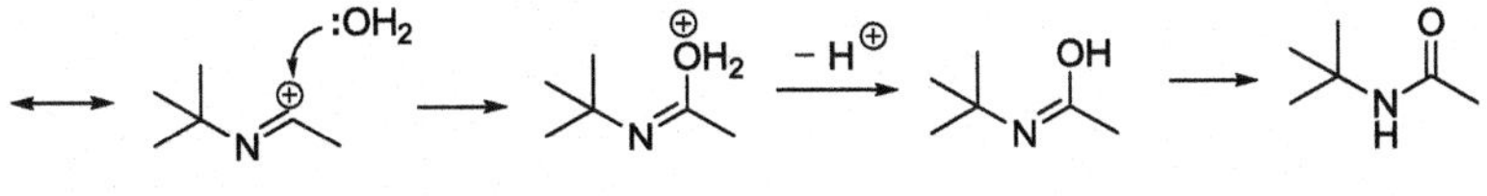

Similarly:

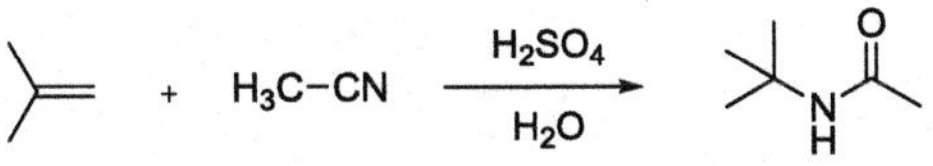

Example 2[3]

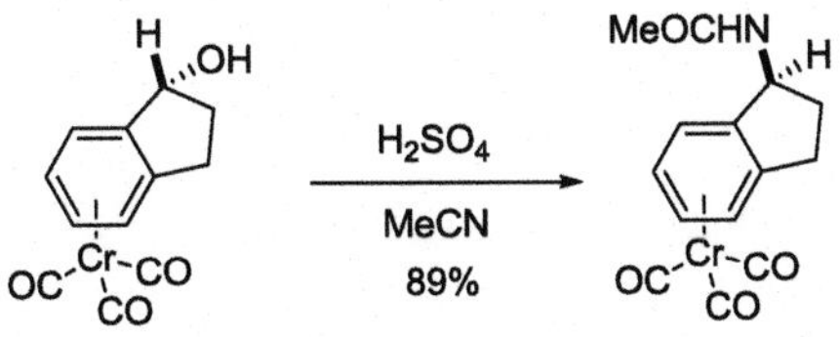

Example 3[4]

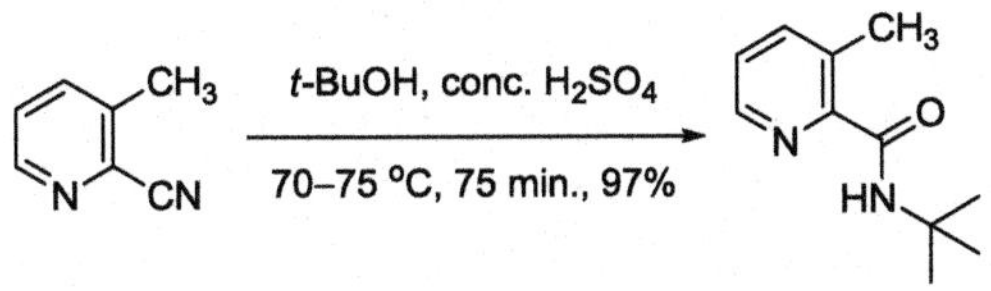

Example 4[5]

J.J. Li, *Name Reactions: A Collection of Detailed Mechanisms and Synthetic Applications*,
DOI 10.1007/978-3-319-03979-4_233, © Springer International Publishing Switzerland 2014

Example 5[6]

Example 6, A cascade of the Prins/Ritter amidation reaction[12]

## References

1. (a) Ritter, J. J.; Minieri, P. P. *J. Am. Chem. Soc.* **1948**, *70*, 4045–4048. (b) Ritter, J. J.; Kalish, J. *J. Am. Chem. Soc.* **1948**, *70*, 4048–4050.
2. Krimen, L. I.; Cota, D. J. *Org. React.* **1969**, *17*, 213–329. (Review).
3. Top, S.; Jaouen, G. *J. Org. Chem.* **1981**, *46*, 78–82.
4. Schumacher, D. P.; Murphy, B. L.; Clark, J. E.; Tahbaz, P.; Mann, T. A. *J. Org. Chem.* **1989**, *54*, 2242–2244.
5. Le Goanvic, D; Lallemond, M.-C.; Tillequin, F.; Martens, T. *Tetrahedron Lett.* **2001**, *42*, 5175–5176.
6. Tanaka, K.; Kobayashi, T.; Mori, H.; Katsumura, S. *J. Org. Chem.* **2004**, *69*, 5906–5925.
7. Nair, V.; Rajan, R.; Rath, N. P. *Org. Lett.* **2002**, *4*, 1575–1577.
8. Concellón, J. M.; Riego, E.; Suárez, J. R.; García-Granda, S.; Díaz, M. R. *Org. Lett.* **2004**, *6*, 4499–4501.
9. Brewer, A. R. E. *Ritter reaction*. In *Name Reactions for Functional Group Transformations*; Li, J. J., Ed.; Wiley: Hoboken, NJ, **2007**, pp 471–476. (Review).
10. Baum, J. C.; Milne, J. E.; Murry, J. A.; Thiel, O. R. *J. Org. Chem.* **2009**, *74*, 2207–2209.
11. Yadav, J. S.; Reddy, Y. J.; Reddy, P. A. N.; Reddy, B. V. S. *Org. Lett.* **2013**, *15*, 546–549.

# Robinson annulation

Michael addition of cyclohexanones to methyl vinyl ketone followed by intramolecular aldol condensation to afford six-membered α,β-unsaturated ketones.

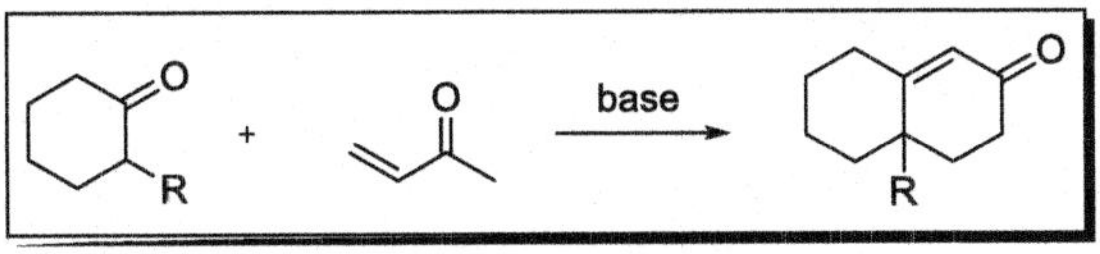

methyl vinyl ketone (MVK)

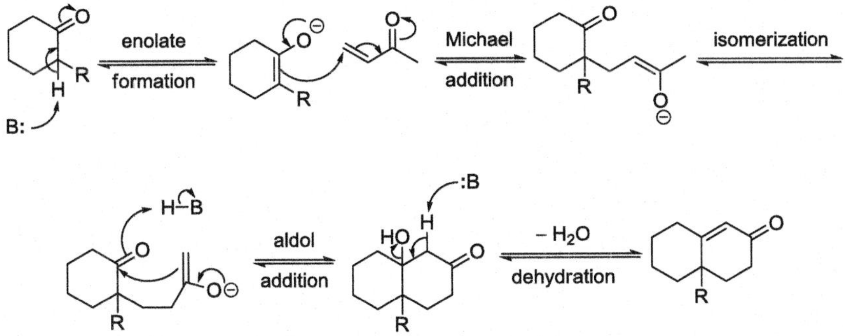

Example 1, Homo-Robinson[7]

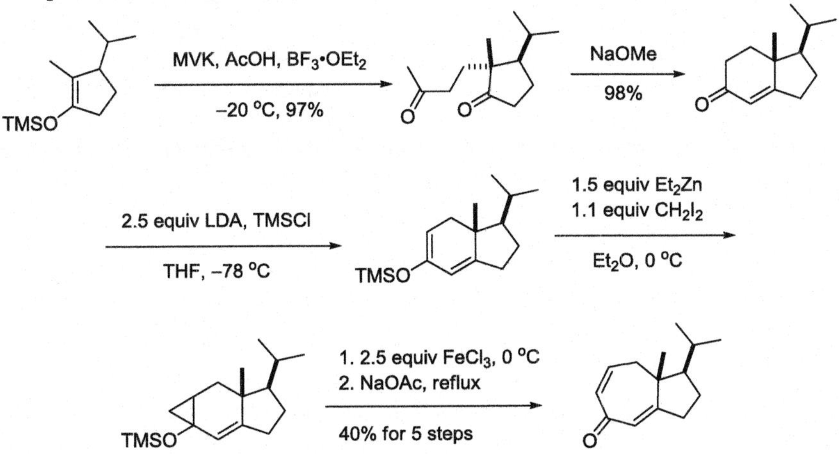

Example 2[8]

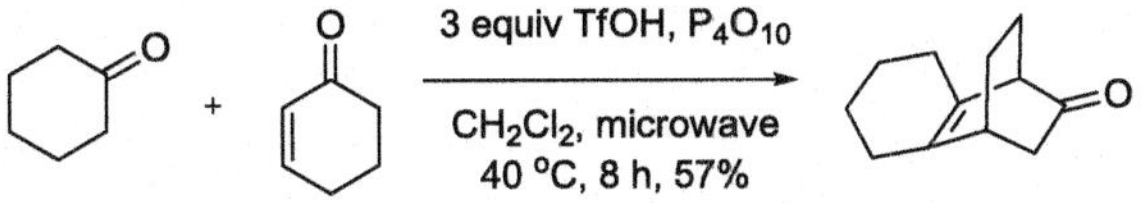

J.J. Li, *Name Reactions: A Collection of Detailed Mechanisms and Synthetic Applications*, DOI 10.1007/978-3-319-03979-4_234, © Springer International Publishing Switzerland 2014

Example 3, Double Robinson-type cyclopentene annulation[9]

Example 4[10]

## References

1.  Rapson, W. S.; Robinson, R. *J. Chem. Soc.* **1935**, 1285–1288.  Robert Robinson used the Robinson annulaton in his total synthesis of cholesterol.  Here is a story told by Derek Barton about Robinson and Woodward: "By pure chance, the two great men met early in a Monday morning on an Oxford train station platform in 1951.  Robinson politely asked Woodward what kind of research he was doing these days; Woodward replied that he thought that Robinson would be interested in his recent total synthesis of cholesterol.  Robinson, incensed and shouting 'Why do you always steal my research topic?', hit Woodward with his umbrella."—An excerpt from Barton, Derek, H. R. *Some Recollections of Gap Jumping,* American Chemical Society, Washington, D.C., **1991**.
2.  Gawley, R. E. *Synthesis* **1976**, 777–794. (Review).
3.  Guarna, A.; Lombardi, E.; Machetti, F.; Occhiato, E. G.; Scarpi, D. *J. Org. Chem.* **2000**, *65*, 8093–8096.
4.  Tai, C.-L.; Ly, T. W.; Wu, J.-D.; Shia, K.-S.; Liu, H.-J. *Synlett* **2001**, 214–217.
5.  Jung, M. E.; Piizzi, G. *Org. Lett.* **2003**, *5*, 137–140.
6.  Singletary, J. A.; Lam, H.; Dudley, G. B. *J. Org. Chem.* **2005**, *70*, 739–741.
7.  Yun, H.; Danishefsky, S. J. *Tetrahedron Lett.* **2005**, *46*, 3879–3882.
8.  Jung, M. E.; Maderna, A. *Tetrahedron Lett.* **2005**, *46*, 5057–5061.
9.  Zhang, Y.; Christoffers, J. *Synthesis* **2007**, 3061–3067.
10. Jahnke, A.; Burschka, C.; Tacke, R.; Kraft, P. *Synthesis* **2009**, 62–68.
11. Bradshaw, B.; Parra, C.; Bonjoch, J. *Org. Lett.* **2013**, *15*, 2458–2461.

# Robinson–Gabriel synthesis

Cyclodehydration of 2-acylamidoketones to give 2,5-di- and 2,4,5-trialkyl-, aryl-, heteroaryl-, and aralkyl-oxazoles.

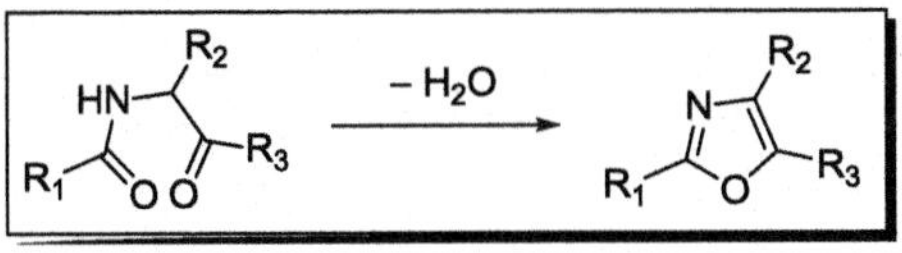

$R_1$, $R_2$, $R_3$ = alkyl, aryl, heteroaryl

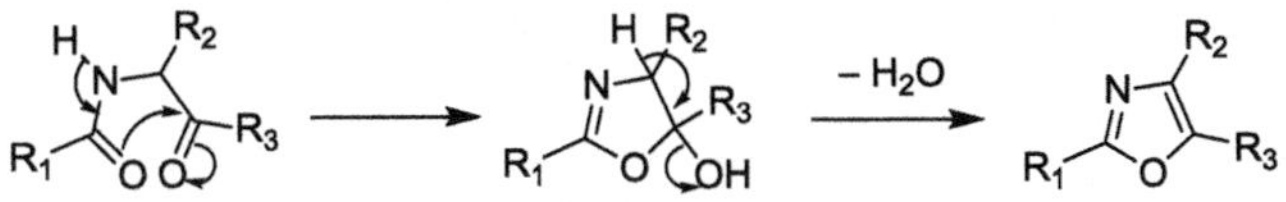

Example 1[3]

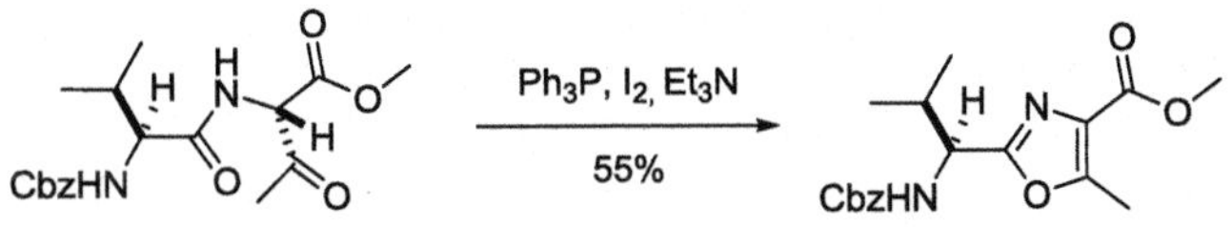

Example 2[4]

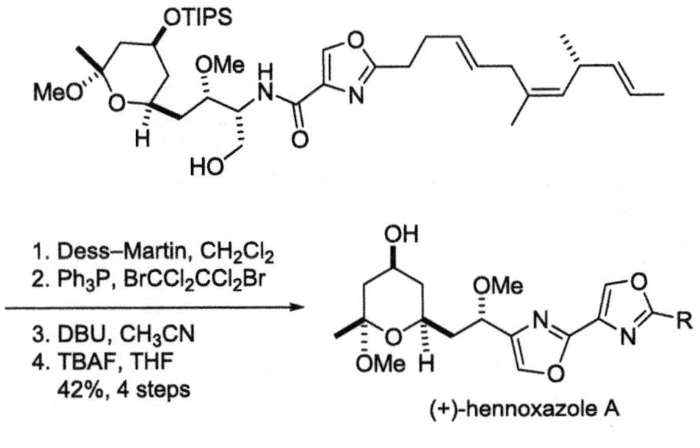

(+)-hennoxazole A

Example 3, Halogen effect[9]

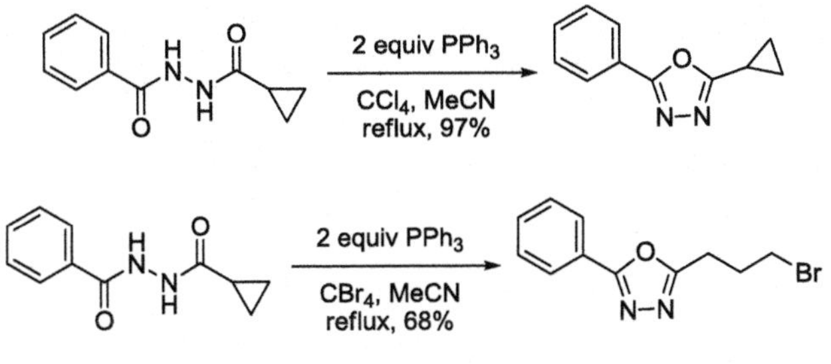

J.J. Li, *Name Reactions: A Collection of Detailed Mechanisms and Synthetic Applications*, DOI 10.1007/978-3-319-03979-4_235, © Springer International Publishing Switzerland 2014

Example 4[10]

Example 5, A cascade Ugi/Robinson-Gabriel reactions[11]

## References

1.  (a) Robinson, R. *J. Chem. Soc.* **1909**, *95*, 2167–2174.  (b) Gabriel, S. *Ber.* **1910**, *43*, 134–138. (c) Gabriel, S. *Ber.* **1910**, *43*, 1283–1287.
2.  Turchi, I. J. In *The Chemistry of Heterocyclic Compounds*, *45*; Wiley: New York, **1986**; pp 1–342. (Review).
3.  Wipf, P.; Miller, C. P. *J. Org. Chem.* **1993**, *58*, 3604–3606.
4.  Wipf, P.; Lim, S. *J. Am. Chem. Soc.* **1995**, *117*, 558–559.
5.  Morwick, T.; Hrapchak, M.; DeTuri, M.; Campbell, S. *Org. Lett.* **2002**, *4*, 2665–2668.
6.  Nicolaou, K. C.; Rao, P. B.; Hao, J.; Reddy, M. V.; Rassias, G.; Huang, X.; Chen, D. Y.-K.; Snyder, S. A. *Angew. Chem. Int. Ed.* **2003**, *42*, 1753–1758.
7.  Godfrey, A. G.; Brooks, D. A.; Hay, L. A.; Peters, M.; McCarthy, J. R.; Mitchell, D. *J. Org. Chem.* **2003**, *68*, 2623–2632.
8.  Brooks, D. A. *Robinson–Gabriel Synthesis*. In *Name Reactions in Heterocyclic Chemistry*; Li, J. J., Ed.; Wiley: Hoboken, NJ, **2005**, 249–253. (Review).
9.  Yang, Y.-H.; Shi, M. *Tetrahedron Lett.* **2005**, *46*, 6285–6288.
10. Bull, J. A.; Balskus, E. P.; Horan, R. A. J.; Langner, M.; Ley, S. V. *Angew. Chem. Int. Ed.* **2006**, *45*, 6714–6718.
11. Shaw, A. Y.; Xu, Z.; Hulme, C. *Tetrahedron Lett.* **2012**, *53*, 1998–2000.

# Robinson–Schöpf reaction

1,4-Diketone condensations with primary amines to give tropinones.

Example 1[5]

J.J. Li, *Name Reactions: A Collection of Detailed Mechanisms and Synthetic Applications*,
DOI 10.1007/978-3-319-03979-4_236, © Springer International Publishing Switzerland 2014

Example 2[9]

## References

1. Robinson, R. *J. Chem. Soc.* **1917**, *111*, 762–768.
2. Paquette, L. A.; Heimaster, J. W. *J. Am. Chem. Soc.* **1966**, *88*, 763–768.
3. Büchi, G.; Fliri, H.; Shapiro, R. *J. Org. Chem.* **1978**, *43*, 4765–4769.
4. Guerrier, L.; Royer, J.; Grierson, D. S.; Husson, H. P. *J. Am. Chem. Soc.* **1983**, *105*, 7754–7755.
5. Royer, J.; Husson, H. P. *Tetrahedron Lett.* **1987**, *28*, 6175–6178.
6. Villacampa, M.; Martínez, M.; González-Trigo, G.; Söllhuber, M. M. *J. Heterocycl. Chem.* **1992**, *29*, 1541–1544.
7. Bermudez, J.; Gregory, J. A.; King, F. D.; Starr, S.; Summersell, R. J. *Bioorg. Med. Chem. Lett.* **1992**, *2*, 519–522.
8. Langlois, M.; Yang, D.; Soulier, J. L.; Florac, C. *Synth. Commun.* **1992**, *22*, 3115–3116.
9. Jarevång, T.; Anke, H.; Anke, T.; Erkel, G.; Sterner, O. *Acta Chem. Scand.* **1998**, *52*, 1350–1352.
10. Amedjkouh, M.; Westerlund, K. *Tetrahedron Lett.* **2004**, *45*, 5175–5177.
11. Eastman, K. J. *Robinson-Schoepf condensation* In *Name Reactions in Heterocyclic Chemistry II*; Li, J. J., Ed.; Wiley: Hoboken, NJ, **2011**, pp 470–476. (Review).

## Rosenmund reduction

Hydrogenation reduction of acid chloride to aldehyde using $BaSO_4$-poisoned palladium catalyst. Without this poisoning, the resulting aldehyde may be further reduced to the corresponding alcohol. The possible by-products are alcohol, ester and alkane.

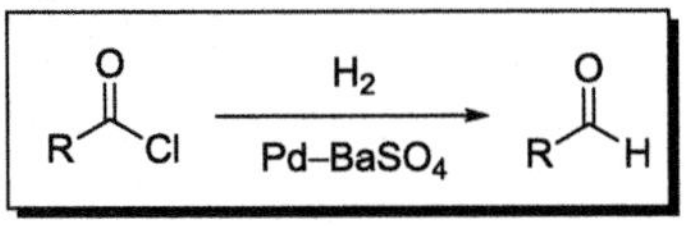

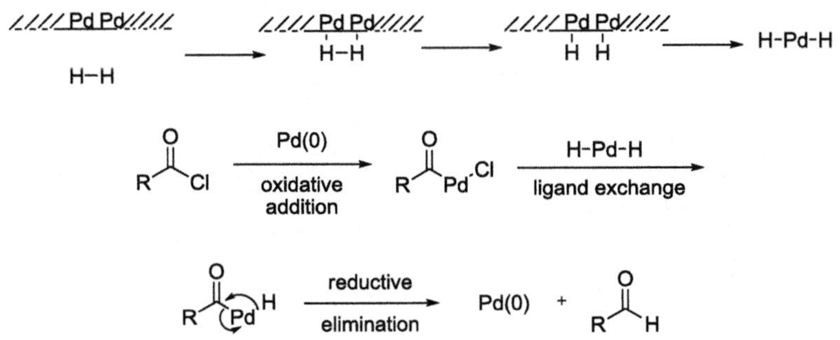

Example 1[4]

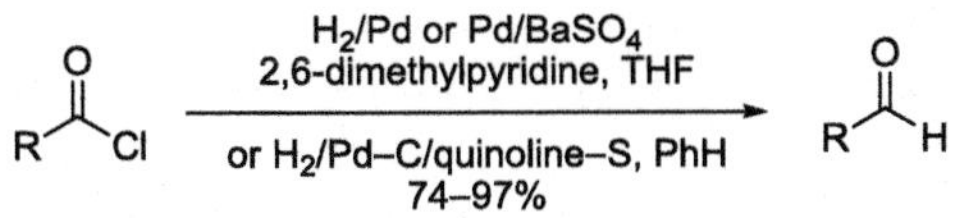

Example 2[6]

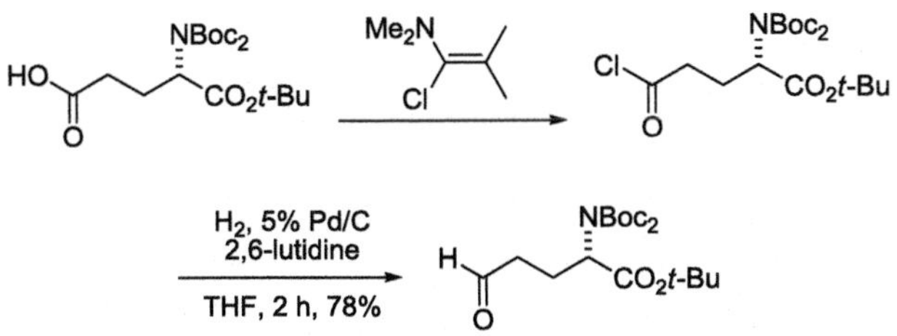

Example 3[9]

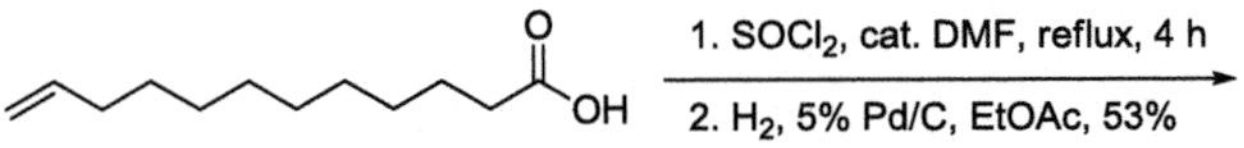

J.J. Li, *Name Reactions: A Collection of Detailed Mechanisms and Synthetic Applications*,
DOI 10.1007/978-3-319-03979-4_237, © Springer International Publishing Switzerland 2014

Example 4[11]

1. (COCl)₂, cat. DMF, EtOAc

2. H₂, Pd/C, quinaldine, EtOAc

50 Kg scale

## References

1.  Rosenmund, K. W. *Ber.* **1918,** *51,* 585–594. Karl Wilhelm Rosenmund was born in Berlin, Germany in 1884. He was a student of Otto Diels and received his Ph.D. In 1906. Rosenmund became professor and director of the Pharmaceutical Institute in Kiel in 1925.
2.  Mosettig, E.; Mozingo, R. *Org. React.* **1948,** *4,* 362–377. (Review).
3.  Tsuji, J.; Ono, K.; Kajimoto, T. *Tetrahedron Lett.* **1965,** *6,* 4565–4568.
4.  Burgstahler, A. W.; Weigel, L. O.; Schäfer, C. G. *Synthesis* **1976,** 767–768.
5.  McEwen, A. B.; Guttieri, M. J.; Maier, W. F.; Laine, R. M.; Shvo, Y. *J. Org. Chem.* **1983,** *48,* 4436–4438.
6.  Bold, V. G.; Steiner, H.; Moesch, L.; Walliser, B. *Helv. Chim. Acta* **1990,** *73,* 405–410.
7.  Yadav, V. G.; Chandalia, S. B. *Org. Proc. Res. Dev.* **1997,** *1,* 226–232.
8.  Chandnani, K. H.; Chandalia, S. B. *Org. Proc. Res. Dev.* **1999,** *3,* 416–424.
9.  Chimichi, S.; Boccalini, M.; Cosimelli, B. *Tetrahedron* **2002,** *58,* 4851–4858.
10. Ancliff, R. A.; Russell, A. T.; Sanderson, A. *J. Chem. Commun.* **2006,** 3243–3245.
12. Britton, H.; Catterick, D.; Dwyer, A. N.; Gordon, A. H.; et al. *Org. Process Res. Dev.* **2012,** *16,* 1607–1617.

# Rubottom oxidation

α-Hydroxylation of enolsilanes.

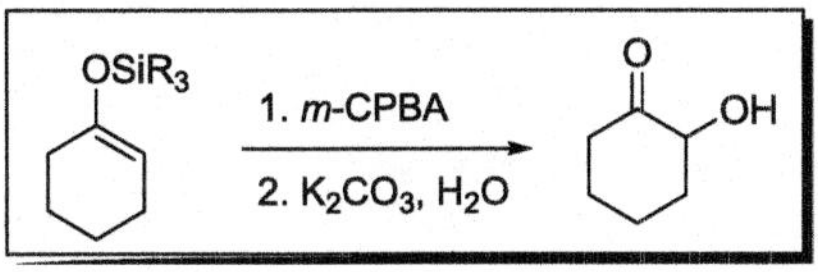

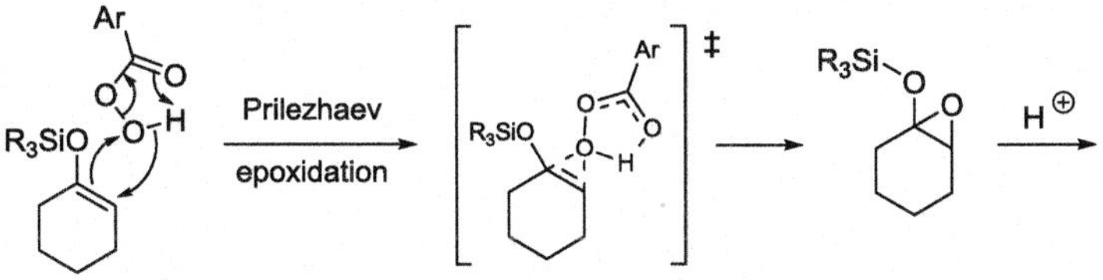

The "butterfly" transition state

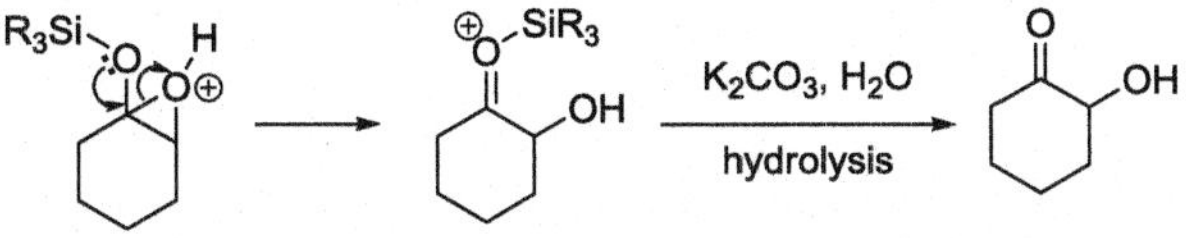

Example 1[2]

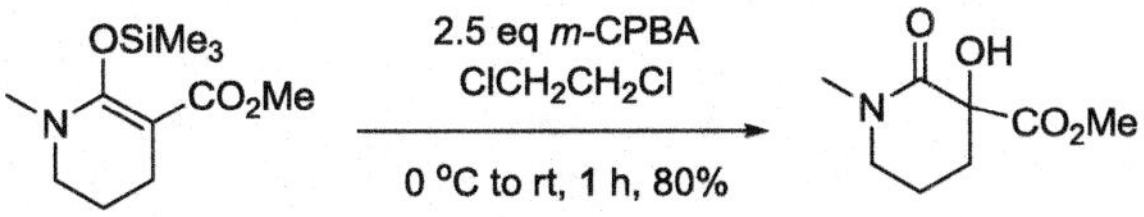

Example 2[3]

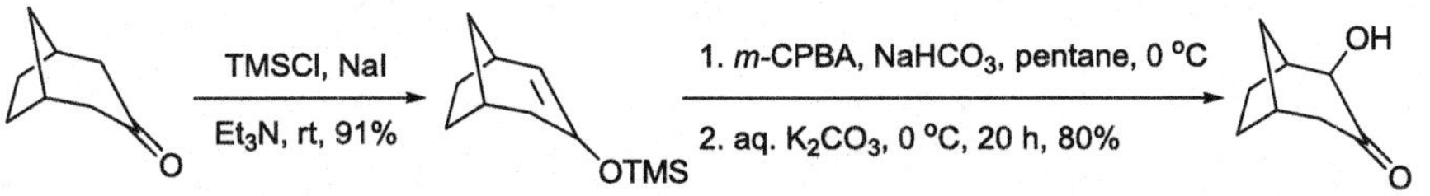

Example 3[4]

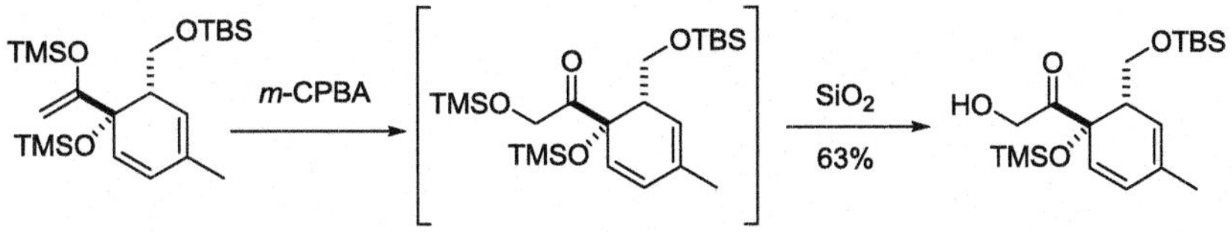

J.J. Li, *Name Reactions: A Collection of Detailed Mechanisms and Synthetic Applications*,
DOI 10.1007/978-3-319-03979-4_238, © Springer International Publishing Switzerland 2014

Example 4[5]

Example 5, Double Rubottom oxidation[11]

## References

1. Rubottom, G. M.; Vazquez, M. A.; Pelegrina, D. R. *Tetrahedron Lett.* **1974**, *15*, 4319–4322. George Rubottom discovered the Rubottom oxidation when he was an assistant professor at the University of Puerto Rico. He is now a grant officer at the National Science Foundation.
2. Andriamialisoa, R. Z.; Langlois, N.; Langlois, Y. *Tetrahedron Lett.* **1985**, *26*, 3563–2366.
3. Jauch, J. *Tetrahedron* **1994**, *50*, 12903–12912.
4. Crimmins, M. T.; Al-awar, R. S.; Vallin, I. M.; Hollis, W. G., Jr.; O'Mahoney, R.; Lever, J. G.; Bankaitis-Davis, D. M. *J. Am. Chem. Soc.* **1996**, *118*, 7513–7528.
5. Paquette, L. A.; Sun, L.-Q.; Friedrich, D.; Savage, P. B. *Tetrahedron Lett.* **1997**, *38*, 195–198.
6. Paquette, L. A.; Hartung, R. E.; Hofferberth, J. E.; Vilotijevic, I.; Yang, J. *J. Org. Chem.* **2004**, *69*, 2454–2460.
7. Christoffers, J.; Baro, A.; Werner, T. *Adv. Synth. Cat.* **2004**, *346*, 143–151. (Review).
8. He, J.; Tchabanenko, K.; Adlington, R. M.; Cowley, A. R.; Baldwin, J. E. *Eur. J. Org. Chem.* **2006**, 4003–4013.
9. Wolfe, J. P. *Rubottom oxidation.* In *Name Reactions for Functional Group Transformations*; Li, J. J., Ed.; Wiley: Hoboken, NJ, **2007**, pp 282–290. (Review).
10. Wang, H.; Andemichael, Y. W.; Vogt, F. G. *J. Org. Chem.* **2009**, *74*, 478–481.
11. Isaka, N.; Tamiya, M.; Hasegawa, A.; Ishiguro, M. *Eur. J. Org. Chem.* **2012**, 665–668.
12. Fujiwara, H.; Kurogi, T.; Okaya, S.; Okano, K.; Tokuyama, H. *Angew. Chem. Int. Ed.* **2012**, *51*, 13062–13065.

# Rupe rearrangement

Acid-catalyzed rearrangement of tertiary α-acetylenic (terminal) alcohols, leading to the formation of α,β-unsaturated ketones rather than the corresponding α,β-unsaturated aldehydes. *Cf.* Meyer–Schuster rearrangement.

Example 1[4]

Example 2[8]

Example 3[9]

J.J. Li, *Name Reactions: A Collection of Detailed Mechanisms and Synthetic Applications*,
DOI 10.1007/978-3-319-03979-4_239, © Springer International Publishing Switzerland 2014

# References

1.  Rupe, H.; Kambli, E. *Helv. Chim. Acta* **1926**, *9*, 672.
2.  Swaminathan, S.; Narayanan, K. V. *Chem. Rev.* **1971**, *71*, 429–438. (Review).
3.  Hasbrouck, R. W.; Anderson Kiessling, A. D. *J. Org. Chem.* **1973**, *38*, 2103–2106.
4.  Baran, J.; Klein, H.; Schade, C.; Will, E.; Koschinsky, R.; Bäuml, E.; Mayr, H. *Tetrahedron* **1988**, *44*, 2181–2184.
5.  Barre, V.; Massias, F.; Uguen, D. *Tetrahedron Lett.* **1989**, *30*, 7389–7392.
6.  An, J.; Bagnell, L.; Cablewski, T.; Strauss, C. R.; Trainor, R. W. *J. Org. Chem.* **1997**, *62*, 2505–2511.
7.  Yadav, J. S.; Prahlad, V.; Muralidhar, B. *Synth. Commun.* **1997**, *27*, 3415–3418.
8.  Takeda, K.; Nakane, D.; Takeda, M. *Org. Lett.* **2000**, *2*, 1903–1905.
9.  Weinmann, H.; Harre, M.; Neh, H.; Nickisch, K.; Skötsch, C.; Tilstam, U. *Org. Proc. Res. Dev.* **2002**, *6*, 216–219.
10. Mullins, R. J.; Collins, N. R. *Meyer–Schuster Rearrangement*. In *Name Reactions for Homologations-Part II*; Li, J. J., Ed.; Wiley: Hoboken, NJ, **2009**, pp 305–318. (Review).
11. Chang, Y.-J.; Wang, Z.-Z.; Luo, L.-G.; Dai, L.-Y. *Chem. Papers* **2012**, *66*, 33–38.

# Saegusa oxidation

Palladium-catalyzed conversion of enol silanes to enones, also known as the Saegusa enone synthesis or the Saegusa–Ito oxidation.

The mechanism is similar to that of the Wacker oxidation.

Regenerating the Pd(II) oxidant:

Larock reported regeneration of the Pd(II) oxidant using oxygen:[4]

Example 1[3]

J.J. Li, *Name Reactions: A Collection of Detailed Mechanisms and Synthetic Applications*, DOI 10.1007/978-3-319-03979-4_240, © Springer International Publishing Switzerland 2014

Example 2[9]

Example 3[10]

Example 4[11]

## References

1. Ito, Y.; Hirao, T.; Saegusa, T. *J. Org. Chem.* **1978**, *43*, 1011–1013. Takeo Saegusa was a professor at Kyoto University in Japan.
2. Dickson, J. K., Jr.; Tsang, R.; Llera, J. M.; Fraser-Reid, B. *J. Org. Chem.* **1989**, *54*, 5350–5356.
3. Kim, M.; Applegate, L. A.; Park, O.-S.; Vasudevan, S.; Watt, D. S. *Synth. Commun.* **1990**, *20*, 989–997.
4. Larock, R. C.; Hightower, T. R.; Kraus, G. A.; Hahn, P.; Zheng, D. *Tetrahedron Lett.* **1995**, 36, 2423–2426.
5. Porth, S.; Bats, J. W.; Trauner, D.; Giester, G.; Mulzer, J. *Angew. Chem. Int. Ed.* **1999**, *38*, 2015–2016. The authors proposed a sandwiched Pd(II) as a possible alternative pathway.
6. Williams, D. R.; Turske, R. A. *Org. Lett.* **2000**, *2*, 3217–3220.
7. Nicolaou, K. C.; Zhong, Y.-L.; Baran, P. S. *J. Am. Chem. Soc.* **2000**, *122*, 7596–7597.
8. Sha, C.-K.; Huang, S.-J.; Zhan, Z.-P. *J. Org. Chem.* **2002**, *67*, 831–836.
9. Uchida, K.; Yokoshima, S.; Kan, T.; Fukuyama, T. *Org. Lett.* **2006**, *8*, 5311–5313.
10. Angeles A. R; Waters, S. P.; Danishefsky S. J. *J. Am. Chem. Soc.* **2008**, *130*, 13765–13770.
11. Lu, Y.; Nguyen, P. L.; Lévaray, N.; Lebel, H. *J. Org. Chem.* **2013**, *78*, 776–779.

## Sakurai allylation reaction

Lewis acid-mediated addition of allylsilanes to carbon nucleophiles. Also known as the Hosomi–Sakurai reaction. The allylsilane will add to the carbonyl compound directly if the electrophile (carbonyl group) is not part of an α,β-unsaturated system (Example 2), giving rise to an alcohol.

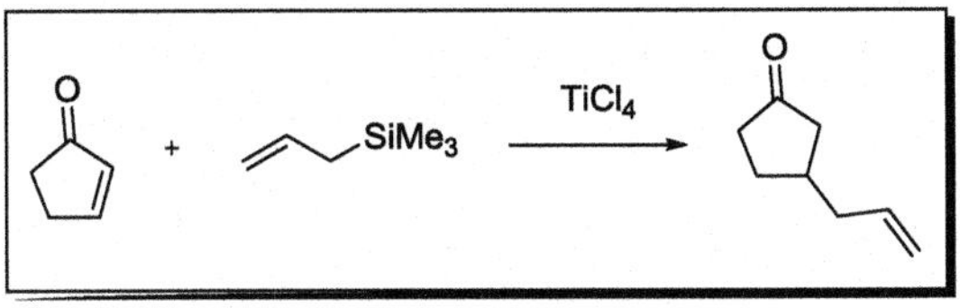

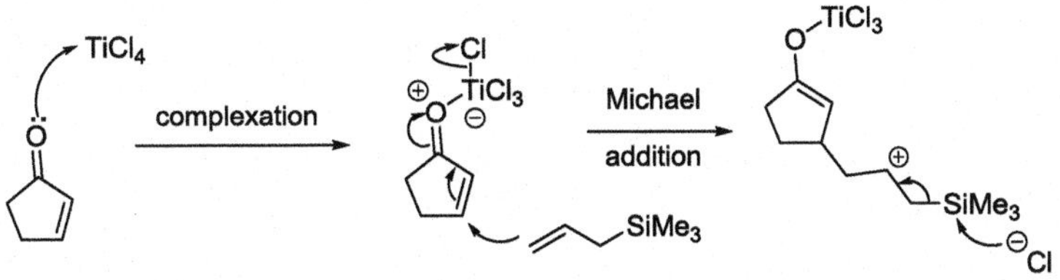

The β-carbocation is stabilized by the β-silicon effect

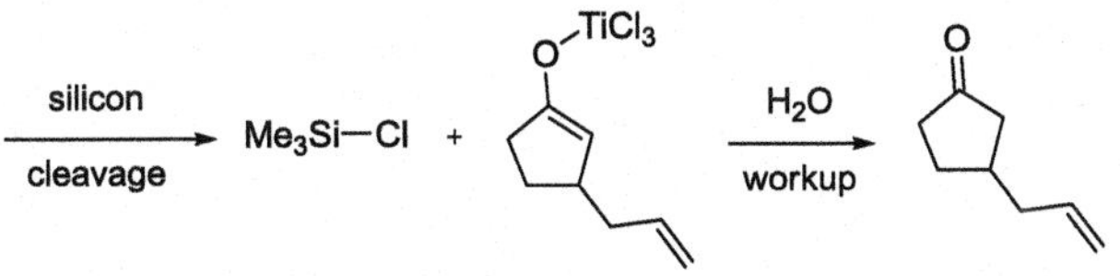

Example 1[2]

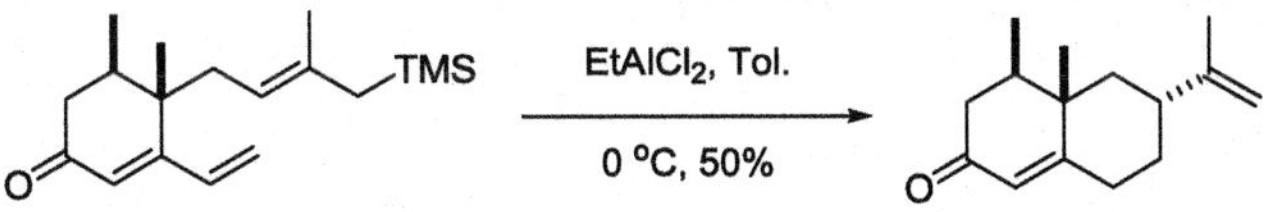

Example 2[6]

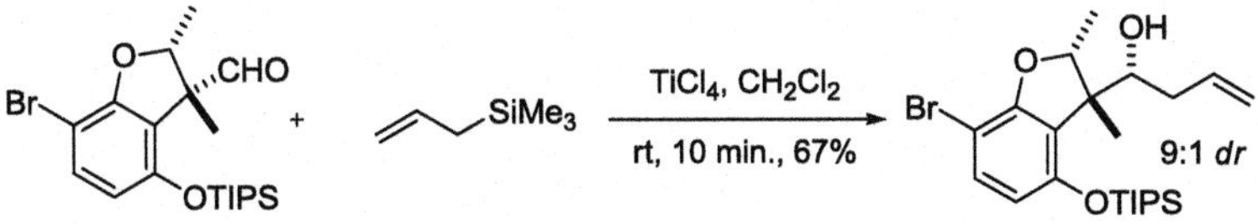

Example 3[9]

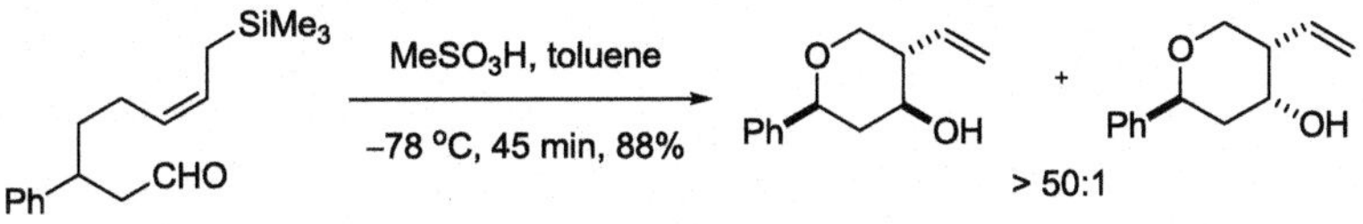

J.J. Li, *Name Reactions: A Collection of Detailed Mechanisms and Synthetic Applications*, DOI 10.1007/978-3-319-03979-4_241, © Springer International Publishing Switzerland 2014

534                                    Name Reactions

## Example 4[10]

## Example 5[11]

## Example 6[12]

## References

1. Hosomi, A.; Sakurai, H. *Tetrahedron Lett.* **1976**, 1295–1298. Hideki Sakurai was a professor at Tohuko University in Japan. This reaction is also known as the Hosomi–Sakurai reaction.
2. Majetich, G.; Behnke, M.; Hull, K. *J. Org. Chem.* **1985**, *50*, 3615–3618.
3. Tori, M.; Makino, C.; Hisazumi, K.; Sono, M.; Nakashima, K. *Tetrahedron: Asymmetry* **2001**, *12*, 301–307.
4. Leroy, B.; Markó, I. E. *J. Org. Chem.* **2002**, *67*, 8744–8752.
5. Itsuno, S.; Kumagai, T. *Helv. Chim. Acta* **2002**, *85*, 3185–3196.
6. Trost, B. M.; Thiel, O. R.; Tsui, H.-C. *J. Am. Chem. Soc.* **2003**, *125*, 13155–13164.
7. Knepper, K.; Ziegert, R. E.; Bräse, S. *Tetrahedron* **2004**, *60*, 8591–8603.
8. Rikimaru, K.; Mori, K.; Kan, T.; Fukuyama, T. *Chem. Commun.* **2005**, 394–396.
9. Jervis, P. J.; Kariuki, B. M.; Cox, L. R. *Org. Lett.* **2006**, *8*, 4649–4652.
10. Kalidindi, S.; Jeong, W. B.; Schall, A.; Bandichhor, R.; Nosse, B.; Reiser, O. *Angew. Chem. Int. Ed.* **2007**, *46*, 6361–6363.
11. Norcross, N. R.; Melbardis, J. P.; Solera, M. F.; Sephton, M. A.; Kilner, C.; Zakharov, L. N.; Astles, P. C.; Warriner, S. L.; Blakemore, P. R. *J. Org. Chem.* **2008**, *73*, 7939–7951.
12. Li, L.; Ye, X.; Wu, Y.; Gao, L.; Song, Z.; Yin, Z.; Xu, Y. *Org. Lett.* **2013**, *15*, 1068–1071.

# Sandmeyer reaction

Haloarenes from the reaction of a diazonium salt with CuX.

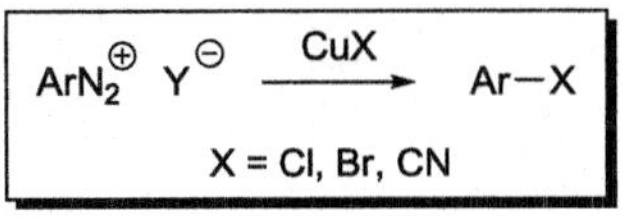

$$ArN_2^{\oplus} \; Y^{\ominus} \xrightarrow{\text{CuX}} Ar{-}X$$

X = Cl, Br, CN

Mechanism:

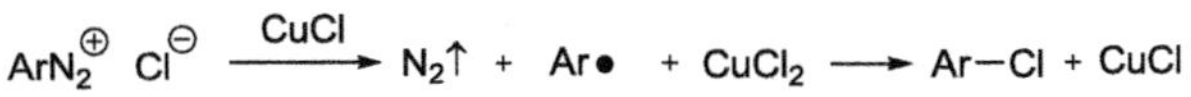

$$ArN_2^{\oplus} \; Cl^{\ominus} \xrightarrow{\text{CuCl}} N_2\uparrow \; + \; Ar\bullet \; + \; CuCl_2 \longrightarrow Ar{-}Cl + CuCl$$

Example 1[4]

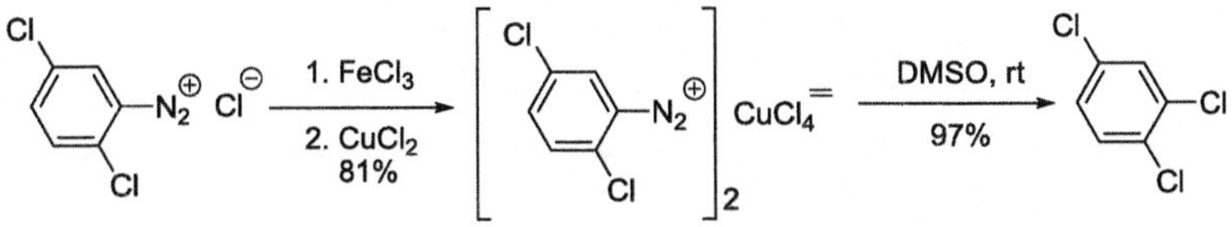

Example 2[7]

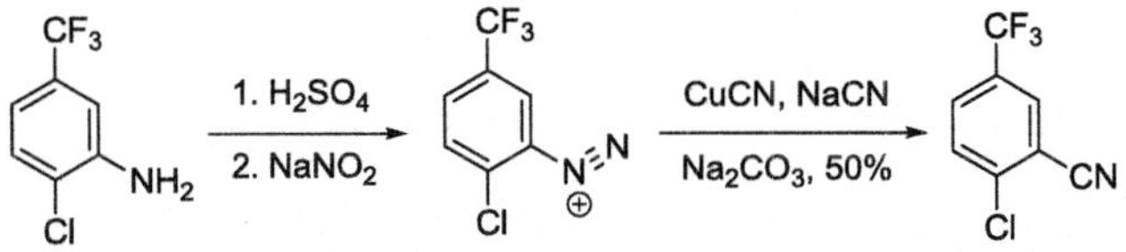

Example 3[8]

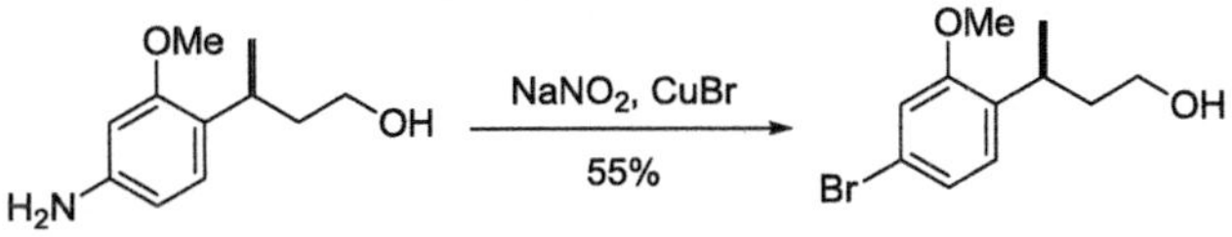

Example 4[9]

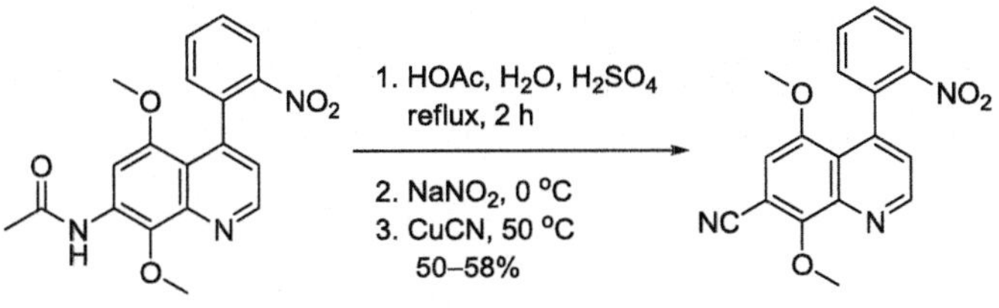

J.J. Li, *Name Reactions: A Collection of Detailed Mechanisms and Synthetic Applications*,
DOI 10.1007/978-3-319-03979-4_242, © Springer International Publishing Switzerland 2014

Example 5[11]

## References

1.  Sandmeyer, T. *Ber.* **1884,** *17*, 1633. Traugott Sandmeyer (1854–1922) was born in Wettingen, Switzerland. He apprenticed under Victor Meyer and Arthur Hantzsch although he never took a doctorate. He later spent 31 years at the company J. R. Geigy, which is now part of Novartis.
2.  Suzuki, N.; Azuma, T.; Kaneko, Y.; Izawa, Y.; Tomioka, H.; Nomoto, T. *J. Chem. Soc., Perkin Trans. 1* **1987,** 645–647.
3.  Merkushev, E. B. *Synthesis* **1988,** 923–937. (Review).
4.  Obushak, M. D.; Lyakhovych, M. B.; Ganushchak, M. I. *Tetrahedron Lett.* **1998,** *39*, 9567–9570.
5.  Hanson, P.; Jones, J. R.; Taylor, A. B.; Walton, P. H.; Timms, A. W. *J. Chem. Soc., Perkin Trans. 2* **2002,** 1135–1150.
6.  Daab, J. C.; Bracher, F. *Monatsh. Chem.* **2003,** *134*, 573–583.
7.  Nielsen, M. A.; Nielsen, M. K.; Pittelkow, T. *Org. Proc. Res. Dev.* **2004,** *8*, 1059–1064.
8.  Kim, S.-G.; Kim, J.; Jung, H. *Tetrahedron Lett.* **2005,** *46*, 2437–2439.
9.  LaBarbera, D. V.; Bugni, T. S.; Ireland, C. M. *J. Org. Chem.* **2007,** *72*, 8501–8505.
10. Gehanne, K.; Lancelot, J.-C.; Lemaitre, S.; El-Kashef, H.; Rault, S. *Heterocycles* **2008,** *75*, 3015–3024.
11. Dai, J.-J.; Fang, C.; Xiao, B.; Yi, J.; Xu, J.; Liu, Z.-J.; Lu, X.; Liu, L.; Fu, Y. *J. Am. Chem. Soc.* **2013,** *135*, 8436–8439.

# Schiemann reaction

Fluoroarene formation from arylamines. Also known as the Balz–Schiemann reaction.

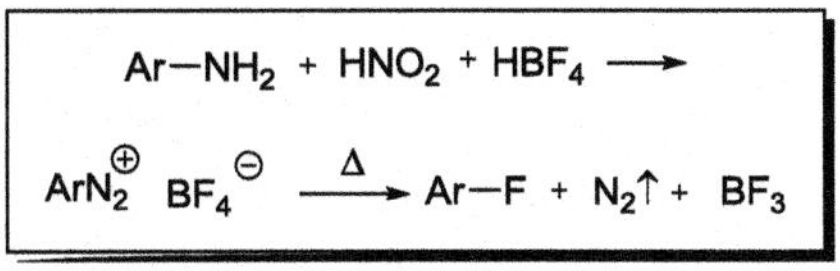

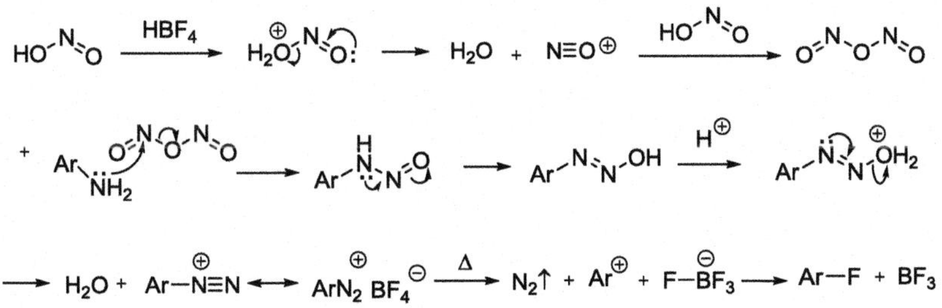

Ar—NH$_2$ + HNO$_2$ + HBF$_4$ ⟶

ArN$_2^{\oplus}$ BF$_4^{\ominus}$ $\xrightarrow{\Delta}$ Ar—F + N$_2$↑ + BF$_3$

Example 1[4]

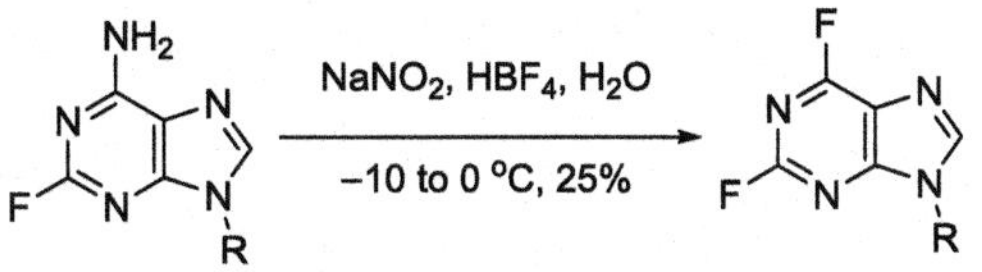

R = 2,3-5-tri-O-acetyl-β-D-ribofuranose

Example 2, Photo-Schiemann reaction[6]

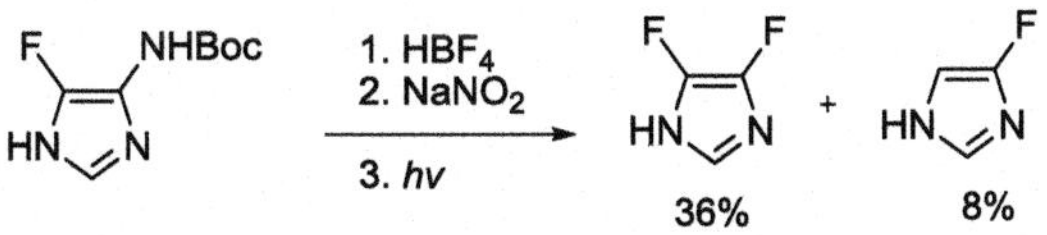

36%   8%

Example 3, Photo-Schiemann reaction[8]

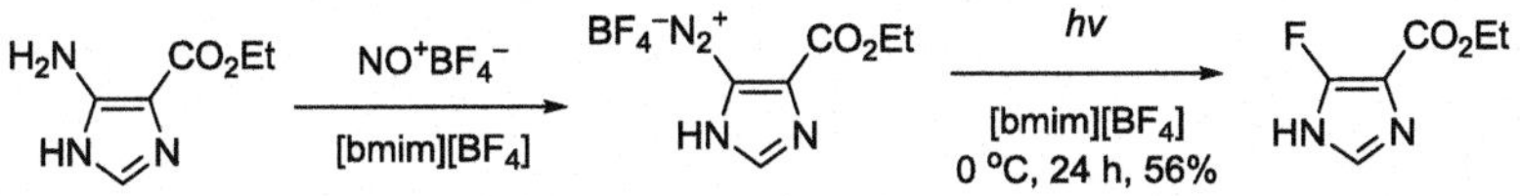

Example 4[10]

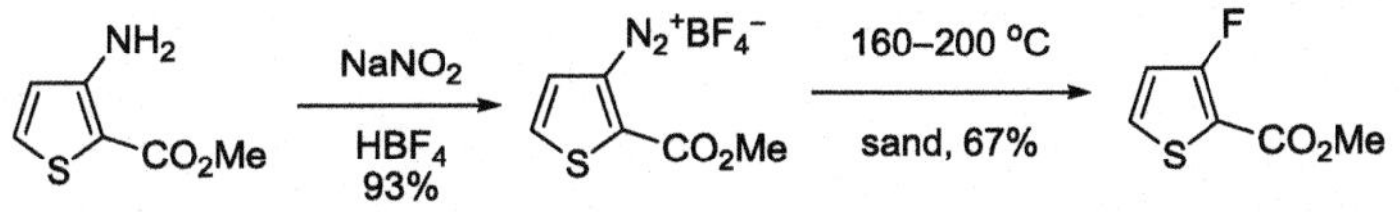

J.J. Li, *Name Reactions: A Collection of Detailed Mechanisms and Synthetic Applications*, DOI 10.1007/978-3-319-03979-4_243, © Springer International Publishing Switzerland 2014

## References

1.  Balz, G.; Schiemann, G. *Ber.* **1927,** *60*, 1186–1190. Günther Schiemann was born in Breslau, Germany in 1899. In 1925, he received his doctorate at Breslau, where he became an assistant professor. In 1950, he became the Chair of Technical Chemistry at Istanbul, where he extensively studied aromatic fluorine compounds.

2.  Roe, A. *Org. React.* **1949,** *5*, 193–228. (Review).

3.  Sharts, C. M. *J. Chem. Educ.* **1968,** *45*, 185–192. (Review).

4.  Montgomery, J. A.; Hewson, K. *J. Org. Chem.* **1969,** *34*, 1396–1399.

5.  Laali, K. K.; Gettwert, V. J. *J. Fluorine Chem.* **2001,** *107*, 31–34.

6.  Dolensky, B.; Takeuchi, Y.; Cohen, L. A.; Kirk, K. L. *J. Fluorine Chem.* **2001,** *107*, 147–152.

7.  Gronheid, R.; Lodder, G.; Okuyama, T. *J. Org. Chem.* **2002,** *67*, 693–720.

8.  Heredia-Moya, J.; Kirk, K. L. *J. Fluorine Chem.* **2007,** *128*, 674–678.

9.  Gribble, G. W. *Balz-Schiemann reaction.* In *Name Reactions for Functional Group Transformations*; Li, J. J., Ed.; Wiley: Hoboken, NJ, **2007,** pp 552–563. (Review).

10. Pomerantz, M.; Turkman, N. *Synthesis* **2008,** 2333–2336.

# Schmidt rearrangement

The Schmidt reactions refer to the acid-catalyzed reactions of hydrazoic acid with electrophiles, such as carbonyl compounds, tertiary alcohols and alkenes. These substrates undergo rearrangement and extrusion of nitrogen to furnish amines, nitriles, amides or imines.

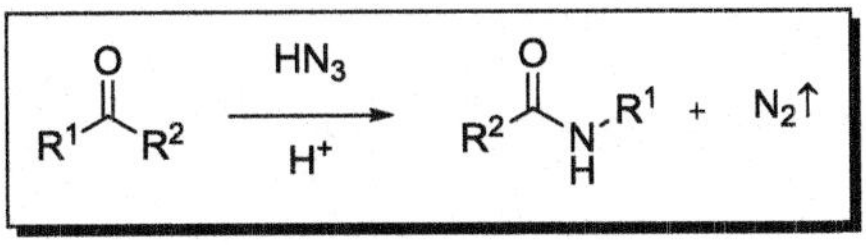

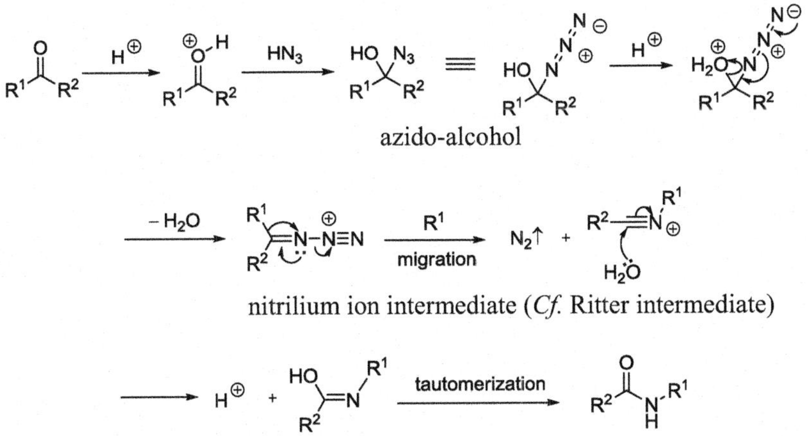

azido-alcohol

nitrilium ion intermediate (*Cf.* Ritter intermediate)

Example 1, A classic example[3]

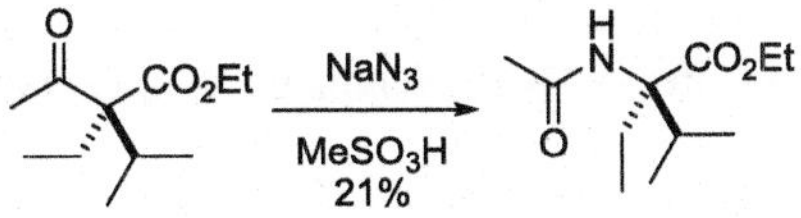

Example 2[5]

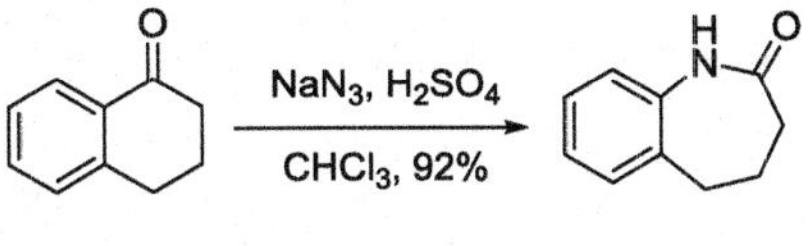

Example 3, Intramolecular Schmidt rearrangement[6]

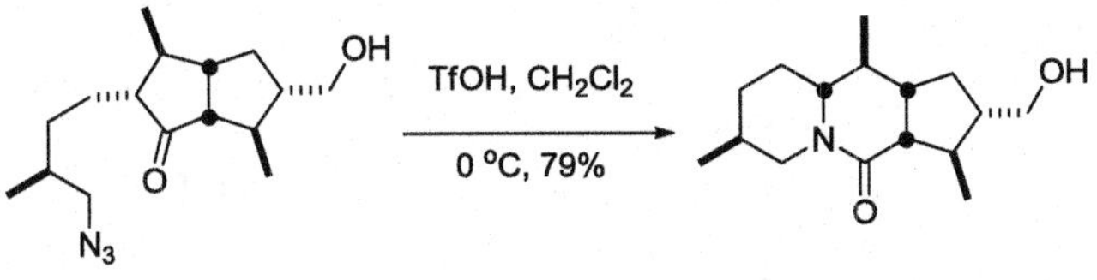

J.J. Li, *Name Reactions: A Collection of Detailed Mechanisms and Synthetic Applications*, DOI 10.1007/978-3-319-03979-4_244, © Springer International Publishing Switzerland 2014

Example 4, Intramolecular Schmidt rearrangement[8]

Example 5, Intermolecular Schmidt rearrangement[9]

Example 6[11]

## References

1. (a) Schmidt, K. F. *Angew. Chem.* **1923**, *36*, 511. Karl Friedrich Schmidt (1887–1971) collaborated with Curtius at the University of Heidelberg, where Schmidt became a Professor of Chemistry after 1923. (b) Schmidt, K. F. *Ber.* **1924**, *57*, 704–706.
2. Wolff, H. *Org. React.* **1946**, *3*, 307–336. (Review).
3. Tanaka, M.; Oba, M.; Tamai, K.; Suemune, H. *J. Org. Chem.* **2001**, *66*, 2667–2573.
4. Golden, J. E.; Aubé, J. *Angew. Chem. Int. Ed.* **2002**, *41*, 4316–4318.
5. Johnson, P. D.; Aristoff, P. A.; Zurenko, G. E.; Schaadt, R. D.; Yagi, B. H.; Ford, C. W.; Hamel, J. C.; Stapert, D.; Moerman, J. K. *Bioorg. Med. Chem. Lett.* **2003**, *13*, 4197–4200.
6. Wrobleski, A.; Sahasrabudhe, K.; Aubé, J. *J. Am. Chem. Soc.* **2004**, *126*, 5475–5481.
7. Gorin, D. J.; Davis, N. R.; Toste, F. D. *J. Am. Chem. Soc.* **2005**, *127*, 11260–11261.
8. Iyengar, R.; Schidknegt, K.; Morton, M.; Aubé, J. *J. Org. Chem.* **2005**, *70*, 10645–10652.
9. Amer, F. A.; Hammouda, M.; El-Ahl, A. A. S.; Abdel-Wahab, B. F. *Synth. Commun.* **2009**, *39*, 416–425.
10. Wu, Y.-J. *Schmidt Reactions.* In *Name Reactions for Homologations-Part II*; Li, J. J., Ed.; Wiley: Hoboken, NJ, **2009**, pp 353–372. (Review).
11. Gu, P.; Sun, J.; Kang, X.-Y.; Yi, M.; Li, X.-Q.; Xue, P.; Li, R. *Org. Lett.* **2013**, *15*, 1124–1127

# Schmidt's trichloroacetimidate glycosidation

Lewis acid-promoted glycosidation of trichloroacetimidates with alcohols or phenols.

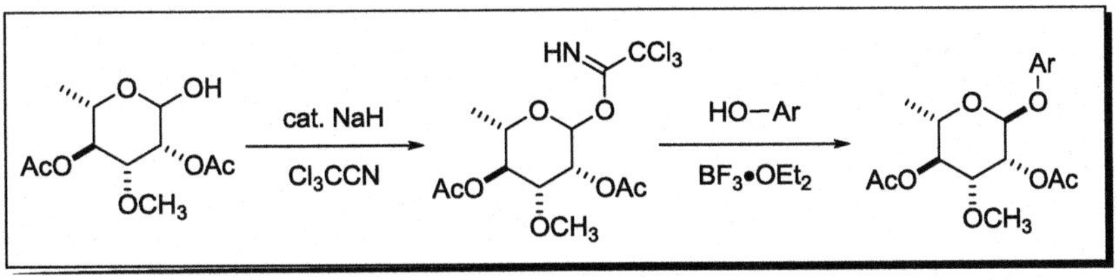

trichloroacetimidate

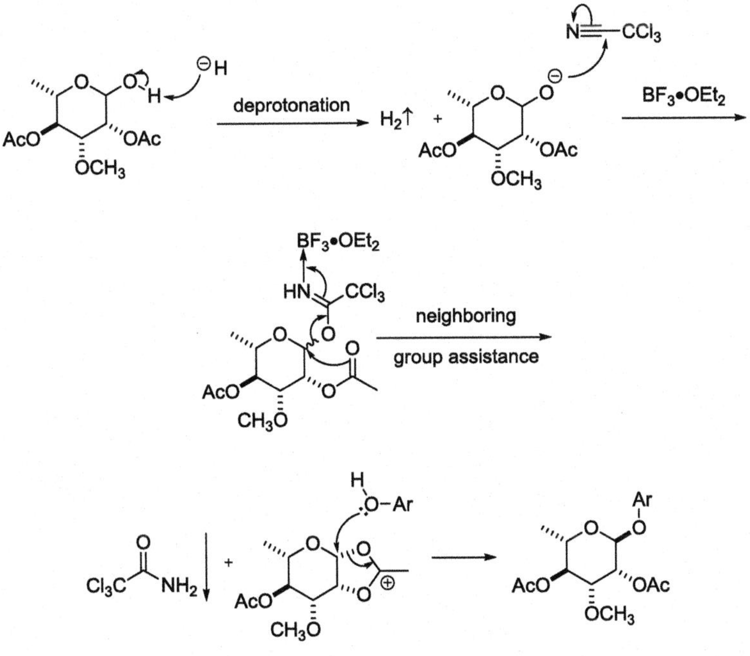

Example 1[5]

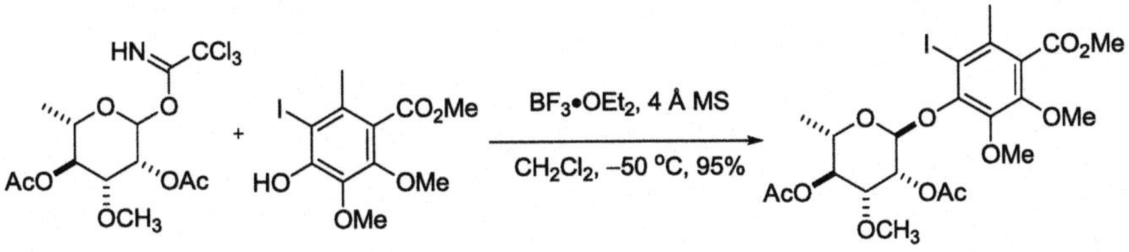

J.J. Li, *Name Reactions: A Collection of Detailed Mechanisms and Synthetic Applications*,
DOI 10.1007/978-3-319-03979-4_245, © Springer International Publishing Switzerland 2014

Example 2[7]

Example 3[9]

40%          +          30%

## References

1.  (a) Grundler, G.; Schmidt, R. R. *Carbohydr. Res.* **1985**, *135*, 203–218.  (b) Schmidt, R. R. *Angew. Chem. Int. Ed.* **1986**, *25*, 212–235. (Review).
2.  Smith, A. L.; Hwang, C.-K.; Pitsinos, E.; Scarlato, G. R.; Nicolaou, K. C. *J. Am. Chem. Soc.* **1992**, *114*, 3134–3136.
3.  Toshima, K.; Tatsuta, K. *Chem. Rev.* **1993**, *93*, 1503–1531. (Review).
4.  Nicolaou, K. C. *Angew. Chem. Int. Ed.* **1993**, *32*, 1377–1385. (Review).
5.  Groneberg, R. D.; Miyazaki, T.; Stylianides, N. A.; Schulze, T. J.; Stahl, W.; Schreiner, E. P.; Suzuki, T.; Iwabuchi, Y.; Smith, A. L.; Nicolaou, K. C. *J. Am. Chem. Soc.* **1993**, *115*, 7593–611.
6.  Fürstner, A.; Jeanjean, F.; Razon, P. *Angew. Chem. Int. Ed.* **2002**, *41*, 2097–2101.
7.  Yan, L. Z.; Mayer, J. P. *J. Org. Chem.* **2003**, *68*, 1161–1162.
8.  Harding, J. R.; King, C. D.; Perrie, J. A.; Sinnott, D.; Stachulski, A. V. *Org. Biomol. Chem.* **2005**, *3*, 1501–1507.
9.  Steinmann, A.; Thimm, J.; Thiem, J. *Eur. J. Org. Chem.* **2007**, *66*, 5506–5513.
10. Coutrot, F.; Busseron, E.; Montero, J.-L. *Org. Lett.* **2008**, *10*, 753–756.
11. Geng, Y.; Kumar, A.; Faidallah, H. M.; Albar, H. A.; Mhkalid, I. A.; Schmidt, R. R. *Angew. Chem. Int. Ed.* **2013**, *52*, 10089–10092.

# Scholl reaction

The elimination of two aryl-bound hydrogens accompanied by the formation of an aryl-aryl bond under the influence of Friedel–Crafts catalysts. *Cf.* Friedel–Crafts reaction.

Example 1[7]

# References

1. Scholl, R.; Seer, C. *Ann,* **1912**, *394*, 111.
2. Olah, G. A.; Schilling, P.; Gross, I. M. *J. Am. Chem. Soc.* **1974**, *96*, 876.
3. Dopper, J. H.; Oudman, D.; Wynberg, H. *J. Org. Chem.* **1975**, *40*, 3398.
4. Rozas, M. F.; Piro, O. E.; Castellano, E. E.; et al. *Synthesis* **2002**, 2399.
5. King, B. T. *J. Am. Chem. Soc.* **2004**, *126*, 15002–15003. (Mechanism).
6. King, B. T. *J. Org. Chem.* **2006**, *71*, 5067–5081. (Mechanism).
7. Pradhan, A.; Dechambenoit, P.; Bock, H.; Durola, F. *J. Org. Chem.* **2013**, *78*, 2266–2274.

J.J. Li, *Name Reactions: A Collection of Detailed Mechanisms and Synthetic Applications*,
DOI 10.1007/978-3-319-03979-4_246, © Springer International Publishing Switzerland 2014

# Shapiro reaction

The Shapiro reaction is a variant of the Bamford–Stevens reaction. The former uses bases such as alkyl lithium and Grignard reagents whereas the latter employs bases such as Na, NaOMe, LiH, NaH, $NaNH_2$, *etc.* Consequently, the Shapiro reaction generally affords the less-substituted olefins (the kinetic products), while the Bamford–Stevens reaction delivers the more-substituted olefins (the thermodynamic products).

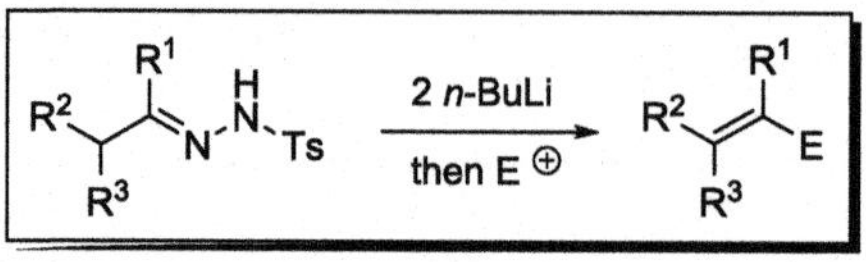
2 n-BuLi
then E ⊕

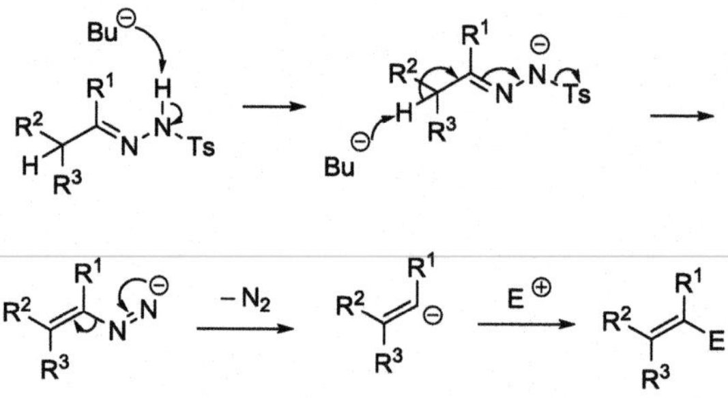
– N2
E ⊕

Example 1[2]

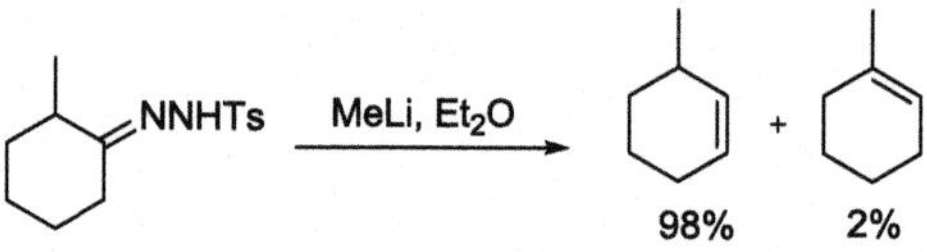
NNHTs
MeLi, Et2O
98%
2%

Example 2[3]

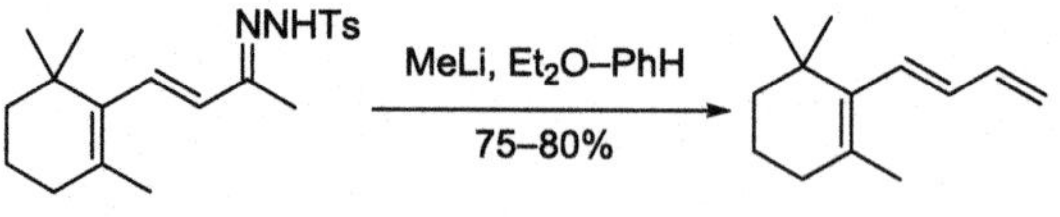
NNHTs
MeLi, Et2O–PhH
75–80%

Example 3[7]

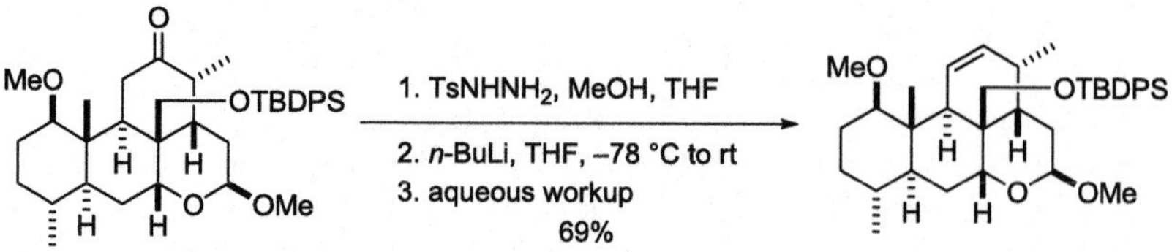
MeO
OTBDPS
OMe
1. TsNHNH2, MeOH, THF
2. n-BuLi, THF, –78 °C to rt
3. aqueous workup
69%

J.J. Li, *Name Reactions: A Collection of Detailed Mechanisms and Synthetic Applications*,
DOI 10.1007/978-3-319-03979-4_247, © Springer International Publishing Switzerland 2014

Example 4[8]

55% yield
one diastereomer

Tris =

Example 5[11]

1. 2.5 equiv n-BuLi, THF
−78 °C, 30 min to 0 °C, 20 min

2. 1.5 equiv NFSI, THF
−78 °C, 30 min to rt, 2 h
70%

NFSI =

# References

1. Shapiro, R. H.; Duncan, J. H.; Clopton, J. C. *J. Am. Chem. Soc.* **1967**, *89*, 471–472. Robert H. Shapiro published this 1967 JACS paper when he was an assistant professor at the University of Colorado. He was denied tenure despite having been immortalized with a reaction named after him.
2. Shapiro, R. H.; Heath, M. J. *J. Am. Chem. Soc.* **1967**, *89*, 5734–5735.
3. Dauben, W. G.; Lorber, M. E.; Vietmeyer, N. D.; Shapiro, R. H.; Duncan, J. H.; Tomer, K. *J. Am. Chem. Soc.* **1968**, *90*, 4762–4763.
4. Shapiro, R. H. *Org. React.* **1976**, *23*, 405–507. (Review).
5. Adlington, R. M.; Barrett, A. G. M. *Acc. Chem. Res.* **1983**, *16*, 55–59. (Review).
6. Chamberlin, A. R.; Bloom, S. H. *Org. React.* **1990**, *39*, 1–83. (Review).
7. Grieco, P. A.; Collins, J. L.; Moher, E. D.; Fleck, T. J.; Gross, R. S. *J. Am. Chem. Soc.* **1993**, *115*, 6078–6093.
8. Tamiya, J.; Sorensen, E. J. *Tetrahedron* **2003**, *59*, 6921–6932.
9. Wolfe, J. P. *Shapiro reaction.* In *Name Reactions for Functional Group Transformations*; Li, J. J., Corey, E. J., eds, Wiley: Hoboken, NJ, **2007**, pp 405–413.
10. Bettinger, H. F.; Mondal, R.; Toenshoff, C. *Org. Biomol. Chem.* **2008**, *6*, 3000–3004.
11. Yang, M.-H.; Matikonda, S. S.; Altman, R. A. *Org. Lett.* **2013**, *15*, 3894–3897.

## Sharpless asymmetric amino-hydroxylation

Osmium-mediated *cis*-addition of nitrogen and oxygen to olefins. Regio-selectivity may be controlled by ligand. Nitrogen sources (X–NClNa) include:

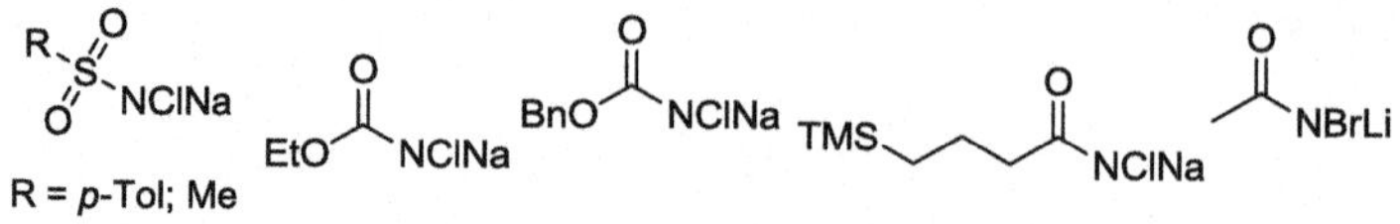

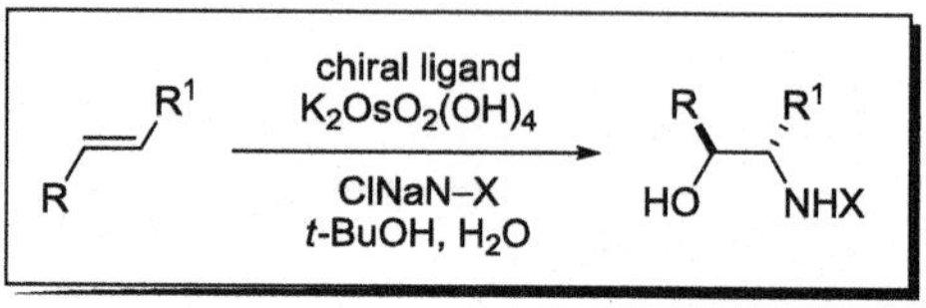

R = *p*-Tol; Me

The catalytic cycle:

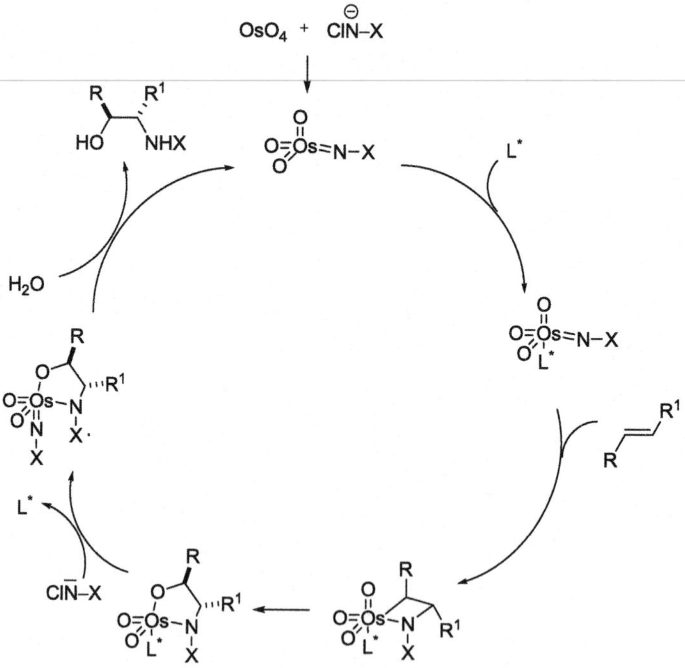

Example 1[1b]

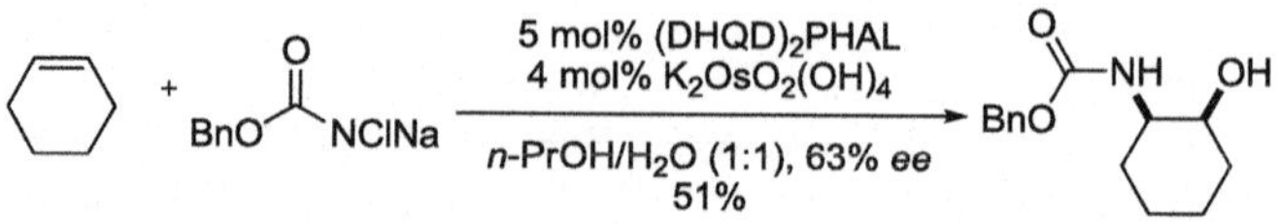

5 mol% (DHQD)$_2$PHAL
4 mol% K$_2$OsO$_2$(OH)$_4$

*n*-PrOH/H$_2$O (1:1), 63% *ee*
51%

J.J. Li, *Name Reactions: A Collection of Detailed Mechanisms and Synthetic Applications*, DOI 10.1007/978-3-319-03979-4_248, © Springer International Publishing Switzerland 2014

(DHQD)$_2$-PHAL = 1,4-bis(9-O-dihydroquinidine)phthalazine:

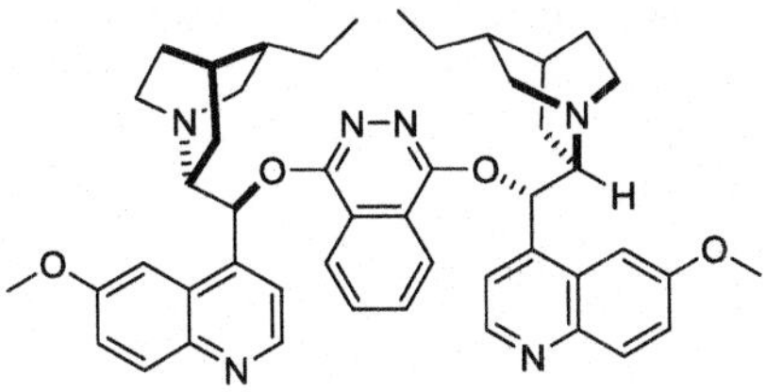

Example 2[2]

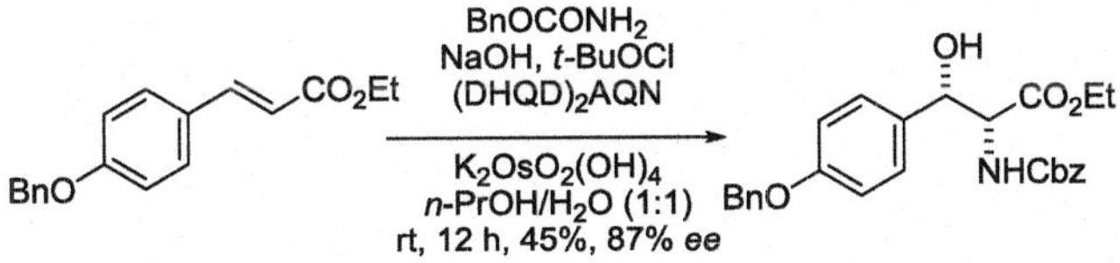

Example 3[6]

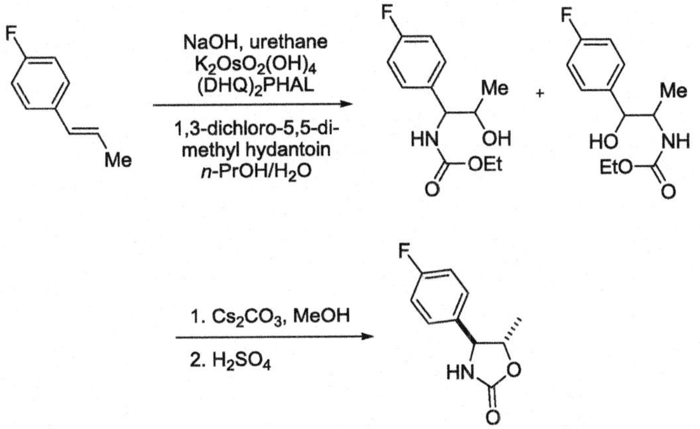

Example 4[13]

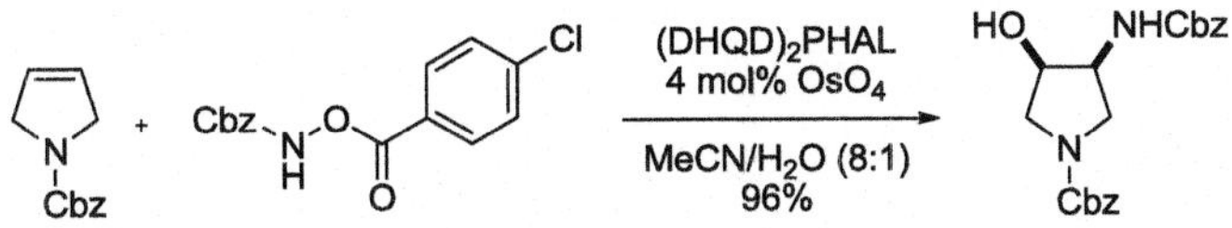

**References**

1.  (a) Herranz, E.; Sharpless, K. B. *J. Org. Chem.* **1978**, *43*, 2544–2548.  K. Barry Sharpless (USA, 1941–) shared the Nobel Prize in Chemistry in 2001 with Herbert William S. Knowles (USA, 1917–) and Ryoji Noyori (Japan, 1938–) for his work on

chirally catalyzed oxidation reactions. (b) Li, G.; Angert, H. H.; Sharpless, K. B. *Angew. Chem. Int. Ed.* **1996,** *35,* 2813–2817. (c) Rubin, A. E.; Sharpless, K. B. *Angew. Chem. Int. Ed.* **1997,** *36,* 2637–2640. (d) Kolb, H. C.; Sharpless, K. B. *Transition Met. Org. Synth.* **1998,** *2,* 243–260. (Review). (e) Thomas, A.; Sharpless, K. B. *J. Org. Chem.* **1999,** *64,* 8379–8385. (f) Gontcharov, A. V.; Liu, H.; Sharpless, K. B. *Org. Lett.* **1999,** *1,* 783–786.

2.   Nicolaou, K. C.; Boddy, C. N. C.; Li, H.; Koumbis, A. E.; Hughes, R.; Natarajan, S.; Jain, N. F.; Ramanjulu, J. M.; Braese, S.; Solomon, M. E. *Chem. Eur. J.* **1999,** *5,* 2602–2621.

3.   Lohr, B.; Orlich, S.; Kunz, H. *Synlett* **1999,** 1139–1141.

4.   Boger, D. L.; Lee, R. J.; Bounaud, P.-Y.; Meier, P. *J. Org. Chem.* **2000,** *65,* 6770–6772.

5.   Demko, Z. P.; Bartsch, M.; Sharpless, K. B. *Org. Lett.* **2000,** *2,* 2221–2223.

6.   Barta, N. S.; Sidler, D. R.; Somerville, K. B.; Weissman, S. A.; Larsen, R. D.; Reider, P. *J. Org. Lett.* **2000,** *2,* 2821–2824.

7.   Bolm, C.; Hildebrand, J. P.; Muñiz, K. In *Catalytic Asymmetric Synthesis;* 2nd edn., Ojima, I., Ed.; Wiley–VCH: New York, **2000,** 399. (Review).

8.   Bodkin, J. A.; McLeod, M. D. *J. Chem. Soc., Perkin 1* **2002,** 2733–2746. (Review).

9.   Rahman, N. A.; Landais, Y. *Cur. Org. Chem.* **2000,** *6,* 1369–1395. (Review).

10.  Nilov, D.; Reiser, O. *Recent Advances on the Sharpless Asymmetric Aminohydroxylation.* In *Organic Synthesis Highlights* Schmalz, H.-G.; Wirth, T., eds.; Wiley–VCH: Weinheim, Germany **2003,** 118–124. (Review).

11.  Bodkin, J. A.; Bacskay, G. B.; McLeod, M. D. *Org. Biomol. Chem.* **2008,** *6,* 2544–2553.

12.  Wong, D.; Taylor, C. M. *Tetrahedron Lett.* **2009,** *50,* 1273–1275.

13.  Harris, L.; Mee, S. P. H.; Furneaux, R. H.; Gainsford, G. J.; Luxenburger, A. *J. Org. Chem.* **2011,** *76,* 358–372.

14.  Kumar, J. N.; Das, B. *Tetrahedron Lett.* **2013,** *54,* 3865–3867.

## Sharpless asymmetric dihydroxylation

Enantioselective *cis*-dihydroxylation of olefins using osmium catalyst in the presence of cinchona alkaloid ligands.

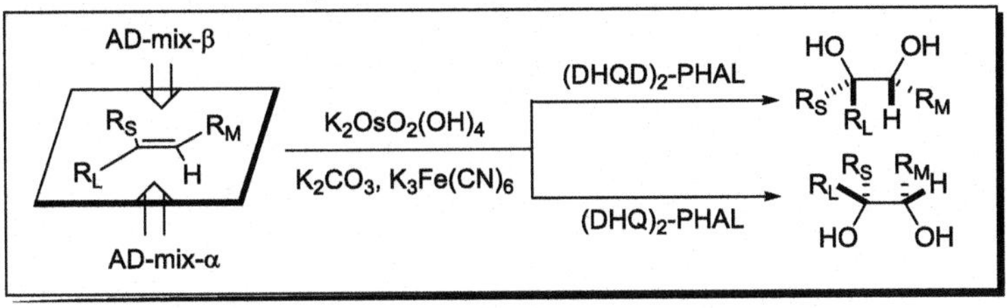

(DHQ)$_2$-PHAL = 1,4-bis(9-*O*-dihydroquinine)phthalazine:

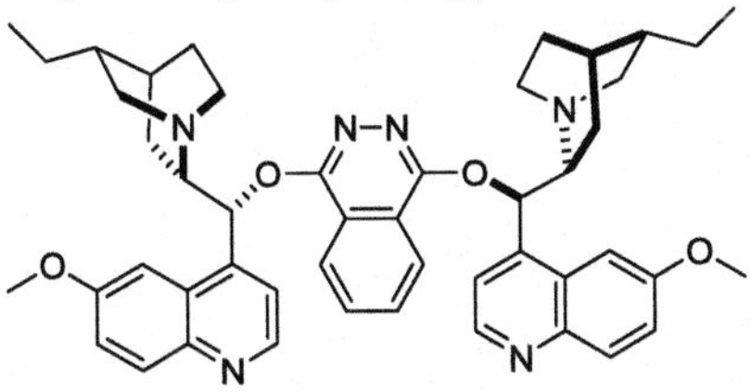

The concerted [3 + 2] cycloaddition mechanism:[5]

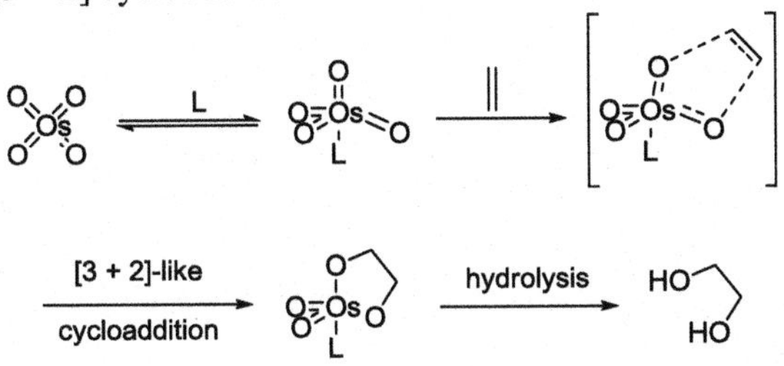

Example 1[2]

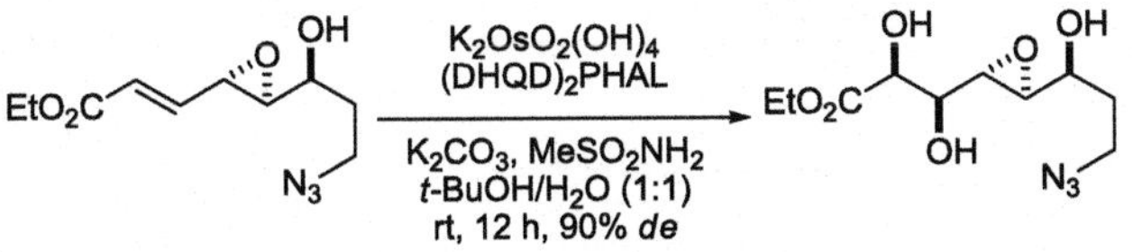

Example 2[4]

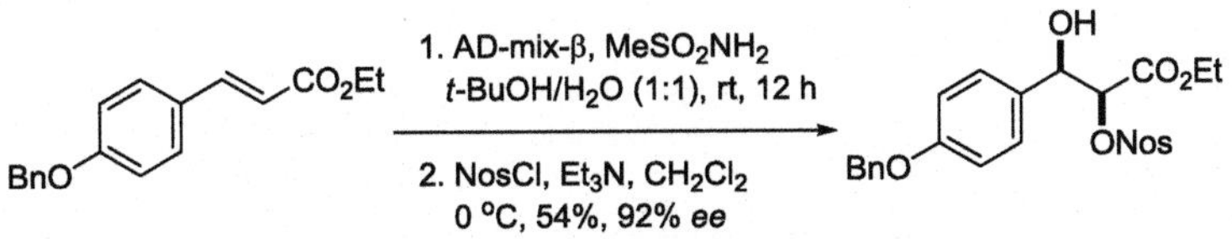

Nos = nosylate = 4-nitrobenzenesulfonyl

J.J. Li, *Name Reactions: A Collection of Detailed Mechanisms and Synthetic Applications*,
DOI 10.1007/978-3-319-03979-4_249, © Springer International Publishing Switzerland 2014

The catalytic cycle: (the secondary cycle is shut off by maintaining a low concentration of olefin):

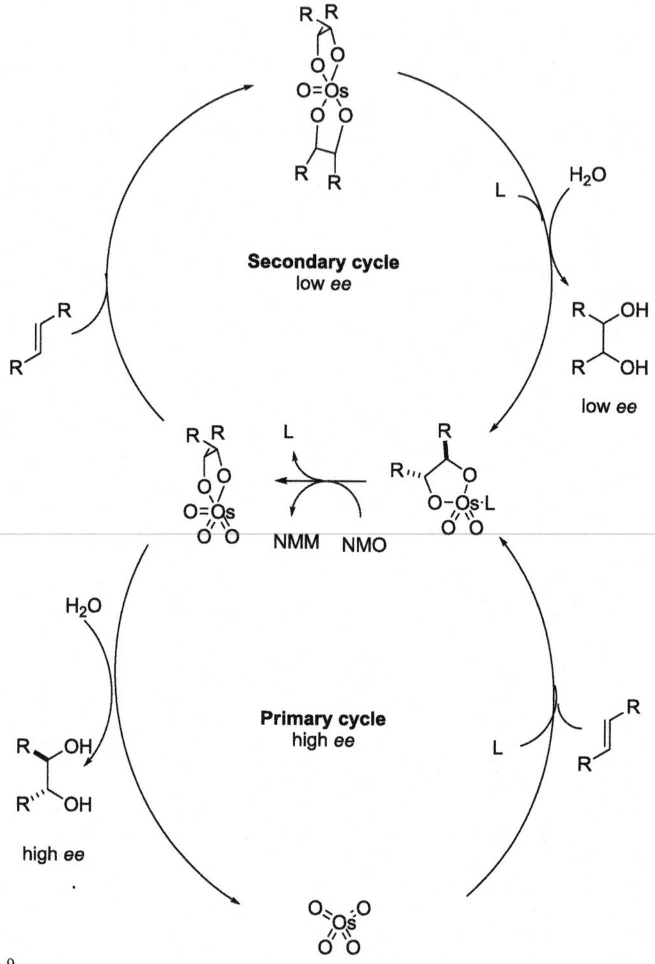

Example 3[9]

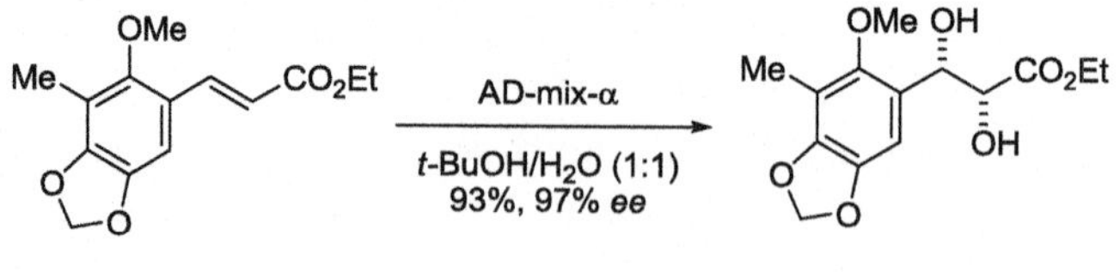

Example 4[10]

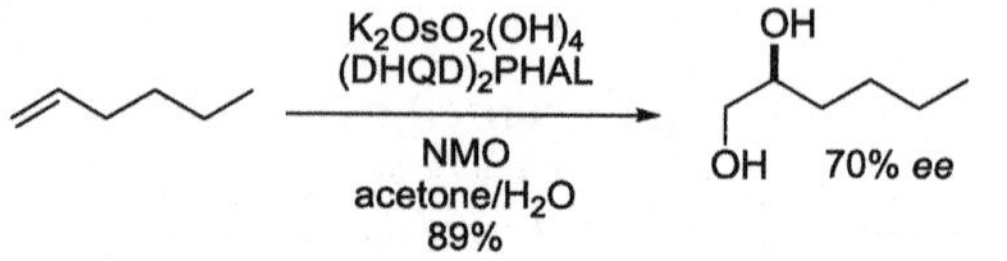

Example 5[13]

## References

1. (a) Jacobsen, E. N.; Markó, I.; Mungall, W. S.; Schröder, G.; Sharpless, K. B. *J. Am. Chem. Soc.* **1988**, *110*, 1968–1970. (b) Wai, J. S. M.; Markó, I.; Svenden, J. S.; Finn, M. G.; Jacobsen, E. N.; Sharpless, K. B. *J. Am. Chem. Soc.* **1989**, *111*, 1123–1125.
2. Kim, N.-S.; Choi, J.-R.; Cha, J. K. *J. Org. Chem.* **1993**, *58,* 7096–7699.
3. Kolb, H. C.; VanNiewenhze, M. S.; Sharpless, K. B. *Chem. Rev.* **1994**, *94*, 2483–2547. (Review).
4. Rao, A. V. R.; Chakraborty, T. K.; Reddy, K. L.; Rao, A. S. *Tetrahedron Lett.* **1994**, *35*, 5043–5046.
5. Corey, E. J.; Noe, M. C. *J. Am. Chem. Soc.* **1996**, *118*, 319–329. (Mechanism).
6. DelMonte, A. J.; Haller, J.; Houk, K. N.; Sharpless, K. B.; Singleton, D. A.; Strassner, T.; Thomas, A. A. *J. Am. Chem. Soc.* **1997**, *119*, 9907–9908. (Mechanism).
7. Sharpless, K. B. *Angew. Chem. Int. Ed.* **2002**, *41*, 2024–2032. (Review, Nobel Prize Address).
8. Zhang, Y.; O'Doherty, G. A. *Tetrahedron* **2005**, *61*, 6337–6351.
9. Chandrasekhar, S.; Reddy, N. R.; Rao, Y. S. *Tetrahedron* **2006**, *62*, 12098–12107.
10. Ferreira, F. C.; Branco, L. C.; Verma, K. K.; Crespo, J. G.; Afonso, C. A. M. *Tetrahedron: Asymmetry* **2007**, *18*, 1637–1641.
11. Ramon, R.; Alonso, M.; Riera, A. *Tetrahedron: Asymmetry* **2007**, *18*, 2797–2802.
12. Krishna, P. R.; Reddy, P. S. *Synlett* **2009**, 209–212.
13. Kamal, A.; Vangala, S. R. *Org. Biomol. Chem.* **2013**, *11*, 4442–4448.

## Sharpless asymmetric epoxidation

Enantioselective epoxidation of allylic alcohols using $t$-butyl peroxide, titanium tetra-$iso$-propoxide, and optically pure diethyl tartrate.

The catalytic cycle:

J.J. Li, *Name Reactions: A Collection of Detailed Mechanisms and Synthetic Applications*,
DOI 10.1007/978-3-319-03979-4_250, © Springer International Publishing Switzerland 2014

The putative active catalyst:

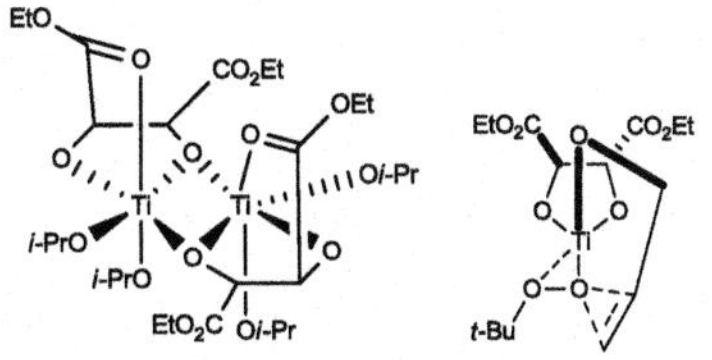

Example 1[3]

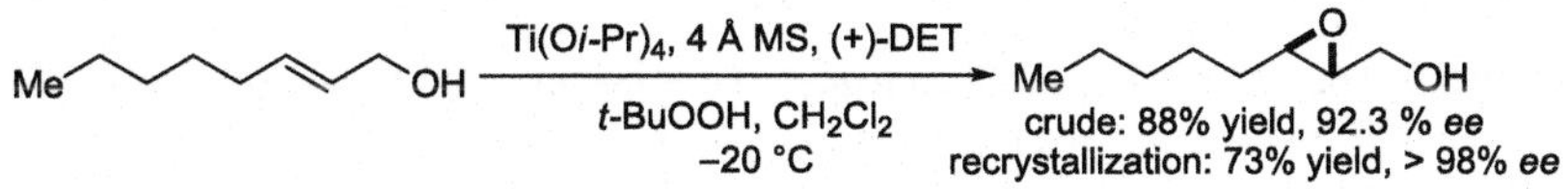

Example 2[3]

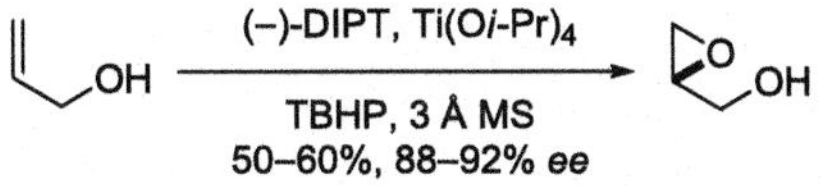

Example 3[11]

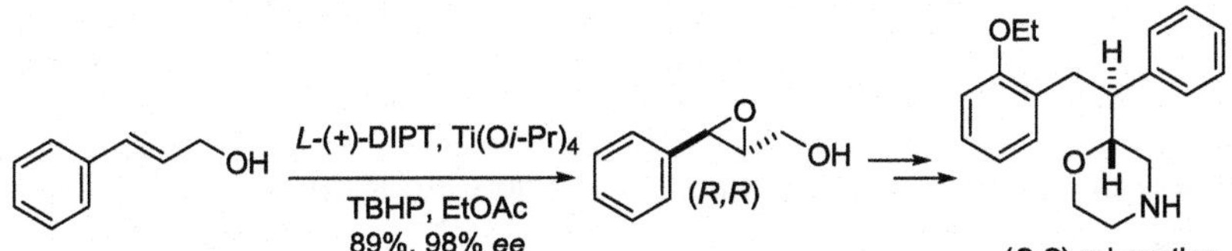

Example 4[12]

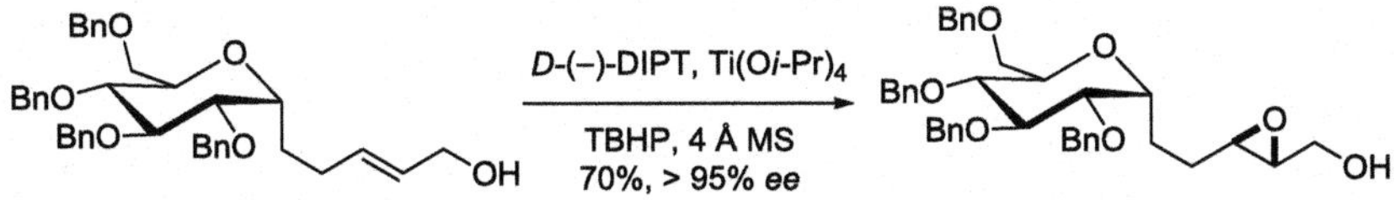

Example 5[14]

$dr = 88:12$

## References

1. (a) Katsuki, T.; Sharpless, K. B. *J. Am. Chem. Soc.* **1980**, *102*, 5974–5976. (b) Williams, I. D.; Pedersen, S. F.; Sharpless, K. B.; Lippard, S. J. *J. Am. Chem. Soc.* **1984**, *106*, 6430–6433. (c) Woodard, S. S.; Finn, M. G.; Sharpless, K. B. *J. Am. Chem. Soc.* **1991**, *113*, 106–113.
2. Pfenninger, A. *Synthesis* **1986**, 89–116. (Review).
3. Gao, Y.; Hanson, R. M.; Klunder, J. M.; Ko, S. Y.; Masamune, H.; Sharpless, K. B. *J. Am. Chem. Soc.* **1987**, *109*, 5765–5780.
4. Corey, E. J. *J. Org. Chem.* **1990**, *55*, 1693–1694. (Review).
5. Johnson, R. A.; Sharpless, K. B. In *Comprehensive Organic Synthesis*; Trost, B. M., Ed,; Pergamon Press: New York, **1991**; Vol. 7, Chapter 3.2. (Review).
6. Johnson, R. A.; Sharpless, K. B. In *Catalytic Asymmetric Synthesis*; Ojima, I., ed,; VCH: New York, **1993**; Chapter 4.1, pp 103–158. (Review).
7. Schinzer, D. *Org. Synth. Highlights II* **1995**, 3. (Review).
8. Katsuki, T.; Martin, V. S. *Org. React.* **1996**, *48*, 1–299. (Review).
9. Johnson, R. A.; Sharpless, K. B. In *Catalytic Asymmetric Synthesis;* 2nd ed., Ojima, I., ed.; Wiley-VCH: New York, **2000**, 231–285. (Review).
10. Palucki, M. *Sharpless–Katsuki Epoxidation.* In *Name Reactions in Heterocyclic Chemistry*; Li, J. J., Ed.; Wiley: Hoboken, NJ, **2005**, 50–62. (Review).
11. Henegar, K. E.; Cebula, M. *Org. Proc. Res. Dev.* **2007**, *11*, 354–358.
12. Pu, J.; Franck, R. W. *Tetrahedron* **2008**, *64*, 8618–8629.
13. Knight, D. W.; Morgan, I. R. *Tetrahedron Lett.* **2009**, *50*, 35–38.
14. Volchkov, I.; Lee, D. *J. Am. Chem. Soc.* **2013**, *135*, 5324–5327.

# Sharpless olefin synthesis

Olefin synthesis from the *syn*-oxidative elimination of *o*-nitrophenyl selenides, which may be prepared using *o*-nitrophenyl selenocyanate and $Bu_3P$, among other methods.

## Example 1[3]

1. $Bu_3P$, *o*-O$_2$NPhSeCN, THF, rt
2. CSA, PhH, rt to 70 °C
3. *m*-CPBA, 2,4,6-collidine
   $CH_2Cl_2$, 0 °C
   42%, 3 steps

## Example 2[6]

1. $Bu_3P$, *o*-O$_2$NPhSeCN MeO
   THF, rt, 6 h, 97%
2. *m*-CPBA, Et$_3$N
   $CH_2Cl_2$, −78 °C
   86%

J.J. Li, *Name Reactions: A Collection of Detailed Mechanisms and Synthetic Applications*,
DOI 10.1007/978-3-319-03979-4_251, © Springer International Publishing Switzerland 2014

Example 3[9]

Example 4[10]

## References

1. (a) Sharpless, K. B.; Young, M. Y.; Lauer, R. F. *Tetrahedron Lett.* **1973**, *22,* 1979–1982. (b) Sharpless, K. B.; Young, M. Y. *J. Org. Chem.* **1975**, *40*, 947–949.
2. (a) Grieco, P. A.; Miyashita, M. *J. Org. Chem.* **1974**, *39*, 120–122. (b) Grieco, P. A.; Miyashita, M. *Tetrahedron Lett.* **1974**, *21,* 1869–1871. (c) Grieco, P. A.; Masaki, Y.; Boxler, D. *J. Am. Chem. Soc.* **1977**, *97*, 1597–1599. (d) Grieco, P. A.; Gilman, S.; Nishizawa, M. *J. Org. Chem.* **1976**, *41*, 1485–1486. (e) Grieco, P. A.; Yokoyama, Y. *J. Am. Chem. Soc.* **1977**, *99*, 5210–5219.
3. Smith, A. B., III; Haseltine, J. N.; Visnick, M. *Tetrahedron* **1989**, *45,* 2431–2449.
4. Reich, H. J.; Wollowitz, S. *Org. React.* **1993**, *44,* 1–296. (Review).
5. Hsu, D.-S.; Liao, C.-C. *Org. Lett.* **2003**, *5*, 4741–4743.
6. Meilert, K.; Pettit, G. R.; Vogel, P. *Helv. Chim. Acta* **2004**, *87,* 1493–1507.
7. Siebum, A. H. G.; Woo, W. S.; Raap, J.; Lugtenburg, J. *Eur. J. Org. Chem.* **2004,** 2905–2916.
8. Blay, G.; Cardona, L.; Collado, A. M.; Garcia, B.; Morcillo, V.; Pedro, J. R. *J. Org. Chem.* **2004**, *69,* 7294–7302. The authors observed the concurrent epoxidation of a tri-subsituted olefin, possibly by the *o*-nitrophenylselenic acid *via* an intramolecular process.
9. Paquette, L. A.; Dong, S.; Parker, G. D. *J. Org. Chem.* **2007**, *72*, 7135–7147.
10. Yokoe, H.; Yoshida, M.; Shishido, K. *Tetrahedron Lett.* **2008**, *49*, 3504–3506.
11. Debnar, T.; Wang, T.; Menche, D. *Org. Lett.* **2013**, *15*, 2774–2777.

# Shi asymmetric epoxidation

An asymmetric epoxidation using a fructose-derived chiral ketone. It is an organocatalyst with Oxone typically used as the primary oxidant.

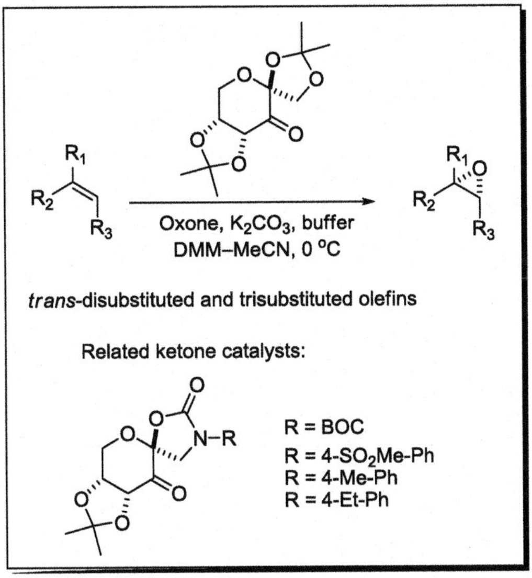

Oxone, $K_2CO_3$, buffer
DMM–MeCN, 0 °C
*trans*-disubstituted and trisubstituted olefins
Related ketone catalysts:
R = BOC
R = 4-$SO_2$Me-Ph
R = 4-Me-Ph
R = 4-Et-Ph


The catalytic cycle:

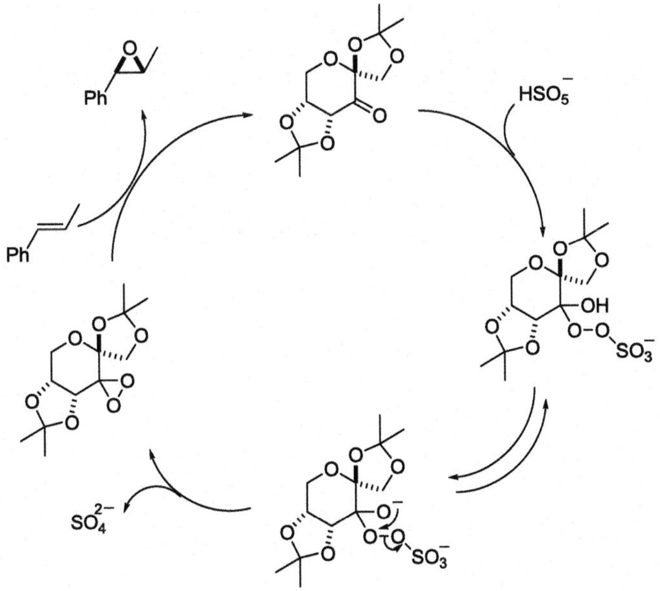

$HSO_5^-$
$SO_4^{2-}$
Ph


J.J. Li, *Name Reactions: A Collection of Detailed Mechanisms and Synthetic Applications*,
DOI 10.1007/978-3-319-03979-4_252, © Springer International Publishing Switzerland 2014

Example 1[6]

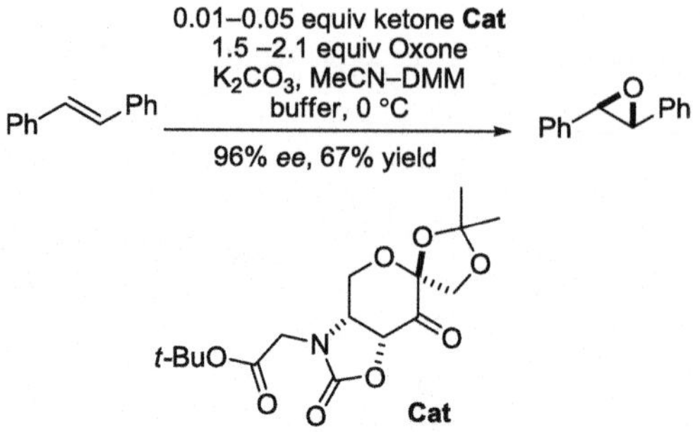

$$Ph\diagup\!\!\!\!\diagdown Ph \xrightarrow[\substack{96\% \ ee, \ 67\% \ yield}]{\substack{0.01-0.05 \ equiv \ ketone \ \textbf{Cat} \\ 1.5-2.1 \ equiv \ Oxone \\ K_2CO_3, \ MeCN-DMM \\ buffer, \ 0 \ °C}} Ph\diagup\!\!\!\!\overset{O}{\triangle}\!\!\!\!\diagdown Ph$$

*t*-BuO

**Cat**

Example 2[7]

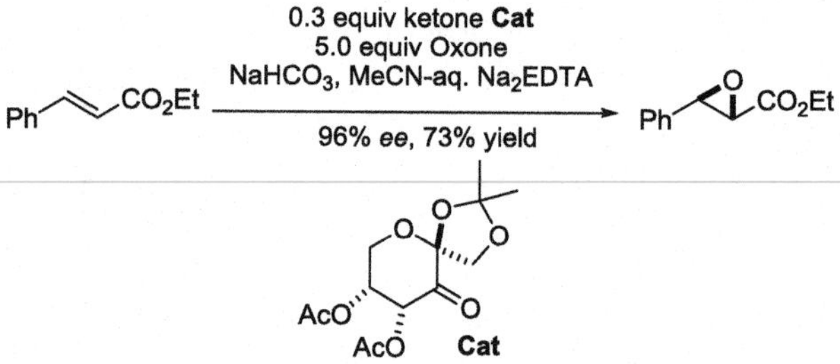

$$Ph\diagup\!\!\!\!\diagdown\diagup\!\!\!\!CO_2Et \xrightarrow[\substack{96\% \ ee, \ 73\% \ yield}]{\substack{0.3 \ equiv \ ketone \ \textbf{Cat} \\ 5.0 \ equiv \ Oxone \\ NaHCO_3, \ MeCN-aq. \ Na_2EDTA}} Ph\diagup\!\!\!\!\overset{O}{\triangle}\!\!\!\!CO_2Et$$

AcO

AcO    **Cat**

Example 3[8]

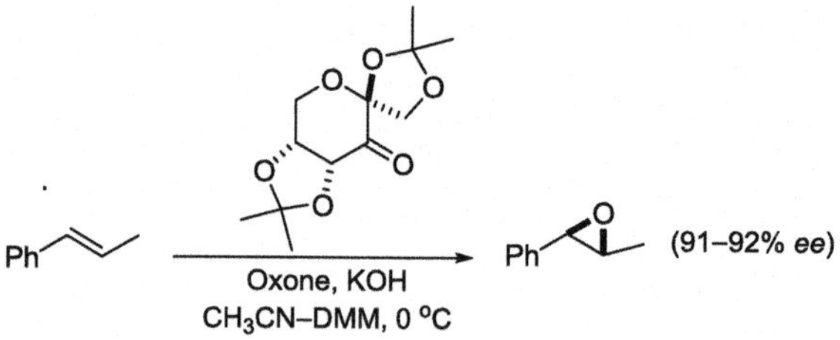

$$Ph\diagup\!\!\!\!\diagdown\diagup \xrightarrow[\substack{Oxone, \ KOH \\ CH_3CN-DMM, \ 0 \ °C}]{} Ph\diagup\!\!\!\!\overset{O}{\triangle}\!\!\!\!\diagdown \quad (91-92\% \ ee)$$

Example 4[8]

$$\xrightarrow[\substack{H_2O_2, \ K_2CO_3 \\ CH_3CN}]{} \quad (96-98\% \ ee)$$

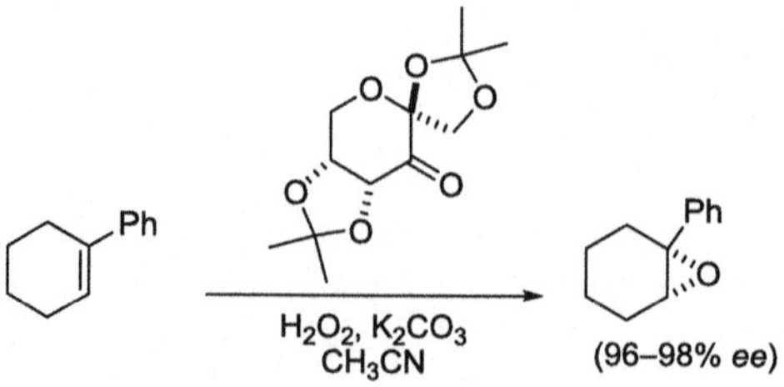

Example 5[9]

Oxone, 83%

> 15 : 1

## References

1.  Wang, Z.-X.; Tu, Y.; Frohn, M.; Zhang, J.-R.; Shi, Y. *J. Am. Chem. Soc.* **1997,** *119,* 11224–11235.
2.  Wang, Z.-X.; Shi, Y. *J. Org. Chem.* **1997,** *62,* 8622–8623.
3.  Tu, Y.; Wang, Z.-X.; Frohn, M.; He, M.; Yu, H.; Tang, Y.; Shi, Y. *J. Org. Chem.* **1998,** *63,* 8475–8485.
4.  Tian, H.; She, X.; Shu, L.; Yu, H.; Shi, Y. *J. Am. Chem. Soc.* **2000,** *122,* 11551–11552.
5.  Katsuki, T. In *Catalytic Asymmetric Synthesis;* 2nd ed., Ojima, I., ed.; Wiley–VCH: New York, 2000, pp 287–325. (Review).
6.  Tian, H.; She, X.; Shi, Y. *Org. Lett.* **2001,** *3,* 715–717.
7.  Wu, X.-Y.; She, X.; Shi, Y. *J. Am. Chem. Soc.* **2002,** *124,* 8792–8783.
8.  Wang, Z.-X.; Shu, L.; Frohn, M.; Tu, Y.; Shi, Y. *Org. Synth.* **2003,** *80,* 9–17; *Coll. Vol. 11,* **2003***,* 183–188.
9.  Julien, C.; Axel, B.; Antoinette, C.; Wolf, D. W. *Org. Lett.* **2008,** *10,* 512–516.
10. Yang, B. V. *Shi Epoxidation.* In *Name Reactions in Heterocyclic Chemistry II*; Li, J. J., Ed.; Wiley: Hoboken, NJ, **2011,** 21–39. (Review).
11. Kumar, V. P.; Chandrasekhar, S. *Org. Lett.* **2013,** *5,* 3610–3613.

## Simmons–Smith reaction

Cyclopropanation of olefins using $CH_2I_2$ and $Zn(Cu)$.

$$CH_2I_2 \; + \; Zn(Cu) \longrightarrow ICH_2ZnI \longrightarrow$$

$$I-CH_2-I \xrightarrow[\substack{Oxidative \\ addition}]{Zn} ICH_2ZnI$$

The Simmons-Smith reagent

$$2 \; ICH_2ZnI \;\rightleftharpoons\; (ICH_2)_2Zn \; + \; ZnI_2$$

$$\longrightarrow \left[ \;\; \right] \longrightarrow \quad + \quad ZnI_2$$

Example 1[2]

$$\xrightarrow[\substack{CH_2I_2, \; Et_2O, \; reflux, \; 36 \; h, \; 90\%}]{Zn/Cu \; [from \; Zn \; and \; Cu(SO_4)_2]}$$

Example 2, An asymmetric version[3]

CONEt₂

OH (1 eq)

OH

CONEt₂

$$\xrightarrow[\substack{CH_2Cl_2, \; 0 \; °C, \; 15 \; h \\ 78\%, \; 94\% \; ee}]{6 \; eq \; Zn/Cu, \; 3 \; eq \; CH_2I_2}$$

MeO                                    MeO          OH

Example 3, Diastereoselective Simmons–Smith cyclopropanations of allylic amines and carbamates[9]

NBn₂

$$\xrightarrow[\substack{CH_2Cl_2, \; rt, \; 1 \; h \\ 92\%, \; > 98\% \; de}]{Et_2Zn, \; CH_2I_2, \; TFA}$$

NBn₂

J.J. Li, *Name Reactions: A Collection of Detailed Mechanisms and Synthetic Applications*,
DOI 10.1007/978-3-319-03979-4_253, © Springer International Publishing Switzerland 2014

Example 4[10]

Example 5[12]

# References

1. Simmons, H. E.; Smith, R. D. *J. Am. Chem. Soc.* **1958**, *80*, 5323–5324. Howard E. Simmons (1929–1997) was born in Norfolk, Virginia. He carried out his graduate studies at MIT under John D. Roberts and Arthur Cope. After obtaining his Ph.D. In 1954, he joined the Chemical Department of the DuPont Company, where he discovered the Simmons–Smith reaction with his colleague, R. D. Smith. Simmons rose to be the vice president of the Central Research at DuPont in 1979. His views on physical exercise were the same as those of Alexander Woollcot's: "If I think about exercise, I know if I wait long enough, the thought will go away."
2. Limasset, J.-C.; Amice, P.; Conia, J.-M. *Bull. Soc. Chim. Fr.* **1969**, 3981–3990.
3. Kitajima, H.; Ito, K.; Aoki, Y.; Katsuki, T. *Bull. Chem. Soc. Jpn.* **1997**, *70*, 207–217.
4. Nakamura, E.; Hirai, A.; Nakamura, M. *J. Am. Chem. Soc.* **1998**, *120*, 5844–5845.
5. Loeppky, R. N.; Elomari, S. *J. Org. Chem.* **2000**, *65*, 96–103.
6. Charette, A. B.; Beauchemin, A. *Org. React.* **2001**, *58*, 1–415. (Review).
7. Nakamura, M.; Hirai, A.; Nakamura, E. *J. Am. Chem. Soc.* **2003**, *125*, 2341–2350.
8. Long, J.; Du, H.; Li, K.; Shi, Y. *Tetrahedron Lett.* **2005**, *46*, 2737–2740.
9. Davies, S. G.; Ling, K. B.; Roberts, P. M.; Russell, A. J.; Thomson, J. E. *Chem. Commun.* **2007**, 4029–4031.
10. Shan, M.; O'Doherty, G. A. *Synthesis* **2008**, 3171–3179.
11. Kim, H. Y.; Salvi, L.; Carroll, P. J.; Walsh, P. J. *J. Am. Chem. Soc.* **2009**, *131*, 954–962.
12. Swaroop, T. R.; Roopashree, R.; Ila, H.; Rangappa, K. S. *Tetrahedron Lett.* **2013**, *54*, 147–150.

# Skraup quinoline synthesis

Quinoline from aniline, glycerol, sulfuric acid and oxidizing agent (e.g. PhNO$_2$).

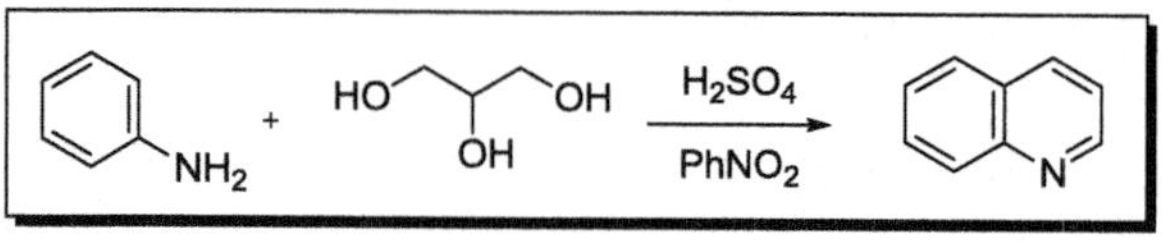

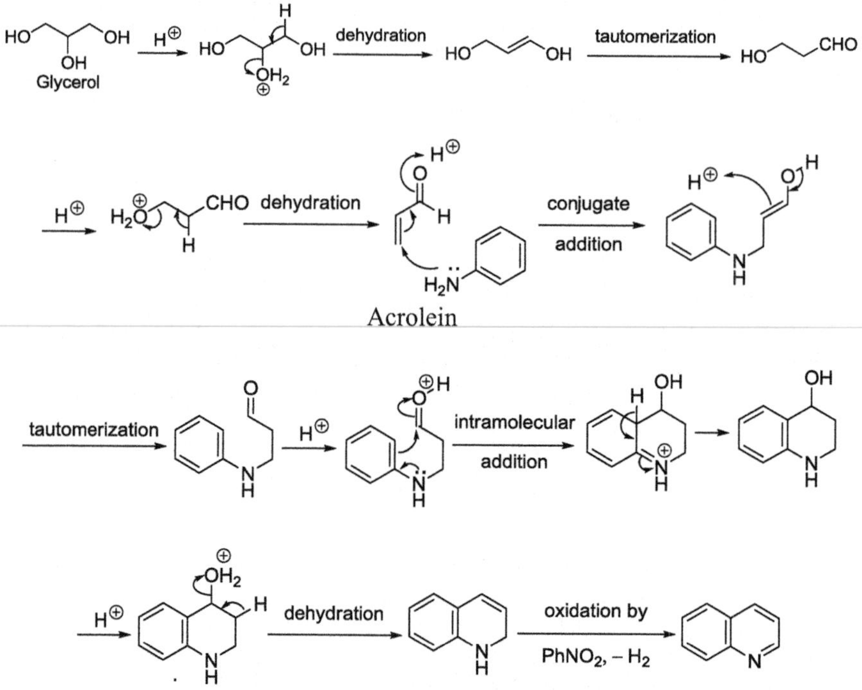

Acrolein

For an alternative mechanism, see that of the Doebner–von Miller reaction.

Example 1[5]

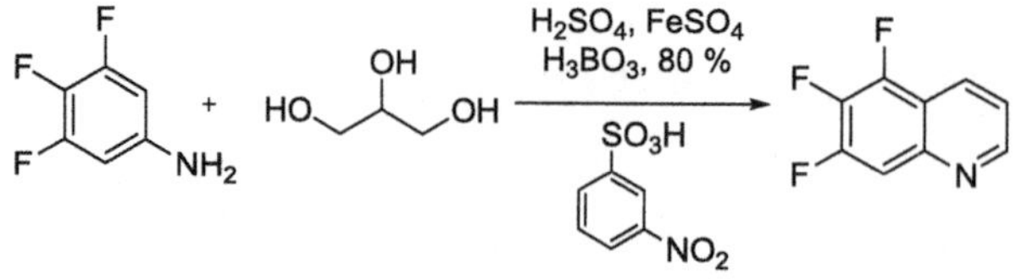

J.J. Li, *Name Reactions: A Collection of Detailed Mechanisms and Synthetic Applications*,
DOI 10.1007/978-3-319-03979-4_254, © Springer International Publishing Switzerland 2014

Example 2[6]

Example 3, A modified Skraup quinoline synthesis[8]

Example 4, A Skraup–Doebner–Von Miller quinoline synthesis[12]

## References

1.  (a) Skraup, Z. H. *Monatsh. Chem.* **1880**, *1*, 316. Zdenko Hans Skraup (1850–1910) was born in Prague, Czechoslovakia. He apprenticed under Lieben at the University of Vienna. (b) Skraup, Z. H. *Ber.* **1880**, *13*, 2086.
2.  Manske, R. H. F.; Kulka, M. *Org. React.* **1953**, *7*, 80–99. (Review).
3.  Bergstrom, F. W. *Chem. Rev.* **1944**, *35*, 77–277. (Review).
4.  Eisch, J. J.; Dluzniewski, T. *J. Org. Chem.* **1989**, *54*, 1269–1274.
5.  Oleynik, I. I.; Shteingarts, V. D. *J. Fluorine Chem.* **1998**, *91*, 25–26.
6.  Fujiwara, H.; Kitagawa, K. *Heterocycles* **2000**, *53*, 409–418.
7.  Ranu, B. C.; Hajra, A.; Dey, S. S.; Jana, U. *Tetrahedron* **2003**, *59*, 813–819.
8.  Panda, K.; Siddiqui, I.; Mahata, P. K.; Ila, H.; Junjappa, H. *Synlett* **2004**, 449–452.
9.  Moore, A. *Skraup Doebner–von Miller Reaction*. In *Name Reactions in Heterocyclic Chemistry*; Li, J. J., Ed.; Wiley: Hoboken, NJ, **2005**, pp 488–494. (Review).
10. Denmark, S. E.; Venkatraman, S. *J. Org. Chem.* **2006**, *71*, 1668–1676. Mechanistic study using $^{13}$C-labelled $\alpha,\beta$-unsaturated ketones.
11. Vora, J. J.; Vasava, S. B.; Patel, Asha D.; Parmar, K. C.; Chauhan, S. K.; Sharma, S. S. *E. J. Chem.* **2009**, *6*, 201–206.
12. Fotie, J.; Kemami Wangun, H. V.; Fronczek, F. R.; Massawe, N.; Bhattarai, B. T. Rhodus, J. L.; Singleton, T. A.; Bohle, D. S. *J. Org. Chem.* **2012**, *77*, 2784–2790.

# Smiles rearrangement

Intramolecular nucleophilic aromatic rearrangement. General scheme:

$$X = S, SO, SO_2, O, CO_2$$
$$YH = OH, NHR, SH, CH_2R, CONHR$$
$$Z = NO_2, SO_2R$$

Mechanism:

spirocyclic anion intermediate (Meisenheimer complex)

Example 1[7]

Example 2, Microwave Smiles rearrangement[9]

J.J. Li, *Name Reactions: A Collection of Detailed Mechanisms and Synthetic Applications*, DOI 10.1007/978-3-319-03979-4_255, © Springer International Publishing Switzerland 2014

Example 3[10]

Example 4[11]

## References

1. Evans, W. J.; Smiles, S. *J. Chem. Soc.* **1935**, 181–188. Samuel Smiles began his career at King's College London as an assistant professor. He later became professor and chair there. He was elected Fellow of the Royal Society (FRS) in 1918.
2. Truce, W. E.; Kreider, E. M.; Brand, W. W. *Org. React.* **1970**, *18*, 99–215. (Review).
3. Gerasimova, T. N.; Kolchina, E. F. *J. Fluorine Chem.* **1994**, *66*, 69–74. (Review).
4. Boschi, D.; Sorba, G.; Bertinaria, M.; Fruttero, R.; Calvino, R.; Gasco, A. *J. Chem. Soc., Perkin Trans. 1* **2001**, 1751–1757.
5. Hirota, T.; Tomita, K.-I.; Sasaki, K.; Okuda, K.; Yoshida, M.; Kashino, S. *Heterocycles* **2001**, *55*, 741–752.
6. Selvakumar, N.; Srinivas, D.; Azhagan, A. M. *Synthesis* **2002**, 2421–2425.
7. Mizuno, M.; Yamano, M. *Org. Lett.* **2005**, *7*, 3629–3631.
8. Bacque, E.; El Qacemi, M.; Zard, S. Z. *Org. Lett.* **2005**, *7*, 3817–3820.
9. Bi, C. F.; Aspnes, G. E.; Guzman-Perez, A.; Walker, D. P. *Tetrahedron Lett.* **2008**, *49*, 1832–1835.
10. Jin, Y. L.; Kim, S.; Kim, Y. S.; Kim, S.-A.; Kim, H. S. *Tetrahedron Lett.* **2008**, *49*, 6835–6837.
11. Niu, X.; Yang, B.; Li, Y.; Fang, S.; Huang, Z.; Xie, C.; Ma, C. *Org. Biomol. Chem.* **2013**, *11*, 4102–4108.

## *Truce–Smile rearrangement*

A variant of the Smiles rearrangement where Y is carbon:

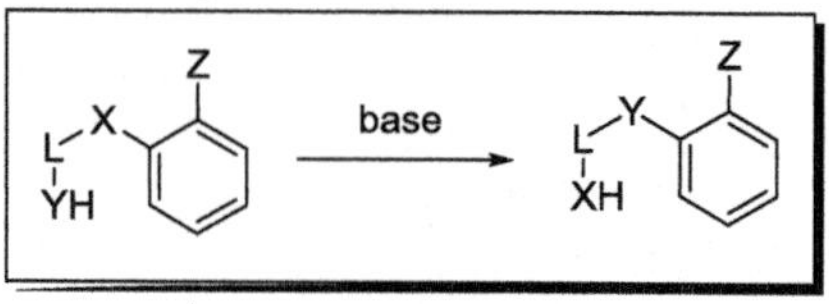

Example 1[6]

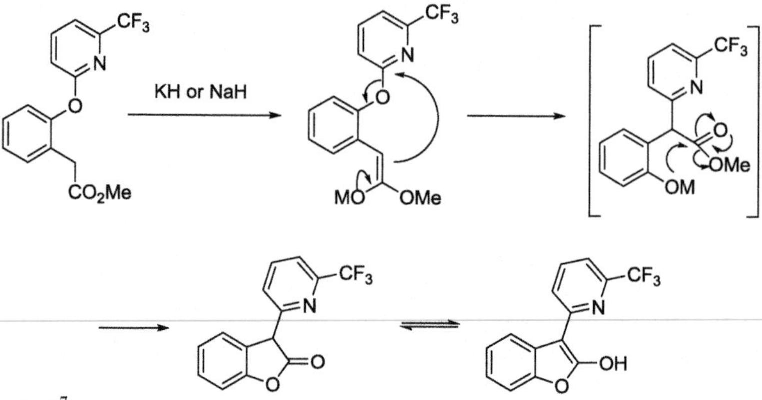

Example 2[7]

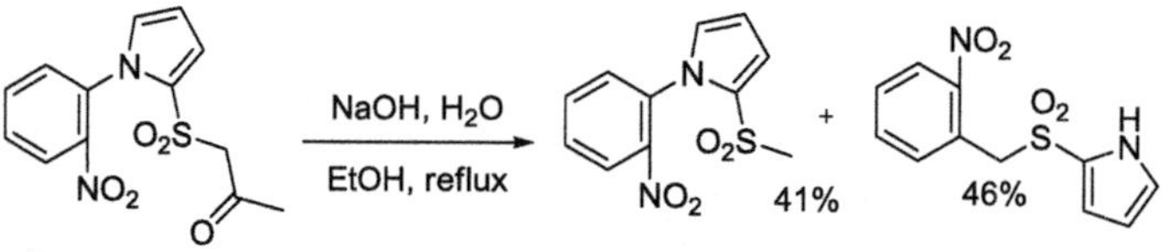

Example 3[8]

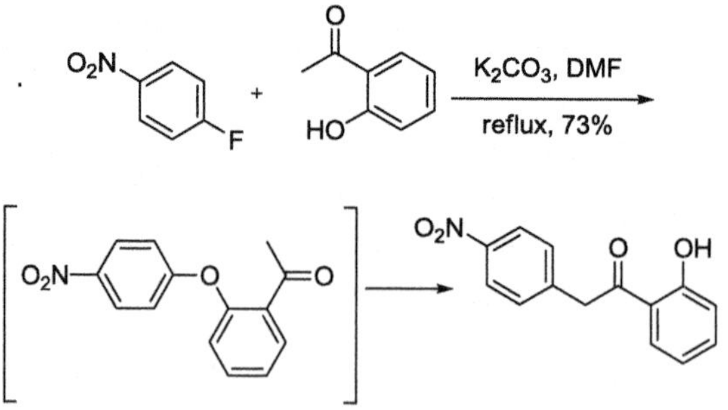

Example 4[10]

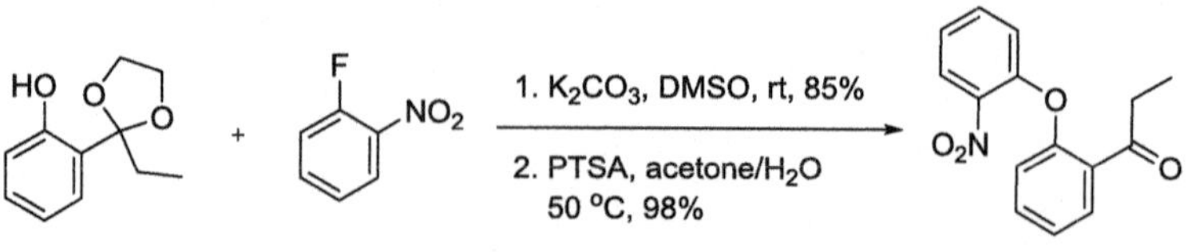

### References

1. Truce, W. E.; Ray, W. J. Jr.; Norman, O. L.; Eickemeyer, D. B. *J. Am. Chem. Soc.* **1958,** *80*, 3625–3629. William E. Truce was a professor at Purdue University.
2. Truce, W. E.; Hampton, D. C. *J. Org. Chem.* **1963,** *28*, 2276–2279.
3. Bayne, D. W; Nicol, A. J.; Tennant, G. *J. Chem. Soc., Chem. Comm.* **1975,** *19*, 782–783.
4. Fukazawa, Y.; Kato, N.; Ito, S.; *Tetrahedron Lett.* **1982,** *23*, 437–438.
5. Hoffman, R. V.; Jankowski, B. C.; Carr, C. S.; Düsler, E. N *J. Org. Chem.* **1986,** *51*, 130–135.
6. Erickson, W. R.; McKennon, M. J. *Tetrahedron Lett.* **2000,** *41*, 4541–4544.
7. Kimbaris, A.; Cobb, J.; Tsakonas, G.; Varvounis, G. *Tetrahedron* **2004,** *60*, 8807–8815.
8. Mitchell, L. H.; Barvian, N. C. *Tetrahedron Lett.* **2004,** *45*, 5669–5672.
9. Snape, T. J. *Chem. Soc. Rev.* **2008,** *37*, 2452–2458. (Review).
10. Snape, T. J. *Synlett* **2008,** 2689–2691.

# Sommelet reaction

Transformation of benzyl halides to the corresponding benzaldehydes with the aid of hexamethylenetetramine (HMTA). *Cf.* Delépine amine synthesis.

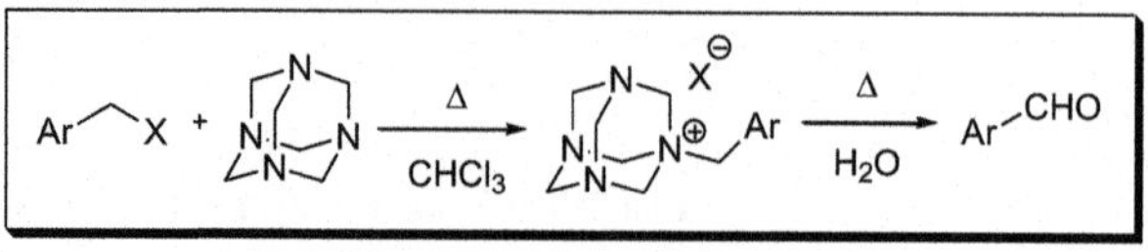
Ar X
Δ
CHCl3
Δ
H2O
Ar CHO

Hexamethylenetetramine (pungent rotten fish smell)

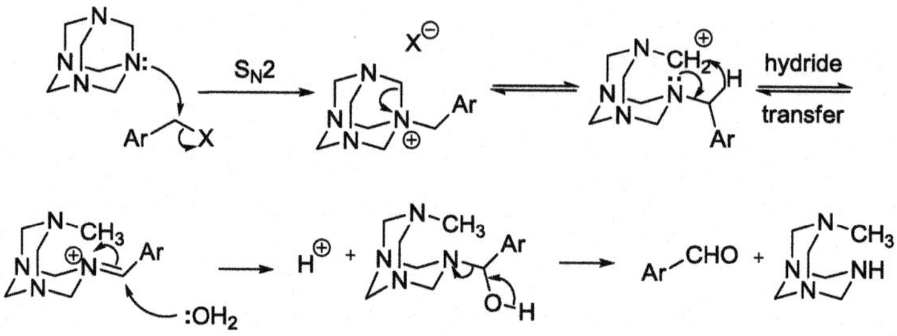
SN2
hydride
transfer
:OH2

hemiaminal

The hydride transfer and the ring-opening of hexamethylenetetramine may occur in a synchronized fashion:

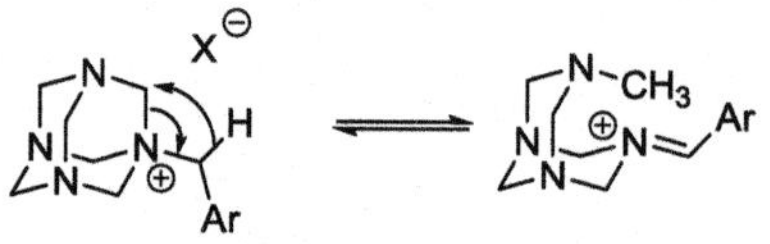

Example 1[3]

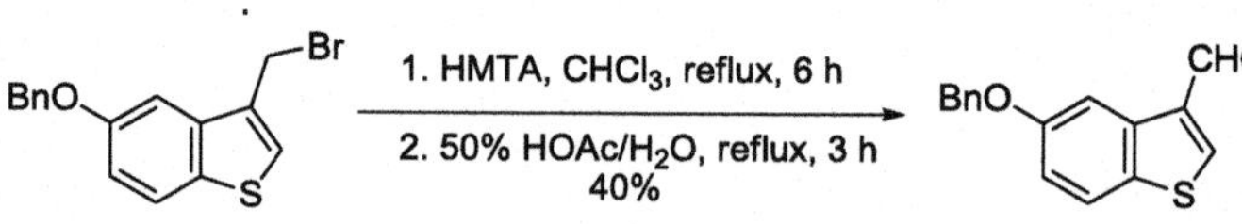
1. HMTA, CHCl3, reflux, 6 h
2. 50% HOAc/H2O, reflux, 3 h
40%

Example 2[4]

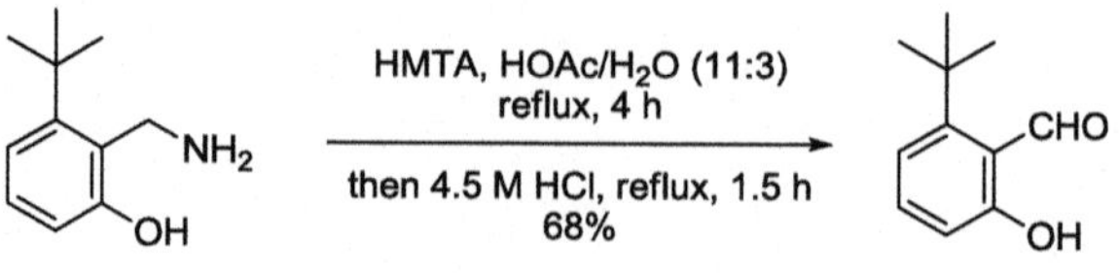
HMTA, HOAc/H2O (11:3)
reflux, 4 h
then 4.5 M HCl, reflux, 1.5 h
68%

J.J. Li, *Name Reactions: A Collection of Detailed Mechanisms and Synthetic Applications*,
DOI 10.1007/978-3-319-03979-4_256, © Springer International Publishing Switzerland 2014

Example 3[7]

Example 4[8]

## References

1.  Sommelet, M. *Compt. Rend.* **1913**, *157*, 852–854. Marcel Sommelet (1877–1952) was born in Langes, France. He received his Ph.D. In 1906 at Paris where he joined the Faculté de Pharmacie after WWI and became the chair of organic chemistry in 1934.
2.  Angyal, S. J. *Org. React.* **1954**, *8,* 197–217. (Review).
3.  Campaigne, E.; Bosin, T.; Neiss, E. S. *J. Med. Chem.* **1967**, *10*, 270–271.
4.  Stokker, G. E.; Schultz, E. M. *Synth. Commun.* **1982**, *12*, 847–853.
5.  Armesto, D.; Horspool, W. M.; Martin, J. A. F.; Perez-Ossorio, R. *Tetrahedron Lett.* **1985**, *26*, 5217–5220.
6.  Kilenyi, S. N., in *Encyclopedia of Reagents of Organic Synthesis*, ed. Paquette, L. A., Wiley: Hoboken, NJ, **1995**, *Vol. 3*, p. 2666. (Review).
7.  Malykhin, E. V.; Shteingart, V. D. *J. Fluorine Chem.* **1998**, *91*, 19–20.
8.  Karamé, I.; Jahjah, M.; Messaoudi, A.; Tommasino, M. L.; Lemaire, M. *Tetrahedron: Asymmetry* **2004**, *15*, 1569–1581.
9.  Göker, H.; Boykin, D. W.; Yildiz, S. *Bioorg. Med. Chem.* **2005**, *13*, 1707–1714.
10. Li, J. J. *Sommelet Reaction*. In *Name Reactions for Functional Group Transformations*; Li, J. J., Ed.; Wiley: Hoboken, NJ, **2007**, pp 689–695. (Review).

# Sommelet–Hauser rearrangement

[2,3]-Wittig rearrangement of benzylic quaternary ammonium salts upon treatment with alkali metal amides *via* the ammonium ylide intermediates.

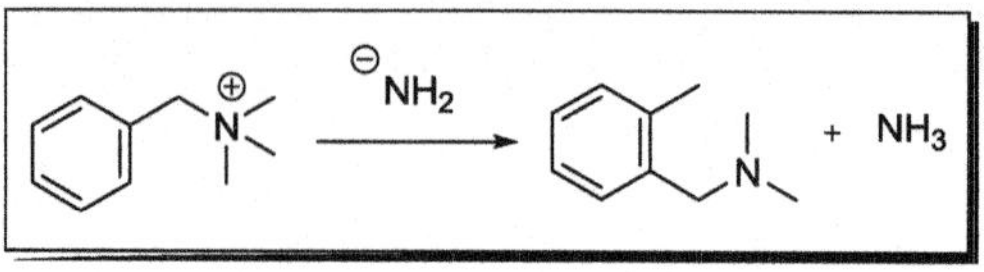

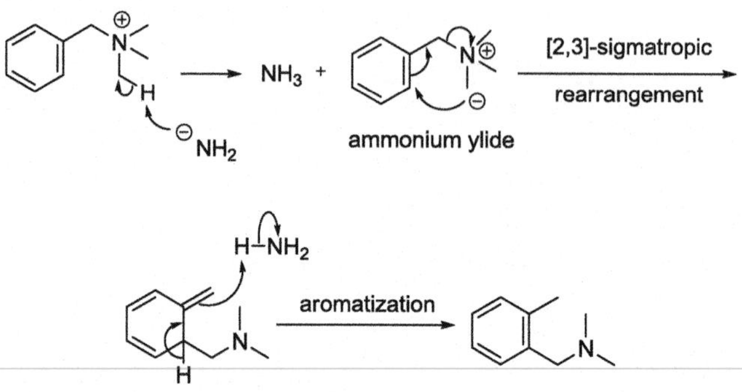

**Example 1[3]**

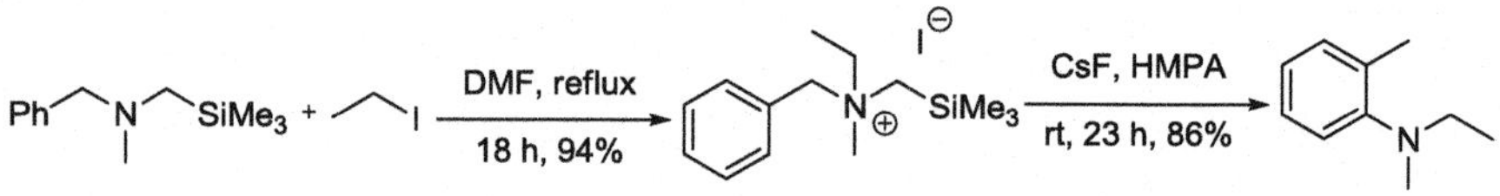

**Example 2[4]**

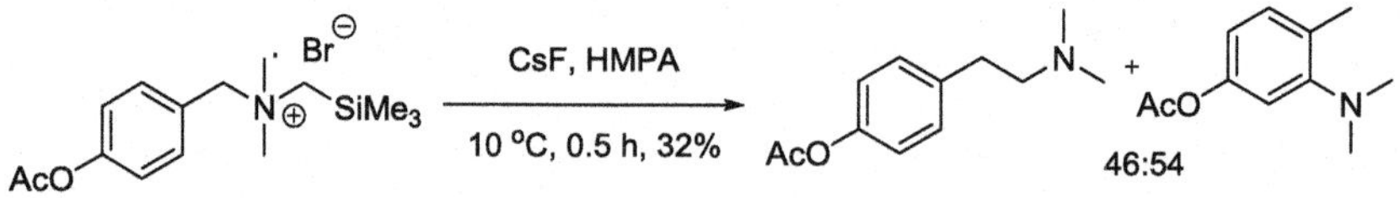

**Example 3[8]**

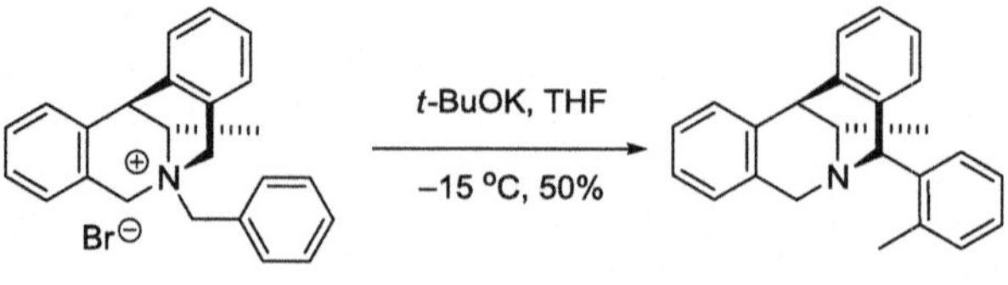

J.J. Li, *Name Reactions: A Collection of Detailed Mechanisms and Synthetic Applications*,
DOI 10.1007/978-3-319-03979-4_257, © Springer International Publishing Switzerland 2014

Example 4[10]

*t*-BuOK, THF

–60 °C, 4 h
57%, > 20:1 *de*
R* = (–)-8-phenylmenthyl

Example 5[12]

BrCN, CH$_2$Cl$_2$

rt, 48 h, 81%

## References

1. (a) Sommelet, M. *Compt. Rend.* **1937**, *205*, 56–58. (b) Kantor, S. W.; Hauser, C. R. *J. Am. Chem. Soc.* **1951**, *73*, 4122–4131. Charles R. Hauser (1900–1970) was a professor at Duke University.
2. Shirai, N.; Sato, Y. *J. Org. Chem.* **1988**, *53,* 194–196.
3. Shirai, N.; Watanabe, Y.; Sato, Y. *J. Org. Chem.* **1990**, *55*, 2767–2770.
4. Tanaka, T.; Shirai, N.; Sugimori, J.; Sato, Y. *J. Org. Chem.* **1992**, *57*, 5034–5036.
5. Klunder, J. M. *J. Heterocycl. Chem.* **1995**, *32*, 1687–1691.
6. Maeda, Y.; Sato, Y. *J. Org. Chem.* **1996**, *61*, 5188–5190.
7. Endo, Y.; Uchida, T.; Shudo, K. *Tetrahedron Lett.* **1997**, *38*, 2113–2116.
8. Hanessian, S.; Talbot, C.; Saravanan, P. *Synthesis* **2006**, 723–734.
9. Liao, M.; Peng, L.; Wang, J. *Org. Lett.* **2008**, *10*, 693–696.
10. Tayama, E.; Orihara, K.; Kimura, H. *Org. Biomol. Chem.* **2008**, *6*, 3673–3680.
11. Zografos, A. L. In *Name Reactions in Heterocyclic Chemistry-II*, Li, J. J., Ed.; Wiley: Hoboken, NJ, 2011, pp 197–206. (Review).
12. Tayama, Eiji; Sato, Ryota; Takedachi, Keisuke; Iwamoto, Hajime; Hasegawa, Eietsu *Tetrahedron* **2012**, *68*, 4710–4718.

## Sonogashira reaction

Pd/Cu-catalyzed cross-coupling of organohalides with terminal alkynes. *Cf.* Cadiot–Chodkiewicz coupling and Castro–Stephens reaction. The Castro–Stephens coupling uses stoichiometric copper, whereas the Sonogashira variant uses catalytic palladium and copper.

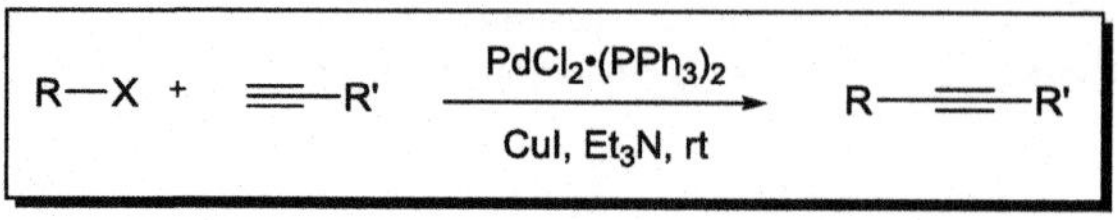

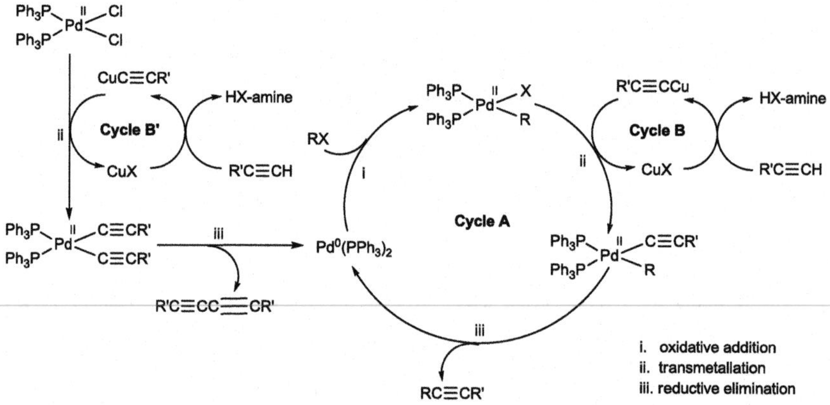

Note that Et$_3$N may reduce Pd(II) to Pd(0) as well, where Et$_3$N is oxidized to the iminium ion at the same time:

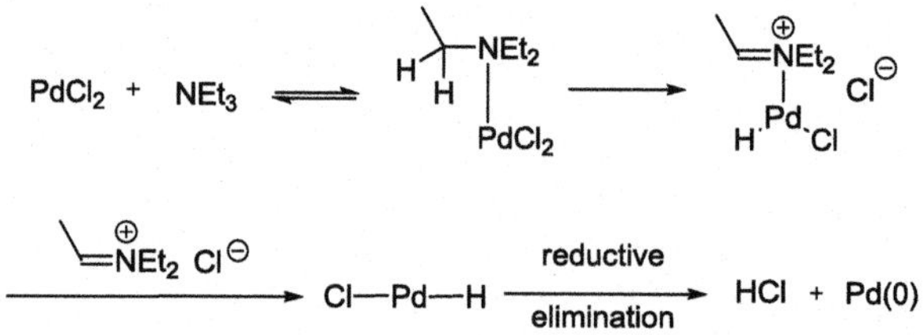

Example 1[2]

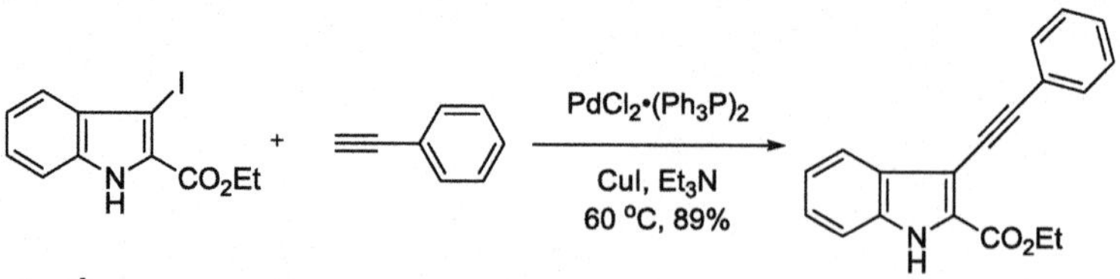

Example 2[3]

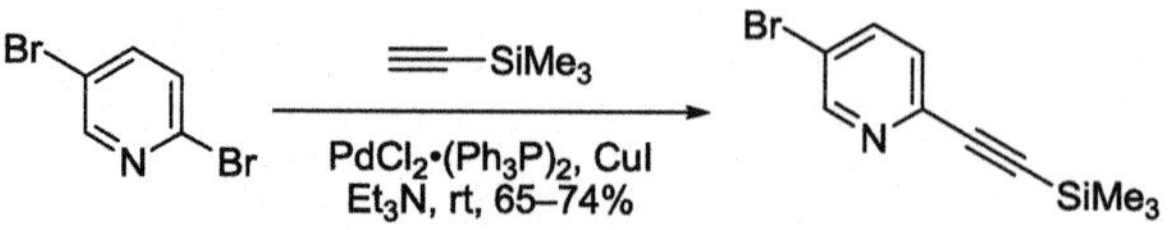

J.J. Li, *Name Reactions: A Collection of Detailed Mechanisms and Synthetic Applications*, DOI 10.1007/978-3-319-03979-4_258, © Springer International Publishing Switzerland 2014

Example 3[8]

Example 4[9]

## References

1.  (a) Sonogashira K.; Tohda, Y.; Hagihara, N. *Tetrahedron Lett.* **1975**, *50*, 4467–4470. Kenkichi Sonogashira was a professor at Fukui University. Richard Heck also discovered the same transformation using palladium but without the use of copper: *J. Organomet. Chem.* **1975**, *93*, 259–263.

2.  Sakamoto, T.; Nagano, T.; Kondo, Y.; Yamanaka, H. *Chem. Pharm. Bull.* **1988**, *36*, 2248–2252.

3.  Ernst, A.; Gobbi, L.; Vasella, A. *Tetrahedron Lett.* **1996**, *37*, 7959–7962.

4.  Hundermark, T.; Littke, A.; Buchwald, S. L.; Fu, G. C. *Org. Lett.* **2000**, *2*, 1729–1731.

5.  Batey, R. A.; Shen, M.; Lough, A. J. *Org. Lett.* **2002**, *4*, 1411–1414.

6.  Sonogashira, K. In *Metal-Catalyzed Cross-Coupling Reactions*; Diederich, F.; de Meijere, A., Eds.; Wiley-VCH: Weinheim, **2004**; *Vol. 1*, 319. (Review).

7.  Lemhadri, M.; Doucet, H.; Santelli, M. *Tetrahedron* **2005**, *61*, 9839–9847.

8.  Li, Y.; Zhang, J.; Wang, W.; Miao, Q.; She, X.; Pan, X. *J. Org. Chem.* **2005**, *70*, 3285–3287.

9.  Komano, K.; Shimamura, S.; Inoue, M.; Hirama, M. *J. Am. Chem. Soc.* **2007**, *129*, 14184–11186.

10. Nakatsuji, H.; Ueno, K.; Misaki, T.; Tanabe, Y. *Org. Lett.* **2008**, *10*, 2131–2134.

11. Gray, D. L. *Sonogashira Reaction*. In *Name Reactions for Homologations-Part II*; Li, J. J., Ed.; Wiley: Hoboken, NJ, **2009**, pp 100–133. (Review).

12. Shigeta, M.; Watanabe, J.; Konishi, G.-i. *Tetrahedron Lett.* **2013**, *54*, 1761–1764.

# Staudinger ketene cycloaddition

Also known as the Staudinger reaction. [2 + 2]-Cycloaddition of ketene and imine to form β-lactams. Other coupling partners for ketenes include: olefin to give cyclobutanone and carbonyl to give β-lactone.

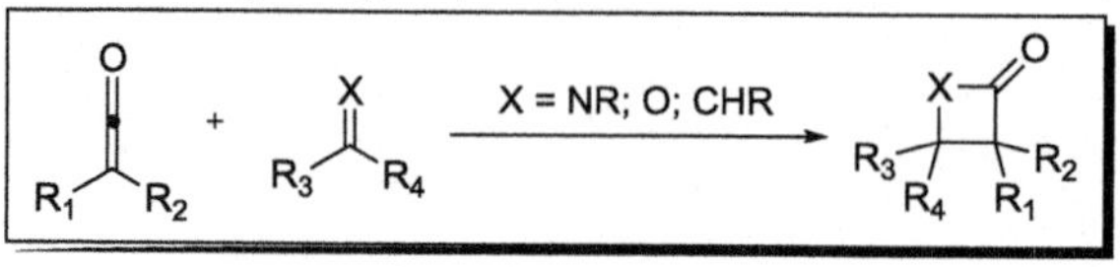

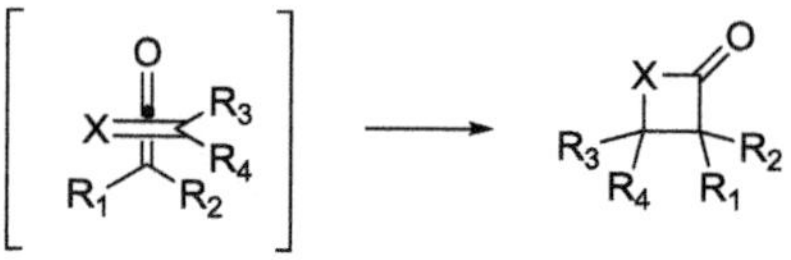

puckered transition state

When X = N:

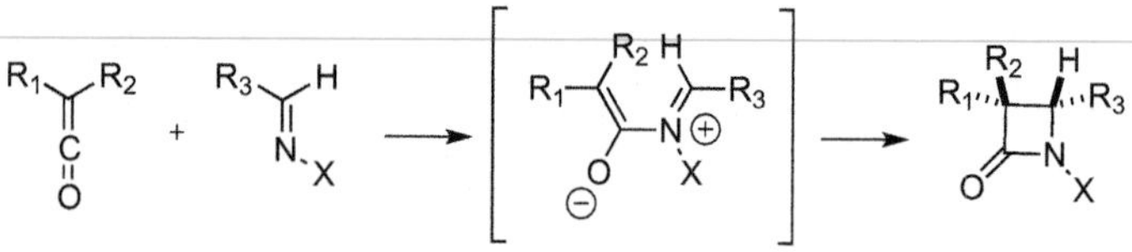

Example 1[6]

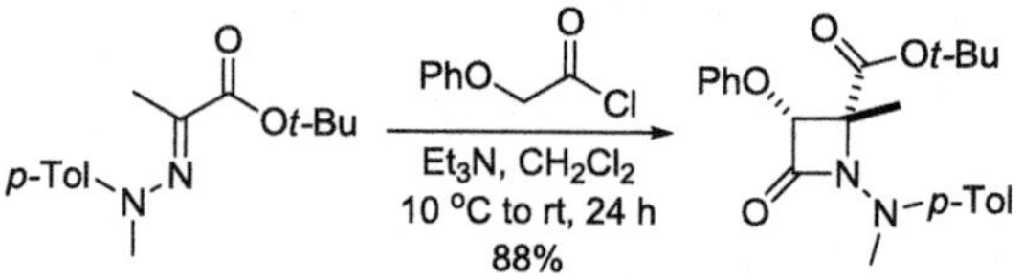

Example 2[7]

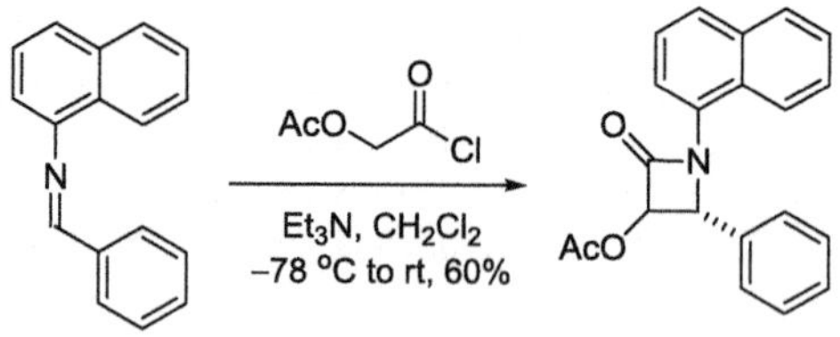

Example 3[9]

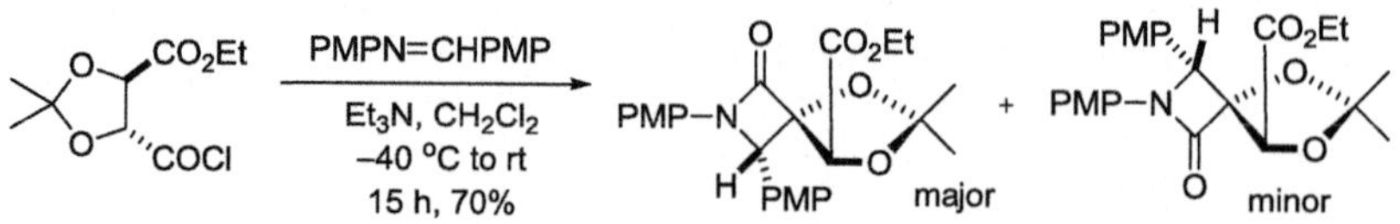

J.J. Li, *Name Reactions: A Collection of Detailed Mechanisms and Synthetic Applications*,
DOI 10.1007/978-3-319-03979-4_259, © Springer International Publishing Switzerland 2014

Example 4[10]

Example 5[11]

## References

1. Staudinger, H. *Ber.* **1907**, *40*, 1145–1146.     Hermann Staudinger (Germany, 1881–1965) won the Nobel Prize in Chemistry in 1953 for his discoveries in the area of macromolecular chemistry.

2. Cooper, R. D. G.; Daugherty, B. W.; Boyd, D. B. *Pure Appl. Chem.* **1987**, *59*, 485–492. (Review).

3. Snider, B. B. *Chem. Rev.* **1988**, *88*, 793–811. (Review).

4. Hyatt, J. A.; Raynolds, P. W. *Org. React.* **1994**, *45*, 159–646. (Review).

5. Orr, R. K.; Calter, M. A. *Tetrahedron* **2003**, *59*, 3545–3565. (Review).

6. Bianchi, L.; Dell'Erba, C.; Maccagno, M.; Mugnoli, A.; Novi, M.; Petrillo, G.; Sancassan, F.; Tavani, C. *Tetrahedron* **2003**, *59*, 10195–10201.

7. Banik, I.; Becker, F. F.; Banik, B. K. *J. Med. Chem.* **2003**, *46*, 12–15.

8. Banik, B. K.; Banik, I.; Becker, F. F. *Bioorg. Med. Chem. Lett.* **2005**, *13*, 3611–3622.

9. Chincholkar, P. M.; Puranik, V. G.; Rakeeb, A.; Deshmukh, A. S. *Synlett* **2007**, *14*, 2242–2246.

10. Cremonesi, G.; Dalla Croce, P.; Fontana, F.; La Rosa, C. *Tetrahedron: Asymmetry* **2008**, *19*, 554–561.

11. Raj, R.; Singh, P.; Haberkern, N. T.; Faucher, R. M.; Patel, N.; Land, K. M.; Kumar, V. *Eur. J. Med. Chem.* **2013**, *63*, 897–906.

12. Tuba, R. *Org. Biomol. Chem.* **2013**, *11*, 5976–5988. (Review).

## Staudinger reduction

Phosphazo compounds (e.g., iminophosphoranes) from the reduction of organic azides using tertiary phosphine (e.g., Ph₃P). Hydrolysis then provides the corresponding amines.

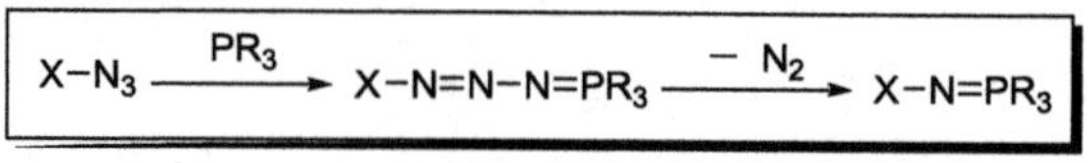

$$X-N_3 \xrightarrow{PR_3} X-N=N-N=PR_3 \xrightarrow{-N_2} X-N=PR_3$$

phosphazide

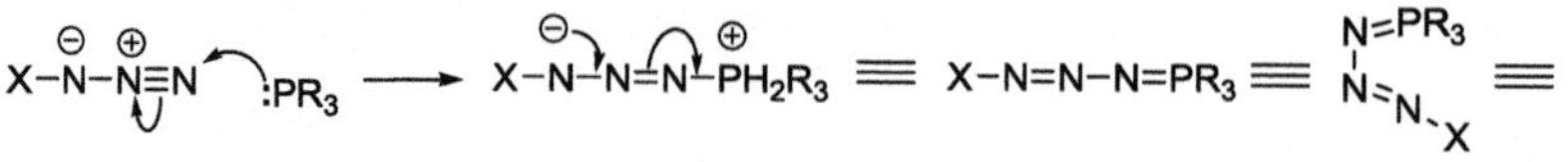

$$X-\overset{\ominus}{N}-\overset{\oplus}{N}\equiv N \quad :PR_3 \longrightarrow X-N\overset{\ominus}{-}N\overset{\oplus}{=}N-PH_2R_3 \equiv X-N=N-N=PR_3 \equiv \overset{N=PR_3}{\underset{N\sim X}{N}}\equiv$$

phosphazide

$$\overset{\oplus}{\underset{N\sim N}{N}}\overset{PR_3}{\underset{X}{\ominus}} \equiv \left[ \overset{N}{\underset{N'\sim N}{N}}\overset{PR_3}{\underset{X}{}} \right]^{\ddagger} \longrightarrow N_2\uparrow + X-N=PR_3 \xrightarrow{H_2O} X-NH_2 + O=PR_3$$

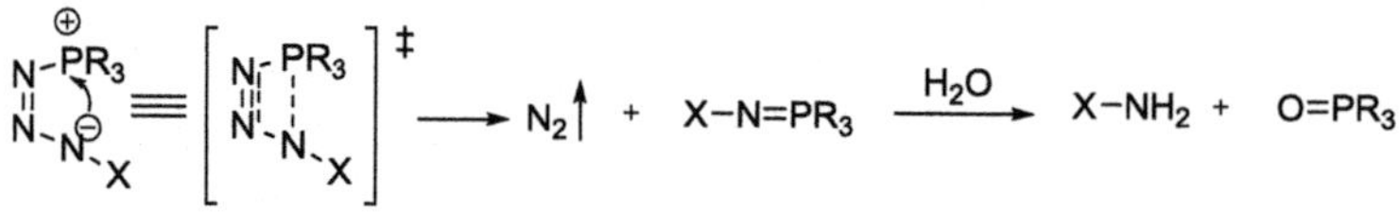

4-membered ring transition state

Example 1[2]

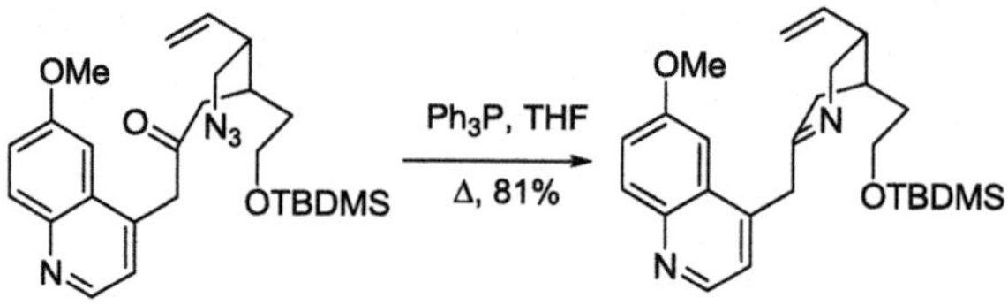

Example 2[3]

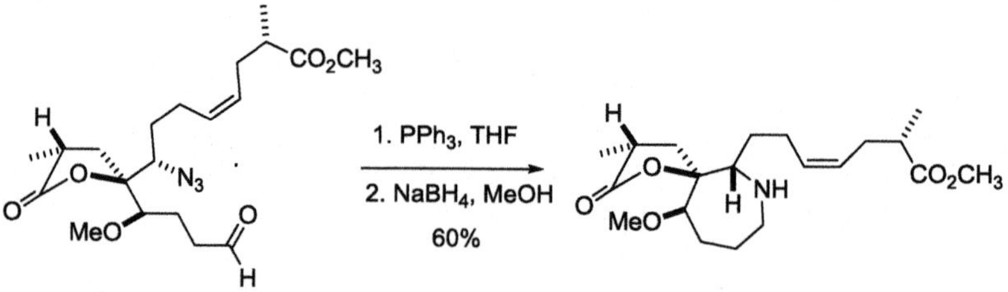

Example 3[4]

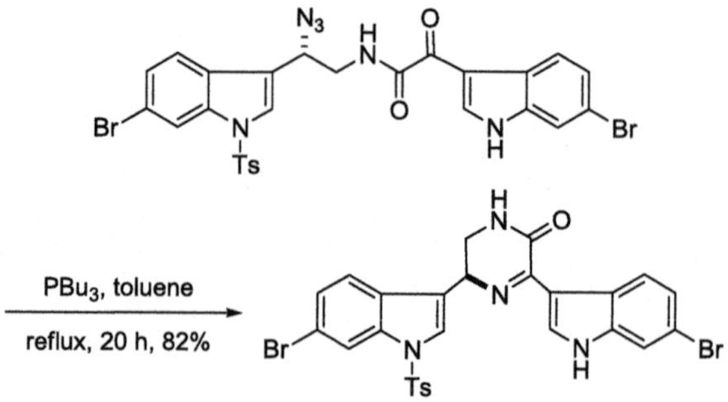

J.J. Li, *Name Reactions: A Collection of Detailed Mechanisms and Synthetic Applications*, DOI 10.1007/978-3-319-03979-4_260, © Springer International Publishing Switzerland 2014

Example 4[8]

Example 5[9]

Example 6, Tandem Staudinger/Aza-Wittig Cyclization[11]

## References

1. Staudinger, H.; Meyer, J. *Helv. Chim. Acta* **1919**, *2*, 635–646.
2. Stork, G.; Niu, D.; Fujimoto, R. A.; Koft, E. R.; Bakovec, J. M.; Tata, J. R.; Dake, G. R. *J. Am. Chem. Soc.* **2001**, *123*, 3239–3242.
3. Williams, D. R.; Fromhold, M. G.; Earley, J. D. *Org. Lett.* **2001**, *3*, 2721–2722.
4. Jiang, B.; Yang, C.-G.; Wang, J. *J. Org. Chem.* **2002**, *67*, 1369–1371.
5. Venturini, A.; Gonzalez, J. *J. Org. Chem.* **2002**, *67*, 9089–9092.
6. Chen, J.; Forsyth, C. J. *Org. Lett.* **2003**, *5*, 1281–1283.
7. Fresneda, P. M.; Castaneda, M.; Sanz, M. A.; Molina, P. *Tetrahedron Lett.* **2004**, *45*, 1655–1657.
8. Li, J.; Chen, H.-N.; Chang, H.; Wang, J.; Chang, C.-W. T. *Org. Lett.* **2005**, *7*, 3061–3064.
9. Takhi, M.; Murugan, C.; Munikumar, M.; Bhaskarreddy, K. M.; Singh, G.; Sreenivas, K.; Sitaramkumar, M.; Selvakumar, N.; Das, J.; Trehan, S.; Iqbal, J. *Bioorg. Med. Chem. Lett.* **2006**, *16*, 2391–2395.
10. Iula, D. M. *Staudinger Reaction.* In *Name Reactions for Functional Group Transformations*; Li, J. J., Ed.; Wiley: Hoboken, NJ, **2007**, pp 129–151. (Review).
11. Kumar, R.; Ermolat'ev, D. S.; Van der Eycken, E. V. *J. Org. Chem.* **2013**, *78*, 5737–5743.

## Stetter reaction

1,4-Dicarbonyl derivatives from aldehydes and α,β-unsaturated ketones and esters. The thiazolium catalyst serves as a safe surrogate for ⁻CN. Also known as the Michael–Stetter reaction. *Cf.* Benzoin condensation.

Example 1, Intramolecular Stetter reaction[2]

J.J. Li, *Name Reactions: A Collection of Detailed Mechanisms and Synthetic Applications*, DOI 10.1007/978-3-319-03979-4_261, © Springer International Publishing Switzerland 2014

Example 2[3]

Example 3[5]

Example 4, Sila-Stetter reaction[9]

## References

1. (a) Stetter, H.; Schreckenberg, H. *Angew. Chem.* **1973**, *85*, 89. Hermann Stetter (1917–1993), born in Bonn, Germany, was a chemist at Technische Hochschule Aachen in West Germany. (b) Stetter, H. *Angew. Chem.* **1976**, *88*, 695–704. (Review). (c) Stetter, H.; Kuhlmann, H.; Haese, W. *Org. Synth.* **1987**, *65*, 26.

2. Trost, B. M.; Shuey, C. D.; DiNinno, F., Jr.; McElvain, S. S. *J. Am. Chem. Soc.* **1979**, *101*, 1284–1285.

3. El-Haji, T.; Martin, J. C.; Descotes, G. *J. Heterocycl. Chem.* **1983**, *20*, 233–235.

4. Harrington, P. E.; Tius, M. A. *Org. Lett.* **1999**, *1*, 649–651.

5. Kikuchi, K.; Hibi, S.; Yoshimura, H.; Tokuhara, N.; Tai, K.; Hida, T.; Yamauchi, T.; Nagai, M. *J. Med. Chem.* **2000**, *43*, 409–419.

6. Kobayashi, N.; Kaku, Y.; Higurashi, K. *Bioorg. Med. Chem. Lett.* **2002**, *12*, 1747–1750.

7. Read de Alaniz, J.; Rovis, T. *J. Am. Chem. Soc.* **2005**, *127*, 6284–6289.

8. Reynolds, N. T.; Rovis, T. *Tetrahedron* **2005**, *61*, 6368–6378.

9. Mattson, A. E.; Bharadwaj, A. R.; Zuhl, A. M.; Scheidt, K. A. *J. Org. Chem.* **2006**, *71*, 5715–5724.

10. Cee, V. J. *Stetter Reaction*. In *Name Reactions for Homologations-Part I*; Li, J. J., Ed.; Wiley: Hoboken, NJ, **2009**, pp 576–587. (Review).

11. Zhang, J.; Xing, C.; Tiwari, B.; Chi, Y. R. *J. Am. Chem. Soc.* **2013**, *135*, 8113–8116.

## Stevens rearrangement

A quaternary ammonium salt containing an electron-withdrawing group Z on one of the carbons attached to the nitrogen is treated with a strong base to give a rearranged tertiary amine.

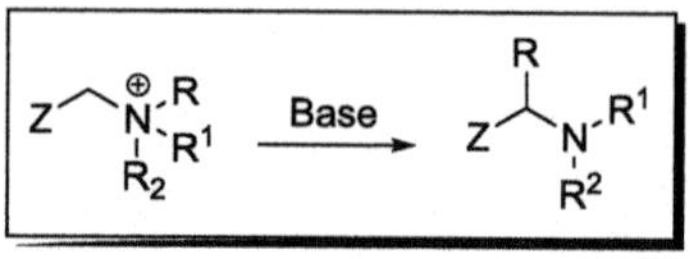

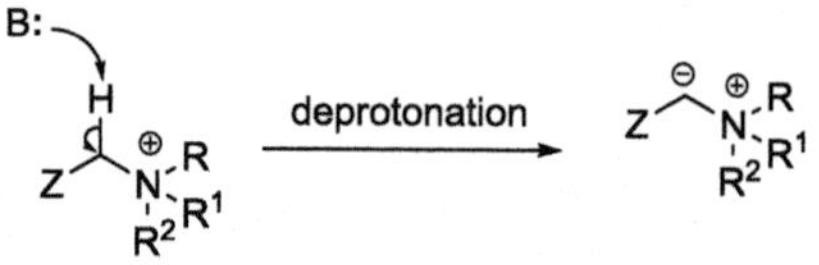

The contemporary radical mechanism:

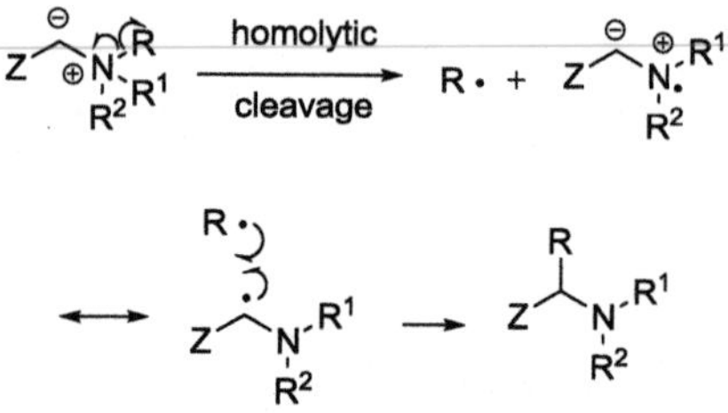

The original ionic mechanism:

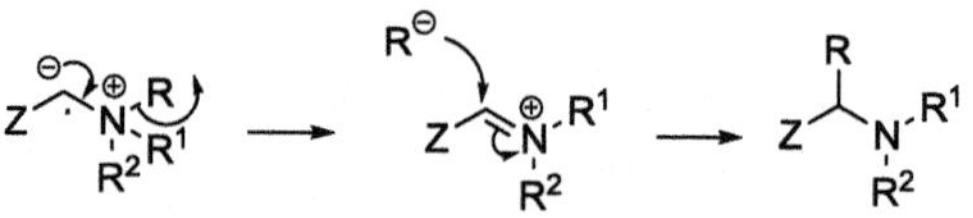

Example 1, Stevens Rearrangement/Reduction Sequence[10]

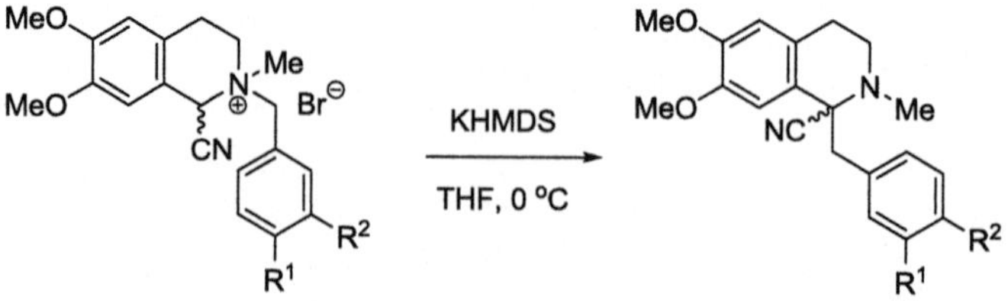

KHMDS

THF, 0 °C

J.J. Li, *Name Reactions: A Collection of Detailed Mechanisms and Synthetic Applications*, DOI 10.1007/978-3-319-03979-4_262, © Springer International Publishing Switzerland 2014

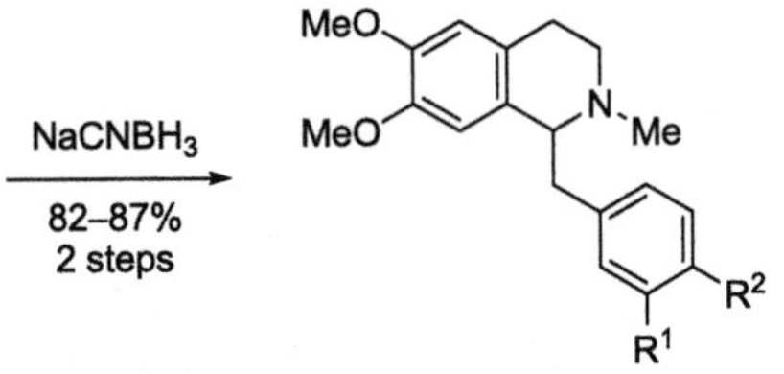

NaCNBH$_3$
→
82–87%
2 steps

## References

1.  Stevens, T. S.; Creighton, E. M.; Gordon, A. B.; MacNicol, M. *J. Chem. Soc.* **1928**, 3193–3197.
2.  Schöllkopf, U.; Ludwig, U.; Ostermann, G.; Patsch, M. *Tetrahedron Lett.* **1969**, *10*, 3415–3418.
3.  Pine, S. H.; Catto, B. A.; Yamagishi, F. G. *J. Org. Chem.* **1970**, *35*, 3663–3665. (Mechanism).
4.  Doyle, M. P.; Ene, D. G.; Forbes, D. C.; Tedrow, J. S. *Tetrahedron Lett.* **1997**, *38*, 4367–4370.
5.  Makita, K.; Koketsu, J.; Ando, F.; Ninomiya, Y.; Koga, N. *J. Am. Chem. Soc.* **1998**, *120*, 5764–5770.
6.  Feldman, K. S.; Wrobleski, M. L. *J. Org. Chem.* **2000**, *65*, 8659-8668.
7.  Kitagaki, S.; Yanamoto, Y.; Tsutsui, H.; Anada, M.; Nakajima, M.; Hashimoto, S. *Tetrahedron Lett.* **2001**, *42*, 6361–6364.
8.  Knapp, S.; Morriello, G. J.; Doss, G. A. *Tetrahedron Lett.* **2002**, *43*, 5797–5800.
9.  Hanessian, S.; Parthasarathy, S.; Mauduit, M.; Payza, K. *J. Med. Chem.* **2003**, *46*, 34–38.
10. Pacheco, J. C. O.; Lahm, G.; Opatz, T. *J. Org. Chem.* **2013**, *78*, 4985–4992.

# Still–Gennari phosphonate reaction

A variant of the Horner–Emmons reaction using bis(trifluoroethyl)phosphonate to give Z-olefins.

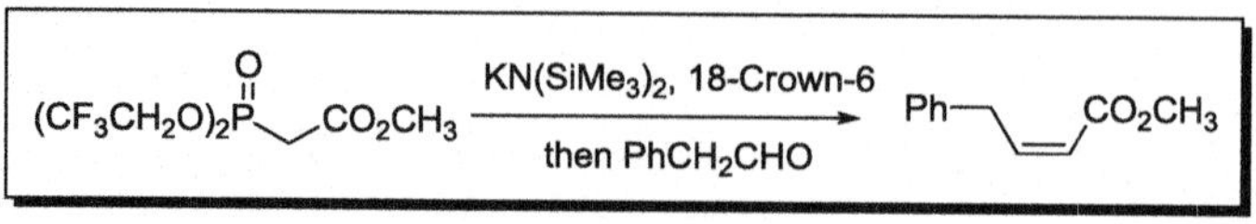

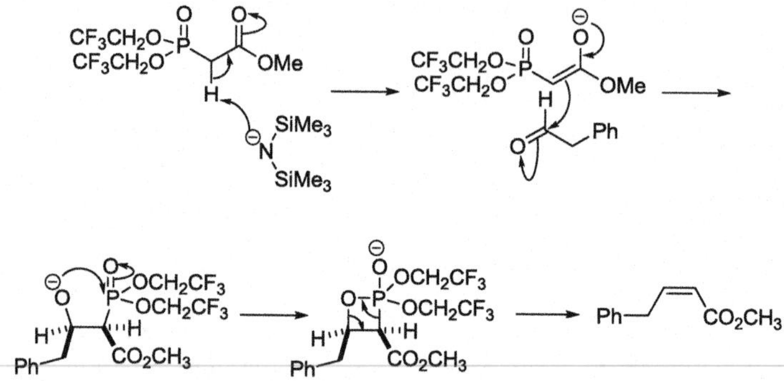

*erythro isomer*, kinetic adduct

Example 1[2]

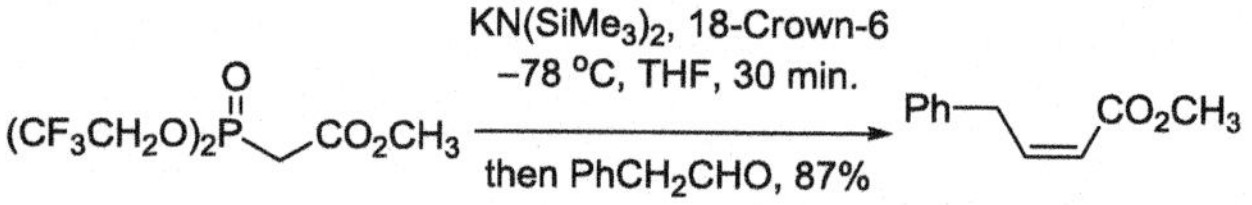

Example 2[3]

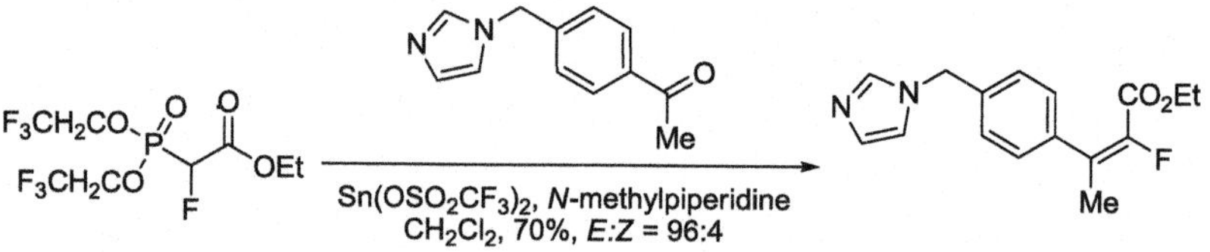

Example 3[4]

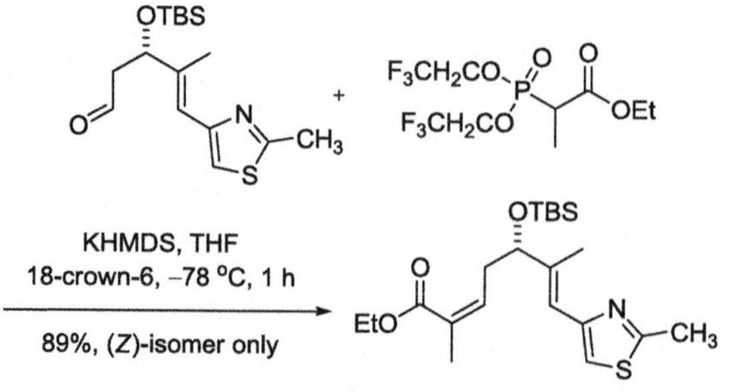

J.J. Li, *Name Reactions: A Collection of Detailed Mechanisms and Synthetic Applications*,
DOI 10.1007/978-3-319-03979-4_263, © Springer International Publishing Switzerland 2014

Example 4[9]

Example 5, An expedient access to Still–Gennari phosphonates[11]

## References

1. Still, W. C.; Gennari, C. *Tetrahedron Lett.* **1983**, *24*, 4405–4408. W. Clark Still (1946–) was born in Augusta, Georgia. He was a professor at Columbia University.
2. Nicolaou, K. C.; Nadin, A.; Leresche, J. E.; LaGreca, S.; Tsuri, T.; Yue, E. W.; Yang, Z. *Chem. Eur. J.* **1995**, *1*, 467–494.
3. Sano, S. Yokoyama, K.; Shiro, M.; Nagao, Y. *Chem. Pharm. Bull.* **2002**, *50*, 706–709.
4. Mulzer, J.; Mantoulidis, A.; Öhler, E. *Tetrahedron Lett.* **1998**, *39*, 8633–8636.
5. Paterson, I.; Florence, G. J.; Gerlach, K.; Scott, J. P.; Sereinig, N. *J. Am. Chem. Soc.* **2001**, *123*, 9535–9544.
6. Mulzer, J.; Ohler, E. *Angew. Chem. Int. Ed.* **2001**, *40*, 3842–3846.
7. Beaudry, C. M.; Trauner, D. *Org. Lett.* **2002**, *4*, 2221–2224.
8. Dakin, L. A.; Langille, N. F.; Panek, J. S. *J. Org. Chem.* **2002**, *67*, 6812–6815.
9. Paterson, I.; Lyothier, I. *J. Org. Chem.* **2005**, *70*, 5494–5507.
10. Rong, F. *Horner–Wadsworth–Emmons reaction*. In *Name Reactions for Homologations-Part I*; Li, J. J., Ed.; Wiley: Hoboken, NJ, **2009**, pp 420–466. (Review).
11. Messik, F.; Oberthür, M. *Synthesis* **2013**, *45*, 167–170.

## Stille coupling

Palladium-catalyzed cross-coupling reaction of organostannanes with organic halides, triflates, *etc.* For the catalytic cycle, see Kumada coupling.

$$R-X \ + \ R^1-Sn(R^2)_3 \ \xrightarrow{Pd(0)} \ R-R^1 \ + \ X-Sn(R^2)_3$$

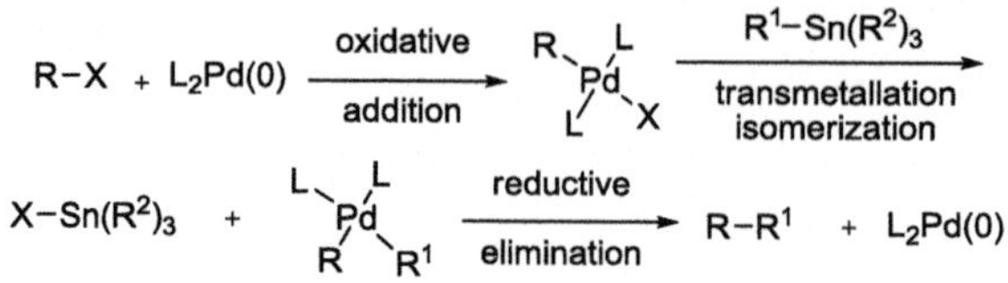

$$R-X \ + \ L_2Pd(0) \ \xrightarrow[\text{addition}]{\text{oxidative}} \ \underset{L}{\overset{R}{\diagdown}}\!Pd\!\underset{X}{\overset{L}{\diagup}} \ \xrightarrow[\substack{\text{transmetallation}\\\text{isomerization}}]{R^1-Sn(R^2)_3}$$

$$X-Sn(R^2)_3 \ + \ \underset{R}{\overset{L}{\diagdown}}\!Pd\!\underset{R^1}{\overset{L}{\diagup}} \ \xrightarrow[\text{elimination}]{\text{reductive}} \ R-R^1 \ + \ L_2Pd(0)$$

Example 1[4]

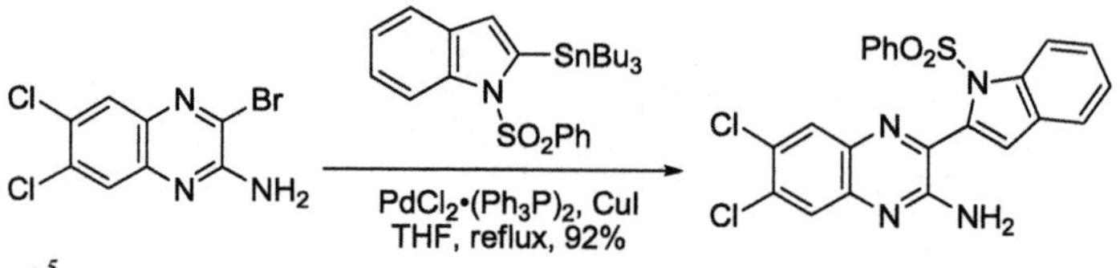

Example 2[5]

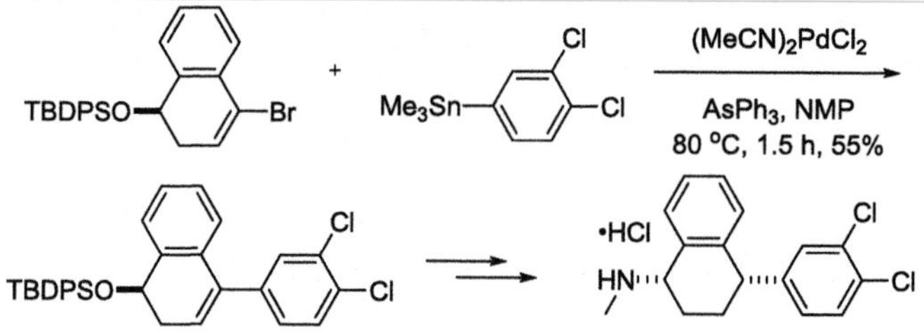

Example 3, π-Allyl Stille coupling[8]

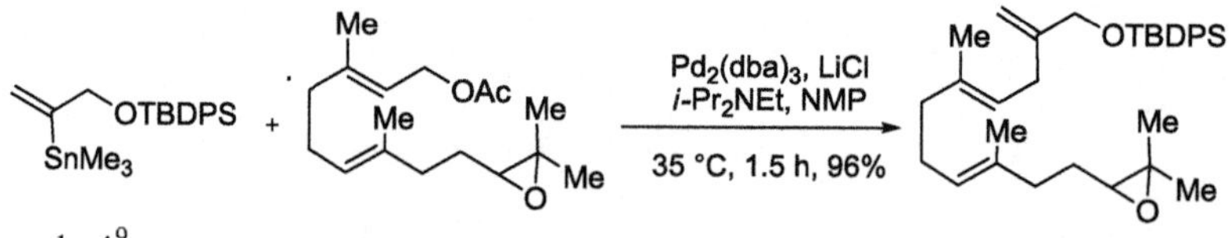

Example 4[9]

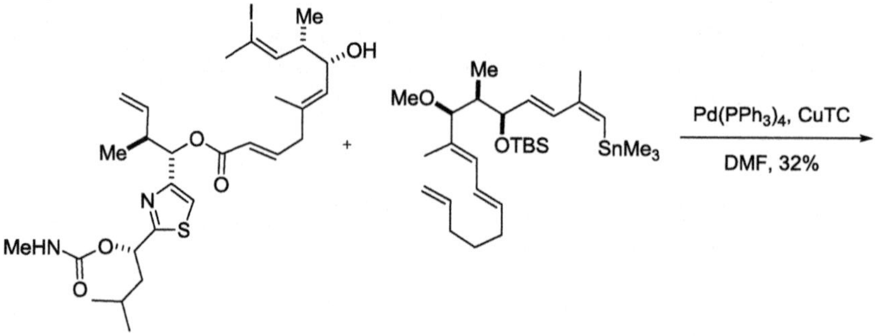

J.J. Li, *Name Reactions: A Collection of Detailed Mechanisms and Synthetic Applications*,
DOI 10.1007/978-3-319-03979-4_264, © Springer International Publishing Switzerland 2014

Example 5[11]

2.5 mol% Pd$_2$(dba)$_3$
10 mol% AsPh$_3$, THF

MW, 150 °C, 23 min.
73%

## References

1. (a) Milstein, D.; Stille, J. K. *J. Am. Chem. Soc.* **1978**, *100*, 3636–3638. John Kenneth Stille (1930–1989) was born in Tucson, Arizona. He developed the reaction bearing his name at Colorado State University. At the height of his career, Stille unfortunately died of an airplane accident returning from an ACS meeting. (b) Milstein, D.; Stille, J. K. *J. Am. Chem. Soc.* **1979**, *101*, 4992–4998. (c) Stille, J. K. *Angew. Chem. Int. Ed.* **1986**, *25*, 508–524.
2. Farina, V.; Krishnamurphy, V.; Scott, W. J. *Org. React.* **1997**, *50*, 1–652. (Review).
3. Duncton, M. A. J.; Pattenden, G. *J. Chem. Soc., Perkin Trans. 1* **1999**, 1235–1249. (Review on the intramolecular Stille reaction).
4. Li, J. J.; Yue, W. S. *Tetrahedron Lett.* **1999**, *40*, 4507–4510.
5. Lautens, M.; Rovis, T. *Tetrahedron*, **1999**, *55*, 8967–8976.
6. Mitchell, T. N. *Organotin Reagents in Cross-Coupling Reactions.* In *Metal-Catalyzed Cross-Coupling Reactions* (2nd edn.) De Meijere, A.; Diederich, F. eds., **2004**, 1, 125–161. Wiley-VCH: Weinheim, Germany. (Review).
7. Schröter, S.; Stock, C.; Bach, T. *Tetrahedron* **2005**, *61*, 2245–2267. (Review).
8. Snyder, S. A.; Corey, E. J. *J. Am. Chem. Soc.* **2006**, *128*, 740–742.
9. Roethle, P. A.; Chen, I. T.; Trauner, D. *J. Am. Chem. Soc.* **2007**, *129*, 8960–8961.
10. Mascitti, V. *Stille Coupling.* In *Name Reactions for Homologations-Part I*; Li, J. J., Ed.; Wiley: Hoboken, NJ, **2009**, pp 133–162. (Review).
11. Chandrasoma, N.; Brown, N.; Brassfield, A.; Nerurkar, A.; Suarez, S.; Buszek, K. R. *Tetrahedron Lett.* **2013**, *54*, 913–917.

## Stille–Kelly reaction

Palladium-catalyzed intramolecular cross-coupling reaction of bis-aryl halides using ditin reagents.

Example 1[6]

## References

1.  Kelly, T. R.; Li, Q.; Bhushan, V. *Tetrahedron Lett.* **1990**, *31*, 161–164.  T. Ross Kelly is a professor at Boston College.
2.  Grigg, R.; Teasdale, A.; Sridharan, V. *Tetrahedron Lett.* **1991**, *32*, 3859–3862.
3.  Iyoda, M.; Miura, M.; Sasaki, S.; Kabir, S. M. H.; Kuwatani, Y.; Yoshida, M. *Heterocycles* **1997**, *38*, 4581–4582.
4.  Fukuyama, Y.; Yaso, H.; Nakamura, K.; Kodama, M. *Tetrahedron Lett.* **1999**, *40*, 105–108.
5.  Iwaki, T.; Yasuhara, A.; Sakamoto, T. *J. Chem. Soc., Perkin Trans. 1* **1999**, 1505–1510.
6.  Yue, W. S.; Li, J. J. *Org. Lett.* **2002**, *4*, 2201–2203.
7.  Olivera, R.; SanMartin, R.; Tellitu, I.; Dominguez, E. *Tetrahedron* **2002**, *58*, 3021–3037.
8.  Mascitti, V. *Stille Coupling.* In *Name Reactions for Homologations-Part I*; Li, J. J., Ed.; Wiley: Hoboken, NJ, **2009**, pp 133–162. (Review).

J.J. Li, *Name Reactions: A Collection of Detailed Mechanisms and Synthetic Applications*,
DOI 10.1007/978-3-319-03979-4_265, © Springer International Publishing Switzerland 2014

# Stobbe condensation

Condensation of diethyl succinate and its derivatives with carbonyl compounds in the presence of bases.

Example 1, Stobbe condensation and cyclization[5]

Example 2[6]

Example 3, Cyclization of the Stobbe product[7]

J.J. Li, *Name Reactions: A Collection of Detailed Mechanisms and Synthetic Applications*,
DOI 10.1007/978-3-319-03979-4_266, © Springer International Publishing Switzerland 2014

Example 4, Two sequential Stobbe condensations[9]

Example 5[11]

## References

1.  Stobbe, H. *Ber.* **1893**, *26*, 2312. Hans Stobbe (1860–1938) was born in Tiehenhof, Germany. He earned his Ph.D. In 1889 at the University of Leipzig where he became a professor in 1894.
2.  Zerrer, R.; Simchen, G. *Synthesis* **1992**, 922–924.
3.  Yvon, B. L.; Datta, P. K.; Le, T. N.; Charlton, J. L. *Synthesis* **2001**, 1556–1560.
4.  Liu, J.; Brooks, N. R. *Org. Lett.* **2002**, *4*, 3521–3524.
5.  Giles, R. G. F.; Green, I. R.; van Eeden, N. *Eur. J. Org. Chem.* **2004**, 4416–4423.
6.  Mahajan, V. A.; Shinde, P. D.; Borate, H. B.; Wakharkar, R. D. *Tetrahedron Lett.* **2005**, *46*, 1009–1012.
7.  Sato, A.; Scott, A.; Asao, T.; Lee, M. *J. Org. Chem.* **2006**, *71*, 4692–4695.
8.  Kapferer, T.; Brückner, R. *Eur. J. Org. Chem.* **2006**, 2119–2133.
9.  Mizufune, H.; Nakamura, M.; Mitsudera, H. *Tetrahedron* **2006**, *62*, 8539–8549.
10. Lowell, A. N.; Fennie, M. W.; Kozlowski, M. C. *J. Org. Chem.* **2008**, *73*, 1911–1918.
11. Webel, M.; Palmer, A. M.; Scheufler, C.; Haag, D.; Muller, B. *Org. Process Res. Dev.* **2010**, *14*, 142–151.
12. Kodet, J. G.; Wiemer, D. F. *J. Org. Chem.* **2013**, *78*, 9291–9302.

## Stork–Danheiser transposition

Treatment of alkoxy-enone (vinylogous ester) with an organometallic (Grignard reagent or organolithium) was followed by treatment with acid to afford another enone where the ketone locates at the enole ther position of the starting material.

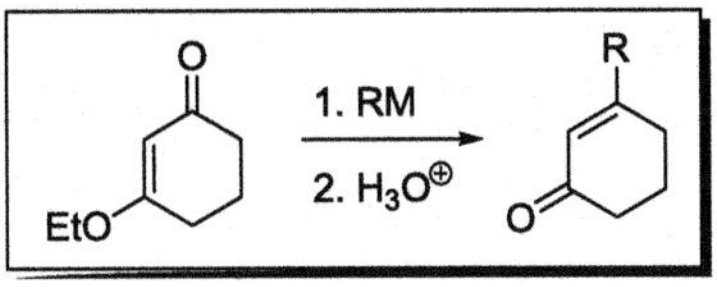

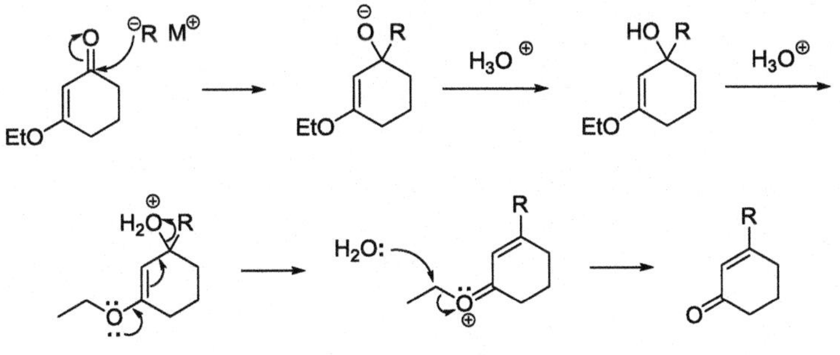

Example 1[2]

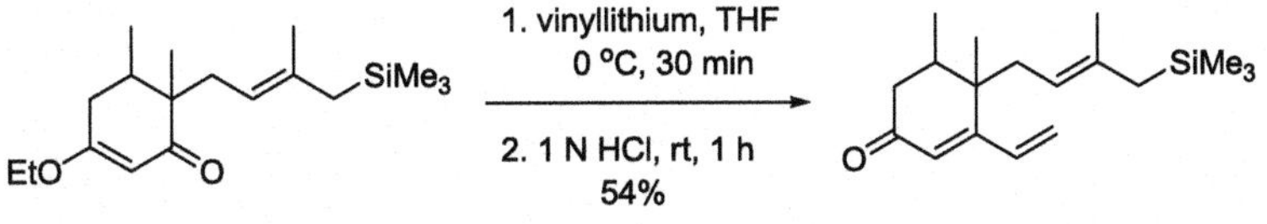

Example 2[6]

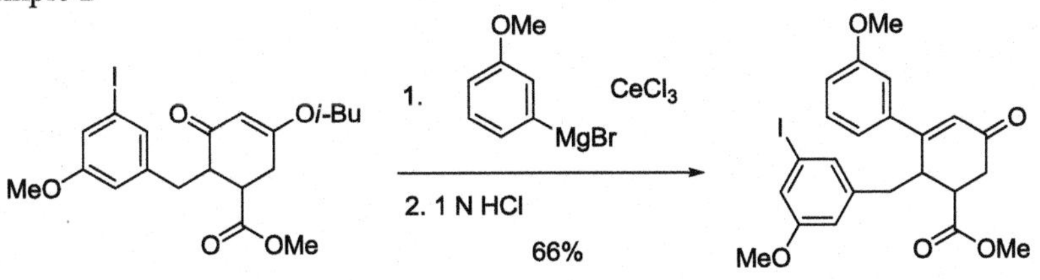

Example 3[7]

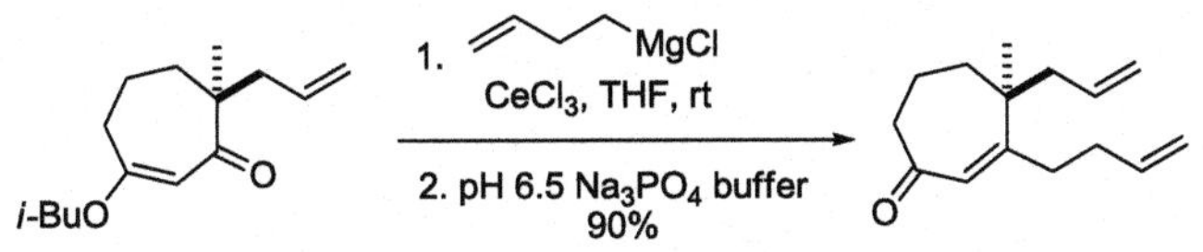

J.J. Li, *Name Reactions: A Collection of Detailed Mechanisms and Synthetic Applications*, DOI 10.1007/978-3-319-03979-4_267, © Springer International Publishing Switzerland 2014

Example 4[9]

## References

1. Stork, G.; Danheiser, R. L. *J. Org. Chem.* **1973,** *38,* 1775.
2. Majetich, G.; Behnke, M.; Hull, K. *J. Org. Chem.* **1985,** *50,* 3615–3618.
3. Kende, A. S.; Fludzinski, P. *Org. Synth.* **1986,** *64.*
4. Liepa, A. J.; Wilkie, J. S.; Winkler, D. A.; Winzenberg, K. N. *Aust. J. Chem.* **1992,** *45,* 759–767.
5. For asymmetric Stork–Danheiser alkylation, see Dudley, G. B.; Takaki, K. S.; Cha, D. D.; Danheiser, R. L. *Org. Lett.* **2000,** *21,* 3407–3410.
6. Grundl, M. A.; Trauner, D. *Org. Lett.* **2006,** *8,* 23–25.
7. Bennett, N. B.; Hong, A. Y.; Harned, A. M.; Stoltz, B. M. *Org. Biomol. Chem.* **2012,** *10,* 56–59.
8. Majetich, G.; Grove, J. L. *Heterocycles* **2012,** *84,* 963–982.
9. Kakde, B. N.; Bhunia, S.; Bisai, A. *Tetrahedron Lett.* **2013,** *54,* 1436–1439.

# Strecker amino acid synthesis

Sodium cyanide-promoted condensation of aldehyde, or ketone, with an amine to afford α-amino nitrile, which may be hydrolyzed to an α-amino acid.

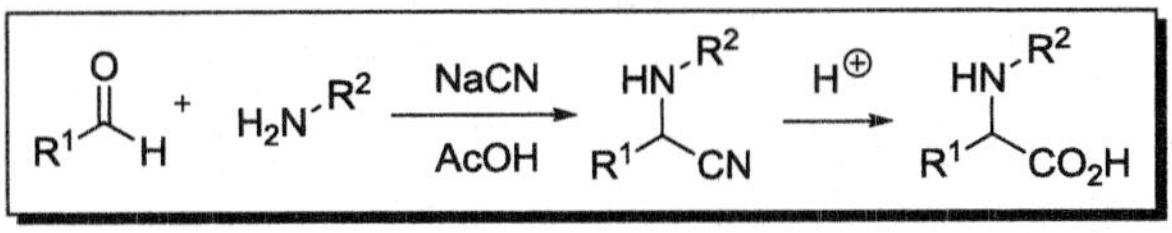

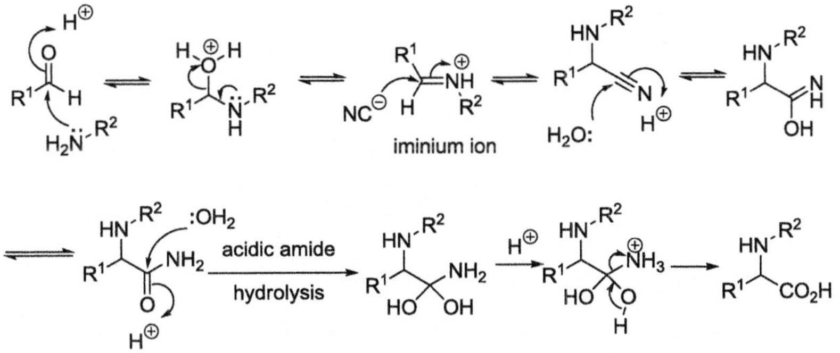

Example 1, Soluble cyanide source[2]

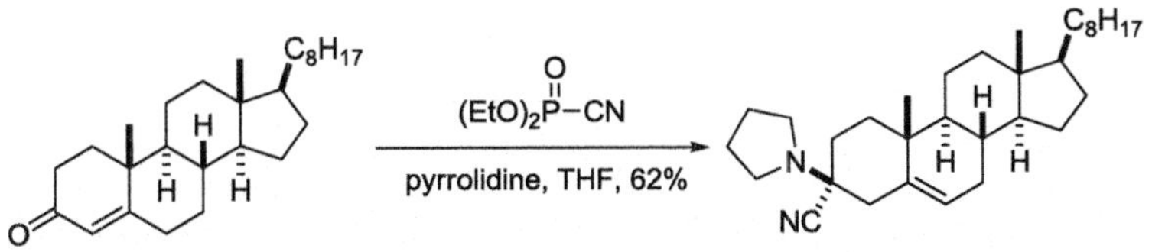

Example 2[3]

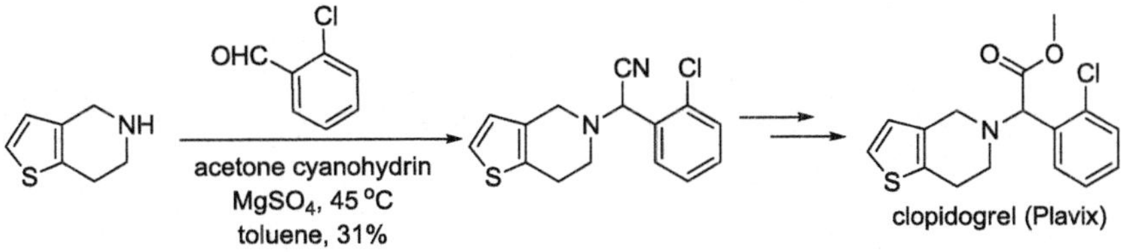

clopidogrel (Plavix)

Example 3[8]

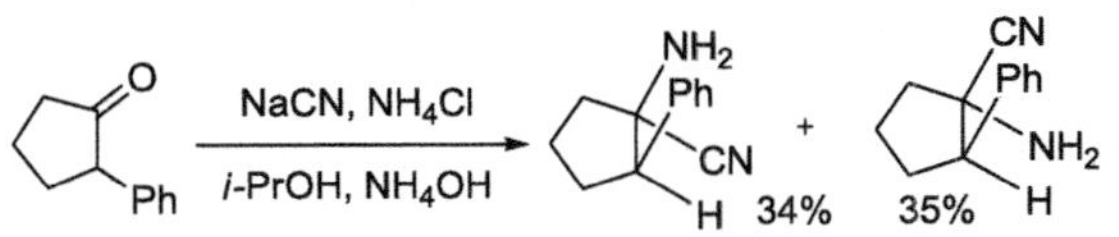

J.J. Li, *Name Reactions: A Collection of Detailed Mechanisms and Synthetic Applications*,
DOI 10.1007/978-3-319-03979-4_268, © Springer International Publishing Switzerland 2014

Example 4[9]

Example 5, Asymmetric Strecker-Type Reaction of Nitrones[11]

## References

1.  Strecker, A. *Ann.* **1850**, *75*, 27–45. Adolph Strecker devised this reaction over 160 years ago. In his paper he described: "The larger crystals of alanine are mother-of-pearl-shiny, hard and crunch between the teeth."
2.  Harusawa, S.; Hamada, Y.; Shioiri, T. *Tetrahedron Lett.* **1979**, *20*, 4663–4666.
3.  Burgos, A.; Herbert, J. M.; Simpson, I. *J. Labelled. Compd. Radiopharm.* **2000**, *43*, 891–898.
4.  Ishitani, H.; Komiyama, S.; Hasegawa, Y.; Kobayashi, S. *J. Am. Chem. Soc.* **2000**, *122*, 762–766.
5.  Yet, L. *Recent Developments in Catalytic Asymmetric Strecker-Type Reactions,* in *Organic Synthesis Highlights V,* Schmalz, H.-G.; Wirth, T. eds.; Wiley–VCH: Weinheim, Germany, **2003**, pp 187–193. (Review).
6.  Meyer, U.; Breitling, E.; Bisel, P.; Frahm, A. W. *Tetrahedron: Asymmetry* **2004**, *15*, 2029–2037.
7.  Huang, J.; Corey, E. J. *Org. Lett.* **2004**, *6*, 5027–5029.
8.  Cativiela, C.; Lasa, M.; Lopez, P. *Tetrahedron: Asymmetry* **2005**, *16*, 2613–2523.
9.  Wrobleski, M. L.; Reichard, G. A.; Paliwal, S.; Shah, S.; Tsui, H.-C.; Duffy, R. A.; Lachowicz, J. E.; Morgan, C. A.; Varty, G. B.; Shih, N.-Y. *Bioorg. Med. Chem. Lett.* **2006**, *16*, 3859–3863.
10. Galatsis, P. *Strecker Amino Acid Synthesis.* In *Name Reactions for Functional Group Transformations*; Li, J. J., Ed.; Wiley: Hoboken, NJ, **2007**, pp 477–499. (Review).
11. Belokon, Y. N.; Hunt, J.; North, M. *Tetrahedron: Asymmetry* **2008**, *19*, 2804–2815.
12. Sakai, T.; Soeta, T.; Endo, K.; Fujinami, S.; Ukaji, Y. *Org. Lett.* **2013**, *15*, 2422–2425.

# Suzuki–Miyaura coupling

Palladium-catalyzed cross-coupling reaction of organoboranes with organic halides, triflates, *etc*. In the presence of a base (transmetallation is reluctant to occur without the activating effect of a base). For the catalytic cycle, see Kumada coupling.

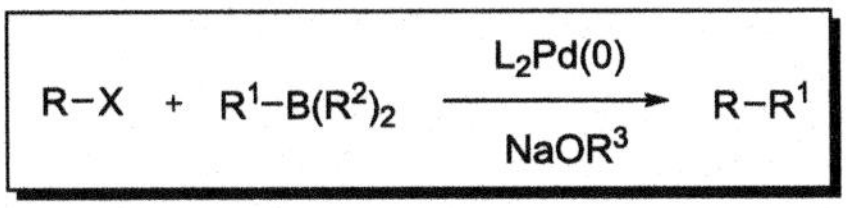

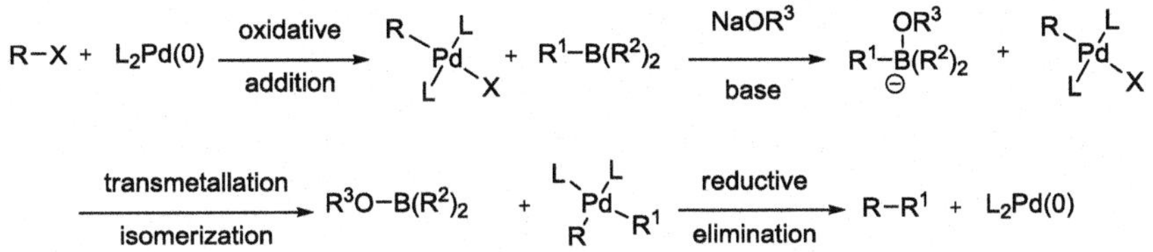

Example 1[2]

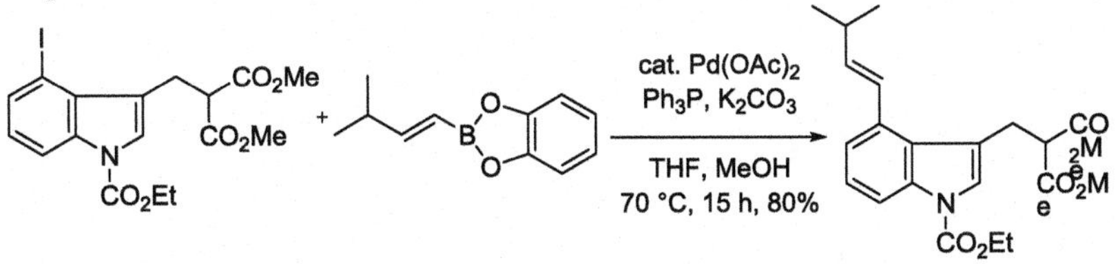

Example 2[4]

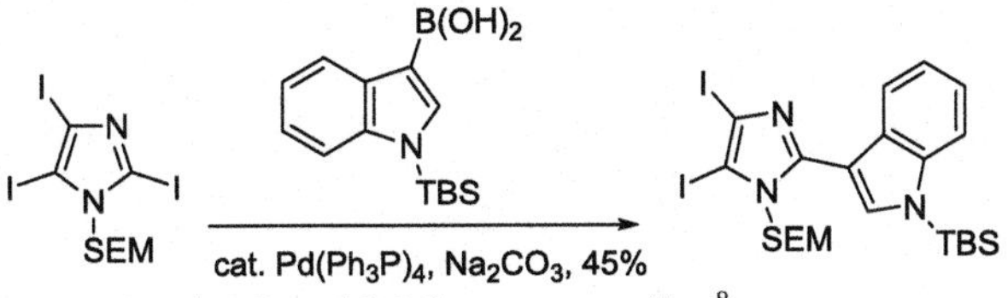

Example 3, Intramolecular Suzuki–Miyaura coupling[8]

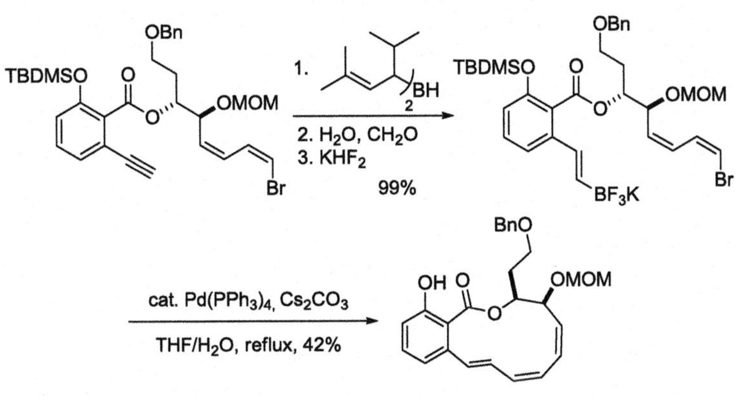

J.J. Li, *Name Reactions: A Collection of Detailed Mechanisms and Synthetic Applications*, DOI 10.1007/978-3-319-03979-4_269, © Springer International Publishing Switzerland 2014

Example 4[9]

Example 5, Nickel-catalyzed Suzuki-Miyaura coupling in green solvents[12]

## References

1.    (a) Miyaura, N.; Yamada, K.; Suzuki, A. *Tetrahedron Lett.* **1979**, *36*, 3437–3440.  (b) Miyaura, N.; Suzuki, A. *Chem. Commun.* **1979**, 866–867.   Akira Suzuki won Nobel Prize in 2010 along with Richard F. Heck and Ei-ichi Negishi "for palladium-catalyzed cross couplings in organic synthesis".
2.    Tidwell, J. H.; Peat, A. J.; Buchwald, S. L. *J. Org. Chem.* **1994**, *59*, 7164–7168.
3.    Miyaura, N.; Suzuki, A. *Chem. Rev.* **1995**, *95*, 2457–2483. (Review).
4.    (a) Kawasaki, I.; Katsuma, H.; Nakayama, Y.; Yamashita, M.; Ohta, S. *Heterocycles* **1998**, *48*, 1887–1901.   (b) Kawaski, I.; Yamashita, M.; Ohta, S. *Chem. Pharm. Bull.* **1996**, *44*, 1831–1839.
5.    Suzuki, A. In *Metal-catalyzed Cross-coupling Reactions*; Diederich, F.; Stang, P. J., Eds.; Wiley–VCH: Weinhein, Germany, **1998**, 49–97. (Review).
6.    Stanforth, S. P. *Tetrahedron* **1998**, *54*, 263–303. (Review).
7.    Zapf, A. *Coupling of Aryl and Alkyl Halides with Organoboron Reagents (Suzuki Reaction).* In *Transition Metals for Organic Synthesis* (2nd edn.); Beller, M.; Bolm, C. eds., **2004**, 1, 211–229. Wiley–VCH: Weinheim, Germany. (Review).
8.    Molander, G. A.; Dehmel, F. *J. Am. Chem. Soc.* **2004**, *126*, 10313–10318.
9.    Coleman, R. S.; Lu, X.; Modolo, I. *J. Am. Chem. Soc.* **2007**, *129*, 3826–3827.
10.   Wolfe, J. P.; Nakhla, J. S. *Suzuki Coupling.* In *Name Reactions for Homologations-Part I*; Li, J. J., Ed.; Wiley: Hoboken, NJ, **2009**, pp 163–184. (Review).
11.   Weimar, M.; Fuchter, M. J. *Org. Biomol. Chem.* **2013**, *11*, 31–34.
12.   Ramgren, S.; Hie, L.; Ye, Y.; Garg, N. K. *Org. Lett.* **2013**, *15*, 3950–3953.

# Swern oxidation

Oxidation of alcohols to the corresponding carbonyl compounds using $(COCl)_2$, DMSO, and quenching with $Et_3N$.

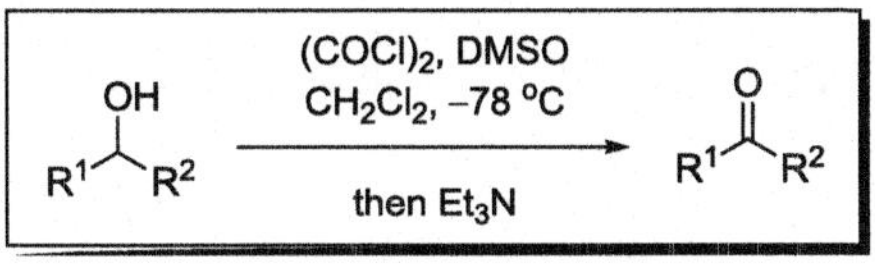

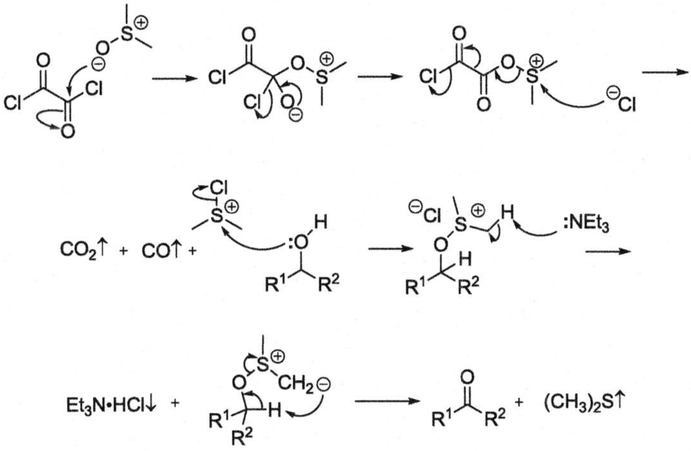

sulfur ylide

Example 1[2]

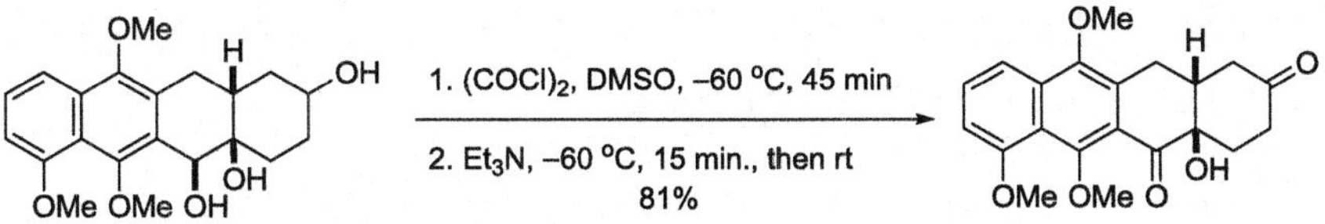

1. $(COCl)_2$, DMSO, $-60\ ^\circ C$, 45 min

2. $Et_3N$, $-60\ ^\circ C$, 15 min., then rt
81%

Example 2[3]

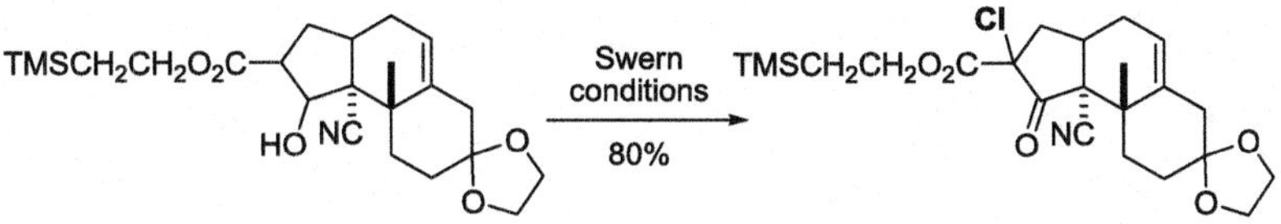

Swern conditions
80%

J.J. Li, *Name Reactions: A Collection of Detailed Mechanisms and Synthetic Applications*, DOI 10.1007/978-3-319-03979-4_270, © Springer International Publishing Switzerland 2014

Example 3[5]

Example 4[7]

## References

1.  (a) Huang, S. L.; Omura, K.; Swern, D. *J. Org. Chem.* **1976**, *41*, 3329–3331. (b) Huang, S. L.; Omura, K.; Swern, D. *Synthesis* **1978**, *4*, 297–299. (c) Mancuso, A. J.; Huang, S. L.; Swern, D. *J. Org. Chem.* **1978**, *43*, 2480–2482. Daniel Swern was a professor at Temple University.
2.  Ghera, E.; Ben-David, Y. *J. Org. Chem.* **1988**, *53*, 2972–2979.
3.  Smith, A. B., III; Leenay, T. L.; Liu, H. J.; Nelson, L. A. K.; Ball, R. G. *Tetrahedron Lett.* **1988**, *29*, 49–52.
4.  Tidwell, T. T. *Org. React.* **1990**, *39*, 297–572. (Review).
5.  Chadka, N. K.; Batcho, A. D.; Tang P. C.; Courtney, L. F.; Cook C. M.; Wovliulich, P. M.; Usković, M. R. *J. Org. Chem.* **1991**, *56*, 4714–4718.
6.  Harris, J. M.; Liu, Y.; Chai, S.; Andrews, M. D.; Vederas, J. C. *J. Org. Chem.* **1998**, *63*, 2407–2409. (Odorless protocols).
7.  Stork, G.; Niu, D.; Fujimoto, R. A.; Koft, E. R.; Bakovec, J. M.; Tata, J. R.; Dake, G. R. *J. Am. Chem. Soc.* **2001**, *123*, 3239–3242.
8.  Nishide, K.; Ohsugi, S.-i.; Fudesaka, M.; Kodama, S.; Node, M. *Tetrahedron Lett.* **2002**, *43*, 5177–5179. (Another odorless protocols).
9.  Ahmad, N. M. *Swern Oxidation.* In *Name Reactions for Functional Group Transformations*; Li, J. J., Ed.; Wiley: Hoboken, NJ, **2007**, pp 291–308. (Review).
10. Lopez-Alvarado, P; Steinhoff, J; Miranda, S; Avendano, C; Menendez, J. C. *Tetrahedron* **2009**, *65*, 1660–1672.
11. Zanatta, N.; Aquino, E. da C.; da Silva, F. M.; Bonacorso, H. G.; Martins, M. A. P. *Synthesis* **2012**, *44*, 3477–3482.

# Takai reaction

Stereoselective conversion of an aldehyde to the corresponding *E*-vinyl iodide using CHI₃ and CrCl₂.

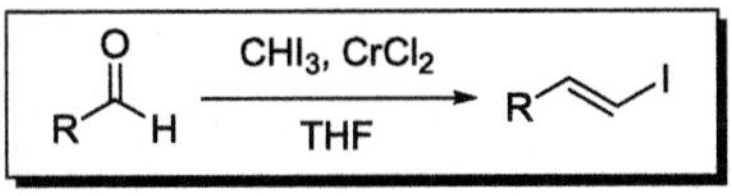

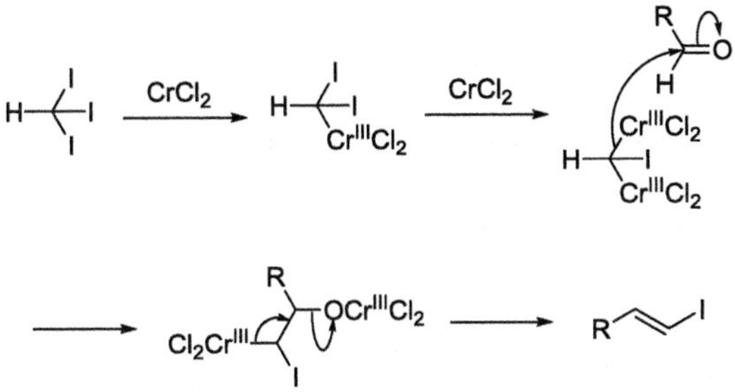

A radical mechanism was recently proposed[10]

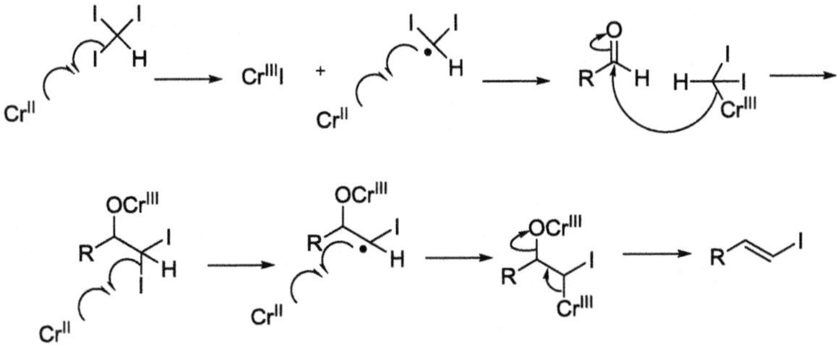

Example 1[2]

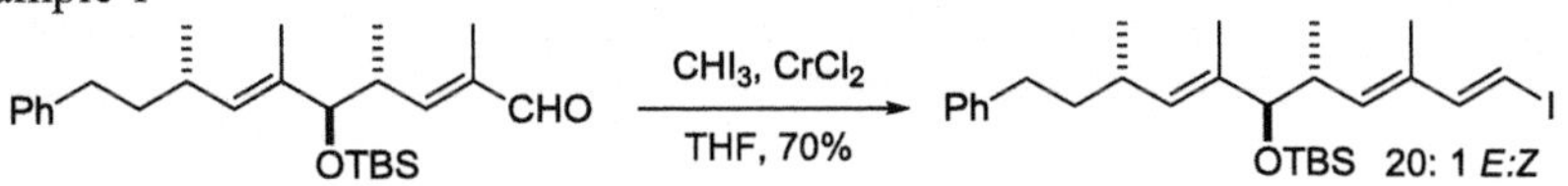

Example 2[3]

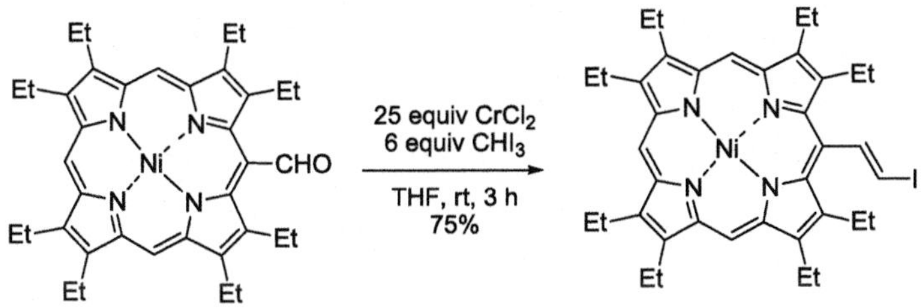

J.J. Li, *Name Reactions: A Collection of Detailed Mechanisms and Synthetic Applications*,
DOI 10.1007/978-3-319-03979-4_271, © Springer International Publishing Switzerland 2014

Example 3[4]

Example 4, A Br/Cl variant[9]

Example 5[10]

Example 5[10]

## References

1. Takai, K.; Nitta, Utimoto, K. *J. Am. Chem. Soc.* **1986,** *108*, 7408–7410.  Kazuhiko Takai was a professor at Kyoto University.
2. Andrus, M. B.; Lepore, S. D.; Turner, T. M. *J. Am. Chem. Soc.* **1997,** *119*, 12159–12169.
3. Arnold, D. P.; Hartnell, R. D. *Tetrahedron* **2001,** *57*, 1335–1345.
4. Rodriguez, A. R.; Spur, B. W. *Tetrahedron Lett.* **2004,** *45*, 8717–8724.
5. Dineen, T. A.; Roush, W. R. *Org. Lett.* **2004,** *6*, 2043–2046.
6. Lipomi, D. J.; Langille, N. F.; Panek, J. S. *Org. Lett.* **2004,** *6*, 3533–3536.
7. Paterson, I.; Mackay, A. C. *Synlett* **2004,** 1359–1362.
8. Concellón, J. M.; Bernad, P. L.; Méjica, C. *Tetrahedron Lett.* **2005,** *46*, 569–571.
9. Gung, B. W.; Gibeau, C.; Jones, A. *Tetrahedron: Asymmetry* **2005,** *16*, 3107–3114.
10. Legrand, F.; Archambaud, S.; Collet, S.; Aphecetche-Julienne, K.; Guingant, A.; Evain, M. *Synlett* **2008,** 389–393.
11. Saikia, B.; Joymati Devi, T.; Barua, N. C. *Org. Biomol. Chem.* **2013,** *11*, 905–913.

# Tebbe reagent

The Tebbe reagent, μ-chlorobis(cyclopentadienyl)(dimethylaluminium)-μ-methylenetitanium, transforms a carbonyl compound to the corresponding *exo*-olefin.

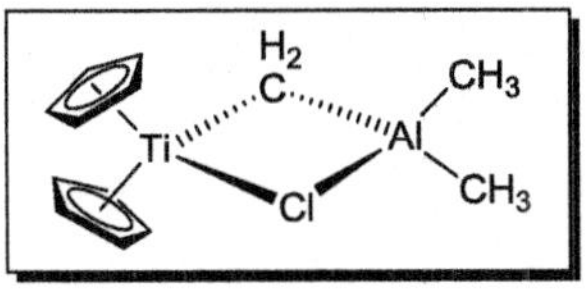

Preparation:[2,6]

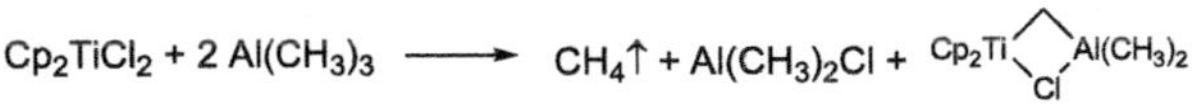

Mechanism:[3]

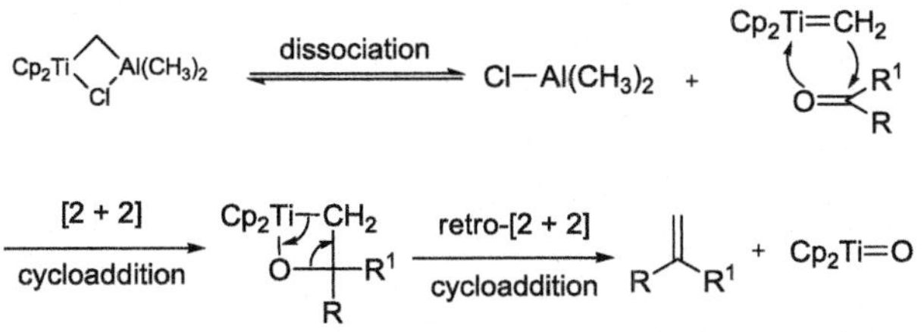

oxatitanacyclobutane                    formation of the strong Ti=O
                                         bond is the driving force.

Example 1, Ketone[2]

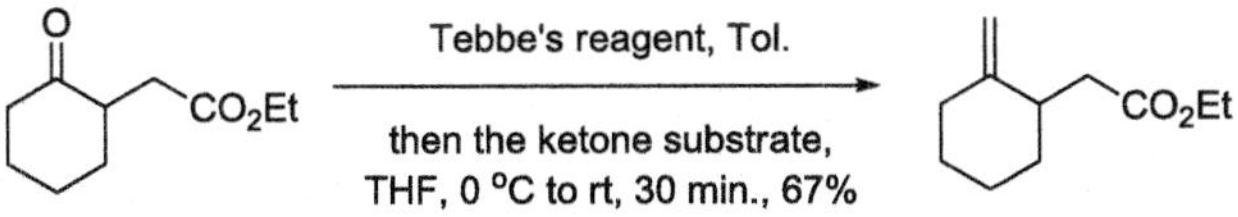

Example 2, Double Tebbe[4]

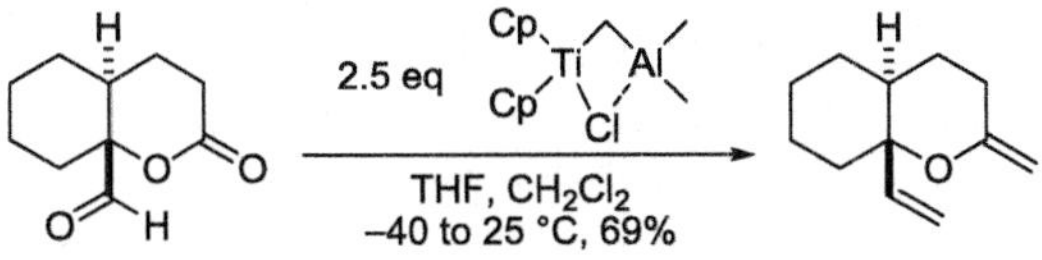

J.J. Li, *Name Reactions: A Collection of Detailed Mechanisms and Synthetic Applications*,
DOI 10.1007/978-3-319-03979-4_272, © Springer International Publishing Switzerland 2014

Example 3, Double Tebbe[5]

Example 4, *N*-Oxide[6]

Example 5, Amide[11]

## References

1.   Tebbe, F. N.; Parshall, G. W.; Reddy, G. S. *J. Am. Chem. Soc.* **1978,** *100*, 3611–3613. Fred Tebbe worked at DuPont Central Research.
2.   Pine, S. H.; Pettit, R. J.; Geib, G. D.; Cruz, S. G.; Gallego, C. H.; Tijerina, T.; Pine, R. D. *J. Org. Chem.* **1985,** *50*, 1212–1216.
3.   Cannizzo, L. F.; Grubbs, R. H. *J. Org. Chem.* **1985,** *50*, 2386–2387.
4.   Philippo, C. M. G.; Vo, N. H.; Paquette, L. A. *J. Am. Chem. Soc.* **1991,** *113*, 2762–2764.
5.   Ikemoto, N.; Schreiber, L. S. *J. Am. Chem. Soc.* **1992,** *114*, 2524–2536.
6.   Pine, S. H. *Org. React.* **1993,** *43*, 1–98. (Review).
7.   Nicolaou, K. C.; Koumbis, A. E.; Snyder, S. A.; Simonsen, K. B. *Angew. Chem. Int. Ed.* **2000,** *39*, 2529–2533.
8.   Straus, D. A. *Encyclopedia of Reagents for Organic Synthesis;* Wiley & Sons, **2000**. (Review).
9.   Payack, J. F.; Hughes, D. L.; Cai, D.; Cottrell, I. F.; Verhoeven, T. R. *Org. Syn., Coll. Vol. 10*, **2004**, p 355.
10.  Beadham, I.; Micklefield, J. *Curr. Org. Synth.* **2005,** *2*, 231–250. (Review).
11.  Long, Y. O.; Higuchi, R. I.; Caferro, T.s R.; Lau, T. L. S.; Wu, M.; Cummings, M. L.; Martinborough, E. A.; Marschke, K. B.; Chang, W. Y.; Lopez, F. J.; Karanewsky, D. S.; Zhi, L. *Bioorg. Med. Chem. Lett.* **2008,** *18*, 2967–2971.
12.  Zhang, J. *Tebbe reagent.* In *Name Reactions for Homolotions-Part I*; Li, J. J., Corey, E. J., Eds., Wiley: Hoboken, NJ, **2009**, pp 319–333. (Review).
13.  Yamashita, S.; Suda, N.; Hayashi, Y.; Hirama, M. *Tetrahedron Lett.* **2013,** *54*, 1389–1391.

# TEMPO oxidation

TEMPO = **Te**tra**m**ethyl **p**entahydropyridine **o**xide. 2,2,6,6-Tetramethylpiperi-dinyloxy is a stable nitroxyl radical, which serves in oxidations as catalyst.

Example 1[4]

Example 2, Trichloroisocyanuric/TEMPO oxidation[5]

J.J. Li, *Name Reactions: A Collection of Detailed Mechanisms and Synthetic Applications*, DOI 10.1007/978-3-319-03979-4_273, © Springer International Publishing Switzerland 2014

Example 3[8]

Example 4[10]

"Ormosil-TEMPO" is a sol-gel hydrophobized nanostructured silica matrix doped
with TEMPO

Example 5[12]

Example 6, TEMPO-mediated aliphatic C–H oxidation with oximes[13]

## References

1.  Garapon, J.; Sillion, B.; Bonnier, J. M. *Tetrahedron Lett.* **1970**, *11*, 4905–4908.
2.  de Nooy, A. E.; Besemer, A. C.; van Bekkum, H. *Synthesis* **1996**, 1153–1174. (Review).
3.  Rychnovsky, S. D.; Vaidyanathan, R. *J. Org. Chem.* **1999**, *64*, 310–312.
4.  Fabbrini, M.; Galli, C.; Gentili, P.; Macchitella, D. *Tetrahedron Lett.* **2001**, *42*, 7551–7553.
5.  De Luca, L.; Giacomelli, G.; Masala, S.; Porcheddu, A. *J. Org. Chem.* **2003**, *45*, 4999–5001.
6.  Ciriminna, R.; Pagliaro, M. *Tetrahedron Lett.* **2004**, *45*, 6381–6383.
7.  Tashino, Y.; Togo, H. *Synlett* **2004**, 2010–2012.
8.  Breton, T.; Liaigre, D.; Belgsir, E. M. *Tetrahedron Lett.* **2005**, *46*, 2487–2490.
9.  Chauvin, A.-L.; Nepogodiev, S. A.; Field, R. A. *J. Org. Chem.* **2005**, *47*, 960–966.
10. Gancitano, P.; Ciriminna, R.; Testa, M. L.; Fidalgo, A.; Ilharco, L. M.; Pagliaro, M. *Org. Biomol. Chem.* **2005**, *3*, 2389–2392.
11. Zhang, M.; Chen, C.; Ma, W.; Zhao, J. *Angew. Chem. Int. Ed.* **2008**, *47*, 9730–9733.
12. Perusquía-Hernández, C.; Lara-Issasi, G. R.; Frontana-Uribe, B. A.; Cuevas-Yañez, E. *Tetrahedron Lett.* **2013**, *54*, 3302–3305.
13. Zhu, X.; Wang, Y.-F.; Ren, W.; Zhang, F.-L.; Chiba, S. *Org. Lett.* **2013**, *15*, 3214–3217.

# Thorpe–Ziegler reaction

The intramolecular version of the Thorpe reaction, which is base-catalyzed self-condensation of nitriles to yield imines that tautomerize to enamine.

## Example 1, A radical-mediated Thorpe–Ziegler reaction[2]

## Example 2[5]

## Example 3[8]

J.J. Li, *Name Reactions: A Collection of Detailed Mechanisms and Synthetic Applications*, DOI 10.1007/978-3-319-03979-4_274, © Springer International Publishing Switzerland 2014

Example 4[9]

Example 5[11]

## References

1.  (a) Baron, H.; Remfry, F. G. P.; Thorpe, Y. F. *J. Chem. Soc.* **1904,** *85*, 1726–1761. (b) Ziegler, K. *et al. Ann.* **1933,** *504,* 94–130. Karl Ziegler (1898–1973), born in Helsa, Germany, received his Ph.D. In 1920 from von Auwers at the University of Marburg. He became the director of the Max-Planck-Institut für Kohlenforschung at Mülheim/Ruhr in 1943. He shared the Nobel Prize in Chemistry in 1963 with Giulio Natta (1903–1979) for their work in polymer chemistry. The Ziegler–Natta catalyst is widely used in polymerization.
2.  Curran, D. P.; Liu, W. *Synlett* **1999,** 117–119.
3.  Dansou, B.; Pichon, C.; Dhal, R.; Brown, E.; Mille, S. *Eur. J. Org. Chem.* **2000,** 1527–1531.
4.  Keller, L.; Dumas, F.; Pizzonero, M.; d'Angelo, J.; Morgant, G.; Nguyen-Huy, D. *Tetrahedron Lett.* **2002,** *43,* 3225–3228.
5.  Malassene, R.; Toupet, L.; Hurvois, J.-P.; Moinet, C. *Synlett* **2002,** 895–898.
6.  Satoh, T.; Wakasugi, D. *Tetrahedron Lett.* **2003,** *44,* 7517–7520.
7.  Wakasugi, D.; Satoh, T. *Tetrahedron* **2005,** *61,* 1245–1256.
8.  Dotsenko, V. V.; Krivokolysko, S. G.; Litvinov, V. P. *Monatsh. Chem.* **2008,** *139,* 271–275.
9.  Salaheldin, A. M.; Oliveira-Campos, A. M. F.; Rodrigues, L. M. *ARKIVOC* **2008,** 180–190.
10. Miszke, A.; Foks, H.; Brozewicz, K.; Kedzia, A.; Kwapisz, E.; Zwolska, Z. *Heterocycles* **2008,** *75,* 2723–2734.
11. Hutt, O. E.; Doan, T. L.; Georg, G. I. *Org. Lett.* **2013,** *15,* 1602–1605.

# Tsuji–Trost reaction

The Tsuji–Trost reaction is the palladium-catalyzed substitution of allylic leaving groups by carbon nucleophiles. These reactions proceed via $\pi$-allylpalladium intermediates.

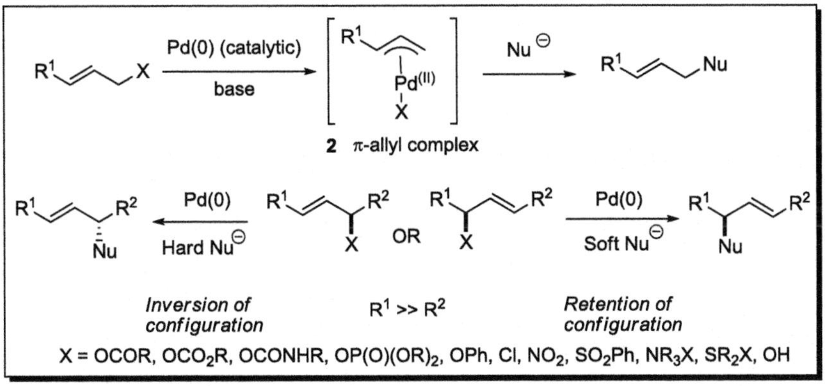

$$X = OCOR,\ OCO_2R,\ OCONHR,\ OP(O)(OR)_2,\ OPh,\ Cl,\ NO_2,\ SO_2Ph,\ NR_3X,\ SR_2X,\ OH$$

The catalytic cycle:

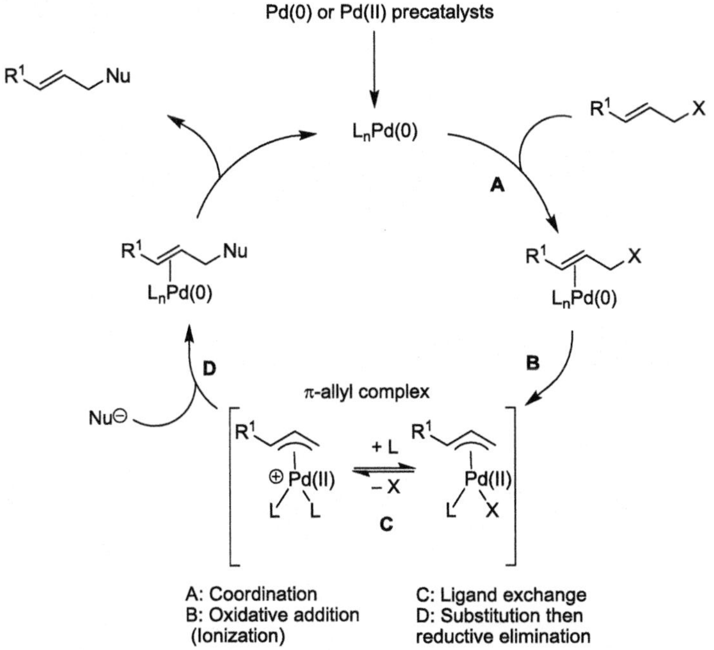

A: Coordination
B: Oxidative addition
(Ionization)

C: Ligand exchange
D: Substitution then
reductive elimination

J.J. Li, *Name Reactions: A Collection of Detailed Mechanisms and Synthetic Applications*, DOI 10.1007/978-3-319-03979-4_275, © Springer International Publishing Switzerland 2014

Example 1, Allylic ether[3]

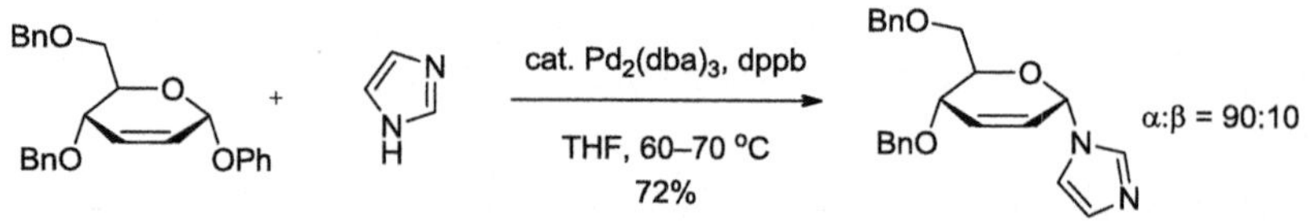
BnO
BnO
OPh
cat. Pd2(dba)3, dppb
THF, 60–70 °C
72%
α:β = 90:10

Example 2, Allylic acetate[3]

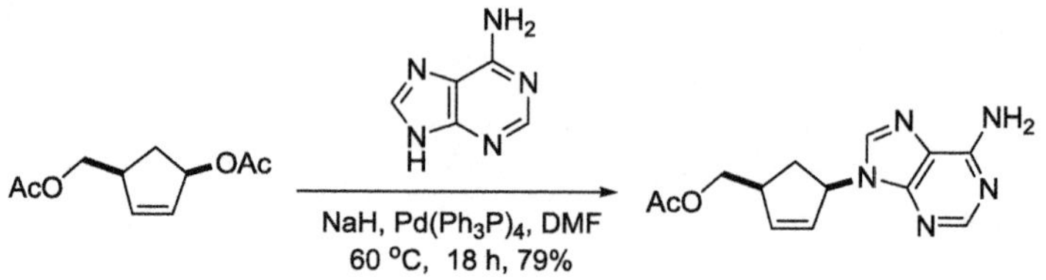
AcO
OAc
NH2
NaH, Pd(Ph3P)4, DMF
60 °C, 18 h, 79%

Example 3, Allylic epoxide[5]

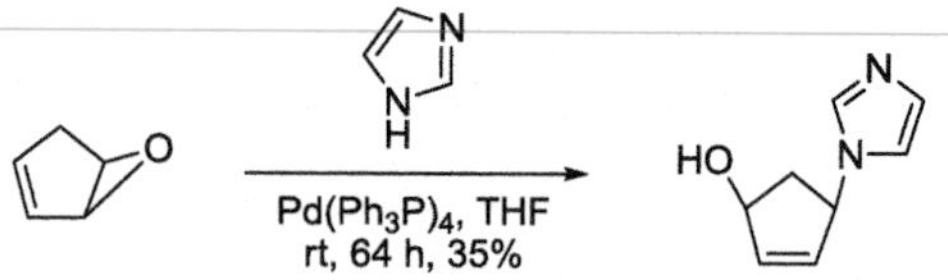
Pd(Ph3P)4, THF
rt, 64 h, 35%
HO

Example 4, Intramolecular Tsuji–Trost reaction[6]

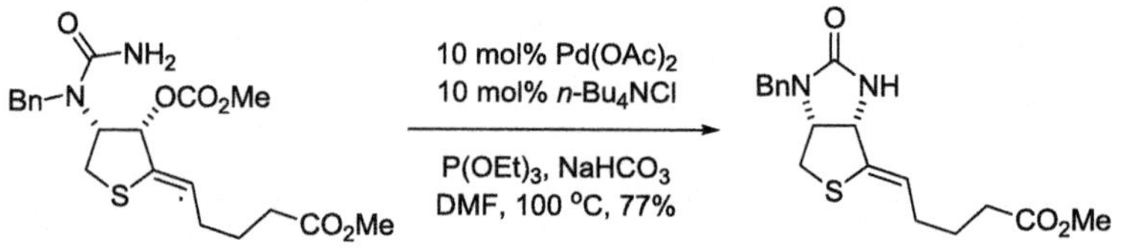
NH2
Bn–N
OCO2Me
CO2Me
10 mol% Pd(OAc)2
10 mol% n-Bu4NCl
P(OEt)3, NaHCO3
DMF, 100 °C, 77%
BnN
NH
CO2Me

Example 5, Intramolecular Tsuji–Trost reaction[7]

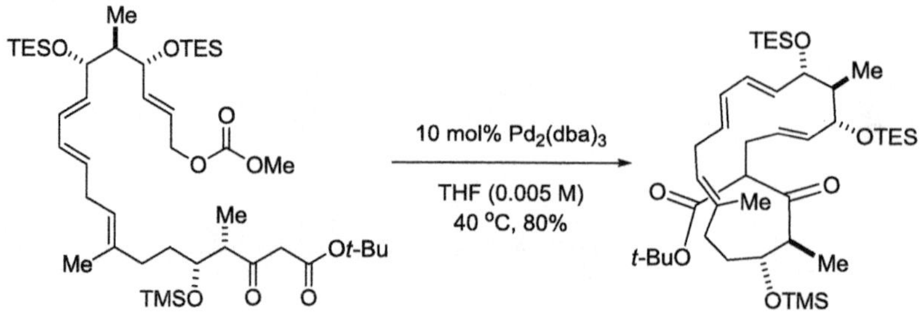
TESO
OTES
Me
OMe
Ot-Bu
TMSO
10 mol% Pd2(dba)3
THF (0.005 M)
40 °C, 80%
t-BuO
OTMS

## Example 6, Asymmetric Tsuji–Trost reaction[8]

## Example 7, Tsuji–Trost decarboxylation–dehydrogenation sequence[12]

## References

1.  (a) Tsuji, J.; Takahashi, H.; Morikawa, M. *Tetrahedron Lett.* **1965**, *6*, 4387–4388. (b) Tsuji, J. *Acc. Chem. Res.* **1969**, *2*, 144–152. (Review). Jiro Tsuji (1927–), now retired, worked at the Toyo Rayon Company in Japan.

2.  Godleski, S. A. In *Comprehensive Organic Synthesis;* Trost, B. M.; Fleming, I., eds.; Vol. *4*. Chapter 3.3. Pergamon: Oxford, 1991. (Review).

3.  Bolitt, V.; Chaguir, B.; Sinou, D. *Tetrahedron Lett.* **1992**, *33*, 2481–2484.

4.  Moreno-Mañas, M.; Pleixats, R. In *Advances in Heterocyclic Chemistry;* Katritzky, A. R., ed.; Academic Press: San Diego, **1996**, *66*, 73. (Review).

5.  Arnau, N.; Cortes, J.; Moreno-Mañas, M.; Pleixats, R.; Villarroya, M. *J. Heterocycl. Chem.* **1997**, *34*, 233–239.

6.  Seki, M.; Mori, Y.; Hatsuda, M.; Yamada, S. *J. Org. Chem.* **2002**, *67*, 5527–5536.

7.  Vanderwal, C. D.; Vosburg, D. A.; Weiler, S.; Sorenson, E. J. *J. Am. Chem. Soc.* **2003**, *125*, 5393–5407.

8.  Trost, B. M.; Toste, F. D. *J. Am. Chem. Soc.* **2003**, *125*, 3090–3100.

9.  Behenna, D. C.; Stoltz, B. M. *J. Am. Chem. Soc.* **2004**, *126*, 15044–15045.

10. Fuchter, M. J. *Tsuji–Trost Reaction.* In *Name Reactions for Homologations-Part I*; Li, J. J., Ed.; Wiley: Hoboken, NJ, **2009**, pp 185–211. (Review).

11. Shi, L.; Meyer, K.; Greaney, M. F. *Angew. Chem. Int. Ed.* **2010**, *49*, 9250–9253.

12. Brehm, E.; Breinbauer, R. *Org. Biomol. Chem.* **2013**, *11*, 4750–4756.

# Ugi reaction

Four-component condensation (4CC) of carboxylic acids, *C*-isocyanides, amines, and carbonyl compounds to afford diamides. Also known as four-component reaction (4CR). *Cf.* Passerini reaction.

## Example 1[2]

## Example 2[5]

J.J. Li, *Name Reactions: A Collection of Detailed Mechanisms and Synthetic Applications*,
DOI 10.1007/978-3-319-03979-4_276, © Springer International Publishing Switzerland 2014

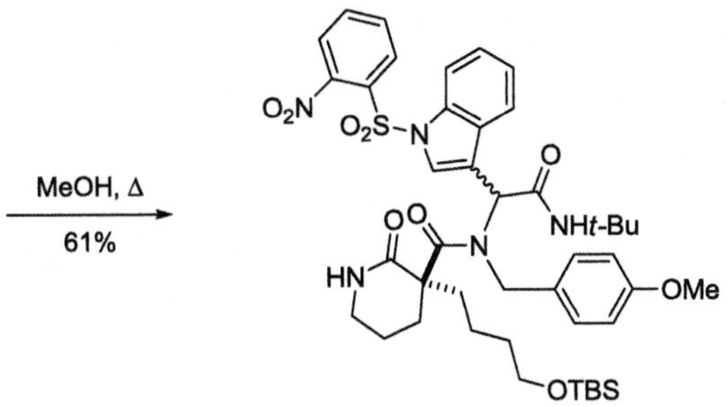

MeOH, Δ
61%

Example 3[7]

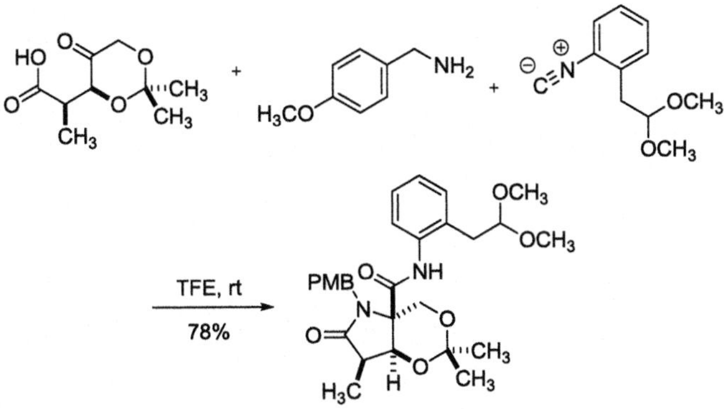

TFE, rt
78%

Example 4[8]

1. MeOH, rt.
2. TBAF, THF
54%, 2 Steps

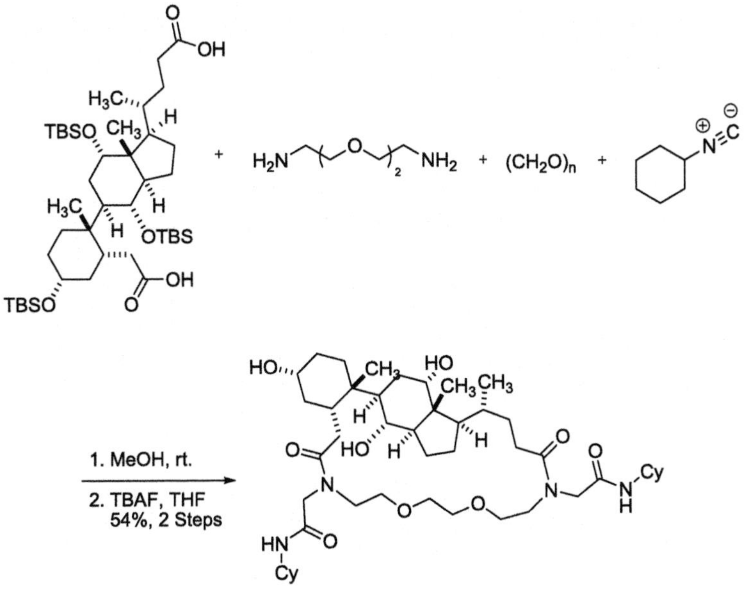

Example 5[11]

**References**

1.  (a) Ugi, I. *Angew. Chem. Int. Ed.* **1962,** *1,* 8–21.; (b) Ugi, I.; Offermann, K.; Herlinger, H.; Marquarding, D. *Liebigs Ann. Chem.* **1967,** *709,* 1–10.; (c) Ugi, I.; Kaufhold, G. *Ann.* **1967,** *709,* 11–28; (d) Ugi, I.; Lohberger, S.; Karl, R. In *Comprehensive Organic Synthesis*; Trost, B. M.; Fleming, I., Eds.; Pergamon: Oxford, **1991,** *Vol. 2,* 1083. (Review); (e) Dömling, A.; Ugi, I. *Angew. Chem. Int. Ed.* **2000,** *39,* 3168. (Review); (f) Ugi, I. *Pure Appl. Chem.* **2001,** *73,* 187–191. (Review).  Ivar Karl Ugi (1930–2005) earned his Ph.D. under the guidance of Prof. Rolf Huisgen.  Since 1962, he worked at Bayer AG, rising through the ranks to director.  But he left Bayer in 1969 to pursue his indendent academic career at the University of Southern California (USC).  In 1973, he moved to the Technische Universität München, where stayed until his retirement in 1999.  Ugi was one of the pioneers in multi-component recations (MCRs).
2.  Endo, A.; Yanagisawa, A.; Abe, M.; Tohma, S.; Kan, T.; Fukuyama, T. *J. Am. Chem. Soc.* **2002,** *124,* 6552–6554.
3.  Hebach, C.; Kazmaier, U. *Chem. Commun.* **2003,** 596–597.
4.  *Multicomponent Reactions* J. Zhu, H. Bienaymé, Eds.; Wiley-VCH, Weinheim, **2005**.
5.  Oguri, H.; Schreiber, S. L. *Org. Lett.* **2005,** *7,* 47–50.
6.  Dömling, A. *Chem. Rev.* **2006,** *106,* 17–89.
7.  Gilley, C. B.; Buller, M. J.; Kobayashi, Y. *Org. Lett.* **2007,** *9,* 3631–3634.
8.  Rivera, D. G.; Pando, O.; Bosch, R.; Wessjohann, L. A. *J. Org. Chem.* **2008,** *73,* 6229–6238.
9.  Bonger, K. M.; Wennekes, T.; Filippov, D. V.; Lodder, G.; van der Marel, G. A.; Overkleeft, H. S. *Eur. J. Org. Chem.* **2008,** 3678–3688.
10. Williams, D. R.; Walsh, M. J. *Ugi Reaction.* In *Name Reactions for Homologations-Part II*; Li, J. J., Ed.; Wiley: Hoboken, NJ, **2009,** pp 786–805. (Review).
11. Tyagi, V.; Shahnawaz Khan, S.; Chauhan, P. M. S. *Tetrahedron Lett.* **2013,** *54,* 1279–1284.

# Ullmann coupling

Homocoupling of aryl iodides in the presence of Cu or Ni or Pd to afford biaryls.

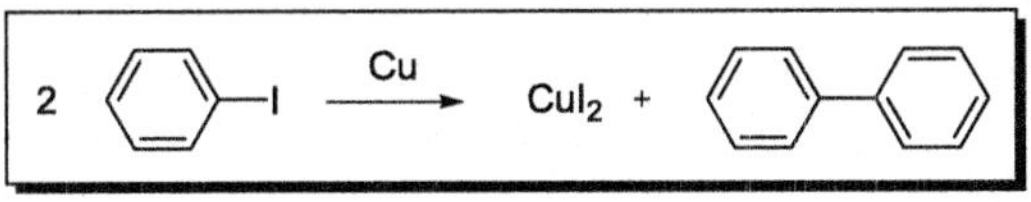

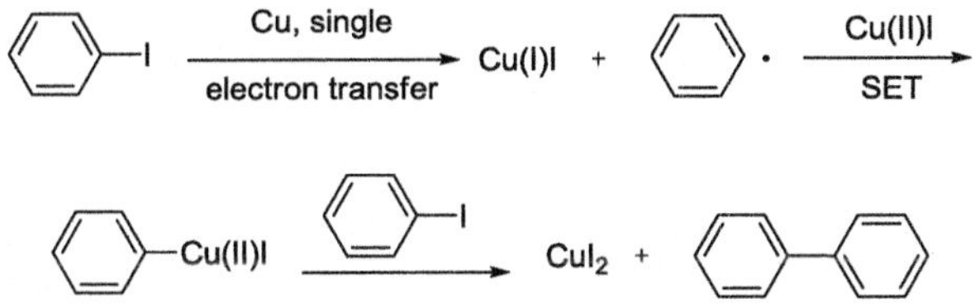

The overall transformation of PhI to PhCuI is an oxidative addition process.

Example 1[3]

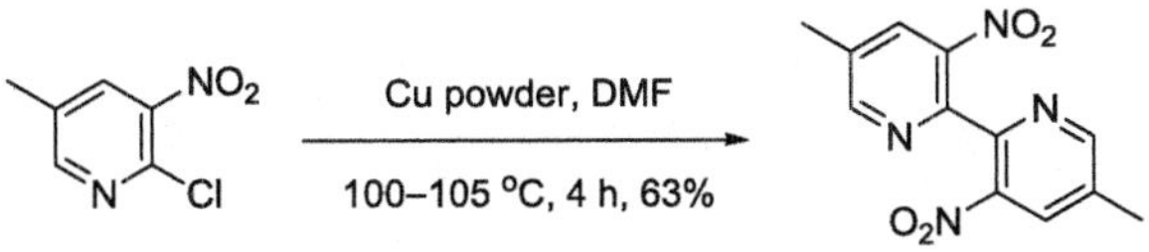

Example 2, CuTC-catalyzed Ullmann coupling (CuTC, Copper(I)-thiophene-2-carboxylate)[4]

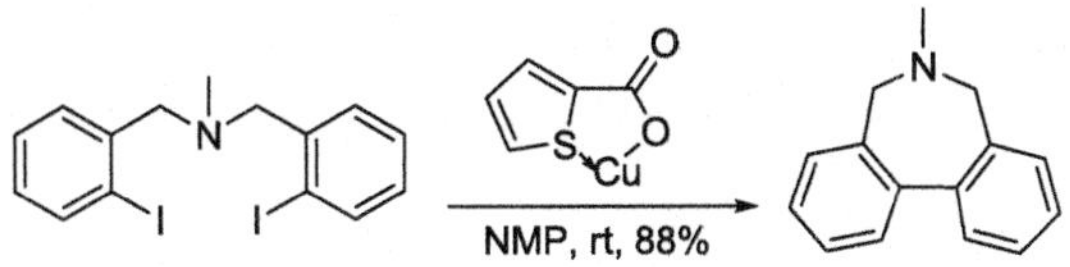

Example 3[5]

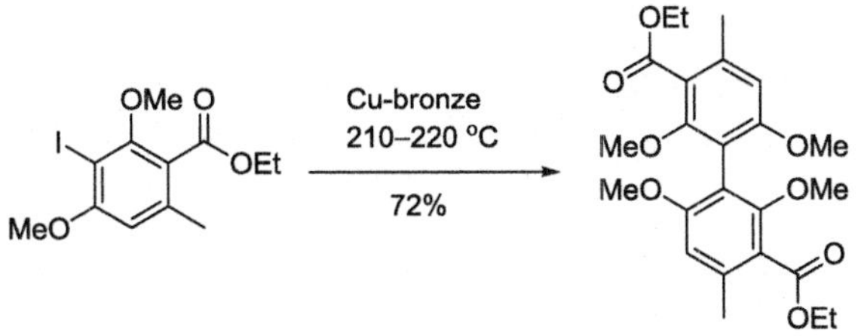

J.J. Li, *Name Reactions: A Collection of Detailed Mechanisms and Synthetic Applications*, DOI 10.1007/978-3-319-03979-4_277, © Springer International Publishing Switzerland 2014

Example 4[8]

Example 5[9]

Example 6[11]

## References

1.  (a) Ullmann, F.; Bielecki, J. *Ber.* **1901**, *34,* 2174–2185. Fritz Ullmann (1875–1939), born in Fürth, Bavaria, studied under Graebe at Geneva. He taught at the Technische Hochschule in Berlin and the University of Geneva. (b) Ullmann, F. *Ann.* **1904**, *332,* 38–81.
2.  Fanta, P. E. *Synthesis* **1974**, 9–21. (Review).
3.  Kaczmarek, L.; Nowak, B.; Zukowski, J.; Borowicz, P.; Sepiol, J.; Grabowska, A. *J. Mol. Struct.* **1991**, *248,* 189–200.
4.  Zhang, S.; Zhang, D.; Liebskind, L. S. *J. Org. Chem.* **1997**, *62,* 2312–2313.
5.  Hauser, F. M.; Gauuan, P. J. F. *Org. Lett.* **1999**, *1,* 671–672.
6.  Buck, E.; Song, Z. J.; Tschaen, D.; Dormer, P. G.; Volante, R. P.; Reider, P. J. *Org. Lett.* **2002**, *4,* 1623–1626.
7.  Nelson, T. D.; Crouch, R. D. *Org. React.* **2004**, *63,* 265–556. (Review).
8.  Qui, L.; Kwong, F. Y.; Wu, J.; Wai, H. L.; Chan, S.; Yu, W.-Y.; Li, Y.-M.; Guo, R.; Zhou, Z.; Chan, A. S. C. *J Am. Chem. Soc.* **2006**, *128,* 5955–5965.
9.  Markey, M. D.; Fu, Y.; Kelly, T. R. *Org. Lett.* **2007**, *9,* 3255–3257.
10. Ahmad, N. M. *Ullman Coupling.* In *Name Reactions for Homologations-Part I*; Li, J. J., Ed.; Wiley: Hoboken, NJ, 2009; pp 255–267. (Review).
11. Chang, E. C.; Chen, C.-Y.; Wang, L.-Y.; Huang, Y.-Y.; Yeh, M.-Y.; Wong, F. F. *Tetrahedron* **2013**, *69,* 570–576.

# van Leusen oxazole synthesis

Formation of 5-substituted oxazoles through the reaction of *p*-tolylsulfonylmethyl isocyanide (TosMIC, also known as the van Leusen reagent) with aldehydes in protic solvents at refluxing temperatures.

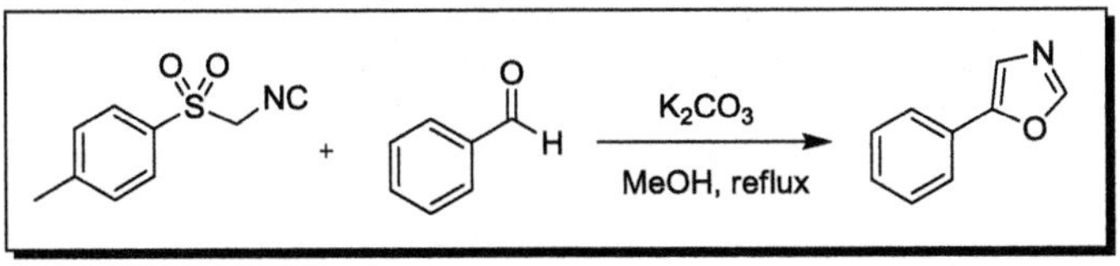

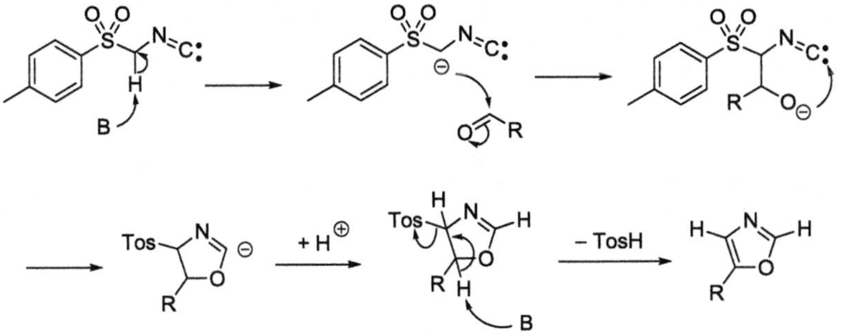

Example 1[3]

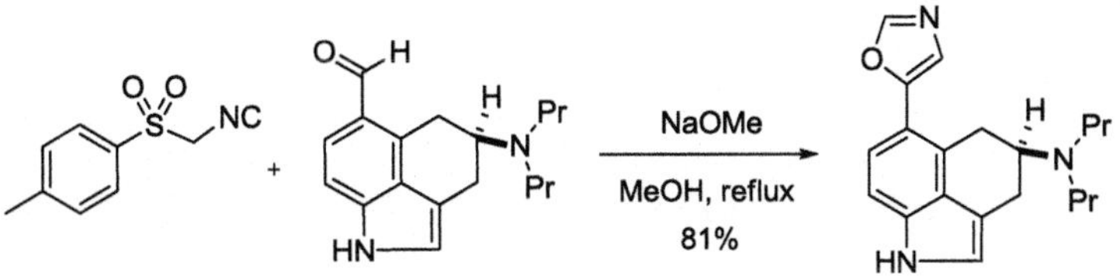

Example 2[5]

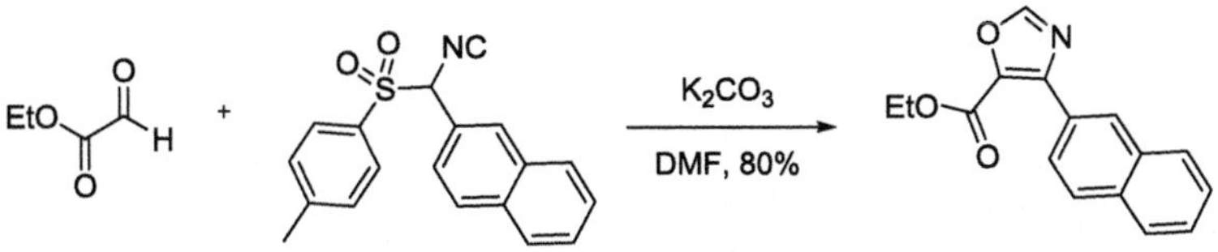

Example 3[9]

J.J. Li, *Name Reactions: A Collection of Detailed Mechanisms and Synthetic Applications*, DOI 10.1007/978-3-319-03979-4_278, © Springer International Publishing Switzerland 2014

Example 4[10]

**References**

1. (a) van Leusen, A. M.; Hoogenboom, B. E.; Siderius, H. *Tetrahedron Lett.* **1972**, *13*, 2369–2381. (b) Possel, O.; van Leusen, A. M. *Heterocycles* **1977**, *7*, 77–80. (c) Saikachi, H.; Kitagawa, T.; Sasaki, H.; van Leusen, A. M. *Chem. Pharm. Bull.* **1979**, *27*, 793–796. (d) van Nispen, S. P. J. M.; Mensink, C.; van Leusen, A. M. *Tetrahedron Lett.* **1980**, *21*, 3723–3726. Van Leusen was a professor at The University Zernikelaan Groningen, The Netherlands.
2. van Leusen, A. M.; van Leusen, D. In *Encyclopedia of Reagents of Organic Synthesis*; Paquette, L. A., Ed.; Wiley: New York, **1995**; *Vol. 7*, 4973–4979. (Review).
3. Anderson, B. A.; Becke, L. M.; Booher, R. N.; Flaugh, M. E.; Harn, N. K.; Kress, T. J.; Varie, D. L.; Wepsiec, J. P. *J. Org. Chem.* **1997**, *62*, 8634–8639.
4. Kulkarni, B. A.; Ganesan, A. *Tetrahedron Lett.* **1999**, *40*, 5633–5636.
5. Sisko, J.; Kassick, A. J.; Mellinger, M.; Filan, J. J.; Allen, A.; Olsen, M. A. *J. Org. Chem.* **2000**, *65*, 1516–1524.
6. Barrett, A. G. M.; Cramp, S. M.; Hennessy, A. J.; Procopiou, P. A.; Roberts, R. S. *Org. Lett.* **2001**, *3*, 271–273.
7. Herr, R. J.; Fairfax, D. J., Meckler, H.; Wilson, J. D. *Org. Process Res. Dev.* **2002**, *6*, 677–681.
8. Brooks, D. A. *van Leusen Oxazole Synthesis.* In *Name Reactions in Heterocyclic Chemistry*; Li, J. J., Ed.; Wiley: Hoboken, NJ, **2005**, pp 254–259. (Review).
9. Kotha, S.; Shah, V. R. *Synthesis* **2007**, 3653–3658.
10. Besselièvre, F.; Mahuteau-Betzer, F.; Grierson, D. S.; Piguel, S. *J. Org. Chem.* **2008**, *73*, 3278–3280.
11. Wu, B.; Wen, J.; Zhang, J.; Li, J.; Xiang, Y.-Z.; Yu, X.-Q. *Synlett* **2009**, 500–504.

# Vilsmeier–Haack reaction

The Vilsmeier–Haack reagent, a chloroiminium salt, is a weak electrophile. Therefore, the Vilsmeier–Haack reaction works better with electron-rich carbocycles and heterocycles.

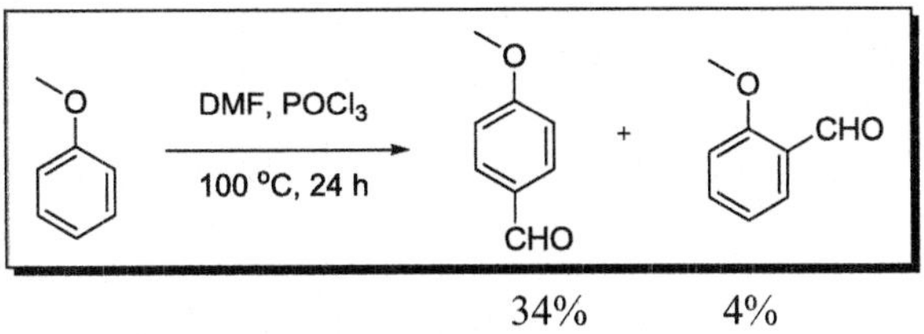

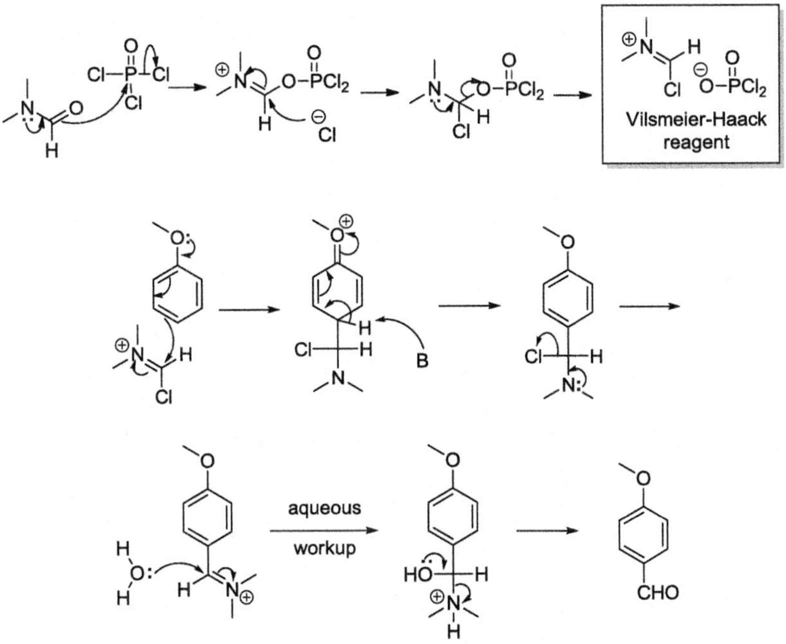

34%          4%

Vilsmeier-Haack reagent

Example 1[2]

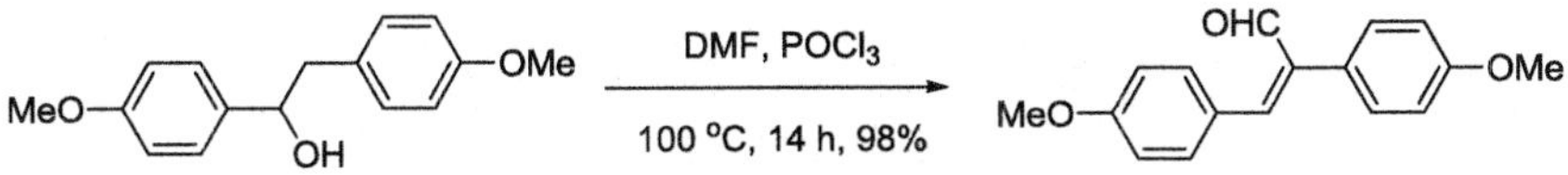

Example 2[3]

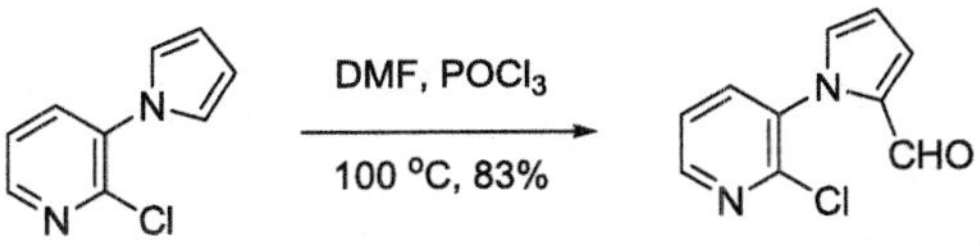

J.J. Li, *Name Reactions: A Collection of Detailed Mechanisms and Synthetic Applications*, DOI 10.1007/978-3-319-03979-4_279, © Springer International Publishing Switzerland 2014

Example 3[9]

Example 4, Reaction outcomes differ as temperature differs[10]

Example 5, An interesting mechanism[11]

## References

1.  Vilsmeier, A.; Haack, A. *Ber.* **1927**, *60*, 119–122. German chemists Anton Vilsmeier and Albrecht Haack discovered this recation in 1927.
2.  Reddy, M. P.; Rao, G. S. K. *J. Chem. Soc., Perkin Trans. 1* **1981**, 2662–2665.
3.  Lancelot, J.-C.; Ladureé, D.; Robba, M. *Chem. Pharm. Bull.* **1985**, *33*, 3122–3128.
4.  Marson, C. M.; Giles, P. R. *Synthesis Using Vilsmeier Reagents* CRC Press, **1994**. (Book).
5.  Seybold, G. *J. Prakt. Chem.* **1996**, *338,* 392–396 (Review).
6.  Jones, G.; Stanforth, S. P. *Org. React.* **1997**, *49*, 1–330. (Review).
7.  Jones, G.; Stanforth, S. P. *Org. React.* **2000**, *56*, 355–659. (Review).
8.  Tasneem, *Synlett* **2003**, 138–139. (Review of the Vilsmeier–Haack reagent).
9.  Nandhakumar, R.; Suresh, T.; Jude, A. L. C.; Kannan, V. R.; Mohan, P. S. *Eur. J. Med. Chem.* **2007**, *42*, 1128–1136.
10. Tang, X.-Y.; Shi, M. *J. Org. Chem.* **2008**, *73*, 8317–8320.
11. Shamsuzzaman, Hena Khanam, H.; Mashrai, A.; Siddiqui, N. *Tetrahedron Lett.* **2013**, *54,* 874–877.

# Vinylcyclopropane–cyclopentene rearrangement

Transformation of vinylcyclopropane to cyclopentene *via* a diradical intermediate.

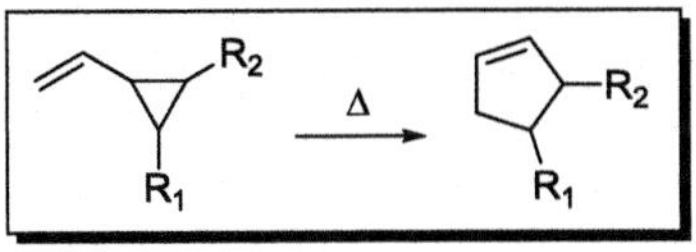
R2
R1
Δ

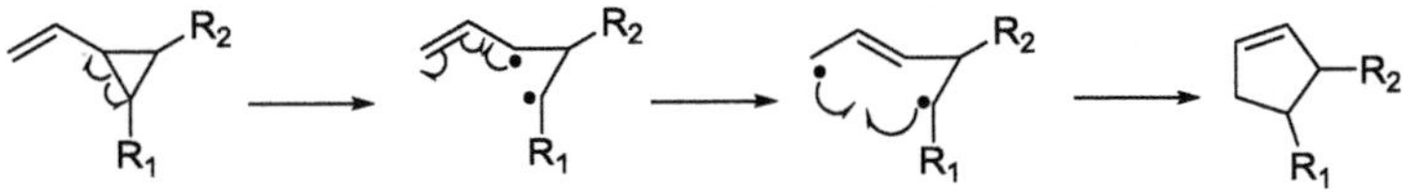
R2
R1

Example 1[1]

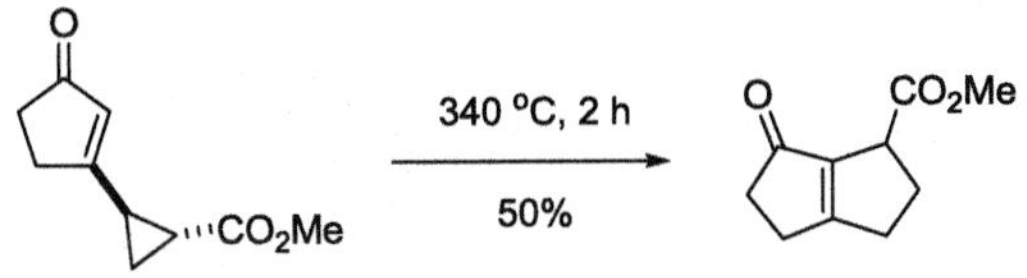
CO2Me
340 °C, 2 h
50%

Example 2[2]

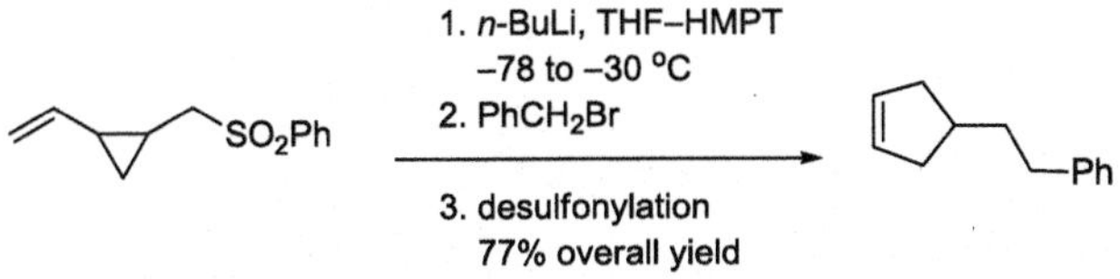
SO2Ph
1. n-BuLi, THF–HMPT
–78 to –30 °C
2. PhCH2Br
3. desulfonylation
77% overall yield
Ph

Example 3[9]

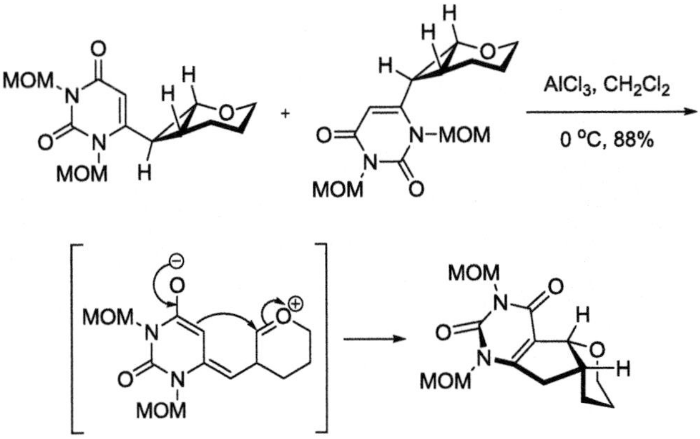
MOM
AlCl3, CH2Cl2
0 °C, 88%

J.J. Li, *Name Reactions: A Collection of Detailed Mechanisms and Synthetic Applications*,
DOI 10.1007/978-3-319-03979-4_280, © Springer International Publishing Switzerland 2014

Example 4[10]

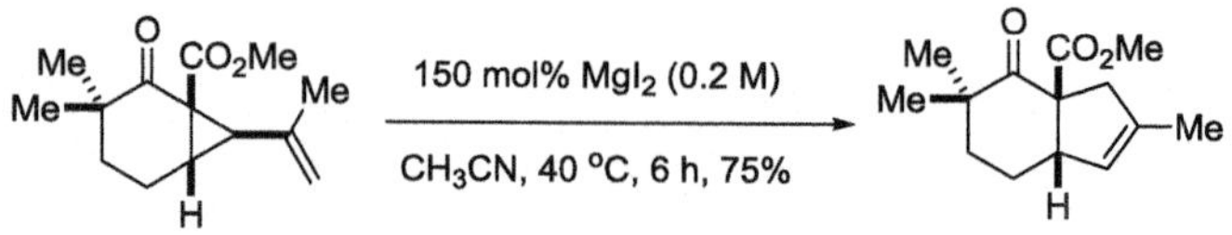

Example 5[11]

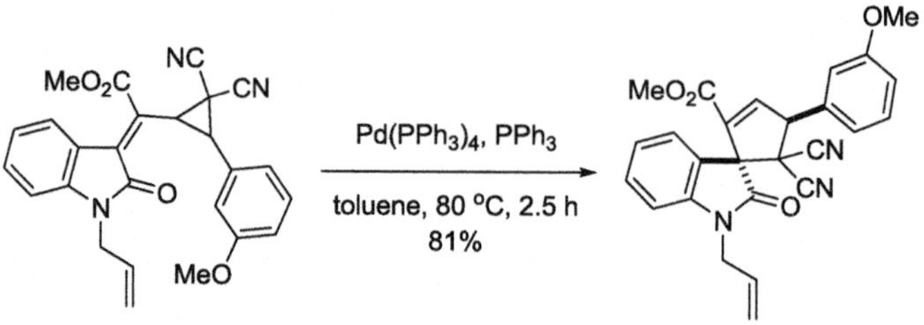

## References

1.  Brule, D.; Chalchat, J. C.; Garry, R. P.; Lacroix, B.; Michet, A.; Vessier, R. *Bull. Soc. Chim. Fr.* **1981**, *1–2,* 57–64.
2.  Danheiser, R. L.; Bronson, J. J.; Okano, K. *J. Am. Chem. Soc.* **1985**, *107,* 4579–4581.
3.  Hudlický, T.; Kutchan, T. M.; Naqvi, S. M. *Org. React.* **1985**, *33,* 247–335. (Review).
4.  Goldschmidt, Z.; Crammer, B. *Chem. Soc. Rev.* **1988**, *17,* 229–267. (Review).
5.  Sonawane, H. R.; Bellur, N. S.; Kulkarni, D. G.; Ahuja, J. R. *Synlett* **1993**, 875–884. (Review).
6.  Hiroi, K.; Arinaga, Y. *Tetrahedron Lett.* **1994**, *35,* 153–156.
7.  Baldwin, J. E. *Chem. Rev.* **2003**, *103,* 1197–1212. (Review).
8.  Wang, S. C.; Tantillo, D. J. *J. Organomet. Chem.* **2006**, *691,* 4386–4392.
9.  Zhang, F.; Kulesza, A.; Rani, S.; Bernet, B.; Vasella, A. *Helv. Chim. Acta* **2008**, *91,* 1201–1218.
10. Coscia, R. W.; Lambert, T. H. *J. Am. Chem. Soc.* **2009**, *131,* 2496–2498.
11. Lingam, K. A. P.; Shanmugam, P. *Tetrahedron Lett.* **2013**, *32,* 4202–4206.

# von Braun reaction

Different from the von Braun degradation reaction (amide to nitrile), the von Braun reaction refers to the treatment of tertiary amines with cyanogen bromide, resulting in a substituted cyanamide.

Example 1[4]

floxetine (Prozac)

Example 2[5]

Example 3[9]

## References

1. von Braun, J. *Ber.* **1907**, *40*, 3914–3933. Julius von Braun (1875–1940), born in Warsaw, Poland, was a Professor of Chemistry at Frankfurt.
2. Hageman, H. A. *Org. React.* **1953**, *7*, 198–262. (Review).
3. Fodor, G.; Nagubandi, S. *Tetrahedron* **1980**, *36*, 1279–1300. (Review).
4. Mody, S. B.; Mehta, B. P.; Udani, K. L.; Patel, M. V.; Mahajan, Rajendra N.. Indian Patent IN177159 (1996).
5. McLean, S.; Reynolds, W. F.; Zhu, X. *Can. J. Chem.* **1987**, *65*, 200–204.
6. Chambert, S.; Thomasson, F.; Décout, J.-L. *J. Org. Chem.* **2002**, *67*, 1898–1904.
7. Hatsuda, M.; Seki, M. *Tetrahedron* **2005**, *61*, 9908–9917.
8. Thavaneswaran, S.; McCamley, K.; Scammells, P. J. *Nat. Prod. Commun.* **2006**, *1*, 885–897. (Review).
9. McCall, W. S.; Abad Grillo, T.; Comins, D. L. *Org. Lett.* **2008**, *10*, 3255–3257.
10. Tayama, E.; Sato, R.; Ito, M.; Iwamoto, H.; Hasegawa, E. *Heterocycles* **2013**, *87*, 381–388.

J.J. Li, *Name Reactions: A Collection of Detailed Mechanisms and Synthetic Applications*, DOI 10.1007/978-3-319-03979-4_281, © Springer International Publishing Switzerland 2014

# Wacker oxidation

Palladium-catalyzed oxidation of olefins to ketones, and aldehydes in certain cases.

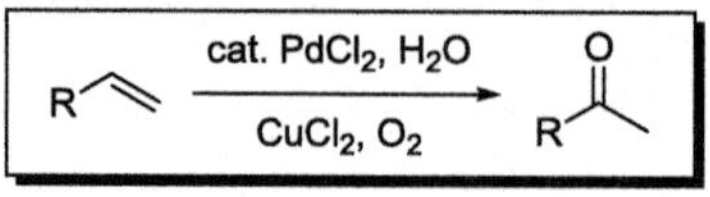

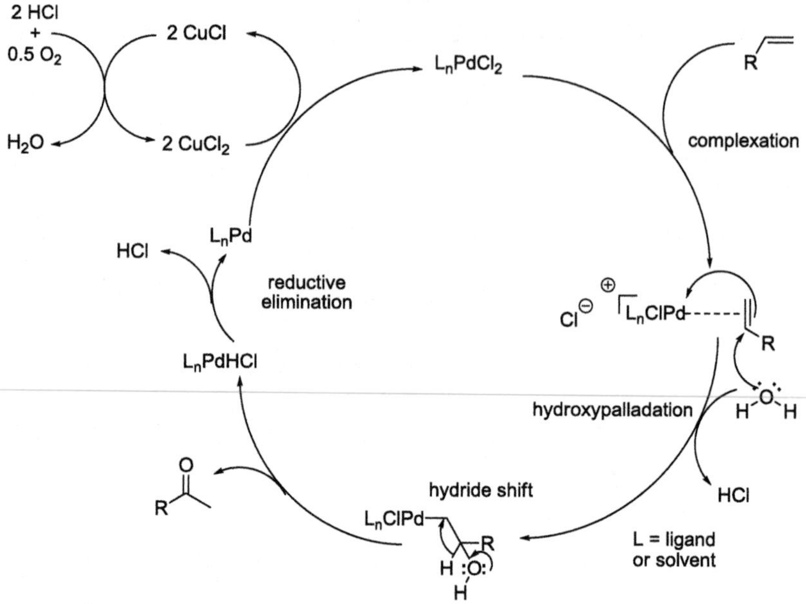

Example 1[5]

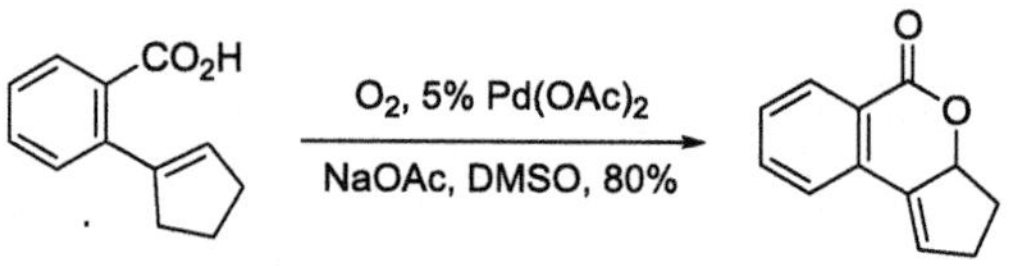

Example 2[7]

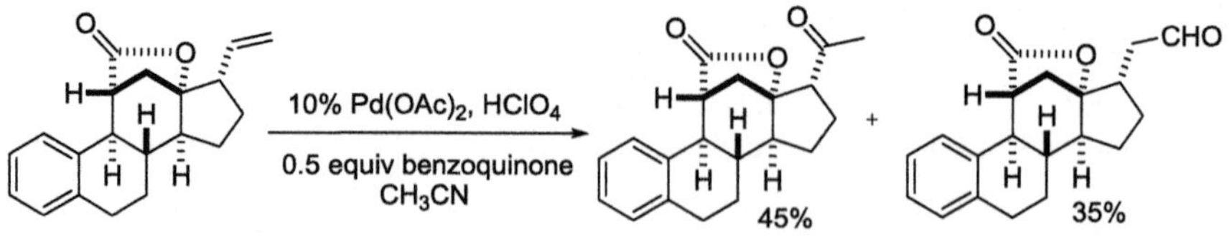

Example 3[9]

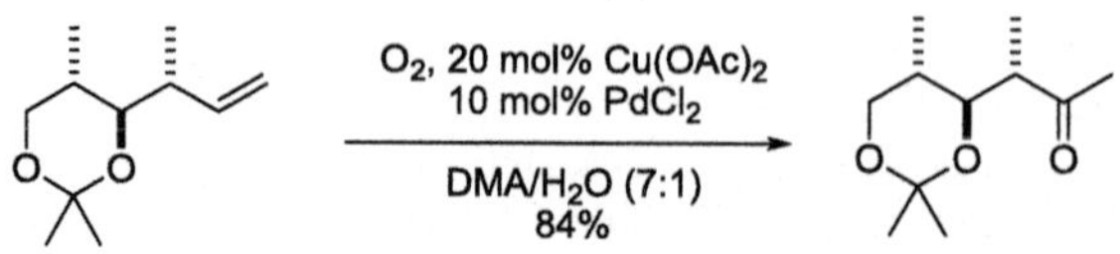

J.J. Li, *Name Reactions: A Collection of Detailed Mechanisms and Synthetic Applications*,
DOI 10.1007/978-3-319-03979-4_282, © Springer International Publishing Switzerland 2014

Example 4[10]

5 eq. PdCl$_2$, air

DMF/H$_2$O (7:1)
81%

12 : 1

CHO

Example 5[10]

5 mol% Pd(Quinox)Cl$_2$
12.5% AgSbF$_6$

12 equiv t-BuOOH (aq)
CH$_2$Cl$_2$, 0.1 M, rt, 66%

Quinox =

## References

1.  Smidt, J.; Sieber, R. *Angew. Chem. Int. Ed.* **1962**, *1*, 80–88.  Wacker is not a person, but a place in Germany where Wacker Chemie developed this process.  Since Hoechst AG later refined the reaction, this is sometimes called Hoechst–Wacker process.

2.  Tsuji, J. *Synthesis* **1984**, 369–384. (Review).

3.  Hegedus, L. S. In *Comp. Org. Syn.* Trost. B. M.; Fleming, I., Eds.; Pergamon, **1991**, *Vol. 4*, 552. (Review).

4.  Tsuji, J. In *Comp. Org. Syn.* Trost. B. M.; Fleming, I., Eds.; Pergamon, **1991**, *Vol. 7*, 449. (Review).

5.  Larock, R. C.; Hightower, T. R. *J. Org. Chem.* **1993**, *58*, 5298–5300.

6.  Hegedus, L. S. *Transition Metals in the Synthesis of Complex Organic Molecule* **1994**, University Science Books: Mill Valley, CA, pp 199–208. (Review).

7.  Pellissier, H.; Michellys, P.-Y.; Santelli, M. *Tetrahedron* **1997**, *53*, 10733–10742.

8.  Feringa, B. L. *Wacker oxidation*. In *Transition Met. Org. Synth.* Beller, M.; Bolm, C., eds.; Wiley–VCH: Weinheim, Germany. **1998**, *2*, 307–315. (Review).

9.  Smith, A. B.; Friestad, G. K.; Barbosa, J.; Bertounesque, E.; Hull, K. G.; Iwashima, M.; Qiu, Y.; Salvatore, B. A.; Spoors, P. G.; Duan, J. J.-W. *J. Am. Chem. Soc.* **1999**, *121*, 10468–10477.

10. Kobayashi, Y.; Wang, Y.-G. *Tetrahedron Lett.* **2002**, *43*, 4381–4384.

11. Hintermann, L. *Wacker-type Oxidations* in *Transition Met. Org. Synth. (2nd edn.)* Beller, M.; Bolm, C., eds., Wiley–VCH: Weinheim, Germany. **2004**, *2*, pp 379–388. (Review).

12. Li, J. J. *Wacker–Tsuji oxidation*. In *Name Reactions for Functional Group Transformations*; Li, J. J., Ed.; Wiley: Hoboken, NJ, **2007**, pp 309–326. (Review).

13. Okamoto, M.; Taniguchi, Y. *J. Cat.* **2009**, *261*, 195–200.

14. DeLuca, R. J.; Edwards, J. L.; Steffens, L. D.; Michel, B. W.; Qiao, X.; Zhu, C.; Cook, S. P.; Sigman, M. S. *J. Org. Chem.* **2013**, *78*, 1682–1686.

# Wagner–Meerwein rearrangement

Acid-catalyzed alkyl group migration of alcohols to give more substituted olefins.

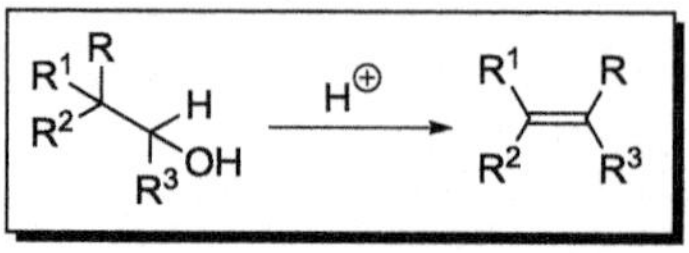

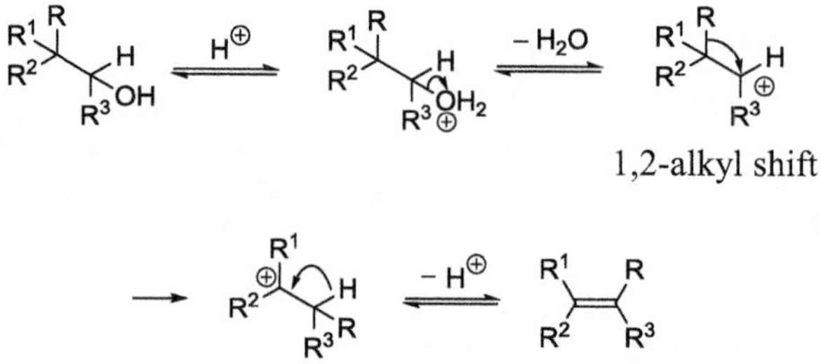

1,2-alkyl shift

Example 1[3]

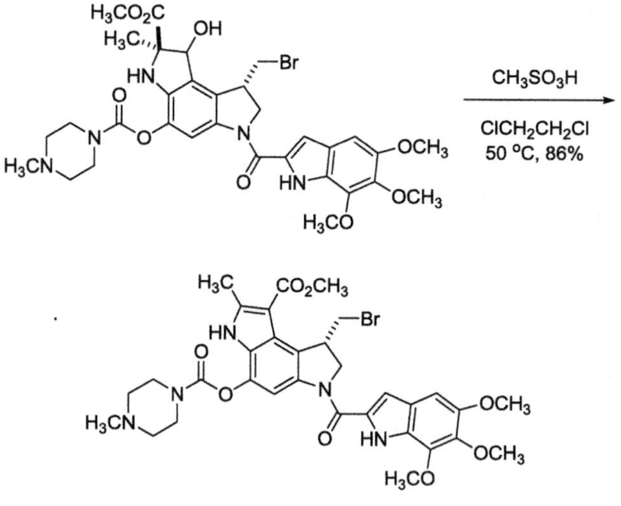

$$CH_3SO_3H$$

$$ClCH_2CH_2Cl$$
50 °C, 86%

Example 2[6]

TFA, $CH_2Cl_2$

72 h, 76%

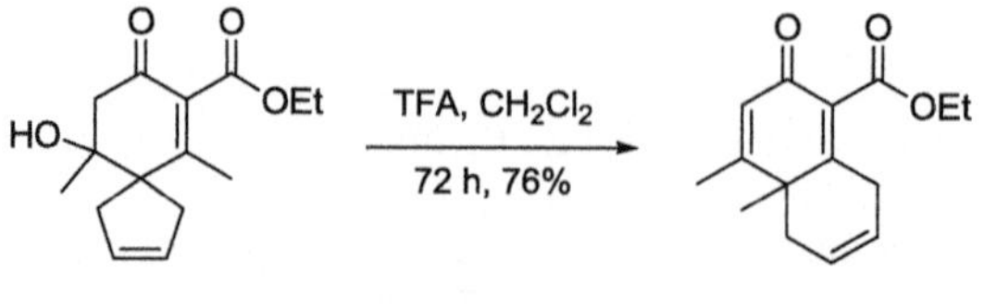

J.J. Li, *Name Reactions: A Collection of Detailed Mechanisms and Synthetic Applications*,
DOI 10.1007/978-3-319-03979-4_283, © Springer International Publishing Switzerland 2014

Example 3[7]

Example 4[9]

## References

1.  Wagner, G. *J. Russ. Phys. Chem. Soc.* **1899**, *31*, 690. Wagner first observed this rearrangement in 1899 and German chemist Hans Meerwein unveiled the mechanism in 1914.
2.  Hogeveen, H.; Van Kruchten, E. M. G. A. *Top. Curr. Chem.* **1979**, *80*, 89–124. (Review).
3.  Kinugawa, M.; Nagamura, S.; Sakaguchi, A.; Masuda, Y.; Saito, H.; Ogasa, T.; Kasai, M. *Org. Proc. Res. Dev.* **1998**, *2*, 344–350.
4.  Trost, B. M.; Yasukata, T. *J. Am. Chem. Soc.* **2001**, *123*, 7162–7163.
5.  Guizzardi, B.; Mella, M.; Fagnoni, M.; Albini, A. *J. Org. Chem.* **2003**, *68*, 1067–1074.
6.  Bose, G.; Ullah, E.; Langer, P. *Chem. Eur. J.* **2004**, *10*, 6015–6028.
7.  Guo, X.; Paquette, L. A. *J. Org. Chem.* **2005**, *70*, 315–320.
8.  Li, W.-D. Z.; Yang, Y.-R. *Org. Lett.* **2005**, *7*, 3107–3110.
9.  Michalak, K.; Michalak, M.; Wicha, J. *Molecules* **2005**, *10*, 1084–1100.
10. Mullins, R. J.; Grote, A. L. *Wagner–Meerwein Rearrangement.* In *Name Reactions for Homologations-Part II*; Li, J. J., Ed.; Wiley: Hoboken, NJ, **2009**, pp 373–394. (Review).
11. Ghorpade, S.; Su, M.-D.; Liu, R.-S. *Angew. Chem. Int. Ed.* **2013**, *52*, 4229–4234.

## Weiss–Cook condensation

Synthesis of *cis*-bicyclo[3.3.0]octane-3,7-dione.  The product is frequently decarboxylated.

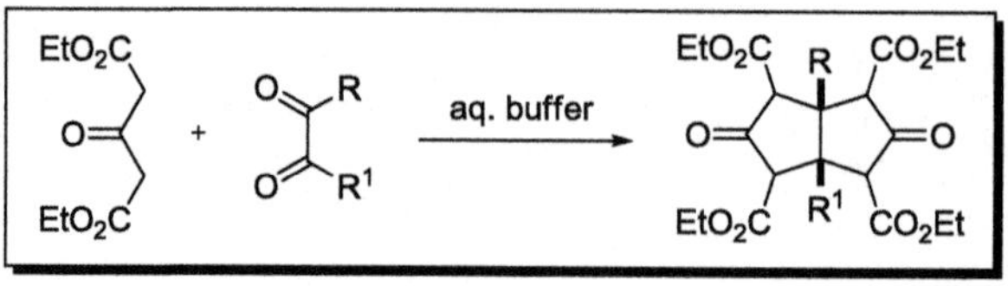

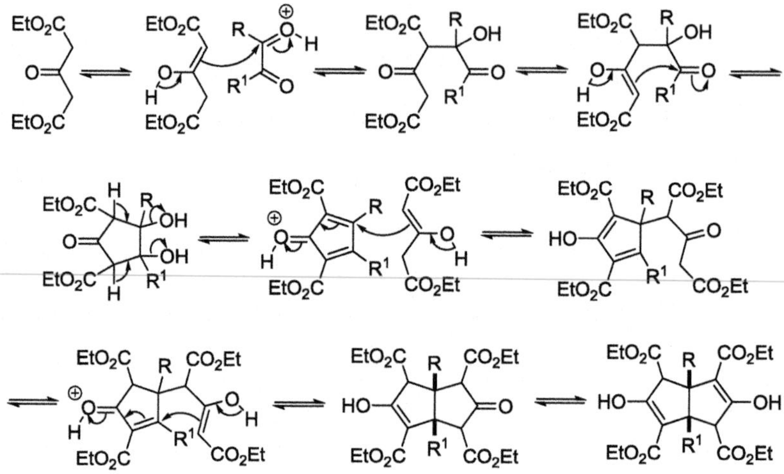

Example 1[2]

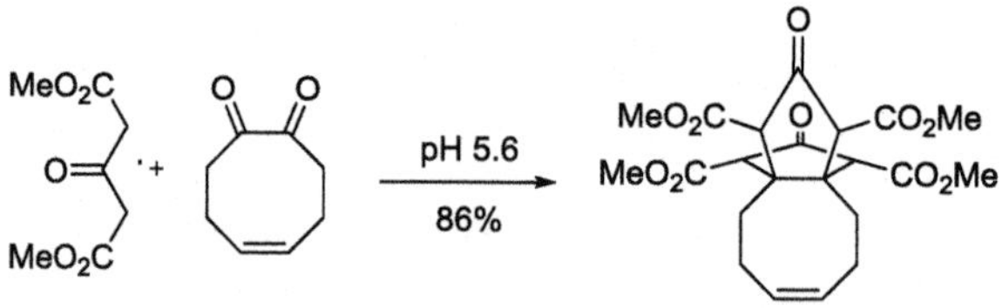

Example 2[3]

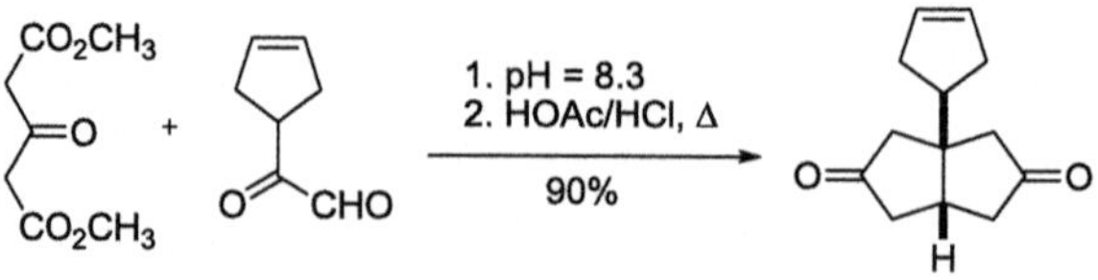

Example 3[4]

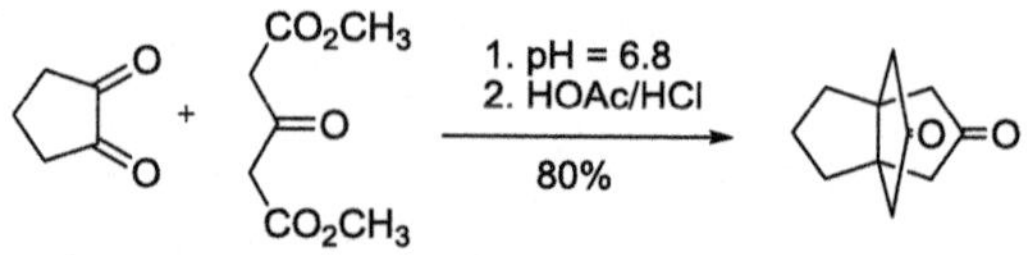

J.J. Li, *Name Reactions: A Collection of Detailed Mechanisms and Synthetic Applications*,
DOI 10.1007/978-3-319-03979-4_284, © Springer International Publishing Switzerland 2014

Example 4[9]

## References

1. Weiss, U.; Edwards, J. M. *Tetrahedron Lett.* **1968**, *9*, 4885–4887. Weiss was a scientist at the National Institutte of Health in Bethesda, Maryland.
2. Bertz, S. H.; Cook, J. M.; Gawish, A.; Weiss, U. *Orga. Synth.* **1986**, *64*, 27–38. James M. Cook is a professor at University of Wisconsin, Milwaukee.
3. Kubiak, G.; Fu, X.; Gupta, A. K.; Cook, J. M. *Tetrahedron Lett.* **1990**, *31*, 4285–4288.
4. Wrobel, J.; Takahashi, K.; Honkan, V.; Lannoye, G.; Bertz, S. H.; Cook, J. M. *J. Org Chem.* **1983**, *48*, 139–141.
5. Gupta, A. K.; Fu, X.; Snyder, J. P.; Cook, J. M. *Tetrahedron* **1991** *47*, 3665–3710.
6. Paquette, L. A.; Kesselmayer, M. A.; Underiner, G. E.; House, S. D.; Rogers, R. D.; Meerholz, K.; Heinze, J. *J. Am. Chem. Soc.* **1992**, *114*, 2644–2652.
7. Fu, X.; Cook, J. M. *Aldrichimica Acta* **1992**, *25*, 43–54. (Review).
8. Fu, X.; Kubiak, G.; Zhang, W.; Han, W.; Gupta, A. K.; Cook, J. M. *Tetrahedron* **1993**, *49*, 1511–1518.
9. Williams, R. V.; Gadgil, V. R.; Vij, As.; Cook, J. M.; Kubiak, G.; Huang, Q. *J. Chem. Soc., Perkin Trans. 1* **1997**, 1425–1428.
10. van Ornum, S. G.; Li, J.; Kubiak, G. G.; Cook, J. M. *J. Chem. Soc., Perkin Trans. 1* **1997**, 3471–3478.
11. Galatsis, P. *Weiss–Cook Reaction*, In Name Reactions for Carbocyclic Ring Formations, Li, J. J., Ed.; Wiley: Hoboken, NJ, 2010, pp 181–196. (Review).

# Wharton reaction

Reduction of α,β-epoxy ketones by hydrazine to allylic alcohols.

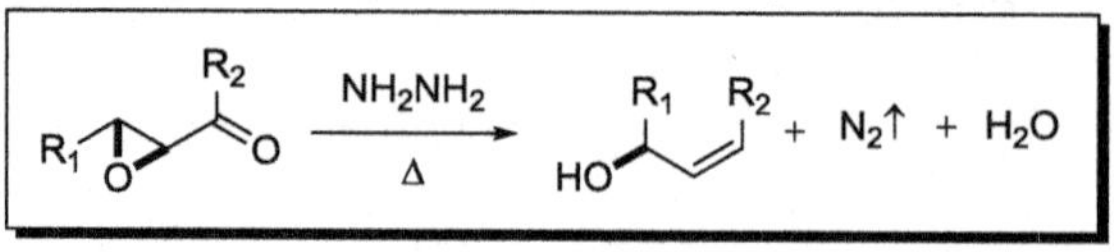

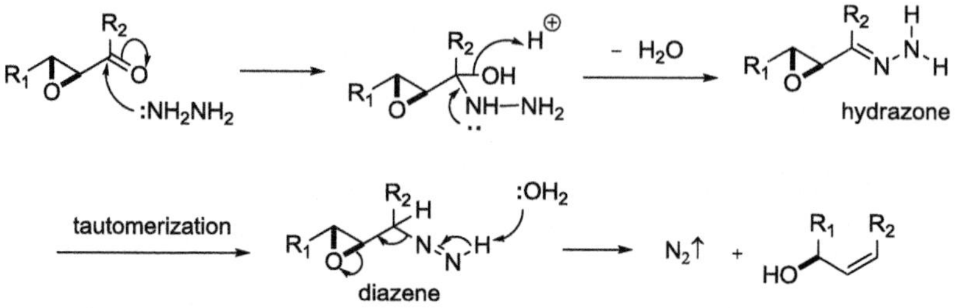

Example 1[5]

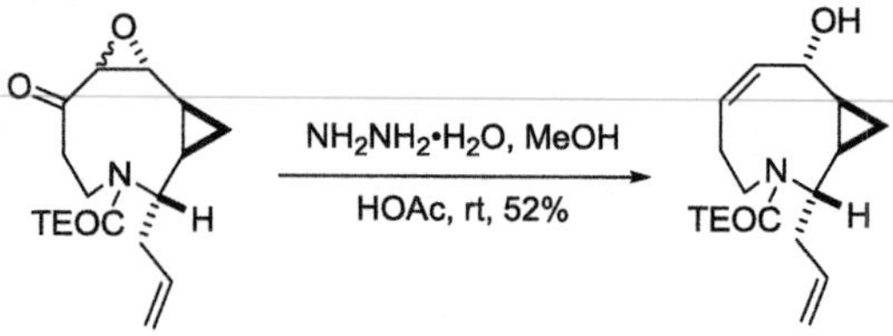

Example 2[6]

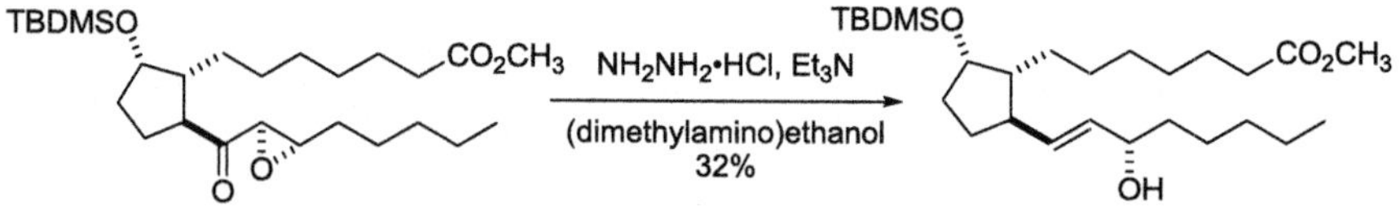

Example 3[7]

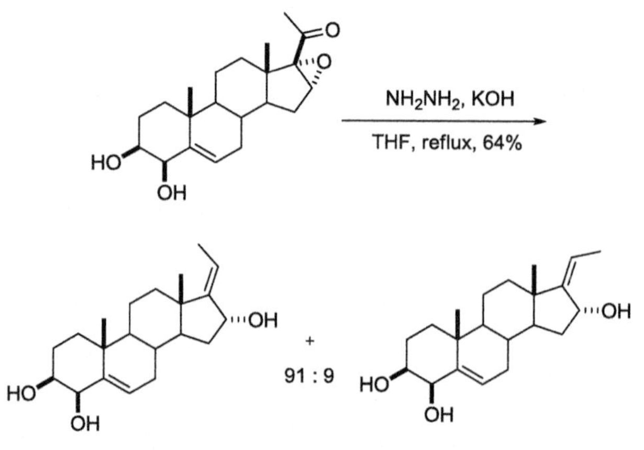

J.J. Li, *Name Reactions: A Collection of Detailed Mechanisms and Synthetic Applications*,
DOI 10.1007/978-3-319-03979-4_285, © Springer International Publishing Switzerland 2014

Example 4[8]

$$NH_2NH_2 \cdot H_2O, MeOH$$
$$HOAc, rt, 59\%$$

Example 5[10]

$$NH_2NH_2 \cdot H_2O, HOAc$$
$$MeOH, 40\%$$

Example 6[11]

1. NaClO, pyr., EtOH
   −10 °C, 20 min, 93%

2. $NH_2NH_2 \cdot H_2O$, $Et_3N$, AcOH
   $CH_3CN$, 30 min, rt, 74%

## References

1. (a) Wharton, P. S.; Bohlen, D. H. *J. Org. Chem.* **1961**, *26*, 3615–3616. (b) Wharton, P. S. *J. Org. Chem.* **1961**, *26*, 4781–4782. Peter S. Wharton earned his Ph.D. at Yale University under the tutelage of Harry H. Wasserman and began his independent academia career at University of Wisconsin at Madison. This was his first paper out of graduate school!
2. Caine, D. *Org. Prep. Proced. Int.* **1988**, *20*, 1–51. (Review).
3. Dupuy, C.; Luche, J. L. *Tetrahedron* **1989**, *45*, 3437–3444. (Review).
4. Thomas, A. F.; Di Giorgio, R.; Guntern, O. *Helv. Chim. Acta* **1989**, *72*, 767–773.
5. Kim, G.; Chu-Moyer, M. Y.; Danishefsky, S. J. *J. Am. Chem. Soc.* **1990**, *112*, 2003–2004.
6. Yamada, K.-i.; Arai, T.; Sasai, H.; Shibasaki, M. *J. Org. Chem.* **1998**, *63*, 3666–3672.
7. Di Filippo, M.; Fezza, F.; Izzo, I.; De Riccardis, F.; Sodano, G. *Eur. J. Org. Chem.* **2000**, 3247–3249.
8. Takagi, R.; Tojo, K.; Iwata, M.; Ohkata, K. *Org. Biomol. Chem.* **2005**, *3*, 2031–2036.
9. Li, J. J. *Wharton Reaction*. In *Name Reactions for Functional Group Transformations*; Li, J. J., Ed.; Wiley: Hoboken, NJ, **2007**, pp 152–158. (Review).
10. Hoye, T. R.; Jeffrey, C. S.; Nelson, D. P. *Org. Lett.* **2010**, *12*, 52–55.
11. Isaka, N.; Tamiya, M.; Hasegawa, A.; Ishiguro, M. *Eur. J. Org. Chem.* **2012**, 665–668.

# Williamson ether synthesis

Ether from the alkylation of alkoxides by alkyl halides. In order for reaction to go smoothly, the alkyl halides are preferred to be primary. Secondary halides work as well sometimes, but tertiary halides do not work at all because $E_2$ elimination will be the predominant reaction pathway.

Example 1, Cyclic etherification[9]

# References

1.  Williamson, A. W. *J. Chem. Soc.* **1852**, *4*, 229–239. Alexander William Williamson (1824–1904) discovered this reaction in 1850 at University College, London.
2.  Dermer, O. C. *Chem. Rev.* **1934**, *14*, 385–430. (Review).
3.  Freedman, H. H.; Dubois, R. A. *Tetrahedron Lett.* **1975**, *16*, 3251–3254.
4.  Jursic, B. *Tetrahedron* **1988**, *44*, 6677–6680.
5.  Tan, S. N.; Dryfe, R. A.; Girault, H. H. *Helv. Chim. Acta* **1994**, *77*, 231–242.
6.  Silva, A. L.; Quiroz, B.; Maldonado, L. A. *Tetrahedron Lett.* **1998**, *39*, 2055–2058.
7.  Peng, Y.; Song, G. *Green Chem.* **2002**, *4*, 349–351.
8.  Stabile, R. G.; Dicks, A. P. *J. Chem. Educ.* **2003**, *80*, 313–315.
9.  Austad, B. C.; Benayoud, F.; Calkins, T. L.; et al. *Synlett* **2013**, *17*, 327–332.

J.J. Li, *Name Reactions: A Collection of Detailed Mechanisms and Synthetic Applications*,
DOI 10.1007/978-3-319-03979-4_286, © Springer International Publishing Switzerland 2014

# Willgerodt–Kindler reaction

Conversion of a ketone to thioamide, with functional group migration.

In Carmack's mechanism,[2] the most unusual movement of a carbonyl group from methylene carbon to methylene carbon was proposed to go through an intricate pathway *via* a highly reactive intermediate with a sulfur-containing heterocyclic ring. The sulfenamide serves as the isomerization catalyst. e.g.:

J.J. Li, *Name Reactions: A Collection of Detailed Mechanisms and Synthetic Applications*, DOI 10.1007/978-3-319-03979-4_287, © Springer International Publishing Switzerland 2014

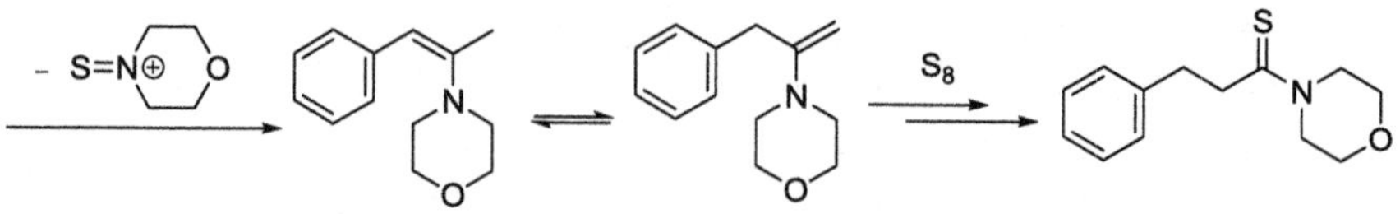

Example 1, The Willgerodt–Kindler reaction was a key operation in the initial synthesis of racemic naproxen (Aleve):[3]

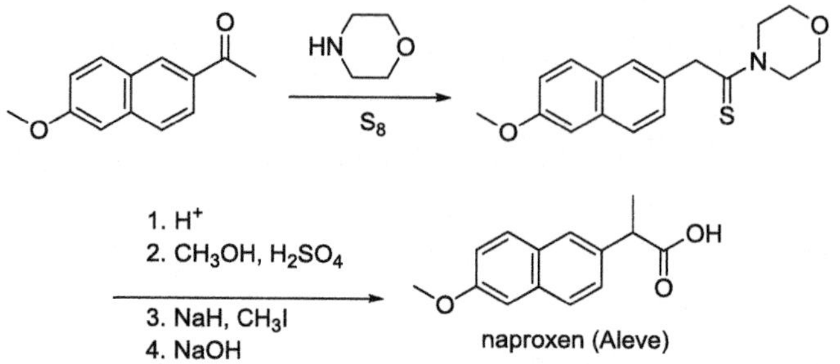

naproxen (Aleve)

Example 2[5]

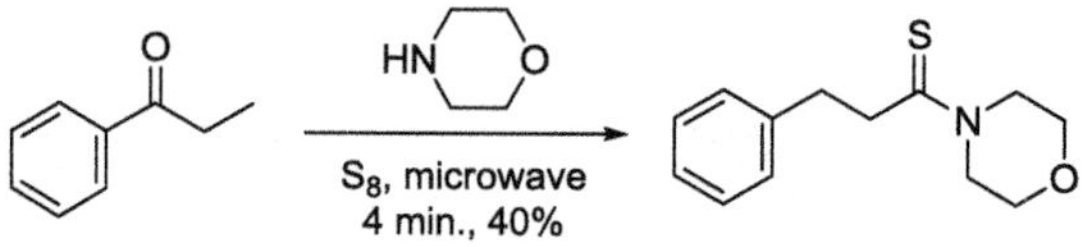

$S_8$, microwave
4 min., 40%

Example 3, A domino annulation reaction under Willgerodt–Kindler conditions:[10]

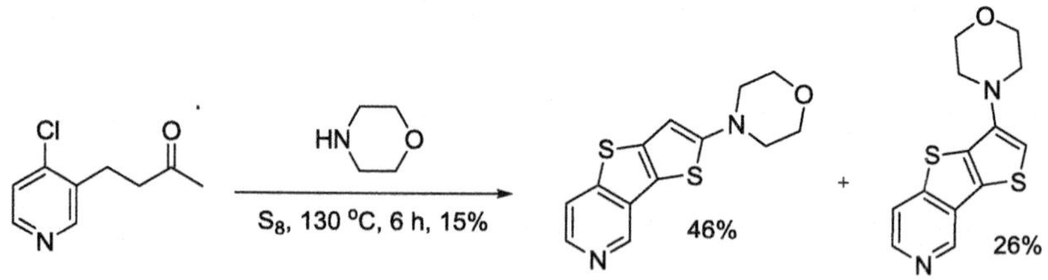

$S_8$, 130 °C, 6 h, 15%          46%          +          26%

## References

1.  (a) Willgerodt, C. *Ber.* **1887**, *20*, 2467–2470.  Conrad Willgerodt (1841–1930), born in Harlingerode, Germany, was a son of a farmer.  He worked to accumulate enough money to support his study toward his doctorate, which he received from Claus.  He became a professor at Freiburg, where he taught for 37 years.  (b) Kindler, K. *Arch. Pharm.* **1927**, *265*, 389–415.
2.  Carmack, M.; Spielman, M. A. *Org. React.* **1946**, *3*, 83–107.  (Review).
3.  Harrison, I. T.; Lewis, B.; Nelson, P.; Rooks, W.; Roskowski, A.; Tomolonis, A.; Fried, J. H. *J. Med. Chem.* **1970**, *13*, 203–205.
4.  Carmack, M. *J. Heterocycl. Chem.* **1989**, *26*, 1319–1323.

5.  Nooshabadi, M.; Aghapoor, K.; Darabi, H. R.; Mojtahedi, M. M. *Tetrahedron Lett.* **1999,** *40*, 7549–7552.

6.  Alam, M. M.; Adapa, S. R. *Synth. Commun.* **2003,** *33,* 59–63.

7.  Reza Darabi, H.; Aghapoor, K.; Tajbakhsh, M. *Tetrahedron Lett.* **2004,** *45*, 4167–4169.

8.  Purrello, G. *Heterocycles* **2005,** *65*, 411–449. (Review).

9.  Okamoto, K.; Yamamoto, T.; Kanbara, T. *Synlett* **2007,** 2687–2690.

10. Kadzimirsz, D.; Kramer, D.; Sripanom, L.; Oppel, I. M.; Rodziewicz, P.; Doltsinis, N. L.; Dyker, G. *J. Org. Chem.* **2008,** *73*, 4644–4649.

11. Eftekhari-Sis, B.; Khajeh, S. V.i; Büyükgüngör, O. *Synlett* **2013,** *24*, 977–980.

# Wittig reaction

Olefination of carbonyls using phosphorus ylides, typically the Z-olefin is obtained.

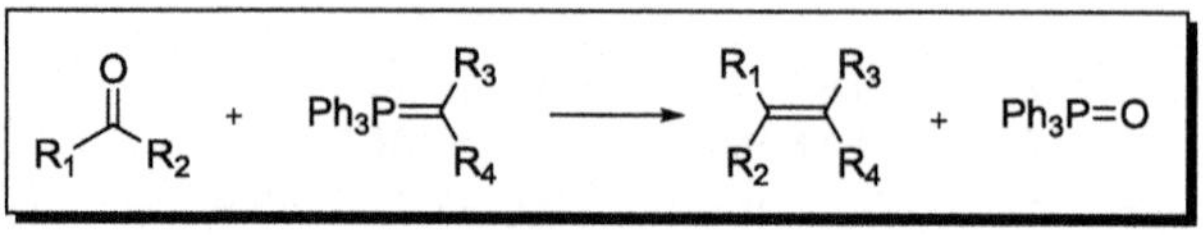

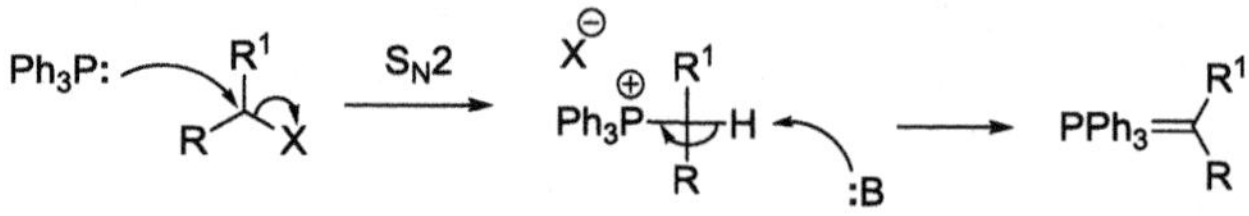

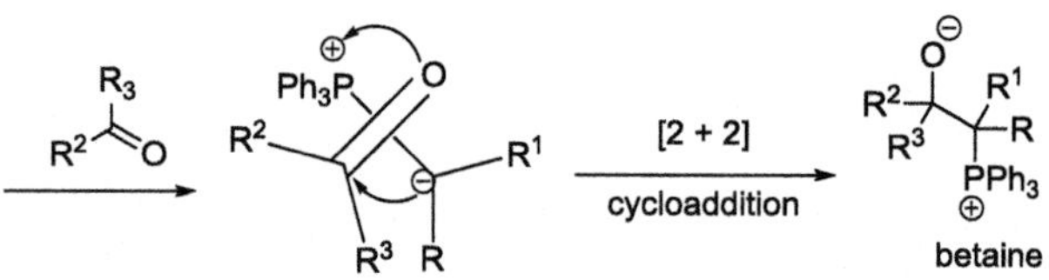

The "puckered" transition state, irreversible and concerted

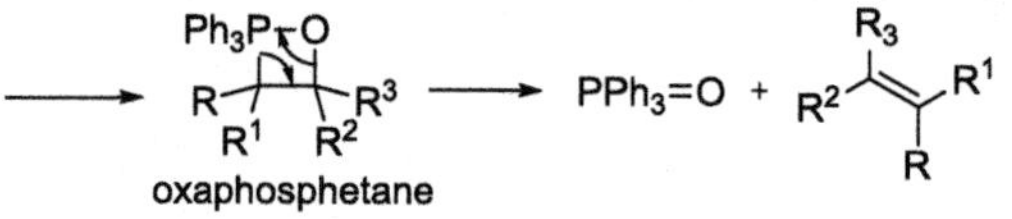

oxaphosphetane

Example 1[3]

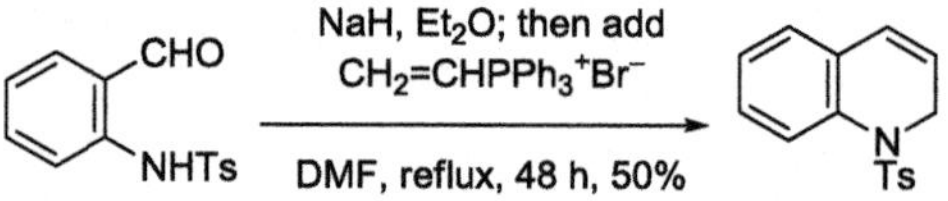

Example 2[4]

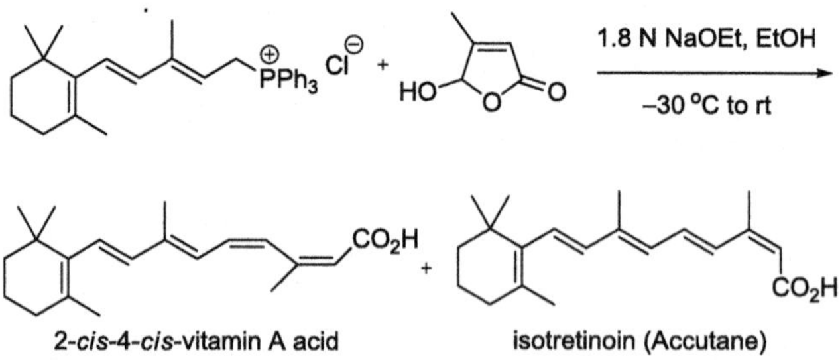

2-*cis*-4-*cis*-vitamin A acid                     isotretinoin (Accutane)

Example 3[5]

$t$-BuO$_2$C                                         CO$_2t$-Bu

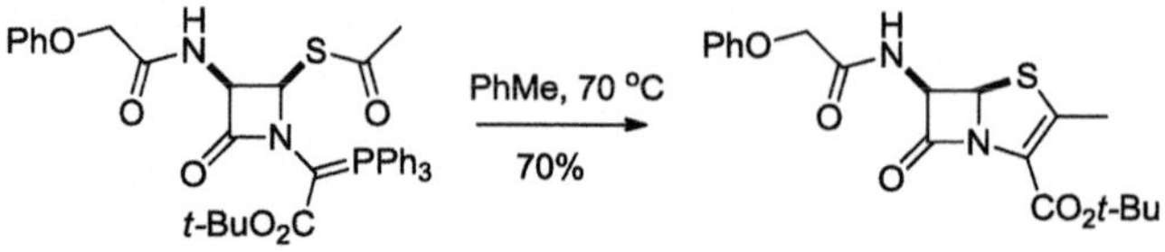

J.J. Li, *Name Reactions: A Collection of Detailed Mechanisms and Synthetic Applications*, DOI 10.1007/978-3-319-03979-4_288, © Springer International Publishing Switzerland 2014

Example 4[9]

Example 5[11]

Z/E = 60:40

## References

1. Wittig, G.; Schöllkopf, U. *Ber.* **1954**, *87*, 1318–1330. Georg Wittig (Germany, 1897–1987), born in Berlin, Germany, received his Ph.D. from K. von Auwers. He shared the Nobel Prize in Chemistry in 1981 with Herbert C. Brown (USA, 1912–2004) for their development of organic boron and phosphorous compounds.
2. Maercker, A. *Org. React.* **1965**, *14*, 270–490. (Review).
3. Schweizer, E. E.; Smucker, L. D. *J. Org. Chem.* **1966**, *31*, 3146–3149.
4. Garbers, C. F.; Schneider, D. F.; van der Merwe, J. P. *J. Chem. Soc. (C)* **1968**, 1982–1983.
5. Ernest, I.; Gosteli, J.; Greengrass, C. W.; Holick, W.; Jackman, D. E.; Pfaendler, H. R.; Woodward, R. B. *J. Am. Chem. Soc.* **1978**, *100*, 8214–8222.
6. Murphy, P. J.; Brennan, J. *Chem. Soc. Rev.* **1988**, *17*, 1–30. (Review).
7. Maryanoff, B. E.; Reitz, A. B. *Chem. Rev.* **1988**, *89*, 863–927. (Review).
8. Vedejs, E.; Peterson, M. J. *Top. Stereochem.* **1994**, *21*, 1–157. (Review).
9. Nicolaou, K. C. *Angew. Chem. Int. Ed.* **1996**, *35*, 589–607.
10. Rong, F. *Wittig reaction* in. In *Name Reactions for Homologations-Part I*; Li, J. J., Ed.; Wiley: Hoboken, NJ, **2009**, pp 588–612. (Review).
11. Kajjout, M.; Smietana, M.; Leroy, J.; Rolando, C. *Tetrahedron Lett.* **2013**, *38*, 1658–1660.

## *Schlosser modification of the Wittig reaction*

Also known as the Wittig–Schlosser reaction. The normal Wittig reaction of non-stabilized ylides with aldehydes gives *Z*-olefins. The Schlosser modification of the Wittig reaction of nonstabilized ylides furnishes *E*-olefins instead.

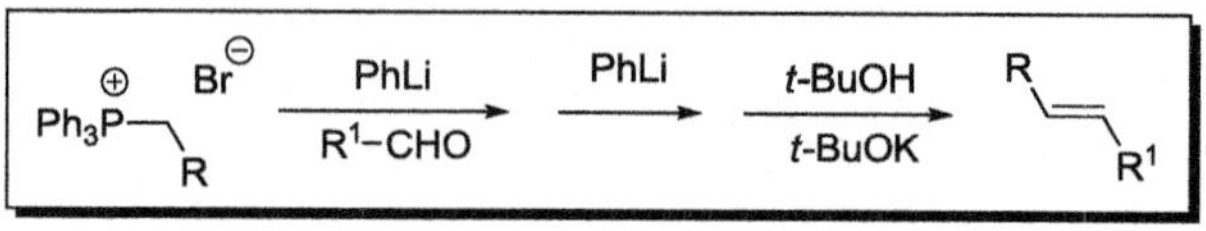

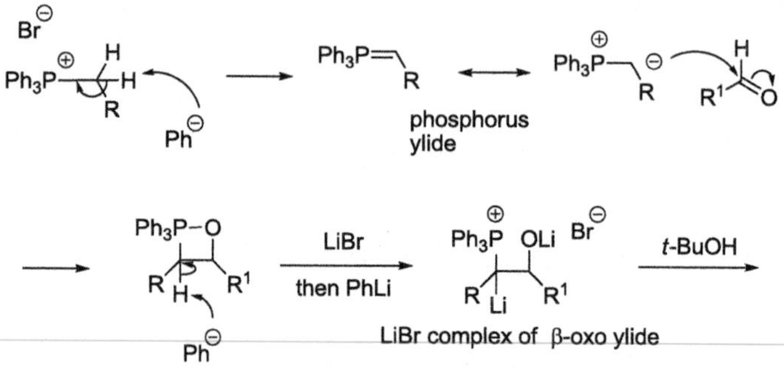

These conditions allow for the *erythreo* betaine to interconvert to the *threo* betaine

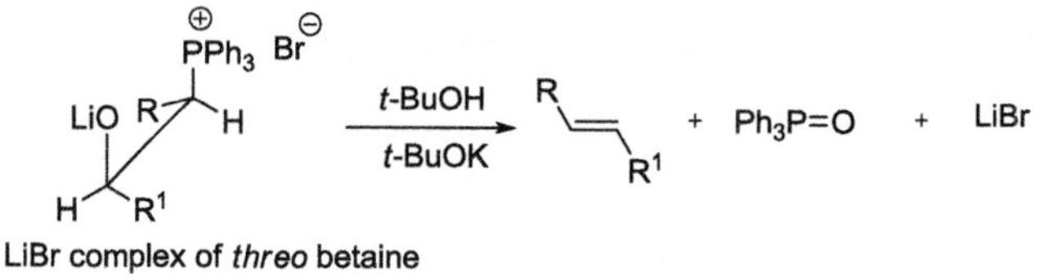

LiBr complex of *threo* betaine

Example 1[6]

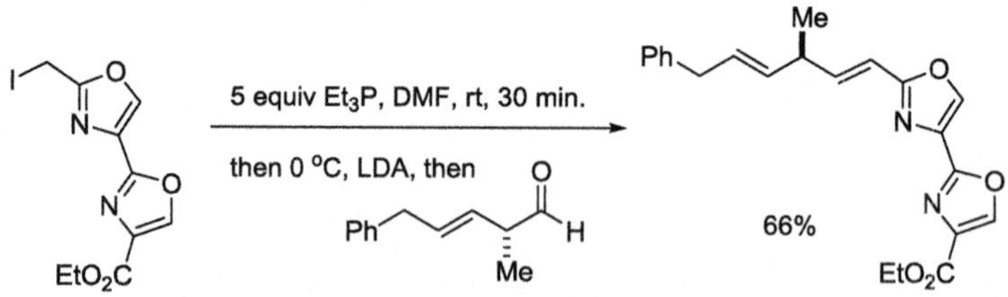

Example 2[10]

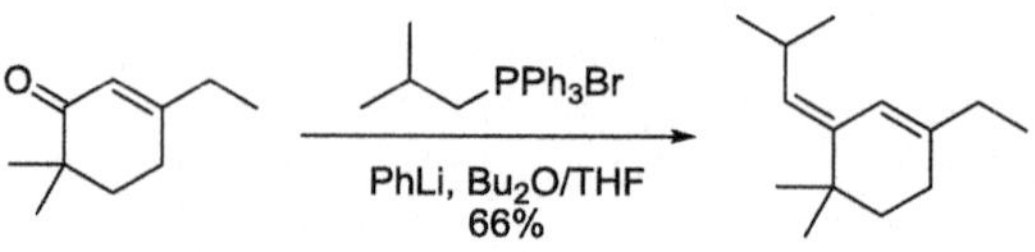

Example 3[11]

## References

1. (a) Schlosser, M.; Christmann, K. F. *Angew. Chem. Int. Ed.* **1966**, *5*, 126. (b) Schlosser, M.; Christmann, K. F. *Ann.* **1967**, *708*, 1–35. (c) Schlosser, M.; Christmann, K. F.; Piskala, A.; Coffinet, D. *Synthesis* **1971**, 29–31. Born in Ludwigshafen on Rhine (Germany), Manfred Schlosser earned his Ph.D. in 1960 under Georg Wittig. He intiailly worked at at the German Cancer Research Center and moved to France to be a professor at the University of Lausanne. Schlosser retired in 2004.
2. van Tamelen, E. E.; Leiden, T. M. *J. Am. Chem. Soc.* **1982**, *104*, 2061–2062.
3. Parziale, P. A.; Berson, J. A. *J. Am. Chem. Soc.* **1991**, *113*, 4595–606.
4. Sarkar, T. K.; Ghosh, S. K.; Rao, P. S.; Satapathi, T. K.; Mamdapur, V. R. *Tetrahedron* **1992**, *48*, 6897–6908.
5. Deagostino, A.; Prandi, C.; Tonachini, G.; Venturello, P. *Trends Org. Chem.* **1995**, *5*, 103–113. (Review).
6. Celatka, C. A.; Liu, P.; Panek, J. S. *Tetrahedron Lett.* **1997**, *38*, 5449–5452.
7. Panek, J. S.; Liu, P. *J. Am. Chem. Soc.* **2000**, *122* 11090–11097.
8. Duffield, J. J.; Pettit, G. R. *J. Nat. Prod.* **2001**, *64*, 472–479.
9. Kraft, P.; Popaj, K. *Eur. J. Org. Chem.* **2004**, 4995–5002.
10. Kraft, P.; Popaj, K. *Eur. J. Org. Chem.* **2008**, 4806–4814.
11. Hodgson, D. M.; Arif, T. *Org. Lett.* **2010**, *12*, 4204–4207.
12. Mikula, H.; Hametner, C.; Froehlich, J. *Synth. Commun.* **2013**, *43*, 1939–1946.

# [1,2]-Wittig rearrangement

Treatment of ethers with bases such as alkyl lithium results in alcohols.

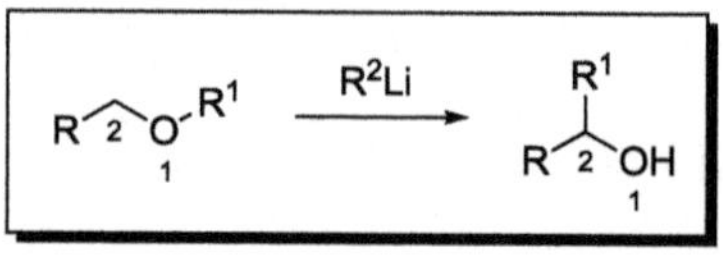

The [1,2]-Wittig rearrangement is believed to proceed via a radical mechanism:

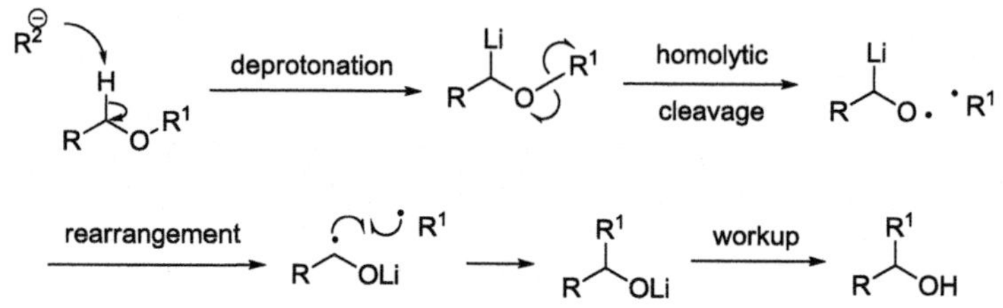

Example 1, Aza [1,2]-Wittig rearrangement[2]

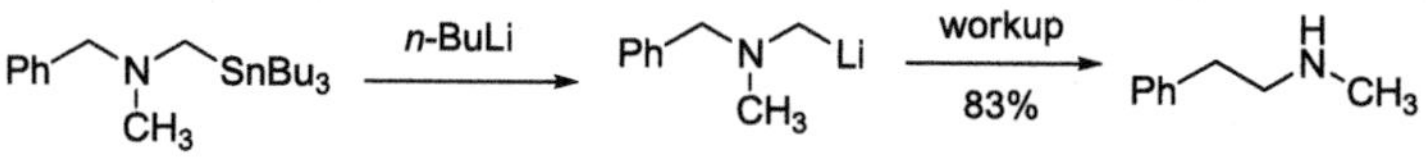

Example 2[3]

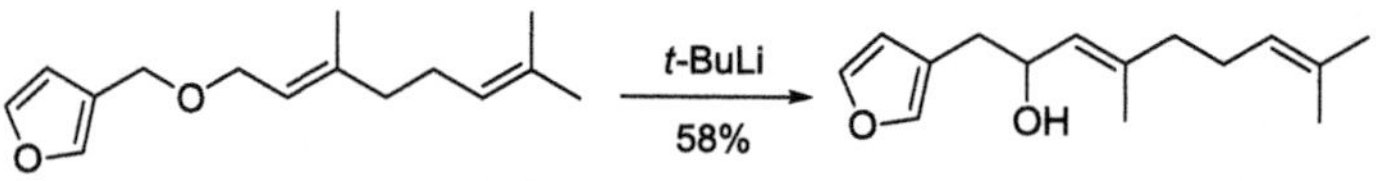

Example 3[4]

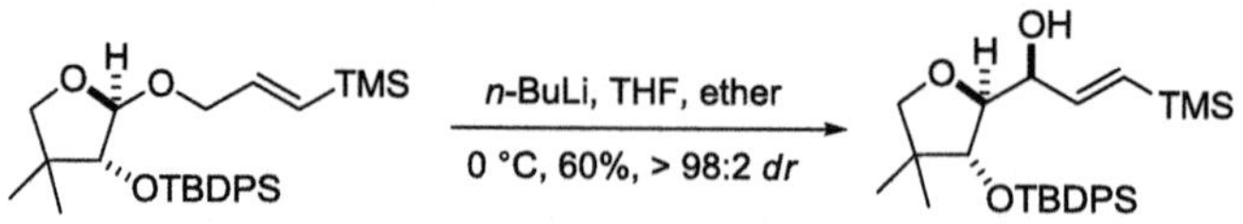

Example 4[6]

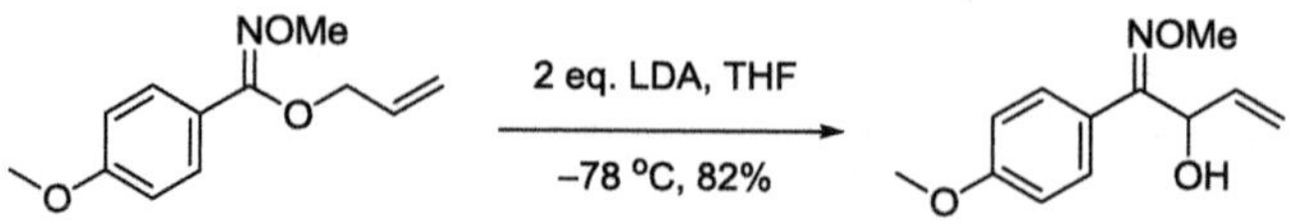

J.J. Li, *Name Reactions: A Collection of Detailed Mechanisms and Synthetic Applications*,
DOI 10.1007/978-3-319-03979-4_289, © Springer International Publishing Switzerland 2014

Example 5[8]

Example 6[9]

Example 7[11]

## References

1  Wittig, G.; Löhmann, L. *Ann.* **1942,** *550,* 260–268.

2  Peterson, D. J.; Ward, J. F. *J. Organomet. Chem.* **1974,** *66,* 209–217.

3  Tsubuki, M.; Okita, H.; Honda, T. *J. Chem. Soc., Chem. Commun.* **1995,** 2135–2136.

4  Tomooka, K.; Yamamoto, H.; Nakai, T. *J. Am. Chem. Soc.* **1996,** *118,* 3317–3318.

5  Maleczka, R. E., Jr.; Geng, F. *J. Am. Chem. Soc.* **1998,** *120,* 8551–8552.

6  Miyata, O.; Asai, H.; Naito, T. *Synlett* **1999,** 1915–1916.

7  Katritzky, A. R.; Fang, Y. *Heterocycles* **2000,** *53,* 1783–1788.

8  Tomooka, K.; Kikuchi, M.; Igawa, K.; Suzuki, M.; Keong, P.-H.; Nakai, T. *Angew. Chem. Int. Ed.* **2000,** *39,* 4502–4505.

9  Miyata, O.; Asai, H.; Naito, T. *Chem. Pharm. Bull.* **2005,** *53,* 355–360.

10  Wolfe, J. P.; Guthrie, N. J. *[1,2]-Wittig Rearrangement.* In *Name Reactions for Homologations-Part II*; Li, J. J., Ed.; Wiley: Hoboken, NJ, **2009,** pp 226–240. (Review).

11  Onyeozili, E. N.; Mori-Quiroz, L. M.; Maleczka, R. E., Jr. *Tetrahedron* **2013,** *69,* 849–860.

## [2,3]-Wittig rearrangement

Transformation of allyl ethers into homoallylic alcohols by treatment with base. Also known as the Still–Wittig rearrangement. *Cf.* Sommelet–Hauser rearrangement.

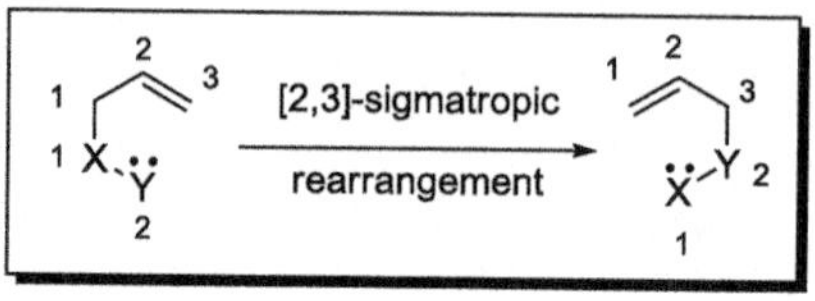
[2,3]-sigmatropic
rearrangement

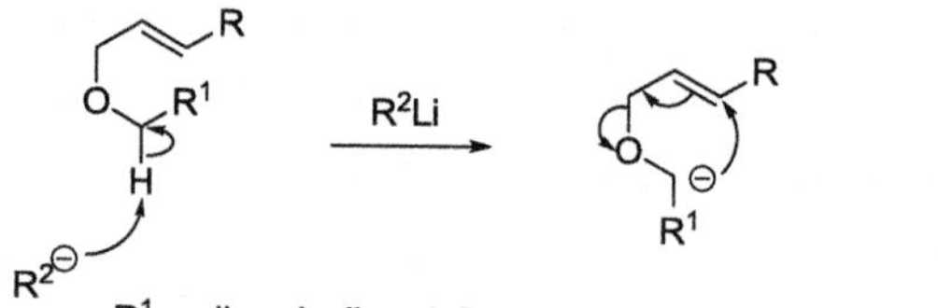
$R^2Li$

$R^1$ = alkynyl, alkenyl, Ph, COR, CN.

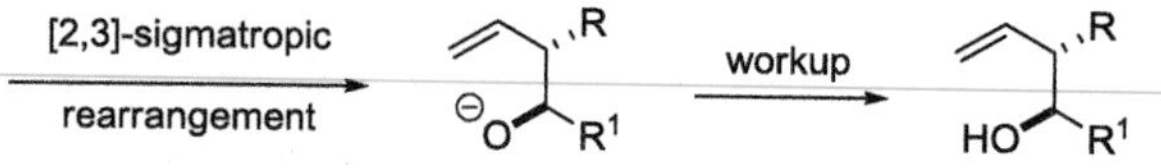
[2,3]-sigmatropic
rearrangement
workup

Example 1[3]

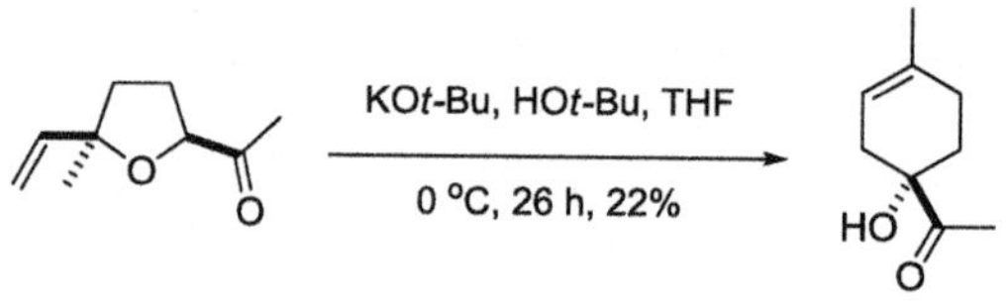
KO*t*-Bu, HO*t*-Bu, THF
0 °C, 26 h, 22%

Example 2[5]

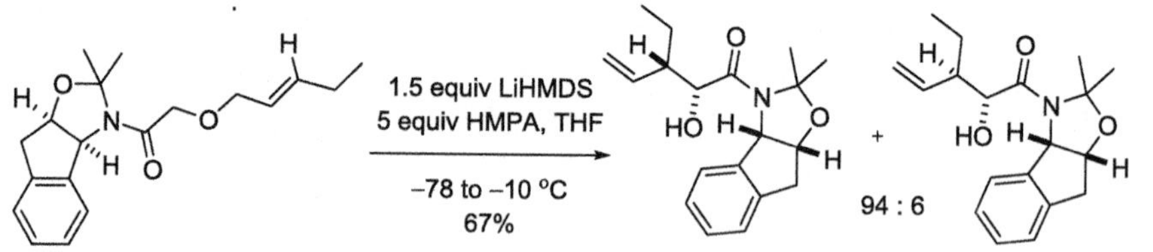
1.5 equiv LiHMDS
5 equiv HMPA, THF
–78 to –10 °C
67%
94 : 6

Example 3[6]

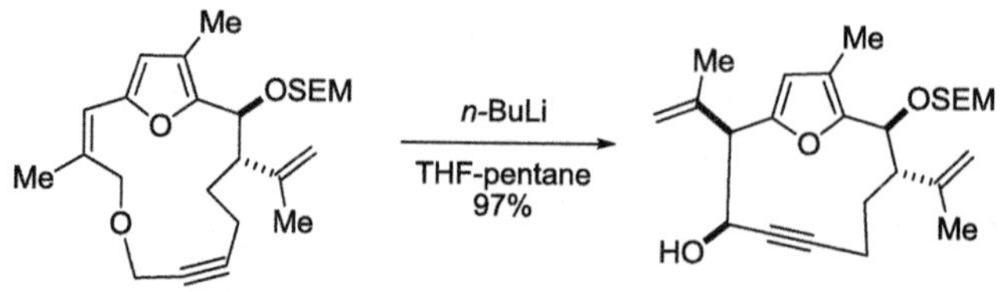
*n*-BuLi
THF-pentane
97%

J.J. Li, *Name Reactions: A Collection of Detailed Mechanisms and Synthetic Applications*,
DOI 10.1007/978-3-319-03979-4_290, © Springer International Publishing Switzerland 2014

Example 4, Tandem Wittig rearrangement/alkylative cyclization reactions[6]

## References

1. Cast, J.; Stevens, T. S.; Holmes, J. *J. Chem. Soc.* **1960**, 3521–3527.
2. Thomas, A. F.; Dubini, R. *Helv. Chim. Acta* **1974**, *57*, 2084–2087.
3. Nakai, T.; Mikami, K.; Taya, S.; Kimura, Y.; Mimura, T. *Tetrahedron Lett.* **1981**, *22*, 69–72.
4. Nakai, T.; Mikami, K. *Org. React.* **1994**, *46*, 105–209. (Review).
5. Kress, M. H.; Yang, C.; Yasuda, N.; Grabowski, E. J. J. *Tetrahedron Lett.* **1997**, *38*, 2633–2636.
6. Marshall, J. A.; Liao, J. *J. Org. Chem.* **1998**, *63*, 5962–5970.
7. Maleczka, R. E., Jr.; Geng, F. *Org. Lett.* **1999**, *1*, 1111–1113.
8. Tsubuki, M.; Kamata, T.; Nakatani, M.; Yamazaki, K.; Matsui, T.; Honda, T. *Tetrahedron: Asymmetry* **2000**, *11*, 4725–4736.
9. Schaudt, M.; Blechert, S. *J. Org. Chem.* **2003**, *68*, 2913–2920.
10. Ahmad, N. M. *[2,3]-Wittig Rearrangement*. In *Name Reactions for Homologations-Part II*; Li, J. J., Ed.; Wiley: Hoboken, NJ, 2009, pp 241–256. (Review).
11. Everett, R. K.; Wolfe, J. P. *Org. Lett.* **2013**, *15*, 2926–2929.

## Wohl–Ziegler reaction

The Wohl–Ziegler reaction is the reaction of an allylic or benzylic substrate with
N-bromosuccinimide (NBS) under radical initiating conditions to provide the corresponding allylic or benzylic bromide. Conditions used to promote the radical
reaction are typically radical initiators, light and/or heat; carbon tetrachloride
($CCl_4$) is typically utilized as the solvent.

N-Bromosuccinimide (NBS) contains a small amount of HBr from the reaction between NBS and moisture. The minute amount of HBr, in turn, reacts with NBS to
provide a low, constant concentration of $Br_2$. Moreover, NBS reacts with the HBr
by-product to produce $Br_2$ and to prevent HBr addition across the double bond.

N-bromosuccinimide (NBS)                    succinimide

Initiation:

2,2'-azobisisobutyronitrile (AIBN)

Propagation:

The bromine radical is now available for the next cycle of the radical chain reaction.

Termination:

J.J. Li, *Name Reactions: A Collection of Detailed Mechanisms and Synthetic Applications*,
DOI 10.1007/978-3-319-03979-4_291, © Springer International Publishing Switzerland 2014

## Example 1[3]

NBS, CCl₄ → NBS, CCl$_4$
reflux, 30 min., 48%

## Example 2[7]

1.2 equiv NBS
0.1 equiv AIBN

solvent free, 60 to 65 °C
89%

## Example 3[8]

NBS, CCl$_4$

reflux, 71%

## Example 4[9]

NBS, AIBN (cat)

CCl$_4$, reflux, 2 h,
95%

## References

1. Wohl, A. *Ber.* **1919**, *52*, 51–63. Alfred Wohl (1863–1939), born in Graudenz, Germany, received his Ph.D. from A. W. Hofmann. In 1904, he was appointed Professor of Chemistry at the Technische Hochschule in Danzig.
2. Ziegler, K.; Spath, A.; Schaaf, E.; Schumann, W.; Winkelmann, E. *Ann.* **1942**, *551*, 80–119. Karl Ziegler (1898–1973), born in Helsa, Germany, received Ph.D. in 1920 from von Auwers at the University of Marburg. He became the director of the Max-Planck-Institut für Kohlenforschung at Mülheim/Ruhr in 1943 and stayed there until 1969. He shared the Nobel Prize in Chemistry in 1963 with Giulio Natta (1903–1979) for their work in polymer chemistry. The Ziegler–Natta catalyst is widely used in polymerization.
3. Djerassi, C.; Scholz, C. R. *J. Org. Chem.* **1949**, *14*, 660–663.
4. Allen, J. G.; Danishefsky, S. J. *J. Am. Chem. Soc.* **2001**, *123*, 351–352.
5. Detterbeck, R.; Hesse, M. *Tetrahedron Lett.* **2002**, *43*, 4609–4612.
6. Stevens, C. V.; Van Heecke, G.; Barbero, C.; Patora, K.; De Kimpe, N.; Verhe, R. *Synlett* **2002**, 1089–1092.
7. Togo, H.; Hirai, T. *Synlett* **2003**, 702–704.
8. Marjo, C. E.; Bishop, R.; Craig, D. C.; Scudder, M. L. *Mendeleev Commun.* **2004**, 278–279.
9. Yeung, Y.-Y.; Hong, S.; Corey, E. J. *J. Am. Chem. Soc.* **2006**, *128*, 6310–6311.
10. Curran, T. T. *Wohl–Ziegler reaction.* In *Name Reactions for Homologations-Part I*; Li, J. J., Ed.; Wiley: Hoboken, NJ, **2009**, pp 661–674. (Review).
11. Tsuchiya, D.; Kawagoe, Y.; Moriyama, K.; Togo, H. *Org. Lett.* **2013**, *15*, 4194–4197.

# Wolff rearrangement

Conversion of an α-diazoketone into a ketene.

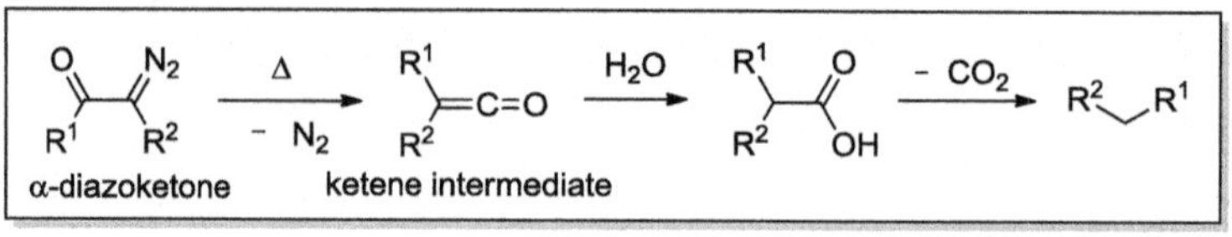

Step-wise mechanism:

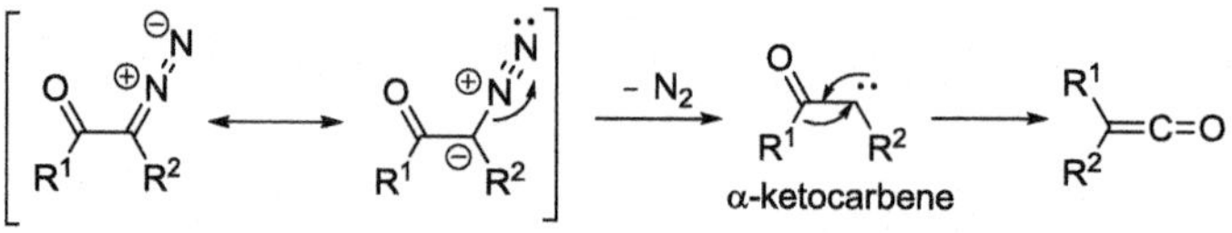

Treatment of the ketene with water would give the corresponding homologated carboxylic acid.

Concerted mechanism:

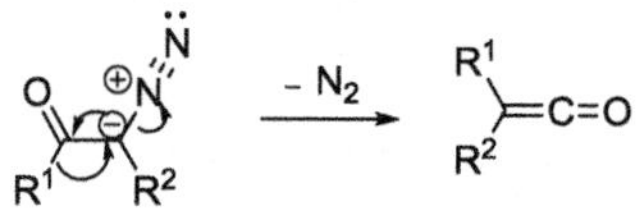

Example 1[2]

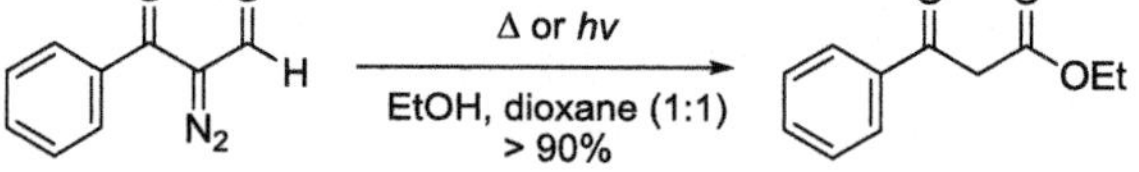

Example 2[3]

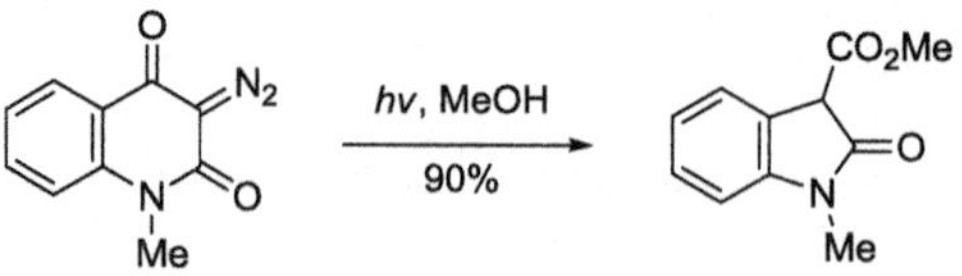

Example 3[4]

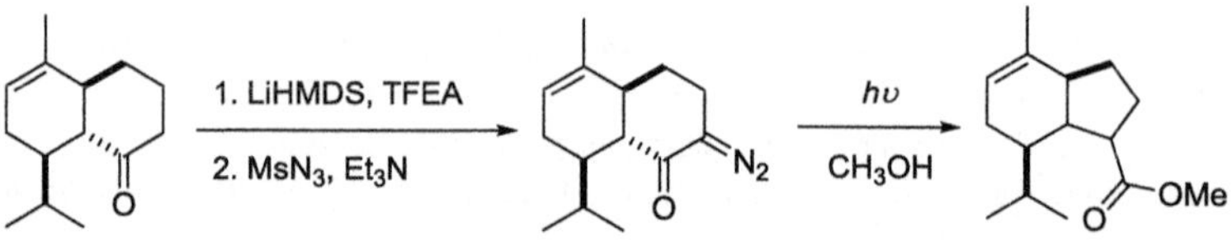

J.J. Li, *Name Reactions: A Collection of Detailed Mechanisms and Synthetic Applications*,
DOI 10.1007/978-3-319-03979-4_292, © Springer International Publishing Switzerland 2014

Example 4[9]

55% — Wolff rearrangement  39% — N-H insertion

Example 5[11]

BuNH₂, PhCHO
toluene, 110 °C
6 h, 80%

## References

1. Wolff, L. *Ann.* **1912**, *394*, 23–108. Johann Ludwig Wolff (1857–1919) earned his doctorate in 1882 under Fittig at Strasbourg, where he later became an instructor. In 1891, Wolff joined the faculty of Jena, where he collaborated with Knorr for 27 years.
2. Zeller, K.-P.; Meier, H.; Müller, E. *Tetrahedron* **1972**, *28*, 5831–5838.
3. Kappe, C.; Fäber, G.; Wentrup, C.; Kappe, T. *Ber.* **1993**, *126*, 2357–2360.
4. Taber, D. F.; Kong, S.; Malcolm, S. C. *J. Org. Chem.* **1998**, *63*, 7953–7956.
5. Yang, H.; Foster, K.; Stephenson, C. R. J.; Brown, W.; Roberts, E. *Org. Lett.* **2000**, *2*, 2177–2179.
6. Kirmse, W. "100 years of the Wolff Rearrangement" *Eur. J. Org. Chem.* **2002**, 2193–2256. (Review).
7. Julian, R. R.; May, J. A.; Stoltz, B. M.; Beauchamp, J. L. *J. Am. Chem. Soc.* **2003**, *125*, 4478–4486.
8. Zeller, K.-P.; Blocher, A.; Haiss, P. *Mini-Reviews Org. Chem.* **2004**, *1*, 291–308. (Review).
9. Davies, J. R.; Kane, P. D.; Moody, C. J.; Slawin, A. M. Z. *J. Org. Chem.* **2005**, *70*, 5840–5851.
10. Kumar, R. R.; Balasubramanian, M. *Wolff Rearrangement*. In *Name Reactions for Homologations-Part II*; Li, J. J., Ed.; Wiley: Hoboken, NJ, **2009**, pp 257–273. (Review).
11. Somai Magar, K. B.; Lee, Y. R. *Org. Lett.* **2013**, *15*, 4288–4291.

## Wolff–Kishner reduction

Carbonyl reduction to methylene using basic hydrazine.

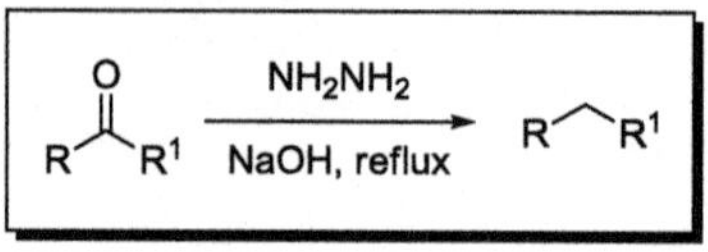

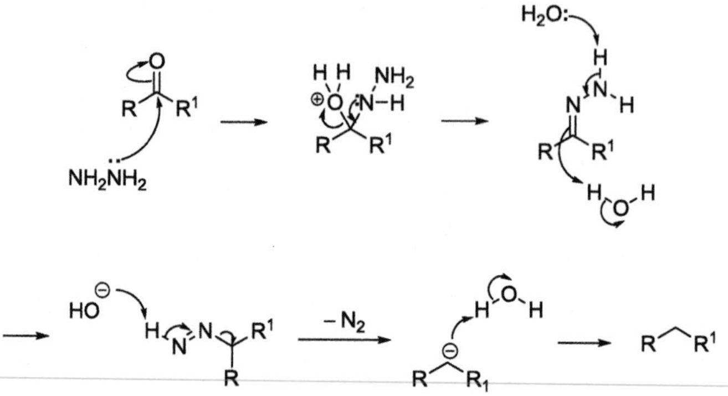

Example 1, The Huang Minlon modification, with loss of ethylene[5]

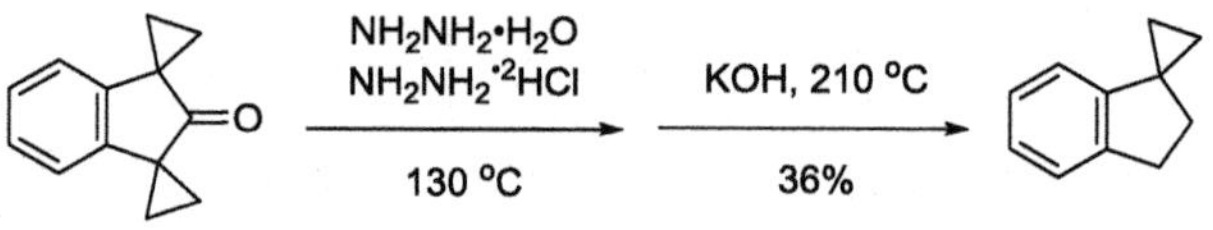

Example 2[7]

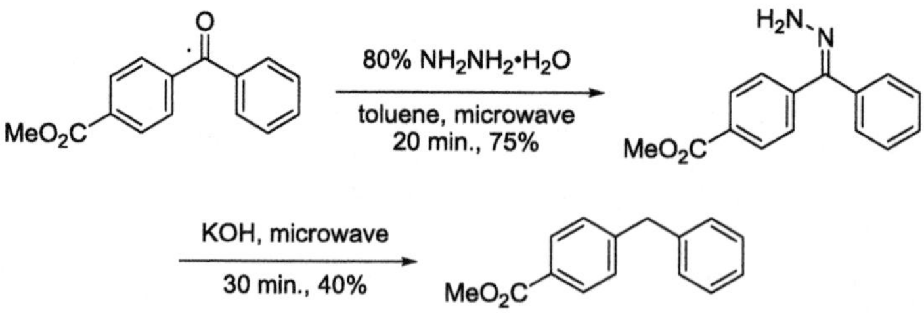

Example 3[8]

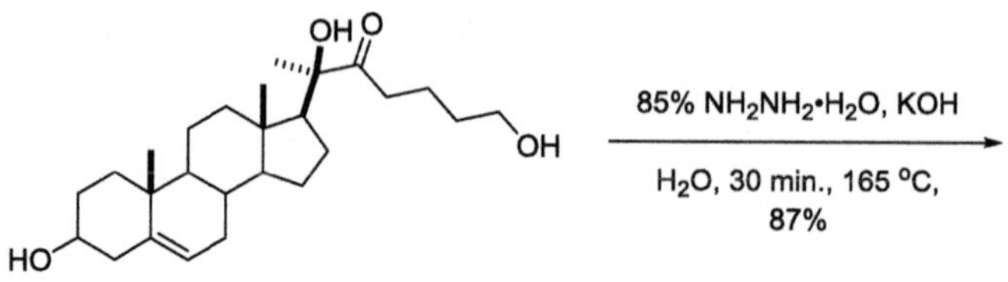

J.J. Li, *Name Reactions: A Collection of Detailed Mechanisms and Synthetic Applications*,
DOI 10.1007/978-3-319-03979-4_293, © Springer International Publishing Switzerland 2014

**Example 4, Huang Minlon modification[10]**

1,5-hydride shift

Example 3[13]

NH$_2$NH$_2$•H$_2$O

diethylene glycol
reflux, 4.75 h, 75%

1. 8 equiv NH$_2$NH$_2$
4 equiv powdered KOH
diethyleneglycol (10 L/kg)
H$_2$O, rt to 143 °C, 2 h
143 to 155 °C, 3.5 h

2. MeCN/H$_2$O
85%

**References**

1. (a) Kishner, N. *J. Russ. Phys. Chem. Soc.* **1911**, *43*, 582–595. Nicolai Kishner was a Russian chemist. (b) Wolff, L. *Ann.* **1912**, *394*, 86. (c) Huang, Minlon *J. Am. Chem. Soc.* **1946**, *68*, 2487–2488. (d) Huang, Minlon *J. Am. Chem. Soc.* **1949**, *71*, 3301–3303. (The Huang Minlon modification).
2. Todd, D. *Org. React.* **1948**, *4*, 378–422. (Review).
3. Cram, D. J.; Sahyun, M. R. V.; Knox, G. R. *J. Am. Chem. Soc.* **1962**, *84*, 1734–1735.
4. Murray, R. K., Jr.; Babiak, K. A. *J. Org. Chem.* **1973**, *38*, 2556–2557.
5. Lemieux, R. P.; Beak, P. *Tetrahedron Lett.* **1989**, *30*, 1353–1356.
6. Taber, D. F.; Stachel, S. J. *Tetrahedron Lett.* **1992**, *33*, 903–906.
7. Gadhwal, S.; Baruah, M.; Sandhu, J. S. *Synlett* **1999**, 1573–1592.
8. Szendi, Z.; Forgó, P.; Tasi, G.; Böcskei, Z.; Nyerges, L.; Sweet, F. *Steroids* **2002**, *67*, 31–38.
9. Bashore, C. G.; Samardjiev, I. J.; Bordner, J.; Coe, J. W. *J. Am. Chem. Soc.* **2003**, *125*, 3268–3272.
10. Pasha, M. A. *Synth. Commun.* **2006**, *36*, 2183–2187.
11. Song, Y.-H.; Seo, J. *J. Heterocycl. Chem.* **2007**, *44*, 1439–1443.
12. Shibahara, M.; Watanabe, M.; Aso, K.; Shinmyozu, T. *Synthesis* **2008**, 3749–3754.
13. Kuethe, J. T.; Childers, K. G.; Peng, Z.; Journet, M.; Humphrey, G. R.; Vickery, T.; Bachert, D.; Lam, T. T. *Org. Process Res. Dev.* **2009**, *13*, 576–580.

# Woodward *cis*-dihydroxylation

*Cf.* Prévost *trans*-dihydroxylation.

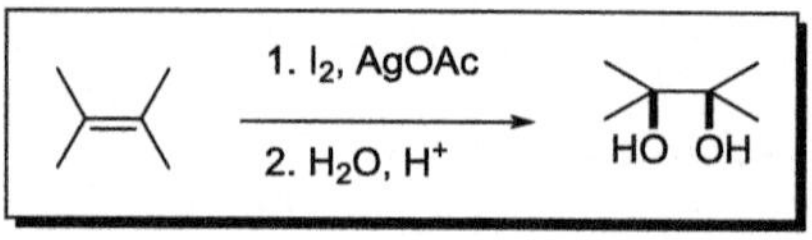

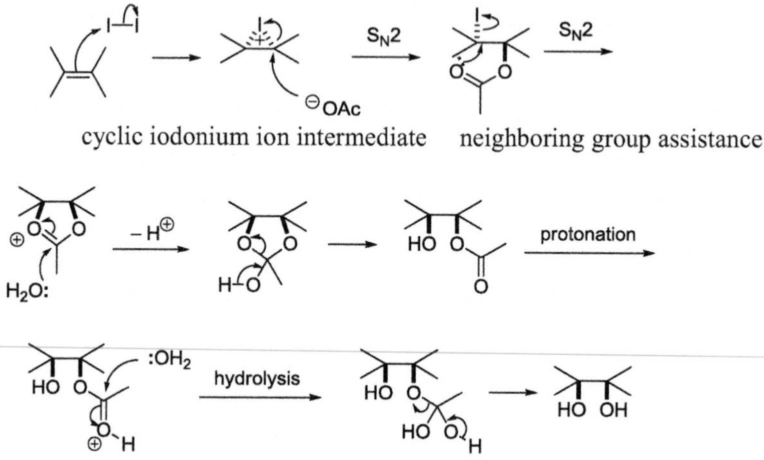

cyclic iodonium ion intermediate    neighboring group assistance

## Example 1[1]

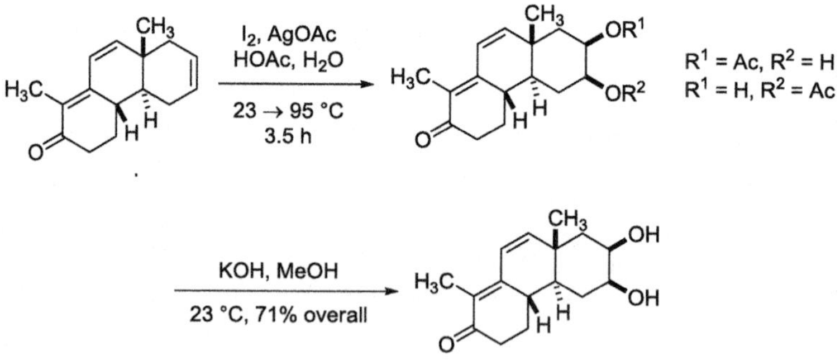

R[1] = Ac, R[2] = H
R[1] = H, R[2] = Ac

KOH, MeOH
23 °C, 71% overall

## Example 2[6]

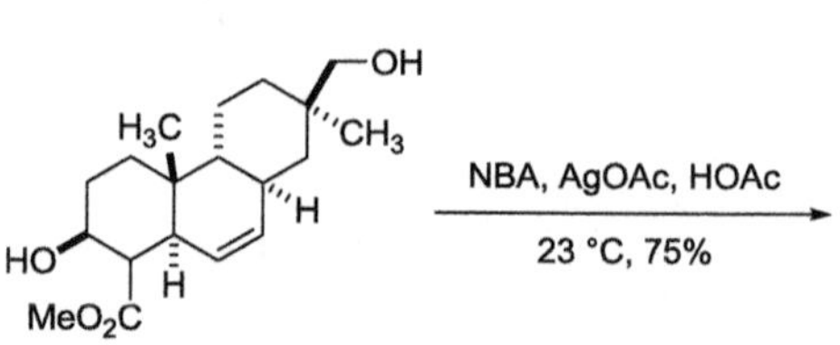

NBA, AgOAc, HOAc
23 °C, 75%

J.J. Li, *Name Reactions: A Collection of Detailed Mechanisms and Synthetic Applications*,
DOI 10.1007/978-3-319-03979-4_294, © Springer International Publishing Switzerland 2014

## References

1. Woodward, R. B.; Brutcher, F. V., Jr. *J. Am. Chem. Soc.* **1958**, *80*, 209–211. Robert Burns Woodward (USA, 1917–1979) won the Nobel Prize in Chemistry in 1953 for his synthesis of natural products.
2. Kirschning, A.; Plumeier, C.; Rose, L. *Chem. Commun.* **1998**, 33–34.
3. Monenschein, H.; Sourkouni-Argirusi, G.; Schubothe, K. M.; O'Hare, T.; Kirschning, A. *Org. Lett.* **1999**, *1*, 2101–2104.
4. Kirschning, A.; Jesberger, M.; Monenschein, H. *Tetrahedron Lett.* **1999**, *40*, 8999–9002.
5. Muraki, T.; Yokoyama, M.; Togo, H. *J. Org. Chem.* **2000**, *65*, 4679–4684.
6. Germain, J.; Deslongchamps, P. *J. Org. Chem.* **2002**, *67*, 5269–5278.
7. Myint, Y. Y.; Pasha, M. A. *J. Chem. Res.* **2004**, 333–335.
8. Emmanuvel, L.; Shaikh, T. M. A.; Sudalai, A. *Org. Lett.* **2005**, *7*, 5071–5074.
9. Mergott, D. J. *Woodward* cis-*dihydroxylation*. In *Name Reactions for Functional Group Transformations*; Li, J. J., Ed.; Wiley: Hoboken, NJ, **2007**, pp 327–332. (Review).
10. Burlingham, B. T.; Rettig, J. C. *J. Chem. Ed.* **2008**, *85*, 959–961.

# Yamaguchi esterification

Esterification using 2,4,6-trichlorobenzoyl chloride (the Yamaguchi reagent).

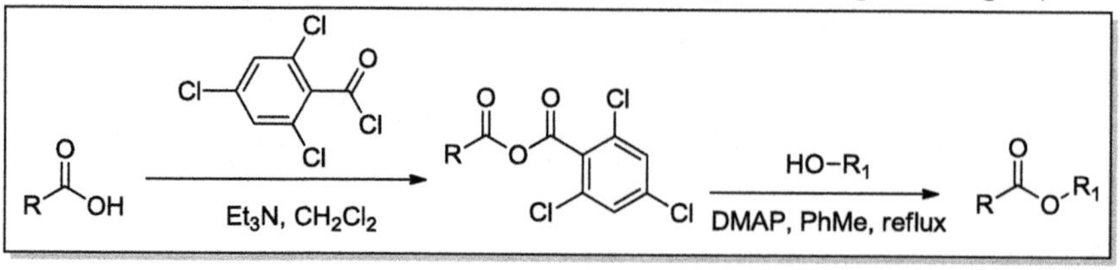

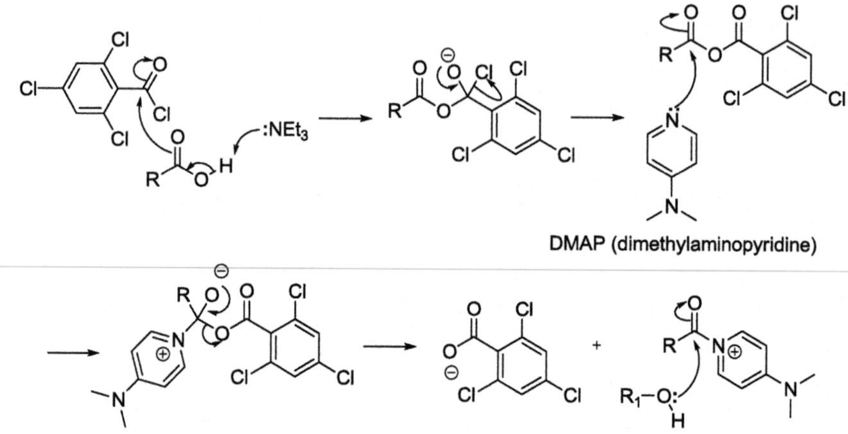

DMAP (dimethylaminopyridine)

Steric hindrance of the chloro substituents blocks attack of the other carbonyl of the mixed anhydride intermediate.

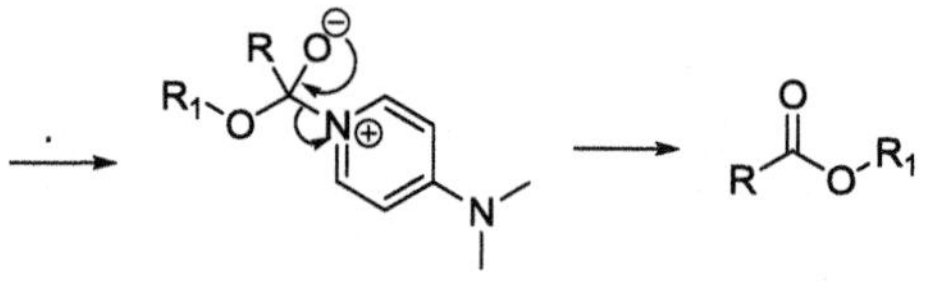

Example 1, Intermolecular coupling[5]

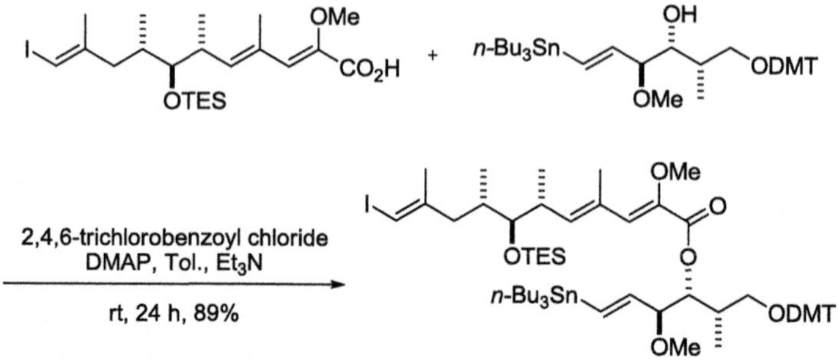

2,4,6-trichlorobenzoyl chloride
DMAP, Tol., Et₃N

rt, 24 h, 89%

J.J. Li, *Name Reactions: A Collection of Detailed Mechanisms and Synthetic Applications*,
DOI 10.1007/978-3-319-03979-4_295, © Springer International Publishing Switzerland 2014

Example 2, Intramolecular coupling[7]

2,4,6-trichlorobenzoyl chloride
Et₃N, THF
_____
then DMAP, toluene, 62%

Example 3, Dimerization[8]

Yamaguchi reagent
_____
125 °C, 6 h, 66%

## References

1. (a) Inanaga, J.; Hirata, K.; Saeki, H.; Katsuki, T.; Yamaguchi, M. *Bull. Chem. Soc. Jpn.* **1979**, *52*, 1989–1993. (b) Kawanami, Y.; Dainobu, Y.; Inanaga, J.; Katsuki, T.; Yamaguchi, M. *Bull. Chem. Soc. Jpn.* **1981**, *54*, 943–944. Masaru Yamaguchi was a professor at the Kyushu University.

2. Richardson, T. I.; Rychnovsky, S. D. *Tetrahedron* **1999**, *55*, 8977–8996.

3. Paterson, I.; Chen, D. Y.-K.; Aceña, J. L.; Franklin, A. S. *Org. Lett.* **2000**, *2*, 1513–1516.

4. Hamelin, O.; Wang, Y.; Deprés, J.-P.; Greene, A. E. *Angew. Chem. Int. Ed.* **2000**, *39*, 4314–4316.

5. Quéron, E.; Lett, R. *Tetrahedron Lett.* **2004**, *45*, 4533–4537.

6. Mlynarski, J.; Ruiz-Caro, J.; Fürstner, A. *Chem., Eur. J.* **2004**, *10*, 2214–2222.

7. Lepage, O.; Kattnig, E.; Fürstner, A. *J. Am. Chem. Soc.* **2004**, *126*, 15970–15971.

8. Smith, A. B. III.; Simov, V. *Org. Lett.* **2006**, *8*, 3315–3318.

9. Ahmad, N. M. *Yamaguchi esterification.* In *Name Reactions for Functional Group Transformations*; Li, J. J., Ed.; Wiley: Hoboken, NJ, **2007**, pp 545–550. (Review).

10. Wender, P. A.; Verma, V. A. *Org. Lett.* **2008**, *10*, 3331–3334.

11. Carrick, J. D.; Jennings, M. P. *Org. Lett.* **2009**, *11*, 769–772.

12. Lu, L.; Zhang, W.; Sangkil Nam, S.; Horne, D. A.; Jove, R.; Carter, R. G. *J. Org. Chem.* **2013**, *78*, 2213–2247

## Zaitsev's elimination rule

$E_2$ thermodynamic elimination gives the more substituted olefin as the major product because it is more stable.

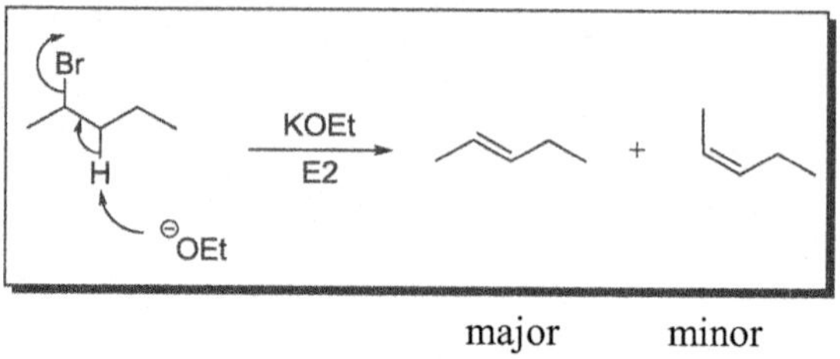
Br
H
⊖OEt
KOEt
E2
+
major
minor

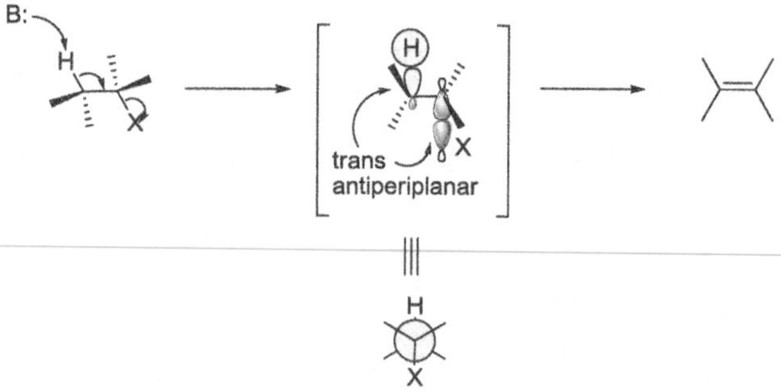
B:
H
X
H
trans
antiperiplanar
X

Example 1[2]

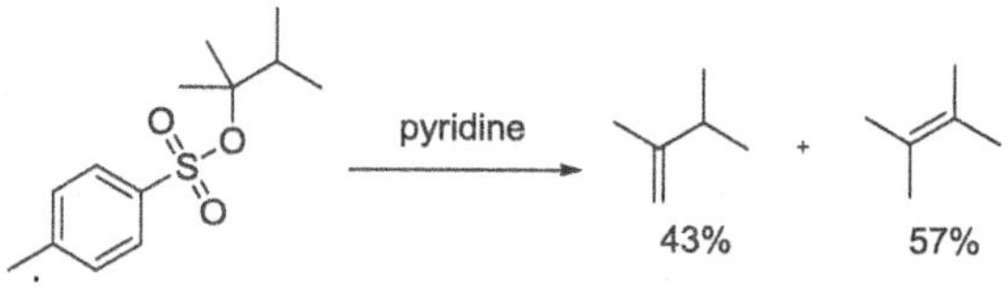
pyridine
43%
57%

Example 2[3]

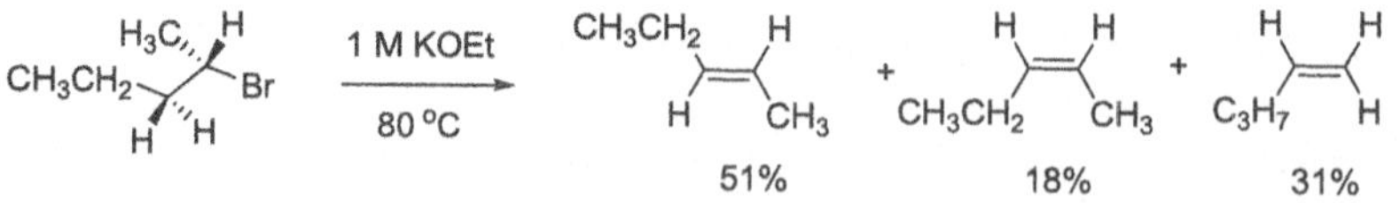
1 M KOEt
80 °C
51%
18%
31%

Example 3[5]

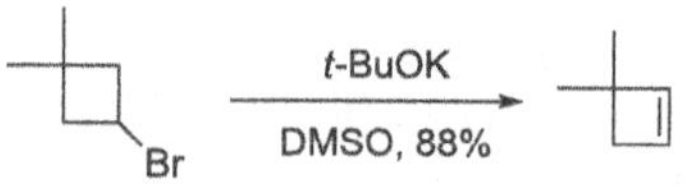
Br
t-BuOK
DMSO, 88%

J.J. Li, *Name Reactions: A Collection of Detailed Mechanisms and Synthetic Applications*,
DOI 10.1007/978-3-319-03979-4_296, © Springer International Publishing Switzerland 2014

Example 4[8]

Zaitsev elimination product

## References

1. Aleksandr Mikhailovich Zaitsev (sometimes spelled as Saytseff, 1841–1910), like Markovnikov, was also a protégé of Aleksandr Mikhailovich Butlerov (1828–1882). But unlike Markovnikov's lack of tact and inability to compromise with administrators, Zaitsev was a skilled politician. He held the position of chair at Kazan' University for over four decades and educated a generation of organic chemistry.
2. Brown, H. C.; Wheeler, O. H. *J. Am. Chem. Soc.* **1956**, *78*, 2199–2210.
3. Chamberlin, A. R.; Bond, F. T. *Synthesis* **1979**, 44–45.
4. Elrod, D. W.; Maggiora, G. M.; Trenary, R. G. *Tetrahedron Comput. Methodol.* **1990**, *3*, 163–174.
5. Larsen, N. W.; Pedersen, T. *J. Mol. Spectrosc.* **1994**, *166*, 372–382.
6. Reinecke, M. G.; Smith, W. B. *J. Chem. Educ.* **1995**, *72*, 541.
7. Guan, H.-P.; Ksebati, M. B.; Kern, E. R.; Zemlicka, J. *J. Org. Chem.* **2000**, *65*, 5177–5184.
8. Guan, H.-P.; Ksebati, M. B.; Kern, E. R.; Zemlicka, J. *J. Org. Chem.* **2000**, *65*, 5177-5184.
9. Hagen, T. J. *Zaitsev Elimination, In Name Reactions for Functional Group Transformations*; Li, J. J., Ed.; Wiley: Hoboken, NJ, **2007**, pp 414–421. (Review).
10. Ramos, D. R.; Castillo, R.; Canle L., M.; Garcia, M. V.; Andres, J.; Santaballa, J. A. *Org. Biomol. Chem.* **2009**, *7*, 1807–1814. (Mechanism).

# Zhang enyne cycloisomerization

Enynes are cycloisomerized regio- and enantio-selectively with a Rh complex with phosphine ligands.

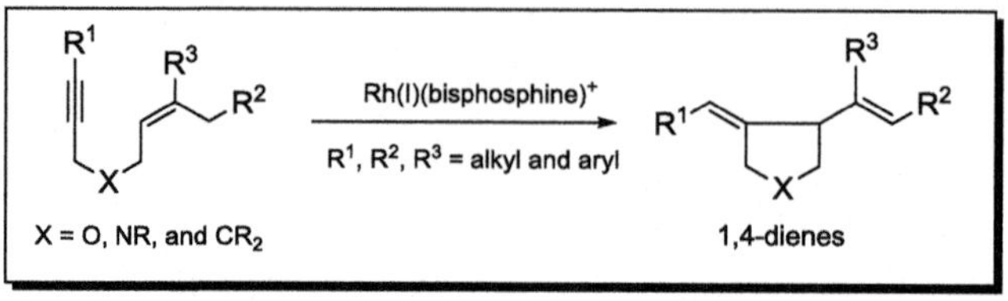

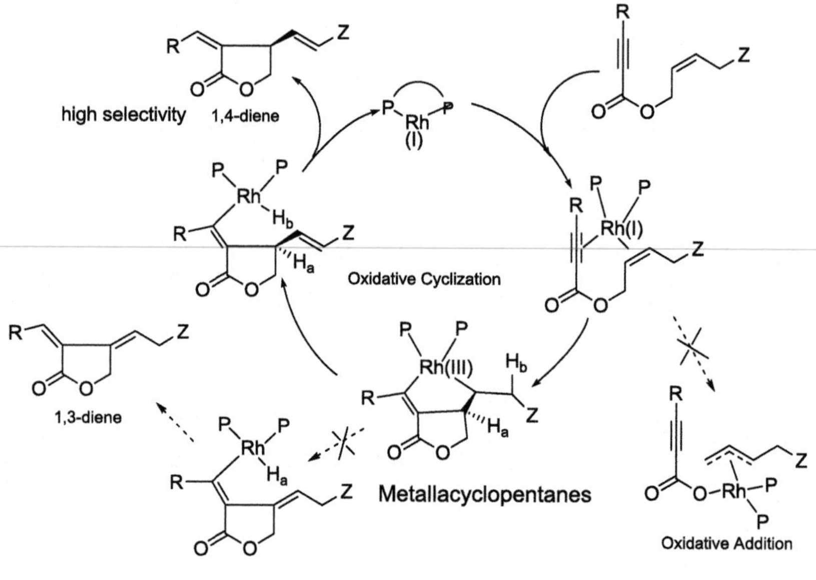

Example 1[3]

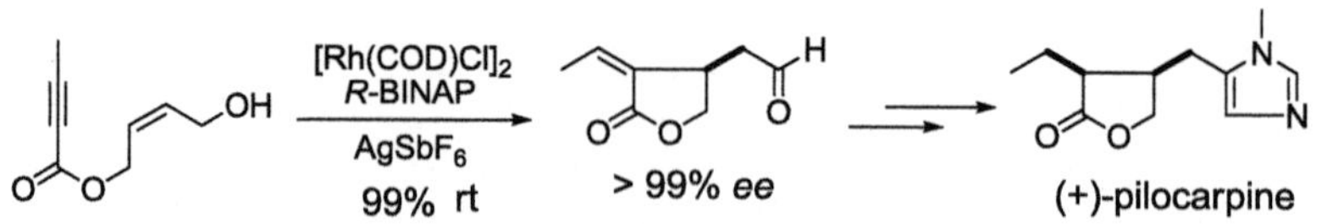

Example 2[4]

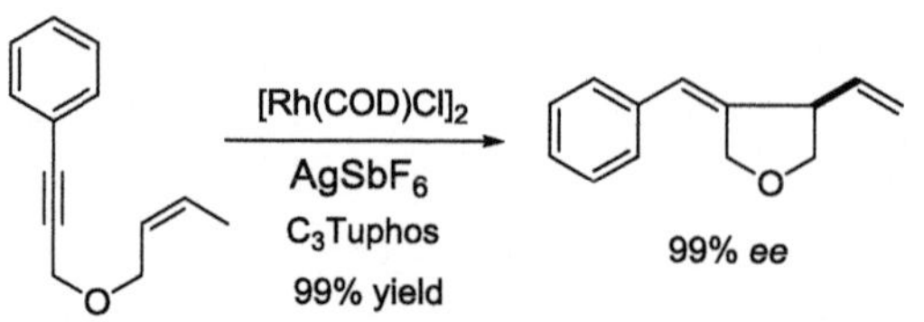

J.J. Li, *Name Reactions: A Collection of Detailed Mechanisms and Synthetic Applications*, DOI 10.1007/978-3-319-03979-4_297, © Springer International Publishing Switzerland 2014

Example 3[5]

COOEt

[Rh(COD)Cl]₂ → 
S-BINAP
AgSbF₆, rt

EtOOC

O N
Bn

91%, >99% ee

Example 4[11]

COOEt

OH

[Rh(COD)Cl]₂ → 
S-BINAP
AgSbF₆, rt

EtOOC CHO

91%, >99% ee

O

## References

1. Cao, P.; Wang, B.; Zhang, X. *J. Am. Chem. Soc.* **2000**, *122*, 6490–6491. Born in 1961, Xumu Zhang studied as an undergraduate in Wuhan University, China. He earned his Ph.D. in 1992 at Stanford under James P. Collman. He began his independent academic career at Pennsyvania State University (1994–2006). Since 2007, he is a distinguished Professor of Chemistry at Rutgers, the State University of New Jersey. In addition to the Zhang enyne cyclization, his has worked in asymmetric hydrogenation, asymmetric hydroformylation and linear selective hydroformylation for pratical synthetic methods. The chiral ligand toolbox developed by Zhang's group includes TangPhos, DuanPhos, Binapine, ZhangPhos, TunePhos, f-binaphane, and YanPhos. His group and industrial partners applied his asymmetric hydrogenation methodologies and completed innovative synthesis of many chiral pharmaceutical intermediates.
2. Cao, P.; Zhang, X. *Angew. Chem. Int. Ed.* **2000**, *39*, 4104–4106.
3. Lei, A.; He, M., Zhang, X. *J. Am. Chem. Soc.* **2002**, *124*, 8198–8199.
4. Lei, A.; He, M.; Wu, S.; Zhang, X. *Angew. Chem. Int. Ed.* **2002**, *41*, 3457–3460.
5. Lei, A.; Waldkirch, J. P.; He, M.; Wu, S.; Zhang, X. *Angew. Chem. Int. Ed.* **2002**, *41*, 4526–4529.
6. Lei, A.; He, M.; Zhang, X. *J. Am. Chem. Soc.* **2003**, *125*, 11472–11473.
7. Tong, X.; Zhang, Z.; Zhang, X. *J. Am. Chem. Soc.* **2003**, *125*, 6370–6371.
8. Tong, X.; Li, D.; Zhang, Z.; Zhang, X. *J. Am. Chem. Soc.* **2004**, *126*, 7601–7607.
9. He, M.; Lei, A.; Zhang, X. *Tetrahedron Lett.* **2005**, *46*, 1823–1826
10. Nicolaou, K. C.; Li, A.; Edmonds, D. J. *Angew. Chem. Int. Ed.* **2006**, *45*, 7086–7088.
11. Nicolaou, K. C.; Li, A.; Edmonds, D. J. *Angew. Chem. Int. Ed.* **2007**, *46*, 3942–3945.
12. Nicolaou, K. C.; et al. *Angew. Chem. Int. Ed.* **2007**, *46*, 6293–6295.
13. Nishimura, T.; Kawamoto, T.; Nagaosa, M.; Kumamoto, H.; Hayashi, T. *Angew. Chem. Int. Ed.* **2010**, *49*, 1638–1641.
14. Corkum, E. G.; Hass, M. J.; Sullivan, A. D.; Bergens, S. H. *Org. Lett.* **2011**, 13, 3522–3525.
15. Jackowski, O.; Wang, J.; Xie, X.; Ayad, T.; et al. *Org. Lett.* **2012**, *14*, 4006–4009.

# Zimmerman rearrangement

Conversion of 1,4-dienes to vinylcyclopropanes under photolysis. Also known as the **Di-π-methane rearrangement**.

1,4-diene                 vinylcyclopropane

diradical              diradical

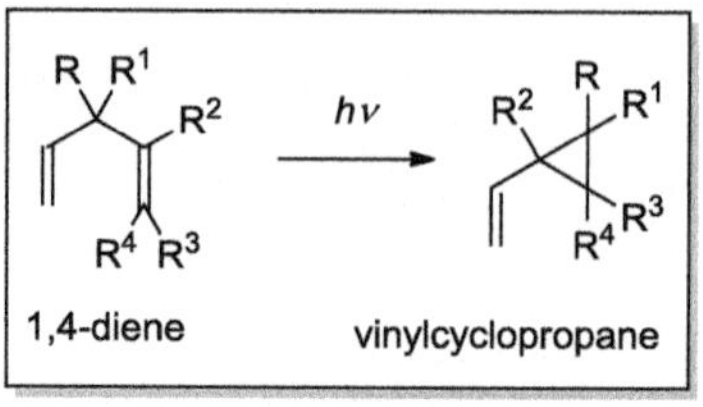

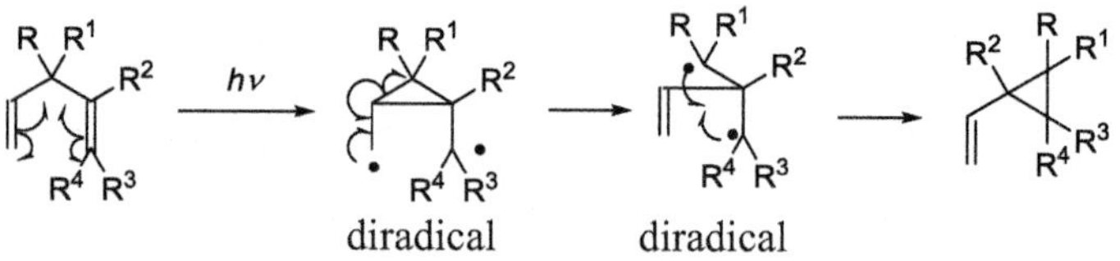

Example 1, Aza-π-methane rearrangement[2]

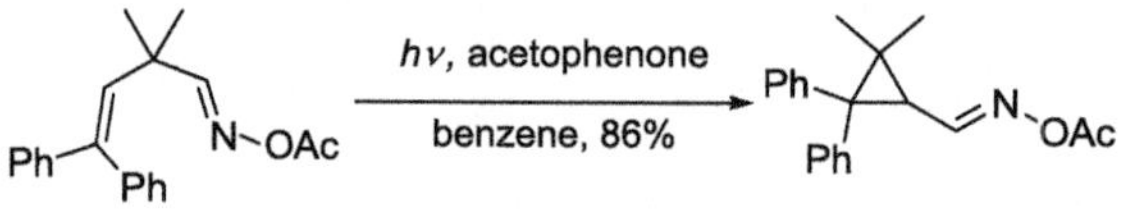

Example 2[4]

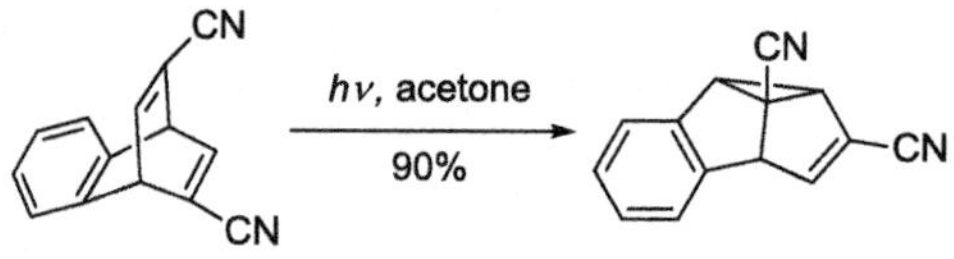

Example 3[8]

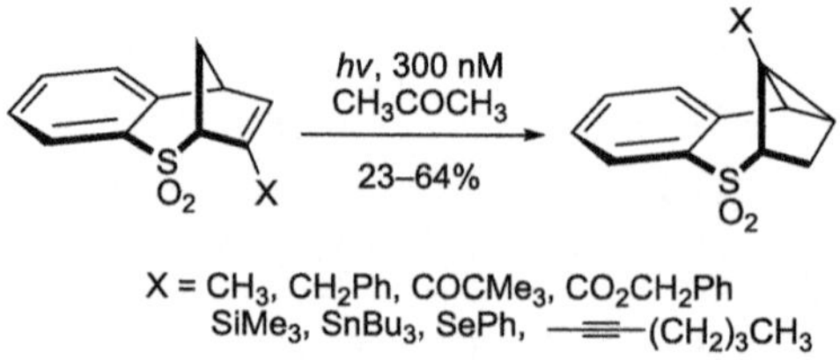

X = CH₃, CH₂Ph, COCMe₃, CO₂CH₂Ph
SiMe₃, SnBu₃, SePh, ——≡—(CH₂)₃CH₃

J.J. Li, *Name Reactions: A Collection of Detailed Mechanisms and Synthetic Applications*, DOI 10.1007/978-3-319-03979-4_298, © Springer International Publishing Switzerland 2014

Example 4, Oxa-π-methane rearrangement[9]

Example 4, Oxa-π-methane rearrangement[10]

## References

1.  (a) Zimmerman, H. E.; Grunewald, G. L. *J. Am. Chem. Soc.* **1966**, *88*, 183–184. Howard E. Zimmerman (1926–2012) was a professor at the University of Wisconsin at Madison. He is also known for the Traxler–Zimmerman trasition state for the asymmetric synthesis. (b) Zimmerman, H. E.; Armesto, D. *Chem. Rev.* **1996**, *96*, 3065–3112. (Review). (c) Zimmerman, H. E.; Církva, V. *Org. Lett.* **2000**, *2*, 2365–2367.
2.  Armesto, D.; Horspool, W. M.; Langa, F.; Ramos, A. *J. Chem. Soc., Perkin Trans. I* **1991**, 223–228.
3.  Jiménez, M. C.; Miranda, M. A.; Tormos, R. *Chem. Commun.* **2000**, 2341–2342.
4.  Ünaldi, N. S.; Balci, M. *Tetrahedron Lett.* **2001**, *42*, 8365–8367.
5.  Altundas, R.; Dastan, A.; Ünaldi, N. S.; Güven, K.; Uzun, O.; Balci, M. *Eur. J. Org. Chem.* **2002**, 526–533.
6.  Zimmerman, H. E.; Chen, W. *Org. Lett.* **2002**, *4*, 1155–1158.
7.  Tanifuji, N.; Huang, H.; Shinagawa, Y.; Kobayashi, K. *Tetrahedron Lett.* **2003**, *44*, 751–754.
8.  Dura, R. D.; Paquette, L. A. *J. Org. Chem.* **2006**, *71*, 2456–2459.
9.  Singh, V.; Chandra, G.; Mobin, S. M. *Synlett* **2008**, 2267–2270.
10. Cox, J. R.; Simpson, J. H.; Swager, T. M. *J. Am. Chem. Soc.* **2013**, *135*, 640–643.

# Zincke reaction

The Zincke reaction is an overall amine exchange process that converts *N*-(2,4-dinitrophenyl)pyridinium salts, known as Zincke salts, to *N*-aryl or *N*-alkyl pyridiniums upon treatment with the appropriate aniline or alkyl amine.

J.J. Li, *Name Reactions: A Collection of Detailed Mechanisms and Synthetic Applications*,
DOI 10.1007/978-3-319-03979-4_299, © Springer International Publishing Switzerland 2014

Example 1[5]

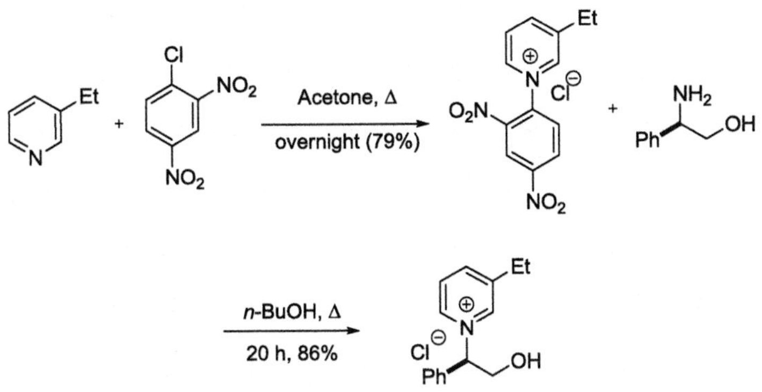

Example 2[6]

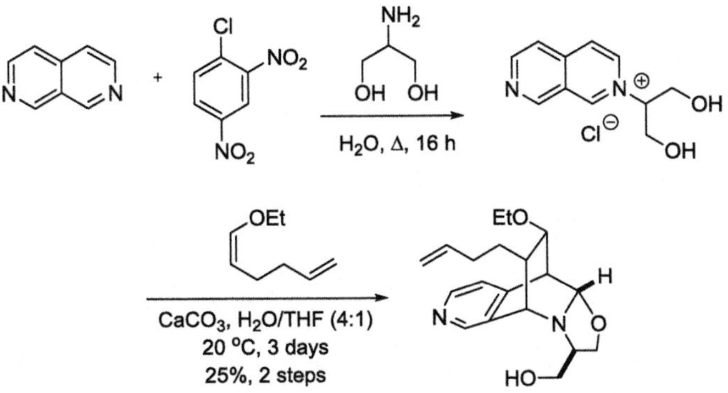

Example 3[9]

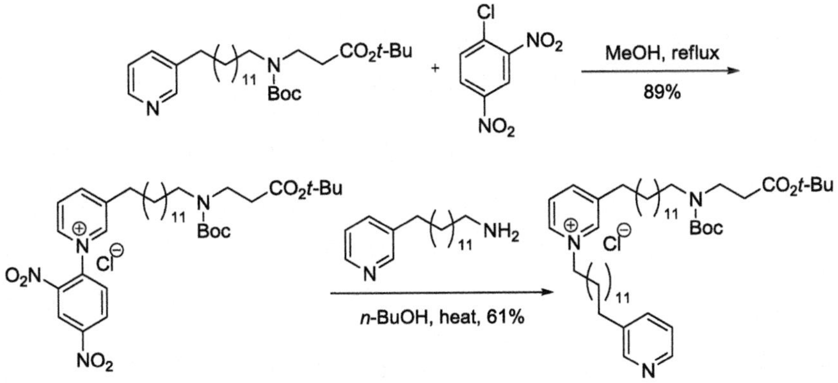

Example 4[10]

Zincke salt

Zincke aldehyde          50-55%

## References

1.  (a) Zincke, Th. *Ann.* **1903**, *330*, 361–374.  (b) Zincke, Th.; Heuser, G.; Möller, W. *Ann.* **1904**, *333*, 296–345.  (c) Zincke, Th.; Würker, W. *Ann.* **1905**, *338*, 107–141.  (d) Zincke, Th.; Würker, W. *Ann.* **1905**, *341*, 365–379. (e) Zincke, Th.; Weisspfenning, G. *Ann.* **1913**, *396*, 103–131.
2.  Epszju, J.; Lunt, E.; Katritzky, A. R. *Tetrahedron* **1970**, *26*, 1665–1673. (Review).
3.  Becher, J. *Synthesis* **1980**, 589–612. (Review).
4.  Kost, A. N.; Gromov, S. P.; Sagitullin, R. S. *Tetrahedron* **1981**, *37*, 3423–3454.  (Review).
5.  Wong, Y.-S.; Marazano, C.; Gnecco, D.; Génisson, Y.; Chiaroni, A.; Das, B. C. *J. Org. Chem.* **1997**, *62*, 729–735.
6.  Urban, D.; Duval, E.; Langlois, Y. *Tetrahedron Lett.* **2000**, *41*, 9251–9256.
7.  Cheng, W.-C.; Kurth, M. J. *Org. Prep. Proced. Int.* **2002**, *34*, 585–588. (Review).
8.  Rojas, C. M. *Zincke Reaction*. In *Name Reactions in Heterocyclic Chemistry*; Li, J. J., Ed.; Wiley: Hoboken, NJ, **2005**, pp 355–375. (Review).
9.  Shorey, B. J.; Lee, V.; Baldwin, J. E. *Tetrahedron* **2007**, *63*, 5587–5592.
10. Michels, T. D.; Rhee, J. U.; Vanderwal, C. D. *Org. Lett.* **2008**, *10*, 4787–4790.
11. Vanderwal, C. D. *J. Org. Chem.* **2011**, *76*, 9555–9567. (Review).

# Zinin benzidine (semidne) rearrangement

Also known as benzidine rearrangement or semidine rearrangement. Acid-promoted rearrangement of hydrazobenzene to 4,4-diaminobiphenyl (benzidine) and 2,4-diaminobiphenyl.

Hydrazobenzene          70% (benzidine)          30% (semidine)

Example 1[9,10]

J.J. Li, *Name Reactions: A Collection of Detailed Mechanisms and Synthetic Applications*, DOI 10.1007/978-3-319-03979-4_300, © Springer International Publishing Switzerland 2014

Example 2, Catalytic Asymmetric Benzidine Rearrangement[9]

CG-50, an acidic resin.

## References

1.  Zinin, N. *J. Prakt. Chem.* **1845,** *36,* 93–107.
2.  Shine, H. J.; Baldwin, C. M.; Harris, J. H. *Tetrahedron Lett.* **1968,** *9,* 977–980.
3.  Shine, H. J.; Zmuda, H.; Kwart, H.; Horgan, A. G.; Brechbiel, M. *J. Am. Chem. Soc.* **1982,** *104,* 5181–5184.
4.  Rhee, E. S.; Shine, H. J. *J. Am. Chem. Soc.* **1986,** *108,* 1000–1006.
5.  Shine, H. J. *J. Chem. Educ.* **1989,** *66,* 793–794.
6.  Davies, C. J.; Heaton, B. T.; Jacob, C. *J. Chem. Soc., Chem. Commun.* **1995,** 1177–1178.
7.  Park, K. H.; Kang, J. S. *J. Org. Chem.* **1997,** *62,* 3794–3795.
8.  Benniston, A. C.; Clegg, W.; Harriman, A.; Harrington, R. W.; Li, P.; Sams, C. *Tetrahedron Lett.* **2003,** *44,* 2665–2667.
9.  Hong, W.-X.; Chen, L.-J.; Zhong, C.-L.; Yao, Z.-J. *Org. Lett.* **2006,** *8,* 4919–4922.
10. Kim, H.-Y.; Lee, W.-J.; Kang, H.-M.; Cho, C.-G. *Org. Lett.* **2007,** *9,* 3185–3186.
11. De, C. K.; Pesciaioli, F.; List, B. *Angew. Chem. Int. Ed.* **2013,** *52,* 9293–9295.

# Index

**A**

Abnormal Beckmann rearrangement, 40
Abnormal Chichibabin reaction, 131
Abnormal Claisen rearrangement, 142
2-Acetamido acetophenone, 104
Acetic anhydride, 62, 192, 229, 468, 486,
 488, 501
Acetone cyanohydrin, 591
Acetonitrile as a reactant, 193
α-Acetylamino-alkyl methyl ketone, 192
Acetylation, 340
Acetylenic alcohols, 123
α,β-Acetylenic esters, 250
Acid chloride, 11, 510, 525
Acid-catalyzed acylation, 325
Acid-catalyzed alkyl group migration, 622
Acid-catalyzed condensation, 155, 157
Acid-catalyzed cyclization, 452
Acid-catalyzed electrocyclic formation of
 cyclopentenone, 424
Acid-catalyzed reaction, 539
Acid-catalyzed rearrangement, 482, 539
Acidic alcohol, 379
Acidic amide hydrolysis, 591
Acidic methylene moiety, 374
Acid-labile acetal, 117
Acid-mediated cyclization, 490
Acid-promoted rearrangement, 217
Acid scavenger, 227
Acrolein, 36, 562
Acrylic ester, 36
Acrylonitrile, 36
Activated hydroxamate, 367
Activated methylene compounds, 344
Activating agent, 486
Activating auxiliary, 507
Activating effect of a base, 593

Activating group, 316, 393, 486
Activation of the hydroxamic acid, 367
Activation step, 357
α-Active methylene nitrile, 279
Acyclic mechanism, 185
2-Acylamidoketones, 521
Acyl-*o*-aminobiphenyls, 413
Acyl anhydride, 260
Acylation, 8, 59, 260, 261, 325, 353, 367, 486
*O*-Acylation, 367
Acyl azides, 188
Acylbenzenesulfonylhydrazines, 369
Acyl derivative, 478
*N*-Acyl derivative, 188
*ortho*-Acyl diarylmethanes, 77
Acylglycine, 229
Acyl group, 260
Acyl halide, 260
Acylium ion, 261, 266, 278, 349
Acyl malonic ester, 289
α-Acyloxycarboxamide, 458
α-Acyloxyketone, 14
α-Acyloxythioether, 501
Acyl transfer, 14, 353, 468, 501
Adamantane-like structure, 478
*cis*-Addition, 546
1,6-Addition/Elimination, 656
1,4-Addition of a nucleophile, 397
Addition of Pd-H, 415
ADDP. *See* 1,1'-(azodicarbonyl)dipiperidine
 (ADDP)
Adduct formation, 407
Adenosine, 411
Aglycon, 246
AIBN. *See* 2,2'-Azobisisobutyronitrile (AIBN)
Air oxidation, 219
Al(O*i*-Pr)$_3$, 386

J.J. Li, *Name Reactions: A Collection of Detailed Mechanisms and Synthetic Applications*, **661**
DOI 10.1007/978-3-319-03979-4, © Springer International Publishing Switzerland 2014

Printed by Printforce, the Netherlands